Methods in Enzymology

Volume 231
HEMOGLOBINS
Part B
Biochemical and Analytical Methods

METHODS IN ENZYMOLOGY

EDITORS-IN-CHIEF

John N. Abelson   Melvin I. Simon

DIVISION OF BIOLOGY
CALIFORNIA INSTITUTE OF TECHNOLOGY
PASADENA, CALIFORNIA

FOUNDING EDITORS

Sidney P. Colowick and Nathan O. Kaplan

Methods in Enzymology

Volume 231

Hemoglobins

Part B

Biochemical and Analytical Methods

EDITED BY

Johannes Everse

DEPARTMENT OF BIOCHEMISTRY
AND MOLECULAR BIOLOGY
TEXAS TECH UNIVERSITY
HEALTH SCIENCES CENTER
LUBBOCK, TEXAS

Kim D. Vandegriff
Robert M. Winslow

DEPARTMENT OF MEDICINE
UNIVERSITY OF CALIFORNIA AT SAN DIEGO
VETERANS AFFAIRS MEDICAL CENTER
SAN DIEGO, CALIFORNIA

ACADEMIC PRESS

A Division of Harcourt Brace & Company

San Diego   New York   Boston   London   Sydney   Tokyo   Toronto

Academic Press, Inc.
525 B Street, Suite 1900, San Diego, California 92101-4495

United Kingdom Edition published by
Academic Press Limited
24–28 Oval Road, London NW1 7DX

International Standard Serial Number: 0076-6879

International Standard Book Number: 0-12-182132-3

PRINTED IN THE UNITED STATES OF AMERICA
94 95 96 97 98 99   EB   9 8 7 6 5 4 3 2 1

Table of Contents

Section I. Preparation, Purification, and Analysis of Human Hemoglobin

Section II. Preparation and Characterization of Nonhuman Hemoglobins

Section III. Hemoglobin Hybrids and Model Hemes

Section IV. Chemically Modified Hemoglobins

Section V. Recombinant Hemoglobin

Section VI. Hemoglobin Stability and Degradation

Section VII. Enzymatic Reactions Catalyzed by Hemoglobin

Section VIII. Xenobiotic Adducts of Human Hemoglobin

Addendum

Contributors to Volume 231

Article numbers are in parentheses following the names of contributors. Affiliations listed are current.

OMOEFE ABUGO (28), *Laboratory of Cellular and Molecular Biology, National Institutes of Health, National Institute on Aging, Gerontology Research Center, Baltimore, Maryland 21224*

A. SEETHARAMA ACHARYA (12, 15), *Division of Hematology, Department of Medicine, Albert Einstein College of Medicine, Yeshiva University, Bronx, New York 10461*

ANDREAS G. ALTEMÜLLER (7), *Fraunhofer-Institut für Biomedizinische Technik, D-6670 St. Ingbert, Germany*

MARK E. ANDRACKI (17), *Department of Biochemistry, University of Iowa, Iowa City, Iowa 52242*

TOSHIO ASAKURA (26), *Department of Pediatrics and Department of Biochemistry and Biophysics, University of Pennsylvania, Philadelphia, Pennsylvania 19104, and Division of Hematology, The Children's Hospital of Philadelphia, Philadelphia, Pennsylvania 19104*

GIORGIO BELVEDERE (39), *Istituto di Ricerche Farmacologiche, "Mario Negri," 20157 Milan, Italy*

RUTH E. BENESCH (16, 31), *Department of Biochemistry and Molecular Biophysics, College of Physicians and Surgeons, Columbia University, New York, New York 10032*

ENRICO BUCCI (10), *Department of Biochemistry, University of Maryland, School of Medicine, Baltimore, Maryland 21201*

STEVEN G. CARMELLA (44), *American Health Foundation, Valhalla, New York 10595*

KEITH W. CHAPMAN (1), *Alliance Pharmaceutical Corporation, San Diego, California 92121*

MARIO CIUFFI (37), *Department of Preclinical and Clinical Pharmacology, "M. Aiazzi Mancini," University of Florence, I-50134 Florence, Italy*

MARGO PANUSH COHEN (5), *Exocell, Inc., Philadelphia, Pennsylvania 19104*

DOUGLAS L. CURRELL (18), *Department of Chemistry and Biochemistry, California State University, Los Angeles, Los Angeles, California 90032*

KELVIN J. A. DAVIES (30), *Department of Biochemistry and Molecular Biology, The Albany Medical College, Albany, New York 12208*

BILLY W. DAY (46), *Departments of Environmental and Occupational, Health and Pharmaceutical Sciences, University of Pittsburgh, Pittsburgh, Pennsylvania 15238*

ALICE DEYOUNG (9), *Department of Medicine, State University of New York at Buffalo, Veterans Administration Medical Center, Buffalo, New York 14215*

KILIAN DILL (35), *Molecular Devices Corporation, Menlo Park, California 94025*

JOHANNES EVERSE (34, 36), *Department of Biochemistry and Molecular Biology, Texas Tech University, Health Sciences Center, Lubbock, Texas 79430*

GIULIO FERMI (22), *MRC Laboratory of Molecular Biology, Cambridge CB2 2QH, England*

PASQUALE FERRANTI (4), *Servizio di Spettrometria di Massa del C.N.R., I-80131 Naples, Italy*

SERGIO FRANCHI-MICHELI (37), *Department of Preclinical and Clinical Pharmacology, "M. Aiazzi Mancini," University of Florence, I-50134 Florence, Italy*

ix

CLARA FRONTICELLI (10), *Department of Biological Chemistry, University of Maryland, School of Medicine, Baltimore, Maryland 21201*

FABIO FUSI (37), *Institute of Pharmacological Sciences, University of Siena, I-53100 Siena, Italy*

GRAZIA GENTILINI (37), *Department of Preclinical and Clinical Pharmacology, "M. Aiazzi Mancini," University of Florence, I-50134 Florence, Italy*

KLAUS GERSONDE (7), *Fraunhofer-Institut für Biomedizinische Technik, and Fachrichtung Medizintechnik der Universität des Saarlandes, D-66386 St. Ingbert, Germany*

CECILIA GIULIVI (30), *Institute for Toxicology, University of Southern California, Health Sciences Campus, Los Angeles, California 90033*

ROY HARRIS (24), *Protein Chemistry Department, Delta Biotechnology, Ltd., Nottingham NG7 1FD, United Kingdom*

STEPHEN S. HECHT (44), *American Health Foundation, Valhalla, New York 10595*

JANE HEIM (28), *Laboratory of Cellular and Molecular Biology, National Institutes of Health, National Institute on Aging, Gerontology Research Center, Baltimore, Maryland 21224*

TIMM-H. JESSEN (22), *Hoechst-AG, General Pharma Research, W-65926 Frankfurt, Germany*

MARIA C. JOHNSON (36), *Blood Research Division, Letterman Army Institute of Research, Presidio of San Francisco, California 94129*

RICHARD T. JONES (21), *Department of Biochemistry and Molecular Biology, Oregon Health Sciences University, School of Medicine, Portland, Oregon 97201*

MAKOTO KATSUMATA (26), *Department of Pathology and Laboratory Medicine, University of Pennsylvania, Philadelphia, Pennsylvania 19104*

MAZHAR KHAN (35), *Desert Analytics, Tucson, Arizona 85719*

TERUYUKI KOMATSU (11), *Department of Polymer Chemistry, Waseda University, Tokyo 169, Japan*

NOBORU H. KOMIYAMA (22), *MRC Laboratory of Molecular Biology, Cambridge CB2 2QH, England*

LAURA D. KWIATKOWSKI (9), *Department of Medicine, State University of New York at Buffalo, Veterans Administration Medical Center, Buffalo, New York 14215*

JACK LEVIN (6), *Department of Laboratory Medicine, University of California, San Francisco, San Francisco, California 94143*

ABRAHAM LEVY (28), *National Center for Research Resources, National Institutes of Health, Bethesda, Maryland 20014*

JOHN S. LOGAN (27), *DNX, Princeton, New Jersey 08540*

DOUGLAS LOOKER (23), *Department of Research and Development, Somatogen, Inc., Boulder, Colorado 80301*

BEN LUISI (22), *MRC Virology Institute, Glasgow GI1 5JR, Scotland*

SHIRLEY L. MACDONALD (19), *National Science Laboratory, Scottish National Blood Transfusion Service, Edinburgh EH1 2QN, Scotland*

VICTOR W. MACDONALD (29), *U.S. Army Medical Research Detachment, National Naval Medical Center, Walter Reed Army Institute of Research, Washington, D.C. 20307*

ANTONIO MALORNI (4), *ICMIB/CNR, and Servizio di Spettrometria di Massa del C.N.R., I-80131 Naples, Italy*

LAURA MANCA (3), *Institute of General Physiology and Biological Chemistry, University of Sassari, I-07100 Sassari, Italy*

JAMES M. MANNING (14, 25), *Department of Biochemistry, Rockefeller University, New York, New York 10021*

MARIO A. MARINI (36), *U.S. Army Medical Research Detachment, National Naval Medical Center, Walter Reed Army Institute of Research, Washington, D.C. 20307*

MICHAEL J. MARTIN (27), *DNX, Millfield, Ohio 45761*

JOSE JAVIER MARTIN DE LLANO (25), *Department of Biochemistry, Rockefeller University, New York, New York 10021*

BRUNO MASALA (3), *Institute of General Physiology and Biological Chemistry, University of Sassari, I-07100 Sassari, Italy*

ANTONY J. MATHEWS (23), *Department of Research and Development, Somatogen, Inc., Boulder, Colorado 80301*

STEVEN L. McCUNE (26), *Department of Biochemistry and Molecular Genetics, The University of Alabama at Birmingham, Birmingham, Alabama 35294*

EVELYN L. McGOWN (35), *Pharmetrix Corp., Menlo Park, California 94025*

J. J. MIEYAL (38), *Department of Pharmacology, Case Western Reserve University, School of Medicine, Cleveland, Ohio 44106*

GLORIANO MONETI (37), *Department of Preclinical and Clinical Pharmacology, "M. Aiazzi Mancini," University of Florence, I-50134 Florence, Italy*

SHARON E. MURPHY (44), *American Health Foundation, Valhalla, New York 10595*

KIYOSHI NAGAI (22), *MRC Laboratory of Molecular Biology, Cambridge CB2 2QH, England*

JUSTIN O. NEWAY (23), *Department of Research and Development, Somatogen, Inc., Boulder, Colorado 80301*

ROBERT W. NOBLE (9), *Department of Medicine and Biochemistry, State University of New York at Buffalo, Veterans Administration Medical Center, Buffalo, New York 14215*

JILL E. OGDEN (24), *Department of Molecular Biology, Delta Biotechnology, Ltd., Nottingham NG7 1FD, United Kingdom*

KENNETH W. OLSEN (33), *Department of Chemistry, Loyola University at Chicago, Chicago, Illinois 60626*

JOSÉE PAGNIER (22), *INSERM, Unité 299, 94275 Le Kremlin-Bicêtre, France*

S. SCOTT PANTER (32), *Department of Neurosurgery, University of California, San Francisco, California and San Francisco General Hospital, San Francisco, California 94143*

DUNCAN S. PEPPER (19), *National Science Laboratory, Scottish National Blood Transfusion Service, Edinburgh EH1 2QN, Scotland*

PIETRO PUCCI (4), *ICMIB/CNR, and Servizio di Spettrometria di Massa del C.N.R., I-80131 Naples, Italy*

M. JANARDHAN RAO (15), *Division of Hematology, Department of Medicine, Albert Einstein College of Medicine, Yeshiva University, Bronx, New York 10461*

MICHAEL P. REILLY (26) *Division of Hematology, The Children's Hospital of Philadelphia, Philadelphia, Pennsylvania 19104*

JOSEPH M. RIFKIND (28), *Laboratory of Cellular and Molecular Biology, National Institutes of Health, National Institute on Aging, Gerontology Research Center, Baltimore, Maryland 21224*

ROBERT I. ROTH (6), *Department of Laboratory Medicine, University of California, San Francisco, San Francisco, California 94143*

RAJENDRA PRASAD ROY (12), *Division of Hematology, Department of Medicine, Albert Einstein College of Medicine, Yeshiva University, Bronx, New York 10461*

HANS HEINRICH RUF (7), *Fraunhofer-Institut für Biomedizinische Technik, and Fachrichtung Medizintechnik der Universität des Saarlandes, D-66386 St. Ingbert, Germany*

THOMAS M. RYAN (26), *Department of Biochemistry and Molecular Genetics, The University of Alabama at Birmingham, Birmingham, Alabama 35294*

MICHELE SAMAJA (39), *Dipartimento di Scienze e Tecnologie Biomediche, Istituto Scientifico San Raffaele, 20132 Milan, Italy*

OLAF SCHNEEWIND (25), *Department of Microbiology and Immunology, University of California School of Medicine,*

Center for the Health Sciences, Los Angeles, California 90024

GIAN PIETRO SGARAGLI (37), *Institute of Pharmacological Sciences, University of Siena, I-53100 Siena, Italy*

PAWAN K. SHARMA (8), *Department of Medicine, University of Tennessee, Chattanooga, Tennessee 37402*

DANIEL SHIH (22), *Department of Biochemistry, Oregon Health Sciences University, Portland, Oregon 97201*

STEPHEN B. SHOHET (2), *Department of Laboratory Medicine, University of California, San Francisco, San Francisco, California 94143*

KULDIP SINGH (46), *Department of Chemistry, University of Missouri-St. Louis, St. Louis, Missouri 63121*

PAUL L. SKIPPER (40, 42, 45), *Division of Toxicology, Massachusetts Institute of Technology, Cambridge, Massachusetts 02139*

D. W. STARKE (38), *Department of Pharmacology, Case Western Reserve University, School of Medicine, Cleveland, Ohio 44106*

GARY L. STETLER (23, 25), *Department of Research and Development, Somatogen, Inc., Boulder, Colorado 80301*

W. G. STILLWELL (42), *Division of Toxicology, Massachusetts Institute of Technology, Cambridge, Massachusetts 02139*

KOLI TAGHIZADEH (45), *Center for Environmental Health Sciences, Massachusetts Institute of Technology, Cambridge, Massachusetts 02139*

JEREMY TAME (22), *Department of Chemistry, University of York, York YO1 5DD, England*

STEVEN R. TANNENBAUM (40), *Department of Chemistry, and Division of Toxicology, Massachusetts Institute of Technology, Cambridge, Massachusetts 02139*

BERNARD J.-M. THEVENIN (2), *Department of Laboratory Medicine, University of California, San Francisco, San Francisco, California 94143*

MARGARETA TÖRNQVIST (43), *Department of Radiobiology, Stockholm University, S-106 91 Stockholm, Sweden*

TIM M. TOWNES (26), *Department of Biochemistry and Molecular Genetics, The University of Alabama at Birmingham, Birmingham, Alabama 35294*

EISHUN TSUCHIDA (11), *Department of Polymer Chemistry, Waseda University, Tokyo 169, Japan*

ANTONIO TSUNESHIGE (13), *Department of Biochemistry and Biophysics, University of Pennsylvania, School of Medicine, Philadelphia, Pennsylvania 19104*

MASSIMO VALOTI (37), *Institute of Pharmacological Sciences, University of Siena, I-53100 Siena, Italy*

SERGE N. VINOGRADOV (8), *Department of Biochemistry, Wayne State University, School of Medicine, Detroit, Michigan 48201*

JOSEPH A. WALDER (17), *Department of Biochemistry, University of Iowa, Iowa City, Iowa 52242*

ROXANNE Y. WALDER (17), *Department of Biochemistry, University of Iowa, Iowa City, Iowa 52242*

MICHAEL T. WILSON (24), *Department of Chemistry and Biological Chemistry, University of Essex, Colchester, CO4 3SQ, United Kingdom*

ROBERT M. WINSLOW (1), *Department of Medicine, University of California, San Diego, San Diego, California 92161*

JOHN S. WISHNOK (41), *Division of Toxicology, Massachusetts Institute of Technology, Cambridge, Massachusetts 02139*

J. TZE-FEI WONG (20), *Department of Biochemistry, Hong Kong University of Science and Technology, Clear Water Bay, Kowloon, Hong Kong*

VAN-YU WU (5), *Exocell, Inc., Philadelphia, Pennsylvania 19104*

HONG XUE (20), *Robertson Institute of Biotechnology, Department of Genetics, University of Glasgow, Glasgow G11 5J5, Scotland*

TAKASHI YONETANI (13), *Department of Biochemistry and Biophysics, University of Pennsylvania, School of Medicine, Philadelphia, Pennsylvania 19104*

LUCILLA ZILLETTI (37), *Department of Preclinical and Clinical Pharmacology, "M. Aiazzi Mancini," University of Florence, I-50134 Florence, Italy*

Preface

Much has happened since "Hemoglobins," Volume 76 of *Methods in Enzymology,* was published in 1981. Methods have been refined and new methods have been devised. At the time Volume 76 went to press, the general feeling in the hemoglobin "community" was that this venerable protein had contributed about all it could to our knowledge of fundamental protein chemistry, and scientists were turning their attention to other proteins.

Three forces have brought hemoglobin back to center stage. First, the expression of human globin genes in *Escherichia coli, Saccharomyces cerevisiae*, and transgenic animals has led to new approaches to hemoglobin chemistry. It is now possible to produce quickly and efficiently site-specific hemoglobin mutants that are being used to test hypotheses on structure–function relationships. The amount of information derived from such studies is staggering. Second, new techniques have been developed that have led to significant advances in the field since 1981. For example, laser photolysis has expanded from nanosecond to femtosecond time domains that are used to explore the very fast events associated with the hemoglobin molecule. Third, hemoglobin has been used in exciting new studies of cell-free oxygen carriers for eventual clinical use, currently being carried out in both academia and industry. Purification and characterization of hemoglobin and chemically or genetically modified hemoglobins are now of critical importance. This new application requires a detailed understanding of the oxygen transport function of hemoglobin and of its interactions with other biological systems.

Thus, what began as a modest effort to update methods described in Volume 76 quickly expanded into two new volumes, 231 and 232. The division of the chapters is somewhat arbitrary, but we settled on methods of biochemical and analytical focus for Volume 231 and those which deal with biophysical methods for Volume 232. As in Volume 76, authors were instructed to emphasize techniques, currently in use in their laboratories, in sufficient detail to allow the interested reader to implement those methods independently.

We wish to thank the authors for their contributions and cooperation and the staff of Academic Press for their assistance. Special thanks are due the Editors-in-Chief of *Methods in Enzymology* for encouraging the preparation of these volumes and to Shirley Light of Academic Press for her expert guidance and support during this work.

JOHANNES EVERSE
KIM D. VANDEGRIFF
ROBERT M. WINSLOW

METHODS IN ENZYMOLOGY

VOLUME LV. Biomembranes (Part F: Bioenergetics)
Edited by SIDNEY FLEISCHER AND LESTER PACKER

VOLUME LVI. Biomembranes (Part G: Bioenergetics)
Edited by SIDNEY FLEISCHER AND LESTER PACKER

VOLUME LVII. Bioluminescence and Chemiluminescence
Edited by MARLENE A. DeLUCA

VOLUME LVIII. Cell Culture
Edited by WILLIAM B. JAKOBY AND IRA PASTAN

VOLUME LIX. Nucleic Acids and Protein Synthesis (Part G)
Edited by KIVIE MOLDAVE AND LAWRENCE GROSSMAN

VOLUME LX. Nucleic Acids and Protein Synthesis (Part H)
Edited by KIVIE MOLDAVE AND LAWRENCE GROSSMAN

VOLUME 61. Enzyme Structure (Part H)
Edited by C. H. W. HIRS AND SERGE N. TIMASHEFF

VOLUME 62. Vitamins and Coenzymes (Part D)
Edited by DONALD B. McCORMICK AND LEMUEL D. WRIGHT

VOLUME 63. Enzyme Kinetics and Mechanism (Part A: Initial Rate and Inhibitor Methods)
Edited by DANIEL L. PURICH

VOLUME 64. Enzyme Kinetics and Mechanism (Part B: Isotopic Probes and Complex Enzyme Systems)
Edited by DANIEL L. PURICH

VOLUME 65. Nucleic Acids (Part I)
Edited by LAWRENCE GROSSMAN AND KIVIE MOLDAVE

VOLUME 66. Vitamins and Coenzymes (Part E)
Edited by DONALD B. McCORMICK AND LEMUEL D. WRIGHT

VOLUME 67. Vitamins and Coenzymes (Part F)
Edited by DONALD B. McCORMICK AND LEMUEL D. WRIGHT

VOLUME 68. Recombinant DNA
Edited by RAY WU

VOLUME 69. Photosynthesis and Nitrogen Fixation (Part C)
Edited by ANTHONY SAN PIETRO

VOLUME 70. Immunochemical Techniques (Part A)
Edited by HELEN VAN VUNAKIS AND JOHN J. LANGONE

VOLUME 71. Lipids (Part C)
Edited by JOHN M. LOWENSTEIN

VOLUME 72. Lipids (Part D)
Edited by JOHN M. LOWENSTEIN

VOLUME 73. Immunochemical Techniques (Part B)
Edited by JOHN J. LANGONE AND HELEN VAN VUNAKIS

VOLUME 199. Cumulative Subject Index Volumes 168–174, 176–194

VOLUME 200. Protein Phosphorylation (Part A: Protein Kinases: Assays, Purification, Antibodies, Functional Analysis, Cloning, and Expression)
Edited by TONY HUNTER AND BARTHOLOMEW M. SEFTON

VOLUME 201. Protein Phosphorylation (Part B: Analysis of Protein Phosphorylation, Protein Kinase Inhibitors, and Protein Phosphatases)
Edited by TONY HUNTER AND BARTHOLOMEW M. SEFTON

VOLUME 202. Molecular Design and Modeling: Concepts and Applications (Part A: Proteins, Peptides, and Enzymes)
Edited by JOHN J. LANGONE

VOLUME 203. Molecular Design and Modeling: Concepts and Applications (Part B: Antibodies and Antigens, Nucleic Acids, Polysaccharides, and Drugs)
Edited by JOHN J. LANGONE

VOLUME 204. Bacterial Genetic Systems
Edited by JEFFREY H. MILLER

VOLUME 205. Metallobiochemistry (Part B: Metallothionein and Related Molecules)
Edited by JAMES F. RIORDAN AND BERT L. VALLEE

VOLUME 206. Cytochrome P450
Edited by MICHAEL R. WATERMAN AND ERIC F. JOHNSON

VOLUME 207. Ion Channels
Edited by BERNARDO RUDY AND LINDA E. IVERSON

VOLUME 208. Protein–DNA Interactions
Edited by ROBERT T. SAUER

VOLUME 209. Phospholipid Biosynthesis
Edited by EDWARD A. DENNIS AND DENNIS E. VANCE

VOLUME 210. Numerical Computer Methods
Edited by LUDWIG BRAND AND MICHAEL L. JOHNSON

VOLUME 211. DNA Structures (Part A: Synthesis and Physical Analysis of DNA)
Edited by DAVID M. J. LILLEY AND JAMES E. DAHLBERG

VOLUME 212. DNA Structures (Part B: Chemical and Electrophoretic Analysis of DNA)
Edited by DAVID M. J. LILLEY AND JAMES E. DAHLBERG

VOLUME 213. Carotenoids (Part A: Chemistry, Separation, Quantitation, and Antioxidation)
Edited by LESTER PACKER

VOLUME 214. Carotenoids (Part B: Metabolism, Genetics, and Biosynthesis)
Edited by LESTER PACKER

Section I

Preparation, Purification, and Analysis
of Human Hemoglobin

[1] Pilot-Scale Preparation of Hemoglobin Solutions*

By ROBERT M. WINSLOW and KEITH W. CHAPMAN

Introduction

Cell-free hemoglobin, suitably modified to confer a physiological oxygen affinity and prolonged plasma retention, is a candidate for a temporary substitute for red blood cells.[1] Such a solution would probably find wide clinical acceptance because the risks of disease transmission from banked human blood are significant,[2,3] and not likely to diminish in the foreseeable future. Although blood substitutes based on hemoglobin have stimulated a considerable amount of research and development for over 50 years, still none is available for clinical use. In fact, in spite of encouraging animal studies, clinical trials have been disappointing, revealing complex biological effects. These problems have made hemoglobin-based red cell substitutes an elusive goal. A serious hindrance to continued progress toward a blood substitute has been the unavailability of high-quality products for further testing and research.[4]

A high degree of purity of hemoglobin for intravenous infusion is critical, because in clinical practice, very large amounts of hemoglobin solution would be used, and even small amounts of contaminants could cause serious side effects. Rabiner[5] showed that by using a better method to lyse red cells, coagulation and renal problems could be reduced substantially. Subsequent studies demonstrated that red cell stromata (membranes) were more toxic than hemoglobin.[6-8] Feola and co-workers[9] believe that even trace amounts of phospholipid contamination of

---

* The opinions expressed herein are the private views of the authors, and are not to be construed as official or as reflecting the views of the U.S. Department of the Army or of the U.S. Department of Defense.

[1] R. M. Winslow, "Hemoglobin-Based Red Cell Substitutes." Johns Hopkins Univ. Press, Baltimore, 1992.

[2] National Institutes of Health Consensus Conference, *JAMA, J. Am. Med. Assoc.* **260**, 2700 (1988).

[3] L. J. Conley and S. D. Holmberg, *N. Engl. J. Med.* **326**, 1499 (1992).

[4] R. M. Winslow, *Transfusion (Philadelphia)* **29**, 753 (1989).

[5] S. F. Rabiner, J. R. Helbert, H. Lopas, and L. H. Friedman, *J. Exp. Med.* **126**, 1127 (1967).

[6] N. I. Birndorf, H. Lopas, and S. Robboy, *J. Lab. Invest.* **25**, 314 (1971).

[7] M. Relihan and M. S. Litwin, *Surgery (St. Louis)* **71**, 395 (1972).

[8] W. Odling-Smee and C. McLarnon, *Ir. J. Med. Sci.* **153**, 385 (1984).

[9] M. Feola, S. C. Simoni, P. C. Canizaro, R. Tran, G. Raschbaum, and F. J. Behal, *Surg., Gynecol. Obstet.* **166**, 211 (1988).

hemoglobin solutions might be toxic, and Lee and co-workers[10] have shown that purified bovine hemoglobin has substantially reduced renal toxicity compared with a crude hemolysate.

Unfortunately, production of highly purified, sterile, endotoxin-free hemoglobin solutions in the academic laboratory is difficult because of the requirement for specialized equipment, sterile techniques, and specially trained personnel. Much of the research in biological systems has been carried out by industry using proprietary materials manufactured to high standards, but such products and experimental results are usually not available to academic investigators.

For these reasons, we undertook the construction of a generic pilot plant at the Letterman Army Institute of Research (LAIR) for the production of purified stroma-free hemoglobin (SFH), hemoglobin A_0, and cross-linked hemoglobin using outdated human blood as a starting material. Our objective for this facility was to be capable of implementing most purification and cross-linking procedures, and to be flexible enough so that changes in the processes could be incorporated as needed, based on results of biological and physicochemical studies.

Materials and Methods

The production facility is housed in four separate, but connected, rooms: one contains the system for generation of pyrogen-free water (PFW), the second is used for reagent preparation, the third contains the filtration assemblies and the bioreactor, and the fourth is a 4° cold room that houses the preparative high-performance liquid chromatography (HPLC) equipment. An additional large walk-in cold room is available to store large tanks of buffers.

The apparatus used for hemoglobin preparation consists of several components, including a manifold for emptying bags of outdated blood; cross-flow filtration devices for washing, lysis, and purification; a bioreactor for cross-linking hemoglobin; a preparative-scale HPLC apparatus; and a hood for final filling of product into sterile bags. The flow loops, including the PFW distribution loops, are closed. All connections are of the sanitary triclamp type and are made in a laminar flow environment.

The cross-flow filtration system (Fig. 1) consists of three pumps (models 30 and 60, Waukesh Fluid Handling, Waukesha, WI), four tanks (DCI Inc., St. Cloud, MN), and four distinct filter packages (AG Technology

---

[10] R. Lee, N. Atsumi, E. E. Jacobs, W. G. Austen, and G. J. Vlahakes, *J. Surg. Res.* **47**, 407 (1989).

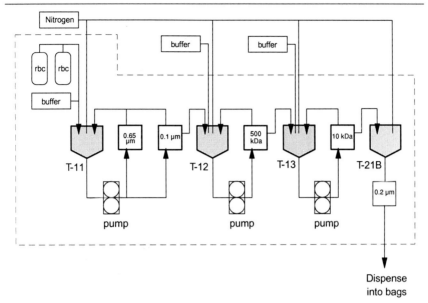

FIG. 1. Hardware diagram for stroma-free hemoglobin production. Stainless steel tanks (T-11, T-12, T-13, and T-21B) are shown shaded. All equipment shown within the dashed line is contained within a class 100 hood. T-11 is a jacketed, temperature-controlled tank. T-21B is a portable tank. The filtration sizes of different filters in the system are indicated. rbc, Red blood cells.

Corp., Needham, MA). All construction materials are of 316L stainless steel with internal finishes of 180 grit. Tubing is flexible and made of reinforced silicone. The hollow fiber filtration cartridges are nonhemolytic, nonpyrogenic polysulfone membranes.

All solutions are chilled after preparation in jacketed stainless-steel 500-liter tanks (Precision Stainless, Inc., Springfield, MO). The red cell wash and lysis tank (T-11) is glycol jacketed for temperature control. The 500- and 10-kDa ultrafiltration loops use double-tube heat exchangers. The PFW is chilled prior to diafiltration of the concentrated hemoglobin. The rotary lobe pumps assure low shear conditions. Pressure profiles are monitored throughout the system and controlled to less than 20 psi. The pH of all solutions is maintained in the physiological range.

Sterility is maintained throughout the process. The cross-flow filtration system is housed in a class 100 environment (Federal Standard 209E). All process tanks, piping, and filters are either steamed in place with pure steam or sanitized chemically with 0.1 N NaOH. Sanitizing solutions are rinsed from the system with PFW. The sterility of the system is verified

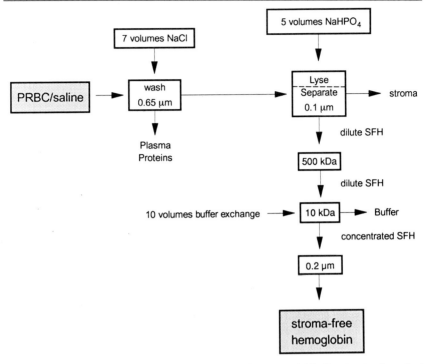

FIG. 2. Flow diagram for the production of stroma-free hemoglobin (SFH). PRBC, packed red blood cells.

prior to processing and throughout the process by sampling for pyrogens using the *Limulus* amebocyte lysate (LAL) test.[11]

All buffer solutions are filtered through 0.2-μm charged filters (Posidyne-Pall, East Hills, NY) whenever a transfer is made into a vessel to reduce the pyrogen load. The final hemoglobin solution is sterile-filtered through 0.2-μm low-protein-binding cartridges (Dominick Hunter, Troy, CA; Posidyne-Pall, or Sartorius, Hayward, CA).

Preparation of Stroma-Free Hemoglobin

A flow diagram for the production of SFH is shown in Fig. 2. Outdated human packed red blood cells (PRBCs) are obtained from Military Blood Collection Centers and Medical Centers. The packed red blood cells are used up to 4 weeks after outdate. All units have tested negative for the human immunodeficiency virus (HIV) and hepatitis B antigen.

[11] E. L. McGown and F. E. Wood, *FASEB J.* **6**, A1339 (1992).

Eighty-four units (about 20 liters) of PRBCs, type A, AB, or O, are pooled using a blood collection set (Saftifilter PLUS, Cutter Biological, Berkeley, CA) with 80 liters of normal saline in a 100-liter jacketed stainless-steel tank (T-11, Fig. 1). The RBCs are washed with 7 volumes of chilled normal saline by diafiltration at a constant volume (80 liters) through three 0.65-μm hollow fiber membrane cartridges (total surface area 69 ft^2; AG Technology, Needham, MA). To verify that plasma is removed completely, the filtrate is assayed for albumin as a marker for plasma proteins. We have found that during the wash, albumin concentrations consistently decrease from more than 2 mg/ml to undetectable levels (<2 μg/ml).

The RBCs are lysed slowly, and the stromata are removed by diafiltration with 5 volumes of chilled 10 mM NaHPO$_4$, pH 7.6, at a constant volume (80 liters) through 0.1-μm hollow fiber membrane cartridges. The filtrate of the 0.1-μm cartridge is directed to a 100-liter stainless-steel tank (T-12). It is then diafiltered at a constant volume (80 liters) through a 500-kDa membrane cartridge to remove any stroma particles. The filtrate (SFH) from the 500-kDa cartridges is then concentrated to 10 g/dl by circulation through three 10-kDa hollow fiber membrane cartridges.

The SFH solution is diafiltered with 6 volumes of Ringer's acetate, pH 7.4, and then transferred from the stainless-steel concentration tank through a 0.2-μm, 10-inch filter (Sartorius code 7 housing) to a 40-liter stainless-steel holding tank (T-21B, Mensco, Freemont, CA). Finally, the SFH product is transferred under sterile conditions into plastic bags and frozen at −80°. Alternatively, the SFH is transferred to a 70-liter bioreactor for cross-linking.

Preparation of Hemoglobin A$_0$

The bench-scale preparation of purified hemoglobin A$_0$ has been described previously.[12] For the large-scale preparation, hemoglobin A$_0$ is isolated chromatographically from SFH by means of a process-scale chromatography system (Biotage Kiloprep 250, Charlottesville, VA), using a strong anion-exchange medium (Accell QMA, Waters, Milford, MA) and an ionic strength gradient. The total process for hemoglobin A$_0$ preparation is divided into three steps: depyrogenation of the chromatography system and buffers, chromatographic separation and collection of hemoglobin A$_0$, and concentration, diafiltration, and sterile filtration of the final product.

[12] S. M. Christensen, F. Medina, R. M. Winslow, S. M. Snell, A. Zegna, and M. A. Marini, *J. Biochem. Biophys. Methods* **17**, 143 (1988).

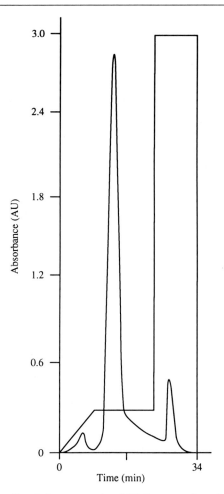

FIG. 3. Elution profile of the preparative HPLC preparation of hemoglobin A_0. The buffers and gradients are described in the text. AU, Absorbance units.

All tanks and equipment used in the process are cleaned with Pyroclean detergent (AlerChek, Portland, ME) to remove pyrogens. Contact surfaces and the column are exposed for at least 45 min in the course of a 22-hr cleaning protocol. The Pyroclean is flushed from the sytem using 1 M sodium acetate until the ultraviolet (UV) absorbance of the effluent is zero. The sodium acetate is flushed from the system with at least 5 column volumes (approximately 25 liters) of PFW. Elution buffer tanks are sprayed with Pyroclean and allowed to stand for 20 min. The tanks are then thoroughly flushed with PFW. A concentrated (3 M) Tris (Sigma Chemical, St. Louis, MO) stock solution, pH 8.1 (at room temperature), is prepared using a 10-liter glass bottle, which is baked at 180° for 4 hr. The concentrated Tris stock is depyrogenated before use with a Pyrosart filter (Sartorius, Hayward, CA).

The elution buffers are prepared by dilution of the concentrated Tris stock solution. Buffer A is composed of 4.9 liters of stock solution diluted to 210 liters with PFW, resulting in a 70 mM Tris buffer. Buffer B is composed of 1.75 liters of stock solution, 3 liters of glacial acetic acid, and 5 liters of 10 N sodium acetate diluted to 105 liters with PFW, resulting in a 50 mM Tris, 500 mM sodium acetate buffer. These solutions are chilled to 4°, elevating the final pH to the working range of 8.5 to 8.7.

The SFH sample (typically 75 ± 5 g in PFW or 10 mM phosphate) is loaded from a sterile 2-liter bag. Tris stock is added to the sample to give a final sample molarity and pH of buffer A.

The elution gradient runs from 0 to 9% buffer B in 9 min, then remains at 9% buffer B for 18 min. The column is then flushed with 100% buffer B for 10 min, and reequilibrated with 100% buffer A for 10 min. The flow rate of each run is 800 ml/min, and the total run time, injection to next injection, is 47 min. The process is monitored at 280 nm using a UV detector (Fig. 3). Fractions are collected by means of an automatic fraction collector installed internally on the system, using elution time windows as a guide. Approximately 50–60% of the total hemoglobin loaded is recovered and collected in a sterile 20-liter bag.

The hemoglobin A$_0$ is diafiltered with 6 volumes of Ringer's acetate, pH 7.3. It is concentrated subsequently to a final concentration of 7–10 g/dl with hollow fiber filters (A/G Technology, Needham, MA) with a nominal molecular weight cutoff of 10 kDa. Finally, it is sterile-filtered into 400- to 1000-ml aliquots in blood bags. These bags are frozen at −80° for future use.

Preparation of Cross-Linked Hemoglobin

The prototype cross-linking agent we have studied most extensively is bis(3,5-dibromosalicyl) fumarate (DBBF). When reacted with deoxyhemoglobin the reagent cross-links between the α chains at lysine-99 residues. The basic methods were described originally by Walder *et al.*,[13-15] Chatterjee *et al.*,[16] Snyder *et al.*,[17] and Tye *et al.*,[18] and later modified by Baxter Healthcare Corp. under contract to the U.S. Army.[19-21] The scheme adapted for our laboratory is shown diagrammatically in Fig. 4.

Twenty-five liters of 10 g/dl SFH or hemoglobin A_0 are transferred to the bioreactor, followed by 15 liters of a 13 : 1 molar equivalence of sodium tripolyphosphate (STP) to hemoglobin (tetramer) at pH 7.4. The solution is mixed slowly at 50–100 rpm.

The solution is deoxygenated at 25° using a 5.3-m^2 capillary membrane oxygenator (Bentley BOS-CM50, Irvine, CA). The N_2/O_2 exchange is accomplished by high nitrogen flow (17 liters/min) and hemoglobin flow (7 liters/min) through the oxygenator. Deoxygenation of 40 liters of hemoglobin solution typically requires 4 to 5 hr, during which time evaporation causes a volume loss of up to 10%. Deoxygenation of hemoglobin is followed spectrophotometrically by monitoring the absorbance at 758 nm. Deoxygenation is considered to be complete when the peak reaches its highest point and is stable for 0.5 hr. The headspace in the bioreactor is purged continually with N_2 at a flow rate of 7 liters/min. Temperature and pH are monitored continuously. Hemoglobin and methemoglobin concentration are monitored spectrophotometrically.

The DBBF used for these studies was the generous gift of Baxter Healthcare Corp. (Round Lake, IL) or was purchased from California State University, Los Angeles, and is suspended in 4 liters of deoxygenated PFW at a 1.9 : 1 molar equivalence to hemoglobin (tetramer). The solution is sparged with N_2 until totally deoxygenated and then transferred to the bioreactor containing the deoxygenated hemoglobin. The cross-

[13] J. A. Walder, R. H. Zaugg, R. Y. Walder, J. M. Steele, and I. M. Klotz, *Biochemistry* **18**, 4265 (1979).
[14] J. A. Walder, R. Y. Walder, and A. Arnone, *J. Mol. Biol.* **141**, 195 (1980).
[15] J. A. Walder, R. Chatterjee, and A. Arnone, *Fed. Proc., Fed. Am. Soc. Exp. Biol.* **41**, 651 (Abstr. 2228) (1982).
[16] R. Chatterjee, E. V. Welty, R. Y. Walder, S. L. Pruitt, P. H. Rogers, A. Arnone, and J. A. Walder, *J. Biol. Chem.* **261**, 9929 (1986).
[17] S. R. Snyder, E. V. Welty, R. Y. Walder, L. A. Williams, and J. A. Walder, *Proc. Natl. Acad. Sci. U.S.A.* **84**, 7280 (1987).
[18] R. W. Tye, F. Medina, R. B. Bolin, G. L. Knopp, G. S. Irion, and S. K. McLaughlin, *Prog. Clin. Biol. Res.* **122**, 41 (1983).
[19] M. K. Bechtel, A. Bagdasarian, W. P. Olson, and T. N. Estep, *Biomater. Artif. Cells Artif. Organs* **16(1–3)**, 123 (1988).
[20] T. N. Estep, M. K. Bechtel, T. J. Miller, and A. Bagdasarian, *Biomater. Artif. Cells Artif. Organs* **16**, 129 (1988).
[21] T. N. Estep, M. K. Bechtel, S. L. Bush, T. J. Miller, S. Szeto, and L. E. Webb, *in* "The Red Cell: Seventh Ann Arbor Conference" (G. Brewer, ed.), p. 325. Alan R. Liss, New York, 1989.

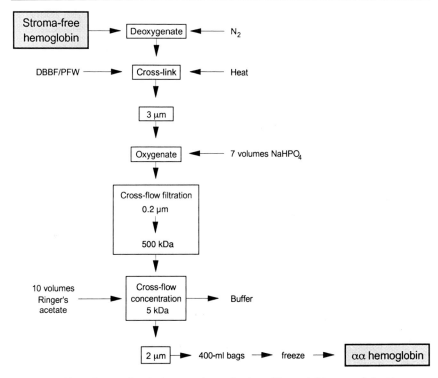

Fig. 4. Flow diagram for the large-scale production of hemoglobin cross-linked between the α chains at lysine-99; DBBF, bis(3,5-dibromosalicyl) fumarate; PFW, pyrogen-free water.

linking reaction occurs at 37° over 3 to 4 hr. The pH is maintained at 6.9 ± 0.1. The reaction is monitored by size-exclusion (tandem SEC3000 and SEC200 Ultraspherogel columns, Beckman, Fullerton, CA) and anion-exchange (POROS QH, Perspective Biosystems, Cambridge, MA) chromatography. At the conclusion of the cross-linking, the pH of the solution is adjusted to 7.37 with 1 M HEPES buffer, pH 9.0. The cross-linked hemoglobin mixture is then heated to 76° for 90 min and cooled as rapidly as possible to 4°.

The coarse non-cross-linked hemoglobin precipitate is removed by passing the solution through a 30-inch, 3.0-μm depth filter. The solution is transferred into 7 volumes of 10 mM $NaHPO_4$ buffer, pH 9.0, that has been chilled thoroughly and oxygenated.

The process continues through the cross-flow filtration system as described previously. The only deviation is that the 0.65-μm hollow fiber membrane cartridges are not used. The filtrate from the 0.1-μm cartridges is directed to the 500-kDa cartridge loop. The filtrate from the 500-kDa loop is directed to the 10-kDa concentration loop. Once the cross-linked hemoglobin has been concentrated to 7 g/dl, a buffer exchange is accomplished by diafiltration of 10 volumes of Ringer's acetate to replace the phosphate buffer. The cross-linked hemoglobin in Ringer's solution is then transferred through a 0.2-μm low-protein-binding 10-inch cartridge into a 40-liter holding tank and stored at 4° until quality control is complete. It is then sterile-filled (Wheaton Unispense II, Millville, NJ) into plastic bags and stored frozen at −80°.

Analytical Methods

Endotoxin is assayed by a kinetic turbidometric method using *Limulus* amebocyte lysate (Associates of Cape Cod, Woods Hole, MA) and a THERMOmax microplate reader with endotoxin software package (Molecular Devices Corp., Menlo Park, CA).[11] The analyses are done in accordance with Food and Drug Administration (FDA) guidelines.[22] Sodium and potassium are measured on the Cobas FARA (Roche Diagnostic System, Montclaire, NJ) using the ion-selective electrode (ISE) module. Albumin is assayed by an immunoturbidometric method using a commercial microalbumin kit (Wako Chemicals USA Inc., Richmond, VA). All other analyses carried out on the Cobas are automated versions of manual spectrophotometric assays.

To measure total organic phosphate, samples are first wet-ashed. Hemoglobin (1.0 ml) is mixed with 0.5 ml of digestion mixture in a Pyrex tube, heated to 180°, then cooled and diluted to 5.0 ml with distilled water. The digestion mixture consists of 5.0 g sodium molybdate, 75 ml H_2O, 75 ml concentrated H_2SO_4 and 100 ml 70–72% perchloric acid. The assay for inorganic phosphate in the oxidized sample is based on the method of Hurst.[23]

Oxygen affinity (P_{50}) is taken as the P_{O_2} at which the optical density change of the solution is half-maximal and is recorded using the Hemox Analyzer (TCS Medical Products Co., Philadelphia). Detailed oxygen equilibrium curves are measured according to the automated device described previously.[24] Hemoglobin concentration is measured by a modification of the Drabkin method as recommended by the International Com-

[22] "Guideline on Validation of the *Limulus* Amebocyte Lysate Test as an End-product Endotoxin Test for Human and Animal Parenteral Drugs, Biological Products and Medical Devices," 1988-201-874/83629. U.S. Govt. Printing Office, Washington, D.C., 1987.

[23] R. O. Hurst, *Can. J. Biochem.* **42**, 287 (1964).

[24] K. D. Vandegriff, F. Medina, M. A. Marini, and R. M. Winslow, *J. Biol. Chem.* **264**, 17824 (1989).

mittee for Standardization in Haematology.[25] Methemoglobin is assayed according to the method of Evelyn and Malloy.[26]

When measuring the concentration of deoxygenated STP-containing hemoglobin, it is necessary to reoxygenate the samples before analysis. The P_{50} of hemoglobin is higher in the presence of STP, and deoxyhemoglobin is thus slower to reoxygenate. Because deoxyhemoglobin has a greater extinction than does cyanomethemoglobin at 630 nm, STP-treated deoxyhemoglobin samples can yield erroneously high methemoglobin results with the Cobas. To ensure complete reoxygenation, we gently agitate each sample under a stream of oxygen for approximately 30 sec prior to methemoglobin analysis.

Results and Discussion

Two significant problems arose when large-scale hemoglobin batches were prepared: denaturation of protein[27] and endotoxin contamination. We found no evidence that ABO blood groups contributed to the purity or final biological reactivity of the solutions. Batches were prepared from only group A, AB, or O cells. The tendency to denature did not differ among these batches, nor did their biological effects when administered to animals.

Characterization of Products

The characteristics of the representative hemoglobin solutions produced in this facility are summarized in Table I. Stroma-free batches were formulated in water or phosphate buffer and used subsequently for preparation of either chromatographically pure hemoglobin A_0 or cross-linked hemoglobin. Methemoglobin concentration was routinely less than 1%, and the solutions could be stored indefinitely at $-80°$. The rabbit pyrogen test was routinely negative. Bacterial culture did not show contamination.

Inorganic phosphate concentration varies somewhat from batch to batch, depending on the persistence of the phosphate buffers used in some stages of production. However, repeated attempts to isolate organic

[25] International Committee for Standardization in Haematology, "Recommendations for Reference Method for Haemoglobinometry in Human Blood," ICSH Standard EP 6/2:1977; J. Clin. Pathol. 31, 139 (1978).

[26] K. A. Evelyn and H. T. Malloy, J. Biol. Chem. 126, 655 (1938).

[27] R. M. Winslow, K. W. Chapman, M. E. Cross, K. Dill, S. McFaul, E. McGown, R. G. Jesse, S. T. Schuschereba, and S. M. Snell, Transfusion (Philadelphia) (submitted for publication).

TABLE I
CHARACTERISTICS OF TYPICAL BATCHES OF HEMOGLOBIN SOLUTIONS[a]

Parameter (units)	Hemoglobin solution		
	Stroma free	A_0	Cross-linked ($\alpha\alpha$)
Formulation	Phosphate	RA	RA
Volume (liters)	5		16
Hemoglobin (g/dl)	7.5	6.4	9.8
Methemoglobin (%)	<1	2	7.5
P_{50} (Torr, 37°, pH 7.4)	—	12.0	30.0
n (Hill)	—	3.0	2.31
P_i (μg/ml)	42	<1	<1
Na^+ (mM)	—	—	150
K^+ (mM)	—	—	3
Cl^- (mEq/liter)	—	—	137
Ca^{2+} (mEq/liter)	—	—	8.4
Osmolarity (mOsm)	—	—	285
Viscosity (cP, 3 rpm)	—	—	8
pH	7.45	7.56	7.56
Sterility	Pass	Pass	Pass
Pyrogen	Negative	Negative	Negative
Endotoxin (EU/ml)	0.02–0.10	0.08–0.10	0.1–0.2
Free iron (μg/ml)	2.8	—	4.2
FPLC			
A_0 (%)	98.6	—	3
Sodium dodecyl sulfate gel			
> 32 kDa (%)	—	—	7
32 kDa (%)	—	—	58
16 kDa (%)	100	—	35
RP-HPLC			
β	53.3	—	46
α	46.7	—	5
$\alpha\alpha$	—	—	49

[a] EU, Endotoxin units; FPLC, fast protein liquid chromatography (Pharmacia, Piscataway, NJ); RP-HPLC, reversed-phase high-performance liquid chromatography; RA, Ringer's acetate.

phosphates by methanol extraction have failed to demonstrate even traces of phospholipids in the final products. Rigorous searches for membrane components have shown no contamination.[28]

Standard chromatographic analysis of the products showed that the solutions contained no unexpected proteins. Isoelectric focusing also showed the material to be fairly homogeneous and yielded only the ex-

[28] B. J. M. Thevenin and S. B. Shohet, this volume [2].

pected bands of hemoglobin and its oxidation products. The reversed-phase HPLC chromatograms of cross-linked hemoglobin revealed that there is about 5 non-cross-linked α chains in a typical batch (see Table I).

Analysis of oxygen equilibrium curves of cross-linked hemoglobin[29] revealed, under physiological conditions (37°, pH 7.4), a Hill's parameter of 2.31–2.4 and a P_{50} of 30 Torr. The P_{50} of the stroma-free hemoglobin under these conditions is 15.5 Torr.

The system described produces relatively large (20–30 liters) lots of purified hemoglobin products for biological testing. The facility is capable of producing injection-grade materials, although our intention is not to achieve full Good Manufacturing Practices compliance (Code of Federal Standards 211).

It is difficult to specify how "pure" a hemoglobin solution needs to be in order to carry out and interpret biological experiments: Hemoglobin contained within red blood cells is already quite pure according to most criteria. In general, there are three types of impurities that must be considered: residual red cell stromal elements, endotoxin, and undesired hemoglobin products. We take the low endotoxin levels in solution to indicate that the solutions are of sufficient purity to permit biological testing. When stored frozen at $-80°$, the SFH, A_0, and cross-linked hemoglobin appear to be quite stable with regard to oxidation.

Our only routine measurement of stromal contamination is the total phosphate assay. Extraction of phospholipid from a hemoglobin solution is extremely difficult. Because of the cost and labor involved in these assays, we have used instead the much simpler total phosphate assay. Although quite sensitive, it does not discriminate the source of the phosphate. Intracellular enzymes constitute a second group of red cell contaminants.[12] We do not know at present whether it is necessary to remove all red cell enzymes from the final product.

The most important problem in the large-scale purification or production of hemoglobin or its derivatives is protein denaturation.[27] Hemoglobin is exquisitely sensitive to the effect of dilution, pressure, shear, temperature, and pH changes. In our experience, these factors can lead to problems involving oxidation, precipitation, heme loss, and subunit dissociation. Thus, it is extremely important that all process steps should be at uniform (and low) temperature, that agitation should not be vigorous, that pH changes should not be abrupt, and that pressures should not be allowed to reach high levels.

Endotoxin removal from hemoglobin solutions can be a vexing problem, because endotoxin is known to bind to many proteins. Previous work

[29] V. W. Macdonald and R. M. Winslow, *J. Appl. Physiol.* **72**, 476 (1992).

from this laboratory[30] indicated that endotoxin and hemoglobin might have a specific interaction, and experiments [unpublished data, (1991)] have attempted to measure the heat of binding of these two macromolecules. Measurement of endotoxin in hemoglobin solutions also can be frustrating. The different types of assays and *Limulus* lysates available may give differing results. We have found the kinetic turbidimetric assay to be the most reliable, and it furnishes the most information about the reaction.

Measurement of endotoxin in hemoglobin solutions is complicated by the fact that hemoglobin interacts with the assay system. The *Limulus* amebocyte lysate assay for bacterial endotoxin is based on the initiation of the LAL coagulation cascade by endotoxin. In the kinetic turbidimetric method, the coagulation process is monitored photometrically as the developing clot scatters light. Hemoglobin causes a concentration-dependent shortening of the time of onset of clot formation.[11] The effect is detectable in the presence of 0.3% SFH and is considerably more severe in cross-linked (DBBF) preparations. Standard curves prepared in water are not appropriate for hemoglobin solutions because they can result in overestimation of endotoxin content by an order of magnitude or more. For reliable results, it is necessary to prepare standard curves containing the same protein at the same concentration and in the same solution conditions as the test samples. The task is considerably simplified in the microplate format because of the large number of samples that can be analyzed concurrently.

Our experience with the large-scale production of hemoglobin solutions for research in blood substitutes has demonstrated that increasing production from bench to pilot scale is complicated by several potential difficulties. Chief among these is denaturation and precipitation of hemoglobin, and endotoxin contamination. The oxidation of both unmodified and chemically modified hemoglobin solutions can be kept at acceptable levels if solutions are scrupulously sterile and denaturing conditions are avoided. Time limits for storage of our solutions at $-80°$ seem to be indefinite.[31]

[30] C. T. White, A. J. Murray, D. J. Smith, J. R. Greene, and R. B. Bolin, *J. Lab. Clin. Med.* **108,** 132 (1986).
[31] G. L. Moore, A. Zegna, M. E. Ledford, J. A. Huling, and R. M. Fishman, *Artif. Organs* **16,** 513 (1992).

[2] Detection of Red Cell Membrane Components in Human Hemoglobin Preparations

By BERNARD J.-M. THEVENIN and STEPHEN B. SHOHET

Introduction

The principle for the detection of membrane contamination in hemoglobin preparations is straightforward: dense membrane particulate components can be separated easily from dilute hemoglobin solutions by density sedimentation. Following separation, detection of membrane components is readily achieved by polyacrylamide gel electrophoresis of the sedimented material, followed by Coomassie blue staining to reveal a protein profile typical of the erythrocyte membrane. If greater sensitivity is desired, silver staining may be substituted for Coomassie blue staining with up to a 100-fold increase in sensitivity. If absolutely necessary, further sensitivity and specificity can be achieved by transfer of the gel onto nitrocellulose and immunostaining with an antibody specific for a membrane protein such as spectrin or band 3.

Method

Red cell membranes are prepared as described in Bennett.[1] Different amounts of total membrane protein are solubilized in sodium dodecyl sulfate (SDS) and submitted to gel electrophoresis on a linear 6–15% SDS–polyacrylamide gel according to Laemmli.[2] The gel lanes are either stained with Coomassie brilliant blue R-250[3] or with silver.[4]

To test for red cell membrane components in a hemoglobin preparation, a 6-ml sample is centrifuged in a TLA 100.3 rotor at 4° for 10 min at ~500,000 g after addition of EDTA to 1 mM and phenylmethylsulfonyl fluoride (PMSF) to 40 μg/ml. A small pellet is obtained from unmodified hemoglobin, and an even smaller pellet from cross-linked hemoglobin.[5] The pellet and the supernatant solution are solubilized in SDS and submitted to gel electrophoresis on a 7.5% SDS–polyacrylamide gel according to Laemmli.[2] The gel is stained with Coomassie blue.

---

[1] V. Bennett, this series, Vol. 96, p. 313.
[2] U. K. Laemmli, *Nature (London)* **227,** 680 (1970).
[3] K. Weber, J. R. Pringle, and M. Osborn, this series, Vol. 26, p. 3.
[4] J. Heukeshoven and R. Dernick, *Electrophoresis (Weinheim, Fed. Repub. Ger.)* **6,** 103 (1985).
[5] R. M. Winslow and K. W. Chapman, this volume [1].

Examples of Typical Results

Figure 1 shows typical Coomassie blue-stained (Fig. 1A) or silver-stained (Fig. 1B) SDS–polyacrylamide gel patterns of various amounts of a red cell membrane preparation. Lanes 1 through 6 (Fig. 1) contain, respectively, 16, 2.7, 0.9, 0.9, 0.3, and 0.1 µg of total membrane protein.

Figure 2 shows a Coomassie blue-stained SDS–polyacrylamide gel of a sample of approximately 100 µg from a preparation of cross-linked hemoglobin (lane 1, Fig. 2)[5] or from the supernatant of a preparation of hemoglobin after ultracentrifugation at 500,000 g (lane 2, Fig. 2); the corresponding pelleted material from ~6 g of the hemoglobin preparation is shown in lane 3 (Fig. 2). No membrane contamination was detectable in these samples.

Figure 3 presents a scanning electron micrograph of the pellet from original hemolysate showing abundant membrane structures and material with only an amorphous pattern.

Discussion

It is comparatively easy to detect red cell membrane or stromal contamination in hemoglobin preparations because membrane fragments or vesicles can be sedimented in the ultracentrifuge from hemoglobin solutions as concentrated as 10 g per 100 ml. Detection of this sedimented membrane contamination with very high sensitivity is then possible using polyacrylamide gel electrophoresis and various staining techniques. As little as 1 µg of total membrane protein can be detected with the Coomassie blue stain; if fastidiously performed, a silver stain can detect as little as 0.05 µg of total membrane protein. Approximately 5 mg of membrane protein is in 1 ml of packed red blood cells, so theoretically this conventional procedure could detect contaminating membranes from as little as 10 nl of red cells, or approximately 100,000 cells. If it is necessary to rule out contamination even below this limit, the technique of Western blotting, for example, with an antispectrin antibody, is suitable for the detection of 0.1 ng of spectrin or membranes from ~1000 cells.[6]

There are some potential pitfalls to the determination of membrane contamination, especially when anticipating the highest sensitivity. These involve primarily the unwanted contamination of bacteria or other proteins in the system and the confounding effects of proteolysis of the membrane proteins at these very low levels. The first of these problems can usually be controlled by prompt, careful handling of the samples, and, if necessary,

[6] H. Towbin, T. Staehelin, and J. Gordon, *Proc. Natl. Acad. Sci. U.S.A.* **76**, 4350 (1979).

A   B

1   2   3   4   5   6

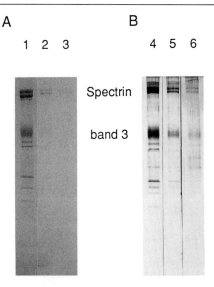

Spectrin

band 3

FIG. 1. SDS–polyacrylamide gel electrophoresis of red cell membrane preparations. Gels are stained with either Coomassie blue (A) or silver (B).

1   2   3

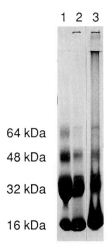

64 kDa

48 kDa

32 kDa

16 kDa

FIG. 2. SDS–polyacrylamide gel electrophoresis analysis of cross-linked hemoglobin. One hundred micrograms of hemoglobin from a cross-linked preparation is shown in lane 1. The supernatant (4 μl) and pellet sedimented from 6 ml of a different sample of 25% cross-linked hemoglobin are shown in lanes 2 and 3, respectively.

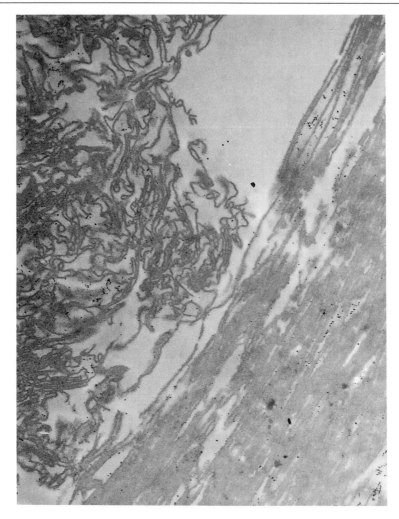

FIG. 3. Scanning electron micrograph of a pellet from a hemolysate before extensive purification. Note the presence of abundant membrane structures as well as amorphous material.

by the use of azide or antibiotic inhibitors of bacterial proliferation. The second problem can be minimized by the use of common protease inhibitors such as EDTA and PMSF.

Finally, the protein pattern of the red cell membrane, with its characteristic high molecular weight (220K–260K) doublet of α- and β-spectrin, is so typical that residual membranes are unlikely to be confused with other particulate contamination (see Fig. 1). Nevertheless, if absolute specificity is required, immunostaining with antispectrin or antiband 3 antibodies as well as morphologic examination in the electron microscope for the typical membrane debris morphology (Fig. 3) can be called on to define unequivocally the nature of the contamination.

[3] Detection of Globin Chains by Reversed-Phase High-Performance Liquid Chromatography

By BRUNO MASALA and LAURA MANCA

Introduction

Reversed-phase high-performance liquid chromatography (RP-HPLC) has greatly influenced and stimulated research in detection, quantitation, and identification of normal and abnormal globin chains in human newborn and adult red blood cells. In fact, RP-HPLC methodology has a number of advantages: it is fast and accurate, minute amounts of material are sufficient for each determination, complete automation is possible, and, most importantly, identification of electrophoretically "silent" mutant hemoglobins due to neutral-to-neutral amino acid substitutions is feasible and reliable.[1-4] It also greatly facilitates the separation and isolation of even microquantities of a mutant globin chain, and of globin proteolytic fragments, for subsequent characterization of the substitution within the abnormal peptide.[1-6] This is possible by means of HPLC techniques for the determination of the amino acid composition at the picomole or subnanomole level,[7,8] as well as microsequencing techniques,[9] fast atom bom-

---

[1] W. A. Schroeder, in "The Hemoglobinopathies" (T. H. J. Huisman, ed.), p. 143. Churchill-Livingstone, Edinburgh, 1986.

[2] F. Kutlar, A. Kutlar, and T. H. J. Huisman, J. Chromatogr. 357, 147 (1986).

[3] T. H. J. Huisman, J. Chromatogr. 418, 277 (1987).

[4] B. Masala and L. Manca, Biophys. Chem. 37, 225 (1990).

[5] F. Kutlar, Y. J. Fei, J. B. Wilson, A. Kutlar, and T. H. J. Huisman, J. Chromatogr. 394, 333 (1987).

[6] J. B. Wilson, H. Lam, P. Pravatmuang, and T. H. J. Huisman, J. Chromatogr. 179, 271 (1979).

[7] B. A. Bidlingmeyer, S. A. Cohen, and T. L. Tarvin, J. Chromatogr. 336, 93 (1984).

[8] K. L. Stone and K. R. Williams, J. Chromatogr. 359, 203 (1986).

[9] J. Y. Chang, this series, Vol. 91, p. 455.

bardment mass spectrometry (FAB mapping),[10,11] electrospray mass spectrometry, and capillary electrophoresis (see elsewhere in this volume).

Different HPLC procedures have been described. Some use cation-exchange systems[12,13] whereas others use the reversed phase with different columns, such as Waters μBondapack C_{18}, Merck LiChrospher 100 CH-8/2, Aquapore RP-300, and Partisil C_{18}.[14-21] These differ markedly in the mobile phase or in the time required for the analysis, which in some cases is labor intensive. Some of these procedures enable the separation of the glycine- and alanine-containing γ chains (the $^G\gamma$ and $^A\gamma$ chains, respectively), which are responsible for the heterogeneity of fetal hemoglobin (Hb F).[14-19] The separation of these chains, which differ only by a methyl group, and the separation of their variants are among the major goals in the study of many aspects of Hb synthesis in perinatal life, and in syndromes characterized by elevated Hb F levels, such as different β-thalassemias (β-thal) and hereditary persistence of fetal hemoglobin (HPFH).

The most significant advance in this field was due to Shelton et al.,[22] who showed that the use of large-pore columns with short hydrocarbon chains (C_4), and simple acetonitrile–water–trifluoroacetic acid gradients, enable the separation and quantitation of the individual globins in a single and relatively fast chromatographic run. With a number of modifications, this basic principle is widely adopted in studies of human globin chain polymorphism in normal and abnormal conditions. Changes in the original procedure also enable the separation of several nonhuman globins.[23]

[10] P. Pucci, P. Ferranti, G. Marino, and A. Malorni, Biomed. Environ. Mass Spectrom. 18, 20 (1989).
[11] G. Marsh, G. Marino, P. Pucci, P. Ferranti, A. Malorni, J. Kaeda, J. Marhs, and L. Luzzatto, Hemoglobin 15, 43 (1991).
[12] S. O. Brennan, Hemoglobin 9, 53 (1985).
[13] H. Wajcman, J. Kister, M. Marden, B. Bohn, Y. Blouquit, J. Descamps, M. Goudemand, C. Poyart, and F. Galacteros, Biochim. Biophys. Acta 1096, 60 (1991).
[14] L. F. Congote, Blood 57, 353 (1981).
[15] L. F. Congote and A. G. Kendall, Anal. Biochem. 123, 124 (1982).
[16] T. H. J. Huisman, C. Altay, B. Webber, A. L. Reese, M. E. Gravely, K. Okonjo, and J. B. Wilson, Blood 57, 75 (1981).
[17] T. H. J. Huisman, B. Webber, K. Okonjo, A. L. Reese, and J. B. Wilson, in "Advances in Hemoglobin Analysis" (S. M. Hanash and G. J. Brewer, eds.), p. 23. Alan R. Liss, New York, 1981.
[18] J. B. Shelton, J. R. Shelton, W. A. Schroeder, and J. De Simone, Hemoglobin 6, 451 (1982).
[19] L. Leone, M. Monteleone, V. Gabutti, and C. Amione, J. Chromatogr. 321, 407 (1985).
[20] V. Baudin and H. Wajcman, J. Chromatogr. 299, 495 (1984).
[21] H. Wajcman, A. Mrad, Y. Blouquit, C. Parmentier, J. Riou, and F. Galacteros, Am. J. Hematol. 32, 294 (1989).
[22] J. B. Shelton, J. R. Shelton, and W. A. Schroeder, J. Liq. Chromatogr. 7, 1969 (1984).
[23] W. A. Schroeder, J. B. Shelton, J. R. Shelton, V. Huynh, and D. B. Teplov, Hemoglobin 9, 461 (1985).

Procedure

The separation of the α- and non-α-globins that are constituents of the Hb tetramer is based on the application of an increasingly hydrophobic environment to the chromatographic column, which is filled with an interacting lipophilic stationary phase. Because of the presence of a constant concentration of trifluoroacetic acid (TFA) in the developers, a pH of 2.0–2.5 is created. Under these conditions, the heme groups are removed from the tetramers, and the different constituent globins dissociate. The order of elution is therefore primarily based on the hydrophobicity of the individual chains, the faster being the less hydrophobic: β, δ, α, $^G\gamma$, and $^A\gamma$. Posttranslationally modified globin subfractions, or genetic variants, are separated as well.

Instrumentation

The basic instrumentation consists of two pumps with a gradient programmer, a sample injector, the chromatographic column, a variable wavelength absorbance monitor (or a single-beam UV monitor equipped with a 220-nm interference filter), a dual-channel recorder, a fraction collector (in case the globins should be further analyzed), and an integrator or a computer (PC)-based chromatography data system. Equipment is also available to automate the application of multiple samples.

The chromatographic column is the analytical Vydac large-pore C_4 column, 4.6 × 250 mm (Cat. #214TP54), manufactured by the Separations Group (P.O. Box 867, Hesperia, CA 92345). It is packed with spheroidal 5-μm silica beads with an average pore size of 300 Å. The same column can also be purchased from Perkin-Elmer (Norwalk, CT; Cat. #N930-1005). Similar columns are supplied by Applied Biosystems (Foster City, CA; the Aquapore BU-300, Cat. #0712-0036) and by Waters-Millipore (Bedford, MA; the Delta-Pak C_4, Cat. #WAT011794), the latter being 3.9 × 150 mm in size. A guard column is not necessary, provided that HPLC-grade reagents are used and that Hb solutions are filtered prior to injection (as described under Sample Preparation).

Larger columns for semipreparative or preparative purposes are available. Examples of these are the C_4 Vydac 10 × 250 mm (Cat. #214TP510) and the larger C_4 Hi-Pore RP304 2.15 × 30 cm (Bio-Rad, Richmond, CA). Several preparative C_{18} columns are also available. Their ability to separate globin chains is often not as satisfactory as that of the analytical ones. Their use is not described here because, as discussed later in detail, the analytical column is also useful for these purposes.

Reagents

Doubly distilled water is recommended to prevent UV-absorbing impurities. In our laboratory high-purity deionized water with a resistance of >15 MΩ per cm is obtained by the Milli-Q system (Millipore), which is fed with a supply of reversed-osmosis purified tap water. Acetonitrile of HPLC quality is necessary and can be purchased from numerous chemical suppliers. Trifluoroacetic acid of reagent grade does not need to be purified.

Sample Preparation

Blood samples collected in anticoagulant are diluted with 4–5 volumes of saline (0.9% NaCl) and washed three times by centrifugation. Lysates are prepared by mixing a volume of the washed red cells with an equal volume of distilled water and 0.4 volumes CCl_4, with occasional stirring. The mixture is centrifuged at 20,000 g for 10 min at 4°, and the clear Hb solution, floating over the stromata, is collected using a pipette.

An optimal separation occurs when a sample containing 80-120 µg globin is injected. Accordingly, Hb concentration should be determined on all samples to be run. After having read the optical density (OD) of the clear Hb solution at 540 nm, the general formula that should be used to obtain the solution Hb concentration is as follows: OD_{540} × 1.14 × dilution = µg/µl, where 1.14 is constant. Depending on the Hb concentration of the blood sample, the collected stroma-free lysate should have a concentration of 8–12 g/dl. Usually 8–12 µl of the above lysate, diluted to 200 µl with developer A (see Preparation of Developers), should have a final concentration of ~0.5 g/dl of Hb, which is equivalent to ~100 µg of globin in 20 µl to be injected. The solution is next filtered to remove particulates. Excellent results are obtained using the syringe-connected 0.45-µm HV filters purchased from Millipore (Cat. #SJHV004 NS).

Quantities ranging from 30 to 200 µg are also suitable, the lower amount being sufficient when the number of different chains in the lysate is expected to be low. The final volume to be injected is not critical because large dilutions (up to 200 µl) have relatively little effect on the pattern of development.

Lysates are used as fresh as possible or are stored at −80°. It is not possible to store Hb solution at 4° for longer than 3–4 weeks. Poor separations or spurious peaks are often observed in outdated blood samples or improperly stored lysates.

Treatment of Diluted Hemoglobin Samples and Globin Powder

It may be necessary to analyze for globin composition from diluted Hb solutions collected after any low- or high-pressure chromatographic isolation procedure. In this case, the solution is dialyzed free of salts and then concentrated by means of the Millipore Centrifugal Ultrafree 10,000 NMWL polysulfone membrane (Cat. #UFC2TGC02). Centrifugation with a swinging rotor producing 1000–2500 g for 50-ml tubes is used as the driving force for filtration.

Heme-depleted globin chains, prepared as acetonic powder,[24] are simply dissolved in water in order to obtain the final 100 μg/15–20 μl concentration and filtered as above.

Preparation of Developers

In individual measuring flasks prepare:
Developer A
    200 ml Acetonitrile and 1 ml TFA, then add water to 1000 ml
Developer B
    600 ml Acetonitrile and 1 ml TFA, then add water to 1000 ml
Pour the developers into the pumping bottles marked A and B, respectively, sonicate for 3 min, and degas with a low stream of helium for 2–3 min.

Developers may be used over a period of 1 week. Sparging with helium should be done each working day and, for a few seconds, between runs. Because developers are volatile, ensure an equal extent of sparging in both developer bottles. Different evaporation patterns for TFA, as well as unbalanced concentration of the developers, may cause baseline drift during recording and reduce reproducibility.

Development of Chromatogram and Recording

The equipment consists of two pumps programmed with a controller, the Vydac C_4 column, an injector, a recorder, and a UV detector. The basic procedure is useful as a primary test on normal or abnormal adult and newborn lysates. Modifications of this procedure for special purposes will be described when appropriate.

After connecting the developer A bottle with the HPLC pump A, and the developer B bottle with pump B:

1. Purge the analyzer at a flow rate of 2–3 ml/min for 5 min with 50% developer B plus 50% developer A with the column disconnected.
2. Stop the pumps and connect the column with the analyzer. With the gradient programmer fixed at 50% developer B plus 50% developer A a flow rate of 1 ml/min should be reached in 2 min.

[24] F. Ascoli, M. R. Rossi Fanelli, and E. Antonini, this series, Vol. 76, p. 72.

3. Connect the recorder, regulate the absorbance monitoring of the effluent at 220 nm and the absorbance units full scale (AUFS) at 0.5, then regulate the chart speed at 20 cm/hr.

4. After having reached the 1 ml/min flow rate as in step 2, equilibrate the column for at least 10 min with the same 50% developer A plus 50% developer B.

5. Reduce the baseline of the recording to 0.0–0.02 AUFS within these 10 min, otherwise continue to equilibrate.

6. Program an 80-min linear gradient from 50 to 60% developer B (with a corresponding decrease in developer A from 50 to 40%) at the same 1 ml/min flow rate.

7. Inject the sample, start the gradient and develop. After completion of the development, continue, at 1 ml/min flow rate, as follows:

8. Run a purging 5-min gradient to reach 90% developer B.

9. Then return to 50% developer B with a 15-min gradient.

10. Reequilibrate until the original absorbance baseline is obtained for at least 10 min. After this, a new chromatography may be started following steps 6 to 10 above.

All procedures are carried out at room temperature, though a thermostatting column device may be necessary if notable temperature variations occur in the laboratory. For periods longer than 2 weeks, the column should be stored in 100% acetonitrile.

Interpretation

Normal Adult Chromatogram

Figure 1A shows the chromatogram of the globin chains in a normal adult lysate. The elution pattern shows the appearance of the most hydrophilic heme group at about 10 min, followed in 22–24 min by a pre-β^A peak, in 28–33 min by the β chain of Hb A, in 34–36 min by the δ chain of the minor Hb A_2, and in 40–44 min by the α chain. The pre-β^A peak consists of a posttranslationally modified protein, presumably containing a β-globin–glutathione adduct as its major component.[25] The small peaks eluting between the δ and α chains are nonglobin proteins.[25]

Normal Newborn Chromatogram

Figure 1B shows the chromatogram of the globin chains in a normal cord blood sample. The $^G\gamma$ and $^A\gamma$ chains of the main Hb F_0 elute com-

[25] W. A. Schroeder, J. B. Shelton, V. Huynh, and J. R. Shelton, *Hemoglobin* **10,** 239 (1986).

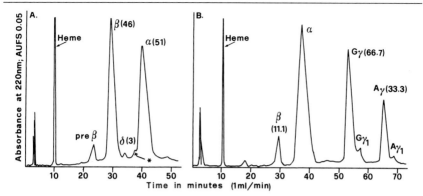

FIG. 1. Normal chromatograms. (A) Adult lysate; (B) newborn lysate. The asterisk indicates nonhemoglobin proteins. The percentage of each globin is in parentheses. In newborn lysate the percentage of β chain refers to the level of Hb A; the percentage of $^{G}\gamma$ and $^{A}\gamma$ chains refers to the composition of Hb F.

pletely separated in 53–57 and 65–69 min, respectively. These chains, which are responsible for the heterogeneity of Hb F, are the products of two nonallelic genes, located on the short arm of chromosome 11 (in the $-^{G}\gamma-^{A}\gamma-$ arrangement) together with the other members of the non-α-globin gene family.[26] The two chains differ solely at residue 136, the $^{G}\gamma$ having a glycine residue and the $^{A}\gamma$ having alanine at that position. As discussed later, the considerably increased retention time of the more hydrophobic $^{A}\gamma$ chain with respect to the $^{G}\gamma$ represents a demonstration as to how the C_4 column works in separating proteins differing in their hydrophobicity. Their $^{G}\gamma_1$ and $^{A}\gamma_1$ subcomponents elute just behind the major chains. These subcomponents, which are present in the minor Hb F_1 fraction, are consistent with enzymatically and posttranslationally acetylated γ-globins.[25] A normal newborn has 70–90% Hb F, 8–12% of which is represented by Hb F_1. The β and α chains should have the same retention time as in an adult sample.

The separation and quantitation of the α-like embryonic ζ chain in normal newborns and in α-thal carriers is also possible.[5,27,28] To do this, continue the development of the chromatogram by programming (in step 6 above) a further linear gradient from 60 to 80% developer B in 60 min

[26] J. L. Slightom, A. E. Blechl, and O. Smithies, *Cell (Cambridge, Mass.)* **21**, 627 (1980).
[27] Y. J. Fei, F. Kutlar, H. F. Harris, II, M. M. Wilson, A. Milana, P. Sciacca, G. Schilirò, B. Masala, L. Manca, Ç. Altay, A. Gurgey, J. M^a. de Pablos, A. Villegas, and T. H. J. Huisman, *Hemoglobin* **13**, 45 (1989).
[28] T. H. J. Huisman, F. Kutlar, and L.-H. Gu, *Hemoglobin* **15**, 349 (1991).

at 1 ml/min flow rate. Next, inject and proceed as in steps 8 to 10. The ζ chain will emerge after about 130 min.

Quantitation and Normal Values in Adult and Newborn

Quantitation of the result requires the calculation of the relative percentages of α and non-α chains. This is done by measuring the peak area of the individual chains. The molecular extinction coefficients of globins are essentially the same at 220 nm. Thus, calculation and comparison of the areas should provide an α/non-α ratio close to unity. Calculation is automatic if the apparatus is equipped with an integrator. Some PC-based software also offers the possibility of acquiring data from different chromatography detectors and a number of data processing and reprocessing facilities. Planimetry is also rapid and accurate.

The percentage of β chain in adults is calculated by the following formula: (pre-β + β) × 100/total chains. In normal adult humans, having 97–98% Hb A, 2–3% Hb A_2, and traces of Hb F, the β chain percentage is about 45–48%.

The percentage of δ chain is equivalent to the percentage of the minor Hb A_2. The average value in normal adult humans is 3.8 ± 0.3% (range 3.2–4.3%).[2] The accuracy of this calculation for diagnostic purposes is, however, less than that obtained by the most successful microchromatographic method of Huisman et al.,[29] or by anion- or cation-exchange HPLC of Hb.[3]

In the newborn, the percentage of β chain is calculated by the following formula: β × 100/total non-α chains. The newborn has about 20% Hb A and 80% Hb F.

In the newborn, as well as in conditions characterized by elevated Hb F levels, the determination of the $^G\gamma$ and $^A\gamma$ chain percentages in the Hb F mixture is of particular importance.

The normal β-chain level obtained on 1000 newborn babies is 20.1 ± 5.5% (range 12.0–34.9%). The composition in terms of γ chain of their 79.9 ± 5.5% Hb F is as follows: $^G\gamma$, 68.9 ± 3.0% (range 59–78%); $^A\gamma$, 31.1 ± 3.0% (range 22–41%).[27,28] These values are indicative of a higher expression of the $^G\gamma$ gene in perinatal life. $^G\gamma$ or $^A\gamma$ chain may be present in abnormal amounts, suggesting the presence of abnormal γ-globin gene arrangements (as discussed under Normal Globin Chains in Abnormal Globin Gene Arrangements).

The level of ζ chain calculated for several normal newborns from different countries and ethnic backgrounds gave values averaging 0.19%

[29] T. H. J. Huisman, W. A. Schroeder, A. N. Brodie, S. M. Mayson, and J. Jakway, J. Lab. Clin. Med. 83, 657 (1975).

(range 0–1.4%), higher values being indicative of abnormal α-globin gene arrangements.[28]

The accuracy of the procedure is excellent. The standard deviation in sets of triplicate assays on the same lysate is within ±0.8 to ±2.2%, depending on the amount of globin injected and their relative level, the baseline of the recorder, etc.

Comments

Under the above-described conditions of development, normal and several abnormal chains can be separated readily. The elution of the different chains is highly reproducible, although retention time may depend on the composition of the developers, the condition of the column, the number of performed runs, the temperature, etc. The complete separation of the globin chains is achieved by increasing from 50 to 60 the percentage of developer B in 80 min. This is equivalent to a 0.125%/min increase of developer B and a 40 to 44% increase of the acetonitrile concentration (i.e., 0.05%/min). Slight variations in the composition of the developers have the effect of altering the globin retention time but not the general elution pattern.[1-4,22] This behavior represents the basic concept to be considered when the separation of a mixture of different globin chains is attempted, because small modifications of the slope of the gradient might considerably improve the separation of a variant polypeptide. When the presence of a variant chain is suspected, it is thus advisable to reduce the increase in developer B (i.e., to 0.09%/min) while increasing the time of development. An example of this will be given in the section on detection of caprine globin chains.

If globins from adult lysates without detectable amounts of Hb F have to be studied, a gradient that saves time and solvents is 50 to 57% developer B in 56 min. Using this gradient, the increase of developer B is the usual 0.125%/min. A higher increase will have the same effect, although the presence of a variant chain may escape observation.

It is evident by comparing many results from the literature that the reproducibility of results from column to column, from equipment to equipment, and from laboratory to laboratory is excellent. In our experience, no significant differences in retention time were observed when using two Vydac columns of different lots in two completely different HPLC equipments, one having a high-pressure and one a low-pressure gradient system. The same column may be used for a large number of analyses with reasonable resolution of peaks. Its life span is increased by the use of high-purity reagents and filtered lysates. It is advisable to reach the running flow rate slowly (in at least 2 min), as well as to apply a 10-

to 15-min gradient whenever it is necessary to purge the column with high concentrations of developer B or to reequilibrate from purging. When a broadening of peaks is observed, or back pressure is higher than usual due to accumulation of particulates at its top surface, the column may be turned upside down to reverse the direction of flow for several hours and then reversed to its original direction. As many as 1500 chromatograms have been run on the same column in our laboratory. The same has been reported by others.[1,3] A guard column may be used to improve the life span, because C_4 silica packing material is available from the column manufacturers.

The amount of globin (30–50 μg) that is necessary for a reliable develop- ment is very small, so that blood collected in a microhematocrit tube or on filter air-dried paper may be used, providing a careful lysate preparation is followed. This may be useful for large testing programs on cord blood samples.

Economic and simple to use C_4 cartridges are commercially available. It is expected that their use would give the same results.

Detection of Abnormal Globin Chains

By April 1991, a total of 585 variant hemoglobins had been described.[30] Most are due to missense mutations leading to single amino acid substitu- tions, whereas a few show more than one substitution; others are deter- mined by globin gene fusion, small deletions, or in-phase insertions, or extensions of the N- or C-terminal regions. From 10 to 20 new variants are described each year.

Separation by RP-HPLC is greatly dependent on the hydrophobicity of the polypeptide. Thus the successful separation of a variant chain depends on the differences in polarity and pK_a value of the amino acid residue introduced into the chain with respect to that deleted from it. Huisman[3] proposed a ranking of the 20 amino acids on the basis of size, pK values at pH <3 (as is the case for the method described here), and increasing polarity. The rank, shown in Table I, suggests that the substitution of a residue by one that precedes it (being more hydrophobic) will increase retention time, whereas the substitution by one that follows (being less hydrophobic) will decrease retention time. Substitutions such as Val → Met little affect retention time because of the nearly equal polarities of the residues involved. As many as 100 different variants described in the literature since 1984 support the validity of the principle,

---

[30] International Hemoglobin Information Center, *Hemoglobin* **15**, 139 (1991). An up-to-date list of Hb variants is published yearly in the third issue of *Hemoglobin* journal.

TABLE I
RANKING OF AMINO ACIDS BASED ON pK VALUES AT pH
<3, POLARITIES, AND SIZES[a]

Rank	Amino acid	Rank	Amino acid
1	Trp	11	Gly
2	Phe	12	Asp
3	Ile	13	Glu
4	Leu	14	Asn
5	Tyr	15	Cys
6	Pro	16	Gln
7	Met	17	Ser
8	Val	18	His
9	Thr	19	Lys
10	Ala	20	Arg

[a] Adapted with permission from Huisman.[3]

indicating that this rank is a helpful guide in the search for substitutions in any globin chain. In a few variants with the Glu → Lys substitution, some with Pro replaced by Thr, Ser, or Arg, and some in which Ala or Thr are replaced by Pro, the observed retention time did not correspond with that expected, indicating that the location of the substitution within the polypeptide could play a prominent role in the chromatographic separation.[3,4]

Figure 2 shows the globin chains of three electrophoretically slow-moving hemoglobins and a silent Hb variant in adult and newborn heterozygotes. In Fig. 2A and B are the lysates of an adult and a newborn, respectively, carriers of the frequently observed Hb S ($\beta^{6\ Glu\to Val}$) and Hb C ($\beta^{6\ Glu\to Lys}$). The β^S chain elutes just behind the β^A, whereas the β^C elutes before it. In both cases elution of the abnormal chain follows the rule suggested by Table I. Figure 2C shows the separation of the abnormal chain of the Hb Lepore–Baltimore, which is one of the three Lepore hemoglobins having a $\delta\beta$ hybrid chain (with a crossover between positions $\delta50$ and $\beta86$).[31] Retention time of the hybrid chain suggests that the structure of the amino-terminal segment of the δ chain, having the Ser → Thr, Thr → Asn, Glu → Ala, and Thr → Ser substitutions at positions 9, 12, 22, and 50 ($\beta \to \delta$), is responsible for its reduced hydrophobicity. Figure 2D shows the separation of the β chain of the silent Hb Hamilton ($\beta^{11\ Val\to Ile}$), which we have found with a particularly high frequency among

[31] B. Masala, L. Manca, J. B. Wilson, B. B. Webber, A. Kutlar, and T. H. J. Huisman, *Hemoglobin* 14, 241 (1990).

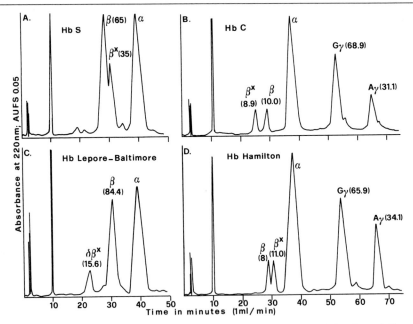

FIG. 2. Separation of variant β-globin chains in heterozygote adults (A and C) and newborns (B and D). The percentage of each globin is in parentheses.

Sardinians.[32-35] This is a typical example of the HPLC detection of a mutant chain that does not change the electrophoretic behavior of the tetramer. The Val → Ile substitution increases retention time.

Figure 3 shows the separation of two α- and two γ-chain variants. In Fig. 3A and B are the Hb G-Philadelphia ($\alpha^{68\ Asn \rightarrow Lys}$) and the Hb J-Sar-

[32] Y. Hattori, F. Kutlar, S. S. Chen, T. H. J. Huisman, P. Demuro, M. Formato, L. Manca, and B. Masala, *Biochem. Genet.* **24,** 669 (1986).

[33] L. Manca, M. Formato, B. Masala, D. Gallisai, and M. Orzalesi, *Hemoglobin* **11,** 161 (1987).

[34] B. Masala, *Hemoglobin* **16,** 331 (1992).

[35] *In vitro* amplification of the $\beta^{Hamilton}$ globin gene showed a guanine-to-adenine transition at the first base of codon 11 [L. Manca, E. Cocco, G. F. Cossu, and B. Masala, *Biochim. Biophys. Acta* **1139,** 17 (1992)], which confirms the Val → Ile substitution detected at the protein level [S. C. Wong, M. A. M. Ali, H. Lam, B. B. Webber, J. B. Wilson, and T. H. J. Huisman, *Am. J. Hematol.* **16,** 47 (1984)]. This observation also supports that the $\beta^{Hamilton}$ mutation is at the level of one of the five methylated CpG dinucleotides that are conserved in the β-globin gene. While further indicating a high mutational rate at that level, the result explains the surprisingly high 1 : 500 prevalence of this mutation among Sardinians [L. Manca, M. Formato, B. Masala, D. Gallisai, and M. Orzalesi, *Hemoglobin* **11,** 161 (1987)].

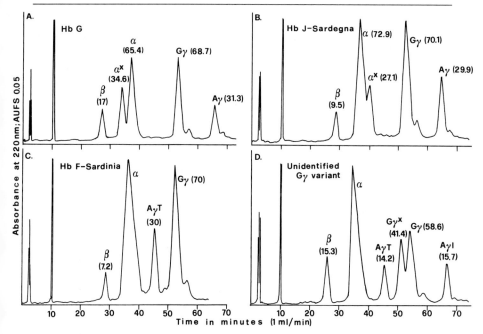

FIG. 3. Separation of variant α- and γ-globin chains in lysates from newborns. The two lower chromatograms refer (C) to a newborn homozygote for the $^{A}\gamma^{T}$ variant chain and (D) to a newborn who is also the carrier of a $^{G}\gamma$ chain variant.

degna ($\alpha^{50\ His\rightarrow Asp}$), respectively. The former is a slow-moving Hb in electrophoresis at alkaline pH and relatively cathodal to Hb A in isoelectric focusing (IEF), whereas the latter is fast moving and relatively anodal.[36,37] Figure 3C and D shows the globin chains of a newborn homozygote and of a heterozygote for the $^{A}\gamma$ globin variant, which is a constituent of the Hb F-Sardinia ($^{A}\gamma^{75\ Ile\rightarrow Thr}$).[30] The Ile → Thr substitution highly reduces retention time of this mutant chain. Better known as the $^{A}\gamma^{T}$ chain (T for Thr at position 75, whereas the normal chain may be thus considered as the $^{A}\gamma^{I}$, I for Ile at the same position), it is present in nearly all populations of the world. Its gene frequency varies from 0.023 in blacks from Ghana to 0.230 in caucasians from Yugoslavia, suggesting a Mediterranean origin of the mutation.[28] Detection and quantitation of this polymorphism may be of importance. Apart from the relevance for genetic and anthropological studies, the high frequency of the $^{A}\gamma^{T}$ allele makes its product a suitable marker for use in evaluating the γ-gene expression being *in cis* or *in trans*

[36] L. Manca, P. Demuro, and B. Masala, *Clin. Chim. Acta* **177,** 231 (1988).
[37] L. Manca and B. Masala, *Hemoglobin* **13,** 33 (1989).

of a particular gene in studies on Hb F synthesis and defects thereof.[38,39] Huisman *et al.*[28] suggested that the level of the $^G\gamma$ chain at birth is not significantly affected by the presence of this anomaly. The occurrence of six $^A\gamma^T$ variants with an additional mutation has been observed.[3,28] The chromatogram in Fig. 3D shows also the presence of an unidentified $^G\gamma$ chain variant.

Quantitation of Abnormal Globin Chains

Generally the percentage of an abnormal β chain is about 50% that of the normal β chain. Lower values are indicative of structural instability of the mutant chain or of inability to associate with the normal counterpart to form tetramers (as is the case with the β^S chain). Similarly, the synthesis under the control of a defective (thalassemic) gene, as well as reduced efficiency in any of the steps in protein synthesis, could affect production of the abnormal chain. The β^{Lepore} chain is an example of a variant produced by a defective gene.

Because the α-globin gene is duplicated, the amount of an α-chain variant should be 20–30% in the heterozygote carrier and 40–60% in the homozygote. Higher or lower percentages are indicative of coinherited α-thal genes or of unstable or undersynthesized globins, respectively. Some mutated α genes (such as the $\alpha^{G-Philadelphia}$) are often linked to an α-thal gene *in cis*.[36]

In the heterozygote newborn, the amount of a $^G\gamma$ variant is about 35% and that of an $^A\gamma$ variant is about 15%. As for other variants, lower values are observed in the case of decreased stability or when the structural gene is defective. Higher values may be observed because of the linkage with a γ-globin gene deletion *in cis*.[28]

Recovery of Globins for Structural Analyses

The use of volatile developers greatly facilitates the recovery of a mutant chain. This is of particular importance for the subsequent characterization of the amino acid substitution.

The eluate containing the abnormal chain is collected in microcentrifuge tubes and evaporated to dryness. This is better achieved in equipment such as the heated Speed Vac Concentrator 100H (Savant Instruments, Farmingdale, NY) connected to a refrigerated condensation trap and a vacuum pump. Lyophilization is also useful. The amount of the recovered

[38] B. Masala, L. Manca, M. Formato, and A. Matera, *Am. J. Hematol.* **21**, 367 (1985).
[39] B. Masala, M. Formato, L. Manca, P. Demuro, D. Gallisai, F. Dore, and M. Longinotti, *Acta Haematol.* **76**, 208 (1986).

component depends on its nature, β-chain variants in adult samples being 15–25% and α-chain variants being 10–18% of the total injected. The 10–50 μg of recovered material, of the 100–200 μg injected, may be sufficient for structural analyses depending on the sensitivity of the procedures. Increasing the sample size excessively, to obtain a larger amount, may result in unfavorable separations and skewing of peaks. It is suggested that whenever a large amount is required several chromatograms be run until sufficient material has been accumulated. Although this procedure is time-consuming, its cost is lower than that of a single run with a highly expensive preparative column.

Detection of Globin Chains of Low-Level Hemoglobins

Quantitation and isolation of globins that are constituents of a low-level Hb fraction are cumbersome and tedious. These procedures require the isolation of the minor Hb by macrochromatography[40] or HPLC,[3] determination of its concentration, and analysis. Contamination by the high-level Hbs may also hinder accurate evaluation and isolation. This is the case for the Hb A_2 variants, which amount 1–2% of the total Hb, and of variants produced under the control of thalassemic genes. Also difficult is the study of the composition of the low percentage of Hb F that is present in the normal adult (often less than 1%) in those affected by syndromes that slightly increase the Hb F level and in patients with a transfusion-dependent β-thal, in which the Hb F is diluted by the Hb A of the donor.

A simple and reliable approach to the determination of the γ chain of low Hb F levels combines IEF to isolate the tetramer with RP-HPLC of globins of the recovered material.[41] The procedure may be applied to investigate genotype/phenotype relationships, to isolate any low-level Hb, and to study abnormal Hbs having globins with the same retention time as normal globins. Any of the different IEF procedures described elsewhere[38,41–43] may be used to isolate Hb F. Lysates, 8–12 g/dl in Hb, are made 0.05% in KCN by addition of small amounts of a 2–3% KCN solution. Lysates are submitted to IEF in slab gels containing ampholytes, which create the suitable 6.7–7.7 pH range. Immediately after focusing, the Hb F band is detached from the support (without fixing and staining)

[40] W. A. Schroeder and T. H. J. Huisman, "The Chromatography of Hemoglobin." Dekker, New York, 1980.
[41] L. Manca and B. Masala, *Hemoglobin* **14**, 517 (1990).
[42] B. Masala and L. Manca, *Clin. Chim. Acta* **198**, 195 (1991).
[43] P. G. Righetti, E. Gianazza, A. Bianchi-Bosisio, and G. Cossu, in "The Hemoglobinopathies" (T. H. J. Huisman, ed.), p. 47. Churchill-Livingstone, Edinburgh, 1986.

with a blade or a spatula and transferred into a microcentrifuge tube. Hb is eluted in a few hours with, depending on the volume of the recovered gel, 150–250 μl of deionized water. The eluate is recovered, filtered, and injected as described. Depending on the Hb concentration of the recovered eluate, the monitoring of the HPLC effluent is from 0.01 to 0.05 AUFS. The hydrophilic ampholytes, eluted from the gel together with Hb, emerge at the solvent front. It is important to proceed with the chromatography within 24 to 36 hr after focusing in order to avoid duplication of peaks or degradation of chains.

Clearly, the initial level of Hb F in whole lysate is essential in determining the recovery of γ chains and the accuracy of calculation. A 15–20% enrichment (as determined by the $\gamma \times 100/\beta + \gamma$ ratio) can be obtained using samples with less than 1%, and a 70–90% enrichment using samples with 6–7% Hb F. A 45-μl lysate (12 g/dl in Hb) sample from an individual with 1% Hb F contains an absolute amount of about 50 μg Hb F. This is usually enough for one HPLC run monitored at a medium–high sensitivity, in spite of the fact that the Hb A contamination may be relevant. Depending on the IEF apparatus, higher amounts of Hb F may be recovered. Because the Hb F_1 fraction is focused away from the γ chains, the recovered γ chains are virtually free of their acetylated subfractions. This simplifies quantitation.

The level of $^G\gamma$ chain in normal adults, without significant alterations of the γ-globin gene structure, seems to vary considerably and is associated with specific chromosomal characteristics.[3] It may range 10–55%, although in some cases it may be as for the newborn (60–75%).

Applications

This method is useful in detecting γ chain composition in any condition characterized by elevated Hb F levels, such as heterozygous β-thal and HPFH due to base changes in the promoter region of the $^G\gamma$- or the $^A\gamma$-globin genes. The example in Fig. 4A refers to a carrier of the cytosine-to-thymine substitution at -158 to the $^G\gamma$ gene linked to the codon 39 nonsense β-thal mutation (CAG $\rightarrow$ TAG). The result indicates a 71% production of the $^G\gamma$ chain because of the presence of the substitution in the promoter of the $^G\gamma$ gene.[27,28]

The example in Fig. 4B, refers to a carrier of the $^A\gamma$-HPFH due to a guanine-to-adenine substitution at -117 to the $^A\gamma$ gene.[44] This syndrome is characterized by 10–15% levels of Hb F of the $^A\gamma$ type. HPLC of enriched Hb F, while supporting the view that the γ gene, which increases

[44] K. G. Yang, T. A. Stoming, Y. J. Fei, S. Liang, S. C. Wong, B. Masala, R. B. Huang, Z. P. Wei, and T. H. J. Huisman, *Blood* **71**, 1414 (1988).

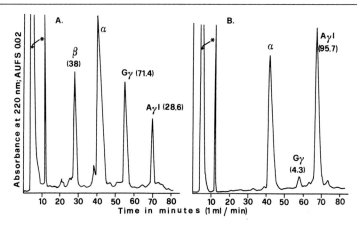

FIG. 4. Separation of globin chains of Hb F recovered by isoelectric focusing gel. As deduced by the percentage of β chain, the recovered Hb F was (A) 62% and (B) nearly 100%. The asterisk denotes the peak of the most hydrophilic ampholytes.

its expression in this syndrome, is the normal $^A\gamma^I$, indicates that some activity of the unaffected $^G\gamma$ gene(s) is also present.

This method has been extensively adopted in the study of the γ-chain heterogeneity of several transfusion-dependent β-thal homozygotes.[45-47] β-Thal is a heterogeneous group of disorders characterized by the partial (β^+-) or total (β^0-) impairment of the β-chain synthesis due to a wide variety of mutations (see Huisman[48] for a complete list of mutations and references). In most cases, the β-chain synthesis is replaced by the synthesis of different amounts of Hb F. The study of Hb F composition in these syndromes is thus of importance because it may suggest the genetic condition causing the thalassemic abnormality. In the Sardinian population (as well as in others), the same β^0 mutation may be linked to either the $-^G\gamma-^A\gamma^I-$ (30% of cases) or the $-^G\gamma-^A\gamma^T-$ (70% of cases) arrangement, whereas most of the β^+ mutations are linked to the $-^G\gamma-^A\gamma^I-$.[39,49] β-Thal homozygotes at birth and in perinatal life show the same $^G\gamma/^A\gamma$ globin chain ratio as the normal newborn, whereas in adult life nontransfused patients show a ratio that clearly appears to be dependent on specific

[45] L. Manca, E. Cocco, D. Gallisai, B. Masala, and J. G. Gilman, *Ann. N.Y. Acad. Sci.* **612,** 485 (1990).

[46] L. Manca, E. Cocco, B. Masala, and J. G. Gilman, *Am. J. Hematol.* **35,** (1990).

[47] L. Manca, E. Cocco, D. Gallisai, B. Masala, and J. G. Gilman, *Br. J. Haematol.* **78,** (1991).

[48] T. H. J. Huisman, *Hemoglobin* **14,** 661 (1990).

[49] B. Masala, L. Manca, D. Gallisai, A. Stangoni, K. D. Lanclos, F. Kutlar, K. G. Yang, and T. H. J. Huisman, *Hemoglobin* **12,** 661 (1988).

TABLE II
γ-Chain Composition of Low Hb F Levels of Severely
Affected β^0-Thalassemia Homozygotes

Number	$^{G}\gamma$ (%)	$^{A}\gamma^{I}$ (%)	$^{A}\gamma^{T}$ (%)
8	52.8 ± 2.9	47.2 ± 2.8	—
24	56.0 ± 3.2	24.4 ± 2.9	19.6 ± 2.5
26	62.2 ± 3.9	—	37.8 ± 3.9

chromosomal characteristics.[49] As shown in Table II, the amount of the $^{A}\gamma^{T}$ globin in severely affected patients examined immediately before a blood transfusion (in order to minimize the possible effect of the donor Hb F) is found to be 20% lower than the $^{A}\gamma^{I}$. A 4-base pair deletion has been found in the promoter region of the $^{A}\gamma^{T}$ gene, which could account for its reduced expression.[45–47] Similar results have been obtained in non-transfusion-dependent patients.[50] It has been suggested that the 4-base pair deletion might be present in the promoter of the $^{A}\gamma^{T}$ gene also in the absence of the β-thal mutation *in cis*.[31]

Normal Globin Chains in Abnormal Globin Gene Arrangements

Detection of globins in newborn samples using RP-HPLC provides information about the occurrence of abnormal γ-globin gene arrangements and expression. As shown by thousands of analyses by several laboratories,[28] normal healthy babies without detectable Hb variants have $^{G}\gamma$ and $^{A}\gamma$ chain percentages ranging between 60–70 and 40–30%, respectively, as the result of the normal $-^{G}\gamma$-$^{A}\gamma$- arrangement on both chromosomes. Significant deviations from these values are indicative of γ-globin gene rearrangements, as directly demonstrated by application of DNA technology.[32,51–53] These arrangements are due to different unequal crossing-over events. Examples are babies with low $^{G}\gamma$ (<50%) or high $^{G}\gamma$ (>80%) levels. These abnormalities may be summarized as follows: the $-^{A}\gamma$-$^{A}\gamma$- and the $-^{G}\gamma$-$^{G}\gamma$- arrangements replacing the $-^{G}\gamma$-$^{A}\gamma$-, the deletion of a γ-globin gene, γ-globin gene triplications, quadruplications, and quintuplications. A few babies have been described as having no $^{G}\gamma$ or $^{A}\gamma$ chains.[52]

[50] L. Manca, D. Gallisai, B. Masala, and J. G. Gilman, *Blood* **72**, 66a (1988).

[51] T. H. J. Huisman, Y. J. Fei, and F. Kutlar, *Hemoglobin* **12**, 699 (1988).

[52] T. Harano, K. Harano, M. Ukita, Y. Wada, A. Hayashi, Y. Ohba, T. Miyaji, F. Kutlar, and T. H. J. Huisman, *Hemoglobin* **12**, 723 (1988).

[53] L. Manca, B. Masala, M. Orzalesi, H. J. Huang, and T. H. J. Huisman, *Hemoglobin* **12**, 741 (1988).

TABLE III
PERCENTAGE OF GLOBIN CHAINS DETECTED IN ABNORMAL γ-GLOBIN
GENE ARRANGEMENTS

Arrangement	Composition (%) of Hb F			β chain (%)
	$^G\gamma$	$^A\gamma^I$	$^A\gamma^T$	
$-^A\gamma^I-^A\gamma^T-$	35.6 ± 3.4	47.6 ± 4.1	17.1 ± 2.6	14.3 ± 5.9
$-^G\gamma-^G\gamma-$	87.4 ± 3.0	12.6^a ± 3.0		19.7 ± 6.7
$-^{GA}\gamma-$	38.3 ± 5.4	61.7^a ± 5.4		34.7 ± 4.6
$-^G\gamma-^{AG}\gamma-^A\gamma-$ or $-^G\gamma-^G\gamma-^A\gamma-$	84.6 ± 2.6	15.4^a ± 2.6		10.6 ± 5.3

^{a} In these arrangements the $^A\gamma$ chain may be of the $^A\gamma^I$ or of the $^A\gamma^T$ type.

As shown in Table III, baby carriers of the $-^A\gamma-^A\gamma-$ arrangement have a low (35–40%) $^G\gamma$ level with normal amounts of β chain. All of them also have about 17% $^A\gamma^T$ and 48% $^A\gamma^I$, thus indicating the duplication being *in cis* to the $^A\gamma^T$ gene (i.e., the $-^A\gamma^I-^A\gamma^T-/-^G\gamma-^A\gamma^I-$ arrangement).[32,53] Babies with the $-^G\gamma-^G\gamma-$ duplication have about 87% $^G\gamma$ and normal levels of β chain. Babies with the $-^{GA}\gamma-$ arrangement have about 38% $^G\gamma$ and increased levels of β chain (35%). This arrangement arose from two types of crossovers between the $^G\gamma$ and the $^A\gamma^I$ genes or between the $^G\gamma$ and the $^A\gamma^T$ genes. Thus its hybrid protein product may be an $^A\gamma^I$ or an $^A\gamma^T$ chain, depending on the $^A\gamma$ allele involved and the position of the crossover.[51,52] As indicated by the high β-chain level, this arrangement has to be considered as a γ-thal. Babies with γ-globin gene triplications ($-^G\gamma-^{AG}\gamma-^A\gamma-$ or $-^G\gamma-^G\gamma-^A\gamma-$) that are considered the counterpart of the above deleted arrangement have 85% $^G\gamma$ and reduced β chain levels, suggesting an overproduction of γ chain under the control of the triplicated γ genes. γ-Globin gene quadruplications ($^G\gamma^G\gamma^G\gamma^A\gamma/^G\gamma^A\gamma$) have been described in Japanese newborns and in Turks,[52,54] and quintuplication ($^G\gamma^G\gamma^G\gamma^G\gamma^A\gamma/^G\gamma^A\gamma$) has been found in blacks.[27,55] All have high $^G\gamma$ levels.

Globin quantitation in these abnormal conditions suggests interesting considerations concerning globin gene expression.[28,52,53,56] It is likely that the $^G\gamma$ level in carriers of the hybrid $^{GA}\gamma$ gene, which is higher than in carriers of the $-^A\gamma^I-^A\gamma^T-$ duplication, depends on the presence of the more active $^G\gamma$ promoter in its 5' block. Similarly, the $^G\gamma$ level in the $-^G\gamma-^{AG}\gamma-^A\gamma-$ arrangement is lower than in the $-^G\gamma-^G\gamma-$ because of the presence of

[54] K. G. Yang, J. Z. Liu, F. Kutlar, A. Kutlar, Ç. Altay, A. Gurgey, and T. H. J. Huisman, *Blood* **68,** 1394 (1986).
[55] Y. J. Fei, K. D. Lanclos, F. Kutlar, E. L. Walker, III, and T. H. J. Huisman, *Blood* **72,** 827 (1988).
[56] K. Harano, T. Harano, F. Kutlar, and T. H. J. Huisman, *FEBS Lett.* **190,** 45 (1985).

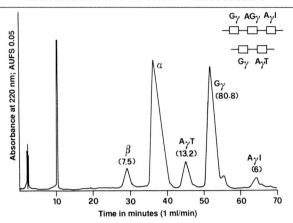

FIG. 5. Separation of globins from a newborn carrier of a γ-globin gene triplication. The chromosomal arrangement of globin genes is also schematized. This may be suggested by the percentages of globins and indicates an $^A\gamma^I$ gene *in cis* to the triplication and an $^A\gamma^T$ gene *in trans* to it.

the $^A\gamma$ promoter in the inserted hybrid $^{AG}\gamma$ gene. The results shown in Fig. 5 suggest that the $^A\gamma^I$ gene is located *in cis* to a triplicated arrangement and that its activity is appreciably decreased (60–70% lower) as compared with the corresponding value in the normal -$^G\gamma$-$^A\gamma$- arrangement. This interpretation is possible due to the presence of the $^A\gamma^T$ marker allele, which is *in trans* to the abnormal arrangement and produces normal levels of $^A\gamma^T$ chain. In quadruplicated and quintuplicated arrangements the activity of the 3' located gene may be even lower.[56]

Globin Chain Synthesis

The excellent resolution of chains provides a suitable tool for globin biosynthesis studies.[15,57,58] These are of importance in detecting carriers of different thalassemic genes, particularly those that do not appreciably alter the hematological phenotype.

[^{3}H]Leucine-labeled chains are prepared as described,[59] separated by the procedure described here, and 1-ml fractions of the eluate are collected and mixed with 10 ml of the scintillation cocktail. Fractions are counted in a liquid scintillation counter, and radioactivity in each peak is determined. Calculation of the α/β ratio makes it possible to assess the presence of

[57] B. P. Alter and D. D. Stump, *Hemoglobin* **11**, 341 (1987).
[58] S. Rahbar and Y. Asmerom, *Hemoglobin* **13**, 475 (1989).
[59] J. B. Clegg and D. J. Weatherall, this series, Vol. 76, p. 749.

α- or β-thal genes or of an unstable variant chain. This method has been extensively used on fetal blood samples for prenatal diagnosis of β-thal.[57]

Detection of Nonhuman Hemoglobins

An extensive study by Schroeder *et al.*[23] shows that slight modifications to the gradient used for human globins permit the separation, quantitation, and recovery of nonhuman chains from 16 species, including common mammalian and avian species. This may be of particular interest because Hb polymorphism is often noticeable in some vertebrates. Separation is achieved simply by adapting the basic procedure to the properties of the globin mixture to be separated. As a general rule, the procedure for human chains should be used when studying a species for the first time, then modifications to the slope of the gradient may be introduced to better separate chains.

Detection of Caprine Globin Chains

Owing to the presence of different allelic and nonallelic α- and β-globin genes, the Hb polymorphism of caprines is quite complex. Five Hbs (A, B, E, G, and H) have been described in domestic sheep (*Ovis aries*),[60,61] four (A, D, D^{Malta}, and E) in goats (*Capra hircus*),[60] three (A, B, and M) in mouflons (*Ovis musimon*),[60,62,63] and two in Barbary sheep (*Ammotragus lervia*),[60] all of them being due to β-globin alleles. To these, specific fetal and embryonic Hbs have to be added, and several species also have an additional nonallelic β^{C} gene, which is linked to the β^{A} but not to the β^{B} gene.[64] The β^{C} gene is responsible for the reversible production of the Hb C, at the expense of the Hb A synthesis, under physiological stimuli such as anemia or hypoxia.[60] Several α-chain variants also contribute to the complication of the polymorphism of caprine Hbs.

As shown in Fig. 6, normal and abnormal globins of a newborn goat are separated by means of a 48–51% developer B gradient in 35 min, followed by a 51–63% developer B gradient in 70 min, both at the 1 ml/min flow rate. We have found this gradient to be the most useful in separating many caprine globin chains.[65] The inset in Fig. 6 shows that the

[60] M. D. Garrick and L. M. Garrick, in "Red Blood Cells of Domestic Mammals" (N. S. Agar and P. G. Board, eds.), p. 165. Elsevier, Amsterdam, 1983.
[61] L. Kilgour, S. C. Dixon, and E. M. Tucker, *Anim. Genet.* **21,** 115 (1990).
[62] S. Naitana, S. Ledda, E. Cocco, L. Manca, and B. Masala, *Anim. Genet.* **21,** 67 (1990).
[63] B. Masala, L. Manca, E. Cocco, S. Ledda, and S. Naitana, *Comp. Biochem. Physiol. A* **100A,** 675 (1991).
[64] K. J. Garner and J. B. Lingrel, *J. Mol. Evol.* **28,** 175 (1989).
[65] L. Manca, B. Masala, S. Ledda, and S. Naitana, *J. Chromatogr.* **563,** 158 (1991).

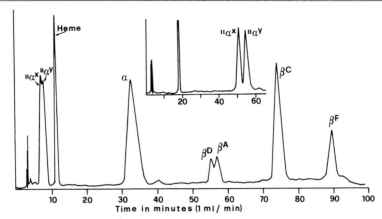

FIG. 6. Separation of globin chains of a newborn goat carrier of two β globins (β^A and β^D) and of two unidentified α-chain variants ($^{II}\alpha^X$ and $^{II}\alpha^Y$). The inset shows the separation of these two variants by means of a shallow gradient.

two most hydrophilic, unidentified α-chain variants, eluting unseparated before the heme group, may be completely resolved by two shallow gradients. The first was 40–42% developer B in 20 min to elute the heme group, the second was 42–47.5% developer B in 44 min to separate the two variants.

Figure 7 shows globins of an anemic sheep of the Hb AB phenotype that is the carrier of three allelic α chains. These are the normal $\alpha^{113\ \text{Leu}}$ chain, and the variants $^I\alpha^{8\ \text{Ala}\ 113\ \text{Leu}}$ and $^{II}\alpha^{113\ \text{His}}$ that are present in different amounts as the result of the particular arrangement and expression of the

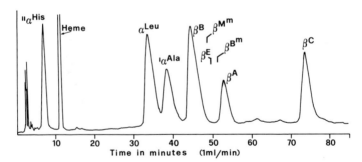

FIG. 7. Separation of globin chains of a newborn sheep carrier of two β globins (β^A and β^B) and of two common α-chain variants ($^{II}\alpha^{113\ \text{His}}$ and $^I\alpha^{8\ \text{Ala}\ 113\ \text{Leu}}$) together with the normal $\alpha^{113\ \text{Leu}}$. The positions of the sheep β^E variant and of mouflon β^{B^m} and β^{M^m} chains are also indicated.

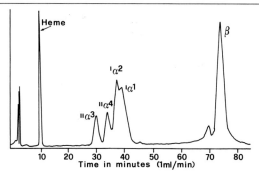

FIG. 8. Separation of the four different α-globin chains of the Italian river buffalo (*Bubalus bubalis*) of the Hb AB phenotype.

sheep α-globin gene complex.[66] The reduced level of the β^A chain with respect to the β^B, which corresponds to the appearance of the β^C chain, is an indication of β^A–β^C-globin gene linkage.

This procedure has been used to assess the presence of the two new β^B chain variants in sheep (temporarily termed the β^E-Sardegna) and in mouflon (the β^M chain).[62,63,65] Retention times of these chains are also indicated in Fig. 7. The mouflon α chain has the same retention time as sheep $\alpha^{113\ \text{Leu}}$.

Detection of Bovine Globin Chains

The same gradient has been used in separating the complex mixture of globins of the Italian river buffalo (*Bubalus bubalis*) and in elucidating the arrangement of its α-globin gene complex.[67] Two different α chains (the $^I\alpha^1$ and the $^{II}\alpha^3$) are resolved in animals described as being homozygotes for the Hb AA phenotype, and two different α chains (the $^I\alpha^2$ and the $^{II}\alpha^4$) are resolved in the Hb BB phenotype. The four chains are found in AB heterozygote animals as shown in Fig. 8. This result is indicative of an α-globin gene complex in which two linked α-globin genes encode different proteins at different levels of efficiency in both the AA and the BB phenotypes. Interpretation of this arrangement, which is very similar to that of horses,[68] had not been possible before using other chromatographic techniques that did not allow the separation of chains having

[66] R. Vestri, E. Pieragostini, F. Yang, P. di Gregorio, A. Rando, and P. Masina, *Br. J. Haematol.* **77,** 110 (1991).

[67] A. Di Luccia, L. Iannibelli, P. Ferranti, L. Manca, B. Masala, and L. Ferrara, *Biochem. Genet.* **29,** 421 (1991).

[68] J. B. Clegg, S. E. Y. Goodbourn, and M. Braend, *Nucleic Acids Res.* **12,** 7847 (1984).

neutral-to-neutral substitutions. The $^{I}\alpha^{1}$ and the $^{II}\alpha^{3}$ chains of the Hb AA river buffalo have been isolated by RP-HPLC and their amino acid sequences have been determined by the FAB overlapping technique.[69,70] Two substitutions distinguish the two chains at positions 130 and 132: $^{I}\alpha^{1}$ has Phe and Asn, respectively, whereas $^{II}\alpha^{3}$ has Leu and Ser. This explains the decreased retention time of the $^{II}\alpha^{3}$ chain. Attempts to separate a newly discovered β chain variant[71] have been unsuccessful.

The Vydac C_4 column has been used to separate two α and two β chains in the snake *Liophis miliaris*.[72]

Conclusions

Apart from the biochemical, physiological, and molecular relevance of studies on new variant Hbs, careful and continued determination of the γ-chain composition of Hb F in the newborn, in (apparently) normal adults, and in syndromes characterized by elevated Hb F levels will certainly continue to extend knowledge of Hb synthesis defects. These findings, in turn, will contribute to the identification of DNA segments as being part of the controlling mechanisms of globin genes and to knowledge of how mutations influence globin gene expression. Future studies may also contribute to elucidation of the largely unknown mechanisms responsible for the switch in the $^{G}\gamma/^{A}\gamma$ chain ratio that occurs after birth. There seems no doubt that HPLC will continue to be most useful in this field of biochemical analysis. Moreover, this methodology may greatly facilitate studies of the Hb polymorphism of vertebrates and the determination of the amino acid sequence of globin chains.

[69] P. Ferranti, A. Malorni, G. Marino, P. Pucci, A. Di Luccia, and L. Ferrara, *Int. J. Mass Spectrom. Ion Processes* **111,** 287 (1991).

[70] P. Ferranti, A. Di Luccia, A. Malorni, P. Pucci, M. Ruoppolo, G. Marino, and L. Ferrara, *Comp. Biochem. Physiol. B* **101B,** 91 (1992).

[71] A. Di Luccia, L. Iannibelli, E. Addato, B. Masala, L. Manca, and L. Ferrara, *Comp. Biochem. Physiol. B* **99B,** 887–892 (1991).

[72] M. S. A. Matsuura, K. Fushitani, and A. F. Riggs, *J. Biol. Chem.* **264,** 5515 (1989).

[4] Structural Characterization of Hemoglobin Variants Using Capillary Electrophoresis and Fast Atom Bombardment Mass Spectrometry

By PASQUALE FERRANTI, ANTONIO MALORNI, and PIETRO PUCCI

Introduction

The determination of abnormal hemoglobins requires investigation of the primary structure of the mutant protein and consists of, in general, a number of sequential analytical and preparative steps. The preliminary requirement to detect abnormal hemoglobins is a fast and sensitive technique capable of detecting the variant and providing useful information for structural characterization. Because the presence of the variant is often associated with mild or severe disease, simple identification methods are needed for clinical diagnosis and genetic counseling. This points out the need for fast and high-resolution techniques to characterize variants.

Over the past years, capillary zone electrophoresis (CZE) of proteins has developed into an analytical technique of considerable power.[1-5] Electrophoretic separations are performed in very small diameter (typically 20–100 μm i.d.) capillary tubes; the small capillary dimensions permit the application of extremely high electric fields (as high as 0.5–1 kV/cm) and yield very rapid and high-resolution separations compared to most electrophoretic methods. Furthermore, CZE combines the instrumental advantages of high-performance liquid chromatography (HPLC) such as detection and quantification with an improved separating power.[6,7]

Globin chains obtained by precipitation of erythrocyte hemolysate in cold acetone can be directly analyzed by CZE in coated capillaries without any prior treatment; each globin migrates at a characteristic rate that depends on charge density and varies with pH and the composition of the background electrolyte, so that even similarly charged hemoglobins are quickly differentiated.[8]

[1] J. W. Jorgenson and K. D. Lukacks, *Science* **222**, 266 (1984).

[2] K. A. Cobb and M. Novotny, *Anal. Chem.* **61**, 2226 (1989).

[3] F. Foret and P. Bocek, *Adv. Electrophor.* **3**, 271 (1989).

[4] H. Ludi and E. Gassmann, *Anal. Chim. Acta* **213**, 215 (1988).

[5] R. M. McCormick, *Anal. Chem.* **60**, 2322 (1988).

[6] P. D. Grossmann, K. J. Wilson, G. Petrie, and H. H. Lauer, *Anal. Biochem.* **173**, 265 (1988).

[7] R. D. Smith, J. A. Olivares, N. T. Nguyen, and H. R. Udseth, *Anal. Chem.* **60**, 436 (1988).

[8] P. Ferranti, A. Malorni, P. Pucci, S. Fanali, A. Nardi, and L. Ossicini, *Anal. Biochem.* **194**, 1 (1991).

For further structural characterization, CZE enzymatic maps of globin chains can be performed so that the variant can be identified by direct comparison of the map with that of the corresponding normal globin chain. These maps complement the reversed-phase high-performance liquid chromatography (RP-HPLC) tryptic mapping systems that are widely used in variant analysis. Map information not only plays a role in the detection, but is also a valid tool for structural identification, particularly in combination with analysis by fast atom bombardment mass spectrometry (FAB/MS).

In FAB/MS, sample ions are directly generated from a liquid solution by bombardment with a high-energy beam of particles (xenon atoms or cesium ions) and mass analysis is carried out using instruments capable of measuring masses up to 12–15 kDa at full sensitivity. Double-focusing magnetic sector mass spectrometers offer, at present, a wide range of capabilities making them ideally suited for structural analysis of peptides and proteins. Based on FAB/MS analysis, a strategy to produce peptide maps of proteins has been developed and termed FAB mapping.[9]

When the FAB mapping procedure is applied to the analysis of hemoglobinopathies, the general strategy consists in the construction of the peptide maps of α, β, and γ normal chains by FAB/MS analysis of the respective tryptic digests.[10] Then, the FAB map of the variant globin is compared to that of the normal chain. The modified peptide(s) exhibit(s) unusual molecular weights, which means that the globin fragment in which the variations took place can be identified.

Sometimes, the mass difference is enough to identify the modification; in other cases, simple protein chemistry procedures (Edman degradation steps, carboxypeptidase B hydrolysis, subdigestion with other proteolytic agents, etc.) carried out on the same peptide mixture and followed by FAB/MS analysis are needed to identify unambiguously the site and the nature of the modification. This strategy has been successfully applied to the analysis of several hemoglobin variants.[11-14]

[9] H. R. Morris, M. Panico, and G. W. Taylor, *Biochem. Biophys. Res. Commun.* **117**, 299 (1983).
[10] P. Pucci, P. Ferranti, G. Marino, and A. Malorni, *Biomed. Environ. Mass Spectrom.* **18**, 20 (1989).
[11] R. De Biasi, D. Spiteri, M. Caldora, R. Iodice, P. Pucci, A. Malorni, P. Ferranti, and G. Marino, *Hemoglobin* **12**, 323 (1988).
[12] P. Pucci, P. Ferranti, A. Malorni, and G. Marino, *Adv. Mass Spectrom.* **11**, 1428 (1989).
[13] F. Frigeri, G. Pandolfi, A. Camera, B. Rotoli, P. Ferranti, A. Malorni, and P. Pucci, *Clin. Chem. Enzymol. Commun.* **3**, 289 (1990).
[14] A. Malorni, P. Pucci, P. Ferranti, and G. Marino, *in* "Mass Spectrometry in the Biological Sciences: A Tutorial" (M. Gross, ed.), p. 325. Kluwer Academic Publishers, New York, 1992.

The only limitation so far reported for FAB mapping is that not all the peptide fragments are observed, due to the suppression phenomena that always occur in FAB/MS analysis of mixtures.[15] Moreover, because the identification of the mutated residues is based on the mass difference between normal and variant peptides, it is straightforward only if the replaced amino acid occurs once in the particular tryptic peptide. A different approach to the problem is the use of different proteolytic enzymes, such as *Staphylococcus aureus* V8 protease, to generate peptide maps that are complementary to the tryptic data.[16]

Coupling capillary electrophoresis data with analysis of enzymatic digests by mass spectrometry allows fast differentiation of hemoglobin variants and can lead to the easy identification of the mutation. This papers reviews some of the results obtained by application of the strategy to the characterization of abnormal hemoglobins, underlining advantages and limitations along with the potential uses in protein structural analysis.

Experimental Conditions

Materials

Hb S, Hb C, Hb F, and patient blood samples are from the Hematology Division of the Ospedale Cardarelli (Naples), from the Division of Hematology, Second Medical School, University of Naples, and from the Department of Hematology, Hammersmith Hospital, London.

Ammonium bicarbonate, sodium hydrogen phosphate, trifluoroacetic acid (TFA), and all reagent-grade and HPLC-grade solvents are from Carlo Erba (Italy). Trypsin (TPCK treated) is from Sigma (St. Louis, MO) and *Staphylococcus aureus* (strain V8) protease is from Boheringer Mannheim.

Apparatus

Capillary Zone Electrophoresis. CZE experiments are performed on a Bio-Rad HPE 100 unit (Richmond, CA) equipped with an on-column spectrophotometric detector (190–380 nm). The apparatus is equipped with a high-voltage power supply (HVPS) capable of delivering up to 12 kV. Sampling and electrophoresis are controlled by a microprocessor.

[15] S. Naylor, S. Findeis, B. W. Gibson, and D. H. Williams, *J. Am. Chem. Soc.* **108,** 6359 (1986).
[16] P. Pucci, P. Ferranti, A. Malorni, and G. Marino, *Biomed. Environ. Mass Spectrom.* **19,** 568 (1990).

Separations are performed in either 20 cm × 25 μm i.d. or 50 cm × 50 μm i.d. capillaries supplied by Bio-Rad mounted in cartridges with an integral flow cell for on-column optical detection. All capillaries have an internal wall covalently coated with a hydrophilic polymer. Detection is carried out using UV absorbance, and the capillary is not thermostatted.

The capillary is filled with the background electrolyte using a 100-μl Hamilton microsyringe. Electropherograms are recorded using an LKB 2210 recorder. Background electrolytes used for CZE are 20 mM sodium phosphate solutions with pH values ranging from 2.5 to 4.5 for globin chain analysis and from 2.5 to 4.0 for tryptic mapping.

HPLC Separations. RP-HPLC of proteins and peptides are performed on a Gilson apparatus equipped with two pumps, Model 302, and a UV detector (Model Holocrome) controlled by an Apple computer. For protein separation a Vydac large-pore (330 Å) C_4 (Cat. #214TP54; 4.6 × 250 mm) column manufactured by the Separations Group (Hesperia, CA 92345) is used; peptide analyses are performed on a μBondapak (4.6 × 300 mm) column purchased from Waters.

Methods

Preparation of Hemolysates and Globin Chains. Hemolysates are prepared from fresh blood (2 ml) anticoagulated with either EDTA or heparin. Red cells are washed twice with 0.9% (w/w) NaCl, hemolyzed by adding water (1 : 4, v/v), and then centrifuged at 10,000 rpm for 30 min.[10] Hemoglobin is converted to globin by precipitation as described by Clegg et al.[17]: the hemolysate is treated with a 20-volume excess of 2% HCl in acetone at −20°, care being taken to remove any trace of acid by washing the precipitate three times with cold (−20°) acetone. Finally the precipitate is dissolved in 10% (v/v) acetic acid, freeze-dried, and stored at 4°, the sample remaining stable for a period of several weeks.

RP-HPLC Purification of Globin Chains. RP-HPLC chromatography of globin chains is performed using a (250 × 4.6 mm) Vydac C_4 column according to the procedure described by Shelton et al.[18] with slight modifications: the elution system is a linear gradient between mixtures of 0.1% trifluoroacetic acid in water and 0.1% TFA in acetonitrile. Mixture A contains 80% of 0.1% TFA in water whereas mixture B contains 40%. Solvents have to be carefully degassed and can then be used over a period of 4–5 days.

The sample size is 500 μg for each run. Globin chains from acetone precipitation are dissolved in 0.1% TFA in water and filtered through 0.45-

[17] J. B. Clegg, M. A. Naughton, and D. J. Weatherall, *J. Mol. Biol.* **19**, 91 (1966).
[18] J. B. Shelton, J. R. Shelton, and W. A. Schroeder, *J. Liq. Chromatogr.* **7**, 1969 (1984).

μm pore size (Cat. #HAWPO1300, Waters) filters. They are then injected into the column equilibrated at 46% of mixture B; after a 2-min hold, a linear gradient from 46 to 58% of mixture B is applied. The flow rate is 1 ml/min. At the end of each run the column is washed with 100% mixture B for 10 min.

Preparation of Enzymatic Digests and Edman Degradation. Tryptic and V8 protease digests are both performed in 0.4% ammonium bicarbonate at pH 8.5, 37°, for 4 hr and at pH 8.0, 40°, for 6 hr, respectively, with an enzyme : substrate ratio of 1 : 50 (w/w). Edman degradation is carried out directly on the enzymatic protein digest using 5% phenyl isothiocyanate in pyridine as coupling agent.[19]

Pyridylethylation of Globin Chains and Peptide Mapping. Reduction and pyridylethylation of globin chains is carried out in 0.3 M Tris-HCl buffer, pH 8.0, containing 6 M guanidinium chloride and 1 mM EDTA, by incubation with a 10-fold molar excess of dithiothreitol over —SH groups at 37° for 2 hr, followed by incubation with a 5-fold molar excess of vinylpyridine (Fluka, Ronkonkoma, NY) (previously distilled under vacuum and stored at −50°) at room temperature for 30 min. Samples are freed from salts and low molecular weight molecules by HPLC chromatography as described above.

Capillary Zone Electrophoresis of Globin Chains. Free zone electrophoresis of globin chains is performed with the polarity of the internal power supply of the instrument set so that the components would migrate toward the detector as cations (i.e., the cathode is at the detector end of the capillary).

Before starting the analysis the capillary and the electrode reservoir at the detector end of the capillary are filled with background electrolyte (BGE), and the inlet-side electrode reservoir is flushed with distilled water. A 10-μl volume of the sample solution in water is then injected into the reservoir just ahead of the capillary inlet. The power supply is turned on, and the sample is introduced into the capillary by electromigration for 5–10 sec at 8 kV. The inlet electrode reservoir is then flushed with BGE, and the power supply is turned back on at 15 μA constant current. Electropherograms are obtained by monitoring absorbance at 206 nm. After a run, the capillary is carefully washed with water to remove uneluted sample.

The free zone electrophoretic peptide maps are obtained by electrophoresis as described above for globin chain analysis at 15 μA constant current in 20 mM sodium phosphate buffer, pH 2.5.

[19] P. Pucci, C. Carestia, G. Fioretti, A. M. Mastrobuoni, and L. Pagano, *Biochem. Biophys. Res. Commun.* **130**, 84 (1985).

RP-HPLC Peptide Mapping. The RP-HPLC tryptic maps are produced by injection of a 100-μg sample onto a 300 × 4.6 mm) C_{18} μBondapak column equilibrated with 5% acetonitrile containing 0.1% TFA in water at a flow rate of 1 ml/min. After a 5-min hold, a linear gradient to 40% acetonitrile is applied over 90 min. The column effluent is monitored at 206 nm. Peaks are identified by FAB/MS analysis on the basis of the molecular weight comparisons with those of the expected tryptic peptides.

FAB/MS Mapping of Globin Chains. FAB/MS spectra are recorded on a VG Analytical ZAB 2SE double-focusing mass spectrometer equipped with a cesium gun operating at 25 keV (2 μA) using glycerol/thioglycerol (1/1) as matrix.

Before the analysis, the mixture of peptides (1–3 nmol) is dissolved in 0.1 M HCl and 0.1–0.3 nmol samples are loaded onto a matrix-coated probe trip. Spectra are recorded on UV-sensitive paper and manually counted. The mass values recorded are assigned to the corresponding peptides along the globin chain sequence on the basis of their molecular weights. All the mass values recorded in FAB mode are shown as monoisotopic masses.

Results and Discussion

Analysis of Globin Chains by CZE

Although still in its early development, high-voltage capillary zone electrophoresis has been shown to be remarkably efficient for the separation of small, charged molecules.[20] Although proteins present some difficulties due to their easy adsorption on the capillary wall, progress has been made, particularly through chemical coating of the capillary wall for the proteins and peptides separated at low pH values.[21,22]

By using coated capillaries, extremely narrow bands can be observed, so that proteins with very subtle differences in charge density can be resolved by manipulating parameters such as buffer composition, pH, capillary dimensions, and applied voltage.

Optimizing Experimental Conditions

Figure 1 shows a typical separation obtained for a sample of globins from a normal subject on a 20-cm capillary in 20 mM sodium phosphate

[20] M. Novotny, *J. Microcolumn Sep.* **2,** 7 (1990).
[21] J. Frenz, S. Wu, and W. S. Hancock, *J. Chromatogr.* **480,** 379 (1989).
[22] J. B. Wilson, H. Lam, P. Pravatmuang, and T. H. J. Huisman, *J. Chromatogr.* **179,** 271 (1979).

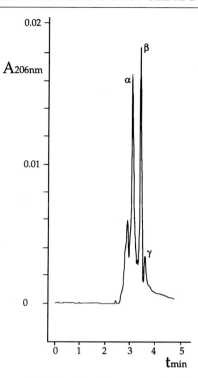

FIG. 1. Electropherogram of globin chains from a normal subject. Globins (10 μg; sampling: electrophoresis at 8 kV for 6 sec) were analyzed in 20 mM sodium phosphate, pH 2.5, 15 μA, in a 20 cm × 25 μm (i.d.) coated capillary. Detection was by absorbance at 206 nm.

buffer, pH 2.5, carried out at constant current (15 μA, 8 kV); at this pH the globin chains migrate with the best resolution.

The more cationic α chain has a mean migration time of $m = 3.23$ min, whereas the β-globin migrates at $m = 3.58$ min. The relative ratio measured by integration was 56 : 44 (assuming for simplicity that the two globins have identical molar extinction coefficients at 206 nm); the minor peak after the β-globin is the γ chain with $m = 3.78$ min. It should be emphasized that the separation requires 5 min, a short time compared with most electrophoretic techniques, which are often very time consuming.

Working with a fixed current rather than a fixed voltage means that much better reproducibility can be obtained. With fixed voltage the pattern is still very reproducible but with marked shifts in retention times from run to run, which can affect peak identification (experiments were not carried out in thermostatted conditions). When working under a constant

TABLE I
MIGRATION TIMES AND REPRODUCIBILITY OF
CZE SEPARATION OF GLOBINS FROM Hb A [a]

Peak	Migration time (min)		
	M	SD	RSD%
α-Globin	3.23	0.059	1.83
β-Globin	3.58	0.061	1.70
γ-Globin	3.78	0.028	0.74

[a] M, Mean; SD, standard deviation; RSD%, relative standard deviation.

current, sequential injections of samples give identical retention times within the standard deviation. Automation of capillary washing and sample introduction steps reduces the problem of reproducibility. Calculated standard deviation values for six runs are shown in Table I.

In the presence of a genetic variant, a modification is expected in the pattern described for normal subjects, resulting in the appearance of abnormal peak(s) and/or the disappearance or decrease of a normal peak. The mobility of the variant peak can facilitate identification of the variant, or, at least, provide valuable information for further structural investigation. Furthermore, quantifying the variant may enable a clinical diagnosis.

If necessary, a much higher resolution can be obtained by increasing the capillary length, as shown in Fig. 2 for the 50-cm capillary: the globin pattern is unchanged but the separation time is longer, the total run time being about 15 min. However, this analysis is still faster than most of the current electrophoretic techniques. The use of larger capillaries is recommended for variants with charges similar to that of normal hemoglobin and therefore not easily detectable with low-resolution techniques.

Globin Chain Tryptic Mapping by CZE

At present the structural characterization of protein variants is based on RP-HPLC separation of the peptides generated by tryptic digestion, followed by identification of the abnormal peptide, amino acid analysis, and peptide sequence determination. By careful standardization of analytical parameters, the HPLC map can be used as the fingerprint of a protein and to detect the presence of its variants.[23]

[23] W. S. Hancock, C. A. Bishop, R. L. Partridge, and M. T. W. Hearn, *Anal. Biochem.* **89,** 203 (1978).

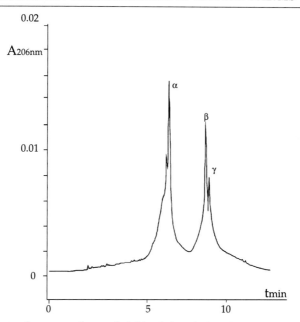

FIG. 2. Electropherogram of normal globin chains obtained in a 50 cm × 50 μm (i.d.) coated capillary. Conditions as in Fig. 1.

Figure 3A shows the RP-HPLC tryptic map of the hemoglobin β chain using a μBondapak column. Peptides were identified by FAB/MS analysis and are numerically labeled according to their position in the protein sequence, so that T-1 corresponds to the N-terminal peptide and T-15 is the C-terminal one. Due to a chymotryptic-like activity of the enzyme a new nontryptic peptides were also found and labeled with the letter C and the respective position in the sequence. Even though the separation was very good, a complete resolution of all the peptides could not be achieved by changing the analytical conditions.

Capillary electrophoresis, with a selectivity different from that of RP-HPLC, offers a convenient alternative for the production of the tryptic map of a protein; Fig. 3B illustrates the CZE map of the β-globin chain. Peptides were identified by running aliquots of the pure HPLC fractions characterized by FAB/MS analysis (values are shown in Table II) and labeled according to the above convention. An analogous map was produced for the α-globin chain (not shown).

The map of a genetic variant can be compared to that of the corresponding normal chain, immediately identifying the globin as an α or β variant. Furthermore, the shift or disappearance of one or more peaks helps to

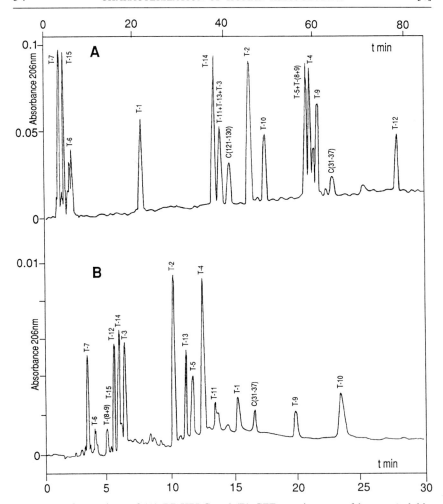

FIG. 3. Comparison of (A) RP-HPLC and (B) CZE tryptic maps of human β-globin. Peptides are marked by a number indicating the position in the protein sequence. See text for details.

locate the modification within a well-defined portion of the protein sequence.

The CZE map of Fig. 3B was performed on a 20-cm capillary with a run time of about 30 min (compared with 90 min required for HPLC), and complete resolution of all peptides should be emphasized. Sample requirement is much less for CZE (a few micrograms, of which only a small fraction is actually injected) than for HPLC. Furthermore, the absence of

TABLE II
MIGRATION TIMES AND REPRODUCIBILITY OF CZE TRYPTIC MAPPING OF β-GLOBIN[a]

Peak	Migration time (min)			m/z	Position in sequence
	M	SD	RSD%		
T-7	3.33	0.051	1.53	412	(62–65)
T-6	3.85	0.066	1.71	246	(60–61)
T-(8 + 9)	4.80	0.060	1.25	1797	(66–82)
T-15	5.20	0.048	0.92	319	(145–146)
T-12	5.40	0.062	1.82	1719	(105–120)
T-14	5.81	0.080	1.37	1149	(133–144)
T-3	6.24	0.072	1.15	1314	(18–30)
T-2	10.16	0.091	0.89	932	(9–17)
T-13	11.14	0.092	0.82	1378	(121–132)
T-5	11.57	0.082	0.71	2058	(41–59)
T-4	12.39	0.074	0.60	1274	(31–40)
T-11	13.41	0.072	0.54	1126	(96–104)
T-1	15.16	0.093	0.61	952	(1–8)
C (31–37)	16.48	0.088	0.53	889	(31–37)
T-9	19.44	0.085	0.43	1669	(67–82)
T-10	23.46	0.091	0.38	1421	(83–95)

[a] See Fig. 3B. M, Mean; SD, standard deviation; RSD%, relative standard deviation; m/z, mass/charge ratio.

gradient elution and of organic solvents in the CZE buffer system results in a much lower background absorbance, and the UV detector can be used at the highest sensitivity setting without significant noise and baseline instability. The low sample tolerance of CZE, however, makes preparative separations difficult: it is doubtful that sufficient material for further structural characterization could be collected from the capillary; peptides might easily be identified by direct interfacing to a mass spectrometer.[24]

Precisely because peptide mapping is essentially a comparison technique, a high degree of reproducibility is required. As was the case in the analysis of whole globins, migration times were reproduced better when the current was kept constant and the injection procedure carefully repeated. Table II shows the mean migration times for six sequential injections, giving a standard deviation always lower than 2%.

Analysis of Common Hemoglobin Variants

The usefulness of CZE to detect modifications in the primary structure of hemoglobin can be demonstrated for the three most common variants:

[24] R. M. Caprioli, W. T. Moore, M. Martin, B. B. DaGue, K. Wilson, and S. Moring, *J. Chromatogr.* **480**, 247 (1989).

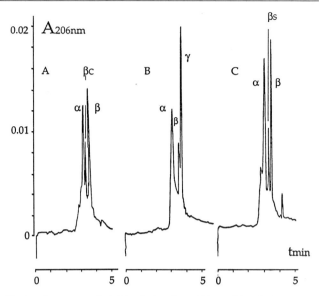

FIG. 4. Electropherogram of globin chains from (A) a subject heterozygous for Hb C, (B) a subject suffering from Cooley's disease, and (C) a subject heterozygous for Hb S. Samples were analyzed as in Fig. 1.

Hb C, Hb F, and Hb S. Because the pathologies related to these variants, particularly to Hb S, are responsible for many worldwide hemoglobinopathies, it is essential for the effectiveness of the proposed procedure to identify these variants selectively.

Figure 4A shows the CZE analysis of globin chains from a subject heterozygous for Hb C. Two of the three peaks, compared with the normal pattern in Fig. 1, were shown to correspond to the α and β chains, respectively. The intermediate peak is, instead, due to the variant β chain: its increased mobility with respect to the β chain is expected on the basis of the amino acid substitution of a glutamic acid residue with lysine. Under the working conditions described, the protein has an extra positive charge, which increases its mobility ($m = 3.28$ min) compared to that of the β normal globin. The variant has a relative amount of 42%, as expected for an asymptomatic C-type globin.

The analysis of an adult subject with Cooley's disease is shown in Fig. 4B. Due to the absence of the normal Hb A, the red cells contain only Hb F, the fetal hemoglobin, and two peaks are therefore detected in the electropherogram: the faster one is the α-globin and the second one ($m = 3.71$ min) is due to the γ-globin of Hb F. A small peak (about 9%

of the γ-globin) is also present at the β position, indicating that the synthesis of the normal hemoglobin is not completely defective.

Figure 4C refers to the electropherogram of a globin sample from an individual heterozygous for Hb S. CZE analysis demonstrates the presence of the variant chain as a peak (m = 3.52 min) just ahead of the β-globin chain, with a relative amount of about 25%. It is worth noting that the amino acid substitution does not produce a net charge difference at pH 2.5. Nevertheless, the variant peak is well resolved from the normal β-globin peak, as was the α chain, indicating that even similarly charged variants can be detected using this procedure.

CZE can therefore differentiate Hb C, Hb S, and Hb F on the basis of their different mobilities, with greater selectivity, higher resolution, and much shorter times than required by conventional electrophoretic and/or HPLC techniques.

Unknown Hemoglobin Variants: Selected Results

Hb O-Arab. The electropherogram shown in Fig. 5 refers to a globin sample from a woman suspected of hosting a genetic variant: two main peaks are present, of which only the first one corresponded to the α chain; the slower peak had a greater mobility (m = 3.47 min) than expected for the β chain, which was absent, suggesting the occurrence of a more cationic β variant. The relative amount α : β variant was 55 : 45, and the absence of the β-chain peak indicated either that the subject was homozygous for the variant or that she was unable to express the β-globin gene for a further genetic defect (i.e., β-thalassemia).

The variant globin was purified using RP-HPLC on a C_4 column; the chromatogram, shown in Fig. 6, confirmed the presence of the variant peak and the normal β peak was absent, so that enough sample could be prepared for structural characterization.

The tryptic CZE map (see Fig. 7) clearly exhibited the typical pattern of a β-globin, except for the absence of the peptide T-13 (peptide 121–132), immediately indicating the protein portion where the modification took place. Because the electropherogram showed a faster anomalous peak, the mutated peptide 121–132 should have shifted to that position due to a modification greatly enhancing its cationic properties, as already seen for the whole protein (i.e., a net negative charge loss, a net positive charge gain, or both). A second slower peak was also present in the map.

To confirm this hypothesis a second aliquot of the tryptic digest was used to perform the FAB map; Fig. 8 shows a comparison of the variant (Fig. 8A) and of the normal (Fig. 8B) β chain; missing is the signal at m/z 1378, corresponding to peptide 121–132, which is always present in the

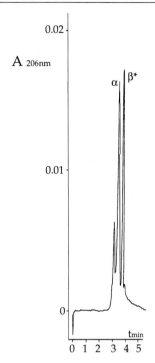

FIG. 5. Electropherogram of globin chains from an O-Arab subject analyzed as described in Fig. 1. The variant (β^*) peak is shown.

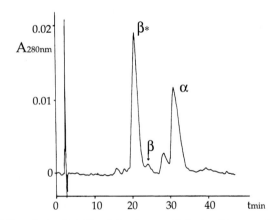

FIG. 6. HPLC analysis of globin chains from an O-Arab subject, performed on a Vydac C_4 column. For experimental conditions, see text. The variant (β^*) peak is shown.

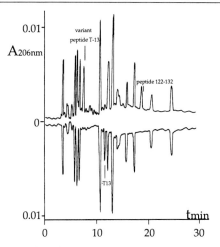

FIG. 7. CZE tryptic mapping of the variant O-Arab globin (top) compared with normal β-globin (bottom).

normal β map. On the other hand, two novel signals are present at *m/z* 1377 and 1249, respectively; the first one was supposed to correspond to the changed T-13 with the N-terminal Glu substituted with Lys, whereas the second one corresponded to peptide 121–132, missing Glu-121. Mass

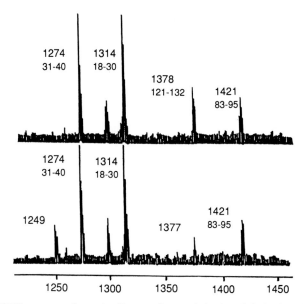

FIG. 8. FAB spectra of tryptic digests of normal (top) and O-Arab variant (bottom) β-globin chains in the region where differences were observed.

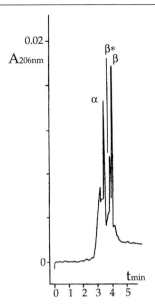

FIG. 9. Electropherogram of globin chains from a San Jose subject analyzed as described in Fig. 1.

value assignments were confirmed by an Edman degradation step followed by FAB/MS analysis according to the FAB mapping procedure. These two observations, coupled with CZE data, identified the variant as Hb O-Arab.[25] In this variant the glutamic acid at position 121 of the β chain is replaced by lysine. This amino acid substitution creates a further hydrolysis site for trypsin, producing two novel peptides: the variant peptide 121–132 (m/z 1377) with lysine at position 121, which is much more cationic in CZE than the normal 121–132. The second peptide is the fragment 122–132 produced by hydrolysis at Lys121.

Hb San Jose and Hb Heathrow: Mapping Using V8 Protease. The second case is presented in Fig. 9. The CZE globin analysis shows both α- and β-globin peaks, as well as an anomalous peak ($m = 3.52$ min) just ahead of the β-globin peak. Integration demonstrated a decrease in the β-globin peak area, suggesting that a relative amount (33%) was a β variant.

The CZE tryptic map of the variant globin overlapped perfectly with that of the β-globin (see Fig. 10) except for the N-terminal peptide T-1

[25] K. Kamel, K. Hoerman, and A. Awny, *Am. J. Phys. Anthropol.* **26,** 107 (1970).

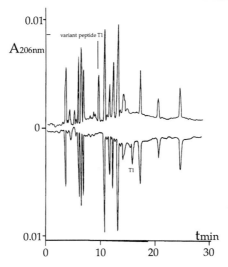

FIG. 10. CZE tryptic mapping of the variant San Jose globin (top) compared with the normal β-globin chain (bottom).

(peptide 1–8); a variant peak appeared, instead, as the consequence of a modification producing a more cationic peptide. The FAB mass spectrum of the tryptic digest is shown in Table III. Note the absence of the N-terminal peptide 1–8 (m/z 952) and the presence of an anomalous peak

TABLE III
FAB/MS ANALYSIS OF TRYPTIC DIGEST OF
VARIANT β-GLOBIN FROM Hb SAN JOSE[a]

Peptide	m/z
1–8 (Gly-7)	880
9–17	932
18–30	1314
31–40	1274
41–59	2058
66–82	1797
67–82	1669
83–95	1421
96–104	1126
105–120	1719
121–132	1378
133–144	1149

[a] See Fig. 10.

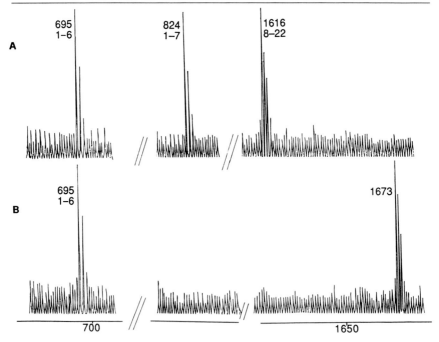

FIG. 11. Partial FAB spectra of V8 protease digest of normal (A) and variant (B) β-globin from Hb San Jose.

at m/z 880. This mass value corresponds to peptide 1–8 with a glutamic acid substituted by glycine: the mass difference between normal and variant peptide is, in fact, -72 mass units. Unfortunately, the N-terminal tryptic peptide contains two glutamic acid residues at positions 6 and 7. Therefore, in this case, it is impossible, on the basis of the tryptic map, to locate the site of the substitution. The particular nature of the C-terminal site of peptide 1–8, Glu-Glu-Lys, in fact, prevents the use of carboxypeptidases to release any amino acids from that particular peptide.

An alternative approach was the construction of a FAB map using V8 protease as proteolytic agent. The V8 protease, in fact, would cleave the protein at the level of Glu-6 or Glu-7, depending on which one had been replaced by Gly, and at Glu-22, yielding peptides 1–6 and 7–22 or 1–7 and 8–22.

The FAB spectrum of the variant globin digested with V8 protease is shown in Fig. 11, compared to the normal β-chain spectrum in the mass region where differences were observed. It should be noted that the signals at m/z 824 (peptide 1–7) and m/z 1616 (peptide 8–22) are absent in the variant spectrum, whereas a new signal was recorded at m/z 1673. This

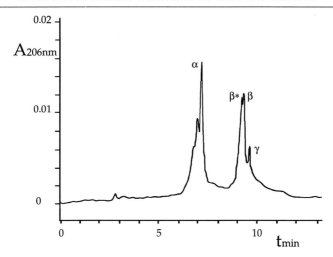

FIG. 12. Electropherogram of globin chains from Hb Heathrow performed using a 50 cm × 50 μm (i.d.) coated capillary.

value corresponds to peptide 7–22, with a glycine residue at position 7 instead of the expected glutamic acid. This hypothesis was easily confirmed by the shift of the signal from m/z 1673 to m/z 1616 following the loss of the N-terminal glycine after a single cycle of Edman degradation. The variant hemoglobin is therefore identified as Hb San Jose a $\beta^{7\ Glu\rightarrow Gly}$ hemoglobin,[13,26] in perfect agreement with CZE analysis. As for Hb S, which exhibits a similar electrophoretic pattern, it is possible using CZE to resolve variant globins that, at low pH, do not present any charge difference from the normal ones.

As shown by the analysis of this case, hydrolysis of globins with V8 protease can be extremely useful and complementary to tryptic data for the identification of mutated amino acids. As a further example of this, identification of a very rare variant (Heathrow) can be considered. Routine electrophoresis of hemoglobin as well as isoelectrofocusing of globin yielded only normal bands. For CZE analysis, a 50-cm-long capillary was chosen, and is always recommended for "silent" variants. Unfortunately, although the variant was obvious from a double-top β-globin peak, it was impossible to achieve complete resolution of the variant peak (see Fig. 12).

A variant peak, also revealed by RP-HPLC (estimated to be about 35%), was eluted after the normal β-globin; the protein was subjected to pyridylethylation followed by tryptic hydrolysis and FAB analysis. The

[26] R. L. Hill, R. T. Swenson, and H. C. Schwartz, J. Biol. Chem. 235, 3182 (1960).

peptide map thus constructed was similar to that of the normal β-globin except that the signal at m/z 1126 (corresponding to peptide T-11) was missing, and a new anomalous signal was detected at m/z 1092. The two substitutions, justifying the difference of 34 mass units, were possibly His[97] → Cys and Phe[103] → Leu/Ile.

However, if a novel cysteine was present, it should be pyridylethylated, giving rise to a peptide at m/z 1197 instead of m/z 1092. Therefore the most likely change appeared to be Phe[103] → Leu/Ile. In order to confirm this, a second map was constructed by means of V8 protease hydrolysis of the tryptic digest. The presence of the peak at m/z 402 (instead of 436, corresponding to the normal peptide 102–104), together with a peak at m/z 709 (corresponding to the normal peptide 96–101) was consistent with the Phe → Leu/Ile substitution. Because it is not possible to distinguish between the isobaric residues Leu and Ile using FAB/MS, the variant tryptic peptide was purified by RP-HPLC; the variant peptide exhibited, in fact, a shorter retention time as compared to the normal peptide T-11.

Amino acid analysis revealed the absence of phenylalanine and the presence of an additional leucine residue, whereas no isoleucine was found, demonstrating that phenylalanine-103 is replaced by leucine, and the variant was identified as Hb Heathrow.[27,28]

Summary and Future Directions

The combination of CZE with FAB/MS analysis according to the FAB mapping procedure couples a potentially fast high-resolution separation method with one of the most powerful techniques for structural characterization. Capillary electrophoresis has the ability to handle very small volume samples (picoliter range) and attomole (10^{-18} mol)-level detection limits reaching separation efficiencies of the order of 10^6 theoretical plates. Previous attempts at CZE of proteins with UV detection have often resulted in broad, tailing peaks due to the absorption of proteins on active sites of the negatively charged capillary wall. However, the introduction of coated capillaries made CZE a fast and sensitive method for the detection of even similarly charged variants; the resolution obtained using this technique is comparable to the well-established electrophoretic and chromatographic procedures and requires smaller samples. Its speed and

[27] J. M. White, L. Szur, I. D. S. Gillies, P. A. Lorkin, and H. Lehmann, *Br. Med. J.* **3**, 665 (1973).
[28] G. Marsh, G. Marino, P. Pucci, P. Ferranti, A. Malorni, J. Kaeda, J. Marsh, and L. Luzzatto, *Hemoglobin* **15**, 43 (1991).

simplicity with respect to the above techniques and the possibility of automation and of development on a micropreparative scale mean the whole procedure is an attractive alternative to conventional systems.

CZE globin chain peptide mapping has been developed so that reproducible and characteristic "fingerprints" can be obtained from low-nanogram-level quantities of purified globin; the sample requirement for peptide mapping is brought down to a level compatible with microscale protein isolation and amino acid analysis.

Direct FAB analysis of peptide mixtures generated by proteolytic digestion of the variant chain avoids, in most cases, the need for chromatographic separation of peptides. Assignment of the mass values to the corresponding peptides generates variant protein FAB maps that do not depend on the chromatographic behavior of the peptides. The choice of trypsin as a proteolytic agent has to be recommended in the case of globin chains because almost all the expected tryptic peptides can be easily detected in FAB mode. However, in many cases, the tryptic map alone cannot lead to the unambiguous identification of the site and of the nature of the genetic mutation. In this case, the use of *S. aureus* V8 protease complements trypsin in that it specifically cleaves polypeptide chains at the level of glutamic acid residues; the narrow specificity of the enzyme, on the other hand, makes the identification of the variated peptides in the spectra very easy, being an alternative approach when data available from tryptic digests do not furnish unambiguous results.

The advantages of the proposed procedure mean that it can be widely used in the diagnosis of hereditary pathologies caused by the presence of mutant proteins, in posttranslational modifications, in protein polymorphism research, and in other fields of protein structure analysis.

[5] Purification of Glycated Hemoglobin

By Margo Panush Cohen and Van-Yu Wu

Introduction

The term glycohemoglobin (glycated hemoglobin) refers to hemoglobin that has been modified postribosomally by the attachment of glucose to the polypeptide chain. This modification, generically designated nonenzymatic glycation, is a condensation reaction between the carbohydrate and a free amino group at the amino terminus or an ε-amino group of a lysine residue. The study of hemoglobin glycation has served as a model system, for it is now clear that many proteins undergo nonenzymatic glycation. The reaction is initiated with attachment of acyclic glucose to a protein amino group via nucleophilic addition, forming an aldimine, also known as a Schiff base. This intermediate product subsequently undergoes an Amadori rearrangement to form a 1-amino-1-deoxyfructose derivative in stable ketoamine linkage. The reaction is slow and continuous, and the extent of glycation is largely dependent on the glucose concentration to which the protein is exposed during its lifetime in the circulation. Other factors that influence the sites and amount of glycation include the accessibility and pK_a of the amino groups within the structure of the protein. Although any free amino group is a potential site for nonenzymatic glycation, these factors, and other microenvironmental considerations such as the availability of nearby catalytic carboxyl groups for acceleration of the Amadori rearrangement, help explain the reactivity of certain free amino groups and the preferential glycation of specific amino acid residues. This review describes the products of the nonenzymatic glycation reaction between glucose and hemoglobin, the glycation of hemoglobin *in vitro,* the purification from erythrocyte lysates of glycohemoglobin species formed *in vivo* or after *in vitro* glycation, and procedures for identification and characterization of glycohemoglobin preparations.

Reaction Products

The reaction between glucose and hemoglobin yields the following products: (1) hemoglobin A_{1c}, which is identical to hemoglobin A_0 except that glucose is linked to the amino-terminal valine residue of the β chain[1-3]; and (2) hemoglobin glycated at other positions on the α and β subunits, collectively called non-A_{1c} glycohemoglobin, or glycated hemoglobin (Fig. 1). Non-A_{1c} glycohemoglobin is identical to hemoglobin A_0 except that glucose is linked to an ε-amino group of one or more lysine residues in the α or β chains. The lysine residues that undergo glycation *in vivo* are, in order of prevalence, Lys-66(β), Lys-61(α), and Lys-17(β).[4] Incubation of [^{14}C]glucose followed by ion-exchange chromatography and peptide mapping has allowed identification of the major sites of lysine glycation

[1] The N-terminal valine residue of the α chain also may undergo glycation *in vivo* or *in vitro,* according to Shapiro *et al.* (Reference 2), but with a much lower prevalence. According to Acharya *et al.* (Reference 3), molecular aspects such as the isomerization potential of the microenvironment determine site selectivity in the glycation of hemoglobin.
[2] R. Shapiro, M. J. McManus, C. Zalut, and H. F. Bunn, *J. Biol. Chem.* **255,** 3120 (1980).
[3] A. S. Acharya, R. P. Roy, and B. Dorai, *J. Protein Chem.* **10,** 345 (1991).
[4] H. F. Bunn, R. Shapiro, M. McManus, L. Garrick, M. McDonald, P. M. Gallop, and K. H. Gabbay, *J. Biol. Chem.* **254,** 3892 (1979).

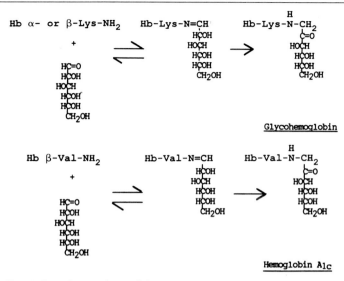

FIG. 1. Reaction products of the nonenzymatic glycation of hemoglobin.

occurring *in vitro*; in order of prevalence, these are Lys-16(α), Lys-66(β), Lys-17(β), Lys-7(α), and Lys-120(β).[2] Formation of these glycohemoglobin adducts, like the N-terminal modification that forms Hb A_{1c}, is increased in diabetic patients with ambient hyperglycemia.[5] However, modification by glucose at these other sites does not usually change the isoelectric point, making the separation of non-A_{1c} glycohemoglobins from Hb A_0 or Hb A_{1c} difficult with traditional electrophoretic or ion-exchange chromatographic methods. In contrast, purification of Hb A_{1c} relies on the change in charge properties conferred by the N-terminal modification, which results in its earlier elution than Hb A_0 from an ion-exchange column. In fact, it was the recognition of this earlier elution that was instrumental in the evolution of a system of nomenclature to describe the N-terminally modified hemoglobins and in their subsequent structural identification.[6-9] Thus, hemoglobins modified at the N terminus of the β chain, which constitute the early eluting species, were designated Hb A_{1a1},

[5] R. L. Garlick, J. S. Mazer, P. J. Higgins, and H. F. Bunn, *J. Clin. Invest.* **71**, 1062 (1983).

[6] D. W. Allen, W. A. Schroeder, and J. Balog, *J. Am. Chem. Soc.* **80**, 1628 (1958).

[7] A. G. Schnek and W. A. Schroeder, *J. Am. Chem. Soc.* **83**, 1470 (1961).

[8] M. J. McDonald, R. Shapiro, M. Bleichman, J. Solway, and H. F. Bunn, *J. Biol. Chem.* **253**, 2327 (1978).

[9] L. M. Garrick, M. J. McDonald, R. Shapiro, M. Bleichman, M. McManus, and H. F. Bunn, *Eur. J. Biochem.* **106**, 353 (1980).

TABLE I
MINOR HEMOGLOBIN DERIVATIVES

Designation	Modification
Hb A_0	None
Hb A_{Ia1}	β-N-Fructose 1,6-bisphosphate
Hb A_{Ia2}	β-N-Glucose 6-phosphate
Hb A_{Ib}	β-N-Carbohydrate
Hb A_{Ic}	β-N-Glucose
GlycoHb	α/β-ε-Lysine glucose

Hb A_{Ia2}, Hb A_{Ib}, and Hb A_{Ic} according to their successive elution profile (see Table I).

Glycation of Hemoglobin in Vitro

Following separation by centrifugation and washing, erythrocytes are lysed in hypotonic solution and the released hemoglobin is incubated in phosphate-buffered saline, pH 7.4, containing 500 mg/dl glucose. The hemoglobin and glucose solutions are sterile filtered before incubation. Incubation time of 4–7 days and a temperature of 25° are preferred to avoid heme precipitation and the possible cross-linking of protein related to advanced glycation end product formation. After dialysis to remove free glucose, the hemoglobin is subjected to ion-exchange and affinity chromatography as described below. Incubation with glucose can be conducted with sodium cyanoborohydride (20 mM) to reduce the Schiff base or can be followed by treatment with sodium borohydride to reduce the ketoamine.[10,11]

Hemoglobin can be reacted in vitro with a number of other hexoses as well as with aldotrioses and with ketoses.[12–15] The Schiff base and stable adducts that are formed with the amino groups are analogous to those formed by the condensation of glucose with hemoglobin. However, the site selectivity of some carbohydrates for particular amino groups in the hemoglobin polypeptide may differ from that of glucose. For example, Acharya et al.[3] have reported that the sites of nonreductive in vitro glyca-

[10] P. J. Higgins and H. F. Bunn, J. Biol. Chem. 256, 5204 (1981).
[11] R. Fluckiger and P. M. Gallop, this series, Vol. 106, p. 76.
[12] H. F. Bunn and P. J. Higgins, Science 213, 222 (1981).
[13] A. S. Acharya and B. N. Manjula, Biochemistry 26, 3524 (1987).
[14] G. Suarez, R. Rajaram, A. L. Oronsky, and M. A. Gawinowicz, J. Biol. Chem. 264, 3674 (1989).
[15] J. D. McPherson, B. Shilton, and D. J. Walton, Biochemistry 27, 1901 (1988).

tion of hemoglobin A with glyceraldehyde are, in order of relative reactivity, Lys-16(α), Val-1(β), Lys-66(β), Lys-82(β), Lys-61(α), and Val-1(α). Lys-59(β) and Lys-120(β) also condense with glyceraldehyde.[16] The sites of reductive glycation of hemoglobin A with glyceraldehyde *in vitro* are, in order of relative reactivity, Val-1(β), Val-1(α), Lys-66(β), Lys-61(α), and Lys-16(α). Like hemoglobin that is nonenzymatically glycated with glucose in the non-A_{1c} position, hemoglobin modified by glyceraldehyde at ε-amino groups of lysine coelutes with hemoglobin A_0 on ion-exchange chromatography[3] (see below).

Purification of Hemoglobin A_{1c}

Erythrocyte lysate is dialyzed against 0.05 M potassium phosphate, pH 6.5, containing 0.01 M KCN and 0.015 M NaCl, and applied to an ion-exchange column (Bio-Rex 70; Bio-Rad Corp, Richmond, CA) equilibrated with the same buffer. Fast-moving hemoglobins (A_{1a}, A_{1b}, and A_{1c}) elute under these conditions[17,18]; the major hemoglobin A_0 peak (which also contains non-A_{1c} glycohemoglobin[19]) is then eluted by making the buffer 1.0 M in NaCl. The fast-moving peak is dialyzed against 0.05 M potassium phosphate, pH 6.6, and is reapplied to an ion-exchange column equilibrated in the same buffer, and then eluted with a linear salt gradient (0.00–0.10 M NaCl). Hemoglobins A_{1a}, A_{1b}, and A_{1c} sequentially elute from the column under these conditions (Fig. 2). If necessary, fractions containing Hb A_{1c} can be reapplied to the column and eluted as a homogeneous peak with the 0.00–0.10 M NaCl salt gradient.

Purification of Non-A_{1c} Glycohemoglobin

Separation of non-A_{1c} glycohemoglobin from (unglycated) hemoglobin A_0 is based on the ability of phenyl boronate in alkaline solution to complex with *cis*-diol groups of sugars. The hemoglobin A_0 peak from ion-exchange chromatography is dialyzed against 0.01 M Tris–acetate buffer, pH 8.5, containing 0.75 g/liter asparagine, 1.25 g/liter taurine, 0.746 g/liter methionine, 0.15 M NaCl, and 10 mM MgCl$_2$, and applied to a column of m-aminophenylboronic acid agarose gel equilibrated with the same buffer. The nonadsorbed fraction (Hb A_0) is eluted with this buffer, and the adsorbed (glycohemoglobin) fraction is then eluted by making the buffer

[16] A. S. Acharya and J. M. Manning, *J. Biol. Chem.* **255**, 1406 (1980).
[17] Y. Enoki, Y. Ohga, A. Kaneko, and H. Kohzuki, *Hemoglobin* **6**, 143 (1982).
[18] Y. Enoki, Y. Ohga, S. Sakata, H. Kohzuki, and S. Shimuzu, *Hemoglobin* **10**, 607 (1986).
[19] E. C. Abraham, M. Stallings, A. Abraham, and R. Clardy, *Biochim. Biophys. Acta* **774**, 335 (1983).

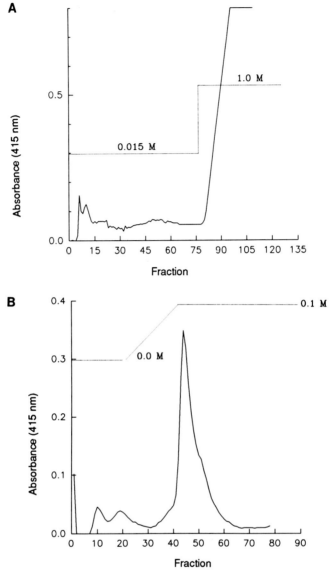

FIG. 2. Purification of hemoglobin A_{1c} on ion-exchange chromatography. About 400 mg of hemoglobin from lysed erythrocytes applied to a 2.6×30 cm column of Bio-Rex 70 equilibrated with $0.01\ M$ KCN and $0.015\ M$ NaCl. (A) Initial column; (B) reapplication of fast-moving peak for separation of Hb A_{1c} from other N-terminally modified hemoglobins.

100 mM in sorbitol lacking $MgCl_2$ and NaCl. If necessary, the glycohemoglobin isolate can be further purified by repassage on phenyl boronate; additional passages may be required because nonspecific adsorption of unglycated species to phenyl boronate can occur, and the phenyl boronate resin may have variable binding affinity and capacity. In an alternative schedule for separation on phenyl boronate of glycated hemoglobin from hemoglobin A_0, which may increase yield, the peak from ion-exchange chromatography is dialyzed against 0.05 M ammonium acetate, pH 9.0, containing 0.05 M KCN and 1 mM $MgCl_2$ and the nonadsorbed fraction (A_0) is eluted with the same buffer. The column is then washed with 0.15 M NaCl/0.05 M Tris, pH 7.4, containing 0.05 M KCN, and the adsorbed (glycated) hemoglobin is eluted with this buffer made 200 mM in sorbitol.

The development of monoclonal antibodies specific to the deoxyfructosyllysine residues in glycohemoglobin allows purification of non-A_{1c} glycohemoglobin from Hb A_0 by immunoaffinity chromatography.[20,21] The monoclonal antibody is immobilized onto Sepharose by coupling to CNBr-activated Sepharose 4B. The column is equilibrated with 0.1 M acetate buffer, pH 8.5, containing 0.15 M NaCl, and nonadsorbed hemoglobin species are eluted with the same buffer. Glycohemoglobin binding to the immobilized monoclonal antibodies is eluted by making the buffer 0.05 M in diethylamine, which brings the pH to 11.5. When the hemoglobin A_0 peak from ion-exchange chromatography is applied, this procedure yields purified non-A_{1c} glycohemoglobin containing the epitope specifically recognized by the antibody; if the applied hemoglobin contains some Hb A_{1c}, those Hb A_{1c} molecules that also are glycated at the ε-amino lysine site will bind to the immunoaffinity column.[21]

HPLC Analysis of Glycohemoglobin Preparations

HPLC is helpful in screening ion-exchange chromatographic fractions to identify those fractions that contain the hemoglobin of interest and are relatively free of other species, thereby facilitating the purification process; HPLC is useful as well as for validation of the purity of the preparation. This is particularly true for the N-terminally modified hemoglobins A_{1a}, A_{1b}, and A_{1c}, in which minor differences in the isoelectric points can be exploited. Most HPLC columns do not separate non-A_{1c} glycohemoglobins from Hb A_0; HPLC columns containing a phenyl boronate resin will selectively bind glycohemoglobin species, but will not distinguish A_{1c}

[20] L. A. Steward, V. Y. Wu, E. Shea, and M. P. Cohen, *J. Immunol. Methods* **140,** 145 (1991).
[21] V. Y. Wu, L. A. Steward, and M. P. Cohen, *Biochem. Biophys. Res. Commun.* **176,** 207 (1991).

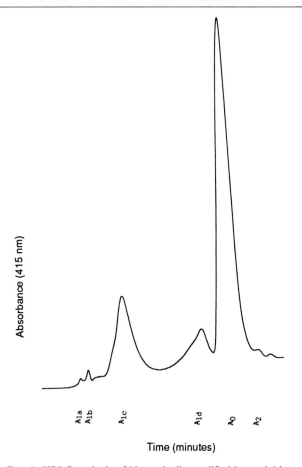

FIG. 3. HPLC analysis of N-terminally modified hemoglobins.

from non-A_{1c} glycohemoglobins. SP-5PW (Waters, Milford, MA), a cation-exchange resin in which a sulfopropyl group is introduced onto a 10-μm hydrophilic rigid resin, affords good separation of N-terminally modified hemoglobins from each other and from Hb A_0 (Fig. 3). The column is preequilibrated with 96% buffer A (0.0124 M NaH_2PO_4, 0.0032 M KCN, 100 ml of acetonitrile added to each liter of buffer; pH titrated to 7.0) and 4% buffer B (0.12 M NaH_2PO_4, 100 ml of acetonitrile added to each liter of buffer; pH titrated to 6.54) and run at a flow rate of 1.2 ml/min for a total of 16 min. The elution is programmed to change the ratios of buffer A to buffer B from 96/4 at 0, 70/30 at 5 min, 30/70 at 10 min, and 0/100 at 11 min. This HPLC program does not detect the presence in any of

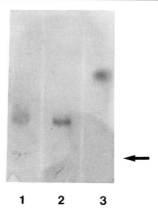

1   2   3

Fig. 4. Agarose gel electrophoresis of non-A_{1c} glycohemoglobin (lane 1), hemoglobin A_0 (lane 2), and hemoglobin A_{1c} (lane 3). Gel is oriented with anode at top.

the peaks (i.e., A_{1a}, A_{1b}, A_{1c}, or A_0) of hemoglobin glycated in the non-A_{1c} position, presumably because the charge modification resulting from glycation of lysine amino groups is insufficient to be registered by this analytic system.

Gel Electrophoretic Analysis

Electrophoretic mobilities of glycohemoglobins and Hb A_0 on SDS-polyacrylamide gels are similar, although some microheterogeneity can be observed, consistent with minor modification of charge and molecular weight as a result of nonenzymatic glycation of N-terminal and/or lysine residues. Agarose gel electrophoresis affords better discrimination; when performed on citrate agarose gels in 0.1 M sodium citrate, pH 6.2, Hb A_{1c} migrates substantially faster than Hb A_0 and non-A_{1c} glycohemoglobin migrates slightly more toward the anode than does Hb A_0, consistent with its slightly greater electronegativity as a result of glycation of lysine amino groups (Fig. 4).

Carbohydrate Analysis

The amount of hydroxymethylfurfuraldehyde (HMF) formed on reaction with thiobarbituric acid[22] can be used for analysis of the carbohydrate content of glycohemoglobin preparations. Because the inherent color of

[22] K. A. Ney, K. J. Colley, and S. V. Pizzo, *Anal. Biochem.* **118**, 294 (1981).

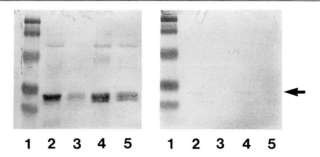

FIG. 5. SDS-gel electrophoresis stained with Amido black (left) and nitrocellulose transfer immunoblotted with LAS34 monoclonal antibody against Hb A_{1c} (right). Lane 1, Molecular weight standard; lane 2, Hb A_{1c}; lane 3, non-A_{1c} glycohemoglobin; lane 4, Hb A_{1c}; lane 5, Hb A_0. A single immunoreactive band (arrow) consistent with hemoglobin A_{1c} monomer is seen in lanes 2 and 4 on the right.

hemoglobin may spuriously elevate absorbance values, samples should be blanked against Hb A_0 subjected to the same procedure. Results are expressed as moles of HMF/mole of hemoglobin, which should have a ratio of at least 1 mol/mol of β-chain monomer for Hb A_{1c} and may have 1–4 mol/mol of tetramer for non-A_{1c} glycohemoglobin. However, HMF release may not be quantitative.[11]

Immunoblotting

Highly specific monoclonal antibodies reactive with glucose in keto-amine linkage with the N-terminal valine residue or with the lysine amino group of hemoglobin can be used for validation of glycohemoglobin preparations and confirmation of their purity. The antibodies must be specific for the relevant glycated epitope in hemoglobin, because cross-reactivity with (unglycated) hemoglobin or with other (glycated) proteins may confound results. Two antibodies described in the literature seem to meet these criteria.[20,23] Hemoglobin A_{1c}, but not Hb A_{1a}, Hb A_{1b}, Hb A_0, or non-A_{1c} glycohemoglobin, will immunoreact with monoclonal antibodies that were raised against a glycated synthetic peptide that comprised several amino acid units corresponding to the N terminus of the hemoglobin β chain. The peptide was glycated *in vitro* at the N-terminal valine position and coupled to an immunogenic carrier. Another monoclonal antibody (LAS34) raised against Hb A_{1c} isolated from normal human erythrocyte lysates shows specific immunoreactivity with Hb A_{1c} (unpublished), and absence of reactivity with Hb A_0 or with hemoglobin glycated in non-A_{1c} positions (Fig. 5). Non-A_{1c} glycohemoglobin, but not Hb A_0, Hb A_{1a},

[23] W. J. Knowles, V. T. Marchesi, and W. Haign, U.S. Pat. 4,727,036 (1988).

Hb A_{1b}, or Hb A_{1c} (unless it also is glycated at a lysine residue) will immunoreact with monoclonal antibodies raised against non-A_{1c} glyco-hemoglobin purified from human erythrocyte lysates.[20]

Acknowledgments

This work was supported in part by grants from the Ben Franklin Fund of the Commonwealth of Pennsylvania and from the Department of Human and Health Services (HL 44767).

[6] Measurement of Endotoxin Levels in Hemoglobin Preparations

By ROBERT I. ROTH and JACK LEVIN

Endotoxin (lipopolysaccharide, LPS) is a glycolipid component of the outer portion of the cell wall of gram-negative bacteria (Fig. 1). Bacterial endotoxin, which is responsible for the fever, disseminated intravascular coagulation, hypotension, and cardiovascular shock that frequently accompany gram-negative septicemia, produces biological and pathophysiological effects at very low concentrations (in the picogram per milliliter range), even in the absence of living bacteria. Because endotoxemia is associated with substantial morbidity and mortality, FDA guidelines require that sufficiently low endotoxin levels be documented to allow human use of parenteral drugs or medical devices. This review will focus on the use of the *Limulus* amebocyte lysate test to measure endotoxin levels in hemoglobin preparations, but will also describe its more general applications for the detection of endotoxin in protein solutions and blood.

Limulus Amebocyte Lysate Test

The *Limulus* amebocyte lysate (LAL) test is an *in vitro* biological assay for endotoxin, based on endotoxin activation of the horseshoe crab coagulation cascade.[1] Lysates of the amebocytes (blood cells) of the North American horseshoe crab *Limulus polyphemus* (or the Japanese horseshoe crab *Tachypleus tridentatus*) contain an endotoxin-activated proteolytic coagulation cascade[2] that is exquisitely sensitive to bacterial endotoxins. This *in vitro* assay has replaced the older rabbit pyrogen test for endotoxin,

---

[1] J. Levin and F. B. Bang, *Bull. Johns Hopkins Hosp.* **115**, 265 (1964).
[2] N. S. Young, J. Levin, and R. A. Prendergast, *J. Clin. Invest.* **51**, 1790 (1972).

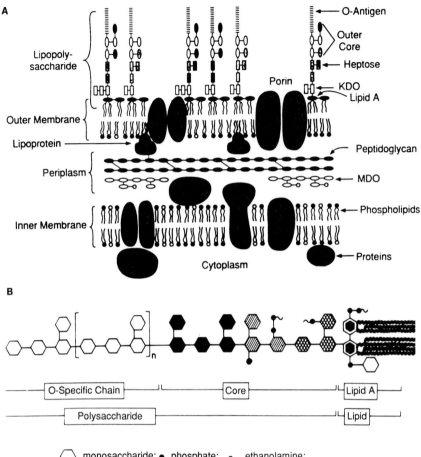

FIG. 1. (A) Schematic representation of the *E. coli* envelope. Lipopolysaccharide is located in the external portion of the outer membrane. Ovals and rectangles depict sugar residues. Circles represent the polar heads of phospholipids. MDO, Membrane-derived oligosaccharides; KDO, 3-deoxy-D-*manno*-octulosonic acid. Reproduced, with permission, from the *Annual Review of Biochemistry,* Vol. 59, © 1990 by Annual Reviews Inc. (B) Schematic structure of lipopolysaccharide (*Salmonella*). Repeating units of the O-specific chain plus core oligosaccharide form the hydrophilic polysaccharide portion of lipopolysaccharide. The hydrophobic lipid A portion consists of a β-1,6-linked D-glucosamine disaccharide backbone substituted with phosphoryl, phosphorylethanolamine, arabinose, and fatty acyl groups. Reproduced with permission from E. Th. Rietschel *et al., Scand. J. Infect. Dis.* (Suppl.) **31,** 10 (1982).

and currently is used to monitor endotoxin in intravenous fluids, parenteral drugs, and medical devices. The LAL test is the most sensitive assay currently available for the detection of endotoxin and can detect crude, native LPS associated with the bacterial cell wall (Fig. 1A), highly purified lipopolysaccharide, and lipid A (Fig. 1B). The minimum concentration of endotoxin detectable by LAL reagents is typically 5–10 pg/ml, although specific batches of LAL may be sufficiently sensitive to quantify endotoxin in the femtogram per milliliter range. Sensitivity also depends on the biological activity of a specific endotoxin.

Activation of the LAL coagulation cascade by endotoxin leads to the subsequent activation of a trypsinlike serine protease, clotting enzyme, which cleaves the clottable protein coagulogen. Proteolytic activation of coagulogen to coagulin by clotting enzyme is followed by polymerization of coagulin, thus resulting in formation of a gel. The rate of this process is directly proportional to the endotoxin concentration,[1,3] and the rate of LAL activation can be used to quantify endotoxin concentrations over a very wide range (e.g., from 1 pg/ml to at least 100 ng/ml).[4] An example of the relationship between rates of gelation and endotoxin concentration is shown in Fig. 2 using a lysate with a lesser sensitivity range.

Three different end points have been utilized for the LAL assay (Table I). First, gelation (formation of a solid gel)[3] is the basis of the gel-clot, pass/fail test that is frequently used for manufacturing and other commercial applications. In this assay, a small volume of test sample (e.g., 50 μl) and a similarly small volume of lysate (or lyophilized reagent) are incubated together at 37° for a fixed amount of time, typically 1 hr, and observed for generation of a solid, translucent gel (Table I and Fig. 3). The lysate has been characterized previously to determine the minimal concentration of LPS required to generate a solid gel within the selected time (cutoff concentration), and therefore the LPS concentration in an unknown sample can be determined to be either greater or less than the gelation cutoff. Samples that gel the lysate can be retested after a series of dilutions to obtain further quantification. In addition, the actual time of gelation can be utilized to provide a more precise measurement of endotoxin concentration. Furthermore, by the recognition of earlier clotting stages (flocculation and increased viscosity)[1] it is possible to use the LAL gelation reaction as a quantitative assay that not only becomes more sensitive but can measure endotoxin over a wider range of concentrations. Second, spectrophotometric measurement of increasing turbidity (related to coagulin polymerization)[3] is a quantitative, automated version of the LAL test that

[3] J. Levin and F. B. Bang, *Thromb. Diath. Haemorrh.* **19,** 903 (1970).
[4] J. F. Cooper, J. Levin, and H. N. Wagner, Jr., *J. Lab. Clin. Med.* **78,** 138 (1971).

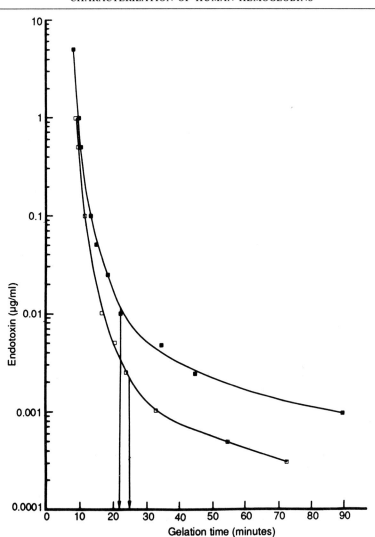

FIG. 2. Gelation of *Limulus* amebocyte lysate by *E. coli* (■) and *Klebsiella* (□) endotoxins. Incubation mixtures containing 0.1 ml *Limulus* lysate and 0.1 ml endotoxin were incubated at 37°. Gelation times are proportional to endotoxin concentration over a wide range of endotoxin concentrations [from J. F. Cooper, J. Levin, and H. N. Wagner, Jr., *J. Lab. Clin. Med.* **78**, 138 (1971), with the permission of the publisher].

makes use of the characteristic lag period and subsequent rate of increase in light scattering, which are endotoxin concentration-dependent (Table I and Fig. 4). An automated method for determination of endotoxin concentration by measurement of increased viscosity also has been de-

TABLE I
PERFORMANCE OF *Limulus* AMEBOCYTE LYSATE TEST FOR QUANTIFICATION OF
ENDOTOXIN CONCENTRATION IN SAMPLES OF SFH

1. To ensure that endotoxin is not introduced into the SFH sample to be assayed, heat all glassware (180°, 3–4 hr), or utilize sterile plasticware previously documented to be endotox-in-free (<10 pg/ml endotoxin, based on currently utilized standards, is usually satisfactory)
2. Transfer 50-μl aliquots of SFH into either endotoxin-free borosilicate glass test tubes or sterile microtiter plate wells
3. Add 50 μl of *Limulus* amebocyte lysate (LAL), the sensitivity of which has been previously determined utilizing a common reference lipopolysaccharide (e.g., *E. coli* 055:B5 or EC-5) for the standard curve
4. Incubate mixtures at 37° and monitor end points of LAL activation as follows:
    (a) For gel-clot assay, observe the mixture for the presence of a solid gel at a standard time, e.g., 1 hr
    (b) For gelation assay with increased sensitivity, observe the mixture at 15- to 30-min intervals for pregelation changes (weak flocculation, heavy flocculation, increased viscosity) and ultimately for gelation time (if endotoxin concentration is sufficient)
    (c) For turbidimetric assay, monitor increase in light scattering nephelometrically or monitor increase in spectrophotometric absorbance (305 nm)
    (d) For chromogenic assay, preincubate the SFH–LAL mixture for a standard time at 37°, e.g., 10–30 min, then add 100–200 μl of chromogenic substrate S-2222 or S-2423 (Kabi Vitrum, Molndal, Sweden), 0.25–0.5 m*M*, pH 7.5–8.0, and monitor amidolytic activity at 405 nm. Absorbance at 405 nm can be determined at a fixed time after addition of chromogenic substrate, e.g., 10–30 min, or kinetic analysis of the rate of increase in absorbance at 405 nm can be performed. Detection of amidolytic activity at 405 nm may be optimized by conversion of hemoglobin to carboxyhemoglobin, to minimize interference by oxyhemoglobin or methemoglobin in the spectrophotometric detection of the enzymatic activity.[30]

scribed.[5] Third, spectrophotometric measurement of the rate of generation of clotting enzyme activity, which is also endotoxin concentration dependent and is typically measured with a chromogenic or fluorogenic substrate for the protease,[6] is the basis for an automated, quantitative assay for the protease that generates the subsequent turbidity (Table I and Fig. 5). If the chromogenic LAL assay is utilized, additional sensitivity for lower concentrations of LPS can be achieved with derivatization of the chromogen with diaminocinnamaldehyde.[7] The concentration of endotoxin in a sample is quantified by determining its LAL reactivity with any of the above described methods and relating LAL activation to that produced by a series of concentrations of a standard endotoxin (i.e., a standard curve).

[5] R. Homma, Y. Takada, I. Karube, K. Kimura, and H. Muramatsu, *Anal. Biochem.* **204,** 398 (1992).
[6] P. Friberger, M. Knos, and L. A. Mellstam, *in* "Endotoxins and their Detection with the *Limulus* Amebocyte Lysate Test" (S. W. Watson, J. Levin, and T. J. Novitsky, eds.), p. 195. Alan R. Liss, New York, 1982.
[7] R. I. Roth, J. Levin, and S. Behr, *J. Lab. Clin. Med.* **114,** 306 (1989).

FIG. 3. Gelation of *Limulus* amebocyte lysate by bacterial endotoxin. *Limulus* lysate initially is a clear, colorless fluid (left). After addition of endotoxin and incubation at 37°, the lysate becomes an opaque, solid gel (right) [from J. Levin, *in* "Blood Cells of Marine Invertebrates" (W. Cohen, ed.), p. 153, Alan R. Liss, New York, © copyright 1985. Reprinted with permission of Wiley-Liss, a division of John Wiley and Sons, Inc.].

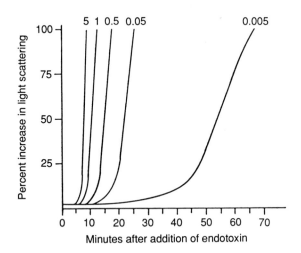

FIG. 4. Increase in turbidity of *Limulus* lysate after addition of bacterial endotoxin. After 0.1 ml *E. coli* endotoxin (0.005–5 μg/ml) was added to 0.9 ml *Limulus* lysate, the mixtures were incubated at 37° and light scattering (turbidity) was measured photofluorometrically. The endotoxin concentration in each incubation mixture is indicated at the top of each curve. At low endotoxin concentrations, there is a substantial lag period followed by a gradual increase in light scattering. At higher concentrations of endotoxin, there is a concentration-dependent decrease in the lag period and increase in the rate of increased turbidity [from J. Levin, P. A. Tomasulo, and R. S. Oser, *J. Lab. Clin. Med.* **75,** 903 (1970), with the permission of the publisher].

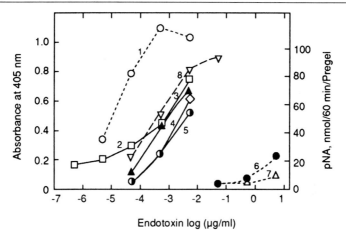

Endotoxin log (µg/ml)

FIG. 5. Increase in amidase activity in *Limulus* lysate after addition of bacterial endotoxin. Preparations of *E. coli* endotoxins (1–5) and *E. coli* polysaccharides (6–8) were added to *Limulus* lysate, followed by the addition of the chromogenic substrate *tert*-butoxycarbonyl-Val-Leu-Gly-Arg-*p*-nitroaniline. Amidase activity (cleavage of the substrate by *Limulus* clotting enzyme) was quantified by determining the absorbance at 405 nm by free *p*-nitroaniline (pNA). Bacterial endotoxins produce a concentration-dependent increase in absorbance at 405 nm, whereas polysaccharides are many orders of magnitude less reactive [from T. Harada-Suzuki, T. Morita, S. Iwanaga, S. Nakamura, and M. Niwa, *J. Biochem. (Tokyo)* **92,** 793 (1982), with the permission of the authors].

Determination of endotoxin concentration in a sample with the LAL test is rapid (typically completed within 1–2 hr) and reproducible [coefficients of variation (CV) for replicate samples using common commercial LAL reagents are ±10%]. The results are generally in good agreement with the standardized *in vivo* assay for endotoxin, the rabbit pyrogen test, and with many other (although less sensitive) *in vitro* assays for endotoxin.[4,8,9] Comparison of *Limulus* amebocyte test assays for endotoxin, performed in our laboratory, with rabbit pyrogen test assays of five preparations of stroma-free hemoglobin* has demonstrated that for three hemoglobin preparations containing ≥100 pg/ml endotoxin according to the LAL test *in vitro,* fever was produced *in vivo,* whereas fever was not elicited by two similarly purified hemoglobin preparations that contained ≤10 pg/ml endotoxin.

---

* Stroma-free hemoglobin (SFH) designates purified hemoglobin that contains negligible amounts of erythrocyte membrane material or other cytosolic components.
[8] J. van Noordwijk and Y. de Jong, *J. Biol. Stand.* **4,** 131 (1976).
[9] P. A. Tomasulo, *in* "Biomedical Applications of the Horseshoe Crab (Limulidae)" (E. Cohen, F. B. Bang, J. Levin, *et al.,* eds.), p. 293. Alan R. Liss, New York, 1979.

Contribution of Endotoxin to *in Vivo* Toxicity of Hemoglobin

Variably pure solutions of hemoglobin have been reported to produce pathophysiological effects when infused into humans. Hypertension and bradycardia have been commonly observed[10,11] and a decrease in glomerular filtration rate and renal plasma flow has been described.[12] Mild prolongations of the partial thromboplastin time also have been described.[11] In some animal studies, preparations of hemoglobin have been shown to produce fever, disseminated intravascular coagulation with resultant thrombosis, and ischemic parenchymal damage.[13,14] Toxicity was due in part to the presence of environmental LPS and in part to the presence of stromal phosphatidylethanolamine and phosphatidylserine, which were shown to produce death in rabbits, in contrast to other tested phospholipids.[13] Some of these changes were observed following administration of hemoglobin reported to be endotoxin free.[14] Similarly, hemoglobin contaminated with membrane phospholipids or LPS caused, *in vitro,* the release of procoagulant activity (tissue factor) and thromboxane from cultured human endothelial cells.[15] Furthermore, major pathological sequelae have been observed after infusion of hemoglobin, in the absence of detectable LPS or phospholipids, and despite the absence of abnormalities in tests of blood coagulation.[16] However, in other animal studies, hemoglobinemia produced by infusion of hemolyzed blood was not associated with laboratory evidence of disseminated intravascular coagulation or with parenchymal organ damage.[17]

Although endotoxin contamination appears to be responsible for some of the *in vivo* toxicity of hemoglobin, there is also evidence that hemoglobin enhances the toxicity of LPS. A dose of bacteria (*Escherichia freundii*), that when administered either alone or in combination with intact red blood cells caused no mortality in dogs, resulted in a 100% mortality when administered in conjunction with a solution of hemoglobin produced by hemolysis of red blood cells.[18] The mechanism of death was believed to be blockade of the reticuloendothelial system with hemoglobin, with

---

[10] W. R. Amberson, J. J. Jennings, and C. M. Rhode, *J. Appl. Physiol.* **1**, 469 (1949).

[11] J. P. Savitsky, J. Doczi, J. Black, and J. D. Arnold, *Clin. Pharmacol. Ther.* **23**, 73 (1978).

[12] J. L. Brandt, N. R. Frank, and H. C. Lichtman, *Blood* **6**, 1152 (1951).

[13] M. Feola, J. Simoni, P. C. Canizaro, R. Tran, and G. Raschbaum, *Surg. Gynecol. Obstet.* **166**, 211 (1988).

[14] D. H. Marks, T. Cooper, and T. Makovec, *Mil. Med.* **154**, 180 (1989).

[15] M. Feola, J. Simoni, D. Fishman, R. Tran, and P. C. Canizaro, *Artif. Organs* **13**, 209 (1989).

[16] C. T. White, A. J. Murray, J. R. Greene, D. J. Smith, F. Medina, G. T. Makovec, E. J. Martin, and R. B. Bolin, *J. Lab. Clin. Med.* **108**, 121 (1986).

[17] J. I. Spector, C. B. Wilson, and W. H. Crosby, *Am. J. Pathol.* **74**, 567 (1974).

[18] M. S. Litwin, C. W. Walter, P. Ejarque, and E. S. Reynolds, *Ann. Surg.* **157**, 485 (1963).

resultant activation of blood coagulation by bacterial endotoxin and production of the generalized Shwartzman reaction. Synergism also was demonstrated in a study in which the combination of SFH and endotoxin produced a 50–100% mortality rate in recipient rabbits at doses of endotoxin and SFH that did not cause death when administered alone.[19] Paradoxically, the incubation of ferrous iron ($FeSO_4$) with bacterial endotoxin prevented the 50% mortality and production of the generalized Shwartzman reaction that had been produced in rabbits by the administration of control preparations of endotoxin.[20] *In vitro*, either SFH or LPS stimulated the generation of procoagulant activity (tissue factor) by both human and rabbit mononuclear cells, with the combination of SFH and *Escherichia coli* LPS producing an additive effect.[21] In these studies, stimulation of procoagulant activity was associated with heme rather than globin.

In summary, it appears that hemoglobin may have inherent pathophysiological effects, some of which may become more manifest if hemoglobin is administered to an already unstable, physiologically compromised recipient. Under such circumstances, e.g., shock or damage to the colon, it would not be surprising that the potential toxicity of hemoglobin would be increased if preparations of hemoglobin are contaminated with bacterial endotoxins or the recipient of hemoglobin is experiencing endotoxemia. In addition, the potential toxicities of hemoglobin and LPS may be synergistic.

Quality Control Monitoring of Hemoglobin Production

LAL testing has been employed for quality control monitoring of the production of chemically modified [bis(3,5-dibromosalicyl) fumarate cross-linked] stroma-free hemoglobin (DBBF-SFH), as described previously.[22] An abbreviated flow scheme for purification of SFH is presented in Fig. 6; results of LAL testing of the production apparatus (prior to hemoglobin production) and of the hemoglobin product at various stages of its purification are presented in Fig. 7. An evaluation for endotoxin with the LAL test is recommended at each step of the production process, because contamination of hemoglobin by environmental endotoxin can occur at any stage. Commonly, endotoxin contamination of hemoglobin preparations occurs as the result of a localized loss of nonpyrogenic (i.e.,

[19] C. T. White, A. J. Murray, D. J. Smith, J. R. Greene, and R. B. Bolin, *J. Lab. Clin. Med.* **108,** 132 (1986).
[20] A. Janoff and W. Zweifach, *J. Exp. Med.* **112,** 23 (1960).
[21] D. J. Smith and R. M. Winslow, *J. Lab. Clin. Med.* **119,** 176 (1992).
[22] R. M. Winslow, K. W. Chapman, and J. Everse, *Biomater. Artif. Cells, Immobilized Biotechnol.* **19,** 503 (1991).

Add 0.9% NaCl to red blood cells
↓
Add NaHPO₄ to red blood cells
↓
Lyse red blood cells
↓
0.1-μm filtration
↓
500-kDa ultrafiltration
↓
10-kDa ultrafiltration/concentration
↓
Deoxygenate hemoglobin
↓
Cross-link hemoglobin with bis(3,5-dibromosalicyl)fumarate (DBBF)
↓
Adjust to pH 9.0 with HEPES
↓
0.1-μm filtration
↓
500-kDa ultrafiltration
↓
10-kDa ultrafiltration/concentration
↓
Buffer exchange to pH 7.4 with Ringer's acetate
↓
Sterile fill cross-linked stroma-free hemoglobin (DBBF-SFH)

FIG. 6. Flow diagram for production of stroma-free hemoglobin; this production scheme is a simplification of that utilized by the Blood Research Division, Letterman Army Institute of Research, The Presidio, San Francisco, California.

endotoxin-free) conditions. An example of a midprocess contamination that was detected by LAL testing of samples at multiple stages is shown (Fig. 7). Because the site responsible for the endotoxin contamination was clearly identified, reestablishment of an endotoxin-free state at the specific step could be approached and accomplished efficiently. Although

SAMPLE	LPS (pg/ml)*
Water rinse of tank	<1
Water rinse of 0.45-μm membrane	<1
NaCl from wash tank	<1
Red blood cells from wash tank	<1
Concentrated red blood cells	<1
Sodium phosphate buffer	<1
Water rinse of 0.1-μm membrane	<1
Water rinse of 10-kDa membrane	<1
Water rinse of concentration tank	<1
Hemoglobin postred blood cell lysis	<1
0.1-μm membrane permeate SFH	10
0.1-μm membrane retentate (stroma)	100**
500-kDa membrane retentate	10
500-kDa membrane filtrate SFH	10
Concentrated SFH	100**
Bagged DBBF-SFH final product	300**

FIG. 7. LAL testing of the production facility and selected steps during the production of an endotoxin-contaminated solution of cross-linked stroma-free hemoglobin (DBBF–SFH). *, Based on an *E. coli* B 055:B5 standard; **, note that relatively high concentrations of endotoxin were detected in the 0.1-μm membrane retentate, in a sample of concentrated SFH, and in the final bagged product. This indicates that breaks in the endotoxin-free technique occurred during the removal of stroma and during the final concentration step.

endotoxin contamination of hemoglobin frequently follows the addition of reagents or sample manipulations (e.g., stabilization of hemoglobin by chemical cross-linking with DBBF or other cross-linking agents), we also have observed that the erythrocytes that constitute the starting material may be contaminated with endotoxin, with subsequent contamination of the purified hemoglobin. Prevention of endotoxin contamination of hemoglobin is far more efficient and economical than attempted removal of endotoxin from the final protein product; therefore, in the following section, we describe several basic procedures to minimize the likelihood of endotoxin contamination.

Depyrogenation Techniques

It is important to recognize that endotoxin has a strong affinity for a wide variety of surfaces and that removal of endotoxin from equipment used for the production of any intravenous product can be technically difficult. Washing of the production apparatus with endotoxin-free solutions (such as those prepared for intravenous use) is often effective for the removal of endotoxin, although extremely large volumes of wash solutions may be required to render systems endotoxin free.** In the case of chromatographic columns, for example, 10–20 or more column volumes of wash solution may be required to generate a LAL-negative system. To make possible the production of endotoxin-free hemoglobin, all glassware and metal containers must be heated at high temperature (180–200° for 3–4 hr) to inactivate endotoxin. Solutions and buffers may be filtered through 10–100 kDa membranes to remove endotoxin, and the erythrocytes used as the starting source of hemoglobin should be LAL tested to establish that no contamination had occurred during procurement of the blood. Sterility is not adequate to ensure lack of endotoxin, because wet, 120° heat is not sufficient to denature endotoxin. Thus, any plasticware used during production of hemoglobin must be LAL tested. In-line filters and membranes can also be a source of endotoxin contamination, and therefore wash solutions should be passed through these sites and subsequently LAL tested prior to use of the apparatus for generation of hemoglobin. When a source of endotoxin contamination is detected, it may be possible to remove endotoxin by means of chemical denaturation. A variety of such treatments have been described (Table II), although the efficacy of these methods has not been reported for hemoglobin production. The use of lipopolysaccharide-binding substances (e.g., polymyxin B or an endotoxin-neutralizing protein), and potentially of other proteins from a recognized family of endotoxin-binding proteins,[23] may provide the basis for removal of endotoxin from solutions when these LPS-binding proteins are immobilized and used to remove endotoxin by affinity binding.

Product quality control requirements of the FDA for intravenous preparations include documentation by LAL testing of adequately low endotoxin concentrations. The acceptable endotoxin concentration threshold is defined according to endotoxin units (EU) per milliliter, because the biological activities, both in the *Limulus* test and *in vivo*, of various endotoxins are widely variable on a weight basis.[24–27] Commercial LAL re-

---

** Because pharmaceutical companies no longer routinely utilize the rabbit pyrogen test, it is technically inappropriate to use the term "pyrogen free."

[23] P. S. Tobias, J. C. Mathison, and R. J. Ulevitch, *J. Biol. Chem.* **263**, 13479 (1988).

[24] M. E. Weary, G. Donahue, F. C. Pearson, and K. Story, Appl. Environ. Microbiol. **40**, 1148 (1980).

[25] M. E. Weary, F. C. Pearson, J. Bohon, and G. Donahue, in "Endotoxins and their Detection with the *Limulus* Amebocyte Lysate Test" (S. W. Watson, J. Levin, and T. J. Novitsky, eds.), p. 365. Alan R. Liss, New York, 1982.

TABLE II
TECHNIQUES FOR DEPYROGENATION OF SOLUTIONS

Technique	Ref.
Ultrafiltration	*a, b*
Polymyxin B	*c, d*
Endotoxin-neutralizing protein	*e*
Acid hydrolysis	*f*
Basic hydrolysis	*g*
Oxidation by hydrogen peroxide	*h*
Alkylation with succinic anhydride	*i*
Barium sulfate adsorption	*j*
Endotoxin–protein dissociation with octyl-β-D-glucopyranoside	*k*
Ultracentrifugation	*l*
Endotoxin removal by LAL gelation	*m*

a K. J. Sweadner, M. Forte, and L. L. Nelson, *Appl. Environ. Microbiol.* **34,** 382 (1977).

b L. W. Henderson and E. Beans, *Kidney Int.* **14,** 522 (1978).

c D. C. Morrison and D. M. Jacobs, *Immunochemistry* **13,** 813 (1976).

d D. M. Jacobs and D. C. Morrison, *J. Immunol.* **118,** 21 (1977).

e N. R. Wainwright, R. J. Miller, E. Paus, T. J. Novitsky, M. A. Fletcher, T. M. McKenna, and T. Williams, *in* "Cellular and Molecular Aspects of Endotoxin Reactions" (A. Nowotny, J. J. Spitzer, and E. J. Ziegler, eds.), p. 315. Elsevier, Amsterdam, 1990.

f O. Luderitz, C. Galanos, V. Lehmann, M. Nurminen, E. T. Rietschel, G. Rosenfelder, M. Simon, and O. Westphal, *in* "Bacterial Lipopolysaccharides" (E. H. Kass and S. M. Wolff, eds.), p. 9. Univ. of Chicago Press, Chicago, 1973.

g M. Niwa, K. C. Milner, E. Ribi, and J. A. Rudbach, *J. Bacteriol.* **97,** 1069 (1969).

h F. A. DeRenzis, *J. Dent. Res.* **60,** 933 (1981).

i J. R. Schenck, M. P. Hargie, M. S. Brown, D. S. Evert, A. L. Yoo, and F. C. McIntire, *J. Immunol.* **102,** 1411 (1969).

j P. S. Reichelderfer, J. F. Manischewitz, M. A. Wells, H. D. Hochstein, and F. A. Ennis, *Appl. Microbiol.* **30,** 333 (1975).

k T. E. Karplus, R. J. Ulevitch, and C. B. Wilson, *J. Immunol. Methods* **105,** 211 (1987).

l R. K. Shadduck, A. Waheed, A. Porcellini, V. Rizzoli, and J. Levin, *Proc. Soc. Exp. Biol. Med.* **164,** 40 (1980).

m F. R. Rickles, J. Levin, E. Atkins, and P. Quesenberry, *in* "Biomedical Applications of the Horseshoe Crab (Limulidae)" (E. Cohen, F. B. Bang, J. Levin *et al.,* eds.), p. 485. Alan R. Liss, New York, 1979.

[26] F. C. Pearson, M. E. Weary, J. Bohon, and R. Dabbah, *in* "Endotoxins and their Detection with the *Limulus* Amebocyte Lysate Test" (S. W. Watson, J. Levin, and T. J. Novitsky, eds.), p. 65. Alan R. Liss, New York, 1982.

[27] R. Homma, K. Kuratsuka, and K. Akama, *in* "Endotoxins and their Detection with the *Limulus* Amebocyte Lysate Test" (S. W. Watson, J. Levin, and T. J. Novitsky, eds.), p. 301. Alan R. Liss, New York, 1982.

agents are available that demonstrate sufficient sensitivity and reproducibility to ensure that hemoglobin products satisfy the FDA requirements. Parenteral formulations such as hemoglobin for resuscitation have an allowable endotoxin concentration (termed the endotoxin limit) that varies according to the volume of infusate and weight of the recipient: endotoxin limit = 5.0 EU/kg/maximal human dose (as prescribed by the FDA).[28]

Technical Issues Related to LAL Testing of Hemoglobin

The following technical considerations concerning endotoxin testing of hemoglobin preparations are important. First, different endotoxins may demonstrate significant differences both in their *in vivo* physiological effects and *in vitro* biological activities. Environmental endotoxins that can cause contamination of hemoglobin are variable in regard to bacterial species, purity (e.g., extent of association of bacterial lipid, protein, and nucleic acids with lipopolysaccharide), and relative biological potencies, as determined by the LAL test, and are not necessarily similar to a standard endotoxin preparation.[26,27]

Second, there are uncertainties in the derivation of the concentration of an unknown endotoxin from a standard endotoxin curve. Standard reference endotoxins (e.g., EC-5) are much more purified than are environmental endotoxins, and their LAL reactivity on a weight basis may be significantly different from that of environmental endotoxins. Nevertheless, for purposes of quality control, we recommend the use of stable, available standard endotoxins such as EC-5 or *E. coli* B 055:B5, which are not totally purified. These endotoxin preparations are structurally and chemically more similar to "native" environmental endotoxins than are synthetic endotoxins, endotoxin partial structures, or lipid A. They are also similar to the gut-derived endotoxins with which infused hemoglobin solutions are likely to interact in patients with major trauma and bacteremia. If necessary, the activity of the endotoxin selected as a standard can be related to the activity of EC-5, the current FDA standard endotoxin.

Third, detection of endotoxin in protein solutions with the LAL test (or other tests for endotoxin) may be complicated by the phenomena of enhancement and inhibition.[29] In studies of cross-linked stroma-free

[28] "Guideline on Validation of the *Limulus* Amebocyte Lysate Test as an End-product Endotoxin Test for Human and Animal Parenteral Drugs, Biological Products, and Medical Devices." Food and Drug Administration, Division of Manufacturing and Product Quality, Office of Compliance, Center for Drug Evaluation and Research, Rockville, MD, 1987.

[29] T. E. Munson, *in* "Endotoxins and their Detection with the *Limulus* Amebocyte Lysate Test" (S. W. Watson, J. Levin, and T. J. Novitsky, eds.), p. 25. Alan R. Liss, New York, 1982.

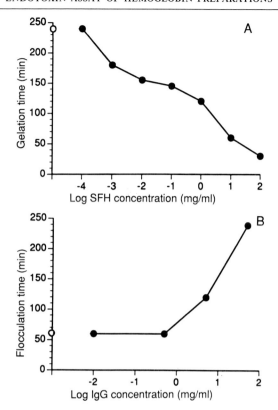

FIG. 8. (A) Enhancement of *Limulus* lysate reactivity for bacterial endotoxin by stroma-free hemoglobin (SFH). The 0.05-ml *Limulus* lysate sample was incubated with 0.05 ml of *E. coli* endotoxin (100 ng/ml) at 37°, in the presence of various concentrations (●) of endotoxin-free stroma-free hemoglobin (100 ng/ml–100 mg/ml). In the absence of added hemoglobin (○), gelation of *Limulus* lysate was observed in 240 min. Gelation time of *Limulus* lysate was decreased by the addition of hemoglobin in a concentration-dependent manner. This results in an apparent increase in the concentration of endotoxin. (B) Inhibition of *Limulus* lysate reactivity for bacterial endotoxin by IgG. The 0.05-ml *Limulus* lysate sample was incubated with 0.05 ml of *E. coli* endotoxin (100 pg/ml) at 37°, in the presence of various concentrations of endotoxin-free IgG (10 μg/ml–50 mg/ml) (●). In the absence of IgG (○) or in the presence of 0.01 or 0.5 mg/ml IgG, flocculation of *Limulus* lysate was observed after 60 min, with gelation at 3 hr. In the presence of 5 or 50 mg/ml IgG, *Limulus* lysate did not form a gel, and flocculation time was increased in a concentration-dependent manner.

hemoglobin, enhanced detection of spiked endotoxin was shown to be directly related to the concentration of hemoglobin (Fig. 8A). This potential enhancement effect, which was not detected in a previous report,[30] should be taken into account. It may be necessary to determine whether

[30] M. Feola, J. Simoni, and P. C. Canizaro, *Artif. Organs* **15,** 243 (1991).

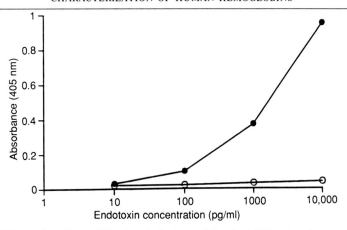

FIG. 9. Detection of bacterial endotoxin in plasma following inhibitor inactivation. Human plasma was spiked with *E. coli* endotoxin (final concentrations from 10 pg/ml to 10 ng/ml), and aliquots then were diluted 10-fold in endotoxin-free 0.15 M NaCl and heated at 65° for 30 min. Other aliquots were untreated. The 0.05-ml plasma samples subsequently were added to 0.05 ml of *Limulus* lysate, incubated at 37° for 30 min, and then chromogenic substrate S-2222 was added. Absorbance of released pNA was measured at 405 nm. Spiked endotoxin was not detected in untreated plasma (○), whereas dilution and heating of plasma (●) resulted in concentration-dependent detection of bacterial endotoxin.

the apparent concentration of endotoxin, which may have been elevated secondary to enhancement, is physiologically relevant when evaluated by other *in vivo* endotoxin assays or by the rabbit pyrogen test, and eventually by administration to humans. Inhibition of detection of endotoxin in the LAL test also is a well-recognized problem with some protein solutions for intravenous use, as illustrated in Fig. 8B for γ-globulin. In the presence of γ-globulin, inhibition of detection of spiked endotoxin was shown to be directly related to IgG concentration. Inhibition also has been shown to interfere with the detection and/or quantification of endotoxin in blood, and, therefore, *in vivo* studies of endotoxemia and hemoglobin infusion cannot be accomplished without treatment of blood to allow the detection of LPS.[31] Endotoxemia can be recognized only after neutralization of plasma inhibitors. LAL detection of LPS in plasma is best achieved using a dilution/heating method for inhibitor neutralization, as shown in Fig. 9. Other plasma treatment methods, e.g., chloroform extraction and acid oxidative techniques, have been evaluated but are not as successful at recovery and detection of endotoxin.[31] Inhibition of detection of endotoxin

[31] R. I. Roth, F. C. Levin, and J. Levin, *J. Lab. Clin. Med.* **116,** 153 (1990).

by plasma proteins is also a well-recognized problem in other assays for bacterial endotoxins.[32]

Acknowledgments

Supported, in part, by U.S. Army Medical Research and Development Command Research Contract MIPR Number 90 MM0535; Research Grant DK 43102 from the NIDDKD, National Institutes of Health; and the Veterans Administration.

[32] J. Levin, P. A. Tomasulo, and R. S. Oser, *J. Lab. Clin. Med.* **75,** 903 (1970).

Section II

Preparation and Characterization of Nonhuman Hemoglobins

[7] Preparation and Characterization of Insect Hemoglobins from *Chironomus thummi thummi*

By Hans Heinrich Ruf, Andreas G. Altemüller, and Klaus Gersonde

Introduction

The lymph of the larvae of the dipteran insect *Chironomus thummi thummi* (CTT) contains a complex mixture of free hemoglobins; these hemoglobins have been named CTT hemoglobins or erythrocruorins.[1-3] In comparison to the tetrameric mammalian hemoglobins, these insect hemoglobins exist as monomers or dimers[1,4,5] and exhibit a high O_2 affinity[6] and a large variation of the Bohr effect[6] (pH dependence of O_2 affinity). The investigation of CTT hemoglobins, which serve as the simplest allosteric model system, has provided a deep insight into the allosteric mechanism[7] (Bohr effect). Many of the physicochemical studies on these hemoglobins have improved the knowledge of the structure–function relationships in O_2-binding heme proteins.[7-9] In this chapter we direct attention to exceptional and selected properties of the CTT hemoglobins. We describe in detail the isolation of large quantities of these hemoglobins using techniques developed in our laboratory.[10-14]

[1] T. Svedberg and J. B. Erikson-Quensel, *J. Am. Chem. Soc.* **56,** 1700 (1934).

[2] R. Huber, H. Formanek, V. Braun, G. Braunitzer, and W. Hoppe, *Z. Elektrochem.* **68,** 818 (1964).

[3] G. Braunitzer and V. Braun, *Hoppe-Seyler's Z. Physiol. Chem.* **340,** 88 (1965).

[4] J. Behlke and W. Scheler, *Eur. J. Biochem.* **3,** 153 (1967).

[5] V. Braun, R. R. Crichton, and G. Braunitzer, *Hoppe-Seyler's Z. Physiol. Chem.* **349,** 197 (1968).

[6] H. Sick and K. Gersonde, *Eur. J. Biochem.* **7,** 273 (1969).

[7] K. Gersonde, H. Sick, M. Overkamp, K. M. Smith, and D. W. Parish, *Eur. J. Biochem.* **157,** 393 (1986).

[8] G. Braunitzer, G. Buse, and K. Gersonde, *in* "Molecular Oxygen in Biology" (O. Hayaishi, ed.), p. 219. North-Holland Publ., Amsterdam, 1974.

[9] K. Gersonde, *Colloq. Ges. Biol. Chem. Mosbach* **34,** 170 (1983).

[10] K. Gersonde, H. Sick, A. Wollmer, and G. Buse, *Eur. J. Biochem.* **25,** 181 (1972).

[11] H. Sick, K. Gersonde, J. C. Thompson, W. Maurer, W. Haar, and H. Rüterjans, *Eur. J. Biochem.* **29,** 217 (1972).

[12] K. Gersonde, H. Twilfer, and M. Overkamp, *Biophys. Struct. Mech.* **8,** 189 (1982).

[13] G. N. LaMar, R. R. Anderson, V. P. Chacko, and K. Gersonde, *Eur. J. Biochem.* **136,** 161 (1983).

[14] A. G. Altemüller, Ph.D. thesis, Universität des Saarlandes, Saarbrücken, Germany (1992).

Properties of CTT Hemoglobins

Polymorphism and Protein Structure

The lymph of the CTT larvae contains at least 12 hemoglobins (CTT Hbs).[15] Each of these CTT Hbs exhibits a unique amino acid sequence, which has been determined by Braunitzer's group,[15] thus the CTT Hbs constitute a typical protein polymorphism.[15] The nomenclature of the CTT Hbs is based on their electrophoretic mobility in polyacrylamide gel electrophoresis (PAGE) at pH 8.9–9.5, beginning the numbering with the CTT Hb band that is nearest to the cathode.[15] Sequence analysis of the primary structures and their comparison reveal a high degree of homology among the CTT Hbs. Therefore, it has been concluded that their evolutionary history originates monophyletically from a common ancestor gene within Insecta.[15]

Svedberg[1] was the first to draw attention to insect hemoglobin, when he determined by ultracentrifugation the molecular mass of CTT Hb as 31,400 Da, which indicates a dimeric state of this hemoglobin. Later, it was discovered that this insect hemoglobin is a mixture of monomeric CTT Hbs (CTTs I, Ia, III, IIIa, and IV) and dimerizing CTT Hbs (CTTs II, VI, VIIa, VIIb, VIII, IX, and X).[4,5] The latter group of proteins constitutes not always dimers, rather their monomer–dimer equilibria depend on *in vitro* conditions, especially on pH and concentration; an increase in pH favors dissociation into monomers.[4] The overall primary structures of the dimers differ from those of the monomers in specific regions. These differences, an elongation of the EF region and two additional negatively charged residues at B6 and G7, have been attributed to dimer formation.[16] The genes for monomeric CTT Hbs are clustered in one chromosomal region and the genes for dimerizing CTT Hbs in a different region.[17] The three-dimensional structure has been resolved only for a monomeric hemoglobin, i.e., CTT III.[18-21] The structure of CTT III closely resembles that of myoglobin.

Oxygen Binding

The hyperbolic nature of the oxygen dissociation curve (ODC) of the hemolymph of CTT indicating a P_{50} value of 0.6 mm Hg at 17° was first

[15] M. Goodman, T. Kleinschmidt, G. Braunitzer, and H. Aschauer, *Hoppe-Seyler's Z. Physiol. Chem.* **364,** 205 (1983).

[16] W. Steer and G. Braunitzer, *Hoppe-Seyler's Z. Physiol. Chem.* **362,** 59 (1981).

[17] H. Tichy, *J. Mol. Evol.* **6,** 39 (1975).

[18] R. Huber, O. Epp, and H. Formanek, *J. Mol. Biol.* **52,** 349 (1970).

[19] R. Huber, O. Epp, W. Steigemann, and H. Formanek, *Eur. J. Biochem.* **19,** 42 (1971).

reported by Fox.[22] The pH dependence of the O_2 affinity (Bohr effect) was first observed in the lymph by Weber.[23] Isolated CTT Hbs exhibit a pH dependence of the O_2 affinity, characteristic for an alkaline Bohr effect with inflection points between pH 7.05 and 7.3, and amplitudes of the Bohr effect curve, which depend on the particular CTT Hb.[6,7,10,24,25] The monomeric CTT Hbs III and IV clearly exhibit a Hill coefficient of one.[10,24] The O_2-binding curves of all isolated dimerizing CTT Hbs also show hyperbolic ODCs.[25,26] The slope of the Bohr effect curve ($-\Delta \log P_{50}/\Delta$pH) for dimers (CTT II, 0.77; CTT X, 4.0)[14,25] is greater than for the monomers (CTT III, 0.28; CTT IV, 0.42).[10,24] The remarkably high value of 4.0 for CTT X is evaluated from a steep pH dependence at pH 7.1–7.2,[25] which correlates to the pH-dependent hyperfine shift of the NMR lines of the methyl protons of the heme.[14] These observations for CTT X have not yet been explained.

Mechanism of Bohr Effect

The comparison of the monomeric CTT Hbs I (nonallosteric), III, and IV allows an insight into the mechanism of the alkaline Bohr effect. This Bohr effect is controlled by a single proton (Bohr proton).[10,11,24,27,28] The Bohr proton-binding site has been assigned by NMR titrations to His-G2.[11] At low pH a salt bridge between the imidazole group of His-G2 and the C-terminal carboxyl group of Met-H22 is formed. This salt bridge stabilizes the T state of the tertiary structure with low ligand affinity. At high pH the dissociation of the Bohr proton induces the transition into the R state with high ligand affinity. The dynamics of the T $\rightleftharpoons$ R transition is reflected in the NMR linewidths of the hyperfine-shifted proton signals, which are significantly broadened around the pK of the Bohr proton.[29] The allosteric interaction between the Bohr proton-binding site and the ligand-binding site at the heme iron has been elucidated using thermodynamics, kinetics, and spectroscopy.[7] The simple nature of the Bohr effect

[20] W. Weber, W. Steigemann, T. A. Jones, and R. Huber, *J. Mol. Biol.* **120**, 327 (1978).
[21] W. Steigemann and E. Weber, *J. Mol. Biol.* **127**, 309 (1979).
[22] H. M. Fox, *J. Exp. Biol.* **21**, 161 (1945).
[23] R. E. Weber, *Proc. K. Ned. Akad. Wet., Ser. C* **66**, 284 (1963).
[24] H. Sick and K. Gersonde, *Eur. J. Biochem.* **45**, 313 (1974).
[25] M. Overkamp, Ph.D. thesis, Rheinisch-Westfälische Technische Hochschule, Aachen, Germany (1980).
[26] D. Zepke, H. Sick, and K. Gersonde, *Biophys. Struct. Mech.* **7**, Suppl., 288 (1981).
[27] K. Gersonde, L. Noll, H. T. Gaud, and S. J. Gill, *Eur. J. Biochem.* **62**, 577 (1976).
[28] G. Steffens, G. Buse, and A. Wollmer, *Eur. J. Biochem.* **72**, 201 (1977).
[29] V. P. Chacko, G. N. LaMar, K. Gersonde, and H. Sick, *Eur. J. Biochem.* **161**, 375 (1986).

as well as the lack of a distal histidine[11,21,30,31] make the CTT Hbs III and IV good model systems for the study of the allosteric mechanism in hemoglobins.

Monomeric deoxyCTTs exhibit a pH-induced $T \rightleftharpoons R$ conformational transition as revealed by the pH-dependent hyperfine shifts of the heme methyl protons, the pK value of which (pK 7.5) reflects the dissociation of the Bohr proton.[32] On the other hand, there is evidence that the electronic structure of the iron–histidine bond is not affected by pH. The absence of any pH dependence has been demonstrated by NMR for the proton hyperfine shifts at the proximal histidine,[32] by resonance Raman studies of the Fe–N^ε stretching mode (N^ε from the proximal His),[33] by EPR for the hyperfine coupling constants of cobalt and N^ε in cobalt CTT IV,[12] and for the on-rates of oxygen binding.[7] Therefore, it is concluded that the pH-induced tertiary structure transition, observed in the deoxy state, is not transmitted to the heme iron. Thus O_2, before it binds to the heme iron, "sees" only one type of binding site, which is not influenced by pH variation, i.e., conformational changes.

The formation of the Fe–O_2 bond in the monomeric oxyCTTs is accompanied by a change from five- to six-coordination of the heme iron. In the liganded state (with O_2, CO, or NO), pH-dependent variations of the electronic structure of the proximal imidazole–heme iron–ligand system become evident. EPR of oxycobalt CTT IV exhibits a pH-dependent hyperfine coupling of the central metal [Co(II)], indicating different electron densities at the metal for the T and R states.[12] Replacement of O_2 by NO results in the transition of the heme complex from hexa- to pentacoordination as shown by EPR[34,35] and resonance Raman studies.[36] Resonance Raman spectroscopy provides further evidence for pH-dependent changes of the cobalt–axial ligand bond in monomeric and dimeric CTT Hbs.[37,38] For CO-ligated CTT Hbs, reciprocal changes in the two iron–axial ligand

[30] N.-T. Yu, H. M. Thompson, H. Mizukami, and K. Gersonde, *Eur. J. Biochem.* **159,** 129 (1986).

[31] D. H. Peyton, G. N. LaMar, S. Ramaprasad, S. W. Unger, S. Sankar, and K. Gersonde, *J. Mol. Biol.* **221,** 1015 (1991).

[32] G. N. LaMar, R. R. Anderson, D. L. Budd, K. M. Smith, K. C. Langry, K. Gersonde, and H. Sick, *Biochemistry* **20,** 4429 (1981).

[33] E. A. Kerr, N.-T. Yu, K. Gersonde, D. W. Parish, and K. M. Smith, *J. Biol. Chem.* **260,** 12665 (1985).

[34] M. Overkamp, H. Twilfer, and K. Gersonde, *Z. Naturforsch., C: Biosci.* **31C,** 524 (1976).

[35] M. Christahl, and K. Gersonde, *Biophys. Struct. Mech.* **8,** 271 (1982).

[36] E. A. Kerr, N.-T. Yu, and K. Gersonde, *FEBS Lett.* **178,** 31 (1984).

[37] N.-T. Yu, H. Mackin-Thompson, D. Zepke, and K. Gersonde, *Eur. J. Biochem.* **157,** 579 (1986).

[38] H. C. Mackin, N.-T. Yu, and K. Gersonde, *Biophys. J.* **51,** 289 (1987).

bonds, induced by the allosteric transition (Bohr effect), are demonstrated by resonance Raman spectroscopy.[39] The off-rate of oxygen binding decreases with increasing pH, reflecting the higher affinity of the R state.[7] The on-rate is pH insensitive, thus the off-rate is responsible for the Bohr effect and for the control of oxygen affinity. Because both allosteric CTT Hbs contain no distal histidine,[11,21,30,31] the pH dependence of the off-rate must be based on a modulation of the proximal histidine–iron interaction, a typical trans effect that is linked to the $T \rightleftharpoons R$ conformational transition.

Heme Rotational Disorder

Heme rotational disorder, i.e., the rotation of the heme around the α,γ-meso axis, was first discovered in CTT Hbs.[40–42] This disorder has been resolved by NMR[13,40–47] and resonance Raman spectroscopy,[48] and has been correlated to the off-rates of O_2 binding[7] and to the Bohr effect.[40,44] Heme rotational disorder has been recognized as a modifier of functions as well as an indicator of heme–protein interactions. It also reflects small changes in the heme pocket due to silent mutations in CTTs I and III.[13,49] Specific contacts between the heme side groups (asymmetry of the heme periphery) and the amino acid side chains that form the protein pocket can explain the differences in the functional properties of the two heme rotational isomers.[7]

A model for the protein–heme interactions with four specific sites (A–D) in the protein pocket has been proposed[7] (see Fig. 1). These sites specifically interact with the methyl and vinyl residues of the porphyrin. A high stability of the protein–heme complex requires the occupation of site C with a vinyl group. At this site, the replacement of vinyl by methyl or hydrogen increases the off-rate for O_2, lowering the oxygen affinity.

[39] K. Gersonde, E. A. Kerr, N.-T. Yu, D. W. Parish, and K. M. Smith, *J. Biol. Chem.* **261,** 8678 (1986).

[40] G. N. LaMar, M. Overkamp, H. Sick, and K. Gersonde, *Biochemistry* **17,** 352 (1978).

[41] G. N. LaMar, D. B. Viscio, K. Gersonde, and H. Sick, *Biochemistry* **17,** 361 (1978).

[42] G. N. LaMar, K. M. Smith, K. Gersonde, H. Sick, and M. Overkamp, *J. Biol. Chem.* **255,** 66 (1980).

[43] W. Ribbing and H. Rüterjans, *Eur. J. Biochem.* **108,** 79 (1980).

[44] W. Ribbing and H. Rüterjans, *Eur. J. Biochem.* **108,** 89 (1980).

[45] D. Krümpelmann, W. Ribbing, and H. Rüterjans, *Eur. J. Biochem.* **108,** 103 (1980).

[46] G. N. LaMar, R. Krishnamoorthi, K. M. Smith, K. Gersonde, and H. Sick, *Biochemistry* **22,** 6239 (1983).

[47] D. H. Peyton, G. N. LaMar, and K. Gersonde, *Biochim. Biophys. Acta* **954,** 82 (1988).

[48] K. Gersonde, N.-T. Yu, E. A. Kerr, K. M. Smith, and D. W. Parish, *J. Mol. Biol.* **194,** 545 (1987).

[49] D. H. Peyton, R. Krishnamoorthi, G. N. LaMar, K. Gersonde, K. M. Smith, and D. W. Parish, *Eur. J. Biochem.* **168,** 377 (1987).

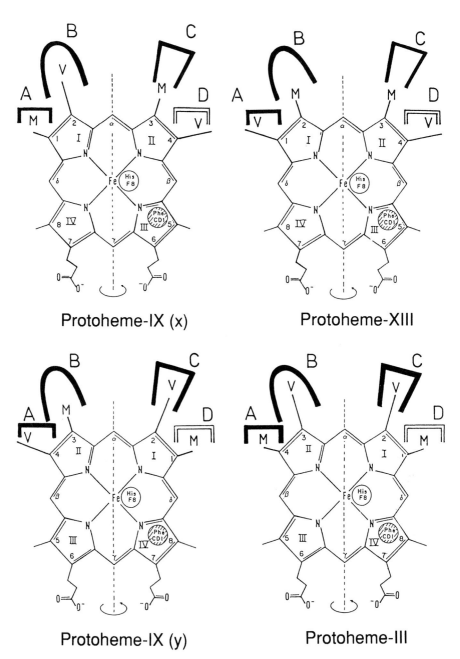

Protoheme-IX (x)

Protoheme-XIII

Protoheme-IX (y)

Protoheme-III

FIG. 1. Model depicting the heme–apoprotein contact sites (A, B, C, and D) and their interaction with the heme rotational components x and y of proto-IX CTT Hbs and with the "symmetric" proto-III and proto-XIII CTT Hbs. The proximal histidine, His-F8, is above the heme plane and the Phe-CD1 is below the heme plane (M, methyl groups; V, vinyl groups). Reproduced with permission from Gersonde et al.[7]

The interaction of vinyl with site C seems to be involved in closing a protein channel, thus hindering the escape of O_2. The pH-dependent modulation of the off-rate (Bohr effect) is most pronounced when site B is occupied by a vinyl group.

Preparation Methods

An improved method for the isolation of larger amounts of monomeric (CTTs I, III, and IV) and dimerizing (CTTs II and X) hemoglobins from the lymph of the larvae of *C. thummi thummi* will be described. This procedure also results in (pre)purified hemoglobin fractions, which can be used for the isolation of other CTT hemoglobins (CTTs V–IX). It is the main objective of this isolation procedure to provide pure and functionally active hemoglobins in the large amounts that are necessary for biophysical, especially spectroscopic, studies.

The extract from the larvae contains proteolytic enzymes, which can degrade the hemoglobins rapidly. To overcome this problem, the homogenization of the larvae and the preparation of the crude hemoglobin extract are carried out under 40% saturation with ammonium sulfate. Furthermore, during preparation, hemoglobin fractions are stored at 4° as precipitates in saturated ammonium sulfate or as lyophilized salt-free powders. In general, the purification steps are scheduled in such a way to minimize the periods that the crude hemoglobin extract is contaminated with potentially active proteases. In addition to proteolytic degradation, the loss of the heme group via dissociation is a problem for the isolation of large amounts of functionally active CTT hemoglobins. Therefore, all preparative steps are carried out at 4° (unless indicated otherwise). The purification protocol consists of three major parts:

The separation into two Hb fractions, i.e., collection of monomeric and dimeric hemoglobins by gel filtration, starting with the crude extract from 1 kg of CTT larvae, takes 8 days. Two more days are required for an additional gel filtration of the overlapping fraction.

The isolation of the monomeric CTTs I, III, and IV is performed by anion-exchange chromatography from the previously mentioned purified monomer fraction. This step takes about 5 days.

The dimerizing CTTs II and X are isolated by cation-exchange chromatography from the purified dimer fraction, followed by anion-exchange chromatography. This step requires at least 2 weeks.

Equipment

Conventional equipment for protein purification employing ammonium sulfate precipitation and liquid chromatography is required. Ion-exchange chromatography is controlled by continuous monitoring of pH and conductivity. Because of the rather large scale of the preparation, buffer containers with a capacity of 25–50 liters and a dialysis setup (beakers of 5 liters) are required. A cooling centrifuge (Beckman J 2-21 M/E) supplied with two fixed-angle rotors (Beckman JA 14 and JA 20) is used. The CTT larvae are homogenized in a rotating blade homogenizer. Conventional liquid chromatography equipment with columns, UV monitor (with a preparative cell of 1–3 mm optical pathlength), and a fraction collector with a capacity of about 5 liters is employed. The eluant flow is driven by gravity. The columns required for chromatography are specified in detail below.

Fast protein liquid chromatography (FPLC) is performed with a Biopilot system (Pharmacia, Piscataway, NJ) consisting of a binary solvent delivery system (controller LCC-500 plus; two pumps P-6000), injection system (motor valve IMV-7, superloop 50 ml), a prefilter (Type 6000), UV monitor at 280 nm (UV-M/1 with flow cell S-2), and fraction collector (Frac-100 or motor valve IMV-8). The system is controlled by a personal computer (AT) with the software FPLC manager (Pharmacia). The HPLC system (Pharmacia/LKB) consists of a solvent conditioner (Type 2156), ternary gradient pump (Type 2249), manual injection valve (Type 2154, Valco) and a dual-wavelength monitor (Type 2141); proteins are monitored at 280 nm, hemoglobins at 405 nm. The system is controlled by a personal computer (AT) with the software HPLC manager (Pharmacia/KLB) in combination with the Nelson software for data acquisition. The inert flow system of the HPLC is made from glass, titanium, or fluorocarbon polymers. All solvents and samples used for the FPLC or HPLC are filtered through 0.22-μm filters.

The hemoglobin assay and the characterization of the proteins are based on spectrophotometry.

Materials

The phosphate buffer salts $NaH_2PO_4 \cdot H_2O$ and $Na_2HPO_4 \cdot 2H_2O$ (GR, Merck) and ammonium sulfate (grade "for biochemistry," Merck, Darmstadt) are required in large quantities. Triton X-100 and NaOH are applied for the hemoglobin assay. All chemicals are used as obtained from the suppliers. Dialysis tubing is Visking 36/32 (Serva, Heidelberg). Larvae from CTT are purchased from a German breeder (Stiels, Brüggen) as 1-kg frozen blocks and are stored below $-20°$.

Separation of Monomeric and Dimerizing Hemoglobins

A schedule for this protocol has been worked out to provide yields of 4–5 g of monomers and 8–12 g of dimers from 1 kg of CTT larvae within 8 days. To follow this schedule requires equipment setup and buffer solution preparation before starting the procedure. The proteolytic degradation of the hemoglobins is minimized by storing the proteins, especially the crude extract of the hemoglobins, as precipitates in saturated ammonium sulfate solution. All steps are carried out at 4° (unless stated otherwise).

Preparation of Buffers

Buffer for gel filtration (0.04 M sodium phosphate, pH 5.5; at least 50 liters) is prepared in batches of 25 liters. About 1.1 liters of 0.04 M Na_2HPO_4 is added to 24 liters of 0.04 M NaH_2PO_4 until the correct pH is achieved (pH 6.05 at room temperature will adjust to pH 5.5 at 4°; cooling down of this buffer to 4° takes at least 1 day).

Homogenization buffer (0.1 M sodium phosphate, pH 6.5; 1 liters) is prepared: about 300 ml of 0.1 M Na_2HPO_4 is added to 700 ml of 0.1 M NaH_2PO_4. The final pH is adjusted by further addition of 0.1 M Na_2HPO_4 or NaH_2PO_4.

Ammonium sulfate solution (saturated, pH 5.8) is prepared. The saturation with ammonium sulfate requires about 700 g/liter. Solid ammonium sulfate should always be present to ensure saturation. The pH value is adjusted and maintained during the salting out of hemoglobin by adding a solution saturated with Na_2HPO_4.

Preparation of Gel Filtration Columns

Sephadex G-50 medium (Pharmacia) is suspended in 0.04 M phosphate buffer, pH 5.5, according to the instructions of the manufacturer. This suspension is poured along a glass rod into the columns and the gel is settled under flow.

Hemoglobin Assay

The hemoglobin concentration is determined employing the alkaline hematin detergent (AHD-575) method.[50,51] In this assay hemoglobin is converted to an alkaline hematin derivative, mostly incorporated in detergent micelles. The stable color of this solution is measured at 575 nm (the extinction coefficient is 6960 M^{-1} cm^{-1}). An aliquot of the hemoglobin solution (usually 20–200 μl) is added to 3 ml of a reagent solution (2.5% Triton X-100 in 0.1 M NaOH; foaming should be avoided when dissolving the Triton X-100 in NaOH; the reagent is stable at room temperature) and mixed. The AHD derivative is formed within 30 sec. Its green color is stable for days and its absorbance is measured at 575 nm against a water

[50] R. Zander, W. Lang, and H. U. Wolf, Clin. Chim. Acta 136, 83 (1984).
[51] H. U. Wolf, W. Lang, and R. Zander, Clin. Chim. Acta 136, 95 (1984).

blank. An average molecular mass of 15,500 Da is used for calculating the concentration (g/liter) of CTT hemoglobins. Compared to the usual hemoglobin assay based on the cyanomet derivative, the AHD-575 method has the advantage of using a nontoxic and stable reagent and of having a colored derivative with high stability.

Schedule of Purification

This schedule assumes that the buffer solutions have been prepared and that the gel filtration columns have been poured and equilibrated. The following time schedule lists the steps of the procedure; the details are described later.

Day 1
    Homogenization of CTT larvae and extraction of hemoglobins
    Salting out of hemoglobins with ammonium sulfate (overnight)
Day 2
    Splitting up of the precipitated hemoglobin into three batches
    Batch 1: Dissolving and first gel filtration; pooling of the Hb fractions
        and salting out
    Batches 2 and 3 remain in saturated ammonium sulfate for storage
Day 3
    Batch 2: Dissolving and first gel filtration; pooling of the Hb fractions
        and salting out
    Batch 1: Dissolving and start of the second gel filtration
Day 4
    Batch 3: Dissolving and first gel filtration; pooling of the Hb fractions
        and salting out
    Batch 1: Finishing the second gel filtration; pooling of the fractions,
        i.e., fractions with monomers, dimers, and mixtures of both hemo-
        globins (overlapping fraction); salting out of the three fractions
Day 5
    Batch 2: Dissolving and start of the second gel filtration
Day 6
    Batch 2: Finishing of the second gel filtration; pooling of the fractions
        of Hb monomers, dimers, and overlapping fractions; salting out of
        each fraction
Day 7
    Batch 3: Dissolving, adding of the overlapping fractions from batches
        1 and 2, and start of the second gel filtration
Day 8
    Batch 3: Finishing of the second gel filtration; pooling of fractions of

Hb monomers, dimers, and overlapping fractions; salting out of each fraction

Despite the high workload on days 2–4, it is necessary to keep the time interval between the first gel filtration and the subsequent salting out as short as possible. Typically 1 kg of CTT larvae provides prepurified monomers (25%), prepurified dimers (56%), and the overlapping fraction with nonseparated hemoglobins (9%). The yields (%) are related to the hemoglobin content of the crude extract after the first precipitation. Further purification of the monomers and dimers is achieved by a third gel filtration on Sephadex G-50, which takes another 2 days for each fraction.

Homogenization of CTT Larvae

A 1-kg block of frozen larvae (containing up to 18 g hemoglobin) is wrapped into a cotton cloth and then broken into small pieces using a hammer. The pieces are transferred into a beaker containing 1 liter of 0.1 M phosphate buffer, pH 6.5. To achieve 40% saturation, ammonium sulfate is added to 221 g/liter of total volume. This mixture is homogenized and thawed in a rotating blade homogenizer at maximum speed for 5–10 min. The homogenate is centrifuged (30 min at 14,000 g; rotor Beckman JA 14 at 12,000 rpm). The reddish-brown turbid supernatant containing the extracted hemoglobins is filtered through fine mesh cloth and poured into dialysis tubing. The pellet contains debris and is discarded.

Salting Out of Hemoglobins

The hemoglobin in the supernatant is salted out by dialysis against a saturated ammonium sulfate solution at pH 5.8. This procedure serves two purposes: the transfer of the hemoglobins into a precipitated state, where proteases are not active, and the concentration of hemoglobin solution between the chromatographic steps. The dialysis bags are suspended in a 5-liter beaker filled with saturated ammonium sulfate solution. An excess of solid ammonium sulfate is added to ensure saturation. This solution is slowly and continuously stirred by a magnetic stirrer. Every hour the pH value is checked and adjusted to pH 5.8 with a saturated Na_2HPO_4 solution. To prevent the precipitated hemoglobins from sticking to the wall of the dialysis tubing, the dialysis bags are kneaded every 15 min during the first 5 hr of dialysis. The dialysis is continued overnight (in the presence of solid ammonium sulfate, otherwise the salting out will be incomplete). Under these conditions hemoglobins are completely precipitated. The suspension of the precipitated hemoglobin is split into three equal batches, each of which is subjected to the following gel filtration steps separately. The salted-out hemoglobin is collected by centrifugation (30 min at 14,000 g; rotor JA 14 at 12,000 rpm). The colorless supernatant is discarded.

Dissolving of Salted-Out Hemoglobins

A minimal amount of 0.04 M phosphate buffer, pH 5.5 (about 100 ml), is added to the sedimented precipitate of one batch and stirred for 30 min to dissolve the hemoglobin. This Hb solution is centrifuged for 2 hr at 31,000 g (rotor JA 20 at 20,000 rpm) to sediment undissolved material. The supernatant, a clear solution of hemoglobin, is collected.

First Gel Filtration

This step should lead to an immediate separation of the hemoglobins from proteases and other proteins. The salted-out hemoglobins are dissolved in about 100 ml of 0.04 M phosphate buffer, pH 5.5, as described above. Volume and hemoglobin concentration of this solution are determined to obtain the hemoglobin content of the crude extract corresponding to a starting value of 100%. To maintain a uniform valence state of the hemoglobin during all chromatographic steps, the hemoglobins are oxidized to the ferric (met) state by the addition of two equivalents of solid $K_3[Fe(CN)_6]$ (43 mg/g of hemoglobin). The resulting methemoglobin solution is applied to a Sephadex G-50 column (5.5 × 50 cm) equilibrated and eluted with 0.04 M phosphate buffer, pH 5.5, at a flow rate of 8 ml/min. The clear and colored hemoglobin fractions are collected and pooled. Volume and hemoglobin content of this pool are determined and the hemoglobin is salted out overnight as described above. The elution of the column is continued to have it ready for the gel filtration the next day.

Second Gel Filtration

This gel filtration, also on Sephadex G-50, is performed at a higher resolution compared to the first one and provides the separation between monomeric and dimeric hemoglobins. A Sephadex G-50 column (10 × 120 cm) is equilibrated and eluted with 0.04 M phosphate buffer, pH 5.5, at a flow rate of 8 ml/min.

The precipitate of the first gel filtration is sedimented and dissolved in about 100 ml of 0.04 M phosphate buffer, pH 5.5, as described above. To achieve an efficient timing, the sample preparation, i.e., dissolution and centrifugation for the first and for the second gel filtration, should be performed in parallel. The hemoglobin solution applied to the gel column should be less than about 150 ml and should contain 2–3.5 g hemoglobin.

Fractions of 16 ml each are collected overnight and pooled the next day as monomers or dimers according to the elution diagram as recorded by the UV monitor at 280 nm. Two clearly separated peaks are eluted, the dimer peak around 1200 ml and the monomer peak around 2400 ml elution volume. There occurs also an overlapping fraction with monomers and dimers not clearly separated. Each pooled fraction is characterized by its volume and its hemoglobin content. Then the hemoglobin is salted out as described above. The overlapping fraction is added to another batch for a repeated separation in the second gel filtration.

Third Gel Filtration

If a further purification of the monomer and dimer fractions, obtained from the second gel filtration, is required, both fractions are rechromatographed on Sephadex G-50. The precipitates of each of the monomer and dimer fractions from all three batches are combined and dissolved as described above (the maximum volume is 250 ml and the maximum hemoglobin concentration 2.3 g/100 ml). Column and conditions are the same as described for the second gel filtration. The typical recovery of hemoglobin is 80%. The yields are about 4.5 g of monomers and 10 g of dimers from 1 kg of CTT larvae.

Salt is removed from the resulting purified monomers and dimers by dialysis against water (check the conductivity). The salt-free hemoglobin solution is frozen at $-20°$ and lyophilized. The lyophilized hemoglobin is stored at $-20°$. Alternatively, salted-out hemoglobins can be stored as wet precipitates at 4°. Stability of hemoglobins is equally high during either type of storage; the choice depends on practical considerations.

Isolation of Monomeric CTT Hemoglobins I, III, and IV

The monomer fraction contains CTT hemoglobins I, III, and IV, which have been separated by anion-exchange chromatography on DEAE-Sephadex A-50 at pH 9.[10-13] Here we describe a new and improved separation using FPLC on a HiLoad Q Sepharose column (1.6 × 20 cm; Pharmacia) with a salt gradient (buffer A, 10 mM Tris-HCl, pH 9.0; buffer B, identical to A plus 250 mM NaCl). Salted-out fractions are extensively dialyzed against buffer A (check the conductivity). Alternatively, lyophilized hemoglobins are dissolved in buffer A. The 2.5-ml sample (up to 40 mg hemoglobin per 2.5 ml) is passed through a 0.22-μm filter (Acro, LC13; Gelman Sciences, Ann Arbor, MI) and is applied to the column. The elution is performed at 0.75 ml/min with a salt gradient that is composed of linear segments (0 ml, 25% B; 7.5 ml, 25% B; 14 ml, 35% B; 14.1 ml, 45% B;

19.5 ml, 55% B; 19.6 ml, 100% B). The elution diagram shows three separate peaks (CTT I at 25% B, CTT III at 33% B, and CTT IV at 49% B). The assignments of these monomeric hemoglobins are based on the R_f values of PAGE (see below). Typical yields of one run are 3.3 mg of CTT I, 10 mg of CTT III, and 13 mg of CTT IV. For scaling up this preparation, the high cost of the separation medium might be limiting.

Isolation of Dimeric CTT Hemoglobins II and X

The dimer fraction contains at least seven hemoglobins (CTTs II, V, VI, VII, VIII, IX, and X). The dimers are isolated by two ion-exchange chromatographic steps: the conventional liquid chromatography on the cation exchanger SP Sephadex C-50 and then the FPLC on the anion exchanger Fraktogel EMD DEAE-650 (M).

Cation-Exchange Chromatography

A column (5.5 × 25 cm) with SP Sephadex C-50 (Pharmacia) is equilibrated with buffer 1 (5 mM bis-Tris-HCl, pH 5.6). The ammonium sulfate precipitate of a dimer fraction is dissolved in buffer 1 as described above and dialyzed against buffer 1 until the dialyzate shows the same conductivity as buffer 1. Alternatively, a lyophilized sample of dimers is dissolved in buffer 1. The undissolved material is spun down. The hemoglobin concentration is adjusted to less than 3 g/100 ml. This Hb solution (maximum 350 ml) is applied to the column and eluted with buffer 1 at a flow rate of 8 ml/min. The resulting eluate 1 contains CTTs VI, VIIb, and VIII, which are not adsorbed to the column. At an elution volume of 600 ml this fraction is completely eluted and then the elution is continued with buffer 2 (5 mM bis-Tris-HCl, pH 7.6). The resulting eluate 2 contains CTTs, VI, VIIb, VIII, and IX. This elution is complete after about 600 ml of buffer 2. Then the residual is eluted with 600 ml of buffer 3 (50 mM bis-Tris-HCl, pH 7.6). This eluate 3 contains CTTs II and X as the major components and in addition CTTs VIIa, VIIb, and VIII as minor components.[52] The relative hemoglobin contents of the three eluates were about 67, 22, and 11%. After washing the column with about 1000 ml of buffer 1, it is ready for the next separation. The three eluates are concentrated

---

[52] The observation that a hemoglobin with an identical apparent electrophoretic mobility is eluted in all three eluates merits a comment. From the electrophoretic mobility, this material was tentatively assigned to CTT VIIb in all three eluates. A definite assignment would require partial sequencing, which was not attempted. The occurrence of possibly the same hemoglobin, CTT VIIb, in all three eluates can be explained by its association with other hemoglobins, presumably by pH-dependent heterologous dimerization reactions.

by ultrafiltration (membrane PM10, Amicon, Beverly, MA), washed with several volumes of water, and lyophilized for storage.

Anion-Exchange Chromatography

The final isolation of CTT II and CTT X from eluate 3 is achieved by anion-exchange chromatography at pH 9. For this step FPLC on a column (1 × 15 cm; Merck Superformance 150-10) containing Fraktogel EMD DEAE-650 (M) (Merck) is applied. The elution is performed at 1 ml/min with a salt gradient that is composed of linear segments (buffer A, 10 mM Tris-HCl, pH 9.0; buffer B, 10 mM Tris-HCl, pH 9.0 with 250 mM NaCl; 0 ml, 15% B; 20 ml, 25% B; 30 ml, 75% B; 60 ml, 100% B). The above-mentioned lyophilized eluate 3 (about 40 mg per run) is dissolved in a minimum volume of buffer A, passed through a 0.22-μm filter (Acro, LC13), and applied to the column. The elution diagram shows two prominent peaks (CTT II at 27% B and CTT X at 98% B) and three minor bands [CTT VIIa (or VI) at 70% B, CTT VIIb at 76% B, and CTT VIII at 83% B]. A further separation or a definite assignment of the three minor bands was not attempted. The assignments are made using the R_f values of PAGE (see below). The yields of one run are 30 mg of CTT II and 7 mg of CTT X.

Characterization of Hemoglobins

The CTT hemoglobins comprise a family of polymorphic proteins that differ in their primary structures, as Braunitzer et al.[15] have shown. The differences in the primary structure lead to altered electrostatic properties and to a variation in spectroscopic and oxygen binding parameters. For an unequivocal assignment, at least a partial sequencing of the isolated hemoglobin must be performed. Electrophoretic mobility, which is the conventional basis of the nomenclature of CTT hemoglobins, is a quite specific parameter, very useful for routine assignments and purity control. Here we introduce high-resolution anion-exchange HPLC as a nondenaturing method for quantitative analysis, accute assignment, and purity control of CTTs. Furthermore, spectroscopy (UV/visible, NMR, ESR, IR, resonance Raman) and oxygen affinity measurements are useful for characterization. In the following discussion we describe a PAGE system for the determination of electrophoretic R_f values of CTT hemoglobins and a HPLC method based on anion-exchange chromatography for the purity check of the isolated hemoglobins.

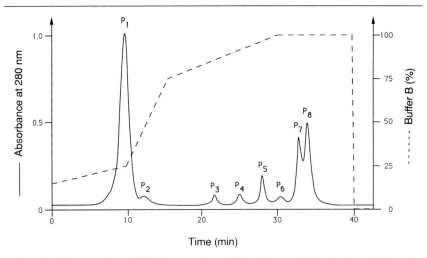

Time (min)

FIG. 2. Anion-exchange HPLC separation of dimeric hemoglobins in eluate 3 from the cation-exchange chromatography containing 0.1 mg of hemoglobin. The elution is performed on a column Mono Q HR 5/5 with a salt gradient as depicted (buffer A, 10 mM Tris-HCl, pH 9.5; buffer B, 20 mM Tris-HCl, pH 9.5, plus 250 mM NaCl; flow rate, 1 ml/min). For further details see the text.

Electrophoretic Mobility (PAGE)

The electrophoretic mobility of the CTT hemoglobins is characterized by their R_f values in a PAGE system according to Maurer.[53] We use 10% slab gels (10 × 10 cm, 1 mm thick) without stacking gel, and the electrophoresis equipment from Biometra model minigel (Göttingen). The gels are prepared from 2.5 ml of solution A [0.46 ml N,N,N',N'-tetramethylenediamine (TEMED) in 100 ml 0.5 M Tris-HCl buffer, pH 8.9], 5 ml of solution B (20 g acrylamide, 0.666 g bisacrylamide in 100 ml H_2O), 2.5 ml of H_2O, and 0.05 ml of solution C [10% (w/v) ammonium persulfate] in the usual way. Solutions A and B are stable for 1 week at 4°, solution C is prepared immediately before use. Electrode buffer is Tris–glycine, which is prepared by a 10-fold dilution of a stock solution (6.0 g Tris and 28.8 g glycine in 1000 ml of H_2O). The hemoglobin samples containing sucrose (a few grains) and bromphenol blue are loaded by a syringe into the sample wells (the maximum volume is 10 μl, containing 3–10 μg protein). The electrophoresis is performed at constant current (starting at 6–8 mA for about 5 min until the hemoglobin has completely migrated into the gel and then continued at 16–18 mA). The separation takes about 30 min and the gel is stained by Coomassie blue in the usual way. The front of the dye should be marked for the determination of R_f values.

[53] H. R. Maurer, "Disk-Elektrophorese." de Gruyter, Berlin, 1968.

Anion-Exchange HPLC

HPLC allows the quick identification and check of the purity of CTT hemoglobins. The final protocol achieves well-isolated peaks in the chromatogram. Each peak is assigned by the electrophoretic R_f value to the CTT hemoglobin. The rate of two monitoring wavelengths, 280 and 405 nm, allows the discrimination of hemoglobin from apoglobin as well as an accurate quantitation by the integration of the chromatograms. We describe the identification and quantitation of those hemoglobins that occur in the monomer fraction and in eluate 3 of the dimer fraction. The described procedure can be adapted to the analysis of other fractions.

HPLC Separation of Dimeric Hemoglobins in Eluate 3. An anion-exchange column Mono Q HR 5/5 (0.5 × 5 cm) from Pharmacia is used for the analytical separation. The lyophilized hemoglobin fraction is dissolved in H_2O to about 1–2 mg/ml and filtered (filter GV-2, Millipore, Bedford, MA). To avoid contamination of the Mono Q column by protein impurities, the hemoglobins are adsorbed to a batch of DEAE-Sepharose fast flow (Pharmacia) and eluted with a minimum amount of 1 M NaCl. Salt is removed by gel filtration on Sephadex G-25 (column 1 × 15 cm), which is equilibrated and eluted with H_2O. The final hemoglobin concentration is about 1 mg/ml for a maximum injection volume of 0.2 ml. The elution is performed at a flow rate of 1 ml/min with Tris-HCl, pH 9.5, with a salt gradient composed of several linear segments (see Fig. 2). This gradient resolves eight peaks (P_1–P_8), three of which (P_2, P_3, and P_7) do not contain heme (see Fig. 2). P_2, P_3, and P_7 do not exhibit an absorbance at 405 nm. The major fractions are P_1 (CTT II) and P_8 (CTT X), which are separated from their apoproteins, P_2 and P_7. The minor fractions P_4–P_6 are tentatively assigned to CTT VIIa (or VI), VIIb, and VIII. The assignments are based on their electrophoretic R_f values. The good correlation of the retention between PAGE and anion-exchange chromatography at pH 9.5 shows that the effective charges of the hemoglobins are similar under both conditions.

HPLC Separation of Monomeric Hemoglobins. Column and sample preparation are as described above for the dimeric hemoglobins. The elution is performed with a buffer system at pH 9.0 with a salt gradient analogous to the separation of the monomers by FPLC (see above).

Acknowledgments

Financial support by the Deutsche Forschungsgemeinschaft and the Fonds der Chemischen Industrie is gratefully acknowledged.

[8] Preparation and Characterization of Invertebrate Globin Complexes

By SERGE N. VINOGRADOV and PAWAN K. SHARMA

The hemoglobins of vertebrates are uniformly intracellular and, with the exception of agnathan hemoglobins, have a heterotetrameric subunit structure. The hemoglobins of the agnathans (jawless cyclostomes), which consist of hagfish and lampreys, occur as homopolymers and exhibit oxygenation-linked monomerization.[1] Furthermore, hemoglobin is ubiquitous in the vertebrates, being absent only in subantarctic fish of the family Chaenichtydae[2] and in larvae of the European eel *Anguilla anguilla*.[3] The picture presented by invertebrate hemoglobins stands in considerable contrast to that for vertebrates. Hemoglobins from invertebrates can be intra- or extracellular and their occurrence is episodic; thus, closely related species living in identical environments may have no hemoglobin, have an intracellular hemoglobin, or have an extracellular hemoglobin.[4] In some cases, intra- and extracellular hemoglobins coexist in the same species.[5] Whenever intracellular hemoglobins are present, they occur in nucleated erythrocytes.[6] Furthermore, invertebrate hemoglobins vary widely in molecular size, ranging from single-chain, myoglobin-like molecules of ~17 kDa to giant heteromultimeric complexes of up to 10,000 kDa.[7]

Structural Diversity of Invertebrate Hemoglobins

Although the majority of intracellular invertebrate hemoglobins are monomeric, many instances of oligomerization to dimers and tetramers and of the formation of heterodimers and heterotetramers are known.[5-10]

[1] T. Brittain, *Comp. Biochem. Physiol. B* **99B**, 731 (1991).
[2] J. C. Hureau, D. Petit, J. M. Fine, and M. Marneux, *in* "Adaptations within Antarctic Ecosystems" (G. A. Llano, ed.), p. 459. Smithsonian Institution, Washington, D.C., 1977.
[3] D. Keilin, *Acta Biochim. Pol.* **3**, 445 (1956).
[4] H. M. Fox, *Proc. R. Soc. London, Ser. B.* **136**, 378 (1949).
[5] R. C. Terwilliger, *Am. Zool.* **20**, 53 (1980).
[6] N. B. Terwilliger, R. C. Terwilliger, and E. Schabtach, *in* "Blood Cells of Marine Invertebrates: Experimental Systems in Cell Biology and Comparative Physiology" (W. D. Cohen, ed.), p. 193. Alan R. Liss, New York, 1987.
[7] S. N. Vinogradov, *Comp. Biochem. Physiol. B* **82B**, 1 (1985).
[8] R. C. Terwilliger and N. B. Terwilliger, *Comp. Biochem. Physiol. B* **81B**, 255 (1985).
[9] C. Bonaventura and J. Bonaventura, "The Mollusca," Vol. 2, p. 1. Academic Press, London, 1983.
[10] A. F. Riggs, *Am. Zool.* **31**, 535 (1991).

Extracellular hemoglobins can be subdivided into four main classes: (1) single-chain, single heme-binding domain, ~17-kDa globin chains, exemplified by the multiple hemoglobins of the insect larvae of *Chironomus*[11]; (2) two-domain, multisubunit hemoglobins consisting of ~35,000 chains, each containing two, linearly connected heme-binding domains, form polymeric aggregates with M_r of 250,000–800,000, found in carapaced branchiopod crustaceans[12]; (3) multi-domain, multisubunit hemoglobins composed of two or more polypeptide chains, each containing from 9 to 20 heme-binding domains, form large structures with M_r in the range of 250,000 to 10,000,000—such molecules occur among the carapaceless branchiopod crustaceans, the planorbid snails, and the clams from the families Astartidae and Carditidae[5,7,10]; (4) hexagonal bilayer hemoglobins and chlorocruorins,* M_r ~3,600,000, characterized by a two-tiered, hexagonal symmetrical electron microscopic appearance, an acidic isoelectric point, and an abnormally low iron content—such molecules are found among the three classes of the annelids (oligochaetes, polychaetes, and leeches) and the vestimentiferans.[7,13-15]

Molecular Diversity of Invertebrate Hemoglobins

The amino acid sequences of invertebrate hemoglobins that had been determined prior to 1987 were of monomeric globin chains of ~17 kDa. An evolutionary tree constructed on the basis of maximum parsimony for 26 sequences, representing the annelids, the arthropods, and the molluscs, 6 plant hemoglobins, and 213 vertebrate globins, showed unequivocally their common origin.[16] The result was consistent with the fossil record and with the hypothesis that the primitive metazoan hemoglobin was monomeric. Since then, over 100 additional amino acid sequences of invertebrate globins have demonstrated that the great diversity in the molecular size and quaternary structure of invertebrate hemoglobins is equalled by a corresponding diversity in the chemical

---

* The prosthetic group, chloroheme, has a 3-formyl instead of a 3-vinyl substitution.

[11] M. Goodman, G. Braunitzer, T. Kleinschmidt, and H. Aschauer, *Hoppe-Seyler's Z. Physiol. Chem.* **354,** 205 (1963).

[12] E. Ilan and E. Daniel, *Comp. Biochem. Physiol. B* **63B,** 303 (1979).

[13] S. N. Vinogradov, O. H. Kapp, and M. Ohtsuki, *in* "Electron Microscopy of Protein" (J. Harris, ed.), p. 135. Academic Press, London, 1982.

[14] S. N. Vinogradov, D. A. Walz, and P. K. Sharma, *Comp. Biochem. Physiol. B* **98B,** 187 (1991).

[15] R. C. Terwilliger and N. B. Terwilliger, *Biol. Soc. Wash. Bull.* **6,** 273 (1985).

[16] M. Goodman, J. Pedwaydon, J. Czelusniak, T. Suzuki, T. Gotoh, L. Moens, D. A. Walz, and S. N. Vinogradov, *J. Mol. Evol.* **27,** 236 (1988).

nature of the constituent globin chains.[17] The latter can be divided into several distinct groups.

1. Multiple, single-chain globins of ~17 kDa; in addition to the 12 globins of *Chironomus*, known examples are the monomeric and polymeric globins of the annelid *Glycera*,[18–21] a bacterial globin (*Vitreoscilla*),[22] the three cytoplasmic hemoglobins of the symbiont-containing clam *Lucina*,[23,23a] the hemoglobin of the parasitic nematode *Trichostrongylus*,[24] the related globinlike host-protective antigen in *Caenorhabditis*,[25] and the "functional" chains of hexagonal bilayer hemoglobins such as those of *Lumbricus* and *Tylorrhynchus*.[26–29]

2. Truncated, single-chain globins (~120 versus 150 residues) of the ciliated protozoans *Paramecium*[30] and *Tetrahymena*,[31] of which the former appears to be polymorphic.[32]

3. Chimeric, ~40-kDa globins in the yeasts *Saccharomyces* and *Candida*[33,34] and in *Escherichia coli*[35] have one N-terminal heme-binding domain and a C-terminal FMN-binding domain.

[17] S. N. Vinogradov, D. A. Walz, B. Pohajdak, L. Moeus, O. H. Kapp, T. Suzuki, and C. N. A. Trotman, *Comp. Biochem. Physiol.* **106B**, 1 (1993).

[18] T. Imamura, T. O. Baldwin, and A. F. Riggs, *J. Biol. Chem.* **247**, 2785 (1972).

[19] P. Simons and J. Satterlee, *Biochemistry* **28**, 8525 (1989).

[20] R. Zafar, L. Chow, M. Stern, S. Vinogradov, and D. Walz, *Biochim. Biophys. Acta* **1041**, 117 (1990).

[21] R. Zafar, L. Chow, J. Scully, P. Sharma, S. Vinogradov, and D. Walz, *J. Biol. Chem.* **265**, 21843 (1990).

[22] S. Wakabayashi, H. Matsubara, and D. E. Webster, *Nature (London)* **322**, 481 (1986).

[23] J. D. Hockenhull-Johnson, M. Stern, P. Martin, C. Dass, D. M. Desiderio, J. B. Wittenberg, S. N. Vinogradov, and D. A. Walz, *J. Protein Chem.* **10**, 609 (1991).

[23a] J. D. Hockenhull-Johnson, M. S. Stern, J. B. Wittenberg, S. N. Vinogradov, O. H. Kapp, and D. A. Walz, *J. Prot. Chem.* **12**, 261 (1993).

[24] M. J. Frenkel, T. A. A. Dolpheide, B. M. Wagland, and C. W. Ward, *Mol. Biochem. Parasitol.* **50**, 27 (1992).

[25] J. Sulston, Z. Du, K. Thomas, R. Wilson, L. Hillier, R. Staden, N. Halloran, P. Green, J. Thierry-Mieg, L. Qiu, S. Dear, A. Coulson, M. Craxton, R. Durbin, M. Berks, M. Metzstein, T. Hawkins, R. Ainscough, and R. Waterston, *Nature (London)* **356**, 37 (1992).

[26] F. Shishikura, J. W. Snow, T. Gotoh, S. N. Vinogradov, and D. A. Walz, *J. Biol. Chem.* **262**, 3123 (1987).

[27] K. Fushitani, M. S. Matsuura, and A. F. Riggs, *J. Biol. Chem.* **263**, 6502 (1988).

[28] T. Suzuki and T. Gotoh, *J. Biol. Chem.* **261**, 9257 (1986).

[29] T. Gotoh and T. Suzuki, *Zool. Sci.* **7**, 1 (1990).

[30] H. Iwaasa, T. Takagi, and K. Shikama, *J. Mol. Biol.* **208**, 355 (1989).

[31] H. Iwaasa, T. Takagi, and K. Shikama, *J. Biol. Chem.* **265**, 8603 (1990).

[32] I. Usuki, A. Hino, and T. Ochiai, *Comp. Biochem. Physiol. B* **93B**, 555 (1989).

[33] H. Zhu and A. F. Riggs, *Proc. Natl. Acad. Sci. U.S.A.* **89**, 5015 (1992).

[34] H. Iwaasa, T. Takagi, and K. Shikama, *Zool. Sci.* **8**, 1134 (1991).

[35] S. Vasudevan, W. Armarego, D. Shaw, P. Lilley, N. Dixon, and R. Poole, *Mol. Cell. Genet.* **226**, 49 (1991).

4. Truncated, chimeric globins (~25–30 kDa) found in the hexagonal bilayer hemoglobins and chlorocruorins[36,37]; these globins are generally heme deficient and are thought to act as "linkers" of dodecameric "functional" globin complexes.[38–40]

5. Multidomain, chimeric globins that consist of linear, covalently bonded ~17-kDa domains (2 to 20) and that aggregate into a variety of structures. Examples comprise the high-oxygen-affinity hemoglobins of the parasitic nematodes *Pseudoterranova*[41] and *Ascaris*,[42,42a] octamers of two-domain globins,[43] and the hemoglobin of the brine shrimp *Artemia*, a dimer of 9-domain chains.[44–46]

Scope

It is obvious from the very brief overview given above that the preparation of invertebrate hemoglobins[46a] ranges from the isolation and purification of polymorphic, single heme-containing polypeptide chains to the isolation of giant heteromultimeric complexes consisting of two or more different subunits. In this review I will describe the isolation and purification of the giant hexagonal bilayer hemoglobins and chlorocruorins of the annelids and their subunits.

Preparation of Native Hexagonal Bilayer Complexes

Because of their high molecular weight, first observed by Svedberg,[47] these molecules can be easily isolated by preparative ultracentrifugation.

[36] T. Suzuki, T. Takagi, and T. Gotoh, *J. Biol. Chem.* **265**, 12168 (1990).
[37] T. Gotoh, T. Suzuki, and T. Takagi, in "Structure and Function of Invertebrate Oxygen Carriers" (S. N. Vinogradov and O. H. Kapp, eds.), p. 217. Springer-Verlag, New York, 1991.
[38] S. N. Vinogradov, M. Mainwaring, S. Lugo, O. H. Kapp, and A. V. Crewe, *Proc. Natl. Acad. Sci. U.S.A.* **83**, 8034 (1986).
[39] O. H. Kapp, A. N. Qabar, M. Bonner, M. Stern, D. A. Walz, M. Schmuck, I. Pilz, J. S. Wall, and S. N. Vinogradov, *J. Mol. Biol.* **213**, 141 (1990).
[40] A. N. Qabar, M. Stern, D. A. Walz, J. Chiu, R. Timkovich, J. S. Wall, O. H. Kapp, and S. N. Vinogradov, *J. Mol. Biol.* **222**, 1109 (1991).
[41] B. Dixon, B. Walker, W. Kimmins, and B. Pohajdak, *Proc. Natl. Acad. Sci. U.S.A.* **88**, 5565 (1991).
[42] I. De Baere, L. Liu, L. Moens, J. Van Beeumen, C. Gielens, J. Richelle, C. Trotman, J. Finch, M. Gerstein, and M. Perutz, *Proc. Natl. Acad. Sci. U.S.A.* **89**, 4638 (1992).
[42a] D. R. Sherman, A. P. Kloek, B. R. Krishnan, B. Guinn, and D. E. Goldberg, *Proc. Natl. Acad. Sci. U.S.A.* **89**, 11696 (1992).
[43] S. Darawshe and E. Daniel, *Eur. J. Biochem.* **201**, 169 (1991).
[44] L. Moens, M. L. Van Hauwaert, K. DeSmet, D. Geelen, G. Verpooten, J. Van Beeumen, S. Wodak, P. Alard, and C. N. A. Trotman, *J. Biol. Chem.* **263**, 4679 (1988).
[45] A. M. Manning, C. N. A. Trotman, and W. P. Tate, *Nature (London)* **348**, 653 (1990).
[46] C. N. A. Trotman, A. M. Manning, L. Moens, and W. P. Tate, *J. Biol. Chem.* **266**, 13789 (1991).
[46a] See also this series, Vol. 76, pp. 43–54.
[47] T. Svedberg and A. Hedenius, *J. Am. Chem. Soc.* **55**, 2834 (1933).

The following procedure, taken from our first studies of the extracellular hemoglobin of *Lumbricus terrestris,* the common North American earthworm,[48,49] has served us well and illustrates the general approach.

Live worms are washed in distilled water, suspended on wire hooks by their posterior ends, and cut at about the seventh body segment (region of the primitive heart) using scissors. The first few drops of blood are collected via mild suction into a 25-ml flask containing a small amount of 0.1 M Tris-HCl buffer, pH 7.0, 1 mM EDTA, and are kept in an ice–water mixture. All subsequent operations are performed in the cold room at 4–7°. The collected blood is filtered through a small plug of glass wool and centrifuged at 20,000 g for ~0.5 hr to remove any particulate matter and cell debris. The clear supernatant hemoglobin solution is then centrifuged for >1 hr at 150,000 g in a Beckman Model L preparative ultracentrifuge or overnight (~18 hr) at 20,000 g at 4°. The supernatant is poured off carefully and the dark red pellet is dissolved in a small amount of Tris-HCl buffer; this step is repeated at least once. The concentrated hemoglobin solution is layered at the top of a preparative column of crosslinked Sepharose CL-6B or CL-4B or Sephacryl S-200 (Pharmacia LKB Biotechnology, Piscataway, NJ 08854), and eluted with the Tris-HCl buffer. The front moiety of the eluted peak is collected, the hemoglobin pelleted by ultracentrifugation, and the pellet dissolved in the desired buffer.

Comments on Method of Preparation

It is imperative that the supernatant hemoglobin solution after low-speed centrifugation and prior to the first ultracentrifugation is clear. If it is not, centrifugation at 20,000 g for 0.5 hr should be repeated as many times as necessary and for longer durations if necessary, until the supernatant solution is clear. The gel filtration step through Sepharose or Sephacryl is not absolutely necessary unless some dissociation of the hemoglobin complex has occurred. In this connection, it should be noted that for some annelid hemoglobins, such as *Amphitrite ornata* hemoglobin,[50] and for some chlorocruorins, e.g., that of *Myxicola infundibulum,*[51] dissociation can occur in the presence of 1 mM EDTA, because of the removal of Ca(II), which is required for maintenance of the structural integrity of the

[48] J. M. Shlom and S. N. Vinogradov, *J. Biol. Chem.* **248,** 7904 (1973).
[49] S. N. Vinogradov, J. M. Shlom, B. C. Hall, O. H. Kapp, and H. Mizukami, *Biochim. Biophys. Acta* **492,** 136 (1977).
[50] E. Chiancone, M. Brenowitz, F. Ascoli, C. Bonaventura, and J. Bonaventura, *Biochim. Biophys. Acta* **623,** 146 (1980).
[51] S. N. Vinogradov, P. Standley, M. Mainwaring, O. H. Kapp, and A. V. Crewe, *Biochim. Biophys. Acta* **828,** 43 (1985).

hexagonal bilayer structure. In such cases, 1–10 mM $CaCl_2$ should be used instead of 1 mM EDTA in the Tris-HCl buffer.

Ideally, the annelid blood should be obtained by its direct removal from annelid blood vessels. This is easily done in the case of the marine polychaete (lugworm) *Arenicola*,[52,53] which has a large dorsal vessel. The worm is held in one hand in a wad of paper towels and is cut longitudinally from the posterior to the anterior end with small, sharp scissors. The exposed internal organs are blotted with a paper towel and the blood is sucked into a micropipette inserted into the blood vessel. A small micropipette of ~200 μl volume can be easily fashioned from the narrow end of a 9-inch Pyrex Pasteur pipette, drawn out over a Bunsen burner flame to a microtip.

Lumbricus hemoglobin is not susceptible to proteolytic attack as judged by two criteria: by the complete absence of alteration(s) in the SDS–PAGE patterns (reduced and unreduced) of the whole blood on standing and by the fact that there is no difference between the SDS–PAGE patterns of whole blood and of the purified hemoglobin.[54]

In cases wherein the properties of the hemoglobin sought are not known and the hemoglobin has to be isolated from homogenized animals, because the animals are small or expensive or the supply of animals is limited, it is prudent to guard against proteolytic attack during the preparation. We routinely include 0.01% (w/v) phenylmethylsulfonyl fluoride (PMSF; Sigma, St. Louis, MO) by adding 1 ml of stock solution (100 mg PMSF in 10 ml acetonitrile) to 100 ml of Tris-HCl buffer. PMSF is effective only against serine proteases and its half-life can vary from ~25 hr at pH 7 at 4° to 110 min at 25°.[55] If the problem of proteolytic degradation cannot be handled by PMSF alone, a cocktail of protease inhibitors, containing PMSF, pepstatin (inhibitor of acid proteases), sodium EDTA (inhibitor of metalloproteases), and leupeptin (a broad-spectrum protease inhibitor), is available from Boehringer Mannheim Corp. (P.O. Box 50414, Indianapolis, IN 46250). Homogenization is carried out in the cold, in a Waring blender, using minimal amounts of more concentrated Tris-HCl buffer in order to compensate for dilution and as quickly as feasible. Such an approach works well in the isolation of hemoglobin from the leech *Macrobdella decora*[39] and of chlorocruorin from *M. infundibulum*[51] and *Eudistylia vancouverii*.[40] An alternative way of liberating the hemoglobin, which works very well for small species such as the aquatic

[52] A. Toulmond, *C. R. Hebd. Seances Acad. Sci.* **270**, 1368 (1970).
[53] S. N. Vinogradov, T. F. Kosinski, and O. H. Kapp, *Biochim. Biophys. Acta* **621**, 315 (1980).
[54] J. M. Shlom, Ph.D. Thesis, Wayne State University, Detroit, MI (1982).
[55] G. James, *Anal. Biochem.* **86**, 574 (1978).

oligochaete *Tubifex,* consists of subjecting the worms placed in a small amount of Tris buffer to one or two freeze–thaw cycles.[56] The homogenized or thawed material is then subjected to the same procedure as described above, omitting filtration through a plug of glass wool, proceeding directly to centrifugation at low speed (20,000 g, 0.5 hr).

Chlorocruorin is found in four families of marine polychaetes. The best preparations of this protein are obtained by homogenization of the plumes of these sessile worms. Good preparations of hexagonal bilayer hemoglobins and chlorocruorins can only be obtained from live animals. Attempts to isolate these molecules from frozen animals have never been successful in our experience.

The annelid hexagonal bilayer molecules prepared by the method described above appear to be physically homogeneous.[48,54] Isoelectric focusing of *L. terrestris* hemoglobin shows the presence of two components with identical SDS–PAGE patterns: the major component has a pI of 5.50 and the minor component a pI of 5.37.[54] The latter component may be due to the minor heterogeneity observed in the amino acid sequence of the monomer subunit.[26]

Lumbricus terrestris hemoglobin is very stable over the range from pH 6 to pH 8[57,58] and will keep for several months when stored in the refrigerator in the Tris-HCl buffer containing 1 mM EDTA. Storage in liquid nitrogen is desirable for longer periods of time. It should be kept in mind that annelid extracellular hemoglobins and chlorocruorins dissociate at pH >8[59] and that alkaline earth cations [Mg(II), Ca(II), and Sr(II)] protect the hexagonal bilayer structure at alkaline pH.[50,56,57,59,60] Thus, interpretation of experiments carried out with extracellular annelid hemoglobins at alkaline pH should take into consideration the possible effect of dissociation.

Extinction Coefficients of Annelid Extracellular Hemoglobins

The absorption spectra of the annelid extracellular hemoglobins appear to be very similar to those of vertebrate hemoglobins and undergo similar

[56] G. Polidori, M. Mainwaring, T. Kosinski, C. Schwarz, R. Fingal, and S. N. Vinogradov, *Arch. Biochem. Biophys.* **233,** 800 (1984).
[57] O. H. Kapp, G. Polidori, M. Mainwaring, A. V. Crewe, and S. N. Vinogradov, *J. Biol. Chem.* **259,** 628 (1984).
[58] M. Mainwaring, S. Lugo, R. Fingal, O. H. Kapp, and S. N. Vinogradov, *J. Biol. Chem.* **261,** 10899 (1986).
[59] S. N. Vinogradov, J. M. Shlom, O. H. Kapp, and P. Frossard, *Comp. Biochem. Physiol. B* **67B,** 1 (1980).
[60] G. Polidori, M. Mainwaring, and S. N. Vinogradov, *Comp. Biochem. Physiol. A* **89A,** 541 (1988).

TABLE I

ABSORPTION MAXIMA AND MILLIMOLAR EXTINCTION COEFFICIENTS
OF *Lumbricus terrestris* AND HUMAN HEMOGLOBINS

Derivative	Lumbricus[a]		Human[b]	
	Wavelength (nm)	E (cm^2 mM^{-1})	Wavelength (nm)	E (cm^2 mM^{-1})
Oxy	577	14.2	577	14.6
	542	13.7	541	13.8
	417	112.7	415	125.0
	345	29.2	344	27.0
	278	51.4	—	—
Deoxy	550	11.4	555	12.5
	430	113.6	430	133
	276	46.3	274	29.2

[a] M. R. Rossi Fanelli, E. Chiancone, P. Vecchini, and E. Antonini, *Arch. Biochem. Biophys.* **141**, 278 (1970).

[b] E. Antonini and M. Brunori, "Haemoglobin and Myoglobin in their Reaction with Ligands." North-Holland Publ., Amsterdam, 1971.

changes on alteration of ligation and/or oxidation state.[61] A comparison of the wavelengths of the absorption bands and the corresponding extinction coefficients (based on heme) for *L. terrestris* and human hemoglobins is shown in Table I.

In practice, it is more convenient to use extinction coefficients derived from dry weight measurements of exhaustively dialyzed hemoglobin and chlorocruorin solutions. Table II lists the values that have been obtained in our laboratory with several hexagonal bilayer molecules.[62-64]

Carbohydrate and Iron Content of Annelid Hemoglobins

The carbohydrate content of *L. terrestris* hemoglobin determined by gas chromatography of the trimethylsilylmethylglycoside derivative was found to be 2.0 ± 0.5 wt%.[65] The ratio of mannose to *N*-acetylglucosamine was 9 : 1. Most of the carbohydrate appears to be located in chain IV of the trimer subunit. No carbohydrate was detected by this method in the

[61] M. C. M. Chung and H. D. Ellerton, *Prog. Biophys. Mol. Biol.* **35**, 53 (1979).
[62] S. N. Vinogradov, unpublished observations (1985).
[63] B. Cameron, *Anal. Biochem.* **35**, 515 (1970).
[64] G. Hanania, A. Yegiayan, and B. Cameron, *Biochem. J.* **98**, 189 (1966).
[65] F. Shishikura, M. Mainwaring, E. Yurewicz, J. Lightbody, D. A. Walz, and S. N. Vinogradov, *Biochim. Biophys. Acta* **869**, 314 (1986).

TABLE II
EXTINCTION COEFFICIENTS OF HEXAGONAL BILAYER MOLECULES BASED
ON DRY WEIGHT MEASUREMENTS

Protein	Wavelength (nm)	E (ml mg^{-1} cm^{-1})	Ref.
Lumbricus hemoglobin	280	2.10 $\pm$ 0.07	56
		2.04 $\pm$ 0.14	40
	540[a]	0.442 $\pm$ 0.013[b]	
Macrobdella hemoglobin	280	2.19 $\pm$ 0.08	39
	540[a]	0.465 $\pm$ 0.016	39
Amphitrite hemoglobin	280	2.09 $\pm$ 0.02	62
Arenicola hemoglobin	540[a]	0.460 $\pm$ 0.018	
Eudistylia chlorocruorin	605	0.472 $\pm$ 0.035	40
	280	2.23 $\pm$ 0.17	40
Sperm whale myoglobin	280	1.70[c]	
		1.74 $\pm$ 0.07	40

[a] Cyanomethemoglobin form.
[b] This value corresponds to an extinction coefficient of 11,200 cm^2/g atom of Fe, based on the spectrophotometrically determined iron content of 0.221 $\pm$ 0.011 wt%,[52] and should be compared to the value of 11.0 $\times$ 10^3 cm^2/g atom of Fe for human hemoglobin.[63]
[c] Calculated using the value 31.2 mM^{-1} cm^{-1} of Hanania *et al.*[64]

hemoglobins of the leech *M. decora*[39] and the polychaete *A. ornata*[62] or the chlorocruorin of *E. vancouverii*.[40] It is thus likely that carbohydrate is present only in the oligochaete hemoglobins.

A recent survey of the literature over the past 30 years on extracellular hemoglobins and chlorocruorins of over 30 species of annelids[14] showed that the range of iron contents was 0.211–0.265 wt%, with a mean of 0.228 $\pm$ 0.013 wt% (N = 28); the range of published heme contents was appreciably broader, 1.83–3.64 wt%, mean = 2.60 $\pm$ 0.38 wt%, with 7 out of the 29 values outside the standard deviation range. Although the greater variation in the heme contents may be due to the variability in the pyridine hemochromogen method used for heme determination in the majority of the published studies, the mean heme content of 2.60 wt% corresponds to an iron content of 0.236 wt%, in excellent agreement with the mean iron content of 0.228 wt%. Experimental values of iron and heme contents outside the ranges of 0.211–0.243 wt% and 2.3–2.7 wt%, respectively, and corresponding to a minimum molecular mass outside the range of 23,000–26,000, should be regarded with caution.

It should be noted here that the iron content of annelid extracellular hemoglobins and chlorocruorins, which is about two-thirds of the normal iron content of vertebrate hemoglobins and myoglobins, reflects the fact that their giant, hexagonal bilayer structure includes some 40 to 50 copies of the truncated, chimeric, heme-deficient globin chains (25–30 kDa), contributing about one-third of the total molecular mass.

Preparation of Dodecameric Complex

In collaboration with S. Gill and J. Wall, we have demonstrated that a stable dodecamer of globin chains is the principal structural and functional subunit of *L. terrestris* hemoglobin[66] and of *E. vancouverii* chlorocruorin.[39] The hemoglobin dodecamer retains the affinity (~12 Torr) and two-thirds of the cooperativity (Hill coefficient n_{max} ~2.1 vs. n_{max} = 3.3) of oxygen binding of the native hemoglobin at neutral pH.[66] The chlorocruorin dodecamer exhibits a more than twofold increase in affinity (47 Torr vs. 112 Torr) and retains slightly less than half of the cooperativity of the native molecule (n_{max} = 1.5 vs. n_{max} = 3.6).[67] These findings imply that there are at least two mechanisms of cooperativity that contribute to the total cooperativity of oxygen binding in the hexagonal bilayer hemoglobins and chlorocruorins, one mediated by intersubunit interactions within the dodecamer and one due to interdodecamer interactions mediated by the chimeric globin linkers.

Preparative isolation of the dodecameric complexes was originally achieved by the following procedure: (1) exposure of the native molecule to 4 M urea at neutral pH (typically dialysis for 4 hr) in the Tris-HCl buffer, 1 mM EDTA; (2) gel filtration at neutral pH on a preparative column of Sephacryl S-200 (Pharmacia) accompanied by the isolation of the peak of dissociated material having the highest molecular weight, i.e., a higher elution volume next to the undissociated molecule; (3) exposure of this fraction to 4 M urea; (4) gel filtration on Sephacryl S-200 at neutral pH and collection of the same peak as before.[66] SDS–PAGE of this material showed it to consist predominantly of the four heme-containing chains[26,27] with <5% of the ~25- to 30-kDa chimeric globin linker chains. Gel filtration measurements indicated that the M_r of the isolated complex was 202,000 ± 15,000 and scanning transmission electron microscopic determinations of the mass of freeze-dried, unstained specimens provided M_r of 202,000 ± 26,000[66] and showed that they consist of three copies each of the four different heme-containing chains I–IV. Variable amounts of the two constituent subunits of the *Lumbricus* hemoglobin dodecamer, the

[66] S. N. Vinogradov, P. K. Sharma, A. N. Qabar, J. Wall, J. A. Westrick, J. Simmons, and S. J. Gill, *J. Biol. Chem.* **266,** 13091 (1991).
[67] J. A. Westrick, S. J. Gill, P. K. Sharma, and S. N. Vinogradov, unpublished observations (1991).

disulfide-bonded trimer of chains II + III + IV, and the monomer subunit (chain I), as well as of the three chimeric globin linker (chains VA, VB, and VI) can also be obtained as by-products of the isolation of the dodecamer. The order of elution from a Sephacryl S-200 column with increased volume, following the dodecamer, is trimer, linker, and monomer subunits, respectively.

We found that the purity of the *Lumbricus* hemoglobin dodecamer complex depends on the duration of contact with the dissociating agent[68]; thus, exposures for less than 1 hr to higher urea concentrations (6 to 8 M), followed immediately by gel filtration, provides a dodecamer subunit virtually free of the linker chains VA, VB, and VI.

The isolation of the dodecamer subunit of *E. vancouverii* chlorocruorin is very similar.[40] In this case, the dodecamer is made up of three copies of a tetramer of at least four different globin chains. The tetramer subunit can also be obtained concurrently with the dodecamer subunit: it consists of disulfide-bonded tetramers and of noncovalent complexes of disulfide-bonded dimers.[40] Interestingly, the oxygen affinity and cooperativity of the tetramer appear to be very similar to that of the dodecamer.[67]

Crystallization of Hexagonal Bilayer Molecules and Dodecamer Subunits

Several types of crystals of *L. terrestris* hemoglobin have been obtained.[69] X-Ray diffraction data to a resolution of 5.5 Å have been collected for type III crystals.[70] The space group is $C222_1$, with one-half of the molecule per asymmetric unit. The observed diffraction symmetry is consistent with *Lumbricus* hemoglobin having a $\boldsymbol{D}_6$ point group symmetry.

Crystals of *Lumbricus* hemoglobin dodecamer and of *E. vancouverii* chlorocruorin suitable for X-ray diffraction analysis have also been obtained.[71] It is possible that solution of the crystal structure of the dodecamer, perhaps by replacement methods, will aid the solution of the complete crystal structure of the *Lumbricus* hexagonal bilayer hemoglobin. In contrast to the *Lumbricus* hemoglobin crystals, the chlorocruorin crystals have orthorhombic symmetry. Measurement of their density suggests that there is one molecule with a mass of 3620 kDa in the *F222* unit cell. The *F222* cell however, appears to contain a pseudocubic *I23* cell that is likely to contain 1/24 of the chlorocruorin molecule, having a mass of 151 kDa. According to the recent findings concerning the quaternary structure of *Eudistylia* chlorocruorin[40] 1/24 of the native molecule should include one-half of a dodecamer, i.e., six ~16-kDa globin chains (96 kDa) and at least

[68] P. K. Sharma and S. N. Vinogradov, unpublished observations (1992).
[69] W. E. Royer, Jr., W. A. Hendrickson, and W. E. Love, *J. Mol. Biol.* **197,** 149 (1987).
[70] W. E. Royer, Jr. and W. A. Hendrickson, *J. Biol. Chem.* **263,** 13762 (1988).
[71] P. Martin, M. Doyle, and B. F. Edwards, private communication (1992).

TABLE III
NOMENCLATURE FOR HEME-CONTAINING CHAINS OF HEXAGONAL BILAYER
HEMOGLOBINS PRESENTLY IN USE AND PROPOSED GENERAL NOMENCLATURE

Hemoglobin source/character	Chain nomenclature[a]				Basis	Ref.
Lumbricus	I	II	III	IV	SDS–PAGE[b]	48
	d	b	a	c	Chromatography	27
Tylorrhynchus	I	IIA	IIC	IIB	Chromatography	76
Strain[c]	A	A	B	B	Aa sequence[d]	77
No. of Cys	2	3	4	3	Aa sequence	—
		Proposed				
	a	A	b	B	—	78

[a] In the majority of annelid hemoglobins chain I corresponds to the monomer subunit and the remaining three chains form a disulfide-bonded trimer subunit.[7]
[b] In order of decreasing mobility.
[c] The amino acid sequences cluster into two groups.[77]
[d] Aa, Amino acid.

two chimeric globin chains of ~27 kDa each to account for the difference (151 − 96 = 55 kDa). These preliminary results imply that the overall stoichiometry in hexagonal bilayer molecules is 12 dodecamers (144 globin chains) linked by 48 chimeric, heme-deficient, ~27-kDa chains. It should be noted that recently another model of the subunit structure of Lumbricus hemoglobin has been proposed.[71a]

Primary Structure of Hexagonal Bilayer Hemoglobins

The complete amino acid sequences of the heme-containing chains of the hexagonal bilayer hemoglobins of the oligochaetes L. terrestris,[26,27] Tubifex tubifex,[72] and Pheretima hilgendorfi,[73] the polychaete Tylorrhynchus heterochaetus,[28] and the vestimentiferan Lamellibrachia[74] have been determined: all four chains in the case of Lumbricus and Tylorrhynchus and single chains in the case of Lamellibrachia (AIII),[75] Tubifex (I), and Pheretima (I). Table III shows the nomenclature of the chains constituting

[71a] D. W. Ownby, H. Zhu, K. Schneider, R. C. Beavis, B. T. Chait, and A. F. Riggs, J. Biol. Chem. 268, 13539 (1993).
[72] M. S. Stern, S. N. Vinogradov, P. K. Sharma, K. Ereifej, and D. A. Walz, Eur. J. Biochem. 194, 67 (1990).
[73] T. Suzuki, Eur. J. Biochem. 185 (1989).
[74] T. Suzuki, T. Takagi, and S. Ohta, Biochem. J. 266, 221 (1990).
[75] It should be noted that the vestimentiferan Lamellibrachia also has a 400-kDa hemoglobin that consists only of four chains. These chains have been sequenced by Suzuki et al. [T. Takagi, H. Iwaasa, S. Ohta, and T. Suzuki, in "Structure and Function of Invertebrate Oxygen Carriers" (S. N. Vinogradov and O. H. Kapp, eds.), p. 245. Springer-Verlag, New York, 1991].

the monomer and trimer subunits of the foregoing hemoglobins, presently in use, as well as a proposed, more general nomenclature.[76-78]

The two heme-deficient linker chains of *Tylorrhynchus* hemoglobin have been sequenced[79] as well as one of the *Lamellibrachia* linker chains.[80] One of the linker chains of *Lumbricus* hemoglobin has been sequenced by Suzuki and Riggs.[81] The linker chains have a characteristic motif of 6 Cys residues separated by 6, 6, 6, 6, and 10 other residues, respectively: $-C-X_6-C-X_6-C-I-X_3-L-X-C-D-G-X_2-D-C-X_2-G-X-D-E-D-X_3-C$. Although this 40-residue motif is not found in any known globin sequence, it corresponds exactly to the seven cysteine-rich repeats of the ligand-binding domain of the low density lipoprotein (LDL) receptors of man and *Xenopus laevis*.[81]

[76] T. Suzuki, T. Furukohri, and T. Gotoh, *J. Biol. Chem.* **260**, 3145 (1985).
[77] T. Gotoh, F. Shishikura, J. W. Snow, K. L. Ereifej, S. N. Vinogradov, and D. A. Walz, *Biochem. J.* **241**, 441 (1987).
[78] T. Gotoh, T. Suzuki, and T. Takagi, *in* "Structure and Function of Invertebrate Oxygen Carriers" (S. N. Vinogradov and O. H. Kapp, eds.), p. 216. Springer-Verlag, New York, 1991.
[79] T. Suzuki, T. Takagi, and S. Ohta, *J. Biol. Chem.* **265**, 12168 (1990).
[80] T. Suzuki, T. Takagi, and S. Ohta, *J. Biol. Chem.* **265**, 1551 (1990).
[81] T. Suzuki and A. F. Riggs, *J. Biol. Chem.* **268**, 13548 (1993).

[9] Fish Hemoglobins

By ALICE DEYOUNG, LAURA D. KWIATKOWSKI,
and ROBERT W. NOBLE

There has been considerable increasing interest in fish hemoglobins in recent years. They exhibit the essential features of mammalian hemoglobins, cooperative ligand binding and heterotropic responses to a variety of ionic species, but they display an astounding variety of functional behaviors. These different properties are of interest as examples of evolutionary adaptation to differing physiological and environmental needs. They also offer valuable systems in which to study specific phenomena in either an exaggerated form or a simplified context. There are components of the hemoglobins of Salmonidae that exhibit neither Bohr effects nor responses to organic phosphates.[1,2] Here

[1] B. Giardina, M. Brunori, I. Binotti, S. Giovenco, and E. Antonini, *Eur. J. Biochem.* **39**, 571 (1973).
[2] H. K. F. Lau, D. E. Wallach, R. R. Pennelly, and R. W. Noble, *J. Biol. Chem.* **250**, 1400 (1975).

cooperativity can be examined without the complexities associated with the simultaneous influences of buffer components on heterotropic interactions. Many fish hemoglobins exhibit exaggerated Bohr effects, commonly known as the Root effect,[3] permitting the study of pH dependencies which are considerably amplified relative to that observed in mammalian hemoglobins. Fish hemoglobins can exhibit widely differing ligand affinities, with the total range reported for different hemoglobins under varying conditions being greater than four orders of magnitude. This would seem to offer an approach to the study of the relationship between protein structure and ligand affinity of the heme prosthetic group. In addition, many fish hemoglobins that exhibit the Root effect lose cooperative ligand binding when they attain their minimum ligand affinity at low pH in the presence of organic phosphates.[4,5] These noncooperative, low-affinity states appear to be excellent models of liganded T states and are ideal for the analysis of the origins of the differences in the ligand affinities of the two extreme quaternary states of hemoglobin, the R and T states.

Preparation and Characterization of Hemoglobin

Considerations for Obtaining Fish

It is preferable to obtain fish hemoglobin from live fish. Because transporting living fish is technically difficult as well as expensive, it is almost always necessary to obtain blood samples from fish in their natural habitat. For this reason, abundantly available, local species are preferred. Some fish are best caught during their spawning season because at this time they travel from the deep waters of rivers and lakes into shallow tributaries, where they are easily netted or trapped. For years this has been the method by which these authors have obtained carp. Fortunately, red cells in the frozen state at liquid nitrogen temperatures can be successfully stored for many years. Several species of fish, such as trout, salmon, and catfish, are farmed commercially. Hatcheries offer a very convenient source of fish, and these may represent a somewhat restricted gene pool, possibly reducing the problems of polyallelism. When fish are not available locally, or when those available will not do, the ability to store red cells for extended periods makes it feasible to collect specimens even in remote locations, as long as a supply of liquid nitrogen is available.

[3] R. W. Root, *Biol. Bull.* (*Woods Hole, Mass.*) **61**, 427 (1931).
[4] A. L. Tan, R. W. Noble, and Q. H. Gibson, *J. Biol. Chem.* **248**, 2880 (1973).
[5] W. A. Saffron and Q. H. Gibson, *J. Biol. Chem.* **253**, 3131 (1978).

Fish Found in Extreme Environments

The properties of fish hemoglobins vary widely and are frequently correlated with fish habitat.[6–9] For this reason it is often important to study fish that live under extreme conditions of, for example, temperature and/or pressure. Bringing deep ocean fish to the surface from abyssal depths potentially involves two drastic changes in environment. There is an enormous reduction in hydrostatic pressure and unless care is taken there may be a considerable increase in temperature. For fish with gas glands and swimbladders, the former change causes catastrophic physical damage and these animals invariably are brought on board ship moribund. However, if such fish are brought slowly to the surface through a water column totally lacking warm layers and are then kept cold, cardiac activity will be retained for several minutes and reasonable blood samples can be obtained. Exposure to warm temperatures is actually more immediately disastrous than is the effect of pressure reduction. This is also true for Antarctic fish.[10] All surface fish should be kept in water at a temperature as close as possible to that from which they were obtained.

Methods for Bleeding

Caudal venipuncture is the preferred method for obtaining blood from fish. This technique is now employed nearly universally and has evolved from the cruder and rarely used method of collecting blood from the severed peduncle. Cardiac puncture continues to be used widely, but penetration of the caudal vein has several advantages. There is less trauma, multiple punctures are possible, and the yields of blood are usually larger. Additionally, the heart can be externally massaged, which facilitates the flow of blood and maintains a higher venous pressure, preventing a premature collapse of the walls of the vessels.

Materials. Before collection, an anesthesia can be administered by immersing the fish in a solution of tricaine methane sulfonate or, alternatively, the animal can be chilled on ice.[11] However, venipuncture is not considered to be a painful procedure. Additionally, if the fish are to be returned to their natural habitat, the exposure to an anesthetic would likely decrease their chances for survival. The sizes of both the needle

[6] D. A. Powers, H. J. Fyhn, U. E. H. Fyhn, J. P. Martin, R. L. Garlick, and S. C. Wood, *Comp. Biochem. Physiol. A* **62A,** 67 (1979).

[7] A. Riggs, *Comp. Biochem. Physiol. A* **62A,** 257 (1979).

[8] J. Sauer and J. P. Harrington, *Comp. Biochem. Physiol. A,* **91A,** 109 (1988).

[9] R. E. Weber and F. B. Jensen, *Annu. Rev. Physiol.* **50,** 161 (1988).

[10] G. N. Somero and A. L. De Vries, *Science* **156,** 257 (1967).

[11] A. Riggs, this series, Vol. 76, p. 5.

and the syringe are important and should be determined according to the sizes of the fish and the vein to be punctured. We have used 18- to 26-gauge needles and 1- to 50-ml syringes. For some large fish with sufficient blood pressure and volume, vacutainers with 20- to 22-gauge needles have proved to be very efficient. In most cases, the smaller vacuum created by a syringe is recommended because of the probability of collapsing the vein due to the draw of the vacutainer. The syringe and needle are normally coated with a solution of heparin. Other anticoagulants, such as citrate, oxalate, and ethylenediaminetetraacetic acid (EDTA), can be used. However, all of these anticoagulants interact with the hemoglobin molecule and it is therefore critical that the red blood cells are washed thoroughly before lysis.[11]

Procedure. The vein to be punctured is found just ventral to the vertebrae located in the peduncle. Starting low in this region, the needle should be inserted underneath the scales at a point directly lateral to or ventral to the vein. After piercing the skin and creating a slight vacuum in the syringe, one should carefully probe until blood is drawn. A second or third collection is often successful at a point anterior to the original entry. The blood is generally diluted with an equal volume of cold 1.0–1.7% NaCl and kept on ice prior to centrifuging and washing the cells. We routinely use 1.0% NaCl for most freshwater fish and 1.3% NaCl for saltwater fish.

Deterring Lysis of Red Blood Cells. Because the erythrocytes of fish are metabolizing, nucleated cells, a glucose-rich medium is necessary to inhibit lysis if blood has to be stored for a significant time before washing and freezing of the cells, e.g., when the blood has to be transported before processing. Instead of mixing the blood with NaCl, a modified Alsever's solution should be used.[12] A 1.0-liter solution contains 20 g D-glucose, 8 g sodium citrate, 0.5 g citric acid, 4.4 g NaCl (for an osmolality equivalent to 1.0% sodium chloride), and Tris, approximately 1.0 g, to raise the pH to 7.5–8.0. Alsever's solution is used only to transport the cells and should not be used in subsequent washings steps.

Prior to washing the erythrocytes, it may be necessary to pass the blood mixture through a cheesecloth if any clots or debris such as scales are present. However, it should be kept in mind that there will be loss due to absorption, and this step should be avoided if possible.

Separation of Erythrocytes from Plasma. Processing the cells consists of (1) centrifuging (2000 *g*, 10 min) the blood mixture, (2) removing the supernatant, and (3) gently stirring the packed cells with 4 volumes of the

[12] J. A. Kolmer, "Clinical Diagnosis by Laboratory Examinations," 2nd ed. Appleton-Century-Crofts, New York, 1949.

appropriate NaCl solution followed by centrifugation. Step (3) is performed three times: two centrifugations should be carried out at 2000 g and the third at 3000 g. If lysis of the cells begins to occur, the number of washes should be reduced to two. The supernatant can be removed by decanting or careful aspiration. The packed cells are poured or pipetted into cryogenic tubes and placed in liquid nitrogen for storage. All handling and centrifuging of the cells and hemoglobin solutions should be carried out between 0° and 4°. These precautions will ensure the highest yield of hemoglobin, the greatest stability of protein, and the least formation of methemoglobin.

Isolation of Hemolysate

Lysis of the cells and removal of stroma and other cellular contents can be accomplished before or after freezing the washed red blood cells. However, hemolysis must be followed quickly by removal of organic phosphates and other ions because this procedure, commonly referred to as stripping, increases hemoglobin stability.

Lysis and Removal of Cellular Debris. Fish red blood cells are nucleated and, consequently, present a problem that mammalian cells do not. When the walls of the nucleus are ruptured, nuclear material is released, forming a large jellylike mass of nucleic acid. A great deal of hemoglobin is trapped in this matrix and separating the hemolysate from this material is difficult and causes enormous waste. Many types of lysis and methods for hemolysate extraction have been described, including using distilled water or organic solvents as hemolytic agents,[13] relying solely on the freezing and thawing process for complete rupture of the cell walls,[14] homogenizing as a means of disrupting the nucleic acid matrix,[11] and using sodium tetraborate in place of NaCl to improve stroma removal.[15] We have experimented with many different techniques and recommend the following lysis procedure. To the thawed cells, 2–3 volumes of 1 mM HCl-Tris, pH 8, are added while vortexing at moderate speed. After incubation on ice for 30 min, 1 M NaCl is added to raise the chloride concentration to at least 0.1 M. The suspension is vortexed and centrifuged at 100,000 g for 30 min at 4°; then the supernatant is decanted. Care must be taken that no cellular debris is decanted with the hemolysate. Recentrifugation is sometimes necessary. Nucleic acids contaminating the hemolysate will slow or completely clog the Sephadex column used prior to stripping. To

[13] W. A. Schroeder and T. H. J. Huisman, "The Chromatography of Hemoglobins." Dekker, New York, 1980.
[14] R. E. Weber, F. B. Jensen, and R. P. Cox, *J. Comp. Physiol. B* **157B,** 145 (1987).
[15] S. Giovenco, I. Binotti, M. Brunori, and E. Antonini, *Int. J. Biochem.* **1,** 57 (1970).

reduce the problems caused by nuclear material, DNA nuclease has been used to lyse chick embryo red cells,[16] but to our knowledge its use with fish erythrocytes has not been reported.

Storage. Storage of hemolysates for more than a few hours should be done only at liquid nitrogen temperatures. All hemoglobin solutions should be kept as cold as possible, but most importantly, we have found that rapid processing through a gel filtration column and then a mixed-bed resin column is highly effective in removing those substances that contribute the most to instability. This procedure will be detailed later in this chapter. The use of EDTA in all solutions that are in contact with the hemoglobin may be beneficial, but this chelating agent is a potential allosteric effector.

Factors Affecting Yield and Quality of Hemoglobin Solutions. Minimizing the loss of hemoglobin during its extraction from the erythrocytes is particularly important for deep ocean fish. These fish are not only difficult and costly to obtain, but the hematocrit and cellular hemoglobin concentrations appear to be significantly lower than those found in temperate and tropical fish. This has also been observed in the cold-adapted fish of the Antarctic Ocean.[17] When compared to human hemoglobin, all fish hemoglobins are more fragile, i.e., more susceptible to autoxidation, denaturation, and precipitation. This greater sensitivity to solution conditions is consistently more evident in the hemoglobins of the abyssal fish. Carbon monoxide is often introduced into the hemoglobin preparation and this can be done at any point. The presence of CO stabilizes the protein by protecting the heme from oxidation and by acting as a reducing agent. Chemical pathways have been postulated for this reduction process[18,19] and the subject of photoreduction is further discussed in this volume.[20] For most investigations, the CO must be removed. This can be accomplished readily and the procedure will be described later in this chapter.

It is common to find that hemoglobin obtained from freshly drawn blood already contains significant amounts of the ferric derivative, i.e., 5% or more. This is not a uniform finding and may result in part from the trauma associated with being caught. However, it is a frequent enough finding that one must be prepared to cycle the hemoglobin through a reduction procedure in order to have consistently fully ferrous material for functional studies. It is virtually impossible to prevent completely the slow formation of methemoglobin when oxygen is present. For this reason

[16] J. L. Brown and V. M. Ingram, *J. Biol. Chem.* **249**, 3960 (1974).
[17] R. M. G. Wells, M. D. Achby, S. J. Duncan, and J. A. Macdonald, *J. Fish Biol.* **17**, 517 (1980).
[18] D. Bikar, C. Bonaventura, and J. Bonaventura, *J. Biol. Chem.* **259**, 10777 (1984).
[19] L. J. Young and W. S. Caughey, *J. Biol. Chem.* **262**, 15019 (1987).
[20] J. Everse, this volume, [34], p. 524.

it is necessary to design experiments to circumvent this problem. As an example, we have started to incorporate an enzymatic reduction system[21] into oxygen equilibrium experiments and have found it to be quite effective to pH 6. However, the general problem of methemoglobin formation can be minimized if some precautions are taken. The presence of metal ions is to be avoided because these greatly accelerate autoxidation. Therefore, reagents with the least possible metal ion contamination should be chosen. In general phosphates seem more likely to contain such contaminants than Tris or bis-Tris buffers. Syringes made entirely of glass should be used. Large syringes, e.g., 30 ml, often require metal adapters for needles or other fittings. These adapters must be replaced frequently because the protective chrome plate wears quickly, exposing the underlying brass. If hemoglobin solutions must be in contact with a metal, the metal should be a high-quality stainless steel.

The aquomet derivative of most hemoglobins is less stable than either the ferrous or the ferric molecule when strong ligands such as cyanide or azide are bound. Autoxidation is therefore frequently a prelude to denaturation and the formation of insoluble material. Both processes are strongly temperature dependent[22] and stability can be greatly enhanced by carrying out experiments below 20°, the temperature at which hemoglobin studies are so frequently performed. Working at 10° or even 15° will sometimes make it possible to take measurements that fail at 20°.

The instability of these proteins is not merely a property of the heme groups but is far more general. As will be mentioned, it also affects the preparation of stable globins. While discussing resistance to chemical change, it is important to note the observation of Goss and Parkhurst that the sulfhydryls of the β chains of carp hemoglobin are readily oxidized.[23] They recommend treatment with dithiothreitol before use.

Removal of Organic Phosphates and Other Ions

As with most vertebrate hemoglobins, organic phosphates in fish red blood cells bind to hemoglobin and modulate ligand affinity, necessitating the removal of these agents for many functional and structural studies.

Procedures. The following removal procedure uses a Dintzis column[11] and is suitable for fish hemoglobins that are stable and soluble under conditions of very low ionic strength. The red cell lysate is passed through a Sephadex G-25 column equilibrated with 1 mM HCl-Tris, pH 8.2, in

[21] A. Hayashi, T. Suzuki, and M. Shin, *Biochim. Biophys. Acta* **310**, 298 (1978).
[22] E. Antonini and M. Brunori, "Hemoglobin and Myoglobin in their Reactions with Ligands." North-Holland Publ., London, 1971.
[23] L. J. Parkhurst and D. J. Goss, *Biochemistry* **23**, 2180 (1984).

order to reduce its ionic strength and to avoid saturating the subsequent deionizing column. The hemoglobin is then deionized by passage through a 1×20 cm ion-exchange column that is equilibrated with deionized H_2O. The column consists of the following resins (from top to bottom):

1 volume of Amberlite IR-120 (ammonium ion form)
1 volume of Amberlite IRA-400 (acetate ion form)
5 volumes of Amberlite MB-1 mixed-bed ion-exchange resin (H^+, OH^-)
0.1 volume of Amberlite IR-120 (H^+ form)

The hemoglobin is passed through the column slowly at a flow rate of about 15 ml/hr. About 50 ml of hemoglobin can be processed through a column of this size. Alternatively, a 1.5×30 cm column of Dowex ion-exchange resins can be constructed as described by Riggs.[11]

Many fish hemoglobins are unstable or insoluble at very low ionic strength, and thus removal of organic phosphate with deionizing columns is inappropriate. For these hemoglobins, a modification of the method of Berman et al.[24] is preferred. This method relies on the fact that organic phosphate binding decreases as the pH becomes more alkaline. The red cell lysate (500 mg) is slowly passed through a Sephadex G-25 (fine) or G-100 column, 2.5×60 cm, equilibrated with 0.1 M NaCl, 0.01 M HCl-Tris, pH 8.4, so that it remains on the column for approximately 5 hr. The assay described by Ames and Dubin[25] is used to test for the presence of phosphate.

Storage. Although the resulting deionized oxyhemoglobin from some species of fish may be relatively stable for up to 1 week at 4°, for most fish hemoglobins the carbonmonoxy derivative is more stable and would be the derivative of choice under these conditions. We recommend rapid freezing and storage of both oxyhemoglobin and carbonmonoxyhemoglobin in liquid nitrogen immediately after preparation to ensure no further changes in composition; slowly drop the hemoglobin into a container of liquid nitrogen so that small beads are formed. Hemoglobins stored this way are stable for years. The amount of beaded hemoglobin needed can be poured from the cryo tubes to thaw, leaving the remaining hemoglobin undisturbed in the frozen state.

Preparation of Derivatives of Fish Ferrous Hemoglobin

In general, the ferrous derivatives of fish hemoglobins are less stable than the human protein and more care in handling is usually required, as discussed previously. The preparation of these derivatives is otherwise

[24] M. Berman, R. Benesch, and R. E. Benesch, *Arch. Biochem. Biophys.* **145,** 246 (1971).
[25] B. N. Ames and D. T. Dubin, *J. Biol. Chem.* **235,** 769 (1960).

straightforward and follows the procedures described by DiIorio[26] for mammalian systems.

Deoxyhemoglobin. For the deoxygenated derivative, most of the oxygen is removed from a solution of oxyhemoglobin by equilibration with humidified, O_2-free nitrogen or by repeated evacuation and exposure to nitrogen or another inert gas. To remove traces of methemoglobin and any residual oxygen, a very small amount of dithionite is added either as a solid or as a concentrated solution so that only very small volumes are needed.

Oxyhemoglobin. Fish deoxyhemoglobin in the absence of dithionite is converted to the oxy derivative on exposure to oxygen. Many fish hemoglobins exhibit very low oxygen affinities under certain conditions. Therefore, to ensure complete saturation with oxygen one must control pH and buffer composition. The appropriate conditions will depend on the hemoglobin being studied.

The oxy derivative can be readily obtained from the carbonmonoxy derivative by exposure of the latter derivative to sufficient light and oxygen. To optimize protein stability this procedure should be carried out above pH 7. In this laboratory the hemoglobin solution at a concentration of ≤ 2 mM in heme is placed in a flask or test tube fitted with a stopper and an inlet and exit needle for oxygen flow. The hemoglobin solution is rocked gently on ice while being both continuously equilibrated with humidified pure oxygen and exposed to a high-intensity, low-heat lamp situated a few centimeters away from the vessel. An ideal fluorescent light source is the Osram 20 W (1200 lumen DWlux-el) electronic light bulb. The usual time for conversion from the CO to O_2 derivative is about 1 hr but depends on the area of the gas–solution interface and the hemoglobin concentration. The absorption spectrum from 650 to 500 nm is checked periodically for complete conversion.

Carbonmonoxyhemoglobin. It is often advantageous to convert red blood cell lysate to the more stable CO derivative before the stripping process or before further processing. The CO derivative is ideal for fractionation of components and is best suited for prolonged storage of the protein in solution and in liquid nitrogen. Converting initially to the CO derivative is also the method of choice for removing methemoglobin from an oxyhemoglobin preparation. Red cell lysate or the oxy- or methemoglobin solution is placed in a flask or tonometer on ice and equilibrated with humidified CO. A tonometer with an attached cuvette is used if carbonmonoxyhemoglobin formation is to be followed spectrally. Once most of the oxygen has been removed, a small amount of solid dithionite or

---

[26] E. E. Di Iorio, this series, Vol. 76, p. 57.

a small volume of concentrated deoxygenated dithionite solution (prepared with 1 mM HCl-Tris, pH 8.2) is added to the hemoglobin solution under continuous CO flow. After a few minutes the hemoglobin solution is removed anaerobically. In this laboratory we use an all-glass syringe that has been deoxygenated. The hemoglobin is then passed through a Sephadex G-25 column in order to remove the products of dithionite oxidation. All columns and buffers are first equilibrated with CO.

Preparation of Fish Ferric Hemoglobins

The aquomet derivative of fish hemoglobin is prepared by reacting the oxy derivative with a threefold molar excess of $K_3Fe(CN)_6$ for about 1/2 hr on ice. Removal of unreacted ferricyanide and ferrocyanide can be accomplished with gel filtration on Sephadex G-25. If the presence of trace amounts of ferro- or ferricyanide will compromise the measurements, as they would, for example, in measurements of paramagnetic susceptibility,[27] then more complete removal with a deionizing column is recommended. However, in such cases, it may be safer to avoid the use of ferricyanide entirely and to oxidize the hemoglobin with a fivefold molar excess of sodium nitrite. Here again, the excess reactant can be removed with a deionizing procedure. Other ferric derivatives are then prepared by adding the appropriate ligand.[26]

Multiple Components and Fractionation

Standard Ion-Exchange Chromatography. For the isolation of the multiple components of a fish hemolysate, it is essential that conditions be chosen that maximize protein stability: (1) reducing any methemoglobin that is present, (2) stripping the protein of organic phosphates, (3) exposing all solutions to carbon monoxide, and (4) performing all procedures at pH values above pH 7, in the cold. Fractionation protocols for three fish hemolysates are outlined below.

Trout Hemoglobin. The hemolysates of two trout, *Salmo gairdneri* and *Salmo irideus,* are composed of four electrophoretically distinct components. The procedures for fractionation involve anion-exchange chromatography using DEAE-Sephadex or DEAE-cellulose. For *S. irideus,* designations I to IV correspond to fractions with increasing anionic mobility and fractions I and IV represent 20 and 65%, respectively, of the total hemoglobin.[28] To isolate the individual components, 1–1.5 g of hemoglobin

[27] R. W. Noble, A. DeYoung, S. Vitale, S. Morante, and M. Cerdonio, *Eur. J. Biochem.* **168,** 563 (1987).
[28] I. Binotti, S. Giovenco, B. Giardina, E. Antonini, M. Brunori, and J. Wyman, *Arch. Biochem. Biophys.* **142,** 274 (1971).

as lysate in 0.1 M Tris-HCl, pH 9.1, is applied to a 2×40 cm DEAE-Sephadex A-50 column equilibrated with the same buffer. Elution of the components is achieved with a linear gradient at a flow rate of 30–40 ml/hr at 4°. The starting buffer is 0.1 M Tris-HCl, pH 9.1, and the final buffer is 0.1 M KH$_2$PO$_4$. In contrast, Lau et al.[2] fractionated the hemoglobins of the North American rainbow trout, S. gairdneri, with a salt gradient using a 2×15 cm DEAE-cellulose column for 500 mg of hemoglobin as lysate. The starting and final buffers are 500 ml each of 0.01 M phosphate, pH 8.0, and 0.01 M phosphate, 0.2 M NaCl, pH 8.0, respectively. Only 250 ml of this gradient are needed for complete elution of the hemoglobin. In the hemolysate of this fish the major hemoglobin components are fractions I and III.

Carp Hemoglobin. Hemolysate from carp, *Cyprinus carpio,* was originally fractionated by Tan et al.[29] into two major and one minor components. We have made several modifications to this procedure. Carp hemolysate is first converted to the CO derivative and put through the reduction and stripping process as already described. Approximately 1 g of deionized hemoglobin is loaded onto a DEAE-cellulose column (2.5×20 cm) equilibrated with 0.01 M HCl-Tris pH 8.2, that is saturated with CO. The flow rate is about 50 ml/hr. Fraction A elutes after 2–3 hr followed by fraction B. The minor fraction, C, is eluted with CO saturated 0.0125 M HCl-Tris, pH 8.2.

Another procedure for fractionating carp hemoglobin components was used by Gillen and Riggs.[30] Phosphate-free hemoglobin solutions are dialyzed against the starting buffer, 0.05 M Tris-HCl, pH 8.7. The hemoglobin (400 mg in 8 ml of this buffer) is applied to a DEAE-Sephadex A-50 column (2.5×50 cm) and eluted with a linear gradient of 1200 ml each of starting buffer and 0.05 M Tris-HCl, pH 8, at a flow rate of 35 ml/hr at 4°.

Fundulus heteroclitus Hemoglobin. The hemoglobin from *Fundulus heteroclitus* has been well characterized by Mied and Powers.[31] The hemolysate is composed of four hemoglobin components, each of a molecular mass of 64,000 Da. The isoelectric points vary from 5 to 9. These components are separated by anion-exchange chromatography on a DEAE-cellulose column (0.9×27 cm) equilibrated with 0.03 M Tris-HCl, pH 8.5, at 4°. Hemoglobin samples are dialyzed against the same buffer and concentrated to a small volume (about 1 ml). Mied and Powers stress the importance of this step in order to obtain resolution of all components because Hb I and Hb II separate without

[29] A. L. Tan, A. DeYoung, and R. W. Noble, *J. Biol. Chem.* **247**, 2493 (1972).
[30] R. G. Gillen and A. Riggs, *J. Biol. Chem.* **247**, 6039 (1972).
[31] P. A. Mied and D. Powers, *J. Biol. Chem.* **253**, 3521 (1978).

fully binding to the resin. A flow rate of 20 ml/hr is used. After elution of the first two components, Hb III and Hb IV can be eluted with a linear gradient consisting of 50 ml each of starting buffer and 0.03 M Tris-HCl, 0.1 M NaCl, pH 8.5.

High-Performance Liquid Chromatography. The literature contains few descriptions of the use of high-performance liquid chromatography (HPLC) for the separation of multiple components of fish hemoglobins in their native, ferrous forms. One such report is that of Duffy *et al.*[32] These investigators isolated the components of the sockeye salmon on a Waters Accell QMA column and subsequently analyzed the purity of the globins using the standard techniques of reversed-phase HPLC. They made the important observation that solution conditions that successfully fractionated the multiple components of the salmon on a standard DE-52 resin were not effective on a Waters Protein-Pak DEAE column, In fact, the isolation of the components on HPLC was eventually performed using a strong anion-exchange medium. We have observed similar, not easily explained differences in the behavior of hemoglobins during standard chromatography and HPLC.

As previously mentioned, conditions for the separation of fish hemoglobins vary widely even when using standard ion-exchange chromatographic methods. Extensive experimentation is almost always necessary to obtain pure samples. HPLC is no exception. Additionally, the conditions that are successful on an analytical column may not succeed on a preparative column, the latter being necessary to obtain enough protein to carry out functional studies. A preparative column is not only expensive but its capacity is limited. One must weigh the benefits of HPLC versus standard ion-exchange chromatography. If separation of hemoglobins with very small charge differences is needed or if maximum recovery of the applied sample is essential, then HPLC should be considered.

We have developed methods for separating the components of some fish hemoglobins as well as an analytical procedure for assaying the purity of previously isolated fractions of carp hemoglobin. Because of the abundance of carp and the large quantities of protein that are often needed for many investigations, these authors have opted to continue to use a standard anion-exchange column at the preparative level. Figure 1 shows an analytical HPLC profile of whole carp ferric hemoglobin. Because the hemes were fully oxidized, all solutions were air equilibrated. However, a similar pH gradient would suffice for the fully ferrous form but the electrophoretic pattern would be shifted toward the right. In the latter case, it is recommended that solutions be exposed to CO.

[32] L. K. Duffy, R. Reynolds, and J. P. Harrington, *J. Chromatogr.* **512**, 291 (1990).

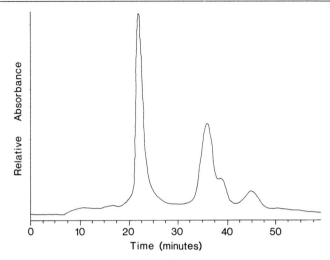

FIG. 1. Profile for the HPLC elution of stripped, fully oxidized carp hemolysate (approximately 5 mg) from a Toso-Haas TSK gel DEAE-5PW anion-exchange column (15 cm × 20 mm i.d.) at 10°. The absorbance was monitored at 406 nm. A 5-ml/min, 60-min linear gradient from 3 mM HCl-Tris, pH 8.2, to 10 mM HCL-Tris, 2.5 mM NaCl, pH 7.55, was used. All solutions contained 10 μM EDTA. The order of elution, from left to right, was carp fractions A, B, and C.

Figure 2 shows the clear separation of the two major components of *Coryphaenoides pectoralis* hemolysate. Several minor components are also evident. As previously discussed in this chapter, precautions such as the exposure of all solutions to CO should be taken to ensure maximal stabilization of the protein.

Preparation of Subunits

Preparation of the isolated subunits of fish hemoglobins has met with very limited success. This is not really surprising given the fact that there are very few vertebrate hemoglobins for which this has been accomplished. Here we are speaking specifically of techniques for obtaining native, heme-containing α and β chains that will recombine spontaneously to form a functionally normal hemoglobin molecule, not the separation of heme-free subunit globins for sequencing purposes. Goss and Parkhurst[33] reported the successful separation of α and β chains of carp hemoglo-

[33] D. J. Goss and L. J. Parkhurst, *Biochemistry* 23, 2174 (1984).

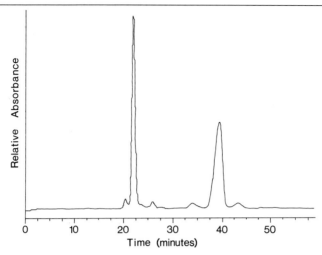

FIG. 2. Profile for the HPLC elution of stripped *C. pectoralis* hemolysate (approximately 3 mg) from a Toso-Haas TSK gel DEAE-5PW anion-exchange column (15 cm × 20 mm i.d.) at 10°. The absorbance was monitored at 420 nm. A 5-ml/min, 60-min linear gradient from 20 mM HCl-Tris, pH 8.3, to 150 mM HCl-Tris, pH 7.65, was used. All solutions contained 10 μM EDTA.

bin. These chains not only recombine to reconstitute carp hemoglobin, but combine with chains derived from human hemoglobin to form hybrid "mermaid" hemoglobins.[34,35]

The isolation procedure is based on the convenient fact that the β chains of carp hemoglobin contain numerous cysteine residues whereas the α chains contain none. A mercurial resin is prepared by reacting *p*-aminophenylmercuric acetate with Affi-Gel 10 (Bio-Rad, Richmond, CA). The chains are dissociated from one another by diluting stock carp hemoglobin (in the CO form) into 0.1 M potassium phosphate, 2.5 M triethylamine, pH 7.0, to a final heme concentration of 100 μM and a triethylamine concentration equal to or greater than 2.2 M. Under these conditions only free carp α chains will fail to react with the mercury resin and their isolation is straightforward.

For the preparation of β chains a similar procedure was reported but a much lower hemoglobin concentration (5 μM) is used. This greatly reduces the amount of undissociated hemoglobin in the system so that most of the material bound to the column are β chains. The β chains are then eluted from the column with 2 mM 2-mercaptoethanol. Although the

[34] T. Causgrove, D. J. Goss, and L. J. Parkhurst, *Biochemistry* **23**, 2168 (1984).
[35] L. J. Parkhurst and D. J. Goss, *Biochemistry* **23**, 2180 (1984).

preparation of the α chain of carp hemoglobin is straightforward and has been used successfully by a number of laboratories, the β-chain preparation procedure has met with more variable success. The source of the difficulty is unclear, but those planning to use such chains should be aware that there are potential problems.

Preparation of Globins

The preparation of globin and the isolation of subunit globins of fish hemoglobin for purposes of sequence analysis are routine procedures and do not differ significantly from those used for mammalian hemoglobins. However, the preparation of globin from which it is possible to reconstitute hemoglobin by the addition of hemin, or which will incorporate other porphyrins, appears to be more difficult with fish hemoglobins than with human and other mammalian hemoglobins. Falcioni et al.[36] have reported the preparation of the globins of two major components of the hemoglobins of the trout, *S. irideus,* and the incorporation not only of protoheme but of meso- and deuteroheme as well. The procedure used for globin preparation was the Fioretti et al.[37] modification of the butanone extraction procedure of Teale[38] and Yonetani.[39] Trout ferric hemoglobin, 1 to 2 mM in heme, is added to an equal volume of 2-butanone. Both solutions must be ice cold. Sufficient 0.1 M HCl is then added to lower the pH of the aqueous phase to 2.8. After 1 min the aqueous phase is separated, extracted with a second volume of cold butanone, and dialyzed against three changes of distilled water. The resulting material is soluble in water (pH 6.5). It is reported to be only partially capable of forming a stoichiometric complex with protoheme, and the hemoglobin thus obtained is not functionally identical to the native protein. Still it is functionally similar and heme incorporation is sufficient to establish the effects of using alternate hemes for reconstitution.

The general applicability of this procedure or the acid–acetone method[40] to the preparation of the globin of other fish hemoglobins is unclear. In our experience with carp hemoglobin, the yield of soluble globin that will recombine with heme has been no better than 1%.

[36] G. Falcioni, E. Fioretti, B. Giardina, I. Ariani, F. Ascoli, and M. Brunori, *Biochemistry* **17,** 1229 (1978).
[37] E. Fioretti, F. Ascoli, and M. Brunori, *Biochem. Biophys. Res. Commun.* **6,** 1169 (1976).
[38] F. W. J. Teale, *Biochim. Biophys. Acta* **35,** 543 (1959).
[39] T. Yonetani, *J. Biol. Chem.* **242,** 5008 (1967).
[40] A. Rossi Fanelli and E. Antonini, *Arch. Biochem. Biophys.* **77,** 478 (1958).

Electrophoretic Characterization

The occurrence of multiple components in fish hemolysates is extremely common.[41–43] A few well-documented exceptions are the hemoglobins of spot (*Leiostomus xanthurus*)[44] and goldfish,[45] which contain single components, and the hemoglobins of Antarctic fish, which express limited polymorphism.[46] However, in many cases analytical electrophoresis reveals a complex pattern of bands. The most frequently used electrophoretic technique is one-dimensional polyacrylamide gel electrophoresis (PAGE).

For fish hemoglobins it is important that the pH of the running buffer be quite alkaline. Harrington *et al.*[41] showed in starch gel[47] that hemolysates of some fish obtained from the Gulf of Mexico contained components whose isoelectric points were significantly above pH 8.6. Because of the position of the origin in starch gels, both anodally and cathodally migrating components can be detected. In PAGE only the anodally migrating components are observed, and it is therefore essential that the pH of the running buffer be above the pI values of the hemoglobins.

Normally, the electrophoretic properties of the ferrous native hemoglobin are of primary interest. Experimental conditions should be optimized to prevent the formation of partially oxidized species and other contaminating derivatives. This can usually be achieved by minimizing sample handling and electrophoresis run time, but the crucial factor is keeping the sample cold, no higher than 10°. If the electrophoresis chamber is not designed to allow continuous cooling, this problem can be overcome by refilling the upper buffer chamber several times during the run with cold buffer, because only this reservoir warms. All runs should be performed in a cold room. Placement of a temperature probe through a hole in the cover simplifies monitoring temperature in the upper buffer.

Polyacrylamide Gel Electrophoresis System. The PAGE system that these authors routinely use is a modification of the method outlined by Bio-Rad Laboratories for their mini-PROTEAN II dual-slab cell. The main advantages of minigels are the ease of preparation, the shorter run times,

[41] J. P. Harrington, M. McEbray, and P. Newton, *Comp. Biochem. Physiol. B* **99B**, 865 (1991).
[42] U. E. H. Fyhn, H. J. Fyhn, B. J. Davis, D. A. Powers, W. L. Fink, and R. L. Garlick, *Comp. Biochem. Physiol. A* **62A**, 39 (1979).
[43] A. Riggs, *in* "Fish Physiology" (W. S. Hoar and D. J. Randall, eds.), p. 209. Academic Press, New York, 1970.
[44] C. Bonaventura, B. Sullivan, J. Bonaventura, and M. Brunori, *J. Biol. Chem.* **251,** 1871 (1976).
[45] A. M. Vaccaro, R. Raschetti, G. Ricciardi, and G. Morpurgo, *Comp. Biochem. Physiol. A* **52A**, 627 (1975).
[46] G. di Prisco, *Comp. Biochem. Physiol.* **90,** 631 (1988).
[47] O. Smithies, *Biochem. J.* **71,** 585 (1959).

and the more efficient dissipation of heat. To obtain the highest resolution, a discontinuous method[48,49] should be applied. Readers are referred to another volume of this series[11] for the gel and buffer compositions used in this laboratory.

Sample Preparation. As little as 1.0 µg of a highly purified hemoglobin is required for detection by common protein stains. Silver staining allows detection of 100-fold smaller amounts. For routine screening, a sample is diluted in the upper or cathode buffer and mixed with an equal volume of 20% sucrose or 50% glycerol[50] to a final protein concentration of approximately 2 mg/ml. If multiple protein components are expected or if one is searching for small levels of contamination in purified samples, greater concentrations of protein should be used. To each sample (50 µl) add 5 µl of bromphenol blue (0.05% in the cathode buffer). To ensure that sulfhydryl groups remain reduced, one may choose to include 50 mM 2-mercaptoethanol in the sample.[51] Again, the samples should be kept cold.

Staining. Staining exclusively for heme proteins is possible and highly sensitive, but this involves the use of reagents that are either known or suspected carcinogens. The problems associated with these chemicals have been dealt with in detail by Riggs.[11] Generally, detecting nonheme proteins is not a factor and numerous staining methods have been described thoroughly in manuals and other publications in which electrophoretic techniques are discussed. However, this laboratory most frequently uses Coomassie brilliant blue G. The procedure is similar to that described by Blakesley *et al.*[52] and has the advantage of fixing the protein and eliminating both the need for lengthy destaining and the use of methanol. Additionally, all but the faintest bands can be seen before removal of the stain. The procedure is as follows: Dissolve 12.5 mg of Coomassie brilliant blue G in 5.0 ml of distilled water. This solution is then diluted into 100 ml of 12.5% (w/v) trichloroacetic acid. The solution is filtered if it is not clear. Depending on the concentration of the applied samples, 1 to 12 hr are required for complete staining. The solution is then replaced two or three times with 5% (v/v) acetic acid. Agitation during staining or destaining is not necessary. We have observed that as long as the gels remain exposed to 5% acetic acid, little fading of the dye occurs.

[48] L. Orstein, *Ann. N.Y. Acad. Sci.* **121**, 321 (1964).
[49] B. J. Davis, *Ann. N.Y. Acad. Sci.* **121**, 404 (1964).
[50] J. F. Robyt and B. J. White, "Biochemical Techniques." Waveland Press, Prospect Heights, IL, 1987.
[51] D. A. Garfin, this series, Vol. 182, p. 425.
[52] R. W. Blakesley and J. A. Boezi, *Anal. Biochem.* **82**, 580 (1977).

Spectral Properties

Extinction Coefficients. For the determination of heme concentrations, we have used extinction coefficients published for human hemoglobin[53]; however the absorption spectra of fish hemoglobins are not identical to those of the human protein. Unfortunately no systematic study of the extinction coefficients has been carried out for any fish hemoglobin. For many years we have also observed that the nature and magnitude of these deviations depend on the type of ligand, pH, presence of organic phosphate, chloride concentration, and other solution conditions. It is also not unexpected that different species of fish and even purified components of a single species exhibit varying spectral characteristics. In 1975, Giardina *et al.*[54] reported that the carbonmonoxy Soret absorption band of trout IV is quite sensitive to both pH and inositol hexaphosphate (IHP). This is not true for its deoxygenated derivative nor for either derivative of trout I. In a study of several Amazonian fish hemolysates, the absorbance maxima and the shapes of their deoxy spectra were found to vary significantly.[55] Their oxy spectra were also found to differ from those of human hemoglobin.

CO Spectrum of Carp Hemoglobin. The carbonmonoxy visible spectra of stripped whole carp hemoglobin and stripped human adult hemoglobin are compared in Fig. 3. These spectra were obtained by first determining the concentrations of the stock solutions by converting them to their respective ferric derivatives. To multiple dilutions of each stock, a few crystals of KCN were added and their spectra recorded in a 1.0-cm pathlength cuvette from 650 to 500 nm. Cyanomethemoglobin is the most stable ferric compound and its spectrum is relatively invariant among the hemoglobins and myoglobins. Although a small blue shift in the carp cyanomethemoglobin spectrum is observed we have assumed that the heme molar extinction coefficient at 540 nm is the same as for human hemoglobin, 11.0×10^3.

To obtain the CO spectra, the methemoglobin was diluted in a CO-saturated pH 7.0 buffer (0.1 M HCl-bis-Tris) to a concentration of 60 μM. Tiny amounts of solid dithionite were added to reduce the oxidized hemoglobin. Several dilutions were made and their spectra were averaged, resulting in the visual comparison in Fig. 3. One can easily see the typical

---

[53] O. W. Van Assendelft, "Spectrophotometry of Hemoglobin Derivatives." Royal Van Gorcum, Assen, The Netherlands, 1970.

[54] B. Giardina, F. Ascoli, and M. Brunori, *Nature (London)* **256,** 761 (1975).

[55] M. Farmer, H. J. Fyhn, U. E. H. Fyhn, and R. W. Noble, *Comp. Biochem. Physiol. A* **62A,** 115 (1979).

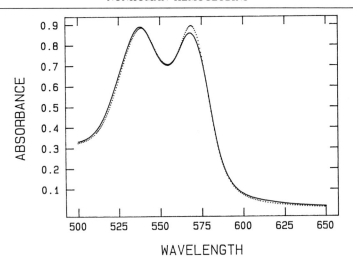

FIG. 3. Spectra of the carbon monoxide derivatives of stripped human hemoglobin (····) and stripped whole carp hemoglobin (—) at 20° in 0.1 M HCl-bis-Tris, pH 7.0. The heme concentration was 60 μM.

blue shift found in many fish hemoglobins, although the extinction coefficient of the β band at 540 nm appears the same for both hemoglobins. In contrast, the extinction coefficient of the α band for carp is much lower than that of the human protein. Variations in spectra can also be found in the Soret region. For the CO derivative of many abyssal fish, we have estimated a maximum molar extinction coefficient to be between 205×10^3 and 210×10^3 as compared to 192×10^3 for human. One must be aware that significant differences in spectra do exist and the values of the human extinction coefficients can only serve as first approximations.

Measurement of Unusual Functional Properties Observed in Fish Hemoglobins

For the most part, measurements of the ligand affinities and kinetics of ligand binding to fish hemoglobins present no special problems and the general techniques used for these measurements will not be discussed here. Instead, the specific problems associated with the measurement of those properties unique to fish hemoglobins will be described.

Fish hemoglobins display a wide array of functional properties and varied relationships between oxygen affinity and the presence of numerous heterotropic effectors, such as protons, chloride, and polyphosphate anions. The properties of these hemoglobins offer examples of molecular evolution and selection that are of interest. Those studying allosteric effects may also find some of these systems very useful. In hemoglobin, cooperative oxygen binding is superimposed on complex effects of protons and anions. There is a component of *S. irideus* hemoglobin that displays neither a Bohr effect nor a sensitivity to organic phosphates,[56] allowing cooperativity to be examined without the simultaneous presence of these other complications. Such a hemoglobin is also of obvious importance in the study of the structures responsible for the heterotropic effects that it lacks. Other differences in the responses of fish hemoglobins to organic phosphates have been reported, ranging from the absence of any effect to significant variations in the relative effects of different organic phosphates. For example, Vaccaro-Torracca *et al.*[57] report that goldfish hemoglobin is less affected by ATP than by GTP, the latter being the predominant organic phosphate in the red cells of this species. Ingermann and Terwilliger measured varying effects of ATP on the adult, mid-, and late-gestation fetal hemoglobins of the sea perch.[58] Other anion effects are also observed. In carp hemoglobin, oxygen affinity is significantly reduced by the addition of chloride ions in the absence of organic phosphates.[59] However, in the presence of IHP (1 mM) the effect is reversed and chloride ions cause an increase in oxygen affinity, presumably due to competition with IHP for the organic phosphate-binding site. This sensitivity to chloride ion concentration has prompted the present authors to control carefully the concentration of buffer chloride ions, adjusting pH by varying other parameters such as the concentrations of uncharged bases of Tris and bis-Tris. Variations on the Bohr effect, in addition to its complete absence, are also found. One component of the hemoglobin of the Amazonian fish, *Pterygoplichthys pardalis*, exhibits a reverse Bohr effect from pH 6 to pH 8.5 in the absence of organic phosphates.[60] That is, its oxygen affinity decreases steadily as the pH is increased. At the other extreme is the Root effect.

Root Effect

The Root effect is an exaggerated Bohr effect, or a pH dependence of ligand affinity. Root[3] first reported an effect of CO_2 that was so large

[56] M. Brunori, *Curr. Top. Cell. Regul.* **9,** 1 (1975).
[57] A. M. Vaccaro-Torracca, R. Raschetti, R. Salvioli, G. Ricciardi, and K. H. Winterhalter, *Biochim. Biophys. Acta* **496,** 362 (1977).
[58] R. L. Ingermann and R. C. Terwilliger, *J. Comp. Physiol.* **144,** 253 (1981).
[59] M. Cerdonio, S. Morante, S. Vitale, C. Dalvit, I. M. Russu, C. Ho, A. DeYoung, and R. W. Noble, *Eur. J. Biochem.* **132,** 461 (1983).
[60] M. Brunori, J. Bonaventura, A. Focesi, Jr., M. I. Goldames-Portus, and M. T. Wilson, *Comp. Biochem. Physiol. A* **62A,** 173 (1979).

as to cause the blood of some marine fishes to be only partially saturated with oxygen when equilibrated with air. Eventually Root and Irving[61] demonstrated that this effect, like the Bohr effect, is related primarily to pH and only indirectly to CO_2 tension. The magnitude of the change in ligand affinity observed over the range from pH 8 to pH 6 varies. It is typically about 100-fold but in some cases the affinities of ligand-binding sites change as much as 10^4-fold.[62]

Measurement of the Root Effect. The definition of the Root effect, a reduction in oxygen affinity so great that the hemoglobin is only partially saturated when equilibrated with air, describes one common and convenient assay for its occurrence. If the spectrum of the hemoglobin is measured as a function of buffer pH when the buffers are equilibrated with air at a fixed temperature, then the pH dependence of fractional saturation can be calculated with reference to the spectra of the fully oxygenated and the fully deoxygenated derivatives. The simplest way to obtain these spectra is to prepare a series of buffers of different pH values and equilibrate them with air at the desired temperature. A stock solution of hemoglobin is then diluted identically into each of these buffers and the spectra of the resultant solutions are recorded. The spectra of the solutions at the higher pH values (>7) in the absence of organic phosphates serve as reference spectra for the fully oxygenated protein. The spectrum of the deoxygenated hemoglobin can be obtained by diluting the stock hemoglobin solution into a deoxygenated buffer and then adding a few crystals of sodium dithionite. There are two difficulties with this procedure. First, as outlined there is the implicit assumption that the spectra of the oxy and deoxy derivatives are independent of pH. This is not in general the case. There is little difficulty in examining the spectrum of the deoxy derivative as a function of pH. However, unless one has the capability of exposing the hemoglobin solution to hyperbaric pressures of oxygen it will not be possible to examine the spectrum of the oxygenated derivative over the entire experimental pH range. Second, pH is not the only solution variable that affects ligand affinity, and the pH dependence of ligand affinity will generally depend on solution composition. Farmer *et al.*[55] compared the magnitudes of the Root effects of a large variety of hemoglobins. The solution that they chose to obtain the maximum possible reduction in oxygen affinity was 100 mM chloride ion containing bis-Tris base and 1 mM IHP, pH 6. The latter is ordinarily the most potent of the organic phosphates. However, it could be argued that the appropriate

[61] R. W. Root and L. Irving, *Biol. Bull.* (*Woods Hole, Mass.*) **84,** 207 (1943).
[62] R. W. Noble, L. D. Kwiatkowski, A. DeYoung, B. J. Davis, R. L. Haedrich, L. T. Tam, and A. F. Riggs, *Biochim. Biophys. Acta* **870,** 552 (1986).

buffer composition is that which most closely approximates the content of the red cells of the organism from which the hemoglobin is derived. If this route is chosen, then there is no universally appropriate buffer composition because the organic phosphate content of red cells varies widely among the various species of fish.

The fundamental deficiency in this method of quantitating the Root effect is the lack of information it offers about the shape of the ligand binding isotherm and the properties of the oxygen-binding sites that are unoccupied at low pH. The measurement of complete oxygen binding isotherms is a partial solution. However, as with the measurement of the Root effect, it is not possible to obtain the spectrum of the fully saturated hemoglobin molecule at low pH. A reference spectrum of the oxygenated derivative is usually obtained in such experiments by adding sufficient solid Tris base to raise the pH to a value that permits complete saturation with oxygen at a partial pressure of 1 atm. However, this procedure again assumes that this spectrum is pH independent, and this is almost certainly not the case. Oxygen binding isotherms also fail to examine the properties of the heme groups that remain unsaturated at the highest oxygen partial pressure available. Were all the sites identical, this would present no problem, but significant site heterogeneity is frequently found, as clearly demonstrated by the hemoglobins of certain deep ocean fish.[62] Therefore, the ligand binding properties of the hemes that one can saturate are not good predictors of the properties of the unoccupied heme groups. This problem can be overcome by using ligands that bind with higher affinity so that all of the ligand-binding sites can be examined. The ideal ligand in this regard is carbon monoxide.

Measurement of CO Binding Equilibria. The equilibria of the reaction of a hemoglobin with CO can be measured as described by Tan *et al.*[29] A tonometer with an attached cuvette is used essentially as described by Nagel *et al.*[63] for measurements of oxygen equilibria. A solution of hemoglobin in the desired buffer is carefully deoxygenated by equilibrating the system with pure N_2. The solution is then titrated anaerobically with dithionite to remove all traces of methemoglobin. The dithionite solution is prepared by dissolving dithionite in deoxygenated 1 mM Tris, pH 8.2. Equilibration at the desired temperature is carried out in the dark by dimming the room lights and covering the tonometer with aluminum foil. The time necessary to reach equilibrium is determined by taking successive spectral measurements until a constant level of saturation is reached. It can vary from 30 to 100 min. There are numerous combinations of cuvette pathlength, hemoglobin concentration, and spectral range that are

[63] R. L. Nagel, J. B. Wittenberg, and H. M. Ranney, *Biochim. Biophys. Acta* **100,** 286 (1965).

suitable. However, we find that when limited amounts of hemoglobin are available and stability is a problem, the use of a 2-mm pathlength, the Soret region of the spectrum, and a hemoglobin concentration of 20 μM in heme equivalents works quite well. As ligand affinity increases. The measurement of CO binding equilibrium becomes more and more difficult because a large fraction of the CO in the system will be bound and a precise determination of CO partial pressure becomes problematical.

Under some conditions certain hemoglobins will be only partially saturated even when exposed to 1 atm of CO. To obtain the fully saturated spectrum, the pH has to be increased. We accomplish this by adding solid Tris to the hemoglobin solution. As discussed in the section on the Root effect, there are difficulties with this procedure. For many fish hemoglobins we have examined, the absorption spectrum for the CO derivative varies with pH and solution conditions. Therefore, the CO spectrum at high pH is an imperfect estimate for the spectrum of the saturated molecule under conditions of low affinity.

Loss of Cooperativity at Minimum Affinity: Occurrence of Liganded T State

The Root effect is associated with a phenomenon that has been of considerable utility in the study of structure–function relationships in hemoglobin. As the affinity of a hemoglobin exhibiting the Root effect varies in response to pH, the cooperativity of ligand binding varies as well. In the presence of organic phosphates maximum cooperativity is typically observed at pH 7. As the pH is lowered to pH 6, there is a marked reduction in ligand affinity accompanied by a reduction in cooperativity. Noble et al.[64] pointed out this apparent loss in cooperativity and suggested that it might be due to an extreme stabilization of the T state such that the hemoglobin remained in what is normally regarded as its deoxy structure even when liganded. Tan et al.[4,65] subsequently demonstrated that at its minimum affinity the properties of the heme groups of carp hemoglobin are insensitive to the presence or absence of ligands on the other heme groups of the molecule. This low-affinity, noncooperative structure seems to satisfy the expectations for a stable T state, and indeed such a phenomenon is predicted by the classical two-state model of Monod, Wyman, and Changeux.[66] Under these conditions the ligand-saturated hemoglobin molecule appears to represent a liganded T state,

[64] R. W. Noble, L.J. Parkhurst, and Q. H. Gibson, J. Biol. Chem. **245**, 6628 (1970).
[65] A. L. Tan and R. W. Noble, J. Biol. Chem. **248**, 7412 (1973).
[66] J. Monod, J. Wyman, and J. P. Changeux, J. Mol. Biol. **12**, 88 (1965).

and this system has been used to compare the properties of liganded T states and R states. Similar behavior has been observed in a variety of other fish hemoglobins.

Difficulties in Accurately Measuring Cooperativity. Demonstrating a true lack of cooperativity requires considerable care. An apparent lack of cooperativity in the equilibrium of ligand binding, i.e., a Hill coefficient of unity, is not sufficient. Differences in the properties of the α and β chains of hemoglobin are frequently observed, which in the absence of cooperativity would result in a Hill coefficient less than one. It is possible for cooperativity to balance such heterogeneity and result in a net Hill coefficient of unity for what is in fact a cooperative system. To prove a lack of cooperativity it is necessary to demonstrate that the properties of the heme groups are insensitive to the presence or absence of ligands on the other hemes of the protein. This is best examined kinetically by using photolysis techniques and measuring the CO recombination reaction following complete and partial photolysis.

Flash photolysis techniques are well documented, beginning with the work of Gibson in 1956.[67] The carbon monoxide and isonitrile derivatives of heme proteins are photosensitive and those ligands are easily and rapidly removed by a pulse of light. The degree of photolysis can be controlled by the integrated intensity of the pulse and its length. Therefore one can produce a solution of fully deoxygenated hemoglobin or one that is randomly liganded. If less than 10% of the ligands is removed, the solution will contain primarily triliganded and fully liganded hemoglobin molecules. This permits a comparison of the kinetics of CO binding to unliganded and partially liganded molecules. Because the affinity changes involved with cooperative ligand binding are always associated with changes in CO combination kinetics, this comparison offers an excellent test for cooperative behavior. This technique is particularly well suited to the study of fish hemoglobins because these proteins exhibit very little tendency to dissociate into $\alpha\beta$ dimers. The dimer–tetramer equilibrium adds to the complexity of the recombination kinetics of mammalian hemoglobins.

Occurrence of Hemoglobins with Vastly Dissimilar Binding Sites

Heme binding site heterogeneity is often observed in hemoglobins and in many cases is probably due to functional differences between the α and β chains. However, site heterogeneity reaches new extremes in some fish hemoglobins.

[67] Q. H. Gibson, *J. Physiol.* (*London*) **134**, 123 (1956).

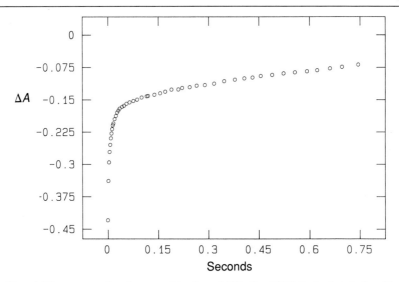

Fig. 4. Time course for the recombination of CO after 100% flash photolysis with *C. pectoralis* in the absence of phosphate at 15°. The change in absorbance (ΔA) was followed at 420 nm. The final buffer composition was 0.1 *M* HCl-bis-Tris, pH 6.1. [CO] = 480 μM and [Hb] = 5 μM in heme equivalents.

In fish hemoglobins the loss of cooperativity at minimum oxygen affinity is frequently associated with a Hill coefficient somewhat below unity, but for these very heterogeneous systems minimum affinities are associated with Hill coefficients below 0.3. In the case of the hemoglobin from *Coryphaenoides armatus,* the CO binding curve at pH 6 in the presence of 1 m*M* IHP is consistent with the presence of two heme populations whose affinities differ by 500-fold.[62] The low-affinity binding sites of these heterogeneous systems exhibit the lowest affinities ever reported for hemoglobins. Again for the hemoglobin of *C. armatus* the low-affinity heme sites bind CO with a P_{50} of 100 mm Hg. This remarkable affinity is associated with equally remarkable kinetic properties. The second-order rate constant for CO combination is about 10^3 M^{-1} sec^{-1}, again some 100-fold lower than the usual rate of CO binding to deoxyhemoglobin. Thus CO binding to the hemoglobin of *C. armatus* at pH 6 in the presence of IHP has two kinetic phases whose rates differ by more than 100-fold. Figure 4 shows the time course of CO recombination for another abyssal fish, *Coryphaenoides pectoralis.* Neither of the two kinetic phases is homogeneous but approximately 80% of each can be fitted to a single rate. Their values differ by two orders of magnitude, the slower being 2.2 × 10^3 M^{-1} sec^{-1}. It should be noted that when the two major hemoglobin components of this fish are isolated and their CO combination reactions are measured, the degree of heterogeneity is amazingly similar, although not identical, to that of the hemolysate. We have made the same observation for the hemoglobin components of another abyssal fish, *Coryphaenoides acrolepis*.

Specific problems arise in the technical aspects of measuring rates of reaction by stopped-flow and flash photolysis for a single hemoglobin possessing simultaneously such disparate binding sites. With both methods, ligand concentration is critical if the kinetic properties of all heme groups are to be examined and enough ligand must be present to saturate all heme populations, including those with extremely low affinities. If the concentration of CO is high enough to saturate the low-affinity heme groups, the reaction of CO with the high-affinity hemes may be so fast that the response or dead times of conventional flash photolysis and stopped-flow systems preclude its precise measurement. One solution is to characterize these two kinetic phases at different CO concentrations. If both kinetic phases are measured at the same high CO concentration, then one must choose a pattern of data collection that permits adequate characterization of both phases. As with measurement of CO equilibria, these kinetic determinations can be seriously perturbed by the presence of significant light. Even the intensity of the monitoring beam should be as low as possible, consistent with precise measurement.

Low-affinity binding sites present a special problem in the measurement of the rates of ligand dissociation. To measure such a process one needs to begin with ligand-saturated molecules and to use the pH jump procedure of Noble *et al.*[64] To measure the rate of oxygen dissociation, hemoglobin in a dilute, high-pH buffer, where its ligand affinity is high, is mixed with a dithionite solution in a more concentrated low-pH buffer in the stopped-flow apparatus. The buffers are chosen such that the final solution after mixing has the desired pH and composition. The rate of CO dissociation from ligand-saturated hemoglobin can similarly be measured with pH jump. In this case the carbonmonoxyhemoglobin is in the dilute, high-pH buffer and the low-pH buffer contains nitric oxide.

It is important to note that when a hemoglobin contains such functionally dissimilar binding sites it presents a special problem in the analysis of cooperativity. At equilibrium it is impossible to measure the binding of ligand to the low-affinity hemes when the high-affinity sites are not fully saturated. The problem persists in the kinetic measurements because of the very slow rate at which the low-affinity hemes bind ligand. Again the reaction of ligands with the low-affinity hemes always follows the saturation of the high-affinity sites. Therefore, it is impossible to determine how the properties of the low-affinity sites respond to the presence or absence of ligand on the high-affinity sites. This means that examining recombination kinetics following full and partial photolysis offers a very incomplete test for cooperative interactions.

Acknowledgments

The work of the authors was supported by research funds from the Veterans Administration and Program Project Grant P01-Hl-40453 from the National Institutes of Health.

[10] Conformational and Functional Characteristics of Bovine Hemoglobin

By CLARA FRONTICELLI and ENRICO BUCCI

Bovine red cells do not contain appreciable amounts of 2,3-diphosphoglycerate[1] (2,3-DPG); however, human and bovine blood have similar oxygen affinity,[2] indicating that either bovine hemoglobin (Hb Bv) has an intrinsically low oxygen affinity, or that its functional characteristics become physiologically efficient through a different mechanism of oxygen affinity regulation. It has been found that this hemoglobin has developed an increased sensitivity to the Cl^- ions of the solvent, which decreases its oxygen affinity to lower levels than those obtainable by human hemoglobin (Hb A) through the interaction with 2,3-DPG.[3,4] These characteristics and the availability of bovine blood have produced an interest in exploring the use of Hb Bv as the basis for an artificial oxygen carrier.

There are two types of Hb Bv, A and B. They differ by three amino acid residues in the β chains in which the substitutions are Gly(A12) → Ser, Lys(AB2) → His, and Lys(GH3) → Asn, from A to B genotype, respectively.[5] The genotype distribution varies and the possible correlation between the frequency of the hemoglobin types and the resistance or susceptibility to several bovine hemoprotozoans has been investigated. Evidence is presented that bovine breeds with a high frequency of A

[1] H. F. Bunn, *Science* **172**, 149 (1971).
[2] E. Bucci, C. Fronticelli, C. Orth, M. Martorana, L. Aebischer, and P. Angeloni, *Biomater., Artif. Cells, Artif. Organs* **16**, 197 (1988).
[3] C. Fronticelli, E. Bucci, and C. Orth, *J. Biol. Chem.* **259**, 10841 (1984).
[4] C. Fronticelli, E. Bucci, and A. Razynska, *J. Mol. Biol.* **202**, 343 (1988).
[5] W. A. Schroeder, J. R. Shelton, J. B. Shelton, B. Robberson, and D. R. Babin, *Arch. Biochem. Biophys.* **120**, 124 (1967).

genotype have a greater resistance to trypanosomal infection,[6] whereas breeds with a high frequency of B genotype have a higher resistance to babesiosis.[7]

Purification of Bovine Hemoglobin

Hereford cows are homozygous for type A hemoglobin and are used for most of the studies reported in the literature. Bovine blood (3 liters) is collected from the slashed neck of slaughtered cows in cold 50 mM sodium citrate and 150 mM NaCl (1 liter). After removal of plasma by centrifugation for 10 min at 4000 rpm the red cells are gently resuspended with twice the volume of cold 0.9% NaCl, centrifuged, and the supernatant removed. This treatment is repeated three times. The washed erythrocytes are hemolyzed by the addition of an equal volume of cold water. Chloroform is added to a 5% (v/v) final concentration and after stirring in the cold for 20 min the sample is centrifuged for 10 min at 10,000 rpm at 5°. The chloroform and the membrane phospholipids sediment to the bottom of the tube. The stroma-free hemolysate is collected and dialyzed against several changes of cold water and stripped from all organic and inorganic ions by repetitive passages through a 2 × 10 cm column containing Dowex MR-3. These preparations contain a maximum of 2% (w/v) methemoglobin. They can be stored at −80° for several months, without an increase in methemoglobin content. However, storing at −20°, as in a household freezer, produces a large amount of ferric hemoglobin. Multiple freezing and thawing of the protein should be avoided because it increases the formation of ferric hemoglobin. Ion-exchange chromatography can be used to obtain a pure sample of bovine hemoglobin. We use Delta-Prep-4000 (Waters, Millford, MA) preparative high-performance liquid chromatography (HPLC) equipment and a gradient made by buffer A (15 mM Tris adjusted to pH 8.3 with acetic acid) and buffer B (15 mM Tris, 200–300 mM sodium acetate, adjusted to pH 7.6 with acetic acid). In a standard preparation, 20 g of hemoglobin equilibrated against buffer A is injected into a 5 × 25 cm glass column (Waters) connected in series with a second similar column. The resin used is a DEAE-Toyopearl 650M (Supelco, Bellefonte, PA), with particle size 40–90 μm. This resin is cost effective and suitable for large-scale chromatography. A typical elution is shown in Fig. 1. The arrow indicates the elution time of methemoglobin when present. This preparation gives a single band in isoelectric focusing on an ampholine PAG plate at pH 5.5–8.5 (Pharmacia, Piscataway, NJ). For the

---

[6] A. D. Banghan and B. S. Blumberg, *Nature (London)* **181,** 1551 (1958).

[7] A. W. Bachmann, R. S. F. Campbell, and D. Yellowlees, *Aust. J. Exp. Biol. Med. Sci.* **56,** 623 (1978).

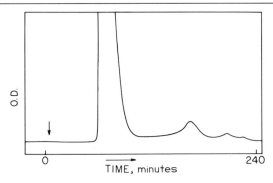

Fig. 1. Elution profile of bovine hemoglobin. The arrow indicates the elution of methemo-globin, when present. Hemoglobin (20 g) was applied to a 5 × 25 cm glass column connected in series with a second similar column. The columns were packed with DEAE-Toyopearl 650M (Supelco). The chromatography was done at 25° using a linear gradient at a flow rate of 40 ml/min. Buffer A: 15 mM Tris at pH 8.3; buffer B: 15 mM Tris, 220 mM sodium acetate at pH 7.6. The elution profile was recorded at 560 nm.

preparation of solutions of purified hemoglobin for *in vivo* use, the protein equilibrated with saline is sterilized by filtration through the Sterifil aseptic system with a 0.45 μm cutoff (Millipore, Bedford, MA). Detoxi-Gel (Pierce, Rockford, IL), with a capacity of 2 mg endotoxin/ml gel, is added in the ratio of 2 ml/10 g of Hb and the suspension is stirred gently overnight. The Detoxi-Gel is removed by filtration through the Sterifil aseptic system. These preparations, when injected into rats in the ratio 60–100 mg/10 g body weight, do not produce side effects except those due to an expansion in vascular volume.[8] Alternatively purified bovine hemoglobin can be purchased from Biopure Inc. (Boston, MA).

Conformational Characteristics

Resistance to Denaturation

Kinetics of denaturation of Hb A and Hb Bv followed spectrophotomet-rically at pH 11.2 at either 2° or 28° indicate a higher resistance of Hb Bv to alkali denaturation.[9] The calculated activation energy was 7.6 kcal/mol for Hb A and 21 kcal/mol for Hb Bv, respectively. No difference was measured on the denaturation of the corresponding globins, indicating that the different alkali resistance was not due to analogous differences

[8] B. K. Urbaitis, A. Razynska, Q. Corteza, C. Fronticelli, and E. Bucci, *J. Lab. Clin. Med.* **117**, 115 (1991).
[9] F. H. Haurowitz, R. L. Hardin, and M. Dicks, *J. Phys. Chem.* **58**, 103 (1954).

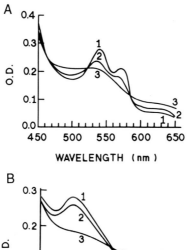

FIG. 2. (A) Absorption spectra or fetal oxyhemoglobin after exposure to 0.2 M acetate buffer, pH 4.51, for 5 hr (1), 39 hr (2), and 62 hr (3). Hemoglobin concentration, 0.5%. (B) Absorption spectra of bovine hemoglobin after exposure to 0.2 M acetate buffer, pH 4.55, for 5 hr (1), 39 hr (2), and 62 hr (3). Adapted from Bucci and Fronticelli.[10]

in the alkali resistance of the corresponding globins. It was proposed that it reflects a difference in the heme–globin complex in the two proteins. Acid denaturation of Hb A and Hb Bv and human fetal hemoglobin (Hb F) after 30 min acid exposure indicated that Hb A and Hb F were stable down to pH 4.8, whereas Hb Bv was stable down to pH 4.1 for at least 24 hr.[10] Whereas Hb A and Hb F formed hemichromogens, Hb Bv at pH 4.5 was quantitatively transformed into methemoglobin, as shown in Fig. 2.

Heme Transfer

Heme transfer was measured by Dr. Ruth Benesch at Columbia University following a technique developed in her laboratory.[11,12] The results, expressed in terms of the equilibrium heme distribution ratio,

[10] E. Bucci and C. Fronticelli, *Boll. Soc. Ital. Biol. Sper.* **37**, 1768 (1961).
[11] R. E. Benesch and S. Kwong, *J. Biol. Chem.* **265**, 14881 (1990).
[12] R. E. Benesch, this volume [31].

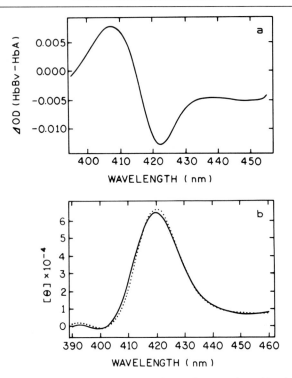

FIG. 3. (a) Differential absorption spectrum in the Soret region of bovine and human oxyhemoglobin. (b) CD spectra of human oxyhemoglobin (——) and bovine oxyhemoglobin (·····). Buffer, 10 mM phosphate at pH 7.2.

$$R = \frac{[\text{methemalbumin}]/[\text{albumin}]}{[\text{methemoglobin}]/[\text{apohemoglobin}]}$$

gave $R = 5.1$ and $R = 0.0071$ for the heme exchange of ferric Hb Bv with human and bovine serum albumin, respectively. These values are slightly higher than for Hb A ($R = 3.4$ and 0.0048)[11] and suggest that in Hb Bv β chains the heme–protein complex is looser than in Hb A β chains.

Optical Spectra

The extinction coefficient of Hb Bv in the cyanomet form at 540 nm is 11,000, as for human hemoglobin.[5] The differential absorption spectrum in the Soret region of the oxy derivatives of purified Hb Bv and Hb A, reported in Fig. 3a, and the corresponding CD spectra shown in Fig. 3b, indicate a 1- to 2-nm shift toward the blue in Hb Bv, which suggests differences in the heme environment in the two proteins.

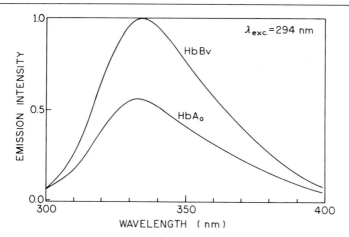

FIG. 4. Tryptophan fluorescence of human and bovine oxyhemoglobin ($\approx$0.3 mg/ml in 10 mM phosphate buffer at pH 7.4; temperature, 20°; excitation wavelength, 294 nm). Measurements were done in our laboratory by Dr. Z. Gryczynski, using right-angle square geometry optics with an SLM-8000 photon counting spectrofluorometer. The polarization of the excitation light was kept horizontal and that of the emission light vertical to eliminate Raman scattering.

Emission Spectra

Probably the best evidence of a conformational difference between human and bovine hemoglobin is given by the tryptophan emission spectra of the two proteins. Emission of tryptophan in heme proteins is largely quenched by energy transfer to a nonemitting heme. The extent of quenching, i.e., of energy transfer, is regulated by the overlap integral between the emission spectrum of tryptophan and the absorption spectrum of heme and by the relative positions and distances of tryptophans and hemes. As presented above, the absorption spectra of human and bovine hemoglobins are very similar, indicating very similar overlap integrals. Figure 4 shows that, after normalizing for protein concentration, the emission of bovine hemoglobin is much more intense than that of human hemoglobin. Work in progress in our laboratory suggests that this is due to a different spatial relationship between tryptophans and hemes.

Chain Separation

The tetrameric Hb Bv has only two cysteines at position βF9, on the protein surface, whereas the βG14 and αG11 cysteines at the $\alpha_1\beta_1$ interface in Hb A are substituted in Hb Bv by valine and serine, respectively.[5] In

Hb A, reaction at acid pH of the cysteines at the $\alpha_1\beta_1$ interface with p-mercuribenzoate allows the separation of the native α and β subunits by ion-exchange chromatography.[13] In Hb Bv these cysteines are absent and alternate methods of subunit separation have to be used. The native α subunits of Hb Bv have been isolated by affinity chromatography using activated thiol-Sepharose 4B.[14] The βF9 cysteines interchange with the thiol groups of the resin and remain covalently bound, whereas the α subunits are eluted from the resin with a low-pH buffer in order to enhance monomer formation. The α- and β-globins can be separated by reversed-phase chromatography using the method of Shelton and Shelton.[15] The buffers used are as follows: buffer A: 400 ml 0.15 M sodium perchlorate, 25 ml methanol, 75 ml acetonitrile, 0.5 ml phosphoric acid, 0.25 ml nonylamine; buffer B: 100 ml 0.15 M sodium perchlorate, 25 ml methanol, 375 ml acetonitrile, 0.5 ml 85% phosphoric acid, 0.25 ml nonylamine. In our laboratory we use a linear gradient from 39% buffer A and 61% buffer B to 100% buffer B in 30 min. For analytical purposes 5–7 μl containing about 1 mg of protein are injected in a C_{18} Bondapak 15–20 μm. For preparative purposes we use a column Beckman Ultrasphere 1 × 25 cm in which up to 20 mg of protein can be injected. Notably in Hb Bv the α chains are eluted soon after the heme and the β chains are last. In Hb A the α chains are eluted last. It is suggested to always confirm the identity of the separate α and β chains by recording the absorption spectrum between 310 and 240 nm. The shape of the spectrum is very sensitive to the different number of tryptophans present in the α chains (one, at position A12) and β chains (two, at positions A12 and C3). If a large number of comparative analyses must be performed it is preferable to use the same stock of eluting buffers. We have observed modifications in the elution time of the α and β chains using different preparations of the elution buffers.

Functional Characteristics

Oxygen Affinity

Hb Bv belongs to a group of mammalian hemoglobins that, in the presence of physiological concentrations of Cl^- ions (100 mM), have an oxygen affinity lower than that of Hb A and are insensitive to 2,3-DPG.[1] The amino acid sequence of Hb Bv, when compared to that of Hb A, shows the presence of a methionine residue at the N termini of the β

[13] E. Bucci and C. Fronticelli, *J. Biol. Chem.* **240**, PC551 (1965).

[14] S. H. De Bruin, J. J. Joordens, and H. S. Rollema, *Eur. J. Biochem.* **75**, 211 (1976).

[15] J. B. Shelton, J. R. Shelton, and W. A. Schroeder, *J. Liq. Chromatogr.* **4**, 1381 (1981).

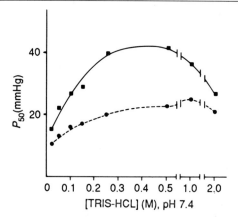

FIG. 5. Effect of Cl⁻ ion concentration on the oxygen affinity of bovine (—) and human (----) hemoglobins in Tris buffer at pH 7.4 and at 37°. Protein concentration, 80–100 mg/ml. From Fronticelli *et al.*[3] Reproduced with permission of the American Society for Biochemistry and Molecular Biology.

chains replacing the dipeptide Val(NA1)-His(NA2) of Hb A.[5] The failure of Hb Bv to respond to 2,3-DPG was explained by the deletion of His(NA2) in the β chains of Hb Bv.[1] The rationale is that the decrease in length by one amino acid of the β subunits increases, in the deoxygenated Hb, the distance between the β-N-terminal residues to about 22 Å,[16] preventing an efficient preferential binding of the 2,3-DPG molecule to deoxyhemoglobin.

Perutz and Imai have compared the oxygen binding isotherms of Hb Bv and Hb A.[17] Analysis in terms of the allosteric constant (Monad, Wyman, and Changeaux model) indicated that the K_R values in Hb Bv and Hb A are similar, whereas the constant K_T and L_0 are lower in Hb Bv. On the basis of these data and model-building considerations, Perutz and Imai proposed that the low oxygen affinity of Hb Bv results from the stabilization of the T structure elicited by the presence in the β chains of the N-terminal methionine residue, which locks the A helix tightly in place, as does organic phosphate in Hb A.

Effect of Chloride Ion and Other Monovalent Anions

In our laboratory we have found that the low oxygen affinity of Hb Bv can be explained by its higher sensitivity to the Cl⁻ ions present in solution.[3] Figure 5 shows the effect of increasing Cl⁻ ions on the oxygen

[16] M. F. Perutz, *Nature (London)* **228**, 726 (1970).
[17] M. F. Perutz and K. Imai, *J. Mol. Biol.* **136**, 183 (1980).

affinity of Hb A and Hb Bv. The positive slopes of the curves indicate that Cl^- ions are released on oxygenation; at high salt concentration the curve is reversed due to the formation of hemoglobin dimers.[18] From the positive slope of the curve the number of the oxygen-linked Cl^- binding sites can be estimated:

$$\Delta Cl^- = d \log p_{1/2}/d \log[Cl^-]$$

where ΔCl^- is the number of Cl^- ions per heme released by the protein on oxygenation and $[Cl^-]$ is the activity of Cl^- ions in solution. From the Data in Fig. 4 it can be calculated that up to a concentration of $[Cl^-] = 0.15\ M$, Hb Bv binds about two Cl^- ions per tetramer more than its oxy derivative. This compares to a value of only one in Hb A.

We have investigated the effect on oxygen affinity of other monovalent ions.[4] Figure 6 shows the oxygen affinity of Hb Bv in the presence of increasing concentrations of halogens, which are molecules with different charge densities in the order $I^- > Br^- > Cl^- > F^-$. The data indicate that the concentration range within which the titration occurred was different for the various salts, as if Hb Bv could discriminate between the halides on the basis of their charge densities, unlike Hb A. This suggests that hydrophobic interactions are relevant to the modulation of oxygen affinity by anions. Notably this effect is observed at 37° and not at 25°.[19] It is known that hydrophobic interactions decrease at low temperature.

Bohr Effect

Figure 7 shows that Hb Bv has a larger Bohr effect than does Hb A.[3] The number of protons released on oxygenation were calculated to be 2.5 per tetramer for Hb Bv and 1.6 for Hb A. The increased preferential binding to deoxyhemoglobin of Cl^- ions and H^+ is consistent. In fact, the electrostatic interaction of Cl^- ions with the positive groups on the protein surface increases their pK values, thereby producing an absorption of protons that are released on oxygenation together with the preferentially bound anions.

Interaction with 2,3-DPG

The sensitivity of the oxygen affinity of Hb Bv to Cl^- ions raised the question of whether the effect of 2,3-DPG could be masked by the Cl^- ions present in the solvent.[3,4] The efficiency of 2,3-DPG and Cl^- ions on a charge basis in Hb Bv is shown in Fig. 8. On the abscissa, the concentra-

[18] G. L. Kellett, *J. Mol. Biol.* **59**, 401 (1971).
[19] E. Bucci, C. Fronticelli, and Z. Gryczynski, *Biochemistry* **30**, 3195 (1991).

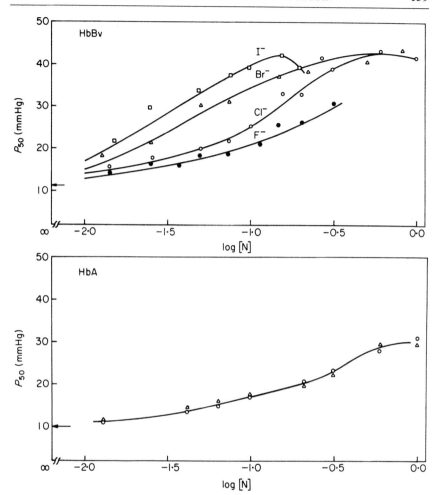

FIG. 6. Oxygen affinity of bovine hemoglobin in 0.15 M HEPES buffer at pH 4.4 at 37°
in the presence of increasing concentrations [N] of halogens, expressed in equivalents/liter.
The different concentrations of halogens were obtained by titrating the Tris buffer to pH
5.5 with the acid form of the halides, prior to adjustment with NaOH to pH 7.4. In this
way, the concentration of the buffer is the concentration of the anions in solution. The
arrow shows the oxygen affinity in water at pH 7.4. From Fronticelli et al.[4] Reproduced
with permission.

tion [N] of 2,3-DPG is expressed in equivalents/liter, assuming that the
five charges of 2,3-DPG are fully protonated at pH 7.4. The data suggest that, in Hb Bv, there are high-affinity sites for anions detectable at
log[N] < −1.0, equally accessible to Cl⁻ and 2,3-DPG, and low-affinity

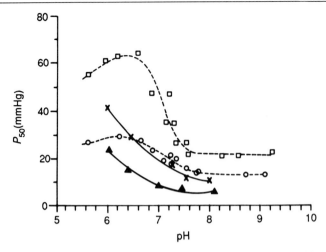

FIG. 7. Bohr effect of human and bovine hemoglobins in the presence of different concentrations of Cl⁻ ions, in bis-Tris or Tris buffer at 37°. Protein concentration, 80–100 mg/ml. □---□, Bovine hemoglobin in 0.15 M Cl⁻; ×—×, bovine hemoglobin in 0.02 M Cl⁻; ○---○, human hemoglobin in 0.15 M Cl⁻; ▲—▲, human hemoglobin in 0.02 M Cl⁻. From Fronticelli et al.[3] Reproduced with permission of the American Society for Biochemistry and Molecular Biology.

sites detectable at log[N] > −1.0, not accessible to 2,3-DPG. The apparent insensitivity of Hb Bv to 2,3-DPG at physiological Cl⁻ ion concentration (0.1 M) is in effect due to a masking of the effect of 2,3-DPG by Cl⁻ ions. In the absence of anions, Hb A and Hb Bv have a very similar oxygen affinity. In the presence of 2,3-DPG, Hb A and Hb Bv also have a similar oxygen affinity, provided Cl⁻ ions are not present in solution.[4]

Hydrophobic Effects and Temperature Dependence of Oxygen Affinity

The oxygen affinity of Hb A and Hb Bv is affected differently by the temperature, as shown by the van't Hoff plot in Fig. 9.[19,20] The estimated values of the intrinsic heat of oxygenation (ΔH^+) at pH 9.0, where the Bohr effect is absent, are −15.7 and −8.1 kcal/mol for Hb A and Hb Bv, respectively. The difference between enthalpies of oxygenation of Hb A and Hb Bv is much larger than the difference between the free energy of oxygen binding, that is, below 1 kcal/heme. This implies a compensation produced by a large oxygen-linked entropy change in Hb Bv, which again is consistent with modifications of the hydrophobic interactions.

[20] A. Razynska, C. Fronticelli, E. Di Cera, Z. Gryczynski, and E. Bucci, *Biophys. Chem.* **38,** 111 (1990).

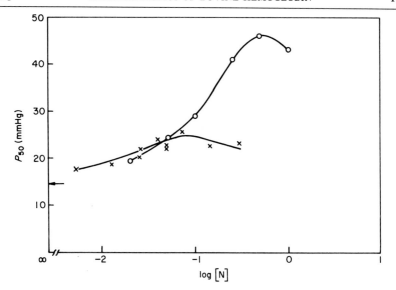

FIG. 8. Oxygen affinity of bovine hemoglobin in 0.15 H HEPES buffer, pH 7.4, at 37°C, in the presence of increasing concentrations of 2,3-DPG ($\times$) or Cl^- ($\bigcirc$). The concentration [N] of the effectors is expressed in equivalents/liter, assuming that 2,3-DPH carries five negative charges. The arrow shows the oxygen affinity in 0.15 M HEPES. Protein concentration, 60 mg/ml. From Fronticelli et al.[4] Reproduced with permission of the American Society for Biochemistry and Molecular Biology.

Consistent with the presence of oxygen-linked hydrophobic effects in Hb Bv are reports from the laboratory of Cordone on the effect of cosolvents on oxygen equilibrium. Vitrano et al.[21] report that in Hb Bv a larger hydrophobic surface than in Hb A is exposed to the solvent during the T-R allosteric transition.

Identification of Chloride Ion Binding Sites

The functional characteristics of Hb Bv point to the presence of a mechanism of oxygen modulation by the Cl^- ions of the solvent, as an alternative to the regulation by 2,3-DPG present in most mammalian hemoglobins. This novel mechanism likely originates from specific characteristics of the Hb Bv amino acid sequence. Hybrids of Hb A and Hb Bv indicate that, in analogy with the allosteric modulation of 2,3-DPG in Hb A, and Cl^- modulation is produced by the β chains of Hb Bv.[22] In a search

[21] E. Vitrano, A. Cupane, and L. Cordone, *J. Mol. Biol.* **180,** 1157 (1984).
[22] C. Fronticelli, *in* "Techniques in Protein Chemistry III" (R. H. Angeletti, ed.), p. 399. Academic Press, San Diego, 1992.

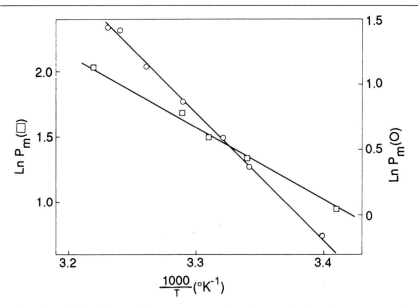

FIG. 9. van't Hoff plots of the temperature dependence of P_m (median O_2 pressure) of human (○) and bovine (□) hemoglobins in 0.1 M borate buffer, pH 9.0. The interpolating lines represent values obtained from global analyses. The left- and right-hand ordinate scales are for bovine and human hemoglobin, respectively. From Bucci et al.[19] Reprinted with permission. Copyright 1991 American Chemistry Society.

for conformational differences between the β chains of Hb A and Hb Bv, hydropathy plots were examined.[23] They indicated the presence of two regions with different hydrophobicity, one comprising the N-terminal end, the other comprising portions of the E helix and EF corner. In these regions of the β chains, comparisons were made between the amino acid sequences of primate and ruminant hemoglobins, of which Hb A and Hb Bv may be considered prototypes. This analysis indicates the presence of constant amino acid substitutions in thue two sets of hemoglobins. On the basis of this observation a model has been proposed for explaining the regulation of oxygen affinity by Cl^- ions, as an alternative to the classic modulation by 2,3-DPG.[23]

    Reaction with methyl acetyl phosphate (MAP) has been used to identify the Cl^- ion binding sites in Hb A and Hb Bv.[24] This study indicates that Cl^- binding sites are present in oxyHb Bv at Met(NA1) and Lys(EF9) in the β chains, in deoxyHb Bv at Val(NA1) in the α chains, and at Lys(EF9)

[23] C. Fronticelli, Biophys. Chem. 37, 141 (1990).
[24] H. Ueno, M. Pospischil, and J. M. Manning, J. Biol. Chem. 264, 12344 (1989).

in the β chains. Notably, in Hb A the α-N-terminal valine is not acetylated by MAP.

Bovine Hemoglobin as Artificial Oxygen Carrier

The presence in Hb Bv of a mechanism of oxygen modulation by the Cl^- ion confers to cell-free solutions of Hb Bv an oxygen affinity similar to that of whole blood, provided physiological concentrations of Cl^- ions (100 mM) are present in the solution.[2] The association of this characteristic with the amino acid sequence of Hb A (20% amino acid substitution, primarily at high-variability sites) and the ready availability of Hb Bv have elicited interest in exploring the use of Hb Bv as an artificial oxygen carrier. There is some concern, however, due to the finding in some cows of the bovine spongiform encephalitis agent, which is thermostable and can remain dormant for many years. Another concern exists as to whether the infusion into humans of large amounts of heterologous protein could trigger allergic reactions, although hemoglobin is known to be a weak antigen.

Although the oxygen affinity of Hb Bv is compatible with efficient oxygen transport *in vivo,* dimerization of the molecule results in a rapid loss of the hemoglobin either through the kidneys or the vascular system. Chemical stabilization of the tetrameric Hb molecule has been explored by acylation with derivatives of dicarboxylic acids conjugated to dibromosalicylic acid. Specific modification at the Lys(EF6) with mono(3,5-dibromosalicyl) fumarate resulted in a product with an oxygen affinity similar to blood, associated to a partial stabilization of the tetrameric hemoglobin.[25] Intramolecular cross-linking of Hb Bv in the oxy or deoxy state with bis(3,5-dibromosalicyl) fumarate results in a product with an oxygen affinity higher or lower, respectively, than that of blood.[2,26] In both cases the hemoglobin is a stabilized tetramer.[8]

Notably, the derivative obtained from cross-linking of the oxyHb Bv with bis(3,5-dibromosalicyl) fumarate is able to deliver oxygen *in vivo* in the rat and in the brains of perfused cats, although it has a $P_{1/2} = 17$ mm Hg, as compared with 27 mm Hg in normal blood.[27]

[25] C. Fronticelli, E. Bucci, A. Razynska, J. Sznayder, B. Urbaitis, and Z. Gryczynski, *Eur. J. Biochem.* **193,** 331 (1990).

[26] C. Fronticelli, T. Sato, C. Orth, and E. Bucci, *Biochim. Biophys. Acta* **874,** 76 (1986).

[27] E. Bucci, C. Fronticelli, A. Razynska, V. Militello, R. Koehler, and B. Urbaitis, *Biomater., Artif. Cells, Artif. Organs* **20,** 243 (1992).

Section III

Hemoglobin Hybrids and Model Hemes

[11] Synthetic Hemes

By EISHUN TSUCHIDA and TERUYUKI KOMATSU

In hemoglobin (Hb) and myoglobin (Mb) the globin chain has a globular and compact conformation, and the protoporphinato-iron (protoheme) group is embedded in the "pocket" of the globular protein. The heme coordinately bonds to the imidazole ring of the F8-His residue within the pocket-forming segment of the globin and becomes the oxygen-binding site.[1-3] The active site is completely surrounded by hydrophobic residues of the globin chain; this aids the formation of an oxygen adduct that is stable to irreversible oxidation through a proton-driven process. Further in the heme pocket, very polar groups within van der Waals distances are present around the coordination site and form a suitable polar environment for oxygen binding.

If the oxygen-binding site, the protoheme, is isolated from Hb and Mb and exposed to oxygen in a solution, the heme complex is immediately and irreversibly oxidized to its ferric [Fe(III)] state and does not act as an oxygen transporter.

The reversible oxygen coordination of Hb and Mb is considered to occur based on the following reasons[4-6]:

1. The heme complex has a five-coordinate structure whose sixth coordination site is vacant, which allows it to also coordinate dioxygen:

$$Fe(II)B + O_2 \rightleftharpoons BFe-O_2 \qquad (1)$$

2. The heme complex is dispersed and diluted to suppress the irreversible oxidation via the μ-dioxo dimer:

$$BFe-O_2 + Fe(II)B \rightarrow BFe-O_2-FeB \rightarrow$$
$$Fe(III)-O-Fe(III) \quad (2)$$

3. The heme complex is surrounded by a hydrophobic environment, causing the proton-driven oxidation to be retarded:

[1] M. F. Perutz, H. S. Muirhead, J. M. Cox, and L. C. Goaman, *Nature* (*London*) **219**, 131 (1986).
[2] B. Shaanan, *Nature* (*London*) **296**, 683 (1982).
[3] G. Fermi, M. F. Perutz, and B. Shaanan, *J. Mol. Biol.* **175**, 159 (1984).
[4] J. H. Wang, *J. Am. Chem. Soc.* **80**, 3168 (1985).
[5] R. D. Jones, D. A. Summerville, and F. Basolo, *Chem. Rev.* **79**, 139 (1979).
[6] E. Tsuchida and H. Nishide, *Top. Curr. Chem.* **132**, 64 (1986).

$$BFe—O_2 + H^+(H_2O) \rightarrow Fe(III)B + HO_2* \qquad (3)$$

In Eqs. (1)–(3), Fe represent heme and B is an axial base such as an imidazole derivative.

Globin protein forms the five-coordinate heme complex and "tucks it away" separately, that is, globin protein protects the heme complex from irreversible oxidation [Eq. (2)] by embedding it separately in the macromolecules and suppresses the proton-driven oxidation [Eq. (3)].

Much research has been directed toward mimicking oxygen transporters like Hb using synthetic heme derivatives.[5–36] Various synthetic hemes have been successful in coordinating oxygen reversibly in organic sol-

[7] C. K. Chang and T. G. Traylor, *Proc. Natl. Acad. Sci. U.S.A.* **70**, 2647 (1973).

[8] J. E. Baldwin and J. Huff, *J. Am. Chem. Soc.* **95**, 5757 (1973).

[9] J. P. Collmam, R. R. Gagne, C. A. Reed, T. R. Halbert, G. Lang, and W. T. Robinson, *J. Am. Chem. Soc.* **97**, 1427 (1975).

[10] J. P. Collmam, J. I. Brauman, K. M. Doxsee, T. R. Halbert, E. Bunnenberg, R. E. Linder, G. N. LaMar, J. D. Gaudio, G. Lang, and K. Spartalian, *J. Am. Chem. Soc.* **102**, 4182 (1980).

[11] J. P. Collmam, J. I. Brauman, J. P. Fitzgerald, P. D. Hampton, Y. Naruta, J. W. Sparapany, and J. A. Ibers, *J. Am. Chem. Soc.* **110**, 3477 (1988).

[12] J. P. Collmam, J. I. Brauman, T. J. Collins, B. L. Iverson, G. Lang, R. B. Pettman, J. L. Sessler, and M. A. Walters, *J. Am. Chem. Soc.* **105**, 3038 (1983).

[13] T. Komatsu, K. Arai, H. Nishide, and E. Tsuchida, *Chem. Lett.*, p. 799 (1992).

[14] J. Geibel, J. Cannon, D. Campbell, and T. G. Traylor, *J. Am. Chem. Soc.* **100**, 3757 (1978).

[15] T. G. Traylor, D. K. White, D. H. Campbell, and A. P. Berzinis, *J. Am. Chem. Soc.* **103**, 4932 (1981).

[16] T. G. Traylor, N. Koga, and L. A. Deardurff, *J. Am. Chem. Soc.* **107**, 6504 (1985).

[17] T. Hashimoto, R. L. Dyer, M. J. Crossley, J. E. Baldwin, and F. Basolo, *J. Am. Chem. Soc.* **104**, 2101 (1982).

[18] J. E. Baldwin, J. H. Cameron, M. J. Crossley, I. J. Dagrely, and T. Klose, *J. Chem. Soc., Dalton Trans.*, p. 1739 (1984).

[19] M. Momenteau, *Pure Appl. Chem.* **58**, 1493 (1986).

[20] D. Lavalette, C. Tetreau, J. Mispelter, M. Momenteau, and J.-M. Lhoste, *Eur. J. Biochem.* **145**, 555 (1984).

[21] M. Momenteau and D. Lavalette, *J. Chem. Soc., Chem. Commun.*, p. 341 (1982).

[22] M. Momenteau, B. Loock, C. Tetreau, D. Lavalette, A. Crisy, C. Schaffer, C. Huel, and J. M. Lhoste, *J. Chem. Soc., Perkin Trans. 2*, p. 249 (1987).

[23] A. R. Battersby, S. A. Barthlomew, and T. Nitta, *J. Chem. Soc., Chem. Commun.*, p. 1291 (1983).

[24] K. S. Suslick, M. M. Fox, and T. J. Reinert, *J. Am. Chem. Soc.* **106**, 4522 (1984).

[25] Y. Uemori and E. Kyuno, *Inorg. Chem.* **28**, 1690 (1989).

[26] C. K. Chang, B. Ward, R. Young, and M. P. Kondylis, *J. Macromol. Sci., Chem.* **25**, 1307 (1988).

[27] J. P. Collmam, J. I. Brauman, B. L. Iverson, J. L. Sessler, R. M. Morris, and Q. H. Gibson, *J. Am. Chem. Soc.* **105**, 3052 (1983).

[28] T. Komatsu, E. Hasegawa, H. Nishide, and E. Tsuchida, *J. Chem. Soc., Chem. Commun.*, p. 66 (1990).

vents, but in aqueous media they have been irreversibly oxidized. Until recently, only Hb and Mb were known oxygen transporters in aqueous media.

In this chapter, procedures used for the preparation and characterization of various synthetic hemes as oxygen transporters are described. The discussion here will be centered not only on the preparation of the synthetic hemes but also on substituted systems of red blood cells and the approach to developing an artificial red cell.

Preparation of Synthetic Hemes

Modified Heme Complexes as Oxygen Carriers

As model compounds for Hb and Mb, many synthetic heme derivatives have been prepared vigorously. Much recent work has been aimed at overcoming the requisites for the reversible oxygen coordination to a heme complex [Eqs. (1)–(3)] and there has been partial success in using aprotic solvents. In aprotic solvents the proton-driven oxidation [Eq. (3)] is excluded, so that the problems of reversible oxygen coordination are how to form the five-coordinate complex as the deoxy state [Eq. (1)] and how to inhibit the irreversible oxidation via dimerization [Eq. (2)]. The first solution oxygenation of a synthetic heme (1) was reported by Chang and Traylor in 1973,[7] simultaneous with the report of oxygenation of another iron macrocycle (2) by Baldwin and Huff,[8] both reactions occurring at low temperature. Several other examples of oxygenation of model compounds were soon reported, and methods of observing reversible oxygenation developed rapidly. Quantitative studies of oxygen binding to model compounds began to appear in 1975. The succesful approach was

[29] T. Komatsu, E. Hasegawa, S. Kumamoto, H. Nishide, and E. Tsuchida, *J. Chem. Soc., Dalton Trans.,* p. 3281 (1991).

[30] E. Tsuchida, T. Komatsu, T. Nakata, E. Hasegawa, H. Nishide, and H. Inoue, *J. Chem. Soc., Dalton Trans.,* p. 3285 (1991).

[31] E. Tsuchida, T. Komatsu, K. Arai, and H. Nishide, *J. Chem. Soc., Dalton Trans.,* p. 2465 (1993).

[32] B. Ward, C. B. Wang, and C. K. Chang, *J. Am. Chem. Soc.* **103,** 5236 (1981).

[33] H. Nishide, Y. Hashimoto, H. Maeda, and E. Tsuchida, *J. Chem. Soc., Dalton Trans.,* p. 2963 (1987).

[34] H. Nishide, H. Maeda, S. Wang, and E. Tsuchida, *J. Chem. Soc., Chem. Commun.,* p. 574 (1985).

[35] E. Tsuchida, S. Wang, M. Yuasa, and H. Nishide, *J. Chem. Soc., Chem. Commun.,* p. 23 (1986).

[36] T. G. Traylor, S. Tsuchiya, D. Campbell, M. Mitchel, D. Stynes, and N. Koga, *J. Am. Chem. Soc.* **107,** 604 (1985).

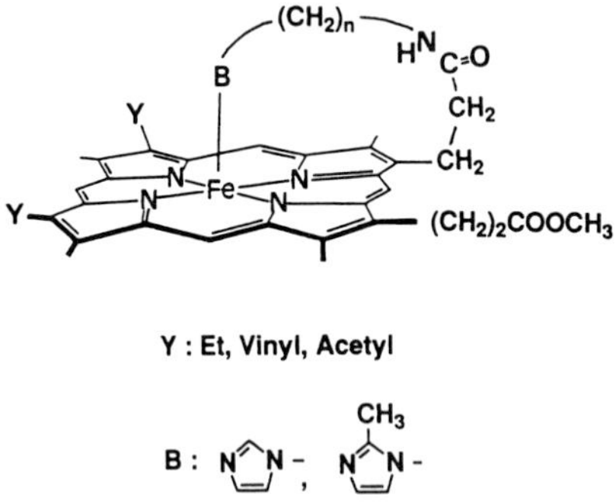

Y : Et, Vinyl, Acetyl

1

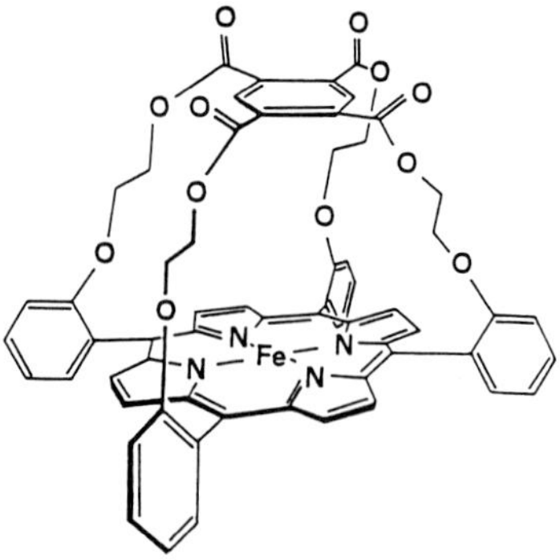

2

an elegant steric modification of porphyrins. Some superstructured hemes have been produced by clever synthetic techniques (3)–(14). In particular, *meso*-tetraphenylporphine (TPP) derivatives sterically encumbered by peripheral substitutes have been widely exploited in the development of porphyrin model system.

A typical example is Collman's 5,10,15,20-tetrakis($\alpha,\alpha,\alpha,\alpha$-*o*-pivalamidophenyl)porphinato-iron (3),[9,27] which has steric bulkiness constructed with the pivalamide groups on one side of the porphyrin plane and with the opposite side of the porphyrin plane unencumbered. The imidazole ligand is allowed to coordinate to the unhindered side of the heme, and the other side remains a pocket for oxygen coordination. Moreover, the bulky pivalamide groups would discourage the dimerization of a μ-dioxo complex. The 3–(1-methylimidazole) complex could bind oxygen reversibly in dry benzene at 25° for over a week. This result shows that a skeleton structure and a special environment around the oxygen-binding site are important to form the reversible oxygen-coordinated complex.

Compound 3 is prepared as follows. Pivaloyl chloride (12 mmol) is allowed to react with 5,10,15,20-tetrakis($\alpha,\alpha,\alpha,\alpha$-*o*-aminophenyl)porphine (1.5 mmol) in tetrahydrofuran (200 ml) containing pyridine (2 mmol) at 5°. The mixture is stirred for 6 hr at room temperature and is separated by column chromatography [silica gel; chloroform/diethyl ether at a ratio of 4/1 (v/v)], affording the corresponding porphyrin (84%). This porphyrin (1 mmol), $FeBr_2$ (20 mmol), and 2,6-lutidine (10 mmol) in dry tetrahydrofuran (200 ml) are refluxed under nitrogen. The reaction is finished after 1 hr. The mixture is then concentrated and chromatographed on a column (basic alumina, chloroform), to give 3 (90%).

In complex **3** the pivalamide groups are believed to provide a distal moiety with a weak hydrogen bonding with a coordinated dioxygen: BFe–O–O···H–NC(=O). In fact, the synthetic hemes with reversible oxygen-coordinating capability often bear amide residues in their substituent groups on the porphyrin plane (amide effect).[9–12,19–22,25–27] Thus there remains a question as to whether the distal amide residues are crucial for reversible oxygen coordination.

A new porphinato-iron derivative **(4)**, 5,10,15-tris[2,6-bis(3,3-dimethyl-butyryloxy)phenyl]-20-[2-(3,3-dimethylbutyryloxy)-6-(5-imidazolylvaleroyl-oxy)phenyl]porphinato-iron (double-sided heme bearing covalently bound axial imidazole, was synthesized.[13,28–31] The eight substituents of **4** are bound to the porphyrin only through ester bonds. Compound **4** forms a stable oxygen adduct reversibly in toluene at 25°. The advantage of the double-sided heme is removal of the complexity of diastereoisomeric properties in the preparation.

Compound **4** is prepared as follows. 5,10,15-Tris[2,6-bis(3,3-dimethyl-butyryloxy)phenyl]-20-[2-(3,3-dimethylbutyryloxy)-6-hydroxyphenyl]porphine is synthesized by condensation of 5,10,15,20-tetrakis[2,6-bis(hydroxy)phenyl]porphine (1.5 mmol) with a 7.2-fold molar excess of 3,3-dimethylbutyryl chloride in dry tetrahydrofuran (200 ml) (10%). This compound (1 mmol) is allowed to react with 5-(imidazolyl)valeroyl chloride (60 mmol) in the presence of triethylamine (60 mmol) in dry acetonitrile at 60° for 6 hr. The mixture is separated by column chromatography [silica gel; chloroform/methanol in a ratio of 20/1 (v/v)] to give the corresponding porphyrin (54%). Iron was inserted by $FeBr_2$ method as described above to afford **4**.

The oxygen-binding affinity of **4** was slightly lower than that of 5,10,15-tris(α,α,α,o-pivalamideophenyl)-β-o-20-(5-imidazolylvaleramidophenyl) porphinato-iron **(5)**,[27] which was structurally similar to **4** except for containing amide groups. This indicates that the amide residue is not crucial for the formation of a reversible and stable oxygen adduct that can serve as a Hb or a Mb model, provided that the porphyrin molecules are modified ingeniously.

Diporphinato-metal derivatives were first synthesized by Ward *et al.*[32] The imidazole complex of diporphinato-copper–iron **(6a)** [M_1 = Cu(II), M_2 = Fe(II)] forms an oxygen-coordinated complex, and its lifetime is fairly long and comparable to that of complex **3** in dry benzene.[33–35] Molecular oxygen coordinates to the heme through an opening of the face-to-face structure of diporphyrin. It was considered that the inert porphinato-copper, tightly linked to the porphinato-iron, protects the coordinated oxygen. Compound **6** was synthesized from 2,7,12,17-tetramethyl-3,13-didecylporphine-8,18-diacetic acid **(7a)**. The *p*-nitrophenyl diester deriva-

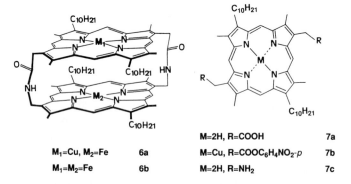

tive **(7b)** of the porphinato-copper and the diamino derivative **(7c)** were derived from **7a** by modifying the previously reported method. Coupling of these compounds in dilute pyridine solution gave **6**.

M=2H, R=COOH	7a
M₁=Cu, M₂=Fe 6a	M=Cu, R=COOC₆H₄NO₂-*p*	7b
M₁=M₂=Fe 6b	M=2H, R=NH₂	7c

Diheme **6b** [M₁ = M₂ = Fe(II)] showed the same coordination, whereas two CO molecules can coordinate to it due to the diiron structure. The coordination equilibrium curve for the **6b**–(1-methylimidazole) complex

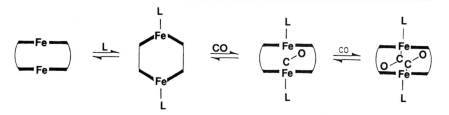

FIG. 1. The structural change of diheme (6b) in the pseudocooperative CO-binding reaction. L, Imidazole derivative.

appeared sigmoidal, whereas the curve for the 6a–(1-methylimidazole) complex was hyperbolic. The pseudocooperative coordination of the synthetic 6b with CO was produced by the triggering of a structural change of the coupled coordination sites (Fig. 1) and, thus, was a good model of the cooperative coordination of Hb, in which the conformational change of the globin protein induces the reactivity change of heme.

The above-mentioned studies on the synthetic heme complexes are of great significance because they have demonstrated steric and environmental effects on oxygen coordination. However, in these studies the oxygen-coordinating ability for these complexes in aqueous media unfortunately was not described.

Characterizations

Many synthetic hemes serving as Hb and Mb models have been well characterized by UV/visible, NMR, ESR, Raman, Mössbauer, and IR spectroscopies, as well as by X ray structural analyses. These model systems have contributed markedly to the understanding of structure–function relationships of the natural system.

O_2 and CO Binding Kinetics

Determination of the equilibrium and kinetic constants of O_2 and CO binding can be easily obtained either by direct photometric titration or by the technique of laser-flash photolysis. In general, reduction to the Fe(II) complex in organic solvents (e.g., toluene and benzene) is carried out by using aqueous $Na_2S_2O_4$ in a heterogeneous two-phase system under anaerobic conditions.[17,20,27,29]

Oxygen and carbon monoxide binding to hemes can be expressed by Eq. (4):

$$Fe(II)B + L \underset{}{\overset{K(O_2)}{\rightleftharpoons}} BFe—L \qquad (4)$$

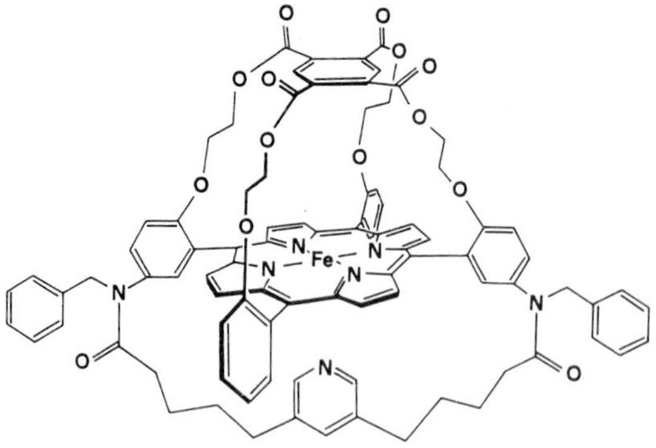

8

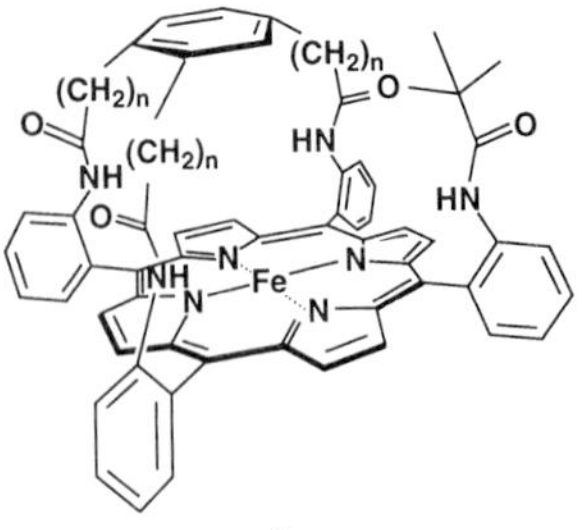

9

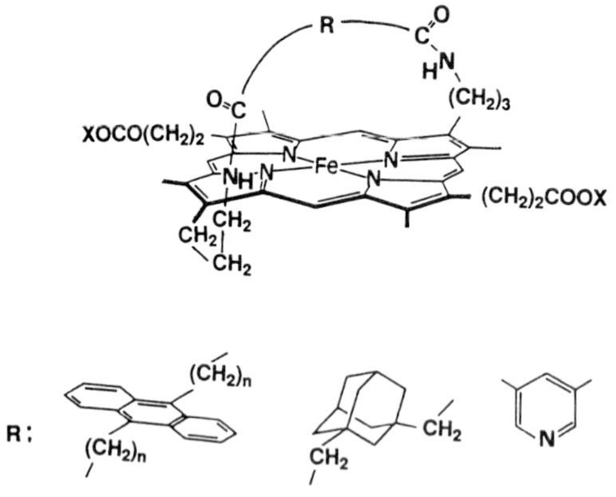

R :

X : -*n*-Bu, -CH₂C₆H₅

10

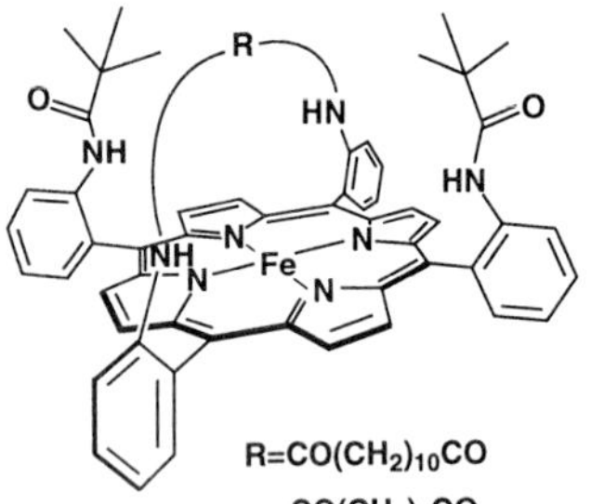

R=CO(CH₂)₁₀CO
CO(CH₂)₈CO
CO(CH₂)₇CO
CO(CH₂)₆CO

11

where L represents the gaseous ligand (O$_2$ or CO). The affinities for O$_2$ and CO were determined from the spectral changes at various partial pressures of the gaseous ligand.

Kinetic measurements are performed by using the laser-flash photolysis technique.[14–16,19,20,26,27,31,36,37] The experiments and data analysis are carried out with USP-500 or TSP-601 (Unisoku Co., Japan). Heme concentrations of 10 μM are used. After flash photolysis, a recombination occurs with k_{obs} given by Eq. (6)

[37] Q. H. Gibson, *Proc. R. Soc. London, Ser. B* **143,** 344 (1955).

$$BFe-L \xrightarrow{h\nu} BFe \underset{k_{off}(L)}{\overset{k_{on}(L)}{\rightleftharpoons}} BFe-L \qquad (5)$$

$$k_{obs} = k_{on}(L)[L] + k_{off}(L) \qquad (6)$$

The gaseous ligand concentrations are always in large excess of the heme concentration, so that the pseudo-first-order approximation can be applied throughout.

The values of k_{on} and k_{off} can be obtained using Eq. (6) at several [L]. However, k_{off} is preferably determined from k_{on}/K, because the k_{off} value obtained from Eq. (6) often contains considerable errors.

$K(O_2)$ is also determined using the competitive rebinding technique. This method is almost a routine tool for studying the binding of O_2 with synthetic heme, even when the oxygenated complex happens to be chemically unstable in the long run with respect to autoxidation.

Photolysis of the BFe–CO in the presence of an appropriate CO and O_2 mixture shows first a rapid recombination to BFe–O_2, followed by a slow return to BFe–CO, due to $K(CO) \gg K(O_2)$ and $k_{on}(CO)[CO] < k_{on}(O_2)[O_2]$ for a wide range of $[O_2]$ and $[CO]$.

$$BFe-CO \xrightarrow{h\nu} BFe \underset{k_{off}(O_2)}{\overset{k_{on}(O_2)}{\rightleftharpoons}} BFe-O_2 \qquad (7)$$

$$\underbrace{\phantom{BFe-CO \xrightarrow{h\nu} BFe \rightleftharpoons BFe-O_2}}_{k_{obs}(slow)}$$

From Gibson's equation [Eq. (8)], a plot of $1/k_{obs}(slow)$ versus $[O_2]/[CO]$ yields a straight line with slope $k_{on}(O_2)/k_{off}(O_2)k_{on}(CO)$.[27,36,37]

$$1/k_{obs}(slow) = \frac{K(O_2)[O_2]}{k_{on}(CO)[CO]} + \frac{1}{k_{on}(CO)[CO]} + \frac{1}{k_{off}(O_2)} \qquad (8)$$

This technique takes advantage of the approximately 10-fold faster rate of O_2 addition to heme compounds as compared to CO and the much greater binding constant of CO to hemes. The $K(O_2)$ derived from the competition method matches that obtained directly under equilibrium conditions.

The kinetic behavior of O_2 and CO binding to the hemes was attributed to the local polarity, solvation, the distal steric hindrance on the coordination site and the proximal pull effect.[7,14–16,19–22,26,27,29,36] The equilibrium and kinetic parameters of O_2 and CO binding to various synthetic hemes are summarized in Table I.

NMR Spectroscopy

The ^1H NMR spectra of all the new metal-free porphyrins have been recorded and used for routine characterization. Further, synthetic hemes in the high- and low-spin Fe(III) forms, as well as the physiologically

TABLE I

O_2 AND CO BINDING RATE CONSTANTS FOR SYNTHETIC HEMES[a]

Heme	O_2			CO			Ref.
	$10^{-7} k_{on}$ (M^{-1} sec^{-1})	$10^{-3} k_{off}$ (sec^{-1})	$P_{1/2}^{a}$ (Torr)	$10^{-6} k_{on}$ (M^{-1} sec^{-1})	$10^{-3} k_{off}$ (sec^{-1})	$10^{4} P_{1/2}^{a}$ (Torr)	
1 (protoheme)[b]	5.3	1.7	2.8	11	250	2.3	15
2-(1-MeIm)[c,d]	—	—	23	0.95	50	54	17
4[c]	6.3	1.2	2.5	13	9.3	0.72	13
5[c]	43	2.9	0.58	36	7.8	0.22	27
9(n = 1)-(1-MeIm)[c,d]	0.22	0.009	0.36	0.58	8.6	15	27
10(n = 7) Anthracene–(1,5-DCIm)[b,e]	6.5	1.0	1.4	6	50	9.2	15
11 (C = 8)-(1-MeIm)[d,f]	0.22	0.002	0.1	0.08	8.2	100	22
12a (C = 12)[f]	36	5	2	35	30	0.9	19
12b (C = 12)[f]	30	40	18	68	69	1.1	19
12c (C = 12)[f]	31	0.62	0.29	40	6.7	0.17	19
3-(1,2-Me₂Im)[c,g]	11	46	38	1.4	140	89	27

[a] $P_{1/2}$: K^{-1}
[b] At 20°, benzene.
[c] At 25°, toluene.
[d] 1-MeIm, 1-Methylimidazole.
[e] 1,5-DCIm, 1,5-Dicyclohexylimidazole.
[f] At 20°, toluene.
[g] 1,2-Me₂Im, 1,2-Dimethylimidazole.

more important Fe(II) complex, have been studied extensively by ^{1}H NMR spectroscopy.

In TPP derivatives, the large isotopic shifts of the pyrrole β protons are very characteristic of oxidation and spin state. Large contact shifts are also observed for coordinated axial ligands, whereas the shifts for the phenyl ring protons are small. The pyrrole β protons of synthetic deoxyhemes (e.g. **4, 5,** and **12**) give rise to four singlets (two protons each), clearly indicating that the Fe(II) is five-coordinated and high spin.[10,13,31]

Momenteau and co-workers showed that changing the attachment mode of both the proximal and distal handles strongly modifies the O$_2$ affinity.[19,20] It was found that the amide derivative **(12a)** had an affinity

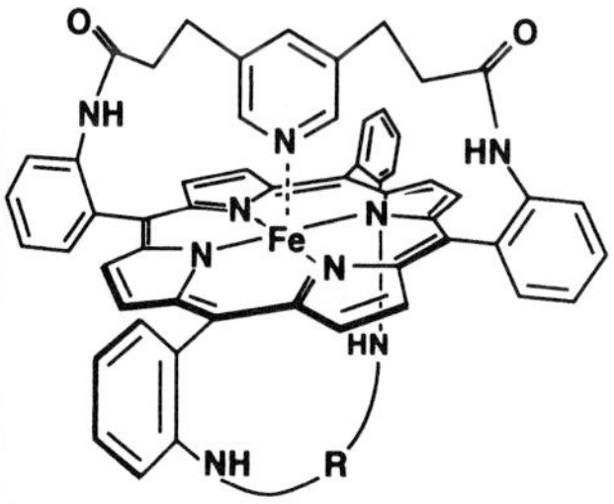

R=CO(CH$_2$)$_{10}$CO
    CO(CH$_2$)$_8$CO
    CO(CH$_2$)$_2$C$_6$H$_4$(CH$_2$)$_2$CO

12a

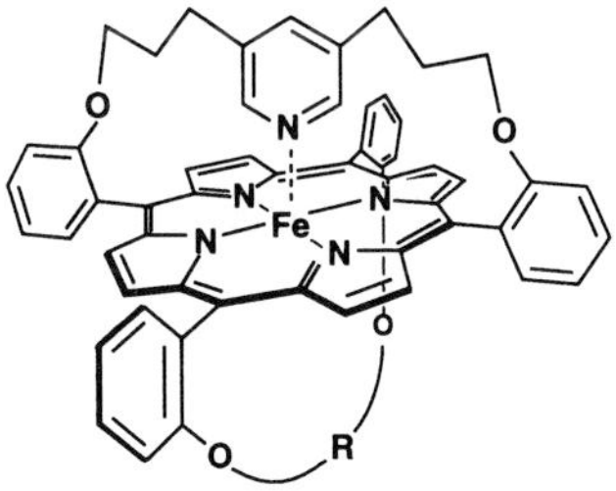

R=(CH$_2$)$_{12}$
    (CH$_2$)$_{10}$
    (CH$_2$)$_4$C$_6$H$_4$(CH$_2$)$_4$
    (CH$_2$)$_3$C$_6$H$_4$(CH$_2$)$_3$

12b

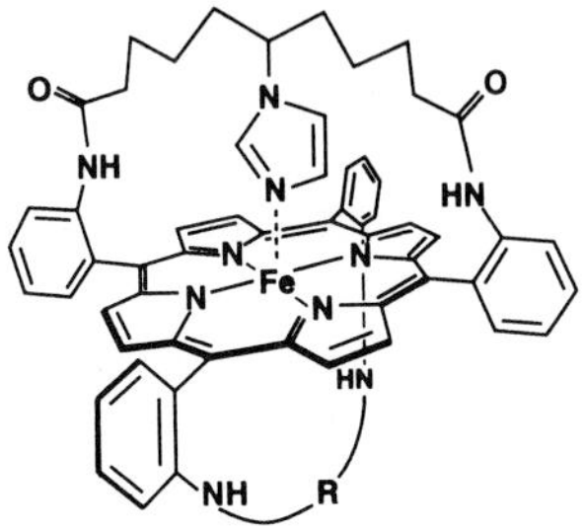

R=CO(CH$_2$)$_{10}$CO

12c

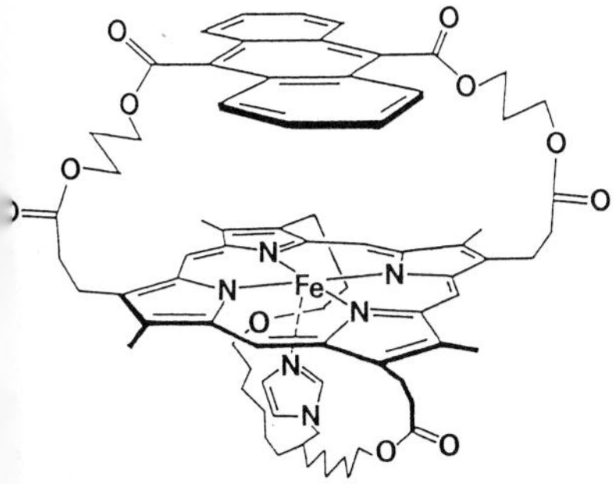

13

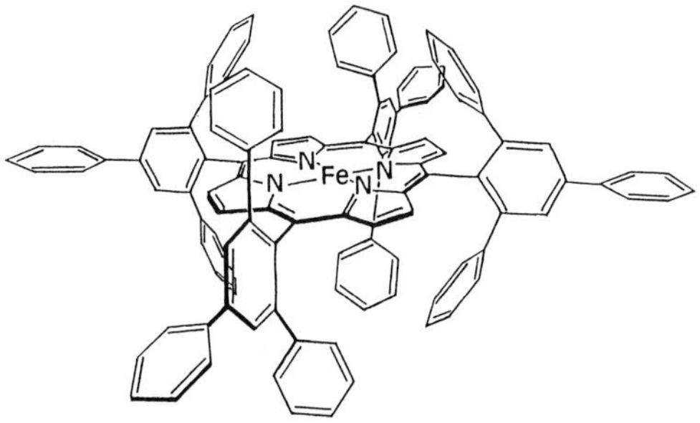

14

for O$_2$ an order of magnitude greater than did the ether analog **(12b)**.[21] This increase in stability of the "amide" oxygenated species was attributed to the presence of the two amide linkages and the possibility of hydrogen bonding with the terminal oxygen atom of the ligated oxygen molecule. The low-temperature ($-27°$) ^{1}H NMR spectra were used to provide evidence for such an interaction.[20] The spectrum of the CO-bound iron(II) amide complex exhibits a C_2 symmetry with the pyrrolic protons appearing as a single resonance. Replacing CO with O_2 gives a large inequivalence of the pyrrolic protons, indicating a preferential orientation of the bent oxygen molecule toward two opposite positions. Examination of amide proton resonances confirms the direct interaction of O_2 with amide groups of the distal chain. Ring current shift calculations from the ${}^1H$ NMR spectrum of the zinc complex allowed an estimate of the position of the distal amide protons and hence the nitrogen atoms. The distance between one of the amide nitrogen atoms and the terminal oxygen atom in a bent configuration may be close to 3 Å. This is consistent with an intramolecular hydrogen bond.

Gerothanassis and co-workers reported the ${}^{17}O$ NMR spectra of the Fe–O_2 moiety of synthetic hemes in organic solvents.[38] The spectra exhibit two well-resolved resonances, in agreement with the end-on structure proposed by Pauling for oxyHb.[39] The sensitivity of ${}^{17}O$ chemical shifts to hydrogen-bonding interactions with distal moieties is demonstrated for the first time. The unusual temperature dependence of the ${}^{17}O$ linewidths of the oxygenated 3–(1-methylimidazole) complex can be attributed to a dynamic equilibrium of two conformers of the Fe–O–O unit with a substantially different electric field gradient tensor at both oxygen sites and a small difference in energy, which is dominated by O_2 pivalamide interactions.

Mössbauer Spectroscopy

Mössbauer spectroscopy has been used extensively to study the electronic nature of Fe(II) hemoproteins as well as of the synthetic hemes. Mössbauer spectra have been recorded for ${}^{57}Fe$-enriched heme complexes under an argon, CO, or O_2 atmosphere.[10,11,30] The spectra were measured with a ${}^{57}Co$ source (3.7 × 10^8 Bq) in a palladium matrix using a Wissel constant-acceleration transducer and then were fitted to Lorentzian curves with a least-squares fitting program. The isomer shifts were referred to the centroid of the Mössbauer spectrum of metallic iron at room temperature. The velocity scale was calibrated by the spectra of metallic iron. Table II contains the Mössbauer parameters for several synthetic hemes.

X-Ray Structural Analyses

Most X-ray studies of CO adducts of heme proteins have suggested a substantial distortion of the Fe–C–O linkage: either a bend or a tilt to the

[38] I. P. Gerothanassis, M. Momenteau, and B. Loock, *J. Am. Chem. Soc.* **111**, 7006 (1989).
[39] L. Pauling, *Nature (London)* **203**, 182 (1964).

TABLE II
MÖSSBAUER PARAMETERS FOR SYNTHETIC HEMES[a]

Heme	Base	Deoxy		O_2		CO		Ref.
		δ	ΔE_Q[b]	δ	ΔE_Q	δ	ΔE_Q	
3	1-MeIm	0.44	1.02	0.27	2.04	0.27	0.27	10
9($n = 1$)	1-MeIm	0.81	2.36	0.23	2.23	—	—	12
9($n = 1$)	1,2-Me$_2$Im[c]	0.92	2.37	0.24	2.32	—	—	12
DS heme[d]	1-HIm[e]	0.44	0.98	0.28	2.07	0.29	0.35	30

[a] Data in mm sec^{-1} at 77 K. δ, Isomer shift.
[b] ΔE_Q, Quadrupole splitting.
[c] 1,2-Me$_2$Im, 1,2-Dimethylimidazole.
[d] DS heme, 5,10,15,20-Tetrakis(2,6-bis(t-butylacetoxy)phenyl)porphinato-iron.
[e] 1-HIm, 1-Hexylimidazole.

heme plane, or both. Such distortion presumably stems from nonbinding interactions of the CO ligand with nearby distal amino acid residues, because CO preferentially binds to Fe in a linear, perpendicular fashion in unconstrained synthetic heme systems. Because O_2 preferentially binds to hemes and to model systems in a bent fashion, it is thought that the distal steric hindrances, hydrogen bonding or porphyrin ruffling, are important factors in O_2/CO discrimination. A number of encumbered porphyrin systems have been prepared in an effort to delineate the structural details of gaseous ligand binding. Unfortunately, as yet, only a few X-ray structures of these encumbered models with bound gaseous ligands have been reported.[12,40–44] Moreover, none of the X-ray structures of the model CO adducts displays the distortion that has been reported to occur in some of the natural systems.

The structure of CO-bound **2**–(1-methylimidazole) and **9**–(1,2-dimethylimidazole) has been determined by single-crystal X-ray diffraction methods.[45,46] The coordinated CO ligand is slightly but detectably distorted

[40] G. B. Jameson, F. S. Molonaro, J. A. Ibers, J. P. Collman, J. I. Brauman, E. Rose, and K. S. Suslick, *J. Am. Chem. Soc.* **102**, 3224 (1980).
[41] G. B. Jameson and J. A. Ibers, *J. Am. Chem. Soc.* **102**, 2823 (1980).
[42] J. W. Sparapany, M. J. Crossley, J. E. Baldwin, and J. A. Ibers, *J. Am. Chem. Soc.* **110**, 4559 (1988).
[43] T. G. Traylor, N. Koga, L. A. Deardurff, P. N. Swepston, and J. A. Ibers, *J. Am. Chem. Soc.* **106**, 5132 (1984).
[44] M. Momenteau, W. R. Scheidt, C. W. Eigenbert, and C. A. Reed, *J. Am. Chem. Soc.* **110**, 1207 (1988).
[45] K. Kim, J. Fettinger, J. L. Sessler, N. Cyr, J. Hughahl, J. P. Collman, and J. A. Ibers, *J. Am. Chem. Soc.* **111**, 403 (1989).
[46] K. Kim and J. A. Ibers, *J. Am. Chem. Soc.* **113**, 6077 (1991).

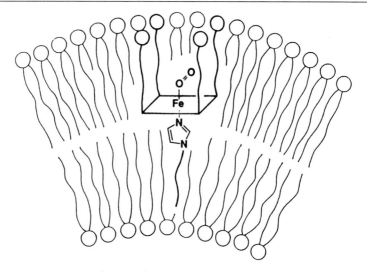

FIG. 2. Phospholipid vesicle-embedded lipid–heme.

from linearity, being both bent and tilted off the axial normal to the porphyrin.

Substituted System of Red Blood Cells

Phospholipid Vesicle-Embedded Lipid–Heme

In Hb and Mb the globin protein protects the heme complexes that
are tucked separately into a hydrophobic domain of the protein. In order
to construct a hydrophobic environment in an aqueous solution, it is
possible to use the macromolecular assembly, "phospholipid vesicle" or
"lipid microsphere," instead of the globin protein.

An amphiphilic heme derivative having four alkylphosphocholine
groups, 5,10,15,20-tetrakis[α,α,α,α-o-(2′,2′-dimethyl-20′-[2″-trimethyl
(ammonioethyl)phosphonatoxy]eicosanamido)phenyl]porphinato-iron,
[lipid–heme (15)][47] was synthesized. Four phosphocholine groups significantly increased the compatibility of the heme complex with a phospholipid and also increased the oxygen-coordinating ability (Fig. 2). With this
heme complex we succeeded in utilizing the hydrophobic region of the
assembly of the phospholipid as the efficient matrix for heme instead of

[47] Y. Matsushita, E. Hasegawa, K. Eshima, and E. Tsuchida, *Chem. Lett.*, p. 1387 (1983).

the globin protein.[48-52] That is, the stereo structure and the amphiphilic property of lipid–heme enhance its compatibility with a phospholipid bilayer and form a very stable assembly of the heme complex. A hydrophobic environment to retard the proton-driven oxidation [Eq. (3)] is constructed around heme in aqueous media by embedding heme in a phospholipid bilayer, and the porphyrin plane of the heme complex is oriented parallel to the bilayer, which prevents the oxidation via μ-dioxo dimers [Eq. (2)].

The lipid–heme is prepared as follows. ω-Benzyloxy-2,2-dimethyleicosanoic acid chloride (10 mmol) is allowed to react with 5,10,15,20-tetrakis ($\alpha,\alpha,\alpha,\alpha$-$o$-aminophenyl)porphine (1.56 mmol) in dry tetrahydrofuran (200 ml) containing pyridine (10 mmol). The mixture is separated by column chromatography [silica gel; benzene/diethyl ether in a ratio of 4/1 (v/v)], affording 5,10,15,20-tetrakis[$\alpha,\alpha,\alpha,\alpha$-$o$-(20'-benzyloxy-2',2'-dimethyleicosanamido)phenyl]porphine (86%). This porphyrin (1 mmol) is debenzylated by $AlCl_3$ (12 mmol) and anisole (12 mmol) in dichloromethane–nitromethane [1/1 (v/v)]; the mixture is cooled with ice–water, followed by extraction with dichloromethane. The extract is concentrated and recrystallized from benzene to give 5,10,15,20-tetrakis[$\alpha,\alpha,\alpha,\alpha$-$o$-[20'-hydroxy-2',2'-dimethyleicosanamido)phenyl]porphine (80%). Reaction with $FeBr_2$ in dry tetrahydrofuran affords 5,10,15,20-tetrakis[$\alpha,\alpha,\alpha,\alpha$-$o$-[20'-hydroxy-2',2'-dimethyleicosanamido)phenyl]porphinato-iron(III) bromide (91%), which is phosphorylated with 2-chloro-2-oxo-1,3,2-dioxaphospholane in dichloromethane, using triethylamine to trap HCl, at room temperature for 12 hr. The resultant phosphate triester is cleaved by an excess of anhydrous triethylamine in dimethylformamide at 60° (24 hr). The red–brown precipitate is collected by filtration and then purified on a gel column (Sephadex LH-60, methanol), giving **15** (89%).

Phospholipid vesicle-embedded complex **15** is easily prepared by modifying the normal method of phospholipid vesicle preparation. Only 50 mol of dimyristoylphosphatidylcholine (DMPC) is enough to solubilize 1 mol of complex **15** in water. Gel permeation chromatography (GPC) (Sepharose CL-4B) and ultracentrifugation of the DMPC/**15** solution shows that complex **15** is completely entrapped within the phospholipid assembly. An

[48] E. Tsuchida, H. Nishide, M. Yuasa, E. Hasegawa, Y. Matsushita, and K. Eshima, *J. Chem. Soc., Dalton Trans.*, p. 275 (1985).

[49] M. Yuasa, K. Aiba, Y. Ogata, H. Nishide, and E. Tsuchida, *Biochim. Biophys. Acta* **860**, 558 (1986).

[50] M. Yuasa, Y. Tani, H. Nishide, and E. Tsuchida, *J. Chem. Soc., Dalton Trans.*, p. 1917 (1987).

[51] M. Yuasa, H. Nishide, and E. Tsuchida, *J. Chem. Soc., Dalton Trans.*, p. 2493 (1987).

[52] E. Tsuchida, H. Maeda, M. Yuasa, H. Nishide, H. Inoue, and T. Shirai, *J. Chem. Soc., Dalton. Trans.*, p. 2455 (1987).

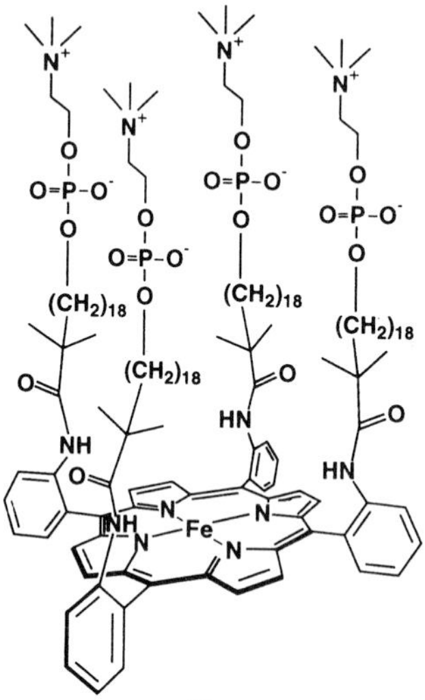

15

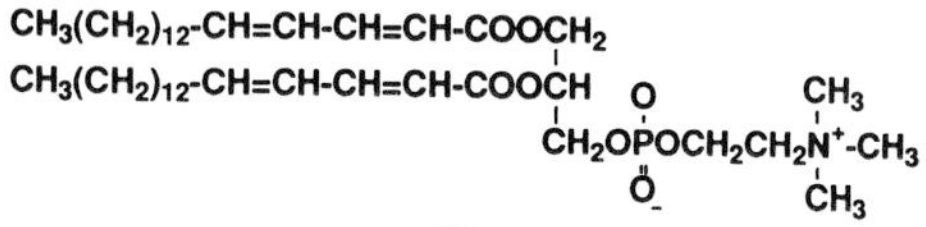

$$CH_3(CH_2)_{12}\text{-}CH=CH\text{-}CH=CH\text{-}COOCH_2$$
$$CH_3(CH_2)_{12}\text{-}CH=CH\text{-}CH=CH\text{-}COOCH$$
$$CH_2OPOCH_2CH_2N^+\text{-}CH_3$$

16

electron micrograph showed that the phospholipid assembly-embedded **15** is a single-walled unilameller vesicle with diameter of ~40 nm.

Using differential scanning calorimetry on the phospholipid vesicle-embedded **15**, the endothermic peak for the gel–liquid crystal phase transition (T_c) was observed at the same temperature as for the corresponding phospholipid vesicle: for example, T_c is 24° both for the DMPC vesicle and for the DMPC/**15** vesicle. This suggests that the orientation of the phospholipid in the DMPC/**15** assembly is equivalent to the DMPC vesicle and that the compatibility of the heme complex with the phospholipid is sufficient to form a stable assembly.

Next we attempted to improve the stability of the phospholipid assembly as the matrix of the hemes, to create a highly oriented physically and mechanically stable structure. To accomplish this, we stabilized the spherical vesicle by polymerization of the phospholipid bilayer.[51–54] The double bond of the phospholipid derivative was rapidly polymerized under UV irradiation, because of its assembled and oriented structure, yielding a covalently bound and very stable lipid bilayer.

The heme complex was embedded in a polymerized vesicle of 1,2-bis(octadecadienoyl)-*sn*-glycero-3-phosphocholine **(16)**. The vesicle

[53] E. Tsuchida, E. Hasegawa, Y. Matsushita, K. Eshima, M. Yuasa, and H. Nishide, *Chem. Lett.*, p. 969 (1985).
[54] E. Tsuchida, H. Nishide, M. Yuasa, E. Hasegawa, K. Eshima, and Y. Matsushita, *Macromolecules* **22**, 2103 (1989).

[16 : 15 = 50 (molar ratio)] was allowed to polymerize under UV irradiation to give poly[1,2-bis(octadecadienoyl)-sn-glycero-3-phosphocholine] vesicle-embedded 15. The degree of polymerization of the polymerized lipid monomer was reduced by the presence of the complex to about one-third to one-fifth of that without the complex.

To fix the heme complex more precisely in the polymer matrix with respect to its orientation, hemes substituted with four alkyl groups that have both a polymerizable double bond and a hydrophilic residue at their top positions (17 and 18) were synthesized. Because of the hydrophilic–lipophilic balance as well as the stereostructure of the heme complexes, they also are highly compatible with the phospholipid matrix and form a stable lipid assembly.[55]

Compounds 17 and 18 are prepared as follows. 5,10,15,20-Tetrakis [$\alpha,\alpha,\alpha,\alpha$-$o$-(20'-hydroxy-2',2'-dimethyleicosanamido)phenyl]porphine (1 mmol) is reacted with muconic acid chloride or itaconic acid chloride (20 mmol) in dry chloroform (100 ml) at 60° for 48 hr. The mixture is separated by column chromatography [silica gel; chloroform/methanol in a ratio of 10/1 (v/v)] to give the corresponding porphyrin (72%). Iron is inserted by refluxing with $FeBr_2$/tetrahydrofuran in the presence of 2,6-lutidine to give 17 and/or 18.

The polymerized phospholipid vesicle-embedded heme complex (e.g., lipid–heme) is prepared as follows. Complex 15 and 1-laurylimidazole [1 : 3 (molar ratio)] are dissolved in methanol (0.5 ml), which is added to a benzene solution (0.1 ml) of 16 [16 : 15 = 50 (molar ratio)]. The solution is evaporated to give a thin film, and then oxygen-free phosphate buffer (30 mM, pH 7.4, 10 ml) is added. The mixture is homogenized by an ultrasonic generator (Nihon Seiki Co., Japan, UP-600; 60 W, 10 min) in an ice–water bath under a nitrogen atmosphere. The prepared phospholipid vesicle-embedded heme solution is incubated at room temperature for a few hours in a quartz cell (pathlength, 10 mm) and then is allowed, under a nitrogen atmosphere, to polymerize for 2 hr at 50° under ultraviolet irradiation with a low-vacuum UV lamp (Riko Kagaku Co., Japan, UVL-32 type; distance between the cell and the lamp is kept at 3 cm) to give the polymerized vesicle-embedded 15 dispersion.

Furthermore, 17 is ligated with an alkylimidazole derivative that possesses a double bond (19). The 19-ligated 17 complex also forms a stable lipid assembly with 16 and it is efficiently copolymerized under UV irradiation in the oriented liquid crystal state to give a polymerized bilayer that is covalently fixed in the matrix through the copolymerization (Fig. 3). The

[55] E. Tsuchida, H. Nishide, M. Yuasa, T. Babe, and M. Fukuzumi, Macromolecules 22, 66 (1989).

m; 6, 12, 18

17

m; 6, 12, 18

18

N—N-(CH₂)₁₀-CH=CH-CH=CH-COONa

19

17/16 copolymerization rate is much faster than that in the homogeneous solution of **17** and **16**, which indicates *in situ* polymerization or fixation in the oriented or ordered structure of the polymerizable double bonds.

The covalent fixation of the porphyrin complexes in the bilayer matrices was confirmed by ^{13}C NMR, GPC, UV–VIS spectroscopy, and elemental analysis.[55]

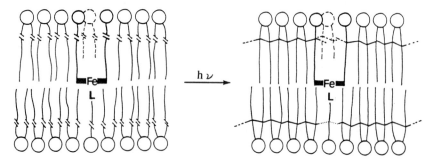

FIG. 3. Copolymerization of phospholipid vesicle-embedded heme, composed of **17, 19,** and **16** (1 : 3 : 50 molar ratio). L, Imidazole derivative.

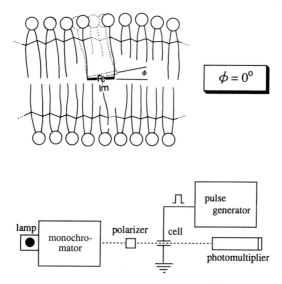

FIG. 4. Orientation of the heme complex in the polylipid bilayer. The transient electric dichromism was measured with the use of a high electric field supplied by an electric field pulse generator with a square-wave pulse with a duration of 1.0 msec. A rotary polarizer was installed between the monochromator and the cell. The electric birefringence of the phospholipid vesicle solution was also measured by the same apparatus by setting an additional analyzer between the cell and the photomultiplier.

The porphyrin plane of the heme complex (**15, 17,** and **18**) is assumed to be oriented nearly parallel to the phospholipid bilayer matrix, which was confirmed by electrooptical measurement (Fig. 4).[56] At first, electric birefringence was measured for the phospholipid vesicles with diameters >100 nm solution; the phospholipid vesicle has a larger reflective index in a direction perpendicular to the electric field, which means that the longer axis of the phospholipid vesicle is aligned parallel to the electric field. Dichromism of the porphinato-iron embedded in the bilayer was monitored under an electric field. The transient absorbance change was much larger when the incident light was polarized parallel with the electric field. The angle for the porphyrin plane of the covalently fixed complex (**17/16**) was also zero, which was not influenced by chemical stimulation such as surfactant addition. From these results, it was concluded that the angle (ϕ) between the porphyrin plane and the phospholipid bilayer is small (Table III). It is assumed that the steric and amphiphilic structure

[56] H. Nishide, M. Yuasa, H. Hashimoto, and E. Tsuchida, *Macromolecules* **20,** 459 (1987).

TABLE III
ORIENTING ANGLE (ϕ) OF HEME PLANE TO
BILAYER MATRIX

Bilayer[a]/heme	ϕ
EYL vesicle/3	19°
EYL vesicle/15	3.9°
EYL vesicle/3 + surfactant	28°
Polylipid vesicle/15	2.6°
Polylipid vesicle/17	1.4°
Polylipid vesicle/17 + surfactant	2.2°

[a] EYL, Egg yolk lecithin; surfactant, Triton X-100.

of the complex, which has four substituent groups built up on the porphyrin plane, keeps the porphyrin plane parallel to the lipid bilayer.

On exposure to oxygen the red and transparent solutions of the macromolecular assembly of the heme complexes with the phospholipids were rapidly and reversibly changed to a brilliant red solution, attributed to the oxygen-coordinated complexes under physiological conditions.

The reduction of the Fe(III) derivative of the heme to the deoxy state [Fe(II)] spontaneously occurred during the polymerization. Complete reduction was confirmed by UV/VIS spectra [λ_{max} = 429, 535, and 562 (sh) nm]. The deoxy solution of the polymerized phospholipid vesicle-embedded 15 (17 or 18) was changed to its oxygen adduct solution on exposure to oxygen (λ_{max} = 422 and 544 nm). The oxygen adduct formation was rapid and reversible, even at remarkably high concentrations of the polylipid vesicle/15, e.g., 20 wt%. The oxygen adduct was slowly degraded to the Fe(III) porphyrin complex and this degradation, i.e., decay of absorbance at 422 nm, obeyed first-order kinetics. The lifetime (half-lifetime) for the reversible oxygen adduct formation was ~1 day under physiological conditions.

The oxygen binding affinity [$P_{1/2}(O_2)$: oxygen pressure at half oxygen binding for the heme] of the polylipid vesicle/15 was determined from the spectral changes at various partial pressures of O_2 using the Hill plot.[48] The O_2 binding affinity of lipid–heme was large enough to apply the Hill plot. An oxygen binding and dissociation equilibrium curve shows that the polylipid vesicle/15 binds oxygen in response to oxygen pressure (Fig. 5). The $P_{1/2}(O_2)$ value of the polylipid vesicle/15 is ~50 Torr at 37° and is close to that of Hb in human blood, but is significantly different from that of Mb. This suggests that the polylipid vesicle/15 has the potential to act as an oxygen carrier under physiological conditions, i.e., to transport

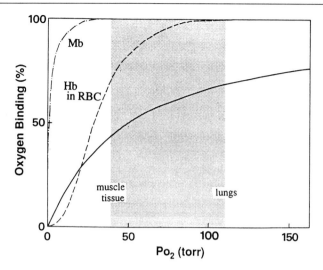

FIG. 5. O_2 binding and dissociation equilibrium curves of the polylipid vesicle/lipid–heme [**15**–1-laurylimidazole–**16** (1 : 3 : 50 molar ratio)] in 30 mM phosphate buffer (pH 7.4, 37°).

oxygen from the lungs [$P(O_2)$; ~110 Torr] to Mb in the tissue [$P(O_2)$; ~40 Torr], as Hb does.

Kinetics of the oxygen binding to the polylipid vesicle-embedded amphiphilic heme were studied by flash photolysis and stopped-flow spectroscopies (Unisoku USP-500 series). Oxygen binding and dissociation of the polylipid vesicle/**15** (**17** or **18**) occurred rapidly and were complete within 1 msec; they are rapid enough to serve as oxygen transporters. Oxygen binding and oxygen dissociation rate constants [$k_{on}(O_2)$, $k_{off}(O_2)$] are summarized in Table IV.

Lipid–Heme/Microsphere

An oil/water (O/W) lipid microsphere emulsified with lipid–heme as surfactant was produced; this gives a red-colored dispersion (lipid–heme/microsphere) and has the ability to bind O_2 reversibly in an aqueous medium (Fig. 6).[57,58]

[57] E. Tsuchida, H. Nishide, T. Komatsu, K. Yamamoto, E. Matsubuchi, and K. Kobayashi, *Biochim. Biophys. Acta* **1108,** 253 (1992).
[58] T. Komatasu, E. Matsubuchi, H. Nishide, and E. Tsuchida, *Chem. Lett.,* p. 1325 (1992).

TABLE IV
OXYGEN BINDING RATE AND EQUILIBRIUM CONSTANTS
FOR PHOSPHOLIPID VESICLE-EMBEDDED HEME

Bilayer/heme	$10^{-4}\,k_{on}\,(M^{-1}\,sec^{-1})$	$k_{off}\,(sec^{-1})$	$10^{-4}\,K\,(M^{-1})$
Polylipid vesicle/**15**	2.3	0.7	1.7
Polylipid vesicle/**17**	1.7	0.8	1.3
DMPC vesicle/**15**	3.7	2.2	1.6
DMPC vesicle/**3**	0.8	0.3	2.5
Hb (R state)[a]	3300	13.1	252
Mb[b]	1000–2000	10–30	33–200

[a] At 20°, pH 7.0. Q. H. Gibson, *J. Biol. Chem.* **245**, 3285 (1970); J. S. Olson, M. E. Anderson, and G. H. Gibson, *ibid.* **246**, 5919 (1971); V. S. Sharma, M. R. Schmidt, and H. M. Ranney, *ibid.* **251**, 4267 (1976).

[b] At 20°, pH 7.0. M. Brunori and T. M. Schuster, *J. Biol. Chem.* **244**, 4046 (1969); F. Antonini and M. Brunori, "Hemoglobin and Myoglobin and their Reactions with Ligands," p. 220. North-Holland Publ., Amsterdam, 1971.

The lipid–heme/microsphere is prepared as follows: **15** and 1-stearyl-imidazole, which is a lipophilic alkyl imidazole derivative with a good compatibility with lipid molecules, are dissolved in triglyceride (TG) [1 : 0.2 : 2~4 (wt ratio)]. Phosphate buffer (pH 7.4, 30 mM, 10 ml) is then added to the mixture. The solution is homogenized with an ultrasonic generator under nitrogen, to give a red-colored O/W emulsion ([heme] = 0.01–30 mM).

Under the scanning and/or transmission electron microscopes, the lipid–heme/microspheres look like spherical particles with diameters of 100–150 nm. The average particle size of the lipid–heme/microsphere was also measured by a dynamic light-scattering method, using a submicron particle analyzer; the diameter of the particle was 110 ± 32 nm.

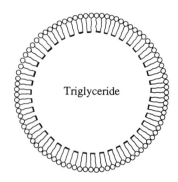

FIG. 6. Lipid–heme/microsphere.

The lipid–heme was reduced to Fe(II) by the addition of a small excess of aqueous ascorbic acid under a nitrogen atmosphere. The solution of the lipid–heme/microsphere was stable and could be stocked for a few months at 25° without precipitation or change of particle size. The viscosity of this emulsion ([heme] = 10 mM) was determined to be 1.2 cP, which was much lower than that of human blood (4.5–5.0 cP).[58]

The most important feature of the lipid–heme/microsphere is that it solubilizes the heme at high concentrations in aqueous medium. That is, the lipid–heme/microsphere suspension (the concentration of TG is 10 wt%) is able to bind O_2 gas up to 23 ml/100 ml medium, which is equal to that of human blood. Furthermore, the lipid–heme/microsphere dispersion is also stable in physiological saline solution or blood plasma and gives a stable O_2 adduct with a $P_{1/2}(O_2)$ of 41 Torr at 37°.

Heme Assembly

The phospholipid vesicle-embedded heme was the first successful example of a totally synthetic O_2 transporting system under physiological conditions. However, less attention has been paid to the molecular assemblies constituted by amphiphilic metalloporphyrins as oxygen carriers in aqueous solution. The authors recently have found that an amphiphilic tetraphenylporphine derivative having four alkyl phosphocholine groups coupled on both sides of the ring plane, 5,10,15,20-tetrakis[2,6-bis(3,3-dimethyl-4-[1-([(2-trimethylammonio)ethoxy]phosphonatoxy)dodecanoxy-carbonyl]butyroyloxy)phenyl]porphinato-iron [octopus heme (20)] forms some self-organized aggregates in aqueous medium.[59]

The synthetic route for 20 is as follows. 1,12-Dodecanediol is monotri-tyrated with tritylchloride and reacted with 3,3-dimethylglutaric acid anhydride, to give the tritylated alkylacid (24%). This compound is allowed to couple with 5,10,15,20-tetrakis[2,6-bis(hydroxy)phenyl]porphine by using N,N-dicyclohexylcarbodiimide and 4-(N,N-dimethylamino)pyridine in dry tetrahydrofuran. The mixture is purified by column chromatography [silica gel; chloroform:acetone 100:1 (v/v)] to yield 8-alkylsubstituted porphyrin (56%). This porphyrin is detritylated with BF_3CH_3OH in dry dichloromethane at room temperature for 1 hr. The reactant is separated by column chromatography [silica gel; chloroform:methanol = 15:1 (v/v)] to afford 5,10,15,20-tetrakis[2,6-bis(3,3-dimethyl(16-hydroxy-4-dodecan-oxycarbonyl)butyroyloxy)phenyl]porphine (75%). Insertion of iron is ac-

[59] T. Komatsu, K. Nakao, H. Nishide, and E. Tsuchida, *J. Chem. Soc., Chem. Commun.* **728** (1993).

complished by using the FeBr$_2$ method and refluxing in dry tetrahydrofuran (85%). The heme is phosphorylated and then the resultant phosphate triesters are cleaved by anhydrous trimethylamine to afford **20** (65%).

20

Octopus heme is dispersed easily in deionized water by vortex mixing ([heme] = 0.1 mM), yielding a transparent red solution. The homogeneous dispersion does not change for several months or longer. The critical micelle concentration (cmc) of octopus heme is ~ 1 μM, as estimated by the Wilhelmy method.[60] Above this concentration the aggregate morphology is clearly elucidated by electron microscopy. The octopus heme in dilute aqueous solution forms the fibrous aggregates.

The morphology of the octopus heme assembly is transformed into totally different structures by adding alkylimidazole and/or phospholipid derivatives. Aggregates of **20** are dispersed with 20-fold molar excess of 1-lauryl-2-methylimidazole by vortex mixing in phosphate buffer (1 mM, pH 7.4), to give spherical vesicles with a diameter of ~ 100 nm. The vesicles constituted by **20**/1-lauryl-2-methylimidazole [1 : 20 (molar ratio)] have the ability to bind oxygen reversibly in an aqueous medium. The molecular assembly composed of complex **20** can mimic the O_2 binding property of Hb.

See Addendum on p. 685.

Approach to Artificial Red Cells

The oxygen-delivering ability of our oxygen carrier was demonstrated *in vivo* by using an exchange transfusion with the polylipid vesicle-embedded lipid–heme or lipid–heme/microsphere suspension to beagles weighing about 8 kg. Under control ventilation, 30 ml/kg (240 ml) of blood was withdrawn from the beagles.[63,64] Then the same amount of the lipid–heme solution was given intravenously. Cardiac output, Hb, lipid–heme concentrations, and blood gases were measured to calculate the amount of oxygen delivered by the lipid–heme *in vivo*.

In the bloodstream, oxygen is mainly shared between the red blood cells and the lipid–heme. The efficiency of each potential oxygen transporter is evaluated by its individual contribution to the total oxygen delivery. First, the withdrawn blood is centrifuged. Red blood cells are separated and the supernatant contains the lipid–heme system. There were no interactions of the lipid–heme system with any blood components. Spectroscopic analysis indicates that the lipid–heme delivers oxygen in the mixture system. The estimated heme concentration in the plasma layer indicates the circulation half-lifetime of the lipid–heme microparticle in the blood stream to be 12 hr. The oxygen consumption (V_{O_2}) by the lipid–heme system was calculated to be 15 ml/min, which is about 20% of the total oxygen consumption.

[60] L. Wilhelmy, *Ann. Phys.* **119**, 177 (1863).
[63] M. Watanabe, K. Kobayashi, T. Ishihara, M. Fukuzumi, H. Nishide, and E. Tsuchida, *Artif. Organs* **14**, 213 (1990).
[64] E. Tsuchida, T. Komatsu, N. Kawai, H. Nishide, T. Kakizaki, and K. Kobayashi, *Artif. Organs Today* **3**, 137 (1993).

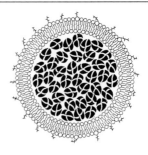

FIG. 9. Stabilized phospholipid vesicle-encapsulated Hb. (a) Polyphospholipid; (b) glycolipid.

As a model of the red blood cell, Hb separated from the red blood cells has been artificially encapsulated with synthetic polymers or phospholipid vesicles. However, vesicle-encapsulated Hb has the following disadvantages: instability of vesicles during freezer storage; aggregation and fusion *in vitro* and probably *in vivo*; leakage of Hb; and clearance of vesicles in the reticuloendothelial system.

In the attempt to prepare a vesicle-encapsulated Hb as an oxygen transporter, our approaches are to stabilize the bilayer membrane by using polymerized phospholipids (**16**) or by introducing a synthetic glycolipid having an oligosaccharide chain.[65] The drastically stabilized vesicles and the excellent performances of vesicle-encapsulated Hb (Fig. 9) are described briefly.

Hb was purified and concentrated to more than 40 g/dl.[66] During the preparation Hb was protected as HbCO in order to prevent metHb formation and/or denaturation of Hb (metHb content was less than 1.0% of total Hb). The Hb vesicle which encapsulated the purified and concentrated Hb with a uni- or bilamellar membrane are prepared by extruding the dispersion of mixed lipid through a membrane filter (final pore size is 0.2 μm).[67] They transport large amounts of oxygen with satisfying rheological properties such as oncotic pressure and solution viscosity. Oxygen-binding affinity [$P_{1/2}(O_2)$] of the vesicle-encapsulated Hb is adjusted so as to exceed

[65] E. Tsuchida, *Biomater., Artif. Cells, Immobilization Biotechnol.* **22,** in press (1994).
[66] H. Sakai, S. Takeoka, Y. Seino, H. Nishide, and E. Tsuchida, *Protein Exp. Purif.* **4,** 563 (1993).
[67] S. Takeoka, K. Terase, H. Yokohama, H. Sakai, H. Nishide, and E. Tsuchida, *J. Macromol. Sci. Pure Appl. Chem.* **31,** 97 (1994).
[68] E. Tsuchida, *Biomater., Artif. Cells, Immobilization Biotechnol.* **20,** 337 (1992).

the ability of oxygen transport of human blood by coencapsulating allosteric effectors in the vesicle. The solution is sterilizable because the diameter of the vesicle-encapsulated Hb is less than 0.2 μm. No change in oxygen-binding affinity and particle size was confirmed during long time storage at 4°.[65,68] The stabilized vesicle-encapsulated Hb can also be stored in a frozen or dried state. The dried vesicle-encapsulated Hb is regenerated by simply adding pure water.

Apart from the artificial red cell, we believe the porphyrin complexes have various potential applications. For example, they will be developed for isolating pure oxygen from air; that is, oxygen-enriching systems containing an oxygen-coordinated complex.[69,70] Other applications include their use as catalysts for superoxide removal in medical systems, selective oxygenation of organic compounds in drug metabolism, photodynamic cancer therapy, and so on.

The combination of a porphinato-metal with a lipid assembly offers possibilities unattainable in other ways. Electron transfer or tunneling via the porphinato-metal complexes[71] and a photochemical hole burning system with highly concentrated memory capacity using the porphyrin moiety[72] have been reported in preliminary studies, and have the potential to be applied as new devices at the molecular level.

[69] E. Tsuchida, H. Nishide, H. Kawakami, and M. Ohyanagi, *J. Phys. Chem.* **92**, 6461 (1988).
[70] H. Nishide, H. Kawakami, T. Suzuki, Y. Azechi, Y. Soejima, and E. Tsuchida, *Macromolecules* **24**, 6306 (1991).
[71] E. Tsuchida, H. Nishide, and M. Kaneko, *J. Phys. Chem.* **90**, 2283 (1986).
[72] E. Tsuchida, H. Ohno, M. Nishikawa, H. Hirooka, and K. Arishima, *J. Phys. Chem.* **92**, 4255 (1988).

[12] Semisynthesis of Hemoglobin

By Rajendra Prasad Roy and A. Seetharama Acharya

This chapter is devoted to the description of chemical approaches of protein engineering that are amenable to the generation of mutants of hemoglobin (Hb). The need and significance of the chemical approaches of protein engineering in this era of oligonucleotide-directed site-specific mutagenesis also have been emphasized. The discussions are limited to aspects relevant to Hb chemistry, and any reader interested in the general principles of some of the approaches discussed here and studies with other proteins is advised to refer to the reviews by Offord.[1-4]

Chemical Synthesis of Proteins

The recombinant methods of protein engineering, although efficient and most of the time quite simple, are limited in their ability to choose from a repertoire of 20 natural amino acids. The need for a higher flexibility, particularly for the introduction of the noncoded amino acids at a given site, is apparent in the recent attempts to develop *in vitro* translation systems. The chemical synthetic approach to systematic variations of protein structure, in principle, is a much more general methodology than is site-directed mutagenesis. It offers a greater flexibility for the site-specific variation of the protein structure: incorporation of noncoded amino acids or introduction of amino acids with spectroscopic probes (NMR, EPR, or fluorophores) or photoactive functional groups, introduction of unnatural bond functionalities in a site-specific fashion, and the introduction of fixed elements of three-dimensional structure are all possible by the chemical approach. However, the total chemical synthesis of a large protein represents more of a challenge. Nonetheless, the pioneering studies of Gutte and Merrifield[5] involving the total chemical synthesis of ribonuclease A (RNase A) has established the feasibility of this approach for the synthesis of proteins. The chemical synthesis of human immunodeficiency virus (HIV) protease[6] and many of its molecular variants demon-

[1] R. E. Offord, *in* "Protein Design and Development of New Therapeutics and Vaccines" (J. B. Hook and G. Poste, eds.), p. 253. Plenum, New York, 1990.
[2] R. E. Offord, "Semisynthetic Proteins." Wiley, Chichester and New York, 1980.
[3] R. E. Offord, *Protein Eng.* 1, 271 (1987).
[4] R. E. Offord, *Protein Eng.* 4, 709 (1991).
[5] B. Gutte and R. B. Merrifield, *J. Am. Chem. Soc.* 91, 501 (1969).
[6] M. Schnolzer and S. B. H. Kent, *Science* 256, 221 (1992).

strates that such an approach indeed provides a viable route for the preparation of the molecular variants. Thus, at this stage the total chemical synthesis is certainly an accessible route for the preparation of variants of medium-size proteins.

Semisynthesis of Proteins

Semisynthesis of proteins is a powerful alternative to total chemical synthesis, in that it provides the same high flexibility of total chemical synthesis to a selected segment of the protein. Semisynthesis, in principle, involves the chemical synthesis of a polypeptide segment of a protein and combines it with the complementary fragment(s) derived from the native protein to generate a functional protein.

Noncovalent Semisynthesis

Semisynthetic studies can be classified broadly as noncovalent and covalent semisyntheses. Noncovalent semisynthesis, as the name implies, involves only the noncovalent interaction of the synthetic segment with the complementary segment from the native protein in generating the active protein. Accordingly, noncovalent semisynthesis is possible only for protein systems for which a fragment complementation has been established. The preparation of RNase S by the subtilisin-mediated[7] limited proteolysis of RNase A represents the first example of the development of such a protein fragment-complementing system. The two polypeptide segments of RNase S are dissociated under acidic conditions and have been isolated in pure form. The isolated fragments alone are inactive. However, when mixed together under physiological conditions, the complementary segments reassociate to generate an enzymatically active complex. Thus, by chemically synthesizing a mutant peptide component and mixing it with the complementary fragment from the wild type, a semisynthetic molecular variant of the fragment-complementing system can be generated.[8-10]

The generation of a fragment-complementing system of proteins by limited proteolysis is the first step in the development of noncovalent semisynthetic approaches of proteins, and this has been established for

---

[7] F. M. Richards and P. J. Vithayathil, *J. Biol. Chem.* **234**, 1459 (1959).

[8] K. Hofmann, J. P. Visser, and F. M. Finn, *J. Am. Chem. Soc.* **91**, 4883 (1969).

[9] F. Scoffone, F. Marchiori, R. Rocchi, G. Vidal, A. Tamburro, A. Scatturin, and A. Marzotto, *Tetrahedron Lett.* **9**, 943 (1966).

[10] M. Pandian, E. A. Padlan, C. Dibello, and I. M. Chaiken, *Proc. Natl. Acad. Sci. U.S.A.* **73**, 1844 (1976).

a number of proteins.[11-13] The region wherein the contiguity of the polypeptide can be interrupted without serious consequences to the three-dimensional structure of the protein is referred to as the "permissible discontinuity region." The existence of permissible discontinuity regions has now been demonstrated in a number of proteins, and this appears to be a general phenomenon. However, the lack of unique procedures (or algorithms) to identify such sites in proteins represents a major limitation to the widespread application of noncovalent semisynthesis in the structure–function relationship studies of proteins.

Covalent Semisynthesis

The noncovalent fragment-complementing system can be converted to the respective covalent forms by chemical (nonenzymatic) and/or enzymatic methods. Religation of the noncovalent complex of cytochrome *c* fragments generated by CNBr is an example of the chemical (nonenzymatic) approach wherein a complex is transformed to the covalent form. Condensation of the unprotected fragments, namely, segments 1–65 and 66–104 of cytochrome *c*, is facilitated by stereochemical orientation of the respective groups (homoserine lactone at position 65 of the cytochrome$_{1-65}$ and the α-amino group of segment 66–104) as a consequence of the complementation of the two fragments to generate a nativelike structure.[14] Covalent semisynthesis of proteins by the enzymatic approach uses the propensity of proteases to work in the reverse direction, namely, the reformation of the peptide bond at the permissible discontinuity site. The subtilisin-catalyzed synthesis of RNase A from RNase S represents the first example of splicing of the discontinuity sites of a protein, and since then protease-mediated semisynthesis has been demonstrated in a number of fragment-complementing systems.[15-18]

It therefore appears reasonable to assume that religation of a peptide bond is possible if the permissible discontinuity site of a protein can be established. The essential structural elements that contribute to the facile

[11] H. Taniuchi, C. B. Anfinsen, and A. Sodja, *Proc. Natl. Acad. Sci. U.S.A.* **58,** 1235 (1967).
[12] D. E. Harris and R. E. Offord, *Biochem. J.* **161,** 21 (1977).
[13] H. Hagenmaier, J. P. Ohms, J. Jahn, and C. B. Anfinsen, *in* "Semisynthetic Peptides and Proteins" (R. E. Offord and C. Di Bello, eds.), p. 23. Academic Press, London, 1978.
[14] C. J. Wallace, P. Mascagni, B. T. Chait, J. F. Collown, Y. Paterson, A. E. Proudfoot, and S. B. H. Kent, *J. Biol. Chem.* **255,** 15199 (1989).
[15] G. A. Homandberg and M. Laskowski, Jr., *Biochemistry* **17,** 5220 (1978).
[16] L. Graf and C. H. Li, *Proc. Natl. Acad. Sci. U.S.A.* **78,** 6135 (1981).
[17] M. Jullierat and A. Homendberg, *Int. J. Pept. Protein Res.* **18,** 335 (1981).
[18] A. Komoriya, G. A. Homendberg, and I. M. Chaiken, *Int. J. Pept. Protein Res.* **16,** 433 (1980).

protease-catalyzed splicing of the discontinuity site appear to be the appropriate stereochemistry of the α-amino and the α-carboxyl groups of the discontinuous region. The stereochemistry at the discontinuity site is afforded by the nativelike conformation of the fragment-complementing systems. Protonation of the α-carboxyl group in the presence of the organic cosolvent is the chemical aspect of the splicing reaction that facilitates the ligation reaction. The propensity of the proteases to work in the reverse direction, i.e., synthetic activity in the presence of the organic solvents (propanediol, butanediol, glycerol, etc.), has been demonstrated using a number of di- and tripeptides. In this chapter we describe the α-globin semisynthetic reaction developed in our laboratory,[19,20] which involves the V8 protease-catalyzed splicing of the complementary segments of α-globin.

Our semisynthetic efforts with the α and β chains of Hb S were initially undertaken as an attempt to translate the original investigations of Homandberg and Laskowski[15] with the RNase S system to the human hemoglobin system. Although an α-globin semisynthetic reaction has been developed, contrary to our initial assumptions, the mechanistic aspects that facilitated this semisynthetic reaction have turned out to be novel and distinct from those of the previously described protease-catalyzed protein semisynthetic reactions.[21,22] However, space does not permit us to discuss the mechanistic aspects in detail here. The major emphasis of this chapter is to describe the methodologies of the α-globin semisynthetic reaction and the reconstitution of the semisynthetic α-globin to semisynthetic Hb.

α-Globin Semisynthetic Reaction

The α-globin semisynthetic reaction is represented schematically in Fig. 1. The Glu^{30}-Arg^{31} peptide bond of the α chain of hemoglobin, which is readily hydrolyzed by V8 protease, has been identified as the permissible discontinuity site.[23] The complementary fragments of α-globin, namely, α^{1-30} and α^{31-141}, can be spliced by the same protease. It may be noted here that this Glu-Arg peptide bond is proximal to the junction of the translational products of exon 1 and exon 2 of the α-globin gene. This is also presented schematically in Fig. 1. It has been suggested that the

[19] A. S. Acharya, Y. J. Cho, and K. S. Iyer, *Prog. Clin. Biol. Res.* **240,** 3 (1987).
[20] R. Seetharam and A. S. Acharya, *J. Cell. Biochem.* **30,** 87 (1986).
[21] K. S. Iyer and A. S. Acharya, *Proc. Natl. Acad. Sci. U.S.A.* **84,** 7014 (1987).
[22] R. P. Roy, K. M. Khandke, B. N. Manjula, and A. S. Acharya, *Biochemistry* **31,** 7249 (1992).
[23] R. Seetharam, A. Dean, K. S. Iyer, and A. S. Acharya, *Biochemistry* **25,** 5949 (1986).

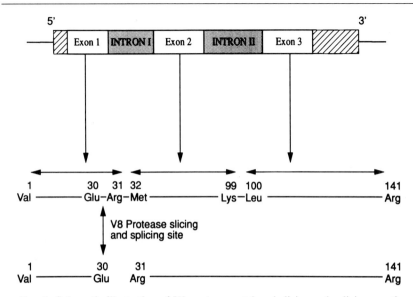

Fig. 1. Schematic illustration of V8 protease-catalyzed slicing and splicing reaction of the α chain at its Glu^{30}-Arg^{31} peptide bond. The permissible discontinuity site is proximal to the junction of the translational products of exon 1 and exon 2 of the α-globin gene.

junctions of translation products of exons in proteins represent regions that are readily accessible for limited proteolysis. Further studies with many proteins will be necessary to establish whether a cross correlation exists between the accessibility of proteins to proteases at sites that represent the junction of the translational products of their exons, and if so, whether the splicing at the same site by the protease in the presence of the organic solvent is a general one.

V8 protease cleaves α-globin selectively and quantitatively at the Glu^{30}-Arg^{31} peptide bond at pH 4.0 and 37° to generate α^{1-30} and α^{31-141}. The same protease can catalyze the splicing of the complementary fragments in a selective and stereospecific fashion to generate intact α^{1-141} (α-globin) in the presence of 30% n-propanol at pH 6.0 and 4°. Thus, this splicing reaction provides a simple methodology to splice a synthetic α^{1-30} (with coded or noncoded amino acids and/or containing NMR- or EPR-sensitive nuclei) to α^{31-141}, yielding the desired semisynthetic α-globin. The mutant α-globin can then be reconstituted with β chain and heme to generate the functional tetramer.[24]

[24] G. Sahni, Y. J. Cho, K. S. Iyer, S. A. Khan, R. Seetharam, and A. S. Acharya, *Biochemistry* **28,** 5456 (1989).

The various steps involved in the semisynthesis of Hb have been classified broadly into five groups for simplicity of the discussion: (1) preparation of the α and β chains, (2) preparation of α^{1-30} and/or α^{31-141} by V8 protease cleavage of α-globin and separation of the complementary fragments by appropriate chromatographic procedures, (3) preparation of mutant α^{1-30} and/or mutant α^{31-141}, (4) V8 protease-catalyzed splicing of mutant α^{1-30} and α^{31-141} and isolation of homogeneous semisynthetic mutant α-globin, and (5) reconstitution of the semisynthetic mutant α-globin with β chain and heme to a tetramer. The first four steps deal with the semisynthesis of α-globin and are presented schematically in Fig. 2.

Semisynthesis of α-Globin

Preparation of α and β Chains of Hb

Hb A, Hb S, and their hydroxymercuribenzoate (HMB)-reacted chains are prepared as described elsewhere in this series.[25] The heme-free globins are obtained by acid–acetone precipitation of the respective chains. Alternatively, the separation of α- and β-globins can also be accomplished by CM-52 chromatography of acid–acetone-precipitated globin in the presence of 8 M urea.[25,26] The total globin ($\alpha + \beta$) obtained from the acid–acetone precipitation of Hb is the starting material for this chromatography. The starting buffer is 5 mM in disodium phosphate, 8 M in urea, and 50 mM in 2-mercaptoethanol, the final pH of which is adjusted to 6.7 with dilute phosphoric acid. The lyophilized sample of the acid–acetone-precipitated globin is dissolved in this buffer at a concentration of 8–10 mg/ml and dialyzed against 50 volumes of the same starting buffer for 2 to 3 hr. Meanwhile, an appropriate amount of preswollen CM-52 (Whatman, Clifton, NJ) is suspended in the starting buffer with occasional mild stirring and then allowed to settle. After 30 min, the buffer at the top is decanted off and fresh buffer is added to make up the original volume. This process is repeated three times. The slurry of the equilibrated resin is packed into a 3 × 10 cm column. The column is equilibrated further with the starting buffer at a flow rate of 45 ml/hr (equivalent to 3 bed volumes). The dialyzed globin sample (100–150 mg of the mixture of α- and β-globin) is loaded onto the equilibrated column. The column is washed with the starting buffer (equivalent to 1 bed volume) to remove any material that does not bind to the column. The chromatographic separation of the α- and β-globins is carried out using a linear gradient of 200 ml each of the starting

[25] E. Bucci, this series, Vol. 76, p. 97.
[26] J. B. Clegg, M. A. Naughton, and D. J. Weatherall, *J. Mol. Biol.* **19**, 91 (1966).

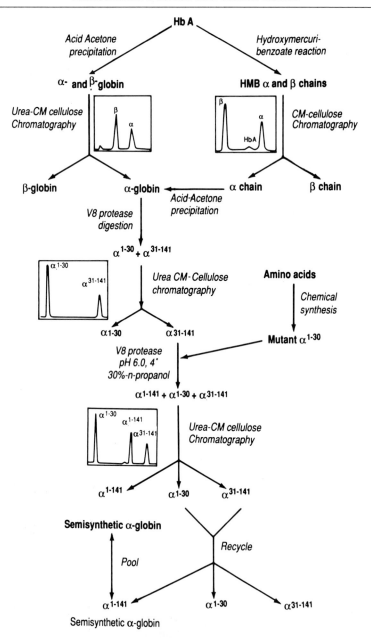

FIG. 2. Protease-catalyzed semisynthesis of α-globin. The various steps involved in the semisynthesis of α-globin are presented as a flow chart. The chromatograms shown in the boxes are not exactly to scale in terms of the positions of the components and are presented merely to provide a general reference.

buffer and 30 mM disodium phosphate containing 8 M urea and 50 mM 2-mercaptoethanol, the pH of which is adjusted to 6.7 with dilute phosphoric acid. The β-globin elutes first from the column. The α-globin elutes almost at the end of the gradient. The fractions containing the two globin samples are pooled separately and dialyzed extensively against 0.1% acetic acid. The dialyzed samples are lyophilized and stored at −20°.

The urea solution should be made fresh before use and, if necessary, should be freed from cyanate by passing the solution through a mixed-bed resin. The pH of the buffers should be adjusted with dilute H_3PO_4 after the addition of urea and 2-mercaptoethanol. The urea–CM-52 chromatography can be used over a wide range of pH and salt concentrations. In our hands, the procedure has worked well for the separation of globins of human and mouse Hb at pH 6.7 and horse Hb at pH 7.0. The same procedure is also useful for the separation of the semisynthetic α-globin and respective α-globin components (described later in this chapter).

*Preparation of Complementary Fragments of α-Globin
by V8 Protease Digestion*

The complementary fragments of α globin needed for the semisynthetic reaction can be prepared by the slicing of either α chain or α-globin. The high selectivity of the digestion at the Glu[30]-Arg[31] peptide bond of the polypeptide chain is seen with both of the forms of the polypeptide chain. Presumably, the presence of heme has very little influence on the conformation of this region of the molecule at pH 4.0. The protein is taken in 10 mM ammonium acetate buffer (pH 4.0) at a concentration of 0.5 mg/ml and digested with V8 protease (1 : 200, w/w) at 37° for 3 hr, by which time the cleavage at the Glu[30]-Arg[31] peptide bond is complete, as observed with the human, horse, and the mouse α-globins.

The specificity of V8 protease cleavage at the Glu[30]-Arg[31] bond at pH 4.0 is very high. This is reflected by the fact that very little cleavage is observed at other peptide bonds of the Glu residues of the polypeptide chain, even when the digestion is continued for 24 hr. This is in contrast to the digestion pattern at pH 7.8, where the cleavage at the peptide bonds of Glu[23] and Glu[27] also occurs on prolonged incubations. In spite of this established high selectivity for the Glu[30]-Arg[31] peptide bond at pH 4.0, it is recommended that investigators follow the kinetics of the digestion of the polypeptide chain when a new α chain (or globin) is subjected to proteolysis with V8 protease. This can be carried out conveniently by analytical reversed-phase high-performance liquid chromatography (RP-HPLC) of the digestion mixture. A typical RP-HPLC profile of a V8 protease digest of human α-globin is shown in Fig. 3. Usually, 100 μl of

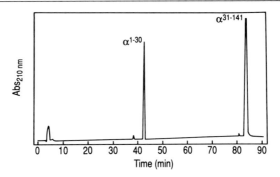

FIG. 3. RP-HPLC of a V8 protease digest of α-globin. The globin digestion was carried out in 10 mM ammonium acetate at pH 4.0 and 37° for 3 hr. An enzyme:substrate ratio of 1:200 (w/w) was used for these digestions; 100 μl (about 50 μg) of the digest was diluted to 1 ml with 0.1% TFA and loaded directly onto an Aquapore RP-300 column (4.6 × 250 mm). A gradient of 5–70% acetonitrile containing 0.1% TFA over a period of 130 min was employed to elute the components. The flow rate was maintained at 1 ml/min throughout the run.

the digestion mixture (containing approximately 50 μg of the protein) is diluted to 1 ml with 0.1% trifluoroacetic acid (TFA) and loaded onto a Vydac C_4 or C_8, RP-300 (4.6 × 250 mm) RP-HPLC column equilibrated with 5% acetonitrile containing 0.1% TFA. The peptides are eluted with a gradient of 5–70% acetonitrile (flow rate, 1 ml/min) in 130 min. The concentration of TFA is maintained at 0.1% throughout the run. The elution pattern is monitored by the absorbance of the effluent at 210 nm. Only two components eluting around 42 and 82 min are seen in this chromatogram. The component that elutes around 42 min is α^{1-30}. Under these chromatographic conditions, the separation of the intact α-globin (α^{1-141}) and α^{31-141} is not achieved. An empirical but reliable method to ascertain the completion of the digestion is to compare the peak area (monitored at 210 nm) of the two components. The ratio of the peak areas of α^{31-141} and α^{1-30} is generally 3.55 to 3.6 on completion of the digestion. However, the absolute purity of the α^{31-141} should be established by tryptic peptide mapping.

The large-scale preparation of α^{1-30} and α^{31-141} from the V8 protease digest of α-globin is carried out conveniently by urea–CM-52 chromatography. Urea–CM-cellulose chromatography was described earlier for the separation of the α- and β-globin from acid–acetone-precipitated Hb. However, a slight modification of the elution system is recommended for the separation of the complementary fragments of α-globin. The lyophilized digest is dissolved in the starting buffer (the same as was used for the separation of globins) and loaded onto the column (without the dialysis

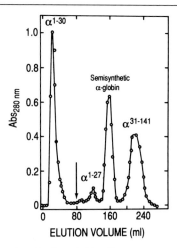

FIG. 4. Purification of semisynthetic α-globin. The urea–CM-52 chromatographic pattern of the semisynthetic reaction mixture of the complementary fragments of α-globin is shown. The lyophilized synthetic reaction mixture was taken in the starting buffer (see text for details) at a concentration of 10 mg/ml and loaded onto the column of CM-52 equilibrated with 5 mM disodium phosphate buffer, pH 6.7, containing 8 M urea and 50 mM 2-mercaptoethanol. The column was washed with the same buffer to elute α^{1-30}. The arrow indicates the position where the gradient elution was started. The gradient was generated from 100 ml each of 25 and 50 mM disodium phosphate buffer, pH 6.7, both buffers containing 8 M urea and 50 mM 2-mercaptoethanol. The column was run at a flow rate of 45 ml/hr and 2-ml fractions were collected. The protein elution was monitored by measuring the absorption of the fractions at 280 nm. The peak marked α^{1-27} represents the small amount of V8 protease-catalyzed hydrolysis of α^{1-30} that occurs during the semisynthetic reaction. The α^{27} is a Glu residue and is consistent with the specificity of V8 protease.

step, because α^{1-30} will escape from the dialysis bag). The α^{1-30} elutes from the column unadsorbed in the wash with the starting buffer. After the elution of α^{1-30}, the column is eluted with a linear gradient generated using 100 ml each of 25 mM and 50 mM disodium phosphate buffer, pH 6.7; both buffers contain 8 M urea and 50 mM 2-mercaptoethanol. The α^{31-141} elutes from the column at a position distinct from that of intact α-globin (see Fig. 4 for details). The fractions containing the peptide α^{1-30} are pooled and freed from the salt using a Sephadex G-25 column (2 × 40 cm) equilibrated and eluted with 0.1% TFA. The fractions containing α^{31-141} are pooled and dialyzed extensively against 0.1% acetic acid. The two complementary segments of α-globin are lyophilized separately and stored at $-20°$. This procedure has been employed successfully by us for the bulk preparation of α^{1-30} and α^{31-141} derived from human and mouse α-globins.

Preparation of Mutant α^{1-30} and/or α^{31-141}

The high selectivity of V8 protease digestion at the Glu[30]-Arg[31] peptide bond of α-globin or the α chain is apparently an unusual feature of the polypeptide chain at pH 4.0. Thus, one could assume that the various mutant human α chains (or globins) will retain this high selectivity unless the mutation introduced perturbs the local conformation of the chain in this region. We have carried out cleavage experiments with two mutant human α chains.[27] The α chain of Hb I contains the α16 Lys → Glu mutation. Even in the presence of this additional Glu residue, the V8 protease digestion was restricted to the Glu-30 site. The α chain of Hb Sealy containing the α47 Asp → His mutation was also sliced at Glu-30 at pH 4.0 in a fashion comparable with that of the wild-type α chain.

The segment α^{1-30} with desired mutations can be synthesized readily using an automated peptide synthesizer. This route is indispensable for introducing noncoded amino acids at a preselected site in this segment. Isosteric substitutions, for example, norvaline in place of valine or norleucine in place of either leucine or isoleucine, could be accomplished readily by chemical synthesis. The total chemical synthesis of α^{31-141} is also possible. However, this may not turn out to be an enjoyable experience. Nonetheless, the segment condensation approaches (chemical or enzymatic) are still viable toward this objective. Development of a new enzyme-catalyzed segment condensation to generate α^{31-141} from two shorter segments, for example, a clostripain-catalyzed condensation of α^{31-92} and α^{93-141}, may provide an alternative. The segment condensation using the oxime resin developed by Kaiser *et al.*[28] also may be attempted. The evaluation of such procedures for the preparation of mutant α^{31-141} is currently in progress in our laboratory.

The major strength of the semisynthetic approach is to introduce non-coded amino acids or amino acids with NMR-sensitive nuclei or a photoactivable group in their side chain in a site-specific fashion. The chemical synthesis of α^{1-30} is certainly one of the best approaches in these cases. Chemical synthesis also could be a viable and possibly simpler method of choice when multiple mutations are desired in this short segment.

The α^{1-30} with multiple mutations also can be prepared from nonhuman α chains that exhibit multiple sequence differences as compared with human α chain. Glu-30 is conserved in most of the nonhuman mammalian α chains. The unusual high selectivity of the Glu[30]-Arg[31] peptide bond is conserved in a number of these α chains. The nonhuman α chains that

[27] G. Sahni, A. K. Mallia, and A. S. Acharya, *Anal. Biochem.* **193**, 178 (1991).
[28] E. T. Kaiser, H. Mihara, G. A. Laforet, J. W. Kelly, L. Walton, M. A. Findeis, and T. Sasaki, *Science* **243**, 187 (1989).

we have tried to date include the ones from monkey, horse, and mouse Hb. These α chains exhibit 4, 18, and 19 sequence differences, respectively.[29-31] For the separation of the mutant α^{1-30} and mutant α^{31-141}, urea–CM-52 chromatography has been found useful just as in the case of the isolation of the complementary segments of human α-globin.

Semisynthesis of α-Globin

Splicing of the complementary segments of α-globin, α^{1-30}, and α^{31-141} (with mutations in one or both the segments) is catalyzed by V8 protease. The splicing reaction is carried out routinely in 50 mM ammonium acetate, pH 6.0, containing 30% n-propanol. However, it should be noted here that this splicing reaction occurs over a wide pH range. The synthetic reaction[24] proceeds at a reasonable rate in the acidic pH region above 4.5, as well as at an alkaline pH up to 8.5. The reaction also shows a considerable degree of flexibility with reference to the concentration of the organic solvent. Although 30% n-propanol was found optimal for the synthetic reaction, an increase in the concentration of the organic solvent to about 60% was not detrimental to the splicing reaction. Thus, a variety of combinations of the concentration of the organic cosolvent and pH can be tried if a complementary fragment containing a mutated residue(s) is found to be insoluble under a given set of conditions. The concentration of the ligating complementary fragments is generally maintained around 1 mM. The reaction appears to proceed with nearly the same overall efficiency even at a concentration of 2 mM of the complementary fragments. However, the splicing reaction is observed at very low levels around a concentration of 0.1 mM.

For a typical splicing reaction the complementary fragments, α^{1-30} and α^{31-141} are mixed at a molar ratio of 1 : 1.2 (carboxyl to amino component) and lyophilized. This lyophilized material is dissolved in an appropriate amount of 100 mM ammonium acetate buffer of desired pH (generally pH 6.0). To this solution n-propanol and water (if needed) are added to get a final concentration of 30% n-propanol and 50 mM ammonium acetate. The final concentration of the complementary fragments is nearly 1 mM. The solution is cooled to 4° in an ice bath for 30 min. An aliquot of V8 protease is added so that the ratio of the enzyme to protein is 1 : 200 (w/w), and the splicing reaction is allowed to proceed at 4° for 48 hr.

The progress of the splicing reaction can be monitored readily by RP-HPLC. This procedure is similar to the one adopted for the analysis of

[29] G. Matsuda, T. Maita, N. Igawa, H. Ota, and T. Miyauchi, *Int. J. Protein Res.* **2**, 13 (1970).
[30] R. C. Ladner, G. M. Air, and J. H. Fogg, *J. Mol. Biol.* **103**, 675 (1976).
[31] R. A. Popp, *J. Mol. Biol.* **27**, 9 (1967).

the slicing reaction of α-globin. The quantitation of synthesis is done by computing the decrease in area of α^{1-30} (42-min peak) and cross-correlating it with the increase in the area of the material eluting around 82 min (α^{31-141} and α-globin do not separate under the present chromatographic conditions, and hence only two peaks are observed). A zero-time aliquot of the reaction mixture serves as a control for the quantitation studies. The splicing reaction reaches an equilibrium in about 48 hr and generally a typical yield of 45–50% is obtained. After equilibrium is reached, the splicing reaction is quenched by diluting the reaction mixture with 0.1% TFA (at least 10-fold) and is lyophilized.

The V8 protease-catalyzed splicing of the complementary fragments at the Glu^{30}-Arg^{31} peptide bond appears to proceed smoothly even when the complementary fragments contain 2, 8, 11, and 19 sequence differences compared with human α-globin. The mutant α^{1-30} of the α-globin of Hb I $\alpha16$ Lys $\rightarrow$ Glu can be spliced readily with the α^{31-141} segment of Hb Sealy ($\alpha47$ Asp α His) to generate a new double mutant of α-globin.[27] Similarly, mouse α^{1-30} containing eight sequence differences compared with that of human α^{1-30}, was spliced with human α^{1-30} to generate mouse–human chimeric α-globin. V8 protease also catalyzed the splicing of human α^{1-30} with mouse α^{31-141} containing 11 sequence differences compared with that of human α^{31-141}, yielding human–mouse chimeric α-globin.

The purification of semisynthetic α-globin from the lyophilized semisynthetic reaction mixture generally is achieved by urea–CM-52 chromatography. The procedure is similar to the one described earlier for the separation of the complementary fragments α^{1-30} and α^{31-141} from the V8 protease digest of α-globin. A chromatographic profile of the splicing reaction mixture of mouse α^{1-30} with human α^{31-141} after 48 hr of the splicing reaction is shown in Fig. 4, as an example. Under these conditions, the order of the elution of the components of the semisynthetic reaction mixture is as follows: α^{1-30}, semisynthetic mouse–human chimeric α-globin and human α^{31-141}. However, this order of elution of semisynthetic α-globin and the complementary fragments need not be true for all the new separations that an investigator wishes to pursue. Besides, other manipulations of the chromatographic system may be required to get a satisfactory separation, depending on the overall charge and nature of the semisynthetic α-globin. As noted earlier, the overall yield of the semisynthetic reaction is about 40–45%. Urea–CM-52 chromatography of the semisynthetic reaction also permits the recovery of the unligated α^{1-30} and α^{31-141}. These complementary fragments thus isolated can be recycled through the semisynthetic reaction to generate another batch of the semisynthetic α-globin.

Reconstitution of Semisynthetic α-globin

There is no unanimity among investigators as to the general procedure of reconstitution of the globins to the tetrameric form. Although the general principles of the original protocol developed by Yip *et al.*[32,33] are, by and large, adhered to in most of the reconstitution experiments, the choice of the buffer and pH has often been the discretion of the individual investigator. It is quite conceivable that the fluctuations in pH or the nature of the buffer ions do not significantly influence the final reconstitution of the tetramer. Nonetheless, it should be noted that a quick survey of the various reconstituted recombinant tetramers of Hb A and Hb S suggest the possible presence of some structural differences between the recombinant Hb A or Hb S and their naturally occurring counterparts. The structural difference between the recombinant Hb S and its counterpart from nature has been suggested by experiments of Adachi and associates.[34] The kinetics of polymerization of recombinant Hb S was found to be distinct from that of native Hb S, although a difference in the oxygen affinity and Hill coefficient was not detected. Similarly, the possibility that the Trp-6(β) Hb S might not have folded properly has been raised by Bihoreau *et al.*[35] Fronticelli and associates[36] have reported differences in the Soret region circular dichroic spectra of recombinant and native Hb A. As discussed by Fronticelli *et al.*[36] this also appears to be true in the case of reconstituted myoglobin. However, these results are not consistent with the studies of Nagai *et al.*[37] describing the preparation of recombinant Hb A. Their investigations demonstrate that the crystal structures of native Hb A and recombinant Hb A are indistinguishable.[37] It is conceivable that either the crystallization process could have acted as a purification step of the recombinant Hb A, or alternatively the solution structure of the recombinant protein is not identical to that of the native protein. These findings suggest that a more stringent comparison of the solution structure of recombinant Hb A/Hb S with their native counterparts should be undertaken to resolve these issues.

[32] Y. K. Yip, M. Waks, and S. Beychok, *J. Biol. Chem.* **248**, 7237 (1972).

[33] Y. K. Yip, M. Waks, and S. Beychok, *Proc. Natl. Acad. Sci. U.S.A.* **74**, 64 (1977).

[34] K. Adachi, E. Rappaport, H. S. Eck, P. Konitzer, J. Kim, and S. Surrey, *Hemoglobin* **15**, 417 (1991).

[35] M. T. Bihoreau, V. Baudin, M. Marden, N. Lacaze, B. Bohn, J. Kister, O. Schaad, A. Dumoulin, S. J. Edelstein, C. Poyart, and J. Pagnier, *Protein Sci.* **1**, 145 (1992).

[36] C. Fronticelli, J. K. O'Donnell, and W. S. Brinigar, *J. Protein Chem.* **10**, 495 (1991).

[37] K. Nagai, B. Luisi, D. Shih, G. Miyazaki, C. Poyart, A. De. Young, L. Kwiatkowski, R. W. Noble, S.-H. Li, and N.-T. Yu, *Nature (London)* **329**, 858 (1987).

We describe below the procedure that is generally used in our laboratory for reconstituting the native or semisynthetic mutant α-globins with the β chain and heme to generate the tetrameric hemoglobin. Most of the previous reconstitution studies of the native globin or recombinant globin have been carried out in the carbonmonoxy form. The reconstituted tetramer is purified in the carbonmonoxy form and subsequently converted to the oxy form by photodissociation. This procedure is likely to generate some amount of oxidized protein as well as some denatured protein. In almost all of these studies, the oxy form of the protein thus generated is not subjected to further purification. We have eliminated this step of photodissociation in our reconstitution studies because we are primarily interested at this stage in generating mutant Hb S for polymerization studies. We are now routinely reconstituting the semisynthetic α-globin in the oxy form in 50 mM Tris-HCl buffer, pH 7.4, through the "alloplex intermediate" pathway. This reconstitution procedure is distinct from the protocol used earlier by us[24] for the preparation of semisynthetic Hb A, which was nearly the same as that described by Nagai et al.[38] The procedure described below has consistently given us reproducible results with the reconstitution of α-globin, semisynthetic α-globin, and the chimeric α-globin. The protocol for the reconstitution of the semisynthetic material is presented schematically as a flow chart in Fig. 5.

Regeneration of the —SH Group of HMB Globins

The native or semisynthetic α-globin (lyophilized material) is dissolved in 50 mM Tris-HCl (pH 7.4) buffer containing 8 M urea, 1 mM ethylenediaminetetraaceticacid (EDTA), and 2 mM dithiothreitol (DTT) to a final protein concentration of 5 mg/ml. This solution is incubated at room temperature for 30 min, after which it is transferred to an ice bath (4°). The HMB-β chain in the oxy form is taken in 50 mM Tris-HCl (pH 7.4) containing 2 mM DTT, 1 mM EDTA, and 1 μg/ml catalase to get a protein solution of 5–10 mg/ml. The chains are incubated at 4° for 1 hr.

Preparation of "Half-Filled" Molecules

The mixture of reduced (thiol) forms of native or semisynthetic α-globin is mixed with the β chains (1.1-fold over α-globin) prepared as described above at 4°. This mixture is diluted with 50 mM Tris-HCl buffer (pH 7.4) containing 1 mM DTT, 1 mM EDTA, and 1 μg/ml

[38] K. Nagai, M. F. Perutz, and C. Poyart, Proc. Natl. Acad. Sci. U.S.A. 82, 7252 (1985).

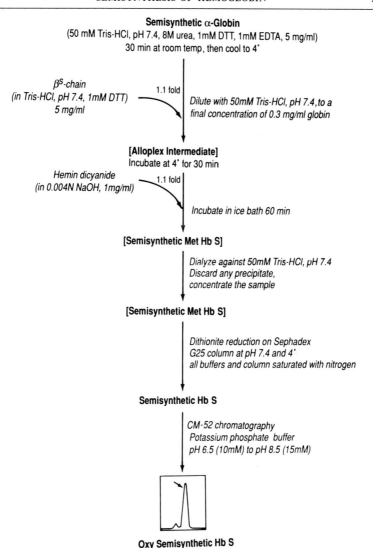

FIG. 5. Reconstitution of semisynthetic α-globin. The flow chart depicts the various steps involved in the reconstitution of the semisynthetic α-globin to tetrameric HbS.

catalase so that the final concentration of the total protein is 0.2–0.3 mg/ml. This dilution also brings down the concentration of urea to less than 0.5 M. The mixture is incubated at 4° for 30 min for the formation of "half-filled" molecules.

Preparation of Hemin Dicyanide

Hemin chloride (25 mg) is dissolved in 1 ml of 0.1 N sodium hydroxide, and 0.3 ml of a 10 mg/ml solution of sodium cyanide is added. The volume is made up to 25 ml with water to obtain a 1.6 mM solution of hemin dicyanide. Proper care should be taken while handling the sodium cyanide. It is a good practice to use gloves to prevent direct contact of sodium cyanide or hemin dicyanide with skin.

Titration of "Half-Filled" Tetramers with Hemin Dicyanide

An appropriate amount of hemin dicyanide solution (representing a 1.1-fold molar excess over the α-globin) is added dropwise to the solution of the "half-filled" tetramer with mild shaking. The solution is allowed to stand for 45–60 min at 4°. The "fully filled" tetramer thus generated is dialyzed extensively aginast 50 mM Tris-HCl buffer (pH 7.4). Some precipitate is formed during the dialysis step. This is centrifuged and discarded. The reconstituted protein is subsequently concentrated by Amicon (Danvers, MA) filtration and subjected to dithionite reduction on a Sephadex G-25 column.

Dithionite Reduction of Reconstituted Tetramer

The reconstitution procedure discussed above yields Hb tetramer in the ferric form. Accordingly, the heme has to be finally reduced to the ferrous state to obtain the functional tetramer. This can be achieved either by an enzymatic method as described by Hayashi et al.,[39] or by the chemical method using sodium dithionite. The sodium dithionite reduction of the ferric form of Hb has been carried out routinely on a Sephadex G-25 (fine) column. This procedure limits the time that the Hb sample will be in contact with dithionite and with the small molecular weight products generated as a result of the dithionite reduction.

The swollen gel (Sephadex G-25) is degassed thoroughly and packed into a column. The column (1.5 × 100 cm) is equilibrated and developed with 50 mM Tris-HCl buffer (pH 7.4). The buffer is bubbled constantly with nitrogen. Dithionite solution (1 ml; 150 mg/ml) prepared anaerobically in the column buffer is loaded gently onto the column and allowed to enter the gel to form a bed of dithionite. The column top is washed with 200 μl of the buffer. Reconstituted Hb (2 ml; 15–20 mg/ml) solution containing 5 μg/ml of catalase is deoxygenated and loaded onto the column. The gel filtration is carried out at a flow rate of 45 ml/hr. The hemoglobin fractions

[39] A. Hayashi, T. Suzuki, and M. Shin, *Biochim. Biophys. Acta* **310,** 309 (1973).

are collected, concentrated by Amicon filtration, and dialyzed against 10 mM potassium phosphate (pH 6.5).

Purification of Semisynthetic Hemoglobin

The oxy form of the reconstituted semisynthetic Hb generated after the dithionite reduction procedure discussed above is purified by CM-52 chromatography using a 0.9 × 30 cm column. The CM-52 resin is equilibrated with 10 mM potassium phosphate, pH 6.5, and packed into a column. The semisynthetic Hb sample dialyzed against the 10 mM phosphate buffer, pH 6.5, is loaded onto the column. Protein loads up to 100 mg can be purified readily on a CM-52 column of the dimension of 0.9 × 30 cm. The chromatogram is developed using a linear gradient of 250 ml each of 10 mM potassium phosphate, pH 6.5, and 15 mM potassium phosphate, pH 8.5. The elution of the protein is monitored at 540 nm. The protein peak representing the semisynthetic Hb is pooled, concentrated by Amicon filtration, and stored at 77 K. The chromatographic elution pattern of mouse α-globin and human–mouse chimeric α-globin (which differs from the mouse α-globin in having one less positive charge) reconstituted with β^s is shown in Fig. 6 as an example. The predominant species in the chromatography is the reconstituted tetramer. The overall yield of this chromatographed semisynthetic Hb (calculated based on the amount of semisynthetic α globin taken for reconstitution) is about 30 to 35%.

Chemical and Functional Characterization of Semisynthetic Hemoglobin

The chemical characterization of semisynthetic hemoglobin is an essential step before undertaking studies of structural and functional consequences of mutation introduced into the molecule. First of all, it is important to ascertain the stoichiometry of the chains in the semisynthetic tetramer. Separation of the globin chains by RP-HPLC is the best way to establish this. It is also important to ascertain that secondary modification of the amino acid side chains of the globins has not occurred during the reconstitution as well as during the dithionite reduction step. This is particularly crucial when the mutant α-globin is reconstituted, because the mutation could conceivably alter the susceptibility of the amino acid side chain in the tetramer to a secondary reaction, for example, oxidation. Information about the presence of such reaction products can be obtained by establishing the molecular mass of the globin chains separated by RP-HPLC. The tetramer can be subjected to this analysis because there is a significant difference in the molecular mass of the two chains.

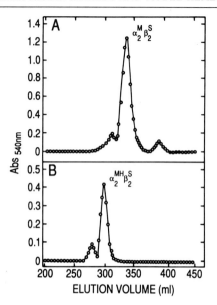

FIG. 6. Purification of reconstituted Hb. The reconstituted tetramer was chromatographed on a CM-52 column (0.9 × 30 cm) employing a linear gradient generated using 250 ml each of 10 mM potassium phosphate buffer (pH 6.5) and 15 mM potassium phosphate buffer (pH 8.5). (A) The chromatogram of mouse α-globin reconstituted with human β^s chain. (B) The chromatogram of the semisynthetic mouse–human chimeric α-globin reconstituted with β^s chain (see text for additional details).

The chain separation is done usually on a Vydac C_4 or C_8 or RP300 (4.6 × 250 mm) column under conditions identical to those described earlier for the separation of α^{1-30} and α^{31-141}. Semisynthetic Hb (50–100 μg) is taken in 1 ml of 0.1% TFA and loaded directly onto the column. The globin chains[40] can also be separated on a Vydac C_4 column (4.6 × 250 mm) employing a gradient of 30–60% acetonitrile containing 0.1% TFA. Visible spectra of the semisynthetic Hb are also recorded to establish that the Soret absorption characteristic of the tetramer has been retained in the semisynthetic Hb. The first derivative spectra of semisynthetic Hb in the region of 350–250 nm also can serve as diagnostic of the quaternary interactions of the Hb fold.[41-44]

[40] S. J. Hoffman, D. L. Looker, J. M. Roehrich, P. E. Cozart, S. L. Durfee, J. L. Tedesco, and G. L. Stetler, *Proc. Natl. Acad. Sci. U.S.A.* **87**, 8521 (1990).
[41] K. Imai, A. Tsuneshige, T. Harano, and K. Harano, *J. Biol. Chem.* **264**, 11174 (1989).
[42] K. Ishimori, K. Imai, G. Miyazaki, T. Kitagawa, Y. Wada, H. Morimoto, and I. Morishima, *Biochemistry* **31**, 3256 (1992).
[43] K. Imai, K. Fushitani, G. Miyazaki, K. Ishimori, T. Kitagawa, Y. Wada, H. Morimoto, I. Morishima, D. Shih, and J. Tame, *J. Mol. Biol.* **218**, 769 (1991).
[44] J. Tame, D. Shih, J. Pagnier, G. Fermi, and K. Nagai, *J. Mol. Biol.* **218**, 761 (1991).

Oxygen affinity measurements of the semisynthetic Hb are an effective approach to evaluate the quality of the preparation. The correct value of the Hill coefficient for semisynthetic Hb ensures retention of the basic architectural principles of the Hb fold. If this is not observed, it is essential to have a control semisynthetic α-globin reconstituted simultaneously with the test sample so that one can be certain that the observed result is indeed a reflection of the consequences of the mutation(s) introduced rather than an artifact of the reconstitution procedures.

Other Semisynthetic Studies of Hemoglobin

The previous semisynthetic manipulations of Hb have been limited mainly to the residues at or near the amino and carboxyl termini of the α and/or β chains. Carboxypeptidase B has been used to prepare desHis $\beta146$, desArg $\alpha141$, and des(Arg $\alpha141$, His $\beta146$) hemoglobins. Similarly, des(His $\beta146$, Tyr $\beta145$) β^s chains have been prepared by the combined action of carboxypeptidase B and carboxypeptidase A of the chain. The method for the preparation of these derivatives of Hb has been described in detail by Kilmartin.[45] However, the feasibility of replacing these residues by a reverse proteolytic approach has not been established and is an area that should be addressed, particularly in view of the fact that these residues are involved in the Bohr effect, and their actual contribution toward the alkaline Bohr effect of Hb[46] remains controversial.

Trypsin-catalyzed formation of the peptide bond at the α-carboxyl group of Arg-141(α) also has been reported.[47] In this case Gly-NH$_2$ (142α) Hb was generated when native Hb was treated with glycine amide in the presence of trypsin. A similar condensation reaction also has been observed with hydrazine. An elegant manipulation of Edman degradation chemistry and carbodiimide-activated coupling at the amino terminus has been made by Gurd[48] and colleagues for the preparation of semisynthetic Hb. A single round of Edman degradation was used to remove the amino-terminal value from the α-globin and the truncated α-globin was reconstituted with β chain. The desVal^1 α-globin also has been coupled chemically with ^{13}C-labeled glycine. The NMR study of the site-specifically ^{13}C-enriched protein enabled the determination of the pK_a of the glycine in the liganded state of the protein. The pK_a of the α-amino group was consistent with the previously reported role of this group in the alkaline Bohr effect, but contrary to what might have been expected from the crystallographic

[45] J. V. Kilmartin, this series, Vol. 76, p. 167.

[46] M. R. Busch, J. E. Mace, N. T. Ho, and C. Ho, *Biochemistry* **30**, 1865 (1991).

[47] K. Nagai, Y. Enoki, and S. Tomita, *J. Biol. Chem.* **257**, 1622 (1982).

[48] S. A. Hefta, S. B. Lyle, M. R. Busch, D. E. Harris, J. B. Matthew, and F. R. N. Gurd, *Proc. Natl. Acad. Sci. U.S.A.* **85**, 709 (1988).

studies, namely, the pK_a value was inconsistent with the presence of a salt bridge of the α-amino group of the α chain to the carboxyl of Arg-141 in the oxy or liganded form of the protein.

Unique Aspects of α-Globin Semisynthetic Reaction

The mechanistic aspects of the V8 protease-catalyzed α-globin semisynthetic reaction are distinct from the other protease-catalyzed splicing of the discontinuity sites of the fragment-complementing systems that generally occur in 90% glycerol. Unlike these systems, the α-globin semisynthetic reaction exhibits a high degree of flexibility in that the truncated segments of the complementary fragments can also be spliced by V8 protease to form the Glu^{30}-Arg^{31} peptide bond. For example, V8 protease catalyzes the condensation of the segments α^{17-30} as well as α^{24-30} with α^{31-40} to generate the respective contiguous segments α^{17-40} and α^{24-40} in much the same way as α^{1-30} and α^{31-141}. Thus, it is clear that single or multiple mutations can be introduced into the segments (α^{1-30} and α^{31-141}) without impairing the proteosynthetic potential of the enzyme. It also has been demonstrated that this splicing reaction is facilitated by the induction of an α-helical conformation in the enzymatically ligated contiguous segment by n-propanol that has been incorporated into the semisynthetic reaction mixture to favor the protonation of the α-carboxylate. This induced α-helical conformation of the nascent contiguous segment acts as the "conformational trap" of the α-globin semisynthetic reaction.

The semisynthetic methods, however, should not be treated as an alternative to site-directed mutagenesis. Both of these approaches have their own strengths and limitations. On the other hand, a combination of these two approaches could provide a much more powerful methodology to delineate long-range communication in proteins. Such studies are particularly pertinent to delineate mechanistic aspects of the integration of quinary structural aspects of deoxy Hb S during the polymerization reaction. For example, let us assume that one would like to investigate the influence of a mutation of the GH corner of the α chain on the dynamics of the microenvironment of His-20(α), a residue of intermolecular contact located at the AB corner of the α chain. Conceivably one could generate the desired mutant α^{1-141} by site-directed mutagenesis and generate α^{31-141} by V8 protease cleavage of this mutant α-globin. (Studies of the identification of other splicing sites in the α^{31-141} region are also currently underway in the laboratory in an attempt to extend the flexibility of the semisynthetic reactions to this region of the molecule.) This could then be ligated with α^{1-30} prepared by the chemical synthesis and containing NMR-sensitive nuclei in His-20. A comparison of the local conformational aspects around

His-20(α) with that of the mutant semisynthetic Hb would help to delineate the long-range mutational effects in bringing out perturbations around the microenvironment of His-20(α). Thus, chemical methods of protein engineering could be coupled with the genetic engineering approaches to generate novel analogs of Hb. Such a combination approach has been developed by Wallace and co-workers for the construction of cytochrome c mutants.[49]

Acknowledgments

This work is supported by National Institutes of Health Grants HL-38655 and a Grant-in-Aid to A. S. Acharya from the American Heart Association, New York City Affiliate. Rajendra Prasad Roy is an Ella Fitzgerald Fellow of the American Heart Association, New York City Affiliate. We wish to thank Dr. Ronald L. Nagel for the facilities extended.

[49] C. J. A. Wallace, G. Guillemete, Y. Hibia, and M. Smith, *J. Biol. Chem.* **266,** 21355 (1991).

[13] Preparation of Mixed Metal Hybrids

By ANTONIO TSUNESHIGE and TAKASHI YONETANI

Introduction

Mixed metal hybrid hemoglobin tetramers, $\alpha(Fe)_2\beta(Mg)_2$, which carry natural protoheme (Fe) in the α subunits and other metallo(proto)porphyrins (Me) in the β subunits, and its complimentary form, $\alpha(Mg)_2\beta(Fe)_2$, have been prepared and utilized in spectroscopic, functional, and structural studies of hemoglobin (Hb). These metalloporphyrins serve as functional and nonfunctional spectroscopic probes. These mixed metal hybrids allow us to observe structural and functional properties of one type of subunit apart from their partner subunits, because properties of the α and β subunits carrying different types of metalloporphyrins are quite distinct. Mixed metal hybrids with combinations of Fe–Co,[1] Fe–Ni,[2] Fe–Zn,[3] and Fe–Mn[4] are typical examples that have been successfully studied.

One principal method available for preparation of mixed metal hybrid Hbs is not only readily performed but is also widely applicable to many metalloporphyrin-containing Hbs. It consists of isolation of α and β sub-

[1] M. Ikeda-Saito, H. Yamamoto, and T. Yonetani, *J. Biol. Chem.* **252,** 8639 (1977).
[2] N. Shibayama, H. Morimoto, and G. Miyazaki, *J. Mol. Biol.* **192,** 323 (1986).
[3] B. M. Hoffman, *J. Am. Chem. Soc.* **97,** 1688 (1975).
[4] M. R. Waterman and T. Yonetani, *J. Biol. Chem.* **245,** 5847 (1970).

units from two parent Hbs containing different prosthetic groups, followed by stoichiometric mixing of two appropriate subunits, which spontaneously form a mixed metal hybrid Hb tetramer (Fig. 1).

Preparation of Metalloporphyrin-Containing Hemoglobin Tetramers

In preparing mixed metal hybrids it is necessary first to prepare metalloporphyrin-substituted Hb tetramers. These are made from apoHb or globin that is prepared by removal of protohemes under well-controlled conditions, followed by incorporation of appropriate metalloporphyrins to form metal-substituted Hb tetramers.

Preparation of ApoHb

There are two well-used methods for preparing apoHb: the acid–acetone method and the acid–butanone method.

Acid–Acetone Method. In this method[5] a one-volume aliquot of approximately 1 mM Hb (per metal basis) is added dropwise into 40 volumes of acid–acetone chilled at $-20°$ with vigorous stirring. The acid–acetone is prepared by mixing 2.5 ml of 2 M HCl and 1 liter of pure acetone. At the end of the addition, the acid–acetone becomes slightly reddish and the protoheme-free Hb precipitates. The suspension is centrifuged at 5000 rpm for 10 min at $-20°$. The precipitate, if white, is separated carefully from the acid–acetone supernatant. Otherwise, the precipitate is resuspended in 10 volumes of chilled acid–acetone, and the procedure is repeated. The white precipitate of apoHb is resuspended in 1 volume of cold distilled water and dialyzed at 4° against 100 volumes of 5 mM bicarbonate and then against 50 mM phosphate buffer, pH 7.4. The dialyzed apoHb solution is ready for reconstitution after removal of any precipitate of denatured globin by centrifugation. Alternatively, after the acid–acetone treatment the colorless precipitate may be resuspended and dialyzed exhaustively against distilled water, centrifuged to remove precipitate, and lyophilized for extended storage. Some reports indicate the use of alkaline borate buffer instead of phosphate as the final dialysis medium in order to prevent formation of large quantities of insoluble precipitates.

Acid–Butanone Method. This method was initially reported by Teale[6] and subsequently modified to apply to heme removal from various

[5] A. Rossi-Fanelli, E. Antonini, and A. Caputo, *Biochim. Biophys. Acta* **30**, 608 (1958).
[6] F. W. Teale, *Biochim. Biophys. Acta* **35**, 543 (1959).

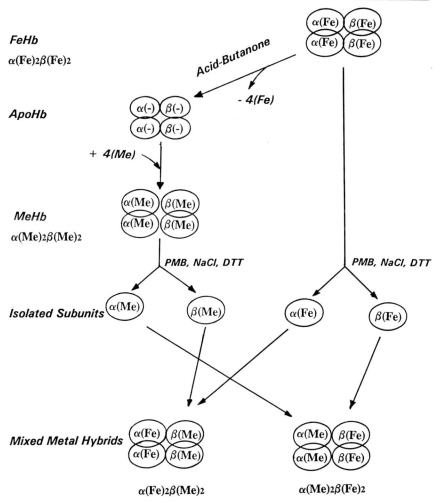

FIG. 1. Flow diagram of the preparation of mixed metal hybrid hemoglobins: $\alpha(Fe)_2\beta(Me)_2$ and $\alpha(Me)_2\beta(Fe)_2$. Fe, Protoheme; Me, metalloporphyrin; PMB, p-hydroxymercuribenzoate; DTT, dithiothreitol.

hemoproteins.[7] OxyFe(II)Hb is converted to MetHb or Fe(III)Hb by addition of a fivefold molar equivalent (per metal) excess of potassium ferricyanide. The mixture is passed through a column of Sephadex G-25 equilibrated with 5 mM phosphate buffer, pH 6.0, to remove ferri- and ferrocyanides. The concentration of the eluate is adjusted to 1 mM

[7] T. Yonetani, *J. Biol. Chem.* **242**, 5008 (1967).

and is acidified to pH 2.5 with dropwise additions of 1 or 0.1 M HCl. The solution is immediately mixed with an equal volume of cold butanone (methyl ethyl ketone) and vortexed for 20 sec. The colored upper layer containing free protoheme in butanone is siphoned off. This process is repeated several times until the lower aqueous phase becomes colorless, which is dialyzed in the cold against 1 mM bicarbonate to remove the dissolved butanone. The dialysis is continued against changes of 1 mM bicarbonate, 1 mM diphosphate, pH 6.3, and 50 mM phosphate, pH 7.0. The dialyzed apoHb solution is centrifuged to remove the precipitate of denatured globin. Because apoHb cannot be stored in a frozen state, it should be used immediately to prepare reconstituted metalloporphyrin-containing Hb.

Recombination of ApoHb with Metalloporphyrin. ApoHb ($\varepsilon_{280 \text{ nm}}$ = 16.2 mM^{-1} cm^{-1}) in diluted neutral buffers such as phosphate or Tris is mixed aerobically with dropwise additions of a 1.1-fold excess of appropriate metalloporphyrin (Porphyrin Products) in dilute NaOH or dimethyl sulfoxide. Because the binding of metalloporphyrins to apoHb is usually accompanied by spectral changes, their stoichiometry is readily determined by titrating apoHb with a metalloporphyrin at an appropriate wavelength. The recombined product is kept at 4° for 30 min, passed through a column of Sephadex G-25 equilibrated with 10 mM phosphate buffer, pH 6.7, and loaded onto a CM-cellulose column equilibrated with 10 mM phosphate buffer, pH 6.7. The adsorbed reconstituted Hb is eluted with a linear gradient of 10 mM phosphate, pH 6.7, and 20 mM phosphate, pH 8.1. The major fraction containing the desired metalloporphyrin-substituted Hb is concentrated to 1-2 mM (per metal) by ultrafiltration and stored at 77 K. Metal ions such as Mn(III) and Co(III) in metalloporphyrin incorporated into apoHb by the above procedure can be reduced to the respective Mn(II) and Co(II) states. However, they are autoxidizable and revert to the original Mn(III) and Co(III) states on reaction with oxygen.

The only exception to this method is the preparation of Co(II)Hb,[8] which can be reversibly oxygenated and was first prepared by Hoffman and Petering.[9] Preparation of functional Co(II)Hb requires anaerobic additions to apoHb of a stoichiometric amount of dithionite-reduced Co(II)-porphyrin in a 10% volume of 50% aqueous pyridine, followed by rapid removal of pyridine and dithionite and purification by the gradient chromatography on CM-cellulose column.[8]

[8] T. Yonetani, H. Yamamoto, and G. V. Woodrow, *J. Biol. Chem.* **249,** 682 (1974).
[9] B. M. Hoffman and D. H. Petering, *Proc. Natl. Acad. Sci. U.S.A.* **67,** 637 (1970).

Preparation of Isolated α and β Subunits

Separation with PMB Treatment

The present method is a modification of the methods of Bucci and Fronticelli[10] and Geraci *et al.*[11] OxyFeHb or carbonmonoxyFeHb or oxy-CoHb, 2 to 3 mM (per metal) in 10 mM phosphate buffer, pH 6.2, and 0.3 M NaCl are incubated at 4° overnight (~12 hr) with 10-fold excess (per tetramer) of p-hydroxymercuribenzoate (PMB), which is dissolved in a minimal volume of 0.1 M NaOH and added slowly dropwise. The Hb–PMB mixture is centrifuged (8000 rpm for 10 min) to remove any turbidity or precipitate, passed through a 4.5 × 30 cm Sephdex G-25 column, equilibrated with 10 mM phosphate buffer, pH 8.0, and adsorbed on the top of a 4.5 × 15 cm DEAE-cellulose (Whatman DE-52; Clifton, NJ) column. A broad band of dilute α-PMB subunits is washed off the column with the same buffer, whereas the β-PMB subunits remain at the top of the column under these conditions. After the α-PMB subunits are fully eluted, the column is washed with 30 mM phosphate buffer, pH 8.3, to remove the undissociated Hb tetramer. When the Hb tetramer is washed off the column, the β-PMB subunits are eluted with 0.1 M phosphate buffer, pH 7.0.

Preparation of PMB-Free Isolated Subunits

The α-PMB subunits[1] are treated with 20 mM dithiothreitol (DTT) (Sigma, St. Louis, MO) and 5 μM catalase (Sigma) for 2 hr and passed through a 10 × 15 cm Sephadex G-25 column, equilibrated with 10 mM phosphate buffer, pH 6.7. The colored fraction of the α-SH subunits is adsorbed on a 2.5 × 3 cm column of CM-cellulose (Whatman CM-52), equilibrated with 10 mM phosphate buffer, pH 6.7. The column is washed with 2 volumes of the same buffer. The α-SH subunits are eluted with 50 mM phosphate buffer, pH 7.4. The β-PMB subunits are incubated with 20 mM DTT and 5 μM catalase for 2 hr and passed through a 10 × 15 cm Sephadex G-25 column, equilibrated with 7 mM Tris buffer, pH 8.0. The colored β-SH fraction is adsorbed on a 2.5 × 7 cm column of DEAE-cellulose (Whatman DE-52), equilibrated with 7 mM Tris buffer, pH 8.0. The column is washed with 2 volumes of the same buffer, and the β-SH subunits are eluted with 50 mM phosphate buffer, pH 7.4.

[10] E. Bucci and C. Fronticelli, *J. Biol. Chem.* **240,** 551 (1965).
[11] G. Geraci, L. J. Parkhurst, and Q. H. Gibson, *J. Biol. Chem.* **244,** 4664 (1969).

The quality and purity of the isolated subunits should be examined by electrophoresis. Various commercially available electrophoresis devices can be used effectively for this purpose. For example, the SepraTek electrophoresis system using Sepraphore III cellulose polyacetate membranes (Gelman) is found to be fast and simple during the DTT treatment to follow the progress of PMB removal. Frequently, incomplete removal of PMB with the DTT treatment may result in a mixture of PMB and SH subunits in isolated subunit preparations. Boyer's spectrophotometric titration of SH groups with PMB[12] is the most convenient method to check the full removal of PMB from the isolated SH subunits. Isolated α-SH and β-SH subunits have one and two titratable SH groups per mole, respectively.

The presence of catalase is found to be essential to prevent oxidative modification of the prosthetic groups and the globin moieties in certain isolated subunits during the DTT treatment.[1] Probably catalase removes hydrogen peroxide produced by DTT.

Approximately equal amounts of isolated α-SH and β-SH subunits are obtainable, with yields of 60–65%. They should be used for reconstitution of mixed metal hybrids as soon as possible. For long-term storage at 77 K, the buffer media of these isolated subunits should be changed from phosphate to Tris (say, 50 mM, pH 7.4) before freezing. We noted, even with this precaution, that some isolated subunits, such as α(Co) subunits, are less stable when frozen.

The α-SH and β-SH subunits, which contain appropriate metalloporphyrins, are incubated individually with 20 mM DTT and 10 μM catalase for 30 min prior to recombination. The α-SH subunits are mixed with an 1.2-fold excess of the β-SH subunits for 1 hr. The mixture is then passed through a 2.5 × 20 cm column of Sephadex G-25, which is equilibrated with 10 mM phosphate buffer, pH 6.7. The colored fraction is then applied to a 2.5 × 5 cm CM-cellulose column, equilibrated with 10 mM phosphate buffer, pH 6.7. Washing of the column with this buffer is continued in order to remove the excess β-SH subunits. The mixed metal hybrids are eluted by a linear gradient chromatography of 10 mM phosphate buffer, pH 6.7 (500 ml), and 20 mM phosphate buffer, pH 8.1 (500 ml). The main fraction, which contains the desired mixed metal hybrids, is concentrated by ultrafiltration. The concentrated mixed metal hybrids can be stored at 77 K for an extended period.

This method of preparation of mixed metal hybrids is well suited for hybrids with metal combinations of Fe(II)–Co(II)[1] and Fe(II)–Fe(III).

---

[12] P. D. Boyer, *J. Am. Chem. Soc.* **76**, 4331 (1954).

Alternate Method of Preparation of Mixed Metal Hybrids

Some metalloporphyrin-substituted Hb tetramers, such as Ni(II)Hb,[2] Zn(II)Hb,[3] Mg(II)Hb,[13] and Mn(III)Hb,[4] some of which often take the T-quaternary structure, are difficult to split into α and β subunits by the method described above. Therefore, an alternate method has been devised. This method consists of isolation of α and β subunits from FeHb, removal of protoheme from the isolated subunits, incorporation of appropriate metalloporphyrins (Me) into apo subunits, and recombination of the α(Me) or β(Me) subunits with β(Fe) or α(Fe) subunits to produce mixed metal hybrids.

Preparation of Isolated α- and β-Apo Subunits

Oxy- or carbonmonoxyFe(II)Hb is first converted to isolated α and β subunits by the method described above. Protohemes of isolated subunits are then oxidized to Fe(III) with a fourfold molar excess of potassium ferricyanide. The isolated Fe(III) subunits are subjected to the acid–butanone treatment, as described previously, to obtain the isolated α-SH and β-SH apo subunits.

Recombination of Isolated Apo Subunits with Metalloporphyrins

This procedure is identical to that for recombination of apoHb with metalloporphyrins, as described previously. It should be noted that because the isolated apo subunits are more unstable than apoHb, extra precaution in handling is advised.

Recombination of Isolated Subunits

The α-SH and β-SH subunits containing appropriate metalloporphyrins are incubated with 20 mM DTT and 10 μM catalase for 30 min at 4° before recombination. The detailed procedure for mixing and subsequent purification with gradient chromatography is the same as described above.

General Comments

Hemoglobin, which is used as the starting material (Fig. 1), should be as fresh as possible and free of metHb, the presence of which is estimated readily from the absorbance around 630 nm.[14] The use of carbonmonox-

---

[13] M. Fujii, X-Y. Zhou, and T. Yonetani, *Biophys. J.* **59**, 610a (1991).
[14] J. V. Kilmartin, K. Imai, R. T. Jones, A. R. Faruqui, J. Fogg, and J. M. Baldwin, *Biochim. Biophys. Acta* **534**, 15 (1978).

yHb rather than oxyHb is an effective way to prevent the formation of metHb during the preparative procedures, particularly during preparation of isolated Fe(II) subunits.[15] However, this procedure is only applicable to Fe(II)- and Ru(II)-containing compounds, because carbon monoxide does not bind other metalloporphyrins.

Mixed metal hybrids $\alpha(Fe)_2\beta(Me)_2$ and $\alpha(Me)_2\beta(Fe)_2$, as described in this chapter, are sometimes called symmetric mixed metal hybrids. Hybrids of this type have been widely used to probe subunit inequivalence. More complex mixed metal hybrids, which are called asymmetric hybrids, have also been prepared. They are tetramers containing protohemes in one, two, or three subunits, and another type of metalloporphyrin in the remaining subunits.[16-18] Such mixed metal hybrids require $\alpha-\alpha$ or $\beta-\beta$ cross-linking of subunits to prevent reversible tetramer–dimer dissociation,[19,20] which leads to scrambling of the desired products of appropriate metal combinations. Preparation methods for these asymmetric mixed metal hybrids have not been firmly established and thus are not discussed in this chapter.

Acknowledgment

We thank the National Heart, Lung, and Blood Institute for support (HL14508).

[15] J. V. Kilmartin, J. A. Hewitt, and J. F. Wooton, *J. Mol. Biol.* **93,** 203 (1975).
[16] S. Miura and C. Ho, *Biochemistry* **24,** 6280 (1982).
[17] S. Miura, M. Ikeda-Saito, T. Yonetani, and C. Ho, *Biochemistry* **26,** 2149 (1987).
[18] T. Inubushi, C. D'Ambrosio, M. Ikeda-Saito, and T. Yonetani, *J. Am. Chem. Soc.* **108,** 3799 (1986).
[19] R. Chatterjee, E. V. Welty, R. Y. Walder, S. L. Pruitt, P. H. Rogers, A. Arnone, and J. A. Walder, *J. Biol. Chem.* **261,** 9929 (1986).
[20] R. Chatterjee, R. Y. Walder, A. Arnone, and J. A. Walder, *Biochemistry* **21,** 5901 (1982).

Section IV

Chemically Modified Hemoglobins

[14] Preparation of Hemoglobin Derivatives Selectively or Randomly Modified at Amino Groups

By JAMES M. MANNING

Introduction

The term "selective or specific modification of a protein" is commonly used to indicate modification of a particular amino acid side chain that may have a special function in the protein because of the particular "region" in which it is located. The formal terms "regioselective" or "regiospecific" are more accurate descriptions of this type of modification. There are many examples of this approach in protein chemistry, including three in this article.

We have assessed the possibility that, in some circumstances, a particular modification of a protein can be made more general in order to obtain information on a class of amino acid side chains, i.e., when the modification is used in a "regiorandom" manner to modify the amino acid side chains that delineate a particular region of the protein. Such an approach, which is qualitative in nature, is designed simply to identify amino acid side chains that participate in a particular function, rather than to attain the quantitative goal of the regioselective approach. On the other hand, the regioselective approach usually has a quantitative goal of modifying completely one particular side chain. Other differences between the regioselective and the regiorandom modification of protein side chains are given in Table I. This approach has been used with hemoglobin to locate the chloride-binding sites along the sides of the central dyad axis, to assess the importance of these sites in governing the oxygen affinity of hemoglobin, and to estimate the contribution of some of the chloride-binding sites to the alkaline Bohr effect.

Four derivatives of hemoglobin involving either regioselective or regiorandom modifications at certain amino groups are described in Scheme I [Eqs. (1)–(4)].

In a previous chapter in this series, the preparation of hemoglobin (Hb) derivatives with uncharged carbamyl groups at the N termini was described.[1] Studies on these derivatives provided answers to two closely related questions, i.e., the mechanism of action of sodium cyanate in

---

[1] J. M. Manning, this series, Vol. 76, p. 159.

TABLE I
DIFFERENCES BETWEEN REGIOSELECTIVE AND REGIORANDOM
MODIFICATION OF PROTEINS

Regioselective
1. Experimental conditions are chosen to obtain modification of most protein molecules at a particular type of side chain, as dictated by the nature of the reagent
2. Special conditions, such as addition of allosteric regulators to block certain sites, may be employed
3. The goal is the *quantitative* modification of one particular amino acid side chain
4. Minor products are removed to obtain a homogeneous protein derivative
5. Total or partial loss of protein function may permit assignment of that function to a particular side chain

Regiorandom
1. Experimental conditions are chosen to limit the extent of modification so that most molecules have one or more of a particular type of side chain modified, but some are unmodified
2. Sometimes buffer anions or cations may be removed if these compete with the site(s) to be identified
3. The goal is a mixture of modified protein molecules with both major and minor sites labeled, so they can be qualitatively identified. The use of a radiolabeled reagent will also permit a measurement of the extent of reactivity of a particular side chain
4. The modified protein derivatives are not separated from one another
5. Retention of protein function is necessary to ensure that protein integrity has been maintained

the inhibition of erythrocyte sickling[2,3] and the identification of Val-1(α) as one of the sites that contributed to the alkaline Bohr effect.[4] The increased oxygen affinity of sickle cells after treatment with sodium cyanate[2] was due primarily to the blocking of a critical salt bridge involving Val-1(α) in carbamylated Hb.[3,4] The deoxy conformation was thus disfavored, leading to an increased oxygen affinity of Hb S and a shift in the equilibrium from the deoxy, sickling form of sickle erythrocytes to the oxy, nonsickling shape. Even though the change in the functional properties of carbamylated Hb could readily explain the physiological effect of sodium cyanate on sickle erythrocytes, for other hemoglobin derivatives there may be no correlation between the biological features conferred by a particular modification and the changes in the functional properties of the derivative.

This chapter is devoted to the preparation of four other types of hemoglobin derivatives with other adducts on their amino groups, i.e., carboxy-

[2] A. Cerami and J. M. Manning, *Proc. Natl. Acad. Sci. U.S.A.* **68,** 1180 (1971).
[3] A. M. Nigen, N. Njikam, C. K. Lee, and J. M. Manning, *J. Biol. Chem.* **249,** 6611 (1974).
[4] J. V. Kilmartin and L. Rossi-Bernardi, *Nature (London)* **222,** 1243 (1969).

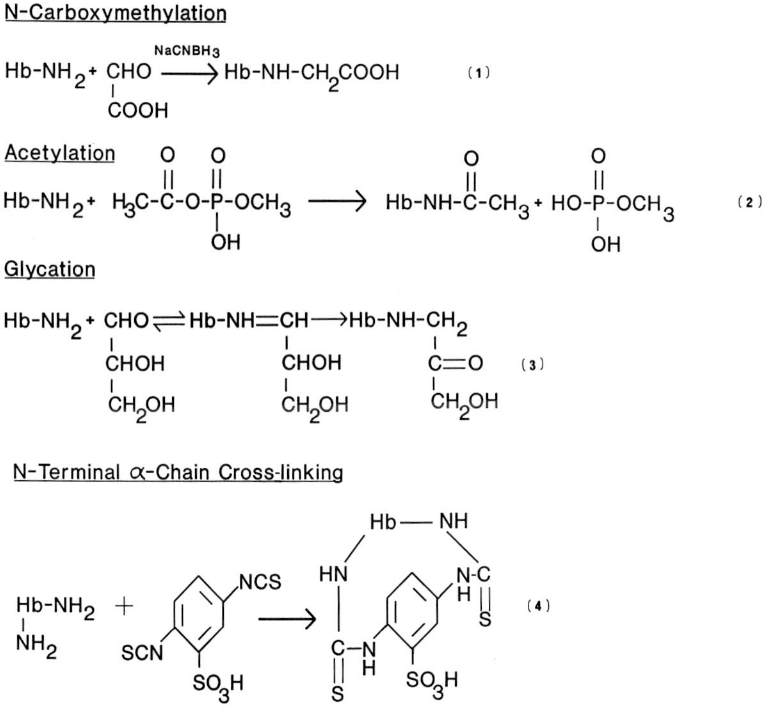

N-Carboxymethylation

$$Hb\text{-}NH_2 + \underset{\underset{COOH}{|}}{CHO} \xrightarrow{NaCNBH_3} Hb\text{-}NH\text{-}CH_2COOH \qquad (1)$$

Acetylation

$$Hb\text{-}NH_2 + \underset{\underset{OH}{|}}{H_3C\overset{\overset{O}{\|}}{C}\text{-}O\text{-}\overset{\overset{O}{\|}}{P}\text{-}OCH_3} \longrightarrow Hb\text{-}NH\text{-}\overset{\overset{O}{\|}}{C}\text{-}CH_3 + HO\text{-}\underset{\underset{OH}{|}}{\overset{\overset{O}{\|}}{P}}\text{-}OCH_3 \qquad (2)$$

Glycation

$$Hb\text{-}NH_2 + \underset{\underset{CH_2OH}{|}}{\underset{CHOH}{|}}{CHO} \rightleftharpoons Hb\text{-}NH\!=\!\underset{\underset{CH_2OH}{|}}{\underset{CHOH}{|}}{CH} \longrightarrow Hb\text{-}NH\text{-}\underset{\underset{CH_2OH}{|}}{\underset{C=O}{|}}{CH_2} \qquad (3)$$

N-Terminal α-Chain Cross-linking

SCHEME I

methyl, acetyl, glycyl, and cross-linked Hb [Eqs. (1)–(4) Scheme I]. The objectives are twofold, i.e., to obtain a derivative with the desired properties and to determine whether any change in the functional properties of the modified hemoglobin can be ascribed to the derivative.

The effect of a given modification on the properties of hemoglobin is dictated by the site(s) to which the adduct is attached, its chemical nature, and the parts of the hemoglobin molecule with which it interacts. A correlation between the nature of the adduct and its effect on hemoglobin function can be made with confidence only if the modifying reagent or the conditions by which it is introduced into the protein do not compromise the functional properties of the hemoglobin, i.e., that they do not adversely affect subunit cooperativity, cause oxidation of the heme, or lead to protein denaturation. In theory, the procedures described here for hemoglobin are applicable to other proteins, but some adjustment of the experimental conditions may be necessary, if the susceptibility of their amino groups, i.e., pK_a values, microenvironment, or the extent to which they interact with solutes, differ significantly from those in hemoglobin.

Both the regioselective and the regiorandom modifications of hemoglobin are described. With either approach there is a functional measurement, i.e.,subunit cooperativity (the Hill coefficient) that permits an independent assessment of the integrity of the protein. Regiorandom chemical modification is different from extensive modification in which as many as possible available side chains of a particular type are modified. For such studies, a high reagent/protein concentration is used for extensive incubation periods. However, protein function can be severely impaired. Regiorandom modification employs low ratios of reagent to protein in order to limit the number of modified side chains for purposes of identification.

Preparation of Hemoglobins and Determination of Functional Properties

It is preferable to use hemoglobin freshly prepared from erythrocytes because commercial preparations are frequently denatured and oxidized to metHb. Human adult hemoglobin A is prepared from whole blood drawn by venipuncture from normal, healthy volunteers into vacutainer tubes containing heparin or EDTA as anticoagulant. Even outdated units of blood can be used as long as extensive oxidation to metHb has not occurred; the initial washing will remove the contents of any lysed erythrocytes together with serum proteins. Pel-Freez Co. (Rogers, Arkansas) has been found to be a reliable source of fresh blood from cows and horses.

Whole blood is centrifuged at 2000 g for 10 min at 4° and, after removal of the supernatant plasma, the cells are washed three times with an equal volume of cold isotonic saline.[1-3] Lysis of the cells is achieved by addition of an equal volume of cold distilled water. After centrifugation at 10,000 g for 10 min at 4°, the lysate is dialyzed against 0.1 N NaCl. For some studies (see random acetylation of Hb below) a mixed anion–cation resin, such as Dowex 501, is used at this step to remove all salts. Even though none of the procedures described herein is so harsh to cause oxidation of the heme, it is, nevertheless preferable to convert the heme to the CO form by saturating the hemoglobin solution, the buffers, and the dialysis medium with CO; cold solutions of these are bubbled with the gas for about 5 min. However, prior to any functional studies that involve the deoxy state of hemoglobin, such as determination of the oxygen equilibrium curve, HbCO must first be converted to oxyhemoglobin by the mild photolysis procedure described in an earlier volume of this series.[1] The concentration of hemoglobin is determined by its absorbance at either 420 or 540 nm, using extinction coefficients also reported in the earlier volume.[1]

If the modification is to be performed with deoxyhemoglobin, a two-armed Erlenmeyer flask that can be flushed with N_2 and then closed to the atmosphere, as described in the earlier volume,[1] is used. Deoxygenation is achieved by bubbling the contents of the flask with nitrogen for 1–2 hr at

25° before the reactants are tipped into the main chamber from the side arms prior to the incubation.

The oxygen equilibrium curves are determined at 37° with a modified Aminco Hem-O-Scan (American Instrument Co., Silver Spring, MD) or a modified TCS Hemox Analyzer (Matheson Gas, East Rutherford, NJ); the P_{50} values are read directly from the plots of percent O_2 saturation versus oxygen pressure. For the Hemox analyzer, two major modifications were made; the gas flow was changed so that it was directed to the top of the stirred hemoglobin solution rather than bubbling in from the bottom, and two Matheson (TCS Medical Products, Huntingdon Valley, PA) FM1050V1A flow meters together with a small humidifier were installed on the N_2 and O_2 gas lines for better control of flow. For estimation of the Hill coefficient (n), the logarithm of the fractional saturation in the range of 40–70% is plotted against the logarithm of the oxygen tension; the slope of the resulting straight line is the n value. After conversion of HbCO to oxyHb by the photolysis procedure, the Hb derivative is concentrated to about 0.1 mM by ultrafiltration in an Amicon cell with a YM10 membrane, and then dialyzed against 0.05 M bis-Tris, pH 7.3. When allosteric modulators, such as 2,3-diphosphoglycerate (2,3-DPG) or chloride, are tested for their effects on modified Hb derivatives, they are added in amounts ranging from 1 to 4 times the Hb concentration. Unmodified Hb or modified hybrid Hb tetramers that have been subjected to the procedures of chromatography, chain separation, and reconstitution described below retain the native properties and full cooperativity of native untreated hemoglobin.

Reductive N-Carboxymethylation of N-Terminal Amino Groups of Hemoglobin

The objective behind studies with this derivative was to provide a stable analog of the CO_2 adduct with hemoglobin (HbNHCOOH).[5,6] The term carboxymethylation used in this section refers to the N-carboxymethylation of amino groupos (HbNHCH$_2$COOH) [Eq. (1)] rather than to the S-carboxymethylation of thiols. The introduction of carboxymethyl (Cm) groups at the N-terminal amino groups of Hb is achieved under mild conditions at neutral pH by treatment of the protein with sodium glyoxylate in the presence of sodium cyanoborohydride,[5] i.e., reductive alkylation with the goal of achieving regioselective modification. Because this

[5] A. DiDonato, W. J. Fantl, A. S. Acharya, and J. M. Manning, *J. Biol. Chem.* **258,** 11890 (1983).
[6] W. J. Fantl, A. DiDonato, J. M. Manning, P. H. Rogers, and A. Arnone, *J. Biol. Chem.* **26,** 12700 (1987).

procedure can, in theory, lead to either mono- or disubstitution by carboxymethyl groups, experimental conditions have been chosen to limit the reaction to the stage of monocarboxymethylation. Hence, the concentration of glyoxylate is lower than the total concentration of protein amino groups, i.e., 1 mM Hb (equivalent to 48 mM in total α- and ε-amino groups) is treated with 10 mM glyoxylate and 100 mM NaCNBH$_3$ at pH 7.2 for 1 hr at 25°. With this ratio of reactants and the choice of neutral pH, the overall extent of reaction favors the preferential monocarboxymethylation of the N-terminal residues. Peptide mapping, amino acid analysis, and X-ray diffraction analysis of the purified carboxymethylated hemoglobin product have indicated the presence of only the monocarboxymethyl derivative of the N-terminal valine for the derivatives Hb$_1$ and Hb$_2$ described below.[5,6] It is very likely that the introduction of a second carboxymethyl group at the same site is sterically unfavorable in hemoglobin. However, other proteins may be able to form dicarboxymethyl derivatives if the microenvironment of the site is conducive to the second substitution.[7]

Separation of Carboxymethylated Hb Derivatives

For purification of the hemoglobin derivatives, the dialyzed or gel-filtered Hb is applied to a column (2.2 × 35 cm) of the anion-exchange cellulose, DE-52 (Whatman, Clifton, NJ), and developed with a linear gradient of 500 ml each of CO-saturated 50 mM Tris–acetate, pH 8.3 and pH 7.3.[5] As shown in Fig. 1, four discrete species are eluted in 85% overall yield. When [^{14}C]glyoxylate is used, the number of carboxymethyl groups per derivative can be calculated. Derivative Hb$_0$, which comprises 9% of the total amount applied, does not contain any ^{14}C label and represents unreacted protein. Hb$_1$, (26% yield) contains about 2 mol of ^{14}C label/Hb tetramer. Hb$_2$ (28% yield) has approximately 3.5 mol of ^{14}C label/Hb tetramer, and peptide mapping shows that it is carboxymethylated on the N-terminal residues of all four chains. The fourth derivative, Hb$_3$ (22% yield), contains 5 to 6 mol of [^{14}C]glyoxylate/Hb tetramer, which represents carboxymethylation on its four N-terminal valine residues as well as some of its lysine residues. Because of this heterogeneity, this derivative is not considered useful for definitive studies.

Preparation of Specifically Carboxymethylated Hybrid Tetramers

To prepare hybrid carboxymethylated tetramers, i.e., those in which the N terminus of either the α or the β chain is carboxymethylated and

[7] T.-P. King, L. Kochoumian, and L. M. Lichtenstein, Arch. Biochem. Biophys. 178, 442 (1977).

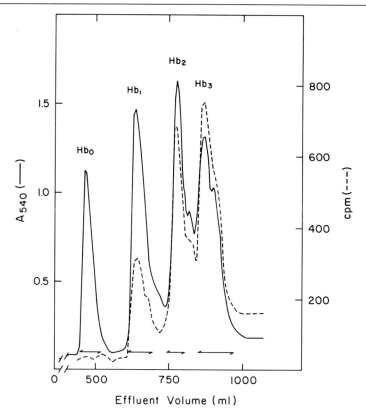

Fig. 1. Separation of specifically carboxymethylated Hb tetramers. The experimental procedures are described in the text and by DiDonato et al.[5]

the other chain has an unblocked N terminus, a strategy that has been employed successfully is recombination of equivalent amounts of carboxymethylated α or β chains with unmodified β or α chains, respectively.[5] It is our experience that this approach is preferable to attempts at modifying either the α or β subunit in the intact tetramer.

Because Hb_2 has its four N-terminal residues carboxymethylated, it is the choice for preparation of the separated carboxymethylated α and β chains as the p-hydroxymercuribenzoate (HMB) derivatives using the same procedures as described in a previous chapter for carbamylated Hb.[1] For example, in the preparation of the carboxymethylated hybrid tetramer $\alpha_2^{Cm}\beta_2$, the HMB-Cm α chain is mixed with an equivalent amount of unmodified HMB β chain at the same pH at which the chains are eluted from the columns. A 300-fold excess of mercaptoethanol (relative to the concentration of original SH groups) is added to remove the HMB groups from chains. After remaining at 4° overnight, the native tetramers are quantitatively regenerated *in situ*. Excess 2-mercaptoethanol is removed by dialysis against several volumes of CO-saturated 10 mM NaCl. The specifically N-carboxymethylated Hb hybrid tetramers prepared in this manner have been shown to be pure and fully functional (see Table II). They can be stored at −80° or in liquid N_2 as the CO derivative.

Effect of Carboxymethylation on Functional Properties of Hb

The hybrid carboxymethylated on the N termini of both α chains, $\alpha_2{}^{Cm}\beta_2$, has an intrinsic oxygen affinity (P_{50} = 12 mm Hg) that is lower than that of the native protein (P_{50} = 7 mm Hg) (Table II). Addition of 2,3-DPG to this hybrid lowers its oxygen affinity about fourfold, to 48 mm Hg. This $\alpha_2{}^{Cm}\beta_2$ hybrid has a reduced response to added chloride because a major chloride-binding site comprising Val-1(α) is blocked; all other chloride-binding sites in this hybrid, especially those involving the β chains and the central cavity, are unmodified and are therefore free to interact with chloride.[5,6]

The hybrid carboxymethylated on the N termini of both β chains, $\alpha_2\beta_2{}^{Cm}$, has an oxygen affinity (P_{50} = 17 Hg) that is also lower than that of the unmodified tetramer (Table II). The presence of the carboxymethyl group protruding into the cleft between the two β chains, as shown by X-ray diffraction analysis,[6] limits but does not prevent further lowering of the oxygen affinity by 2,3-DPG (maximum P_{50} = 25 Hg). The addition of chloride to $\alpha_2\beta_2{}^{Cm}$ results in a significant lowering of the oxygen affinity due mainly to the binding of the chloride anion to the region around the N terminus of the α chain, which is free in $\alpha_2\beta_2{}^{Cm}$.

The hybrid carboxymethylated on all four N-terminal residues, $\alpha_2{}^{Cm}\beta_2{}^{Cm}$, has an oxygen affinity that is considerably lower (P_{50} = 37 mm Hg) than the additive effects of this modification at the N termini of the individual α and β chains (Table II). The addition of 2,3-DPG to this hybrid results in a further lowering of the oxygen affinity (P_{50} = 50 mm Hg). Both $\alpha_2\beta_2{}^{Cm}$ and $\alpha_2{}^{Cm}\beta_2{}^{Cm}$ retain some response to 2,3-DPG, perhaps because the Cm group does not fully occupy the DPG site. Because the hybrid $\alpha_2{}^{Cm}\beta_2{}^{Cm}$ already has two of its major anion-binding regions covalently occupied with a negatively charged anion, further lowering of the oxygen affinity by chloride is marginal.

For all of the hybrids studied (except for $\alpha_2{}^{Cm}\beta_2{}^{Cm}$ at low oxygen tensions), the degree of cooperativity is unaffected by the modification; n values remain at 2.4 (Table II). This finding argues against significant distortion of subunit contacts after introduction of the negatively charged carboxymethyl group.

Acetylation of Hemoglobin

In theory, acetylation should be capable of providing information on the function of amino groups in hemoglobin. However, many of the acetylating agents, such as acetic anhydride or acetyl chloride, are so potent that acetylation with these reagents is difficult to control, devoid of any specificity, and prone to cause denaturation of proteins. Therefore, Kluger and Tsui[8] designed and synthesized methyl acetyl phosphate (MAP), a mild acetylating agent linked to a monoanionic phosphate group. This combination provided for binding of the reagent to a particular site on the protein, prior to acetylation of a nearby amino group by the other end [Eq. (2), Scheme I]. We find that this acetylating agent, when used in a regioselective fashion, is very specific for one of the most avid anion-binding sites of hemoglobin, the DPG cleft between the two β chains.[9] It has proved useful in reducing the gelation of sickle hemoglobin[10] to decrease the high density of sickle cells, and in elucidating the major and minor oxygen-linked chloride-binding sites in Hb when used in a regiorandom manner.[11]

Treatment of Intact Human Erythrocytes with Methyl Acetyl Phosphate (Regioselective)

Methyl acetyl phosphate is prepared essentially by the modified method of Kluger and Tsui.[8,10] The product, dimethyl acetyl phosphate, is collected by distillation under vacuum (0.02 Torr) and then treated with NaI in dry acetone; this product is recrystallized from methanol/ether. Elemental analysis is consistent with the correct structure for MAP. For preparation of [^{14}C]MAP[10] to facilitate identification of acetylated sites, [^{14}C]acetyl chloride is mixed with unlabeled acetyl chloride. The use of a radiolabeled reagent is essential for the regiorandom modification, described below, so that all of the unmodified hemoglobin (as shown in Fig. 3; see later) and unlabeled chymotryptic/tryptic peptides can be removed. Selective acetylation of some amino groups in the DPG cleft of Hb A or Hb S[9–12] is achieved by treating either normal or sickle erythrocytes with MAP (25 mM) in 50 mM HEPES buffer, pH 7.5, containing 0.1 M NaCl at 37°. Erythrocytes are permeable to MAP, but the intracellular

[8] R. Kluger and W.-C. Tsui, *J. Org. Chem.* **45**, 2723 (1980).
[9] H. Ueno, M. A. Pospischil, J. M. Manning, and R. Kluger, *Arch. Biochem. Biophys.* **244**, 795 (1986).
[10] H. Ueno, L. J. Benjamin, M. A. Pospischil, and J. M. Manning, *Biochemistry* **26**, 3125 (1987).
[11] H. Ueno and J. M. Manning, *J. Protein Chem.* **11**, 177 (1992).
[12] H. Ueno, M. A. Pospischil, and J. M. Manning, *J. Biol. Chem.* **264**, 12344 (1989).

concentration attained is not known. After 2 hr in a shaking water bath, the cells are washed with isotonic saline to remove the unreacted MAP. Lysis is achieved by treating the cells with an equal volume of distilled water. The hemolysate is dialyzed against 10 mM potassium phosphate, pH 5.85, prior to chromatography.

Separation of Acetylated Human Hemoglobin Products (Regioselective)

Separation of the acetylated hemoglobin derivatives is achieved by cation-exchange chromatography on carboxymethyl cellulose (2 × 11 cm) with a linear gradient of 150 ml each of 10 mM potassium phosphate, pH 5.85, and 15 mM potassium phosphate, pH 7.50. The elution profile for human deoxyHb A in Fig. 2 shows the appearance of three major modified fractions, in addition to some unmodified hemoglobin. Using [^{14}C]MAP, the number of acetyl groups per tetramer can be calculated. Thus, the modified species in peaks b_1, b_2, and b_3 of Fig. 2 are a mixture of components derivatized at either two, four, or six of the three susceptible sites, respectively. For example, it is likely that the derivative in the least modified Hb peak, which is labeled b_1 in Fig. 2, has its two acetyl groups

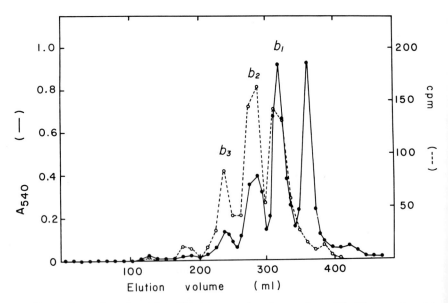

FIG. 2. Separation of acetylated Hb tetramers from human deoxy Hb. The experimental procedures are described in the text and by Ueno and Manning.[11]

per tetramer distributed among Val-1, Lys-82, and Lys-144 of the β chain, within the DPG cleft.[9,10]

Regiorandom Acetylation of Bovine Hemoglobin with MAP

In these studies,[11] bovine Hb was used, because it is more responsive to anions than is human Hb. The random acetylation of the amino groups of bovine Hb to identify sites involved in the binding of chloride is performed with isolated Hb. In this case, the hemolysate is passed through a mixed anion/cation desalting column[11] to remove chloride from the hemoglobin. Deoxygenated or oxygenated bovine hemoglobin (4 μmol) is treated with MAP (20 μmol) for 1–2 hr at 25°. Under these conditions, about 80% of the protein molecules is modified and 20% is free; peptide mapping indicates that 14 of the 24 amino groups per $\alpha\beta$ dimer of bovine Hb are acetylated to varying extents. Just as in the regioselective approach wherein the sites of acetylation are distributed among the susceptible amino groups in the DPG cleft, Hb acetylated in the regiorandom fashion has its modified sites distributed throughout the protein. However, a given hemoglobin tetramer has only a few of its functional amino groups acetylated. Therefore, a large amount of hemoglobin functionality is retained, i.e., cooperativity, n = 2.0, (Table II).

Separation of Acetylated Bovine Hemoglobin Products (Regiorandom)

It is not necessary to separate these acetylated products from each other but only from the unmodified Hb, which is characterized by a 540-nm absorbance and the absence of radioactivity[11,12] (Fig. 3). The modified chains contain multiple sites of acetylation, all of which have been determined by peptide mapping.[11] The strategy for peptide mapping in regiorandom modification is different from that usually employed in regioselective modification. First, the acetylated α-globin and acetylated β-globin (Fig. 3) are separately treated with trypsin (2% by weight) at 37° overnight. This digest is further treated with chymotrypsin (4% by weight) at 37° overnight. Because not all of the sites are modified in every protein molecule in the regiorandom approach, the use of both enzymes is advantageous. Trypsin will not digest an acetylated lysine residue whereas it will cleave the corresponding unlabeled lysine in another molecule. Therefore, these peptides are separable during HPLC. Chymotryptic cleavage is used primarily to hydrolyze at susceptible sites on either side of the acetylated lysine sites, thereby generating smaller peptides that are more readily purified. The use of both enzymes permits assignment of the acetylated site if a peptide has one modified and one unmodified lysine.

TABLE II
FUNCTIONAL PROPERTIES OF MODIFIED HEMOGLOBIN DERIVATIVES

Hemoglobin derivative	Source	P_{50} (mm Hg)	Hill coefficient (n)	Site of modification
Carboxymethylated Hb				
$\alpha_2\beta_2$ (unmodified)[a]	Human	7	2.4	—
$\alpha_2^{Cm}\beta_2$	Human	12	2.4	Val-1(α)
$\alpha_2\beta_2^{Cm}$	Human	17	2.4	Val-1(β)
$\alpha_2^{Cm}\beta_2^{Cm}$	Human	37	2.4	Val-1(α), Val-1(β)
Acetylated Hb				
$\alpha_2\beta_2$ (unmodified)[a]	Human	10	2.3	—
Ac-$\alpha_2\beta_2$ (selective)	Human	28	2.3	Val-1(β), Lys-82(β), Lys-144(β)
Ac-$\alpha_2\beta_2$ (random)	Bovine	44	2.0	See Ueno and Manning[11]
Glycated Hb				
$\alpha_2\beta_2$ (unmodified)[a]	Human	8	2.5	—
$\alpha_2^{DHP}\beta_2$	Human	5	2.5	Val-1(α) + Lys[b]
$\alpha_2\beta_2^{DHP}$	Human	10	2.5	Val-1(β) + Lys[b]
Cross-linked Hb				
(α-DIBS-α)β_2	Human	6	2.5	Val-1(α)–Val-1(α)

[a] The values for unmodified hemoglobin samples vary somewhat for different derivatives because the determinations were made on different instruments under somewhat different conditions, which are described in the individual references. For a given derivative, the precision of the measurement is ±1 mm Hg.
[b] The glycated hybrids contain modifications at Val-1(β), Lys-16(α), Lys-59(β), Lys-82(β), and Lys-120(β).

The amount of radiolabel in each isolated purified peptide is an estimate of the amount of acetylation either in deoxy- or oxyHb. Comparison of the extent of reaction at each site as a function of the presence or absence of O_2 permits the identification of those amino groups that are acetylated in an "oxygen-linked" manner. Five such sites over the entire hemoglobin molecule have been identified.[11] Molecular modeling studies indicate these sites are located either in or at the ends of the molecular dyad axis of Hb to form an intramolecular chloride channel (see Discussion below and Ueno and Manning[11]).

Effect of Acetylation on Oxygen Affinity

The oxygen affinity of Hb is lowered to 28 mm Hb by regioselective acetylation (Table I). The oxygen affinity of Hb acetylated in a regiorandom but extensive fashion is very low (44 mm Hg), which is probably due mainly both to the acetylation of amino groups at the DPG-binding

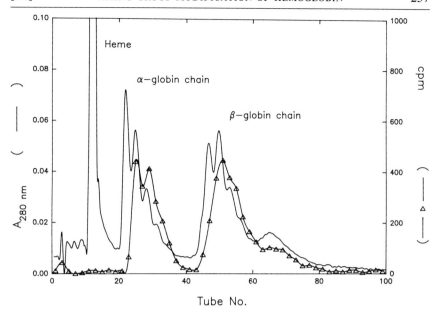

FIG. 3. Separation of acetylated globin chains on HPLC. The experimental procedures are described by Ueno and Manning.[11] The line without symbols represents the absorbance at 280 nm; Δ, radioactivity. For the α-globin acetylated chains, all the tubes containing radioactivity were pooled and analyzed; the same procedure was followed for the β-globin chains. Thus, the unacetylated globin chains are removed and the acetylated mixture is analyzed.

site and along the sides of the dyad axis of hemoglobin (see Discussion below and Ueno and Manning[11]).

Effect of Acetylation on Alkaline Bohr Coefficient

Studies on human HbA had previously shown that chloride-dependent and chloride-independent sites contribute to the alkaline Bohr coefficient.[13–16] Val-1(α) is primarily responsible for the chloride-dependent alkaline Bohr coefficient, and the chloride-independent alkaline Bohr coefficient is contributed predominantly by His-146(β).[13] Our results on the

[13] M. F. Perutz, J. V. Kilmartin, K. Nishikura, J. H. Fogg, P. J. G. Butler, and H. S. Rollema, *J. Mol. Biol.* **138,** 649 (1980).
[14] C. Fronticelli, E. Bucci, and C. Orth, *J. Biol. Chem.* **259**(10), 841 (1984).
[15] C. Ho and I. M. Russu, *Biochemistry* **26,** 6299 (1987).
[16] G. Fermi and M. F. Perutz, *in* "Atlas of Molecular Structures in Biology. 2. Haemoglobin and Myoglobin" (D. C. Phillips and F. M. Richards, eds.), pp. 3–101. Oxford Univ. Press (Clarendon), Oxford, 1981.

influence of chloride on the alkaline Bohr coefficient of extensively acetylated bovine Hb reinforce these conclusions because we can distinguish
between the relative contributions of amino or imidazole groups in the
same molecule, because MAP reacts with Val-1(α) but not with His-
146(β).[11] Hence, in the presence of 0.1 M chloride, unmodified bovine
hemoglobin has an alkaline Bohr coefficient of 0.48, a value consistent
with that previously reported for bovine Hb[14] and close to that found for
human HbA.[13] In low concentrations of chloride, the Bohr coefficient is
reduced to a value of 0.20, which is 40% of the total Bohr coefficient; this
part of the alkaline Bohr coefficient is likely contributed by His-146(β),[13]
to which chloride does not functionally bind. Furthermore, this part of
the Bohr coefficient is not affected by acetylation with MAP in low concentration of chloride (Bohr coefficient, 0.19). That part of the chloride-
dependent alkaline Bohr coefficient blocked by acetylation (0.48 −
0.28 = 0.20; 40%) is assigned primarily to Val-1(α), because it is the major
site of acetylation. The residual chloride-induced alkaline Bohr coefficient
of acetylated Hb (0.28 − 0.19 = 0.09; 20%) is probably due to other
functional amino side chains, probably Lys-81(β), that are not completely
acetylated in a fraction of the tetramers. Thus, the major contributors to
the alkaline Bohr coefficient in bovine Hb are Val-1(α) (40%; chloride
dependent), Lys-81(β) (20%; chloride dependent), and His-146(β), based
on the studies of others[13] and because it is not acetylated by MAP (40%;
chloride independent).

Glycation of Amino Groups in Hemoglobin

Two types of nonenzymatic glycation are described.[17–22] These reactions are referred to as reductive or nonreductive (nonenzymatic) alkylation depending on the presence or absence of sodium cyanoborohydride,
respectively. In general, reductive alkylation [analogous to the reaction
in Eq. (1)] is a faster reaction than the nonreductive type [Eq. (3)]. In the
latter, the sites of reaction are dictated by several factors, including the
ease with which certain amino groups form a Schiff base with the aldehyde
donor and the efficiency with which nearby side chains of the protein,
either in the adjacent sequence or in the overall three-dimensional struc-

[17] A. S. Acharya, L. G. Sussman, and J. M. Manning, *J. Biol. Chem.* **260**, 6039 (1985).
[18] A. S. Acharya, L. G. Sussman, W. M. Jones, and J. M. Manning, *Anal. Biochem.* **136**, 101 (1984).
[19] A. S. Acharya, L. G. Sussman, and J. M. Manning, *J. Biol. Chem.* **258**, 2296 (1983).
[20] N. Mori and J. M. Manning, *Anal. Biochem.* **152**, 396 (1986).
[21] Y. Bai, H. Ueno, and J. M. Manning, *J. Protein Chem.* **8**, 299 (1989).
[22] N. Mori, Y. Bai, H. Ueno, and J. M. Manning, *Carbohydr. Res.* **189**, 49 (1989).

ture, facilitate the subsequent transformation of the Schiff base into a ketoamine adduct, i.e., the Amadori rearrangement. In reductive glycation, sodium cyanoborohydride is included at the inception of the reaction so that the Schiff base between the aldehyde and the susceptible amino group is reduced as soon as it is formed. These reactions have been performed with several sugar aldehydes, including glycoaldehyde, glyceraldehyde, and glucose. The first two compounds form Schiff base adducts very rapidly because they are straight-chain compounds, but glucose exists predominantly in the ring conformation, and therefore is much less reactive than glycoaldehyde or gyceraldehyde. If glucose were substituted for glyceraldehyde in the following description, the experimental conditions would require longer incubation periods and/or higher concentrations of glucose to attain an extent of modification comparable to that obtained with glyceraldehyde.

Nonenzymatic Glycation of Hb A with Glyceraldehyde

When this procedure is performed with intact red cells, the conditions for physiological nonenzymatic glycation are mimicked.[17-19] Erythrocytes are resuspended in an equal volume of isotonic phosphate-buffered saline, pH 7.4, and incubated at 37° for 3 hr with 10 mM [^{14}C]glyceraldehyde. After centrifugation and washing with phosphate-buffered saline, the treated erythrocytes are resuspended in 1 M phosphate buffer, pH 6.0, to their original volume of whole blood (the high-ionic-strength buffer is needed for the subsequent reduction with NaBH$_4$). This treatment promotes lysis of the cells and the glycated Hb is isolated, as described below. These glycated adducts represent a mixture of various species, which are reasonably stable after the Amadori rearrangement, so that their intrinsic properties can be studied. However, they are slowly hydrolyzed and can be stabilized by reduction with NaBH$_4$.[18]

Reductive Glycation with Glyceraldehyde

In this procedure, sodium cyanoborohydride is present at the beginning of the reaction.[17] Purified Hb A (1 mM) is treated with 10 mM glyceraldehyde and 20 mM NaCNBH$_3$ at 37° for 30 min at 25°. Excess reagents are then removed by gel filtration on Sephadex G-25 (2.2 × 35 cm); this column is equilibrated and eluted either with 50 mM Tris–acetate, pH 8.5, or with 10 mM phosphate buffer, pH 5.85, depending on the type of ion-exchange chromatography to be used for separation of the variously glycated species, as described next.

Chromatography of Glycated Hb

The chromatographic separations[19] of hemoglobin glycated either in the absence or presence of NaCNBH$_3$ are shown in Figs. 4 and 5, respectively. During cation-exchange chromatography on CM-52, derivatives glycated on the N termini of α or β chains either reductively or nonenzymatically elute before derivatives glycated on ε-amino groups of lysine residues, which coelute with unmodified Hb. Presumably, the pK_a values of glycated N-terminal groups of hemoglobin subunits are considerably lower than those of ε-glycated lysine residues.

For Hb glycated nonreductively, the modifications are on the N terminus of Val-1(β) and on the ε-amino groups of Lys-16(α), Lys-59(β), Lys-82(β), and Lys-120(β). Derivatives glycated on the N termini of the β chains elute before derivatives containing glycated lysine residues (Fig. 4); however, the first component can contain both types of modification.[19] A possible mechanism for nonenzymatic glycation (Amadori rearrangement) is described below.

For Hb reductively glycated with glyceraldehyde, two dihydroxypropylated (DHP) derivatives elute before unmodified Hb (Fig. 5); DHP Hb A, which is modified on all four N-terminal residues of the α or β chains,

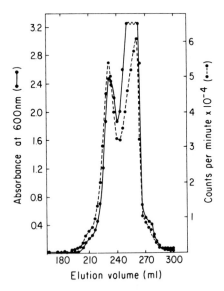

FIG. 4. Separation of nonreductively glycated Hb. The experimental procedures are described in the text and by Acharya *et al.*[19] The first radiolabeled component contains predominantly Hb tetramers modified on the N termini of the chains. The second radiolabeled component contains Hb tetramers modified on the ε-amino groups of Lys residues.

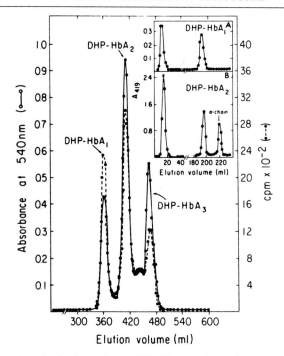

FIG. 5. Separation of reductively glycated Hb. The experimental procedures are described in the text and by Acharya *et al.*[19] Dihydroxypropylated Hb A$_1$ (DHP-HbA$_1$) is modified at its 4-N-terminal group; the inset shows no free α-chain in this sample. DHP-HbA$_2$ is modified at the N termini of its two β chains; the inset shows free α chain in this sample, which is described more fully by Acharya *et al.*[17]

elutes before DHP Hb A$_2$, which is modified on two of the N-terminal residues. Derivatives modified on the ε-amino groups of lysine elute later, together with unmodified Hb A. The locations of these modifications are summarized in Table II.

Effect of Glycation on Oxygen Affinity

As shown in Table II, modification of the N terminus of the α chain by reductive glycation with glyceraldehyde has little effect on the oxygen affinity of the hybrids, whereas reductive glycation at the N terminus of the β chain leads to a slight decrease in the oxygen affinity of Hb A. This type of modification of the α-amino groups has little influence on cooperativity ($n = 2.5$).

Cross-Linking of Hemoglobin at N-Terminal Residues of α Chains

This modification is of the regioselective type. 2,5-Diisothiocyanato-benzene sulfonate (DIBS) forms several cross-linked products with hemoglobin A [Eq. (4)].[23] The derivative that is cross-linked between the N-terminal residues of the α chains is formed in highest yield; its isolation, properties, and plasma retention time are described here. It is likely that formation of this component with hemoglobin is favored by the size and geometry of the cross-linking agent. Kavanaugh et al.[24] reported cross-linking of hemoglobin A between the N termini of both β chains with a larger related compound.

Hemoglobin in 0.1 M potassium phosphate, pH 7.2 (200 μM in the deoxygenated state, usually a total of 3–5 μmol) is treated at 25° with a 10-fold molar excess of the cross-linking agent DIBS (Trans World Chemical, Rockville, MD), which is dissolved in cold buffer just before use. After 15 min the reaction is terminated by further incubation with a 30-fold molar excess of glycylglycine for 15 min at 25°. The solution is then dialyzed at 4° against 50 mM Tris–acetate, pH 8.3. The amount of methemoglobin is ≤5%.

The cross-linked hemoglobin sample (200–250 mg) is applied to a Whatman DE-52 column (2 × 30 cm) and eluted with a linear gradient of 50 mM Tris–acetate from pH 8.3 to pH 6.3 (500 ml of each) (Fig. 6). The recovery of hemoglobin from the column is 80–95%. The major component, labeled B in Fig. 6, is eluted just after unmodified Hb A, labeled A in Fig. 6, and represents about 50% of the total material. This component is very stable at neutral pH, i.e., no hydrolysis of the cross-link is found in neutral aqueous solution at room temperature for several months. However, the cross-link is slowly hydrolyzed at acidic pH with elevated temperature, as described below, and in the presence of excess nucleophiles.

When component B is analyzed by gel filtration, either in the presence or absence of 1 M MgCl$_3$, practically all of the material elutes with a molecular weight of 64,000,[19] suggesting that the cross-link is between like subunits, either α–α or β–β, rather than between α–β subunits. An α–β cross-linked tetramer would have dissociated into 32,000-Da dimers during gel filtration in 1 M MgCl$_2$.[25]

The amount of subunit cross-linking of component B has been determined by SDS–PAGE to be 50%, i.e., two cross-linked and two free

[23] L. R. Manning, S. Morgan, R. C. Beavis, B. T. Chait, J. M. Manning, J. R. Hess, M. Cross, D. L. Currell, M. A. Marini, and R. M. Winslow, *Proc. Natl. Acad. Sci. U.S.A.* **88**, 3329 (1991).

[24] M. P. Kavanaugh, D. T.-B. Shih, and R. T. Jones, *Biochemistry* **27**, 1804 (1988).

[25] A. G. Kirshner and C. Tanford, *Biochemistry* **3**, 291 (1964).

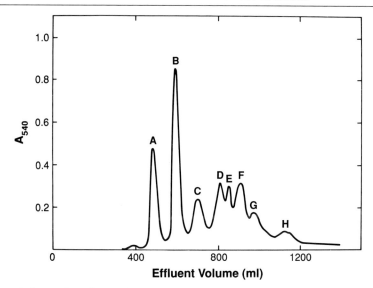

FIG. 6. Separation of cross-linked Hb A. The experimental procedures and the peaks are described in the text and by Manning *et al.*[23]

subunits per tetramer. The number of DIBS moieties per cross-linked hemoglobin tetramer can be approximated from the ratio of absorbance at 295 nm to that at 540 nm. For unmodified hemoglobin, this ratio is 1.42. Values exceeding this ratio are due to the presence of DIBS on the protein. Dividing this value by 15, the extinction coefficient of DIBS ($mM^{-1} cm^{-1}$), and by the hemoglobin concentration, calculated from its absorbance at 540 nm, provides an estimate of moles of DIBS per mole of hemoglobin tetramer. However, because the absorbance at 295 nm is a shoulder on the main protein absorbance band, the calculation gives only a rough estimate. For component B, the amount of DIBS is calculated to be 1.0 ± 0.2 per Hb tetramer.

Several lines of evidence indicate that it is the N-terminal residues of the α chains of component B that are cross-linked by DIBS.[23] Thus, when component B is subjected to reversed-phase HPLC, α chains with a molecular weight of 32,000 and un-cross-linked β chains of molecular weight of 16,000 are isolated. The identity of the α chains was established by amino acid analysis. The molecular weight of the α–α cross-linked globin chains was established by mass spectrometry. These findings were confirmed by N-terminal sequence analysis. Prior to Edman degradation, the DIBS cross-linked α–α chains were first treated for 16 hr with 5% (v/v) acetic acid at 56° to hydrolyze the cross-link and thus permit sequencing; otherwise, no sequence information would be obtainable. Subsequent SDS–gel electrophoresis showed that most of the cross-link (80%) had been cleaved by such treatment. Sequence analysis of the cleaved hemoglobin chains indicated the presence of Leu as the first residue, indicating that the original N-terminal Val had been removed together with the DIBS cross-link through cyclization by the overnight treatment with acetic acid; this chemistry is analogous to that in the Edman degradation.

Oxygen Affinity of Cross-Linked Hb

This hemoglobin derivative, containing a single DIBS cross-link between the N-terminal amino acids of the α chains, was found to have a somewhat increased affinity (P_{50} = 6 mm Hg) (Table I) and full cooperativity (n = 2.5). Its plasma retention time in rats was found to be significantly longer than that of carboxymethylated Hb.[23]

Discussion

Studies on these hemoglobin derivatives have led to some interesting insights into the properties of hemoglobin.[26]

N-Carboxymethylation

As shown by X-ray diffraction analysis,[6] the negatively charged carboxymethyl moiety at Val-1(α) interacts with Ser-131(α). The carbon dioxide hemoglobin adduct (carbamino derivative) displays the same interaction. Because this latter adduct with hemoglobin is very labile and hence difficult to study, the carboxymethyl derivative as a stable analog can be used to obtain information on the naturally occurring carbamino adduct.

The low oxygen affinity of the derivative carboxymethylated on its β chain ($\alpha_2\beta_2{}^{Cm}$), is consistent with the known avidity of anions for the cleft between the two β chains and the consequent lowering of the oxygen affinity of Hb by reduction of the positive charge within this cleft. Further studies on the molecular basis for this phenomenon may be facilitated through studies on this derivative.

The low oxygen affinity of N-carboxymethylated Hb makes it a good starting material for the preparation of a blood substitute.

[26] J. M. Manning, *Adv. Enzymol. Relat. Areas Mol. Biol.* **64**, 55 (1991).

Acetylation

Specific and limited acetylation of sickle cell Hb by MAP leads to a reduction in the gelation of this protein under anaerobic conditions.[10] As a result, the abnormally high density of sickle cells is decreased.[27]

Regioselective acetylation lowers the oxygen affinity of Hb by neutralization of positively charged side chains within the DPG cleft, analogous to the results with $\alpha_2\beta_2{}^{Cm}$ described above. This finding is consistent with the generalization that any reduction in the net positive charge in the 2,3-DPG-binding site leads to a lower oxygen affinity, as found earlier with Hb carbamylated at the N terminus of the β chain[3] and for Hb Raleigh, which is acetylated at Val-1(β), and for Hb Providence.[28]

Molecular modeling has shown that the oxygen-linked chloride-binding sites labeled by MAP in a regiorandom fashion are aligned in a channel along the molecular dyad axis connecting the two well-known anion-binding sites, i.e., the regions comprising Val-1(α) and the 2,3-DPG-binding site.[11] The dyad axis of Hb may be an important site for controlling its oxygen affinity. This central cavity is known to be more open in the deoxy conformation than in the oxy conformation. Any factor that hinders this constriction, such as the binding of chloride or other effectors, may lower the oxygen affinity considerably.

Comparisons of the alkaline Bohr effect of unmodified Hb with Hb acetylated in a regiorandom fashion in the presence or absence of chloride indicate agreement with the conclusions of Perutz and colleagues[13,16,29] regarding the contributions of the chloride-dependent sites [Val-1(α)] and the chloride-independent site [His-146(β)] to the alkaline Bohr coefficient.

Glycation

The gelation of hemoglobin S within sickle cells is significantly reduced by nonenzymatic glycation with glyceraldehyde. This effect is mainly contributed by glycation of Lys-16(α).[18]

The intramolecular catalysis involved in the Amadori rearrangement may involve imidazole groups of His residues in close proximity to the Schiff base,[17–22] either in the primary structure or in the three-dimensional structure of the protein. The mechanism is supported by studies on the relative extent of glycation of model peptides and of tetrameric human and horse Hb[20–22]; the latter Hb contains a Gln-2(β) residue instead of a

[27] H. Ueno, E. Yatco, L. J. Benjamin, and J. M. Manning, *J. Lab. Clin. Med.* **120,** 152 (1992).
[28] C. Bonaventura and J. Bonaventura, in "Biochemical and Clinical Aspects of Hemoglobin Abnormalities" (W. S. Caughey, ed.), p. 647. Academic Press, New York, 1980.
[29] M. F. Perutz, "Mechanisms of Cooperativity and Allosteric Regulation in Proteins," p. 12. Cambridge Univ. Press, Cambridge, 1990.

His-2(β) residue (found in human Hb) and the extent of glycation by glucose at Val-1(β) is significantly reduced.

Cross-Linking

The specific cross-linking of Hb between the N termini of its α chains leads to a prolongation of its plasma survival time,[23] because the α–α cross-linked tetramer cannot dissociate into $\alpha\beta$ dimers, which would be rapidly cleared from the circulation.

Glycolaldehyde is useful for obtaining intertetrameric cross-linking[30]; the site of cross-linking in Hb by this reagent is not yet established.

Acknowledgements

The author is indebted to many talented colleagues with whom he collaborated in various aspects of this work. These include Seetharam Acharya, Adelaide Acquaviva, Arthur Arnone, Yasuo Bai, Ronald Beavis, Ted Bella, Lennette Benjamin, Robert Buzolich, Brian Chait, Michael Cross, Douglas Currell, Alberto Di Donatao, Wendy J. Fantl, John Hess, Wanda Jones, Ronald Kluger, Lois Manning, Mario Marini, Javier Martin de Llano, Sheila Morgan, Nobuhiro Mori, Maria Pospischil, Leslie Sussmann, Hiroshi Ueno, Robert Winslow, and Edward Yatco. The author thanks Dr. William Agosta and Dr. Ronald Kluger for discussions that led to the term ''regiorandom.''

This work was supported in part by NIH Grant HL-18819 and by the U.S. Army Contract DAMD1788C8169.

[30] L. R. Manning and J. M. Manning, *Biochemistry* **27**, 6640 (1988).

[15] Amidation of Basic Carboxyl Groups of Hemoglobin

By M. Janardhan Rao and A. Seetharama Acharya

Crystallographic studies of human hemoglobin A (Hb A) and solution studies on the Bohr effect of the protein have implicated unique roles for one or more side-chain carboxyl groups (the β-carboxyl of Asp and/or the γ-carboxyl of Glu residues) in the structure and function of the protein. The formation and breakage of a salt bridge between the imidazole of His-146(β) and the β-carboxylate of Asp-94(α) have been suggested to play a major role in the alkaline Bohr effect of Hb.[1-3] The side-chain carboxylates of Glu-22(β), Asp-23(α), Glu-43(β), Asp-73(β), and Glu-121(β) of sickle cell hemoglobin (Hb S) have been implicated as part of the intermolecular

[1] M. F. Perutz, *Nature (London)* **228**, 726 (1970).

contact regions of the polymer of the deoxy protein.[4-6] The contribution of some of these acidic amino acid residues in the quinary interaction of Hb S polymer may be simply a consequence of the higher pK_a as compared to those of other side-chain carboxyl groups of the protein. Site-selective chemical modification of Hb A and Hb S was sought as an approach to establish the presence or absence of side-chain carboxyl groups with high pK_a and to evaluate their crucial structure–function role. The design of this chemical strategy was based initially on an assumption that the structural or functional role of one or more of the acidic amino acid residues of Hb A and/Hb S is a consequence of the basic nature of their carboxyl group and that the existing chemical methodologies could be refined to obtain a selective modification of the basic carboxyl groups of the proteins. Indeed, both these concepts have now been established as a result of our amidation studies of Hb A and Hb S over the years.[7-10] It has been established that as a result of their basic nature, the γ-carboxyl group of Glu-43(β) and of Glu-22(β) as well as the β-carboxyl group of Asp-47(β) are readily accessible for carbodiimide-mediated coupling reactions around neutral pH. In addition, the γ-carboxyl group of Glu-43(β) has been demonstrated to contribute significantly to the alkaline Bohr effect.[11] The primary emphasis of this chapter is to discuss the development of procedures for the site-selective amidation of carboxyl groups of Hb A and Hb S and the use of this chemical reaction as an *in situ* probe of the oxy–deoxy conformational changes at the $\alpha_1\beta_2$ interface of mutant hemoglobins as compared with that of Hb A. A brief discussion of the amidation chemistry is also included in this chapter. However, readers interested in detailed mechanistic aspects of the amidation of carboxyl groups of proteins in general or alternate procedures for modification of the carboxyl groups of proteins should refer to other extensive reviews.[12-16]

[2] M. F. Perutz, J. V. Kilmartin, K. Nishikura, J. H. Fogg, and P. J. G. Butler, *J. Mol. Biol.* **138**, 649 (1980).

[3] M. F. Perutz, A. M. Gronenborn, G. M. Clore, J. H. Fogg, and D. T.-B. Shih, *J. Mol. Biol.* **183**, 491 (1985).

[4] B. C. Wishner, K. B. Ward, E. E. Lattman, and W. E. Love, *J. Mol. Biol.* **98**, 179 (1975).

[5] E. A. Padlan and W. E. Love, *J. Biol. Chem.* **260**, 8272 (1985).

[6] E. A. Padlan and W. E. Love, *J. Biol. Chem.* **260**, 8280 (1985).

[7] R. Seetharam, J. M. Manning, and A. S. Acharya, *J. Biol. Chem.* **258**, 14810 (1983).

[8] A. S. Acharya and R. Seetharam, *Biochemistry* **24**, 4885 (1985).

[9] A. S. Acharya and L. Khandke, *J. Protein Chem.* **8**, 231 (1989).

[10] M. J. Rao and A. S. Acharya, *J. Protein Chem.* **10**, 129 (1991).

[11] M. J. Rao and A. S. Acharya, *Biochemistry* **31**, 7231 (1992).

[12] A. N. Glazer, *Annu. Rev. Biochem.* **39**, 101 (1970).

[13] S. Bauminger and M. Wilchek, this series, Vol. 70, p. 151.

General Aspects of Amidation Chemistry

The early attempts to modify the carboxyl groups of proteins involved esterification using methanolic HCl[17-20] or diazo compounds[21-23] and reduction of the esterified carboxyl group to the corresponding primary alcohol using either LiAlH$_4$ or LiBH$_4$. Diborane has also been recommended for the reduction of carboxyl groups in proteins without prior esterification.[24]

Riehm and Scheraga[25] introduced 1-cyclohexyl-3-(2-morpholinoethyl) carbodiimide metho-p-toluene sulfonate, a water-soluble carbodiimide, to modify the carboxyl groups of proteins in an aqueous medium. The reaction was carried out at pH 4.5 so that most of the protein carboxyl groups would exist in their protonated state. The carbodiimide-mediated modification of carboxyl groups of proteins was refined further by Hoare and Koshland[26,27] by incorporating nucleophiles in the system so that isopeptide linkages were generated at the side-chain carboxyl groups. In these studies, carbodiimides, such as 1-cyclohexyl-3-(2-morpholinoethyl)carbodiimide, 1-benzyl-3-dimethylaminopropylcarbodiimide, and 1-ethyl-3-dimethylaminopropylcarbodiimide, were used as carboxyl group activating reagents and glycine methyl ester was used as the nucleophile. The use of denaturant (8 M urea) in the amidation mixture has been advocated to obtain a quantitative modification of carboxyl groups of protein. Urea unfolds the protein, thereby making the carboxyl groups of the protein readily accessible for activation with carbodiimide.

The amidation of carboxyl groups is efficient around pH 4.5. The activation of the carboxyl group to form the O-acylisourea (activated carboxyl) involves the protonated species of the carboxyl group (Fig. 1). The pK_a of the β-carboxyl of Asp and the γ-carboxyl of Glu is around pH 4.8 and hence around pH 4.5 the side-chain carboxyl groups of the

[14] B. F. Erlanger, this series, Vol. 70, p. 85.
[15] T. Borsos and J. J. Langone, this series, Vol. 74, p. 161.
[16] K. L. Carraway and D. E. Koshland, Jr., this series, Vol. 25, p. 616.
[17] H. Fraenkel-Conrat and H. S. Olcott, *J. Biol. Chem.* **161**, 259 (1945).
[18] J. P. Riehm and H. A. Scheraga, *Biochemistry* **4**, 772 (1965).
[19] A. S. Acharya and P. J. Vithayathil, *Int. J. Pept. Protein Res.* **7**, 207 (1975).
[20] P. J. Vithayathil and F. M. Richards, *J. Biol. Chem.* **236**, 1380 (1961).
[21] A. C. Chibnall, J. L. Mangan, and M. W. Rees, *Biochem. J.* **68**, 114 (1958).
[22] M. S. Doscher and P. E. Wilcox, *J. Biol. Chem.* **236**, 1328 (1961).
[23] T. G. Rajagopalan, W. H. Stein, and S. J. Moore, *J. Biol. Chem.* **241**, 4295 (1966).
[24] M. Z. Atassi and A. F. Rosenthal, *Biochem. J.* **111**, 593 (1969).
[25] J. P. Riehm and H. A. Scheraga, *Biochemistry* **5**, 99 (1966).
[26] D. G. Hoare and D. E. Koshland, Jr., *J. Am. Chem. Soc.* **88**, 2057 (1966).
[27] D. G. Hoare and D. E. Koshland, Jr., *J. Biol. Chem.* **242**, 2447 (1967).

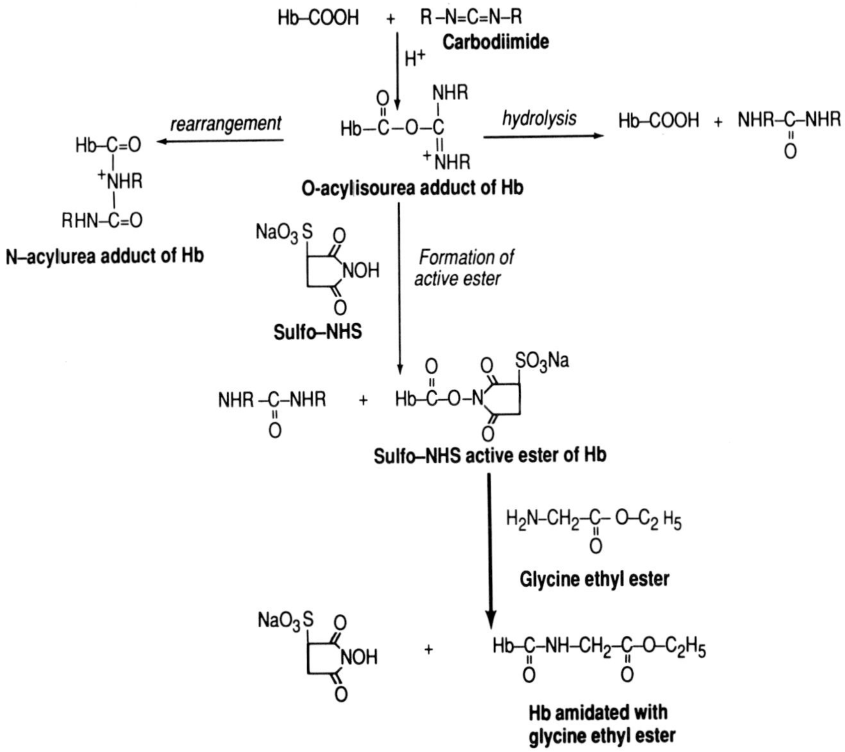

Hb–COOH   +   R–N=C=N–R

Carbodiimide

H^+

$$Hb-\overset{\overset{\displaystyle O}{\|}}{C}-O-\overset{\overset{\displaystyle NHR}{|}}{\underset{\underset{\displaystyle +NHR}{\|}}{C}}$$

O-acylisourea adduct of Hb

rearrangement ←

hydrolysis →   Hb–COOH   +   NHR–$\overset{\displaystyle O}{\underset{\displaystyle \|}{C}}$–NHR

$$Hb-C=O \\ \overset{\displaystyle |}{+NHR} \\ \overset{\displaystyle |}{RHN-C=O}$$

N-acylurea adduct of Hb

Sulfo–NHS

$$NaO_3S \quad O$$ NOH

Sulfo–NHS

Formation of active ester

NHR–$\overset{\displaystyle O}{\underset{\displaystyle \|}{C}}$–NHR   +   Hb–$\overset{\displaystyle O}{\underset{\displaystyle \|}{C}}$–O–N$\underset{O}{\overset{O}{\bigg\langle}}$ SO$_3$Na

Sulfo–NHS active ester of Hb

$$H_2N-CH_2-\overset{\overset{\displaystyle }{}}{\underset{\underset{\displaystyle O}{\|}}{C}}-O-C_2H_5$$

Glycine ethyl ester

$$NaO_3S \quad O$$ NOH   +   Hb–$\overset{\displaystyle O}{\underset{\displaystyle \|}{C}}$–NH–CH$_2$–$\overset{\displaystyle O}{\underset{\displaystyle \|}{C}}$–O–C$_2H_5$

Hb amidated with glycine ethyl ester

FIG. 1. Schematic representation of the amidation reaction of carboxyl groups of Hb: the amidation reaction of carboxyl groups of Hb in the presence of EDC, sulfo-NHS, and GEE is represented. Activation of a carboxyl group of Hb is initated by carbodiimide and results in the formation of an activated carboxyl group (O-acylisourea adduct). The activated carboxyl group can undergo hydrolysis or rearrangement in the absence of a rescuing agent or nucleophile. The sulfo-NHS traps the activated carboxyl group as sulfosuccinimidyl ester, inhibiting formation of the N-acylurea adduct. The sulfosuccinimidyl ester possesses higher stability compared with that of the O-acylisourea adduct, hence the amidation reaction is channeled either through the amidation with nucleophile or its hydrolysis back to the free carboxyl group. The latter can be reactivated by EDC.

proteins are expected to be in their protonated state. Thus the choice of this low pH ensures that most, if not all, of the side-chain carboxyl groups of the protein are present in the protonated state. If the "activated carboxyl group" (O-acylisourea) is not attacked by an amine to form an amide bond (i.e., amidation), it is subjected to hydrolysis to regenerate carboxyl groups. These can be recycled by the carbodiimide activation process. Alternatively, the O-acylisourea can also rearrange to a stable N-acylurea adduct. The partitioning of the O-acylisourea of protein carboxyl groups through the hydrolytic pathway and the rearrangement as a function of pH have not been addressed. The hydrolysis rate constant of O-acylisourea at pH 4.75 is of the order of 2–3 sec^{-1}. Thus, the activated carboxyl groups are readily deactivated[27] if they are not subjected to the nucleophilic attack. Accordingly, the activated carboxyl group needs to be trapped quickly to achieve the derivatization of the carboxyl group. Thus, the use of a high concentration (1 to 2 M) of nucleophile is recommended in the amidation studies of proteins.

One of the unique advantages of the amidation procedure is the flexibility to use a variety of nucleophiles, thereby providing an opportunity to modulate the hydrophobicity/hydrophilicity of the microenvironment of the carboxyl groups and also to investigate the consequence of such perturbations on the structure of the protein. Armstrong and McKenzie[28] converted the Glu and Asp residues of proteins to Gln and Asn, respectively, by using ammonium chloride in the amidation procedure. Wilchek et al.[29,30] have modulated the amidation procedure further in such a way that the modified proteins still retain the negative charge in the microenvironment of the protein carboxyl group, but with a stereochemistry that is distinct from that of the original carboxyl group. Glycine N-phthalimidomethyl or L-alanylglycine N-phthalimidomethyl esters have been used as nucleophiles instead of glycine methyl ester. The N-phthalimidomethyl group of the nucleophile serves as a protecting group for the glycine carboxyl group during the amidation reaction. However, the protecting group can be removed readily after the completion of the amidation reaction by incubating the modified protein in 0.5 M piperidine for 30 hr at 4°. Nitrotyrosine ethyl ester was used as the nucleophile in the amidation procedure by Lacombe et al.[31] This nucleophile has a unique advantage in that the nitro group can be reduced to an amino group under mild conditions to generate aminotyrosine[32] after the amidation reaction. The pK_a of the amino group of amino tyrosine is around 5.0 and thus could serve as a selective anchoring site for tagging bifunctional reagents such as difluoronitrobenzene or other photoactivable reagents monofunctionally at low pH.

The emphasis of most of the early amidation studies of proteins has been the derivatization of all the accessible carboxyl groups of the proteins. In subsequent studies, particularly designed to achieve limited modification of carboxyl groups of proteins, the denaturants are not used and

[27] J. McD. Armstrong and H. A. McKenzie, *Biochim. Biophys. Acta* **147**, 93 (1967).
[29] M. Wilchek, A. Frensdorff, and M. Sela, *Arch. Biochem. Biophys.* **113**, 742 (1966).
[30] M. Wilchek, A. Frensdorff, and M. Sela, *Biochemistry* **6**, 247 (1967).
[31] G. Lacombe, N. V. Thiem, and B. Swynghedauw, *Biochemistry* **20**, 3648 (1981).
[32] M. Sokolovsky, J. F. Riordan, and B. L. Vallee, *Biochem. Biophys. Res. Commun.* **27**, 20 (1967).

limited concentrations of the activating reagent, i.e., carbodiimide, are employed. Thus, only the exposed, very reactive carboxyl groups of proteins are expected to derivatize. Amidation of trypsinogen and trypsin[33] and ribonuclease A (RNAse A) using 1-ethyl-3-[3'-(dimethylamino)propylcarbodiimide] (EDC) and glycinamide or glycine methyl ester as nucleophiles represents early attempts to obtain limited modification of the carboxyl groups of proteins.

General Procedural Aspects of Amidation of Hb A/Hb S with Glycine Ethyl Ester

The major objective of our initial amidation studies of Hb A and Hb S has been to establish the presence of side-chain carboxyl groups in the protein with a pK_a high enough so that these could be derivatized selectively around the neutral pH region and so that homogeneous amidated products of Hb could be isolated to establish the structural and functional consequences of the lack of a negative charge at a specific site. However, the low pH (4.5) used by Hoare and Koshland[27] to amidate the protein carboxyl groups is not appropriate to the amidation studies of Hb. At a pH as low as 4.5, denaturation of Hb is expected to occur. Accordingly, an intermediate pH of 6.0 and higher has been chosen for studies with Hb. The original procedure of Hoare and Koshland[27] involves the use of a pH-state, and the same procedure is used at pH 6.0. Nonetheless, the reaction conditions also have been standardized now so that the amidation reactions are carried out in 4-morpholinoethanesulfonic acid (MES) buffers, giving a higher flexibility for the amidation studies of Hb. Nonetheless, both procedures are described below because both procedures are expected to have specific advantages, depending on the experimental design.

Amidation of Hemoglobin in KCl Using a pH-Stat

The original procedure of Hoare and Koshland using a pH-stat has been adapted for amidation of Hb[7-9] with slight modification. The amidation reaction is carried out in 0.1 M KCl at pH 6.0, and urea is not used during the amidation reaction. EDC has been chosen for activating the carboxyl groups of Hb. Other water-soluble carbodiimides may have specific advantages. However, these aspects have not been investigated so far. But it is worthwhile to study the reaction with other carbodiimides, especially to establish whether the site selectivity of the amidation can be influenced by the nature of the carbodiimide. Generally, all of the

---

[33] T. M. Radhakrishnan, K. A. Walsh, and H. Neurath, *Biochemistry* **8**, 4020 (1969).

amidation studies are carried out at a Hb concentration of 0.5 mM. The concentration of Hb may be varied according to the need of experimental design. A typical amidation reaction mixture is 0.5 mM in Hb, 100 mM in glycine ethyl ester, and 10 or 20 mM in EDC. Radioactive (^{14}C)-labeled product, (either from NEN Research Products, Boston, MA, or American Radiochemical, St. Louis, MO) glycine ethyl ester (GEE) is used as the tracer to simplify the quantitation of the amidation reaction. Hb (1.0 mM) is dialyzed overnight at 4° against 0.1 M KCl adjusted to pH 6.0 with HCl. All the reagents are prepared in 0.1 M KCl as stock solutions and their pH is readjusted to 6.0 using either KOH (1 M) or HCl (1 M). Appropriate aliquots of Hb sample and nucleophile solution (GEE) are taken together and diluted with 0.1 M KCl (pH 6.0) so that the final concentration of Hb and nucleophile are 0.5 and 100 mM, respectively. The amidation reaction is initiated by introducing an appropriate aliquot of a freshly prepared concentrated solution of EDC in 0.1 M KCl (pH 6.0). The EDC solution prepared is used within 3 to 5 min after preparation. A final concentration of 20 mM EDC generally is used for the amidation reaction. The pH of the solution will change as a consequence of the amidation reaction, and this is maintained at 6.0 by using 0.01 M HCl. The concentration of HCl used for the titration is adjusted so that the total volume of HCl used to maintain the pH at 6.0 during the amidation reaction is not more than 5% of the initial reaction mixture. After the desired period of reaction, the amidated protein sample is desalted on a column of Sephadex G-25 (superfine, 0.9 × 30 cm, flow rate 15–20 ml/hr) equilibrated and eluted with 100 mM phosphate buffer, pH 6.8. The quantitation of nucleophile (radioactivity) incorporated into Hb provides an estimate of the amidation reaction. The reaction period and the concentration of EDC can be adjusted appropriately to obtain the desired amount of amidation reaction. A reaction period of 60 min with 20 mM EDC results in an average modification of 3 groups per mole of tetramer. The peptide mapping of amidated sample coupled with the isolation of the amidated peptides and their amino acid sequencing must be carried out to establish the selectivity of the reaction. Such investigations have established that more than 70% of the amidation reaction is on Glu-43(β), identifying the side-chain carboxyl group of this residue as the most basic carboxyl group of Hb A/Hb S. Small amounts of modifications are observed in Asp-47(β) and Glu-22(β). The pK_a values of the carboxyl groups of these two residues are, apparently, intermediate to those of Glu-43(β) and the rest of the carboxyl groups of the protein.

Amidation of Hemoglobin in MES Buffer

Amidation of the carboxyl groups of Hb using the pH-stat (in 0.1 M KCl) is a rather cumbersome procedure and is certainly not convenient if one wishes to pursue the detailed kinetics of the amidation of carboxyl groups of Hb. Accordingly, the feasibility of amidating Hb in MES buffer has been evaluated so that the use of the pH-stat could be avoided. The amidation reaction can be carried out in MES buffer of desired pH. This indeed simplifies the protocol for amidation of Hb.[10] The amidation reactions are carried out routinely in 20 mM MES buffer of desired pH. As discussed above, a typical amidation reaction mixture contains 0.5 mM Hb, 100 mM GEE (with appropriate amounts of [^{14}C]GEE), and 10 mM EDC. A concentrated solution of Hb (1.0 mM or higher) is dialyzed for overnight at 4° against MES buffer, pH 6.0. A stock solution of GEE also is prepared in 20 mM MES buffer, pH 6.0, and the pH of the solution is readjusted to 6.0 with alkali. To an aliquot of dialyzed Hb sample, an appropriate amount of the nucleophile is added and diluted with MES buffer so that the final concentration is 0.5 mM in Hb and 100 mM in GEE. The amidation reaction is initiated by introducing EDC. The stock solution of EDC also is prepared in MES buffer. The concentration of EDC stock solution is such that the volume of the solution added to initiate the amidation reaction is less than 2% of the total reaction volume. The amidation reaction is carried out routinely at room temperature and allowed to proceed for the defined period to achieve the desired extent of amidation. After incubation, the amidation mixture is desalted on a column of Sephadex G-25 (superfine, 0.9 × 30 cm) equilibrated and eluted with phosphate-buffered saline (PBS), pH 7.4. The amount of nucleophile (radioactivity) incorporated into protein is estimated to quantitate the amidation reaction. The peptide mapping of the amidated sample demonstrated that the modification reaction is predominantly on the γ-carboxyl group of Glu-43(β), just as in the case when the reaction is carried out in 0.1 M KCl. Thus the use of MES buffer in place of 0.1 M KCl and the pH-stat does not appear to have influence on the site selectivity of the amidation of Hb S.

Side Reactions during Amidation of Hb around Neutral pH:
  CM-Cellulose Chromatographic Investigation

Although the amidation of Hb A and Hb S at pH 6.0 demonstrated a high degree of selectivity for the amidation of the γ-carboxyl group of Glu-43(β) of Hb S, chromatography of the amidated Hb S (modified at pH 6.0 for 1 hr in the presence of 20 mM EDC and 100 mM GEE) on CM-cellulose showed that the product is heterogeneous (Fig. 2C). The amidated product contained at least five distinct chromatographic components. All of these components elute at positions after that of the unmodified Hb, demonstrating that these derivatives have an increased net positive charge compared to that of the parent molecule. The elution position

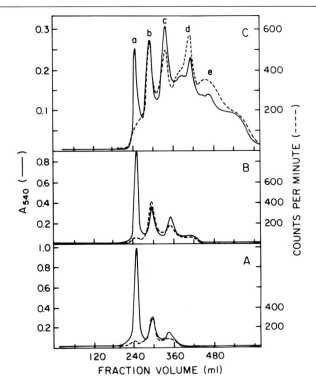

FIG. 2. Chromatography of amidated Hb S. Hb S (0.5 mM) amidated with GEE (100 mM) in the presence of 5, 10, or 20 mM EDC is passed through a Sephadex G-25 column (2.2 × 30 cm) equilibrated and eluted with 100 mM phosphate buffer, pH 6.8, containing 1 mM EDTA. The derivatized protein (50 mg) free from the reagents is chromatographed on a CM-52 cellulose column (0.9 × 30 cm) equilibrated with 15 mM phosphate buffer, pH 6.0. The protein was eluted with a linear gradient generated from 500 ml each of 15 mM phosphate buffer, pH 6.0, and 50 mM phosphate buffer, pH 8.3, both buffers containing 1 mM EDTA. (A) 5 mM EDC, (B) 10 mM EDC, and (C) 20 mM EDC. The components are identified as a–e, in the order of their elution from the column; the position of a corresponds to that of unmodified Hb S.

of the component represented by peak a in Fig. 2C corresponds to that of the unmodified Hb S. After 1 hr of the amidation reaction (on an average, 2.8 groups derivatized) less than 15% of Hb S remains unmodified. All of the components (b–e, Fig. 2C) eluting after the position of Hb S are radioactive. Component b had nearly 2 mol of GEE per mole of tetramer, suggesting that it is a disubstituted derivative of Hb S. Components c, d, and e had 1.6, 2.4, and 2.8 mol of GEE per mole of tetramer, respectively. Thus the increasing retardation of components c–e on the CM-cellulose column does not directly correlate with the extent of their amidation (i.e., the loss of the negative charges of the protein under the physiological conditions) as reflected by the GEE incorporated. Presumably, the heterogeneity of the amidated product reflects side reactions such as N-acylurea adduct formation, which could also contribute toward increasing the net negative charge of the protein. Consistent with this explanation, a control sample of Hb S incubated with only 20 mM EDC at pH 6.0 for 60 min showed the presence of chromatographic components that eluted at positions corresponding to those of components c, d, and e.

In an attempt to suppress the side reactions, the amidation reaction was carried out in the presence of 5 and 10 mM EDC. Nearly 1 mol of GEE per mole of tetramer was incorporated into Hb S after 1 hr of amidation reaction in the presence of 10 mM EDC. Chromatographic analysis of this amidated product showed the presence of one major radioactive component that eluted at the position of component b (Fig. 2B). The amount of this derivative in this sample is nearly 20%. The overall yield of component b is, however, higher than that seen with the amidated product obtained using 20 mM EDC even though the total amount of amidation in this sample is nearly threefold lower than that in the 20 mM EDC-reacted material. Thus the use of a lower concentration of EDC resulted in the suppression of the generation of components d and e, besides lowering the overall amidation reaction. The results demonstrate that the amidation reaction becomes more specific in the presence of 10 mM EDC as a consequence of the suppression of the side reactions that result in the generation of products that elute after component b from CM-52 columns. The chromatographic behavior of the sample amidated in the presence of 5 mM EDC (Fig. 2A) for 1 hr (with an incorporation of 0.5 groups per mole) is comparable to that of 10 mM EDC-reacted material except that component b is lower (about 12%) and the relative amounts of components eluting after b are reduced further. Thus the yield of the disubstituted derivative (component b) present in an amidated sample of Hb A is not directly proportional to the extent of amidation, although under all these conditions the amidation reaction shows a high selectivity for the modification of Glu-43(β). Clearly, as the concentration of the EDC increases, the side reactions become predominant. Presumably, these reaction products represent the stable N-acylurea adducts generated by the rearrangement of the O-acylisourea adduct. Apparently, although the attempt to limit the amidation by increasing the pH to 6.0 (as compared to pH 4.8 employed by Hoare and Koshland[26,27]) has succeeded in having a selective amidation at Glu-43(β), this manipulation also has resulted in an increase in the rearrangement of the O-acylisourea adduct to stable adducts. Alternatively, the microenvironment of Glu-43(β) or that of some other carboxyl group that is activated by EDC is such that it facilitates the rearrangement of the respective O-acylisourea adducts. Accordingly, attempts were directed to determine whether the incorporation of N-hydroxysulfosuccinimide (sulfo-NHS) in the amidation reaction mixture would inhibit the rearrangement of the O-acylisourea adduct and facilitate the isolation of a homogeneous disubstituted derivative of Hb with the amidation on Glu-43(β) in quantitative yields that are consistent with the extent of amidation.

Quantitative Modification of O-Acylisourea Intermediates by Nucleophiles in Presence of Sulfo-NHS

Staros and associates[34,35] have demonstrated recently that the efficiency of the carbodiimide-mediated amidation of carboxyl groups of proteins at a given concentration of EDC is increased considerably in the presence of sulfo-NHS. The sulfo-NHS apparently rescues the activated carboxyl group as stable sulfo-NHS ester (Fig. 1). It has been suggested that the higher degree of stability of the sulfo-NHS ester of the protein carboxyl groups, as compared with that of their corresponding O-acylisourea intermediates, decreases the loss of the activated carboxyl through hydrolysis. Thus the relative efficiency of the EDC-activated carboxyl group to get channeled through the nucleophilic attack pathway (amidation) is increased in the presence of sulfo-NHS. Nonetheless, it is also conceivable that the higher stability of the sulfo-NHS ester could either inhibit or at least decelerate the formation of the N-acylurea adduct from the O-acylisourea adduct. If the sulfo-NHS can completely inhibit the rearrangement reaction, the amidation reaction, in principle, is expected to reflect directly the propensity of the side-chain carboxyl group to remain as a protonated species.

The introduction of sulfo-NHS as the rescuer of the activated carboxyl group resulted in a significant increase in the extent of modification of the basic carboxyl group(s) of Hb. These results are consistent with the results of Staros and colleagues.[34,35] The influence of the presence of 2 mM sulfo-NHS on the extent of amidation of Hb with various concentrations of EDC is shown in Table I. It may be seen that even with a fourfold increase in the concentration of EDC (from 5 to 20 mM), the rescuing efficiency (the increase in the amidation in the presence of the reagent compared with the control) of sulfo-NHS remained nearly the same. A study of the rescuing influence of sulfo-NHS as a function of the nucleophile concentration demonstrated that the rescuing efficiency of sulfo-NHS decreases as the concentration of nucleophile increases from 10 to 500 mM. This observation confirms the higher stability of the sulfo-NHS ester of the

[34] J. V. Staros, R. W. Wright, and D. M. Swingle, *Anal. Biochem.* **156**, 220 (1986).
[35] P. S. R. Anjaneyulu and J. V. Staros, *Int. J. Pept. Protein Res.* **30**, 117 (1987).

TABLE I
INFLUENCE OF SULFO-NHS ON AMIDATION OF Hb[a]

Carbodiimide concentration (mM)	Average number of carboxyl groups of Hb amidated (mol/mol)		Rescuing efficiency of sulfo-NHS[b]
	Presence of sulfo-NHS	Absence of sulfo-NHS	Presence/absence
5	0.58	0.27	2.10
10	0.98	0.50	1.95
20	1.58	0.74	2.13

[a] Amidation of oxyHb (0.5 mM) was carried out at pH 6.0 and 23° in 20 mM MES buffer using 5, 10, and 20 mM EDC for activation of the carboxyl groups and 100 mM GEE as the nucleophile. The reaction period was 15 min. At each concentration of the carbodiimide, the amidation was carried out both in the presence and in the absence of 2 mM sulfo-NHS.
[b] The rescuing efficiency of sulfo-NHS is the increase in the amidation in the presence of sulfo-NHS as compared to that in the absence of sulfo-NHS.

carboxyl group of Hb. As the concentration of the nucleophile increases, the activated carboxyl group (O-acylisourea) is readily subjected to the nucleophilic attack and thus channeled through the amidation pathway much in the same way when sulfo-NHS is added to the system. At a concentration of 600 mM of the nucleophile, the rescuing efficiency of sulfo-NHS decreases to a value of 1.2. Thus, at these high concentrations of the nucleophile, all the activated carboxyl groups are subjected to the nucleophilic attack at nearly the same rate in the presence and absence of sulfo-NHS.

The site selectivity of amidation of carboxyl groups of Hb S by GEE in the presence of sulfo-NHS has also been established. The site selectivity of the amidation reaction remained nearly the same as that in the absence of sulfo-NHS. Thus the increased amidation reaction in the presence of sulfo-NHS does not represent the trapping of an active ester of a new carboxyl group that was previously being hydrolyzed, it reflects only an efficient trapping and channeling of activated Glu-43(β) through the amidation pathway.

Inhibition of Side Reactions during Amidation of Hb in the Presence of sulfo-NHS

The quantitation of the amidation reaction in the presence of sulfo-NHS (as reflected by the incorporation of the nucleophile) only demonstrates the increased efficiency of the nucleophilic attack. However, it does provide an insight as to whether the EDC-mediated side reactions are also eliminated in the presence of sulfo-NHS. The CM-cellulose chromatogram of the sample of Hb amidated in the presence of sulfo-NHS is considerably simpler compared to that of the sample prepared in the absence of sulfo-NHS. The sample contained only three components, one eluting at the position of the unmodified Hb, a second component eluting after the Hb peak has been identified as the disubstituted derivative with the amidation exclusively on Glu-43(β). Accordingly, this component is designated as DiGEE-Hb A. A third component containing 4 mol of nucleophile per mole of Hb and eluting after the disubstituted derivative from the CM-cellulose column was also present (see Fig. 3 for a chromatogram of the amidated sample). The chromatographic study thus confirms that the EDC-mediated side reactions are inhibited in the presence of sulfo-NHS.

Large-Scale Preparation of DiGEE-Hb A

For the large-scale preparation of disubstituted derivative, Hb (Hb A or Hb S) amidated with 10 mM EDC in the presence of 2 mM sulfo-NHS and 100 mM GEE at pH 6.0 for 15 min at room temperature is taken as the starting material. The amidation reaction mixture is first desalted on a Sephadex G-25 column equilibrated and eluted with PBS buffer, pH 7.4. The desalted amidated sample, generally in 5 to 8 ml, is dialyzed overnight against 250 volumes of 10 mM potassium phosphate buffer, pH 6.0, containing 1 mM EDTA. On a 0.9 × 30 cm column we normally chromatograph up to 250 mg of the protein. The column is equilibrated to pH 6.0 with 10 mM potassium phosphate buffer containing 1 mM EDTA. The chromatogram is developed by employing a linear gradient generated from 250 ml each of 10 mM potassium phosphate buffer, pH 6.0, and 50 mM potassium phosphate buffer, pH 8.5, both containing 1 mM EDTA. For the chromatography of 1 g of amidated Hb A, we have generally used a preparative (2 × 30 cm) column of CM-52. For these bigger columns we use a linear gradient generated by 1 liter each of 10 mM potassium phosphate buffer, pH 6.0, and 50 mM potassium phosphate buffer, pH 8.5, both containing 1 mM EDTA, as the initial and final buffers, respectively. A typical large-scale preparative chromatogram of the amidated sample is shown in Fig. 3. Three chromatographically distinct components are present in this amidated sample. Nearly 60–65% of the load eluted at the position of Hb A. The second major component of the amidated sample elutes after the position of Hb A (fraction B), and it accounts for nearly 25% of the protein loaded on the column. The major radioactive component (amidated Hb A) contains 2 mol of nucleophiles per mole of tetramer.

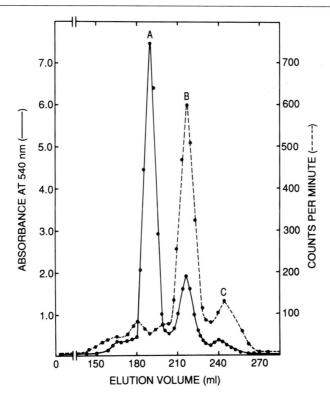

FIG. 3. Preparation of DiGEE-Hb A: Hb A amidated (approximately 150 mg) with GEE (100 mM) in the presence of 10 mM EDC and 2 mM sulfo-NHS at pH 6.0 for 15 min was isolated by desalting the reaction mixture on a Sephadex G-25 column equilibrated and eluted with PBS, pH 7.4. This sample was dialyzed extensively against the starting buffer (10 mM phosphate buffer, pH 6.0). The sample was loaded onto the column of CM-cellulose (0.9 × 30 cm) equilibrated with 10 mM phosphate buffer, pH 6.0. The protein was eluted with a linear gradient of 10 mM phosphate buffer, pH 6.0, to 50 mM phosphate buffer, pH 8.5 (250 ml each). Both buffers contained 1 mM EDTA. Elution of protein from the column was followed by measuring the absorbance of the fractions at 540 nm. A 200-μl aliquot of each fraction was used to measure [^{14}C]GEE incorporated into each of the components. Peak A corresponds to the position of unmodified Hb A and has very little radioactivity associated with it. Peak B contains 2 mol of GEE per mole of protein (DIGEE-Hb A), whereas peak C contains 4 mol of GEE per mole of protein (TETRAGEE-Hb A).

Based on the radioactive [^{14}C]GEE incorporation, fraction B is designated as a disubstituted derivative of Hb A. The disubstituted Hb A-containing fractions are pooled, concentrated, and dialyzed extensively against 10 mM phosphate buffer, pH 6.0, and repurified on a CM-52 column using either the same conditions as used in the first purification step or smaller analytical columns (for 100- to 200-mg loads).

The chemical characterization of the DiGEE-Hb A involves separation of the modified β chain (no modification is detected in the α chain), tryptic peptide mapping, and amino acid sequencing of the amidated peptide. These techniques are discussed elsewhere in the series and are not discussed here. These structural studies of the derivative have confirmed the homogeneity of the disubstituted derivatives of Hb A and Hb S.

Functional Properties of DiGEE-Hb A

The oxygen affinity of DiGEE-Hb A was determined to ascertain whether the amidated protein has maintained normal quaternary structure and cooperativity of oxygen binding. The P_{50} of Hb A was decreased from 10.3 to 8.3 Torr on amidation of Glu-43(β) at pH 7.0 and 37°. The modification of side-chain carboxyl group of Glu-43(β) increases the oxygen affinity with very little influence on the cooperativity. The Hill coefficient of DiGEE-Hb A was nearly same as that of Hb A (n = 2.7 for control Hb A and 2.55 for DiGEE-Hb A). The oxygen affinity of the derivative was modulated by 2,3-diphosphoglycerate (2,3-DPG) and chloride in a fashion comparable with that seen with Hb A. These functional studies suggest that the overall quaternary structure of Hb undergoes very little change on amidation of Glu-43(β).

Amidation of Proteins as an Approach to Establish pK_a Values of Basic Carboxyl Groups

As discussed in the introduction, the protonated form of the carboxyl group is the reactive species for activation with carbodiimide. Therefore the amidation studies of Hb discussed above imply that the γ-carboxyl group of Glu-43(β) of Hb exists as a protonated species around pH 6.0 at room temperature. Presumably, the pK_a of the γ-carboxyl of Glu-43(β) is higher than the other side-chain carboxyl groups of the protein. Because, in the presence of sulfo-NHS, the activated carboxyl group is channeled exclusively though the nucleophilic pathway, the rate of activation of the carboxyl group (i.e., the rate of amidation) is expected to correlate directly with the concentration of the protonated carboxyl group. Thus, if the kinetics of amidation of Hb is studied as a function of pH, the plot of the rate of amidation of a specific carboxyl group versus the respective pH, for example, Glu-43(β), is expected to reflect its ionization behavior. Identification of the carboxyl groups of the protein that are reactive around neutral pH will help to establish the identity of the basic carboxyl groups of the protein, whereas the study of the kinetics of amidation of these basic carboxyl group(s) as a function of pH will help to determine the apparent pK_a of the individual basic carboxyl group of the protein.

Determination of pK_a of γ-Carboxyl Group of Glu-43(β) of Hb

The study of chemical reactivity of the ionizable functional group of a protein as a function of pH is a direct approach to establish the pK_a of such a group if the reactivity is specific for the ionized or un-ionized state of that functional group. The protonated form of a carboxyl group is the reactive species involved in the formation of the activated carboxyl (Fig. 1). The protonated state of the carboxyl group is determined by the pH of the medium. Therefore, if the amidation reaction is carried out at different pH values, the rate of amidation at a given pH should reflect the concentration of the protonated carboxyl group at that pH. Such a study has been now carried out with Hb using a MES buffer system. The study of amidation of oxyHb S as a function of pH, in the range of 5.5 to 7.5, has shown that the pseudo-first-order rate of amidation decreases as the pH is increased. However, the site selectivity of amidation of the carboxyl groups of Hb remained nearly the same in this pH range. The primary site of reaction was Glu-43(β) at all these pH values. A plot of the chemical reactivity of Glu-43(β) as a function of pH,[10] in the region of 5.5 to 7.5, indeed corresponded to a titration curve of an ionizable group (Fig. 4) with an apparent pK_a of 6.35. Therefore, we have proposed that the apparent pK_a of the γ-carboxyl group of Glu-43(β) is around 6.35.

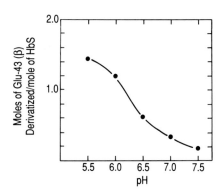

FIG. 4. Reactivity of the γ-carboxyl group of oxyHb S as a function of pH. Amidation of Hb S (0.5 mM) with GEE (100 mM) was carried out at pH 6.0 and 23° using 10 mM EDC and 2 mM sulfo-NHS at various pH values for 20 min. The relative distribution of the amidation at Glu-43(β) and Glu-22(β) was established by tryptic peptide mapping of the unfractionated amidated sample.

Kinetics of Amidation of Glu-43(β) of Hb A and Hb Chesapeake

The high chemical reactivity of the Glu-43(β) γ-carboxyl is apparently a reflection of the unique conformational aspects of the microenvironment of this residue. Glu-43(β) is a part of the $\alpha_1\beta_2$ interface of Hb. This region of the molecule is expected to be significantly distinct in Hb Chesapeake, in which Arg-92(α) is mutated to Leu. The chemical reactivity of Glu-43(β) of Hb Chesapeake has been compared with that of Hb A to determine whether the mutation (i.e., local conformation) could influence the ionization behavior of the residue. The reactivity of Glu-43(β) in Hb Chesapeake has been found to be distinct from that of the Glu-43(β) γ-carboxyl group in Hb A. Pertuz and his associates have invoked a salt bridge between the Glu-43(β) γ-carboxyl group with the guanidino group of Arg-92(α).[36] Mutation of Arg-92(α) to a Leu in Chesapeake could thus influence the intrinsic reactivity of Glu-43(β). Comparison of the pseudo-first-order kinetics of amidation of Hb S and Hb Chesapeake at pH 7.5 demonstrated that the rate of amidation of Glu-43(β) of Hb Chesapeake is nearly 30% higher than that of the same residue in Hb A. The increased chemical reactivity of Glu-43(β) of Hb Chesapeake could be a consequence of the increased pK_a of the carboxyl group, presumably as a result of the increase in the hydrophobic character of the microenvironment. The loss of the hydrophilic guanidino group from the $\alpha_1\beta_2$ interface region with the concomitant introduction of the hydrophobic side chain of Leu should result in an increase in the overall hydrophobicity of the region.

Sensitivity of Chemical Reactivity of Glu-43(β) on Oxy–Deoxy Conformational States of Hb

Glu-43(β) is a residue of the $\alpha_1\beta_2$ interface of Hb, and the conformation of this interface undergoes significant changes on deoxygenation of the molecule. This could conceivably influence the chemical reactivity of Glu-43(β). If the guanidino group of Arg-92(α) indeed forms a salt bridge in the deoxy conformation of the protein, the reactivity of Glu-43(β) should decrease on deoxygenation of the protein. Accordingly, this aspect of the chemical reactivity of Glu-43(β) of Hb has been investigated. To study the reactivity of Glu-43(β) of the protein in the deoxy conformation, a solution of Hb (0.5 mM) in MES buffer (20 mM) and containing 100 mM GEE and 2 mM sulfo-NHS is deoxygenated by passing helium gas continuously for 40–45 min (without dipping into the protein solution) for

[36] G. Fermi and M. F. Perutz, in "Hemoglobin and Myoglobin. Atlas of Molecular Structure in Biology" (D. C. Phillips and F. M. Richards, eds.), p. 101. Oxford Univ. Press (Clarendon), Oxford, 1981.

complete conformational isomerization of the protein to the deoxy state. The conformational transition to the deoxy state is confirmed spectrophotometrically as discussed by Benesch et al.[37] All reagent concentrations and the reaction conditions are the same as those used for the investigations of amidation reactions of oxyHb. The EDC solution is introduced anaerobically to the above deoxygenated Hb sample. The passing of helium gas is continued during the entire period of amidation reaction. Experiments should be rejected if the metHb amount at the end of the reaction is more than 5%. The pseudo-first-order rate of amidation of Hb A in the deoxy conformation increases by nearly 2.3-fold at pH 7.0^{11} (Fig. 5). Nonetheless, the site selectivity of the amidation remains the same. The extent of acceleration of the amidation of Glu-43(β) varies as a function of pH in the range of 5.5 to 7.5. At pH 5.5 and pH 6.0, the chemical reactivity of Glu-43(β) in the oxy and deoxy conformations is nearly the same. The maximum increase in the chemical reactivity is observed at pH 7.0. This suggests that the apparent pK_a of the Glu-43(β) γ-carboxyl group in the deoxy conformation is around $7.0.^{11}$

The increase in the apparent pK_a of the Glu-43(β) γ-carboxyl from 6.35 to 7.0 as a consequence of the conformational transition of Hb from oxy to deoxy conformational states makes this a potential alkaline Bohr group. This can be confirmed independently by a study of the oxy–deoxy transition-dependent proton titration of the amidated derivative of Hb A with a selective modification on both Glu-43(β) and comparing it with that of unmodified Hb. The comparison of proton titration of oxy to deoxy conformational transition of Hb A and DiGEE-Hb A has confirmed the contribution by the Glu-43(β) γ-carboxyl group to the alkaline Bohr effect of Hb A. The contribution is maximal at pH 7.0, and it accounts for about 30% of the Bohr effect of Hb A at this pH.[11]

Flexibility of Amidation Reactions

One of the unique strengths of the amidation procedure is the flexibility to use different amines as nucleophiles, so that the microenvironment and stereochemical aspects of the carboxyl group can be probed. The amidation of carboxyl groups of Hb has also been investigated with other nucleophiles, particularly with amino sugars [glucosamine (GU) and galactosamine (GA)] to establish whether the selectivity of the modification of carboxyl groups of Hb is influenced by the nature of the nucleophile. The amidation conditions and reagent concentrations are the same as those used in the amidation reaction using GEE as the nucleophile, except that

[37] R. E. Benesh, R. Benesch, and S. Young, *Anal. Biochem.* **55**, 245 (1973).

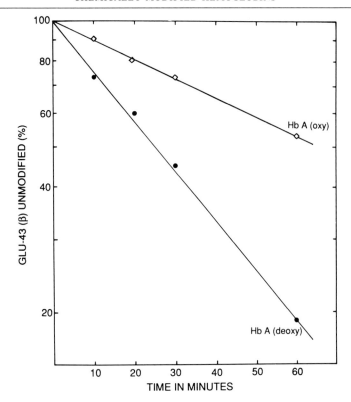

FIG. 5. Influence of the oxy–deoxy conformational transition of Hb A on the chemical reactivity of Glu-43(β). Kinetics of the amidation of Hb A in the oxy and deoxy conformations were carried out at pH 7.0. The modification was quantitated by the incorporation of GEE into Hb. The site selectivity of the reaction to Glu-43(β) was confirmed by tryptic peptide mapping. The amount of Glu-43(β) of Hb A that is not derivatized is plotted as the percentage unmodified as a function of time. The deoxygenation of Hb A increases the reactivity of Glu-43(β) by nearly 2.3-fold.

the reactions have to be carried out for longer periods to get similar levels of modifications. Nonetheless, the selectivity of Glu-43(β) for amidation with these amines is considerably higher than that of Glu-22(β), much in the same way that had been observed earlier with GEE.

The rescuing efficiency of sulfo-NHS for the amidation of Glu-43(β) and Glu-22(β) with galactosamine as the nucleophile has also been calculated and compared with that obtained when GEE is used as the nucleophile. The extent of reaction at Glu-22(β) increased only marginally in the presence of sulfo-NHS, whereas modification of Glu-43(β) increased nearly threefold. The rescuing efficiency of sulfo-NHS for Glu-43(β) is

2.8 when GA is used as the nucleophile, whereas it is 2.0 for GEE. On the other hand, the rescuing efficiency of sulfo-NHS for the Glu-22(β) site is nearly 1.1. This implies that even in the absence of sulfo-NHS almost all of the activated Glu-22 are channeled through the isopeptide bond formation (amidation) pathway. Apparently very little of the O-acylisourea of Glu-22(β) rearranges to the corresponding N-acylurea adduct under the present experimental conditions. On the other hand, in the absence of sulfo-NHS a considerable amount of the O-acylisourea adduct of Glu-43(β) appears to rearrange to the corresponding N-acylurea adduct. Presumably, the microenvironment of Glu-43(β) is more hydrophobic in character relative to that of Glu-22(β), and this favors the rearrangement of the activated carboxyl group (O-acylisourea) of Glu-43(β) to the N-acylurea. This concept is consistent with the earlier analysis that the microenvironment of Glu-43(β) is hydrophobic in character.

The pseudo-first-order kinetics of amidation of Hb with GA at any given pH (in the range of 5.5 to 7.0) is linear. Dependency of chemical reactivity of Glu-43(β) on the pH again suggests an apparent pK_a of 6.35 for its γ-carboxyl. However, the reactivity of Glu-22(β) remained the same in the pH range of 6.0 to 7.0, suggesting that the pK_a of Glu-22(β) is considerably lower than that of Glu-43(β). The relative increase in amidation of Glu-43(β) on deoxygenation is nearly the same with both nucleophiles.

Amidation of carboxyl groups of Hb using methylamine (MA) as nucleophile has also been carried out under conditions similar to those used for the galactosamine system. The selectivity of the amidation of carboxyl groups remains the same as with the GEE or GA systems, i.e., more than 85% of the amidation was at Glu-43(β). The procedure used for the preparation of DiGEE-Hb A also can be employed to isolate the disubstituted derivative of Hb modified at its Glu-43(β) with galactosamine and methylamine.

Influence of Chemical Nature of Substituent at Glu-43(β) on Functional Properties of Disubstituted Derivatives of Hemoglobin

Changes in the functional properties of Hb on the amidation of Glu-43(β) can be a consequence of the structural changes of the protein. A comparison of the functional properties of Hb, amidated with nucleophiles that differ considerably in size and chemical nature, can provide new insights into this structural aspect. Oxygen equilibrium properties of Hb S and its disubstituted derivatives with GEE, GA, and MA have been determined to ascertain whether O_2 affinity and the cooperativity of the molecule are altered to the same degree after chemical modification of the

γ-carboxyl group of Glu-43(β).[38] The amidation of the Glu-43(β) γ-carboxyl group with various nucleophiles increases the oxygen affinity of the protein to the same degree with all three derivatives. The Hill coefficient of disubstituted derivatives of Hb S (DiGEE-Hb S, DiGA-Hb S, and DiMA-Hb S) is nearly the same and is also close to that of unmodified Hb S. The amidation of Glu-43(β) by any of these nucleophiles has very little influence on the cooperativity of the protein. Thus, the slight increase in O_2 affinity of Hb appears to be a direct consequence of the removal of the negative charge at this site, but the amidation alone has a limited influence on the overall conformation and the long-range interactions of the protein.

General Applications of Amidation Chemistry to Structural Studies of Hemoglobin

The identification of Glu-43(β) as the alkaline Bohr group and its location at the $\alpha_1\beta_2$ interface make this residue and the disubstituted Hb A ideal candidates for further investigations of the oxy–deoxy transition-mediated conformational changes of the $\alpha_1\beta_2$ interface of Hb. Selective chemical modification of Glu-43(β) in fact provides an unique opportunity to introduce "structural probes" for monitoring the conformational aspects of the $\alpha_1\beta_2$ interface of the protein during conformational transition. ^{13}C, ^{15}N, ^{19}F, or photoaffinity labels can be introduced at a specific site, which can be useful for mapping the complementary site by the cross-linking approach.

The sensitivity of the chemical reactivity of Glu-43(β) to the oxy–deoxy conformational states of Hb provides a unique opportunity to probe the conformational aspects of the $\alpha_1\beta_2$ interface of Hb variants. A particularly interesting mutant in this respect is Hb S. The relative reactivity of Glu-43(β) in the oxy to deoxy conformational states appears to be sensitive to mutations at distant sites. The relative increase in the chemical reactivity of Glu-43(β) that occurs on deoxygenation of Hb appears to decrease as a consequence of the mutation of Glu-6(β) to Val. We have seen recently that the chemical reactivity of Glu-43(β) of Hb A increases nearly 2.3-fold at pH 7.0. However, the value for Hb S is around 1.9 (M. J. Rao and A. S. Acharya, unpublished results, 1993) under identical conditions. This observation is consistent with the implication of Glu-43(β) as an intermolecular contact site of deoxyHb S, and the amidation of Hb S at

---

[38] M. J. Rao and A. S. Acharya, *Blood* **78**, 198a (1991).

Glu-43(β) increases the solubility of the deoxy protein.[39] The sensitivity of the chemical reactivity of Glu-43(β) to the conformational isomerization of the molecule thus appears to be influenced by long-range mutational effects. In summary, the chemical reactivity of Glu-43(β) and its sensitivity to the conformational isomerization of the protein can be considered as an intrinsic chemical reporter of the conformational aspects of the $\alpha_1\beta_2$ interface of the protein. In combination with the flexibility of site-directed mutagenesis, permitting the generation of mutant Hb, this chemical approach provides a simple procedure to probe the existence of long-range communication between the mutated site and the $\alpha_1\beta_2$ interface, particularly the microenvironment of Glu-43(β). Although the use of the amidation reaction to establish the pK_a of the basic carboxyl group is restricted so far to the studies with Hb, this simple yet powerful chemical approach should prove useful in other protein systems as well.

Acknowledgments

This research was supported by Grant HL-38655 from the National Institutes of Health and a Grant-in-Aid from the American Heart Association. Our thanks are extended to Dr. L. Khandke, who had initiated some of the early studies of the side reactions that occur during the amidation of Hb S. The authors wish to thank Dr. Ronald L. Nagel for the facilities extended to them. The help of Drs. P. Nacharaju, R. P. Roy, and R. E. Hirsch in the preparation of the manuscript is also gratefully acknowledged.

[39] A. S. Acharya, L. Khandke, B. T. Chait, and M. J. Rao, in "Molecular Conformation and Biological Interactions" (P. Balaram and Ramaseshan, eds.), p. 269. Indian Acad. Sci., Bangalore, 1991.

[16] Bis(pyridoxal) Polyphosphates as Specific Intramolecular Cross-Linking Agents for Hemoglobin

By Ruth E. Benesch

Reagents that can form intramolecular cross-links between the two $\alpha\beta$ dimers in the hemoglobin molecule are of interest from several points of view. First, they are necessary for the stabilization of intermediates in the ligand-binding reaction (or, rather, models for them) of the kind $\alpha\alpha^+\beta^+\beta^+$ or $\alpha^+\alpha^+\beta\beta^+$, so that the properties of the intermediates can be studied.[1] Second, for the same reason, they are necessary for the

[1] S. A. Fowler, J. Walder, A. DeYoung, L. D. Kwiatkowski, and R. W. Noble, *Biochemistry* **31**, 717 (1992).

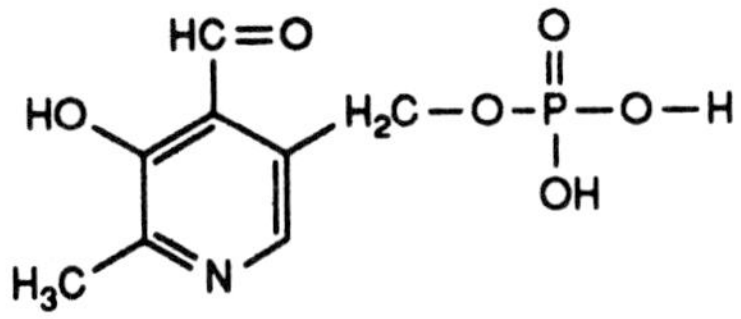

PLP

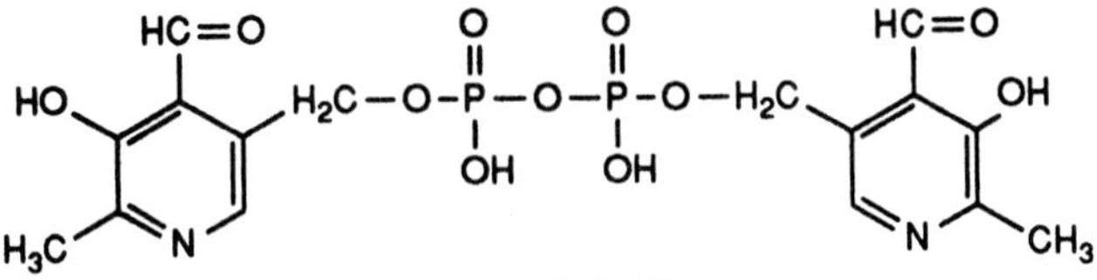

(bis-PL)P$_2$

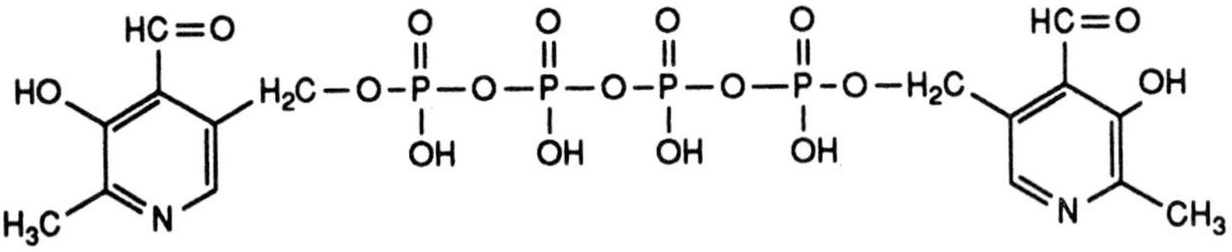

(bis-PL) P$_4$

SCHEME I

investigation of hemoglobins containing three different chains, for example, $\alpha^A\beta^S\alpha^X\beta^S$.[2] Last, because extracorpuscular hemoglobin is excreted rapidly owing to glomerular filtration of $\alpha\beta$ dimers, stabilization of the tetramer by intramolecular cross-links is necessary for use of hemoglobin as a blood substitute.[3]

Two kinds of cross-linking agents and their reactions with human and bovine hemoglobin will be discussed here: (1) 2-nor-2-formylpyridoxal 5'-phosphate (NFPLP), i.e., pyridoxal phosphate (PLP), in which the methyl group in the 2-position of the pyridine ring is replaced by an aldehyde group,[4] and (2) two members of a series of compounds containing two pyridoxal rings connected at the 5'-position by a number of phosphate residues,[5] i.e., bis(pyridoxal) diphosphate [(bisPL)P$_2$] and bis(pyridoxal) tetraphosphate [(bisPL)P$_4$] (see Scheme I). All three compounds show a

[2] R. E. Benesch, S. Kwong, and R. Benesch, *Nature (London)* **299,** 231 (1982).
[3] R. Bolin and F. DeVenuto, *Prog. Clin. Biol. Res.* **122,** 1 (1983).
[4] A. Pocker, *J. Org. Chem.* **38,** 4295 (1973).
[5] R. E. Benesch and S. Kwong, *Biochem. Biophys. Res. Commun.* **156,** 9 (1988).

remarkable specificity for entrance into the central cavity between the β chains in deoxyhemoglobin, but they differ drastically in their relative affinity for the human and bovine proteins.

Synthesis of Reagents

The synthesis of NFPLP has been described in an earlier volume of this series.[6] (bisPL)P$_2$ and (bisPL)P$_4$ are prepared by the anion-exchange method of Michelson[7] with the modifications for pyridoxal compounds described by Fukui *et al.*[8-10] This method involves, first, conversion of pyridoxal phosphate to pyridoxal diphosphate β-diphenyl ester by reaction with diphenyl chlorophosphate, followed by displacement of diphenyl phosphate by another anion. The anions used, in the form of their tributyl-ammonium salts, are PLP in the preparation of (bisPL)P$_2$ and pyrophosphate for the synthesis of (bisPL)P$_4$. In the latter case an excess of pyridoxal diphosphate β-diphenyl ester is used in order to effect substitution at both ends of the pyrophosphate molecule.

(a) General Precautions. Pyridoxal compounds are sensitive to both light and oxygen. Flasks, columns, etc. should be wrapped in aluminum foil, and aqueous reagents, such as column eluents, should be deoxygenated by bubbling nitrogen or argon through them. Solvents must be dry, especially the pyridine used for the anion exchange. It is important that the synthesis be completed promptly, i.e., in 3–4 days, including purification of the products.

(b) Tributylammonium Salt of PLP. 0.8 g (3 mmol) of PLP (Sigma, St. Louis, MO) and 0.7 ml (3 mmol) of tributylamine (TBA; Aldrich, Milwaukee, WI) are refluxed with 30 ml of chloroform for ~1 hr until a clear solution is obtained. This is evaporated *in vacuo* at room temperature and the residue is dried by evaporation of two 10-ml portions of benzene, followed by vacuum dessication over P$_2$O$_5$.

(c) TBA Salt of Pyrophosphate. 390 mg (1 mmol) of TBA pyrophosphate (Sigma, labeled 1.5 mol TBA per mole pyrophosphate) and 0.25 ml (1 mmol) of TBA are refluxed with 10 ml of chloroform for 10 min, followed by drying, first with benzene and then *in vacuo* over P$_2$O$_5$ as above. It should be noted that when the Sigma product is used directly, without the above treatment, no (bisPL)P$_4$ is formed in the subsequent step.

[6] R. Benesch and R. E. Benesch, this series, Vol. 76, p. 147.
[7] A. M. Michelson, *Biochim. Biophys. Acta* **91,** 1 (1964).
[8] S. Shimomura and T. Fukui, *Biochemistry* **17,** 5359 (1978).
[9] M. Tagaya and T. Fukui, *Biochemistry* **25,** 2958 (1986).
[10] M. Tagaya, K. Nakano, and T. Fukui, *J. Biol. Chem.* **260,** 6670 (1985).

(d) *Pyridoxal Diphosphate β-Diphenyl Ester.* 0.53 g (2 mmol) of PLP and 0.95 ml (4 mmol) of TBA are refluxed with 20 ml of chloroform until all is dissolved (∼1 hr). After evaporation *in vacuo,* the residue is dried with two portions of benzene. It is then dissolved in 15 ml of chloroform and 0.2 ml of TBA and 0.52 ml of diphenyl chlorophosphate (Kodak, Rochester, NY) are added and the mixture is stored for 3 hr at room temperature under anhydrous conditions. It is then evaporated *in vacuo* at room temperature, followed by the addition of 30 ml of ice-cold ether. After stirring the mixture for 30 min at 0°, the yellow ether is decanted and the residue is dried, first with two portions of benzene and then *in vacuo* over P_2O_5.

(e) Anion Exchange

(1) (bisPL)P_2. The 3 mmol of TBA–PLP prepared in (*b*) are dissolved in 2 ml of dry pyridine (dried over KOH and distilled just before use) and the solution is added to the 2 mmol of pyridoxal diphosphate β-diphenyl ester prepared in (*d*). The mixture is stored overnight at room temperature under anhydrous conditions.

(2) (bisPL)P_4. The 1 mmol of TBA pyrophosphate from (*c*) is dissolved in 3 ml of dry pyridine and the solution is added to 4 mmol of pyridoxal diphosphate β-diphenyl ester. The mixture is stored for 3 hr at room temperature and then overnight at 4° under anhydrous conditions.

(f) Isolation of Products. For both compounds, the mixture is evaporated *in vacuo* to remove the pyridine. The residue is dissolved in 25 ml of oxygen-free water and extracted with ether. After removal of residual ether from the aqueous layer, it is diluted to ∼200 ml with oxygen-free water and applied to a 1.5 × 17 cm column of Dowex-1 that has been washed with water at room temperature. After washing the column with 200 ml of 0.01 N HCl, it is eluted with 1.5 liters of a linear gradient of LiCl from 0 to 0.05 M, both in 0.01 N HCl for (bisPL)P_2 and from 0 to 0.1 M LiCl for the tetraphosphate, at a flow rate of ∼400 ml/hr. Fractions are monitored by UV spectra after 20-fold dilution of 0.1-ml samples with phosphate buffer (pH 7).

In the case of the diphosphate, the only products eluted by the gradient are PLP and (bisPL)P_2, with the former preceding and well separated from the latter. The relevant fractions are pooled, neutralized to about pH 7 with 1 N LiOH, and concentrated to a small volume by flash evaporation below 30°. The compound is precipitated by the addition of an equal volume of ethanol, followed by about 25 volumes of acetone. LiCl remains in solution in the acetone. The (bisPL)P_2 is filtered, washed with acetone and then with ether, dried *in vacuo* over P_2O_5, and stored, protected from light and moisture, at −20°. The yield is ∼20%. The UV spectrum shows

$\varepsilon/^{max}_{392} = 9.400\ M^{-1}$ with a shoulder between 330 and 340 nm. Thin-layer chromatography (TLC) on silica gel 60 F-254 in butanol–pyridine–water (6 : 4 : 3) shows a single component with $R_f = 0.49$.

In the preparation of (bisPL)P$_4$, five well-separated pyridoxal compounds are eluted by the gradient. Their identification and their formation as a result of both hydrolysis and self-condensation of pyrophosphate are discussed in Benesch and Kwong.[11] The third component is (bisPL)P$_4$. It is isolated and stored as described above for the diphosphate. In some preparations the compound, with an R_f of 0.48 in the above TLC system, was contaminated with a slow-moving impurity. This was easily removed by chromatography on silica gel using 60% methanol–40% CHCl$_3$ as the solvent. The UV spectrum shows maxima at 392 and 324 nm. Solutions are standardized using $9.400\ M^{-1}$ for the extinction coefficient at 392 nm or on the basis of their phosphate content determined by the method of Ames and Dubin.[12] The yield of pure compound is 10–20% based on pyrophosphate.

Preparation of Cross-Linked Hemoglobins

The reaction of deoxyHb with the bis(pyridoxal) polyphosphates is carried out as described previously for NFPLP.[6] The yield of cross-linked hemoglobin may be determined by passage of the reaction mixture through Sephadex G-100 SF in 1 M MgCl$_2$. Under these conditions un-cross-linked Hb is completely dissociated into $\alpha\beta$ dimers and a complete separation of the dimers from the cross-linked tetramers is effected.

The cross-linked tetramers are purified by chromatography on IDA Sepharose (Chelating Sepharose fast flow; Pharmacia, Piscataway, NJ) in 0.05 M bis-Tris buffer, pH 6.5, using a gradient of NaCl from 0 to 0.025 M in the same buffer for elution. The penultimate component is the cross-linked tetramer. The location of the cross-link was determined by chromatography of the globin hydrolyzate and identification of the fluorescent products.[13] For human hemoglobin these were pyridoxylvaline, pyridoxylvalylhistidine, and pyridoxyllysine, whereas for bovine Hb only pyridoxylmethionine and pyridoxyllysine were found in equimolar amounts. In all cases the label was found only in the β-chain globin. Thus these reagents connect the N-terminal residue of one β chain to a lysine (probably residue 82 in the human sequence and 81 in the bovine sequence) on the other one.

This identification of the binding site provides a reasonable explanation for the dramatic difference in the affinity of the three cross-linking re-

[11] R. E. Benesch and S. Kwong, *J. Protein Chem.* **10**, 503 (1991).
[12] B. N. Ames and D. T. Dubin, *J. Biol. Chem.* **235**, 769 (1960).
[13] R. E. Benesch, R. Benesch, R. D. Renthal, and N. Maeda, *Biochemistry* **11**, 3576 (1972).

TABLE I
CROSS-LINKING OF HUMAN
AND BOVINE HEMOGLOBIN[a]

	Cross-linked tetramer (% of total Hb)	
Reagent	Human Hb	Bovine Hb
NFPLP	70	8
(bisPL)P$_2$	21	47
(bisPL)P$_4$	80	84

[a] One mole of reagent per Hb tetramer was used for these experiments and the products were separated on Sephadex G-100 as described in the text.

TABLE II
OXYGENATION PARAMETERS OF NATIVE
AND CROSS-LINKED HEMOGLOBINS

Hemoglobin[b]	P_{50} (mm Hg)	Hill coefficient
Human	4.0	3.0
(bisPL)P$_4$ (XL human)	20	1.8
Bovine	11	2.9
(bisPL)P$_2$ (XL bovine)	41	1.8

[a] Hemoglobin (50 μM) in 0.05 M bis-Tris and 0.1 M Cl$^-$, 1 mM EDTA, pH 7.3, at 20°.[14]
[b] XL, Cross-linked.

TABLE III
STABILITY OF CROSS-LINKED HEMOGLOBINS[a]

Hemoglobin[b]	Initial rate of autoxidation (% Hb$^+$/hr)	Temperature for 50% denaturation (°C)
Human	1.3	43
(bisPL)P$_4$ (XL human)	4.0	51
Bovine	3.3	45
(bisPL)P$_2$ (XL bovine)	6.0	53

[a] Measurements were made as described by Yang and Olsen[15] and White and Olsen[16] with slight modifications.[11]
[b] XL, Cross-linked.

TABLE IV
EFFECT OF INTRAMOLECULAR CROSS-LINKS ON HEME–GLOBIN LINKAGES[a]

Hemoglobin[b]	Time for transfer of one heme/tetramer (min)	Equilibrium heme distribution ratio R
Human	23	3.5
(bisPL)P$_4$ (XL human)	145	0.32
Bovine	17	4.0
(bisPL)P$_2$ (XL bovine)	230	0.52

[a] The measurements were made as described elsewhere in this volume,[18] using human serum albumin as the heme acceptor.
[b] XL, Cross-linked.

agents, NFPLP, (bisPL)P$_2$, and (bisPL)P$_4$ for the two hemoglobins, as illustrated in Table I.

The distance between the two aldehydes in NFPLP is fixed at about 6 Å, which corresponds closely to the gap between β1 and β82 in human Hb but is too short to connect β1 to β81 in the bovine protein, because these two residues are at least 11 Å apart. (bisPL)P$_2$, on the other hand, wherein the aldehydes are about 12 Å apart, can bridge the bovine Hb cavity very well and indeed much more effectively than in the human Hb cavity. Finally, in (bisPL)P$_4$ the two pyridine rings can assume a stacked conformation,[5] imparting the flexibility to bridge both distances with high efficiency.

Properties of Cross-Linked Hemoglobins

It can be seen from Table II that the introduction of these cross-links at the DPG-binding site causes a large decrease in the oxygen affinity. The oxygenation remains cooperative, although the Hill coefficient is substantially decreased.

The rate of oxidation to the ferric form is increased in the cross-linked hemoglobins, about twofold for the bovine protein and threefold for the human one (Table III).[11,15,16]

By contrast, the cross-links increase the resistance of the hemoglobins to heat denaturation, because the temperature for half-denaturation is raised by 8° as a result of the cross-link (Table III). A similar and even more

[14] R. Benesch, G. Macduff, and R. E. Benesch, Anal. Biochem. 11, 81 (1965).
[15] T. Yang and K. W. Olsen, Biochem. Biophys. Res. Commun. 163, 733 (1989).
[16] F. L. White and K. W. Olsen, Arch. Biochem. Biophys. 258, 51 (1987).

dramatic increase in heat stability, resulting from a different intramolecular cross-link, has been reported by Bellini *et al.*[17]

Finally, the stability of the heme–globin linkage is increased very substantially by these cross-links with respect to both the rate and the equilibrium of heme transfer (Table IV).[18] This property is, of course, very desirable in connection with the use of these hemoglobins as blood substitutes, because an early step in catabolism of extracellular hemoglobin is removal of the heme.

[17] A. Bellelli, R. Ippoliti, A. Brancaccio, E. Lendaro, and M. Brunori, *J. Mol. Biol.* **213**, 571 (1990).
[18] R. E. Benesch, this volume [31].

[17] Preparation of Intramolecularly Cross-Linked Hemoglobins

By ROXANNE Y. WALDER, MARK E. ANDRACKI, and JOSEPH A. WALDER

Hemoglobin (Hb), although normally a tetramer, readily undergoes dissociation to form $\alpha\beta$ dimers [Eq. (1)]. The reaction is oxygen linked,

$$\alpha_2\beta_2 \rightleftharpoons 2\alpha\beta \qquad (1)$$

occurring to a much greater extent with oxyHb than with deoxyHb. By cross-linking the molecule intramolecularly this process can be blocked. This has become important both for basic structure–function studies of Hb and for the development of Hb derivatives useful as blood substitutes.

Preparation of mixed metal and valency hybrids has made it possible to isolate stable species representing intermediate ligation states of Hb; these have yielded important new insights into the linkage between ligand binding and the structural and energetic changes that underlie cooperativity.[1-3] However, with native Hb only the two symmetric intermediates can be isolated:

$$\begin{matrix} \beta & \beta \\ {}^*\alpha & {}^*\alpha \end{matrix} \quad \text{and} \quad \begin{matrix} {}^*\beta & {}^*\beta \\ \alpha & \alpha \end{matrix}$$

[1] F. R. Smith and G. K. Ackers, *Proc. Natl. Acad. Sci. U.S.A.* **82**, 5347 (1985).
[2] B. Luisi and N. Shibayama, *J. Mol. Biol.* **206**, 723 (1989).
[3] S. A. Fowler, J. A. Walder, A. DeYoung, L. D. Kwiatkowski, and R. W. Noble, *Biochemistry* **31**, 717 (1992).

(The asterisks indicate a ligated or an oxidized subunit.) The remaining six intermediate species are asymmetric:

$$\begin{array}{cccccc} {}^*\beta\beta & \beta\beta & {}^*\beta\beta & {}^*\beta\ \beta & {}^*\beta{}^*\beta & {}^*\beta\ \beta \\ \alpha\alpha & {}^*\alpha\alpha & {}^*\alpha\alpha & \alpha{}^*\alpha & {}^*\alpha\ \alpha & {}^*\alpha{}^*\alpha \end{array}$$

These six intermediates cannot be prepared due to disproportionation of the tetramer [Eq. (2)].

$$2(\alpha\beta^*)(\alpha\beta) \rightleftharpoons (\alpha\beta^*)_2 + (\alpha\beta)_2 \tag{2}$$

During purification, either by chromatography or electrophoresis, the equilibrium in Eq. (2) becomes displaced completely to the right as the two symmetric species are separated. This reaction can be blocked by cross-linking the two $\alpha\beta$ dimers. With such derivatives all eight intermediate species can be isolated. Obviously, for such purposes it is essential that the cross-link not significantly perturb the allosteric properties of the molecule.

Cross-linking is also important from the standpoint of developing Hb derivatives to serve as blood substitutes. The $\alpha\beta$ dimers having a molecular weight of only 32,000 are readily filtered from the circulation by the kidneys. As a result, unmodified Hb has a very short plasma half-life ($\approx$90 min), and the massive hemoglobinuria that results poses the risk of renal injury. It is essential to cross-link the molecule intramolecularly to prevent this from occurring. A decrease in oxygen affinity, to compensate for the absence of 2,3-diphosphoglycerate (DPG) outside of the red cell, is also desirable to enhance the off-loading of oxygen to the tissues. Although not as critical as for detailed structure–function studies, it is desirable in this case also to maintain a high level of cooperativity.

A number of bifunctional reagents have been described that react selectively with Hb at the DPG-binding site to cross-link the β subunits. The first such compound reported, 2-nor-2-formylpyridoxal 5'-phosphate reacts specifically with deoxyHb to cross-link lysine-82 of one β subunit to the amino-terminal valine residue of the second β chain.[4] The preparation and properties of this derivative were described by Benesch and Benesch.[4] Subsequently, bifunctional analogs of aspirin also were shown to modify Hb selectively at the DPG site.[5,6] In this case the reaction occurs preferentially with oxyHb, and the predominant site of cross-linking is

[4] R. Benesch and R. E. Benesch, this series, Vol. 76, p. 147.
[5] J. A. Walder, R. H. Zaugg, R. Y. Walder, J. M. Steele, and I. M. Klotz, *Biochemistry* **18**, 4265 (1979).
[6] J. A. Walder, R. Y. Walder, and A. Arnone, *J. Mol. Biol.* **141**, 195 (1980).

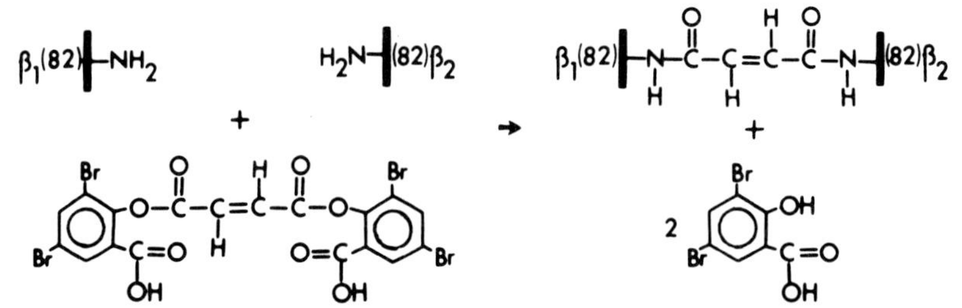

FIG. 1. Reaction of bis(3,5-dibromosalicyl) fumarate with oxyHb to cross-link Lys-82(β_1) and Lys-82(β_2).

between the two lysine-82(β) residues.[6] The parent compound in this series is bis(3,5-dibromosalicyl) fumarate (DBBF) (Fig. 1).

Later we discovered that if the reaction with DBBF is carried out under deoxygenated conditions a new derivative, cross-linked between the α chains, is formed.[7] The site of cross-linking proved to be between Lys-99(α_1) and Lys-99(α_2), very near the center of the Hb molecule. The yield of this derivative is markedly increased in the presence of polyanions that bind within the DPG site and block side reactions from occurring in this region. The preparation of these two cross-linked Hbs with DBBF is described in detail in this chapter. Both derivatives are now used extensively for basic studies of Hb function. The properties of the $\alpha\alpha$ cross-linked derivative (see below) also make it an attractive candidate for use as a blood substitute.[8]

Synthesis of Bis(3,5-dibromosalicyl) Fumarate

3,5-Dibromosalicylic acid (5 g, 0.017 mol) is weighed into a stoppered 200-ml round-bottom flask and is stirred with 100 ml of benzene, freshly distilled and dried over molecular sieves. Two equivalents, 0.034 mol, of N,N-dimethylaniline (4.28 ml) are added with stirring. The mixture suddenly changes to a milky slurry. The flask is chilled in an ice bath and fumaryl chloride (0.008 mol, 0.91 ml) is added dropwise. After all of the fumaryl chloride is added, the flask is allowed to come to room temperature and is stirred for 1 hr. In that time, the milky precipitate becomes soluble. The reaction mixture is then poured into cold H_2O containing 1.1 equivalents of HCl and is stirred for 15 min. The product is extracted into ethyl

[7] R. Chatterjee, E. V. Welty, R. Y. Walder, S. L. Pruitt, P. H. Rogers, A. Arnone, and J. A. Walder, J. Biol. Chem. 261, 9929 (1986).
[8] S. R. Snyder, E. V. Welty, R. Y. Walder, L. A. Williams, and J. A. Walder, Proc. Natl. Acad. Sci. U.S.A. 84, 7280 (1987).

acetate using a separatory funnel. The ethyl acetate layer is dried with Na$_2$SO$_4$ and is filtered into a 500-ml round-bottom flask. The ethyl acetate is evaporated, and the compound is recrystallized twice from boiling ethyl acetate–acetone. Acetone is added in limiting amounts sufficient to achieve solubility of the compound (mp 216–217°). The same procedure can be used to synthesize a number of related cross-linking agents in which the span of the binding group is altered.[6]

Preparation of Hb

The following method is used in the authors' laboratory to prepare Hb from outdated blood, but many other procedures are available. In this and the subsequent preparation of the Hb derivatives described below, gloves should be worn to avoid possible contact with blood-borne pathogens.

One unit of outdated blood is transferred to two 500-ml bottles and centrifuged in a Beckman JA-10 rotor at 5000 rpm for 10 min at 4°. The plasma and white cells are removed by aspiration and the packed red cells are washed three times with an equal volume of cold 0.9% NaCl. Four volumes of H$_2$O are added to the washed red cells and lysis occurs during a 1-hr incubation at 4° with stirring. The lysate is centrifuged at 41,000 rpm in a Ti45 rotor for 90 min at 4°. The supernatant is concentrated to about 1.5 mM using an Amicon (Danvers, MA) PM10 membrane and quickly frozen by dropping into liquid nitrogen. Any residual DPG that may be present does not interfere with the subsequent cross-linking reactions and is removed in the chromatographic purification of the derivatives.

Preparation of Hb Cross-Linked between Lys-82(β_1) and Lys-82(β_2)

There are 44 lysine residues and the four N-terminal amino groups of the α and β chains of Hb at which DBBF can potentially react. With oxyHb DBBF reacts with a high degree of selectivity to symmetrically cross-link the two lysine-82(β) residues. This derivative (2–3 g) can be readily prepared with the following procedure.

Hemoglobin (4 g, 62 μmol) is first exchanged into 50 mM bis-Tris buffer (pH 7.2) by gel filtration over a Sephadex G-25 column (5 × 30 cm) and is then concentrated by ultrafiltration with an Amicon PM10 membrane to 1 mM. DBBF (44 mg, 62 μmol) is then added to the solution as a solid and the reaction mixture (62 ml) is allowed to incubate for 2 hr at 37° in a shaking water bath. Glycine (15.5 ml of a 1 M solution adjusted to pH 8.8 with NaOH) is then added to quench the reaction. The extent of modification can be readily determined at this point either by analytical isoelectric focusing or SDS–polyacrylamide gel electrophoresis.[5,6]

The cross-linked derivative is next purified by chromatography on a DEAE-Sepharose column (5 × 15 cm) first equilibrated with 0.2 M glycine (pH 8.0). The pH and concentration of the glycine buffer used to stop the reaction are chosen to give these same conditions so that the sample can be applied directly to the column. The column is run at a flow rate of 100 ml/hr. After the Hb is applied, the column is washed with 2 volumes of 0.2 M glycine (pH 8.0) and the product is then eluted with a 0.03–0.06 M NaCl gradient in the same buffer (2.5 liters). The product is generally visible as a discrete band on the column following the residual unmodified Hb. Derivatives that are multiply modified (typically about 15% of the reaction mixture) elute subsequently as a diffuse peak. Fractions corresponding to the derivative cross-linked between the two $\beta 82$ residues are pooled, concentrated, and the product is stored frozen in liquid nitrogen.

Preparation of Hemoglobin Cross-Linked between Lys-99(α_1) and Lys-99(α_2) with DBBF

The preparation of this Hb is similar to the derivative cross-linked between the β subunits except that the reaction is carried out under deoxygenated conditions, and inositol hexaphosphate (IHP) is added to block competing reactions within the β cleft. DPG, inositol hexasulfate, and other polyanions can be used for the same purpose. The concentration of IHP typically employed is 5 mM, i.e., a fivefold excess over that of Hb. A 0.1 M stock solution of IHP is prepared by dissolving the Na$^+$ salt in water and titrating the pH of the solution to 7.2 with acetic acid. By decreasing the oxygen affinity of Hb, the addition of IHP also facilitates the removal of oxygen from the solution.

The Hb solution is first deoxygenated by purging the reaction vessel with a stream of humidified N_2. The temperature of the solution is maintained at 25° or below to minimize autoxidation of the Hb, which occurs most rapidly at intermediate partial pressures of oxygen.[9] After 1 hr the reaction vessel is transferred to a 37° water bath and purging with N_2 is continued for 30 min. An equimolar concentration of DBBF is then added as a solid and the reaction is allowed to proceed for 2 hr. Glycine is then added to a final concentration of 0.2 M to terminate the reaction as above. Before addition to the reaction mixture, the glycine solution should first be deoxygenated. After 15 min the reaction flask is transferred to an ice

---

[9] H. F. Bunn and B. G. Forget, "Hemoglobin: Molecular Genetics and Clinical Aspects," p. 641. Saunders, Philadelphia, 1986.

bath, and once the temperature is equilibrated the solution is rapidly reoxygenated. The product is then purified by chromatography on DEAE-Sepharose using the same buffer system (0.2 M glycine, pH 8.0) as described for the separation of the $\beta\beta$ cross-linked derivative. The column is developed with a linear gradient from 0.015 to 0.035 M NaCl. IHP, as expected, binds very tightly to the column and is removed to greater than one part in 500 as determined by phosphate analysis.

Properties of Cross-Linked Hemoglobins

The isoelectric point of both of the cross-linked Hbs is decreased due to the loss of the positive charges of the modified amino groups when converted to amide linkages (Fig. 2). The decrease in the isoelectric point of the $\alpha\alpha$ cross-linked derivative when in the liganded form is anomalously small (compare lanes 5 and 6, Fig. 2). We have attributed this effect to an increase in the pK_a of Glu-101(β) in oxyHb, which becomes somewhat buried from the solvent due to close proximity to the cross-linked bridge.[7] In deoxyHb this residue is more exposed to the solvent and its pK_a is not

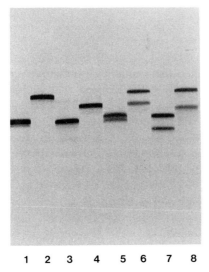

1 2 3 4 5 6 7 8

FIG. 2. Isoelectric focusing gels of the 99α and 82β cross-linked Hbs. Lane 1, oxyHb A; lane 2, deoxyHb A; lane 3, oxyHb 99α; lane 4, deoxyHb 99α; lanes 5 and 6, comixture of Hb A and Hb 99α under oxygenated and deoxygenated conditions, respectively; lanes 7 and 8, comixture of Hb A and Hb 82β under oxygenated and deoxygenated conditions, respectively. Hbs were deoxygenated during electrophoresis by dithionite. From Chatterjee et al.[7]

affected by the cross-link. As a result the expected decrease in isoelectric point due to the loss of two positively charged residues is observed (lane 6, Fig. 2), as seen for the $\beta\beta$ derivative in both the liganded and deoxy forms (lanes 7 and 8, Fig. 2).

Both of the cross-linked Hbs are highly cooperative. The oxygen affinity of the derivative cross-linked between the two lysine-82(β) residues is slightly increased. Shibayama et al.[10] reported that there is a minor impurity also cross-linked between the β chains, but at another site, which cannot be separated from the Lys-82(β_1)–Lys-82(β_2) derivative by ion-exchange chromatography. It can be removed by exploiting a difference in the reactivity of Cys-93(β) of the two derivatives toward the sulfhydryl reagents. This results in an increase in the observed Hill coefficient from about 2.5 to 3.0 and a substantial decrease in K_1, the first stoichiometric constant for the binding of O_2. The P_{50}, which is decreased by about 40% compared to Hb A, remains unchanged. For very detailed structure function studies, this minor Hb should be removed.

The derivative cross-linked between Lys-99(α_1) and Lys-99(α_2) has unique electrophoretic and chromatographic properties, making it easier to purify. As noted above, the isoelectric point of the Hb in the oxy form is decreased much less than that expected from the loss of the two positive charges of the modified amino groups. For this reason the salt concentration required to elute the Hb from DEAE-Sepharose is much lower than that for the derivative cross-linked between the β chains.

The reaction of DBBF at Lys-99(α_1) is very specific for the deoxy quaternary structure. In the liganded form this site is completely inaccessible to the reagent. However, after the reaction has occurred and the dibromosalicylic acid groups are eliminated, the cross-linked Hb is readily able to adopt the liganded conformation. The Hill coefficient is between 2.9 and 3.1. The oxygen affinity of this derivative is decreased, making it a strong candidate for use as a blood substitute. Under physiological conditions the P_{50} is increased slightly more than twofold, giving rise to oxygen transport characteristics very similar to those of whole blood. A further advantage of cross-linking is that the thermostability of the protein is increased, making it possible to heat solutions of the Hb under deoxygenated conditions above 70° for prolonged periods to destroy any viruses that may be present. This derivative is currently undergoing clinical trials and may prove to be the first widely used blood substitute.

[10] N. Shibayama, K. Imai, H. Hirata, H. Hiraiwa, H. Morimoto, and S. Saigo, *Biochemistry* **30**, 8158 (1991).

[18] Derivatives of Hemoglobin Prepared by Reaction with Aryl or Alkyl Isothiocyanates

By Douglas L. Currell

The reaction of protein amino groups with aryl isothiocyanates to form thiourea derivatives, the Edman reaction, has frequently been exploited in the sequencing of proteins. The basis for this application lies in the reaction of aryl isothiocyanates with the amino terminus of the protein. Reaction with the reagent depends on the nucleophilic character of the protein amino group, and as a result, the protonated form is unreactive. Thus, the specificity of the reaction can be controlled to some extent by manipulation of the pH, e.g., reaction at the amino termini is favored by maintenance of the pH near 7.3, because at this pH most of the ε-amino groups of the protein are protonated. Maintenance of the pH near neutrality can also minimize competing hydrolysis reactions of the reagent, which is subject to both acid and base catalysis. Another competing reaction in the case of alkyl isothiocyanates is alkylation of nucleophilic sulfhydryl groups through displacement of the —NCS group. This reaction is unlikely for aryl isothiocyanates because nucleophilic aromatic substitution is extremely slow unless the —NCS group is activated by the presence of electron-withdrawing groups in the 2- and 4-positions.

Synthesis of Isothiocyanates

In general, isothiocyanates are synthesized from the corresponding amino compounds by reaction with thiophosgene according to modifications of the method of Maddy.[1] Amino compounds and thiophosgene are available from Aldrich Chemical Co. (Milwaukee, WI). The following general syntheses are provided by TransWorld Chemicals (Rockville, MD).

Alkyl Isothiocyanates

A three-necked flask, equipped with an overhead stirrer, an addition funnel, and a thermometer, is charged with 2 liters of water, 500 ml of dichloromethane, 201.6 g (2.4 mol) of sodium bicarbonate, and 88 ml (1.15 mmol) of thiophosgene. The mixture is surrounded by an ice bath and

---

[1] A. H. Maddy, *Biochim. Biophys. Acta* **88**, 390 (1964).

METHODS IN ENZYMOLOGY, VOL. 231

stirred vigorously while 1.0 mol of amine (or a dichloromethane solution of the amine) is added dropwise at 5° to 10°. The reaction is followed by monitoring the CO_2 gas evolution. When the CO_2 evolution ceases, the lower layer is separated, dried with magnesium sulfate, stripped of solvent, and vacuum distilled.

Isothiocyanatobenzenesulfonic and Isothiocyanatobenzoic Acids

A three-necked flask, equipped with an overhead stirrer, addition funnel, and thermometer, is charged with 2.9 liters of water, 210 g (2.5 mol) of sodium bicarbonate, and 88 ml (1.15 mol) of thiophosgene. The mixture is cooled to 10° and a solution of 1.0 mol of the aminobenzenesulfonic or aminobenzoic acid in 500 ml of 10% sodium bicarbonate is added dropwise at a rate that will produce a moderate gas evolution. When the addition is complete, the mixture is stirred until the evolution of gas ceases and then is cautiously rendered acidic with 10% hydrochloric acid. The product is filtered, washed once with cold water, air dried, and then recrystallized from acetone. If the sodium salt of the acid is the preferred product, the method of Maddy can be utilized, which omits the acidification with HCl.[1]

4-Isothiocyanatobenzenesulfonamide

A three-necked flask, equipped with an overhead stirrer and a thermometer, is charged with 1.7 liters of water, 415 ml of 37% hydrochloric acid, and 172 g (1.0 mol) of sulfanilamide. The solution is cooled to 10° and 77 ml (1.0 mol) of thiophosgene is added in one portion. Stirring is continued for 3 hr. The precipitated solid is then filtered, washed with cold water, air dried, and recrystallized from acetone.

All of the isothiocyanate compounds must be stored in a desiccator to prevent hydrolysis by atmospheric water.

Reaction Conditions for Modification

The optimum time for the reaction of oxyhemoglobin with isothiocyanates can be determined by following the change of oxygen affinity with time and by following the course of the reaction with analytical high-performance liquid chromatography (HPLC) until a relatively homogeneous product is obtained.[2] The optimum ratio of isothiocyanate to hemoglobin can be determined by studying the effect of various ratios at a constant reaction time. In general, a ratio of reagent to oxyhemoglobin

[2] D. L. Currell, B. Law, M. Stevens, P. Murata, C. Ioppolo, and F. Martini, *Biochem. Biophys. Res. Commun.* **102**, 348 (1981).

tetramer of 10 : 1 and a reaction time of 3 to 5 hr at room temperature have proved useful. The reaction can be terminated by passing the reaction mixture through a mixed-bed ion-exchange resin or by addition of a large excess of glycylglycine.[3] If these conditions lead to the formation of a small amount of methemoglobin, the mixture can be reduced by the method described by Mills et al.[4]

The pH can be maintained by an automatic titrator or a buffer that contains no nucleophilic groups that would react with the reagent. The Goode buffers are frequently used because they meet this criterion. However, Bohn et al.[5] have reported that the Goode buffers have an effect on the oxygen affinity. Bis and bis-Tris buffers suffer the same disadvantage, because HCl must be used in the pH adjustment. In general a pH of 7.0 to 7.3 has been used and leads to specific modification of the N termini. Nothing is gained by adopting lower pH values, and using higher pH values could lead to modification of the ε-amino groups of lysines.

The isothiocyanate reagent is added as a buffered aqueous solution or suspension, prepared immediately before use to minimize hydrolysis reactions. Benzenesulfonates are more soluble than benzoates, which are more soluble than uncharged reagents such as benzenesulfonamides. Solubility of the latter can be improved by solution in solvents such as dimethyl sulfoxide.

In contrast to pyridoxal phosphates,[6] the reaction sites for the monoisothiocyanates are the four amino termini, independent of the state of ligation, although the rate of reaction is faster for deoxyhemoglobin than for oxyhemoglobin. The chain cross-linking reactions with diisothiocyanates[7,8] are dependent on ligation state, because the distance between the chains differs for oxy- and deoxyhemoglobin.

A rough estimate of the number of reaction sites can be obtained by utilizing the absorption at 295 nm shown by the isothiocyanate reagents and the hemoglobin derivatives.[7]

Hybrid tetramers in which only the α chains or only the β chains are modified have been prepared by separation of the chains of the completely reacted tetramer and recombination of the reacted chains with the unre-

[3] J. M. Manning, this series, Vol. 76, p. 161.
[4] A. Hayashi, T. Suzuki, and M. Shin, Biochim. Biophys. Acta 310, 309 (1973); F. C. Mills, M. L. Johnson, and G. K. Ackers, Biochemistry 15, 5350 (1976).
[5] B. Bohn, J. Kistler, M. Marden, and C. Poyart, Nouv. Rev. Fr. Hematol. 29(5), 341 (1987).
[6] R. Benesch and R. E. Benesch, this series, Vol. 76, p. 147.
[7] L. R. Manning, S. Morgan, R. C. Beavis, B. T. Chait, J. M. Manning, J. R. Hess, M. Cross, D. L. Currell, M. A. Marini, and R. M. Winslow, Proc. Natl. Acad. Sci. U.S.A. 88, 3329 (1991).
[8] M. P. Kavanaugh, D. T.-B. Shih, and R. T. Jones, Biochemistry 27, 804 (1988).

acted chains.[9] Another technique for synthesizing the hybrid reacted at the α chain only is to favor reaction at the α-chain termini by carrying out the reaction of deoxyhemoglobin in the presence of a large excess of inositol hexaphosphate (Sigma Chemical, St. Louis, MO), which blocks reaction at the termini of the β chains.[7] Sodium tripolyphosphate (Alfa Morton-Thiokol, Danvers, MA) can be substituted for the inositol hexaphosphate. The reaction time is much shorter under these conditions (15 min) and the reaction can be terminated by addition of a large excess of glycylglycine.[3,7]

To determine whether the isothiocyanate has reacted with the β^{93} sulfhydryls of hemoglobin, a titration of the sulfhydryls is carried out with p-chloromercuribenzoic acid, according to the method of Boyer.[10] Titration of the hemoglobin derivatives prepared with aryl isothiocyanates consistently shows the presence of approximately two sulfhydryls, indicating no reaction with the β^{93} sulfhydryls.

Properties of the Isothiocyanate-Modified Hemoglobins

The structures of the isothiocyanates studied are shown in Fig. 1. The isothiocyanates can be classified according to the nature of the charge: (1) negatively charged reagents, 2-, 3-, and 4-isothiocyanatobenzenesulfonic acids (2-, 3-, and 4-ICBS) and 4-isothiocyanatobenzoic acid (4-ITCB), (2) uncharged reagents, 4-isothiocyanatobenzenesulfonamide (4-ICSA), and (3) positively charged reagents, 3-(diethylamino)propyl isothiocyanate (DEAPI).

A summary of the properties of the isothiocyanate derivatives is given in Table I. In general, negatively charged reagents produce a protein modified at the amino termini of the the chains with a reduced oxygen affinity over a wide pH range, a decreased but still significant cooperativity and alkaline Bohr effect, a greatly reduced response to 2,3-diphosphoglycerate, and an oxygen affinity independent of chloride concentration at pH 7.3 and above.[11,12] Hb-2- and Hb-4-ICBS showed some dependence on chloride concentration at pH values below 7.[12] The decrease in oxygen affinity was rationalized as due to interaction of the negatively charged moiety of the ICBS or ITCB compound with the positively charged amino

[9] A. Bellelli, R. Ippoliti, D. Currell, S. G. Condo, B. Giardina, and M. Brunori, *Eur. J. Biochem.* **161,** 329 (1986).

[10] P. D. Boyer, *J. Am. Chem. Soc.* **76,** 4336 (1954).

[11] D. L. Currell, D. M. Nguyen, S. Ng, and M. Hom, *Biochem. Biophys. Res. Commun.* **106,**1325 (1982).

[12] R. Ippoliti, D. L. Currell, E. Lendaro, A. Bellelli, M. Castagnola, M. Bolognesi, and M. Brunori, *Biophys. Chem.* **37,** 293 (1990).

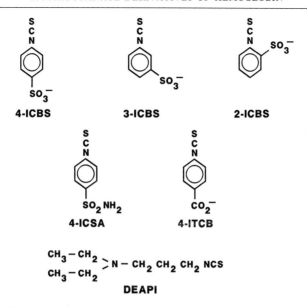

FIG. 1. Isothiocyanates used to prepare hemoglobin derivatives.

group of the lysine-82(β) of the opposite β chain with a resultant stabilization of the deoxy conformation.[2] Computer modeling studies indicated that such an interaction is possible for the 4-substituted compounds but not for the 2- or 3-substituted compounds.[12] This latter observation explains the greater reduction in oxygen affinity of the 4-substituted, negatively charged compounds. Support for this idea is given by the slight increase in oxygen affinity of the uncharged 4-ICSA.[11] The absence of response to chloride of the hemoglobin derivatives reacted with the negatively charged compounds is reasonable because reaction is at or near the oxygen-linked chloride-binding sites.[11] The reduction in the Bohr effect is probably due to the absence of the chloride-dependent Bohr effect.[13] The contribution of the lysine-82(β) to the Bohr effect as suggested by Perutz et al.[14] would be maintained by the above proposed interaction between the sulfonate and the protonated lysine-82(β) in the deoxy conformation. By contrast, hemoglobin reacted with the uncharged 4-ICSA shows an almost unchanged response to the allosteric effectors, chloride, the proton, and 2,3-diphosphoglycerate. This latter difference was exploited in a study of the identity of the dissociation-linked chloride-binding

[13] G. U. van Beek, E. R. P. Zuiderweg, and S. H. de Bruin, Eur. J. Biochem. 99, 379 (1979).
[14] M. F. Perutz, J. V. Kilmartin, K. Nishikura, J. H. Fogg, P. J. G. Butler, and H. S. Rollema, J. Mol. Biol. 138, 649 (1980).

TABLE I

EFFECT OF 2,3-DIPHOSPHOGLYCERATE ON
OXYGEN AFFINITY OF HEMOGLOBIN REACTED
WITH ISOTHIOCYANATE DERIVATIVES[a]

	$\log P_{50}{}^b$ (mm Hg)	
Hemoglobin derivative	No DPG	With DPG $(2.5 \times 10^{-4}\ M)$
Hb A	0.52(2.4)	0.95(2.8)
Hb–4-ITCB	1.01(1.8)	1.15(2.1)
Hb–4-ICBS	0.99(1.7)	1.03(1.8)
Hb–3-ITCB	0.85(1.5)	0.85(1.8)
Hb–2-ICBS	0.76(2.5)	0.78(1.9)
Hb–4-ICSA	0.42(2.1)	0.62(2.4)
Hb–DEAPI[c]	−0.27	0.31
Hb A[c]	0.08	0.95

[a] At 20°; Hb concentration, $5 \times 10^{-5}\ M$ (tetramer); 0.05 M bis-Tris buffer, pH 7.3; Cl⁻ concentration, 0.05 M.
[b] The n values are in parentheses; determined by the spectrophotometric method of A. Rossi Fanelli and E. Antonini, *Arch. Biochem. Biophys.* **77**, 478 (1958).
[c] In HEPES buffer, 0.05 M, no chloride, pH 7.3; R. E. Fan and D. L. Currell, *FASEB J.* **48**, A353 (1988).

sites.[15] That the chloride dependence of the dimer–tetramer assembly for hemoglobin reacted with 4-ICBS and 4-ICSA resembled that of unreacted hemoglobin was taken to indicate a difference in the oxygen-linked and dissociation-linked chloride-binding sites.

The reaction of oxyhemoglobin with the tertiary amine, DEAPI, produces a derivative with a high oxygen affinity, a reduced Bohr effect, and a normal dependence of the oxygen affinity on chloride concentration below 0.1 M, but absent above 0.1 M, indicating disruption of the low-affinity binding sites.[16] Response to 2,3-diphosphoglycerate was still present.

One possible application for the isothiocyanates is in the treatment of sickle cell anemia. The 4-ICSA would seem to be a particularly good candidate because it produces a slight increase in oxygen affinity at pH

[15] D. L. Currell, M. Gattoni, and E. Chiancone, *Biochim. Biophys. Acta* **871**, 316 (1986).
[16] R. E. Fan and D. L. Currell, *FASEB J.* **48**, A353 (1988).

7.3, with maintenance of cooperativity, and has a normal response to the allosteric effectors. In addition, its hydrolysis product is sulfanilamide.

Another potential application lies in the area of blood substitutes. To this end diisothiocyanato compounds, which cross-link the chains, have been studied.[7,8,17]

[17] J. M. Manning, this volume [4].

[19] Hemoglobin Polymerization

By Shirley L. MacDonald and Duncan S. Pepper

Introduction

A reagent to cross-link hemoglobin (Hb) has long been thought desirable, and is needed to stabilize the tetramer (64 kDa) by intramolecular cross-links and also to produce oligomers (<500 kDa) and polymers (>500 kDa) by intermolecular cross-links. Intramolecular cross-links between tetramer subunits prevent dissociation into excretable dimers (32 kDa).

Intermolecular cross-links are necessary both to reduce colloid osmotic pressure and to improve the circulatory half-life of the molecule as a cell-free solution. By judicious cross-linking to form oligomers ($n = 2$, 128 kDa; $n = 3$, 192 kDa; etc.) it is possible to increase the half-life to approximately 40–46 hr, obtained in a total exchange using baboons,[1] and 25 hr in a 90–95% exchange transfusion in rats,[2] which is clinically more useful than the 2–4 hr seen using pure tetramer.[1,2] However, carrying out oligomerization too far, producing polymers with molecular masses >500 kDa, is undesirable because the large aggregates may have altered surface and charge characteristics; these alterations could lead to flow changes in the microcirculation[3] and may well acquire additional untoward toxicities, such as reticuloendothelial system (RES) blockade[3] or stimulation of an immunological response to the altered aggregate surfaces.[3]

The chemical problem of producing a reagent that will cross-link mainly intramolecularly, and only slightly intermolecularly, is not a trivial one, and considerable effort has been expended over many years in designing "smart" cross-linkers that will target a specific site and covalently cross-

[1] S. A. Gould, L. R. Sehgal, A. L. Rosen, H. L. Sehgal, and G. S. Moss, *Ann. Surg.* **211,** 394 (1990).

[2] F. DeVenuto and A. Zegna, *J. Surg. Res.* **34,** 205 (1983).

[3] P. J. Scannon, *CRC Crit. Care Med.* **10,** 261 (1982).

link the Hb subunits while simultaneously effecting a decrease in oxygen affinity. Such molecules include 2-nor-2-formylpyridoxal 5'-phosphate (NFPLP),[4] bis(pyridoxal) phosphate (bis-PLP),[5–7] and bis(3,5-dibromosalicyl) fumarate (DBBF).[8–10]

When designing a cross-linker it is necessary also to pay attention to the problem of oxygen affinity. In some conditions, the physical process of cross-linking is likely to be accompanied by an undesirable increase in the oxygen affinity,[11] and it is usually only when the cross-linking molecule is designed to incorporate an oxygen affinity modifier that the oxygen affinity is maintained or reduced.[4–10] In practice, the cross-linker tends to increase, and the modifier to decrease, the oxygen affinity, so the net effect is a balance between the two effects. However, these combined or dual-purpose molecules are difficult to synthesize and, with the exception of DBBF, are not readily commercially available.

We therefore sought a reliable two-step procedure in which the cross-linking is performed by a simple molecule and the oxygen affinity is modified by a separate simple molecule. The reagents had to be available commercially at low cost, and the procedures had to possess potential for scaleup from the laboratory to production. Initial approaches to this strategy were based on the use of glutaraldehyde as the cross-linker and pyridoxal phosphate (PLP) as the oxygen affinity modifying molecule.[2,12] Glutaraldehyde is a reactive molecule, producing rapid intermolecular, as well as limited intramolecular, cross-linking. The result is a mixture of cross-linked (stabilized) tetramer, unstabilized tetramer and, a wide range of higher molecular weight species.

Even when a soluble product is obtained of suitable MW distribution, it remains unstable on storage[13] or heating (pasteurization). It has been

[4] R. Benesch and R. E. Benesch, this series, Vol. 76, p. 147.
[5] R. E. Benesch and S. Kwong, Biochem. Biophys. Res. Commun. 156, 9 (1988).
[6] R. E. Benesch and S. Kwong, J. Protein Chem. 10, 503 (1991).
[7] P. E. Keipert, A. J. Anderson, S. Kwong, and R. E. Benesch, Transfusion (Philadelphia) 29, 768 (1989).
[8] J. A. Walder, R. H. Zaugg, R. Y. Walder, J. M. Steele, and I. M. Klotz, Biochemistry 18, 4265 (1979).
[9] R. Chatterjee, E. V. Welty, R. Y. Walder, S. L. Pruitt, P. H. Rogers, A. Arnone, and J. A. Walder, J. Biol. Chem. 261, 9929 (1986).
[10] N. Shibayama, K. Imai, H. Hirata, H. Hiraiwa, H. Morimoto, and S. Saigo, Biochemistry 30, 8158 (1991).
[11] K. Kothe and K. Bonhard, Surg., Gynecol. Obstet. 161, 563 (1985).
[12] L. R. Sehgal, A. L. Rosen, S. A. Gould, and G. S. Moss, Transfusion (Philadelphia) 23, 158 (1983).
[13] M. A. Marini, G. L. Moore, S. M. Christensen, R. M. Fishman, R. G. Jesse, F. Medina, S. M. Snell, and A. I. Zegna, Biopolymers 29, 871 (1991).

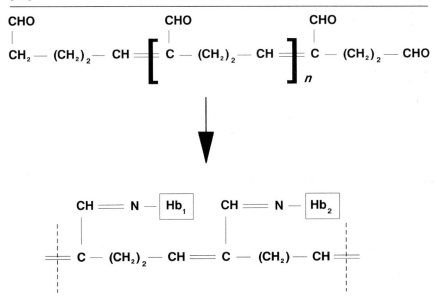

FIG. 1. Reaction between aldehyde groups of glutaraldehyde and amino groups of hemoglobin at neutral pH. The reaction is rapid, nonspecific, and leads to large numbers of polymers of high molecular mass ($>1 \times 10^3$ kDa).

found that glutaraldehyde is toxic at even low doses, and the possibility of reversible reactions releasing free glutaraldehyde *in vivo* cannot be discounted.[13] Much effort has been made to moderate the glutaraldehyde reaction, either by adding modifiers such as lysine,[12] working with low Hb concentrations so as to favor tetramer stabilization over polymer formation,[14] and/or using physically separate compartments (dialysis cartridges) to slow the reaction between Hb and glutaraldehyde.[15]

In retrospect, it seems that the glutaraldehyde molecule is unsuitable for a number of reasons: it is a dialdehyde (too reactive), it is too long a molecule (facilitating intertetramer cross-links), and it lacks any intrinsic or functional specificity, leading to uncontrolled polymerization (Fig. 1). For all these reasons, therefore, we chose to work instead with the glycolaldehyde–pyridoxal phosphate system to provide dual control of cross-linking (MW) and oxygen affinity (P_{50}). The use of glycolaldehyde to polymerize proteins has been known in the literature for some time. Initial

[14] J. C. Hsia, U.S. Pat. 4,857,636 (1989).
[15] L. R. Sehgal, R. E. DeWoskin, G. S. Moss, A. L. Rosen, and H. L. Seghal, U.S. Pat. 4,826,811 (1989).

studies involved RNase[16] and Hb.[17] However, it was not known if the reaction could be controlled under production conditions or whether oxygen affinity could be regulated to a theoretically clinically useful value. A final consideration for clinical products is the need to reduce the residual un-cross-linked tetramer (excretable dimer) to a minimum. Ideally, this would be <2%, and preferably <20%. In this regard, glutaraldehyde is also disappointing because overpolymerization and gelling tend to coexist with high residual un-cross-linked tetramer of greater than 20%, thereby presenting a doubly unsatisfactory situation.

Preliminary ranging studies showed that glycolaldehyde could be used in the concentration range 0.3–1.0%, with Hb solutions in the concentration range 1.0–16.0 g/dl, without excessive polymerization. Fine tuning of reaction parameters was then sought by varying the pH, temperature, and time of the reaction between pH 6–9, 4°–40°, and 2–20 hr. A useful optimum combination was then further evaluated using 0.3% (w/v) glycolaldehyde, 12 g/dl deoxyHb (pH 8.5, 30°), and carrying out the reaction for 2 hr. This gave >80.0% cross-linked product with <2.0% material having a MW of 500 kDa, a P_{50} (unmodified) of 16.0 mm Hg (37°, pH 7.4), and a methemoglobin (metHb) level of <5.0%. A crucial factor was the chemical compatibility with the next reagent in the process, PLP. Because both reagents use the same chemistry (aldehyde + amine → Schiff base) it might be expected *a priori* that they would mutually interfere. We have discovered that this is not the case. Glycolaldehyde and PLP can be used sequentially in either order or together, without interference, and thus it must be concluded that they do not compete with each other for the same amino acid —NH_2 groups within the Hb chains. For practical reasons, it is convenient to perform the glycolaldehyde reaction first, either during the deoxygenation step or following deoxygenation. The PLP is then added toward the end of the cross-linking period, and the two reactions proceed concurrently for an additional 45 min. The glycolaldehyde reaction can be performed either with oxyHb or deoxyHb or a mixture of both. However, the final P_{50} values are somewhat higher if this is carried out mainly or wholly in the deoxy form. The PLP modification step requires anaerobic conditions and is therefore better performed toward the end of the glycolaldehyde reaction. Both reactions are terminated by the addition of a reducing agent (we prefer sodium borohydride but other agents can be used), which serves both to terminate polymerization by converting excess aldehydic groups to primary alcohols and to

[16] A. S. Acharya and J. M. Manning, *Proc. Natl. Acad. Sci. U.S.A.* **80,** 3590 (1983).
[17] W. J. Fantl, L. R. Manning, H. Ueno, A. DiDonato, and J. M. Manning, *Biochemistry* **26,** 5755 (1987).

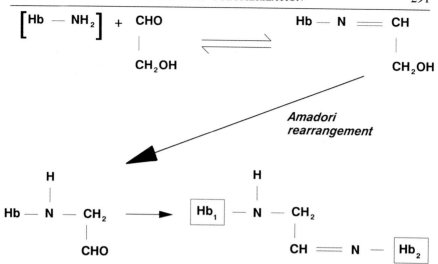

FIG. 2. Reaction between the monoaldehyde glycolaldehyde and amino groups of Hb. The Schiff base formed undergoes the Amadori rearrangement to generate a new aldehyde group capable of cross-linking. Addition of sodium borohydride will reduce these Schiff base linkages to stable secondary amine linkages and reduce excess aldehydic groups to primary alcohols, preventing further cross-linking.

stabilize the reversible Schiff bases by reduction to stable secondary amines. An additional advantage of sodium borohydride is that metHb is reduced to <2.0%. No change in the MW distribution is seen on prolonged storage or heating, other than the precipitation of free dimer at temperatures above 60°.

Chemistry of Glycolaldehyde Reaction

It is known that glycolaldehyde, a simple two-carbon α-hydroxyaldehyde, can undergo a simple univalent reaction with pendant unionized α- or ε-amino groups on proteins.[17] The resultant Schiff base contains an unsaturated double bond, which can migrate via the Amadori rearrangement to the end (α-OH) carbon, thus producing a new unsaturated aldehydic oxygen (Fig. 2). Thus, a key feature of the reagent is that the second aldehyde group only appears when the molecule has reacted with a first amino group. This step is probably reversible. If a second amino group is not in close proximity to the new aldehyde group, the intermediate will revert to its earlier form and dissociate from the protein molecule, leaving it effectively unchanged. Over a period of time, only the cross-links that are sterically favored by the close proximity of two amino groups will accumulate. The remarkably short (two-carbon) chain of glycolaldehyde also dictates that few such cross-links will be formed, and therein must lie the explanation for the slow reaction and remarkable specificity of glycolaldehyde compared with glutaraldehyde. The close physical proximity of the tetramer subunits of Hb naturally tend to favor intra- over intermolecular cross-links. However, by random statistical movement it is inevitable that occasionally two Hb molecules will collide and react such that a cross-link will form on the outer surface between two adjacent but separate molecules. The extent of this "side reaction," which produces oligomers, is directly proportional to the concentration of Hb and can be controlled or eliminated by reducing the Hb concentration. However, this is not practical below approximately 1 g/dl Hb and so a compromise of 6–12 g/dl is chosen to give a workable volume and an acceptable molecular weight distribution of oligomers. From a purely practical point of view, this is also desirable because it reduces the colloid osmotic pressure (COP) from approximately 40 mm Hg to 25 mm Hg for a 14 g/dl solution of Hb, producing an isooncotic preparation suitable for clinical use. This feature also can be used to monitor the progress of the reaction because COP is quickly and easily measured on small-scale samples. A more direct (but slower) measure of the degree of oligomerization can be obtained by fast protein liquid chromatography (FPLC) gel filtration (molecular sieve chromatography). However, it is essential to use 1 M MgCl$_2$ as the solvent to provide the chaotropic conditions necessary to dissociate and so quantify un-cross-linked dimers, which otherwise associate noncovalently with other Hb dimers and chromatograph with an apparently higher molecular weight.

Therefore, in summary, combined steric and kinetic mechanisms operate to provide a controlled and quasi-specific, predominantly intramolecular cross-link together with a restricted amount of oligomerization, which, together with the ability concurrently to perform PLP modification, make glycolaldehyde a very practical reagent for Hb derivatization.

Polymerization Methods

Hemoglobin solutions used for all polymerization protocols should contain as little red cell-derived stromal lipid as possible. This is generally, and rather misleadingly, known as stroma-free hemoglobin. In practice, we have found that hemoglobin solutions prepared from washed, lysed human red cells by hollow fiber ultrafiltration (pore size, 0.2 μm) contain <0.5 μg/ml phospholipid when assayed for total phosphorus content.

Polymerization with Glutaraldehyde

Glutaraldehyde is a widely used reagent for polymerizing hemoglobin, although the precise mechanisms by which glutaraldehyde reacts with proteins remain poorly understood. Commercial preparations of aqueous 25% glutaraldehyde contain unstable derivatives of the aldehyde in the form of compounds that readily revert to glutaraldehyde.[18] At acidic pH (commercial solutions have a pH of 3.0) glutaraldehyde is in equilibrium with its cyclic hemiacetal and polymers of cyclic hemiacetal. For cross-linking purposes the pH is raised to neutrality, whereupon the dialdehyde is transformed into unsaturated aldehyde polymers (Fig. 1). As pH is further increased (n), increases and precipitation may occur.[18]

At neutral pH, the major reactive species (α,β-unsaturated aldehyde) form labile intermediate Schiff bases with amino groups of the protein. This reaction involves mainly lysine residues as well as the α-amino groups and the sulfhydryl group of cysteine.[12] There is evidence to suggest that in all polymerized species, the β chain is extensively modified, with some α-chain involvement also.[13]

The concentration of the Hb is believed not to be critical, similar results being obtained with 2–12 g/dl,[13] although others[14] report that a Hb concentration of 1 g/dl favors tetrameric stabilization over oligomer formation. Reaction temperature is also now considered to be of limited significance,[13] although published reports generally carry out the polymerization at 4°,[2,12] presumably in an attempt to inhibit the formation of metHb.

Coupling Procedure

The pH of the hemoglobin solution is adjusted to pH 6.8–7.0 with 1 M Tris (1 M Na$_2$CO$_3$ is a suitable alternative). A 25% aqueous solution of pure glutaraldehyde (grade 1; Sigma Chem. Co., St. Louis, MO) is diluted to 2.1 g/dl with deionized water. To achieve a 7:1 molar excess glutaraldehyde:Hb, 7 ml of this diluted stock is added per 100 ml of 14 g/dl Hb. Increasing the glutaraldehyde concentration above a 7:1 molar ratio (glutaraldehyde:Hb) simply leads to gel formation, whereas lower concentrations (<1:4 molar excess) produce lesser degrees of cross-linking.[19] The glutaraldehyde should be added slowly with continual mixing. Alternatively, the reaction may be slowed to some degree by use of a kidney dialysis cartridge (molecular weight cutoff 10,000) to prevent too rapid a reaction.[15]

Progress may be monitored directly by the fall in colloid osmotic pressure (Osmomat 050, Gonotec, Berlin, Germany) or by gel filtration on either FPLC or HPLC. Measurement of COP is rapid and provides immediate assessment of polymerization, whereas gel filtration is time consuming and therefore retrospective in nature. The COP of a 12 g/dl un-cross-linked Hb solution is approximately 45 mm Hg, and when this

[18] P. Monzan, P. Germain, and M. Marzarguil, *Biochimie* **57**, 1281 (1975).
[19] D. H. Marks, G. L. Moore, F. Medina, and G. Boswell, *Mil. Med.* **153**, 44 (1988).

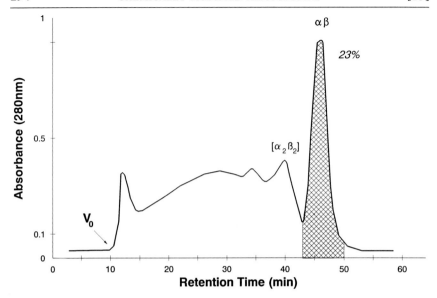

FIG. 3. FPLC gel filtration profile (Superose 12, Pharmacia) of glutaraldehyde-polymerized hemoglobin. The buffer contains 1 M $MgCl_2$, which brings about dissociation of unpolymerized tetramer (hatched area), representing 23% total protein load. There is a considerable amount of material at the void volume V_0 ($>1 \times 10^6$). Elution details are described in the text.

reaches 25 mm Hg, the reaction may be terminated by addition of lysine[12] to a final concentration of 50 mM (5 ml 1 M lysine in 50 mM phosphate buffer, pH 7.4/100 ml 14 g/dl Hb). Alternatively, excess glutaraldehyde may be removed simply by dialysis against saline.[2,12] It has been reported that dialysis is preferable because addition of lysine increases heterogeneity by producing lysylhemoglobin products.[19]

The final product is a complex mixture of unreacted Hb dimers and tetramers as well as various Hb species reacted with glutaraldehyde at different sites and to different degrees (Fig. 3). If pyridoxalation has been carried out previously, various species of pyridoxalated Hb will also be included. In the production of a blood substitute, optimum values for polymerization are necessarily a tradeoff between oxygen affinity and degree of polymerization. As the latter rises, P_{50} falls to values that conceivably could prevent hemoglobin from off-loading oxygen *in vivo*. It is worth pointing out that polymerization of deoxygenated pyridoxalated Hb has been reported to yield a product with a slightly higher P_{50} (22 mm Hg) than that of the oxyhemoglobin derivative (19 mm Hg). The Hill coefficient (n) is also increased under these anaerobic conditions (2.2 vs. 1.9).[13]

Although molecular heterogeneity is not necessarily a disadvantage for the functional properties of an emergency resuscitation fluid, it is a severe problem in quality control procedures. Should the product lead to undesirable side effects *in vivo,* it would be very difficult to define the nature and origin of the problem. The nonspecific nature of the cross-linking leads to an overall rise in oxygen affinity, thereby reducing the effectiveness of any pyridoxalation reaction that has previously been performed.

The nature of glutaraldehyde–Hb cross-links appears to be inherently unstable.[13] When a sample aged 1 month at 4° was subjected to anion-exchange chromatographic analysis, a very different elution profile was obtained. The Schiff bases that constitute glutaraldehyde cross-links are assumed to be capable of rearranging or dissociating, even on short-term storage.

Polymerization with Glycolaldehyde

Glycolaldehyde is a cross-linking agent of latent reactivity in which new aldehyde functions are generated following the Amadori rearrangement of Schiff bases formed with protein amino groups (Fig. 2). The reagent has been used successfully to cross-link carboxymethylated Hb.[17,20] Our primary interest was in maximizing the degree of cross-linking, while avoiding the buildup of high molecular weight polymers, and combining the reaction with pyridoxalation as a means of decreasing oxygen affinity.

Coupling Procedure

Hemoglobin solution (1 liter, 12 g/dl, pH 6.4) in normal saline is transferred to an air-tight, jacketed multiport reaction vessel connected to a circulating water bath. The solution is deoxygenated to <2% oxyHb at 20° using either a nitrogen gas- or 1% sodium dithionite-perfused hydrophobic hollow fiber cartridge (see General Methods). The reaction vessel head space is perfused with nitrogen gas. The pH is then adjusted to 8.5 with degassed 1 M Na_2CO_3 and the temperature increased to 30°. Glycolaldehyde is added to a final concentration of 0.3% (30.1 ml of 10% aqueous degassed stock solution) and mixed gently with a magnetic stirrer. Polymerization is monitored by the decrease in COP, which typically will fall from 45 mm Hg to 25 mm Hg over 90 min. If the solution is not to be pyridoxalated, the cross-linking reaction may be terminated at this stage by the addition of 85.7 ml of 1.56% sodium borohydride (10 mM NaOH) to give a 20 : 1 molar excess $NaBH_4$: Hb. This should be added slowly

[20] L. R. Manning and J. M. Manning, *Biochemistry* 27, 6640 (1988).

and with mixing to prevent any denaturation. It is also advisable to allow venting of the hydrogen gas produced to avoid buildup of pressure.

It is also possible to cross-link Hb solution that is fully oxygenated. The reaction rate is appreciably faster under aerobic conditions but the oxygen affinity of the final product is higher than that found following anaerobic cross-linking (14 and 16 mm Hg, respectively). FPLC reversed-phase chromatography of globin chains (see General Methods) indicates that both α and β chains are involved extensively in oxy-cross-linked Hb, whereas β chains are predominantly involved in deoxy-cross-linked Hb. This presumably results in less distortion of the tetramer per se and allows greater access to the 2,3-DPG-binding site, leading to a higher final P_{50} if pyridoxalation is carried out subsequently.

Glycolaldehyde Coupling Combined with Pyridoxalation

In this procedure,[21] the pH of the glycolaldehyde-polymerized Hb solution is adjusted to pH 7.2 with 2 M NaH$_2$PO$_4$ (degassed) and the temperature is reduced to 20–25°. Pyridoxal phosphate (Sigma) is then added to a final 6 : 1 molar excess PLP : Hb (64.5 ml degassed 4.08% PLP in 1 M Tris-HCl, pH 7.2; PLP is first dissolved in 1 M Tris and then adjusted to pH 7.2 with concentrated HCl). At this pH and temperature the glycolaldehyde cross-linking rate is reduced significantly. After 45 min, 85.7 ml of 1.56% sodium borohydride (10 mM NaOH) is added as above to terminate the cross-linking and stabilize PLP–Hb covalent bonds. After 45 min (or overnight at 4°) the polymerized, pyridoxalated Hb is dialyzed extensively against either isotonic saline or deionized water to remove excess reactants and red cell-derived low molecular weight contaminants. We find a hollow fiber kidney dialysis cartridge (Cobe Centrisystem; surface area 0.9 m^2) to be the most efficient means of dialyzing 1-liter batches of Hb solution (Hb flow rate, 500 ml/minute; saline/water, 1 liter/minute; 2–3 hr). An internal transmembrane pressure of 8–10 psi allows simultaneous concentration to achieve a normal COP of 25 mm Hg.

If wet heat treatment (see General Methods) is to be used to remove un-cross-linked Hb, this is best carried out immediately after sodium borohydride reduction. The COP will fall in accordance with the amount of un-cross-linked Hb that has precipitated.

A typical FPLC molecular weight profile (see General Methods) following cross-linking of Hb (12 g/dl) with glycolaldehyde is shown in Fig. 4. The un-cross-linked fraction is 27% and there is very little polymeric Hb with a molecular weight >500,000. The amount of stabilized tetramer

[21] S. L. MacDonald and D. S. Pepper, Eur. Pat. Appl. 9,104,011.3 (1991).

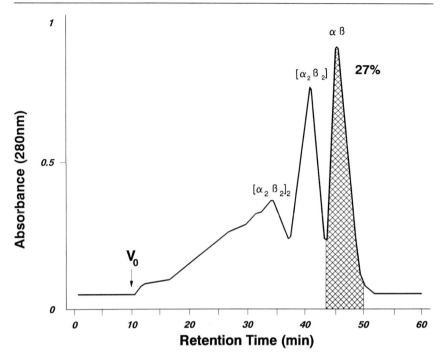

FIG. 4. FPLC gel filtration profile (Superose 12, Pharmacia) of glycolaldehyde-polymerized hemoglobin. As in Fig. 3, the hatched area represents unpolymerized material (27%). There is little high MW material (<5%) at V_0 (>1 × 10^6).

produced, relative to higher molecular weight oligomers, is a function of Hb concentration (Table I), and our final choice of 12 g/dl is a compromise between the optimal Hb concentration (<4 g/dl) for minimizing polymer formation and the constraints of an efficient production process.

TABLE I

EFFECT OF Hb CONCENTRATION ON MOLECULAR MASS PROFILE
FOLLOWING GLYCOLALDEHYDE POLYMERIZATION

Hb (g/dl)	Molecular mass distribution (% total protein)			
	>1 × 10^3 kDa[a]	500–128 kDa	64 kDa	32 kDa
4.0	0	28.2	49.1	22.7
11.0	0	52.6	25.3	22.1
14.0	15.4	51.7	13.7	19.2

[a] Void volume of gel filtration column >1 × 10^3 kDa.

OPTION 1

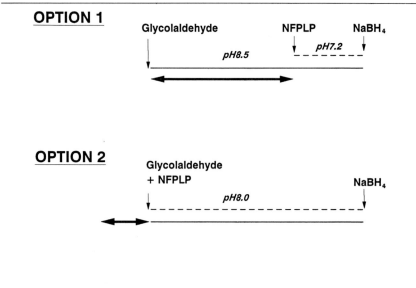

OPTION 2

OPTION 3

FIG. 5. Experimental design options when polymerizing with glycolaldehyde and the combined intratetramer cross-linker/oxygen affinity modifier NFPLP. ↔, Deoxygenation; —, glycolaldehyde polymerization; - - -, NFPLP modification.

Glycolaldehyde Coupling in Combination with Combined Cross-Linker/ Oxygen Affinity Modifiers

Glycolaldehyde–NFPLP. In an attempt to further reduce the un-cross-linked fraction following glycolaldehyde polymerization we substituted the PLP derivative NFPLP in the above reaction process.[21,22] NFPLP selectively introduces a cross-link between β chains.[4] Figure 5 shows the flexibility with which NFPLP and glycolaldehyde cross-linking may be carried out. NFPLP and glycolaldehyde cross-linking may be performed sequentially in either order or simultaneously with little difference in the characteristics of the final product. We find optimal results are obtained when NFPLP is added in 2 : 1 molar excess NFPLP : Hb (0.056 g NFPLP/

[22] S. L. MacDonald and D. S. Pepper, *Biomater., Artif. Cells, Immobilized Biotechnol.* **19,** A424 (1991).

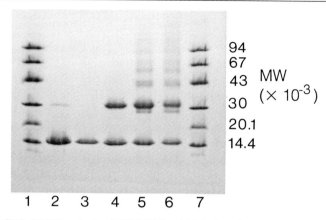

FIG. 6. SDS–PAGE analysis of NFPLP/glycolaldehyde-polymerized hemoglobin. Lanes 1 and 7 contain low MW markers (Pharmacia). Lane 2 contains stroma-free hemoglobin contaminated with red cell-derived nonheme protein. Lane 3 contains stroma-free hemoglobin from which contaminants have been removed by ion-exchange chromatography. Lane 4 contains stroma-free hemoglobin modified by NFPLP. Lanes 5 and 6 contain NFPLP-modified hemoglobin that has been polymerized with glycolaldehyde (30 and 60 min, respectively).

100 ml 7 g/dl Hb solution). Glycolaldehyde is 0.3% (3.09 ml 10% aqueous stock solution/100 ml 7 g/dl Hb solution). Sodium borohydride should be added in a 20 : 1 molar excess to Hb (5 ml 1.56% sodium borohydride in 10 mM NaOH/100 ml 7 g/dl Hb).

Figure 6 is an SDS–PAGE (8–25% gradient gel) analysis of Hb that has first been cross-linked intramolecularly with NFPLP and then further polymerized by glycolaldehyde (option 3 in Fig. 5). Typically, the un-cross-linked fraction is reduced further to 15–20% from 20–25% obtained after the glycolaldehyde–PLP reaction. The final P_{50} is 30–34% mm Hg (37°, pH 7.4), higher than the 24–26 mm Hg typically found after pyridoxalation for reasons that are not entirely clear. Hill coefficients are 1.6–1.8.

Glycolaldehyde–DBBF. DBBF introduces a cross-link between α chains in deoxyHb[8–10] and may also be combined with glycolaldehyde cross-linking. It is necessary to include 5 mM inositol hexaphosphate (IHP) in the reaction mixture in order to protect β chains from DBBF modification, which would have the effect of increasing oxygen affinity. Preliminary experiments suggest that slightly better results are obtained if the DBBF modification is carried out prior to glycolaldehyde modification, although satisfactory results have been obtained when DBBF is substituted for NFPLP in any of the three options in Fig. 5.

DBBF cross-linking is carried out on deoxyHb that has first been dialyzed into 10 mM bis-Tris, pH 7.2, and to which IHP has been added to a concentration of 5 mM (0.46 g/100 ml 7 g/dl Hb solution). DBBF is added to a final molar excess 1.5 : 1.0 DBBF : Hb (0.1 g DBBF/100 ml 7 g/dl Hb, dissolved first in 5 ml 1 M bis-Tris, pH 7.2, and pH adjusted slowly to 7.2 by dropwise addition of 1 M NaOH). The reaction proceeds for 3 hr at 37°. The temperature is then reduced to 30° and pH adjusted to 8.5 with 1 M Tris. Glycolaldehyde is then added to a final concentration of 0.3% (10% aqueous stock solution) and the reaction proceeds for 1 hr. The glycolaldehyde reaction is then terminated by addition of sodium borohydride (5 ml 1.56% NaBH$_4$ in 10 mM NaOH/100 ml 7 g/dl Hb) to a final molar excess 20 : 1 NaBH$_4$: Hb. Reduction proceeds for 30 min followed by dialysis against isotonic saline.

Following preliminary experiments we have obtained a final un-cross-linked fraction of 19%, with no buildup of polymer >500 kDa, and a P_{50} of 30.0 mm Hg (37°, pH 7.4). The Hill coefficient is 1.8–2.2. It should be emphasized that the syntheses of these combined cross-linkers/oxygen affinity modifiers are complex and NFPLP is not yet available commercially, although DBBF may now be purchased from Sigma Chem. Co. (product no. D0292).

Polymerization with Oxidative Ring-Opened Sugars

The aldehydic products of oxidative ring-opened mono- and oligosaccharides[14,23] represent a class of cross-linking reagents of which oxidized raffinose has proved to be particularly useful with regard to Hb polymerization. Oxidative ring opening (e.g., by sodium periodate) produces two aldehyde groups for each sugar monomer in the oligosaccharide. Therefore, raffinose (a trisaccharide) theoretically yields six aldehyde groups. At a low Hb concentration (1 g/dl) and a molar excess of 20 : 1 oxidized raffinose : Hb, the cross-links are almost exclusively intratetrameric, and if modified in the deoxy conformation, the oxygen affinity attains physiological levels (29 mm Hg) without additional modification. The final product is 95–98% stabilized tetramer with no oligomer formation.[14]

Conformationally stabilized, polymerized tetrameric Hb may be produced by increasing the Hb concentration to 4 g/dl and adding oxidized–raffinose in molar excess of 2 : 1 oxidized raffinose : Hb.[14] The formation of intertetramer cross-links then produces a final product in which 70–75% exists in the MW range 128,000–500,000. Only 1–2% remains as unstabilized tetramer. Because the tetramers are already conformationally

[23] J. C. Hsia, *Biomater., Artif. Cells, Immobilized Biotechnol.* **19,** A402 (1991).

stabilized in the deoxy conformation, all the above species should have the same oxygen affinity.

Preparation of Periodate-Oxidized Raffinose and Other Sugars

The sugars (200 mg; glucose, sucrose, raffinose, maltose, maltotriose, maltopentose, maltohexose) in 6 ml distilled water are treated with a fixed molar ratio of solid sodium periodate/saccharide at 20°. The molar ratio of sodium periodate/saccharide is 2.2 for pyranoside and 1.1 for furanoside. After 1 hr, the reaction mixture is cooled in an ice–water bath and sodium bisulfite solution ($NaHSO_3$) is added with vigorous mixing until the precipitated iodine redissolves to yield a colorless solution. The bisulfite acts as a reducing agent, destroying excess periodate, and this reaction is complete when all iodine has been converted to soluble, colorless iodide ion. The pH is adjusted immediately to pH 6.0–6.5 by addition of 6 N NaOH. The resulting oxidized sugar is diluted to give a final concentration of 20 mg/ml, filtered (0.45-μm type GS membrane filter, Millipore, Bedford, MA), and stored at 4°. Theory would suggest that the oxidized raffinose is more stable when stored at pH 6.0–6.5, rather than pH 8.0 as suggested in the original patent.[14]

Preparation of Hemoglobin Covalently Stabilized in Deoxy Conformation by Oxidized Raffinose

A 350-ml solution of Hb (1 g/dl) in 0.1 M phosphate buffer, pH 8.0, is converted to deoxyHb under vacuum for approximately 4 hr at 20°. Sodium dithionite (0.1 mmol in 0.3 ml degassed buffer) is added to the Hb solution and allowed to react for 5 min to ensure complete deoxygenation. Oxidized raffinose (1.08 mmol in 20 ml previously degassed buffer, pH 8.0) is added under vacuum with vigorous mixing and the solution is stirred for 4 hr at 20°. The molar excess oxidized raffinose : Hb is 20 : 1. The reaction mixture is cooled in an ice–water bath and sodium borohydride (15.0 mmol in 5 ml degassed 1 mM NaOH) is added under a nitrogen blanket so as to terminate the cross-linking reaction and to stabilize the Schiff bases by reduction to secondary amines. The reduction proceeds for 45 min.

FPLC gel filtration shows that 95% of the final product is stabilized in the deoxy conformation with <5% dissociating into dimers. The P_{50} is approximately 27 mm Hg (pH 7.4, 37°) and the Hill coefficient is 1.0–1.5.[14] Measurements in a 30% hypervolemic rat model have shown that the plasma half-life is approximately 4–5 hr.[14]

Preparation of Hemoglobin Stabilized in Deoxy Conformation and
Polymerized by Oxidized Raffinose

In this preparation[14] a 77-ml solution of stroma-free Hb (4 g/dl in 0.1 M phosphate buffer, pH 8.0) is deoxygenated under vacuum for approximately 4 hr at 20°. Sodium dithionite (0.1 mmol in 0.3 ml degassed buffer) is added to the Hb solution and allowed to react for 5 min. Oxidized raffinose (0.95 mmol in 20 ml previously degassed buffer; 2 : 1 molar excess oxidized raffinose : Hb) is added under vacuum and the reaction proceeds for 6 hr. The reaction mixture is cooled in an ice–water bath and sodium borohydride (9.5 mmol in 4 ml degassed 1 mM NaOH) is added under a nitrogen blanket. Reduction proceeds for 45 min.

FPLC gel filtration shows that >98% of the Hb is stabilized and cross-linked, with 20–25% being the stabilized tetramer (64 kDa), 50–55% being two, three, and four tetramers combined (128–256 kDa), and 20–25% being pentamers and higher MW polymers combined.[14] The P_{50} is 32 mm Hg (37°, pH 7.2) with a Hill coefficient of 1.6.[14] This product has a plasma half-life of 7–8 hr in a 30% hypervolemic rat model.[14] Both of the oxidized raffinose-derivatized hemoglobins can be converted to the CO derivative and can be pasteurized by heating for 10 hr at 60°.

General Methods

Techniques for Physical Characterization
of Polymerized Hemoglobin Solution

FPLC Gel Filtration. A FPLC Superose 12 column (Pharmacia) is preequilibrated with 50 mM Tris-HCl, 1 M MgCl$_2$, pH 7.0. The buffer is filtered through a 0.2-μm filter and degassed prior to use. Approximately 1 mg of hemoglobin (in 100 μl of elution buffer) is loaded and eluted at a flow rate of 0.3 ml/min. Absorbance is monitored at 280 nm. On-line integration provides a measure of peak areas. Under these conditions, tetramers lacking an intratetramer cross-link will dissociate into dimers and elute at a position corresponding to 32 kDa. The column is calibrated using Pharmacia low molecular weight standards. This analytical-scale technique provides a quantitative means of assessing the extent of polymerization and an assay of the unpolymerized fraction.

Hydrophobic Interaction Chromatography. Hydrophobic interaction chromatography (HIC) represents an alternative to gel filtration in MW fractionation of complex, polymeric Hb solutions. In combined high-ionic strength and chaotropic conditions, hydrophobic regions are effectively unmasked, due to "screening" of ionic interactions, and tetramers lacking

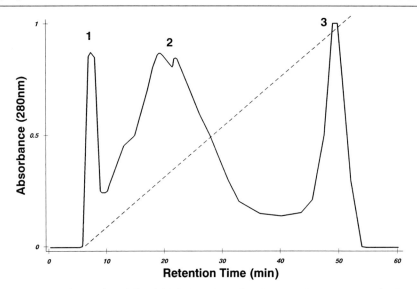

FIG. 7. Elution profile of glycolaldehyde-polymerized hemoglobin adsorbed at high ionic strength onto Fractogel butyl 650 (Merck) and eluted at low ionic strength. The three peaks contain hemoglobin separated according to increasing degrees of hydrophobicity. Elution details are described in the text.

an intratetramer cross-link dissociate, exposing hydrophobic dimer inter-faces. Polymeric Hb may be adsorbed onto hydrophobic ligands of the gel matrix and correspondingly desorbed at low ionic strength.

A Fractogel butyl 650 (Merck) column (10 × 0.5 cm) is equilibrated (1 ml/min) in eluant A (50 mM phosphate, 10% ammonium sulfate, 2 M NaCl, pH 7.2). Eluant B is deionized water. The sample is diluted to 10 mg/ml in eluant A, and 100 μl (1 mg) is chromatographed with a 45-min linear gradient to 100% eluant B (flow rate, 0.5 ml/min). Separation is monitored at 280 nm. Three distinct peaks are eluted (Fig. 7), which, when dialyzed and FPLC gel filtered as described, comprise material of the following composition (Table II). The least hydrophobic Hb (peak 1) consists almost entirely of stabilized tetramer (68 kDa), representing 10% of the original protein load. Peak 2 (68% original protein) is of intermediate hydrophobicity and contains dimeric and nontetrameric Hb of higher molecular mass. Peak 3 (25% original protein) contains predominantly hydrophobic, high molecular mass material. Therefore, HIC does not achieve a clear separation on the basis of molecular mass alone and, although too complex for production-scale removal of un-cross-linked Hb, does provide a very useful method complementary to gel filtration in the small-scale

TABLE II
MOLECULAR MASS PROFILE FOLLOWING HYDROPHOBIC INTERACTION
CHROMATOGRAPHY OF GLYCOLALDEHYDE-POLYMERIZED Hb-SOLUTION

Solution	Molecular mass distribution (% total protein)			
	$>1 \times 10^3$ kDa	500–128 kDa	68 kDa	32 kDa
Start material	<5.0	49.5	20.0	25.5
Peak 1	0	0	96.0	4.0
Peak 2	0	48.5	16.0	35.5
Peak 3	8.2	77.1	7.7	7.0

preparation of material for molecular studies into the nature of complex, polymeric Hb solutions.

SDS–PAGE. The purity of the starting material (stroma-free Hb solution) is monitored by reducing SDS–PAGE (Phastsystem: Pharmacia). The hemoglobin sample is diluted to approximately 1 mg/dl in sample buffer [10 mM Tris-HCl, 1 mM EDTA, 1.25% sodium dodecyl sulfate (SDS), 2.5% dithiothreitol (DTT), pH 8.0], heated to the boiling point for approximately 4 min, resolved on 8–25% gradient Phastgels (Pharmacia), and stained with Coomassie blue. Gels may be scanned using a PhastImage Gel Analyser (Pharmacia) to provide a quantitative estimate of nonheme protein. Electrophoresis of polymerized hemoglobin will provide a measure of the relative proportions of dimer, trimer, and higher MW cross-linked species, although we consider FPLC gel filtration to be a better means of quantitative assessment.

Hemoglobin, Methemoglobin, and Oxyhemoglobin Assay. Hb, metHb, and oxyHb concentrations are measured using an OSM 3 Hemoximeter (Radiometer), which calculates the parameters on the basis of absorbances at six wavelengths between 535 and 670 nm.

Oxygen Affinity Measurement. Oxygen affinity is measured using a Hemox Model B Analyzer (TCS Medical Products). The operating principle is based on dual-wavelength spectroscopy, whereby log/ratio recording at 560 and 570 nm is used to measure optical absorbance changes during sample deoxygenation. The buffer was Hemox buffer, pH 7.4 (TCS, HS-500), although 50 mM bis-Tris, 0.1 M NaCl (pH 7.4) is a satisfactory alternative. Measurements typically are conducted at 37°.

FPLC Globin Chain Analysis. Reversed-phase chromatography of a complex mixture of polymers, oligomers, and un-cross-linked Hb does not yield good peak resolution. In order to investigate which globin chains are primarily involved in such a reaction mixture it is advisable first to isolate stabilized tetramer by gel filtration in 50 mM Tris-HCl, 1 M MgCl$_2$, pH 7.0, as previously described. Fractions containing polymerized tetra-

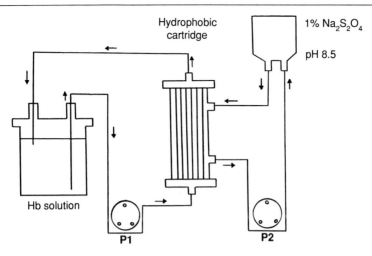

FIG. 8. Experimental design for deoxygenation of hemoglobin using sodium dithionite, pH 8.5. The hemoglobin and dithionite compartments remain physically separated due to the hydrophobic nature of the hollow fiber cartridge and a 5-psi positive internal pressure, although oxygen passes freely from the hemoglobin to the dithionite compartment. Both hemoglobin and dithionite circulate by means of peristaltic pumps (P1 and P2). The head space of the reaction vessel is flushed with nitrogen gas.

mer are then pooled, dialyzed against deionized water, and diluted to approximately 0.5 mg/ml in eluant A (0.3% trifluoroacetic acid, 39% acetonitrile in water). A FPLC ProRPC 5/10 reversed-phase column is equilibrated with eluant A and approximately 50 μg of hemoglobin chromatographed by means of a 45-min linear gradient (0.2 ml/min) to 100% eluant B (45% acetonitrile in water). Separation of heme and globin chains is monitored at 280 nm. The column is calibrated for elution positions of α- and β chains using stroma-free hemoglobin, $\beta + \beta$ chains using NFPLP cross-linked tetramer, and $\alpha + \alpha$ chains using DBBF cross-linked tetramer. The acetonitrile gradient is shallow (39–45%) and care should be taken to prevent evaporation of acetonitrile and consequent shifting of relative globin chain peak positions.

Deoxygenation of Hemoglobin Solution. We have found it most efficient to remove oxygen from Hb solutions prior to polymerization (or modification by PLP, NFPLP, or DBBF) using a method whereby 1 liter of 1% sodium dithionite (pH 8.5) is pumped in a closed circuit through the exterior compartment of a hydrophobic hollow fiber cartridge (Liqui-Cel Module, surface area 0.4 m^2; Hoechst Celanese, Charlotte, NC or Cell-Pharm Hollow Fiber Oxygenator OXY-10, surface area 1 m^2; UniSyn Fibertec, San Diego, CA). The hemoglobin solution is pumped through the inside of the hollow fibers (Fig. 8). As long as the hydrophobic nature

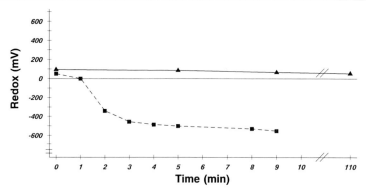

FIG. 9. Comparison of redox potential in hemoglobin solution (▲—▲) and deionized water (■---■) during deoxygenation with 1% sodium dithionite, pH 8.5, in a hydrophobic hollow fiber cartridge. The redox potential does not fall below +60 mV, due to the buffering effect of the protein.

of the polypropylene fiber matrix is maintained, the two solutions do not come into direct physical contact, although oxygen passes freely from the hemoglobin to the dithionite compartment. Maintenance of a slight positive pressure (5 psi) inside the fibers further discourages any leakage. Monitoring of redox potential in the hemoglobin solution indicates any leakage of dithionite into the hemoglobin solution. We have found that the redox potential of the hemoglobin solution rarely falls below +60 mV from a starting value of +120 mV (Fig. 9). This is presumably due to the buffering effect of the protein, although, in our experience, glycolaldehyde cross-linking can proceed satisfactorily until redox potential falls below −100 mV. Under the following conditions (Hb solution flow rate, 500 ml/min; sodium dithionite flow rate, 500 ml/min; internal pressure approximately 5 psi or 34 kPa), 1 liter of 12 g/dl hemoglobin may be deoxygenated to <2% oxyHb within approximately 30 min. This compares with 2–3 hr required when nitrogen gas replaces sodium dithionite solution. Hydrophobicity gradually deteriorates on repeated use, although we have performed up to 15 separate runs using the same cartridge without appreciable loss in performance.

Removal of Excretable Dimer from Polymerized Hemoglobin Solutions

We have assessed a number of methods for the removal of the residual unpolymerized fraction. Unsuccessful techniques include ammonium sulfate precipitation, ultrafiltration, and haptoglobin adsorption. On a small scale, hydrophobic interaction chromatography is successful in separating stabilized tetramer and oligomeric Hb from the remainder of the reaction mixture, but yield (35%) and complexity make this unrealistic for scaleup. To date, only two methods have proved successful in removing dimer to ≤2% and possess potential for scaleup to process.

Large-Scale Gel Filtration. Analytical-scale FPLC gel filtration in 1 M MgCl$_2$ may be used to prepare dimer-free polymerized Hb in 1- to 2-mg quantities as previously described. Scaleup to 1–2 g of Hb may be achieved by equilibrating a 90 × 2.5-cm column packed with preswollen Sephadex G-150 (Pharmacia) with 50 mM Tris-HCl, 1 M MgCl$_2$, pH 7.0 (3 M NaCl may be substituted for 1 M MgCl$_2$ with only slight loss of resolution). Hemoglobin (1–2 g in 10 ml elution buffer containing 1 g glucose) is applied to the column and eluted at a flow rate of 0.7 ml/min. Absorbance is monitored at 280 nm.

Wet Heat Treatment. Either CO-liganded or (under strictly anaerobic conditions) polymerized Hb may be heated to 74–76° without denaturation occurring.[24,25] Hemoglobin that is not cross-linked has a lower thermal stability and so will denature and precipitate. At temperatures greater than 76° polymerized Hb will also denature.

This step is most conveniently carried out immediately following sodium borohydride reduction of polymerized, pyridoxalated Hb when the oxyHb content approaches zero. The temperature of the reaction vessel is increased to 76° and great care should be taken to ensure that adequate mixing prevents any part of the vessel contents exceeding 76°. This temperature is maintained for 90 min, during which period the unpolymerized fraction will precipitate and subsequently may be removed by a graded sequence of filters (5.0–0.1 μm). The glycolaldehyde cross-link has proved stable under these conditions and there is no further buildup of higher MW polymers (unpublished observations, 1992).

If a viral inactivation step also is required the temperature may be reduced to 60° (10 hr) following precipitation of dimer. Strictly anaerobic conditions must be maintained throughout.[24] Wet heat treatment therefore offers several advantages over gel filtration for a production-scale process. It is much faster (<2 hr), can be combined with a viral inactivation step, and offers greater potential for scaleup.

Conclusions

Glycolaldehyde is a very useful, flexible, reliable, and cost-effective option in the spectrum of chemical reagents available for the modification

[24] T. N. Estep, U.S. Pat. 4,831,012 (1989).
[25] T. N. Estep, M. K. Bechtel, T. J. Miller, and A. Bagdasarian, *Biomater., Artif. Cells, Artif. Organs* **16**, 129 (1988).

of Hb solutions. We also have found it to be remarkably compatible with a variety of additional chemical derivatives (NFPLP, DBBF) as well as the oxy and deoxy forms of Hb. Finally, one may realistically contemplate the use of glycolaldehyde in the scaleup of Hb modification and processing on a large scale for the production of therapeutic, clinical grade material.

Acknowledgments

We would like to thank Dr. Chris Prowse for helpful comments during the preparation of the manuscript. We are also grateful for Dr. J. Bakker (Dutch Red Cross, Amsterdam) for the supply of NFPLP and Dr. Kim Vandegriff (Letterman Army Institute of Research, San Francisco) for the supply of DBBF.

[20] Preparation of Conjugated Hemoglobins

By Hong Xue and J. Tze-Fei Wong

Hemoglobin is potentially a useful erythrocyte substitute for carrying oxygen from the lungs to body tissues. For this purpose a distinct disadvantage is the small size of the protein, which causes its rapid excretion through kidneys, leading to a short plasma half-life as well as possible damage to the kidneys.[1-3] Accordingly a number of approaches have been developed to increase its plasma half-life in order to fulfill the requirements of an oxygen-carrying blood substitute. These approaches include encapsulation, polymerization, and covalent conjugation to a carrier molecule. The advantages of a conjugate stem from the wide range of choices with respect to the chemical nature and size of the carrier, as well as the types of linkages that may be constructed to link the hemoglobin to the carrier. The conjugation approach (Fig. 1), first introduced in the form of a dextran–hemoglobin conjugate,[4-9] has now been extended to include other

[1] M. Relihan and M. S. Litwin, *Surgery (St. Louis)* **71,** 395 (1972).
[2] A. S. Haupt, R. Ochs, and G. E. Schubert, *Urol. Res.* **10,** 1 (1982).
[3] S. C. Tam and J. T. Wong, *J. Lab. Clin. Med.* **111,** 189 (1988).
[4] S. C. Tam, J. Blumenstein, and J. T. Wong, *Proc. Natl. Acad. Sci. U.S.A.* **73,** 2128 (1976).
[5] J. E. Chang and J. T. Wong, *Can. J. Biochem.* **55,** 398 (1977).
[6] J. Blumenstein, S. C. Tam, J. E. Chang, and J. T. Wong, *Prog. Clin. Biol. Res.* **19,** 205 (1978).
[7] S. C. Tam, J. Blumenstein, and J. T. Wong, *Can. J. Biochem.* **56,** 981 (1978).
[8] J. T. Wong, *Biomater., Artif. Cells, Artif. Organs* **16,** 237 (1988).
[9] H. Xue, X. F. Wu, and J. T. Wong, *Artif. Organs* **16,** 427 (1992).

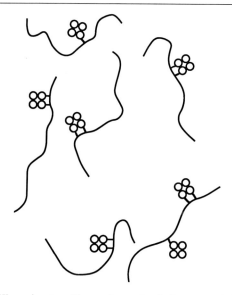

FIG. 1. Carrier–Hb conjugates. The carrier molecule is represented by a random chain and the Hb is represented by a tetramer of globular subunits. Each tetramer may be linked to the carrier through single or multiple linkages, depending on the mode of construction.

carriers, for example, hydroxyethyl starch,[10] inulin,[11] polyvinylpyrrolidone,[12] and polyethylene glycols.[13–17]

Important landmarks in the development of an oxygen-carrying blood substitute include the preparation of stroma-free hemoglobin, which avoids major damage to the kidneys,[18] and the first complete erythrocyte replacement with a blood substitute using stroma-free hemoglobin[19] or a perfluorocarbon emulsion.[20] However, on account of its short residence

[10] J. E. Baldwin, B. Gill, H. Taegtmeyer, and J. P. Whitten, *Tetrahedron* **37**, 1723 (1981).
[11] K. Iwasaki, K. Ajisaka, and Y. Iwashita, *Biochem. Biophys. Res. Commun.* **113**, 513 (1983).
[12] K. Schmidt, *Klin. Wochenschr.* **57**, 1169 (1979).
[13] K. Ajisaka and Y. Iwashita, *Biochem. Biophys. Res. Commun.* **97**, 1076 (1980).
[14] K. Iwasaki and Y. Iwashita, *Artif. Organs* **10**, 411 (1986).
[15] Y. Iwashita, K. Iwasaki, A. Yabuki, A. Gonsho, T. Okami, and K. Kosaka, *in* "International Symposium on Stabilized Hemoglobin" (M. Hori and Y. Nose, eds.), p. 5. ISAO Press, Cleveland, OH, 1987.
[16] M. Leonard, J. Neel, and E. Dellacherie, *Tetrahedron* **40**, 1581 (1984).
[17] M. Leonard and E. Dellacherie, *Biochim. Biophys. Acta* **791**, 219 (1984).
[18] S. F. Rabiner, J. R. Helbert, H. Lopas, and L. H. Friedman, *J. Exp. Med.* **126**, 1127 (1967).
[19] G. S. Moss, R. DeWoskin, A. L. Rosen, H. Levine, and C. K. Palani, *Prog. Clin. Biol. Res.* **19**, 191 (1978).
[20] R. P. Geyer, *Prog. Clin. Biol. Res.* **19**, 1 (1978).

time in the body, stroma-free hemoglobin cannot support the recovery of an animal from a complete blood replacement without the restoration of blood within hours after the replacement. Use of a perfluorocarbon emulsion suffers from a different shortcoming. Because of the insufficient solubility of oxygen in an emulsion under room air pressure, an atmosphere of pure or enriched oxygen is essential to the survival of the blood-replaced animal. In this regard, it is noteworthy that a carrier–hemoglobin conjugate using dextran as carrier allowed the first essentially complete erythrocyte replacement that was followed by survival and spontaneous recovery of the animal under room air.[6,7] This illustrates the important potential of the carrier–hemoglobin conjugation approach in the development of blood substitutes.

The present chapter describes the preparation of several forms of carrier–hemoglobin conjugates that are being extensively investigated as blood substitutes.

Dextran–Hemoglobin

Dextran, a clinical plasma expander, is a biocompatible polymer of mostly α-1,6-linked glucose units. Conjugation between this polymer and human hemoglobin can be achieved by different synthetic methods.

Synthesis by Alkylation

In this method,[4,5] the dextran (Dx) is first derivatized with cyanogen bromide and diaminoethane to contain a free amino group, which is acylated with bromoacetyl bromide. The bromoacetyl function in turn alkylates the sulfhydryl of the β93 cysteine on Hb:

$$Dx + CNBr + diaminoethane \rightarrow aminoethyl\text{-}Dx$$
$$Aminoethyl\text{-}Dx + bromoacetyl\ bromide \rightarrow Dx\text{-}NHCOCH_2Br$$
$$Dx\text{-}NHCOCH_2Br + HS\text{-}Hb \rightarrow Dx\text{-}NHCOCH_2\text{-}S\text{-}Hb$$

In a typical preparation, 1.5 g of cyanogen bromide is dissolved in 15 ml of acetonitrile and is added to 10 g of dextran (MW 20,000) in 375 ml of water. The pH is maintained at 10.8 for 5 min by the addition of 1 M NaOH; it is then lowered to about 2.0–2.5 with concentrated HCl. After stirring for 1 min, 15 ml of diaminoethane is added along with sufficient HCl to prevent the pH from exceeding 9.5. The final pH is adjusted to 9.5. After standing overnight at 4°, the mixture is thoroughly dialyzed against distilled water using a Millipore (Marlborough, MA) Pellicon dialyzer and is lyophilized. The aminoethyl-Dx so obtained is dissolved in 250 ml of 0.1 M sodium phosphate, pH 7.0, and 15 ml of bromoacetyl

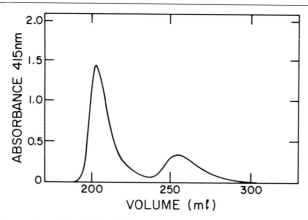

FIG. 2. Separation of Dx–Hb and Hb by gel filtration. The chromatogram illustrates a partial formation of Dx–Hb during synthesis of the conjugate. (After Chang and Wong.[5])

bromide is added through a Pasteur pipette with a finely drawn capillary tip, accompanied by vigorous stirring over a period of 2 hr. Throughout, the pH is maintained at 7.0 with the use of a pH-stat and addition of 1 M NaOH. Afterward the mixture is dialyzed thoroughly against distilled water and is lyophilized to yield about 7 g of the Dx–NHCOCH$_2$Br (Br–dextran). The bromine content of the Br–dextran is in the range of 9–11 glucose residues per bromine atom.

To couple hemoglobin to dextran, 3.3 g of Br–dextran is dissolved in 100 ml of 6% hemoglobin solution in 0.1 M sodium bicarbonate, pH 9.5. The coupling reaction is allowed to proceed with constant mixing at 4°. To determine the yield of Dx–NHCOCH$_2$-S-Hb (Dx–Hb), 0.1 ml of the reaction mixture is applied to a Sephadex G-75 column (3 × 87 cm) equilibrated with 0.05 M phosphate buffer, pH 7.5, and eluted with the same buffer, at a flow rate of 40 ml/hr. Hemoglobin content of the eluant fractions is determined by absorbance at 415 nm, and the proportions of the faster migrating Dx–Hb peak and the slower migrating Hb peak are given by the areas under these peaks (Fig. 2). After 2 days the formation of the Dx–Hb conjugate is essentially complete.

A more convenient procedure for separating Dx–Hb from free Hb is provided by electrophoresis on a cellulose acetate membrane (Gelman, Ann Arbor, MI; 5.7 × 12.7 cm) in 50 mM barbiturate–HCl buffer, pH 8.6. After electrophoresis at a constant current of 10 mA for 30 min, free hemoglobin migrates to the anode distinctly ahead of the Dx–Hb conjugate (Fig. 3).

The half-clearance time of Dx–Hb from canine plasma is 2.4 days (Fig. 4). Allowing for loss of oxygen-carrying function on account of

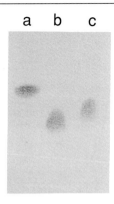

FIG. 3. Electrophoresis of Dx–Hb. Lane a, Free Hb; lane b, Dx–Hb (with MW 40,000 Dx); lane c, Dx–Hb (with MW 20,000 Dx) (anode is at the top).

oxidation of the hemoglobin moiety to methemoglobin, this gives a functional plasma half-life of 1.9 days,[7] which represents the most prolonged half-life of a hemoglobin-based blood substitute so far observed.

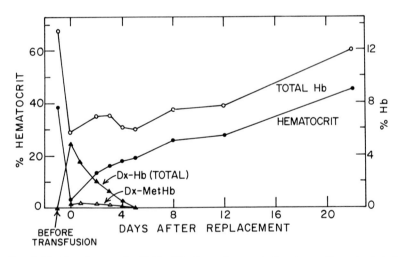

FIG. 4. Blood replacement with Dx–Hb: the replacement of canine erythrocytes to below a 2% hematocrit by Dx–Hb containing 6% Hb moiety. The half-life of Dx–Hb in circulation was 2.4 days. However, a minor fraction of the Dx–Hb underwent oxidation to Dx–metHb, which does not carry any oxygen. Subtracting this fraction gave a half-life for functional Dx–Hb of 1.9 days. This half-life was sufficiently prolonged to allow the animal to recover spontaneously under room air from the blood replacement, maintaining the total Hb (hematocrit Hb plus Dx–Hb) at an adequate level for life support.

Right-Shifting by Periodated Inositol Tetrakisphosphate

The oxygen dissociation curve of hemoglobin is right-shifted inside human erythrocytes by 2,3-diphosphoglycerate, and inside avian erythrocytes by inositol phosphates. Accordingly, right-shifting of the free hemoglobin to increase its P_{50} (half-saturation tension) for oxygen might be advantageous before it is employed as a blood substitute, because this would allow the unloading of more oxygen at the prevalent oxygen tension in the blood. Covalent phosphopyridoxalation[21] of the deoxy form of Hb brings about right-shifting. However, deoxygenation of large quantities of Hb to convert to the deoxy form is cumbersome and prone to methemoglobin formation. Accordingly, it is particularly convenient to have available right-shifting modifiers that could be covalently attached to Hb under aerobic conditions. Periodated inositol tetrakisphosphate (oxyIP$_4$) satisfies this requirement.[8,9] Phytase from wheat bran degrades inositol hexakisphosphate (IP$_6$, or phytate), which is an abundant constituent of cereal brans, to the lower phosphates IP$_5$, IP$_4$, IP$_3$, . . . , and eventually to free inositol. IP$_5$ is the major regulator of hemoglobin affinity for oxygen in most species of avian erythrocytes, and IP$_4$ is the major regulator in ostrich erythrocytes.[22] From the IP$_4$ stage onward, vicinal hydroxyls will be present in the molecule, which could be periodated into the dialdehyde, or oxy, form.[23] The dialdehyde function of oxyIP$_4$ enables its covalent attachment to Hb or Dx–Hb through reductive alkylation. The oxyIP$_4$-modified Dx–Hb has a P_{50} for oxygen of 23 mm Hg at 37°, and oxyIP$_4$-modified Hb has a P_{50} of 55 mm Hg. The right-shifting by oxyIP$_4$ is extensive, because the *in vivo* P_{50} of unmodified Dx–Hb is about 7 mm Hg and that of unmodified Hb is about 15 mm Hg, as estimated from the venous oxygen tension of blood containing half-saturated forms of these blood substitutes.[6]

Inositol tetrakisphosphate is isolated by chromatography on a Dowex-1 column[23] from a partial digest of sodium phytate by wheat bran phytase.[9,24] To convert to the oxy form, 10 g of IP$_4$ in 292 ml of 0.6 M HIO$_4 \cdot 2H_2O$ is incubated in the dark at 26° for 16 hr. The reaction mixture is neutralized to pH 5.2 with 5 N KOH, kept at 0° for 10 min, and centrifuged to remove KIO$_4$. Ascorbic acid is added to the supernatant to a concentration of 1 M. After 10 min, 3 volumes of ethanol are added

[21] R. E. Benesch, S. Yung, T. Suzuki, C. Bauer, and R. Benesch, *Proc. Natl. Acad. Sci. U.S.A.* **70**, 2595 (1973).
[22] R. Isaacks, D. Harkness, R. Sampsell, J. Adler, S. Roth, C. Kim, and P. Goldman, *Eur. J. Biochem.* **77**, 567 (1977).
[23] R. V. Tomlinson and C. E. Ballou, *Biochemistry* **1**, 166 (1962).
[24] F. G. Peers, *Biochem. J.* **53**, 102 (1953).

and the mixture is kept at 4° for 1 hr. The precipitate is recovered by centrifugation, triturated, washed with ethanol, ethanol/ether (1 : 1), and ether, and air dried. The yield is about 7 g oxyIP$_4$, which is stable for at least 1 year at −12°.

To attach the oxyIP$_4$ to Dx–Hb (or Hb) by means of reductive alkylation, 5 g of oxyIP$_4$ in 680 ml of 0.05 M sodium acetate, pH 5.2, is added to 1 liter of Dx–Hb (or Hb) containing 6% hemoglobin moiety, in 0.05 M bis-Tris buffer, pH 7.4. The final pH of the solution is adjusted to 7.4. After a 20-min incubation at 0°, 100 ml of 0.5 M dimethylamine borane in water is added. The solution is incubated for a further 2 hr at 0°. Afterward sodium bicarbonate is added to 0.2 M and the pH is adjusted to 9.5. Mercaptopropionate is added to 16 mM to remove excess bromine from the dextran prior to dialysis in a Millipore Pellicon dialyzer.

Synthesis by Dialdehyde

This synthesis[4] is an adaptation of the method for coupling proteins to cellulose.[25] Ten milliliters of a 12% aqueous solution of sodium periodate is added to 100 ml of a 10% aqueous solution of dextran, and the mixture is left overnight in the dark at 4°. A 3% solution of sodium bisulfite is added until the mixture turns brown and then, once again, colorless. The mixture is dialyzed against distilled water to yield the dextran dialdehyde solution. It is then added to 2 volumes of 3% stroma-free hemoglobin in 0.3 M sodium bicarbonate buffer, pH 9.5; coupling of hemoglobin to dextran is allowed to proceed overnight at 4°. The Dx–Hb complex formed is separated from uncoupled hemoglobin by means of chromatography on a Sephadex G-75 column.

Coupling of Hb to Dx-dialdehyde is pH dependent.[26] When coupling is performed by dissolving 100 mg Dx-dialdehyde in 1 ml of 0.6 M sodium borate buffer and mixing with 1.8 ml of 10% Hb at 6°, many labile imine linkages are formed at pH < 9.6, and the conjugates have a high molecular weight, ranging to above 100,000. At higher pH, the major product has a lower molecular weight range (70,000 > MW > 100,000) and likely consists of a 1 : 1 complex between Dx and Hb, which is only slowly converted to higher molecular weight forms. When this conjugate is formed at pH 9.8 and reduced at pH 7.2 for 30 min with excess NaBH$_4$ (2 mol per mole of initial aldehyde) dissolved in 1 mM NaOH, only the α chain of Hb is found to be modified by Dx.[26] Coupling of Hb to Dx-dialdehyde also proceeds much more rapidly at higher pH, requiring less than 1 hr for

[25] C. Flemming, A. Gabert, and P. Roth, *Acta Biol. Med. Ger.* **30**, 177 (1973).
[26] E. Dellacherie, F. Bonneaux, P. Labrude, and C. Vigneron, *Biochim. Biophys. Acta* **749**, 106 (1983).

completion at pH 10 and only 1.5 hr at pH 9.7, but 6 hr at pH 9.5 and 23 hr at pH 9.1. When prepared at pH 9.75, the oxygen P_{50} for Dx–Hb is 10.1 mm Hg when the conjugate is allowed to form for 1 hr prior to $NaBH_4$ reduction, 9.5 mm Hg when allowed to form for 4 hr, and 8.1 when allowed to form for 18 hr.

Hydroxyethyl Starch–Hemoglobin

Like dextran, hydroxyethyl starch (Hs) is a useful plasma expander and therefore a potentially attractive carrier for a blood substitute.[10] To prepare for conjugation to Hb, the Hs is first converted to aminoethyl-Hs by a modification of the method described for dextran.[4] In a typical preparation, 1.5 g of cyanogen bromide is dissolved in 15 ml of acetonitrile and added to 500 ml of 2% Hs solution. The pH of the solution is maintained at 10.8 for 5–10 min by the addition of 1 M NaOH solution. The pH is then lowered to 2.0–2.5 with concentrated HCl, and 10 ml of diaminoethane is added along with additional HCl to prevent the pH from exceeding 9.5. The final pH is adjusted to 9.5 and the solution is allowed to stand overnight at 4° before being dialyzed against deionized water. The ratio of cyanogen bromide/diaminoethane to Hs can be varied, allowing the synthesis of aminoethyl-Hs in which from 7 to 20% of the glucose residues in the starting polymer are substituted.

Aldehyde-substituted Hs is prepared by reaction of the aminoethyl-Hs with glutaraldehyde. In a typical reaction 500 ml of dialyzed solution of aminoethyl-Hs is treated with 2 g of sodium bicarbonate to give a solution 2% in Hs and approximately 0.05 M in bicarbonate. Then 5 ml of 50% glutaraldehyde solution is added to the solution, which is stirred at room temperature for 2 hr before dialysis.

Hemoglobin is employed as a freeze-dried solid under carbon monoxide. This is reconstituted under argon using deoxygenated deionized water at 4° to give a solution with approximately 2.5 g Hb per ml. In a typical reaction 500 ml of dialyzed solution of the aldehyde-substituted Hs is treated with sodium bicarbonate to give 500 ml of solution approximately 2% in Hs and 0.1 M in bicarbonate. Hemoglobin solution (25 ml) is added and the reaction mixture is stirred at room temperature for 4 hr, after which time gel filtration on Sephadex G-150 indicates that no unbound Hb remains. Sodium borohydride (1.0 g) is then added to the solution, which is stirred for a further 2 hr at room temperature. The Hs–Hb is dialyzed using an Amicon (Danvers, MA) ultrafiltration unit with a 100,000 molecular weight cutoff cartridge to enable the removal of any trace of unbound Hb. Glucose (10 g) is added to the solution prior to freeze-drying and storage under carbon monoxide at 4°.

Hs–Hb also can be synthesized from Hs-dialdehyde as follows. Using a modification of the periodate method described for dextran,[4] 0.03 equivalents of Hs are dissolved in 250 ml of water and treated with 0.028 mol of sodium periodate for 12 hr at 5° in the dark. The solution is dialyzed until ion free. The percent oxidation may be determined using a colorimetric method.[27] The solution is buffered to pH 8.0 by addition of sodium bicarbonate, cooled to 5°, and treated with 5 g of carbonylated Hb. The reaction is allowed to proceed for 18 hr at room temperature when gel filtration indicates complete binding of hemoglobin. The solution is dialyzed against 1% ammonium carbonate and freeze-dried in the presence of glucose.

Inulin–Hemoglobin

Inulin (In) is yet another polysaccharide that has been employed as a carrier for prolonging the plasma half-life of hemoglobin.[11] To synthesize the In–Hb conjugate, inulin is first succinylated by reacting with succinic anhydride in N,N-dimethylformamide at 100° for 2 hr. Subsequently the succinylated inulin is linked to N-hydroxysuccinimide at room temperature overnight using dicyclohexylcarbodiimide as condensation agent in N,N-dimethylformamide. Pyridoxalated Hb prepared under anaerobic conditions[21] is allowed to react with a 10-fold molar excess of the N-hydroxysuccinimide-activated inulin in 0.1 M Tris buffer, pH 7.0, at 4° for 1 hr to yield In–Hb, which is purified with an Amicon PM30 membrane filter until the unreacted inulin and other low molecular weight compounds are removed.

By controlling the succinic anhydride/inulin ratio, the number of N-hydroxysuccinimide-activated succinyl groups on the inulin can be varied. A low density of such groups gives rise to a 82,000 MW In–Hb conjugate, whereas higher densities produce cross-linked In–Hb ranging up to above 300,000 MW. The 82,000 MW In–Hb conjugate has a P_{50} for oxygen of 17.3 mm Hg at 37° compared to the P_{50} of 14.3 mm Hg for free Hb. The plasma half-life for In–Hb is 21 hr in rats, in contrast to a half-life under the same conditions of only 3 hr for free Hb.

Polyethylene Glycol–Hemoglobin

A conjugate between polyethylene glycol (PEG) and pyridoxalated Hb has been prepared and investigated as a blood substitute.[14] Its preparation depends on reaction between Hb and active succinimidyl ester groups on the PEG.

[27] C. S. Wise and C. L. Mehltretter, *Anal. Chem.* **30**, 174 (1958).

Synthesis of Polyethylene Glycol Bis(succinimidyl Succinate)

PEG (200 g; 0.059 mol, average 3400 MW; Nippon Oil and Fats Co. Ltd., Tokyo, Japan) is dissolved in 200 ml of dimethylformamide at 100°, and 15 g of succinic anhydride (0.15 mol) is added. The mixture is stirred for 3 hr at 100°. The dimethylformamide solution is cooled to room temperature and poured into 1 liter of ethyl ether. The resulting PEG ester of succinic acid is filtered through a glass filter and washed with ethyl ether. The ester is then dried under vacuum conditions at 40°. The weight of the product is about 197 g (93% yield).

To activate the succinyl groups on PEG, 197 g of the PEG ester of succinic acid (0.055 mol) is dissolved in 200 ml of dimethylformamide, after which 13 g of N-hydroxysuccinimide (0.11 mol) and 23 g of dicyclohexylcarbodiimide (0.22 mol) are added. The solution is stirred vigorously overnight at 30°. The precipitate of dicyclohexylurea is filtered out and the filtrate is poured into 1 liter of ethyl ether. The polyethylene glycol bis(succinimidyl succinate) formed is isolated, washed with ethyl ether repeatedly, and dried under vacuum conditions at 40°. The weight of the product is about 196 g, representing a yield from PEG of 87%.

The purity and the degree of imidylation of polyethylene glycol bis(succinimidyl succinate) may be estimated by nuclear magnetic resonance using tetramethylsilane as standard (0 ppm) and chloroform-d_1 as solvent.

Conjugation of Pyridoxalated Hemoglobin to Polyethylene Glycol

The following procedure is carried out at 4°. Stroma-free Hb is prepared from a hemolysate of outdated human red blood cells by toluene extraction followed by filtration through a 0.22-μm membrane filter.[12] Then 100 ml of a 0.25 mM Hb solution in 0.1 M sodium phosphate, pH 7.4, is deoxygenated by bubbling argon through the solution for 1 hr. Afterward 13.3 mg of pyridoxal 5-phosphate 2-hydrate (0.05 mmol) is added and the reaction mixture is stirred for 30 min; 9.5 mg of sodium borohydride (0.25 mmol) is then added. After 30 min 0.95 g (0.25 mmol) of polyethylene glycol bis(succinimidyl succinate) is added and the reaction is continued for 1 hr. Argon is bubbled into the solution throughout the reaction. Subsequently 100 ml of the solution is concentrated to ~15 ml by means of ultrafiltration on an Amicon XM100 membrane. An electrolyte solution is then added and the concentration process repeated. By repeating this concentration procedure three times, unreacted PEG and other low molecular weight compounds are removed. During the reaction of the hemoglobin with polyethylene glycol bis(succinimidyl succinate), the oxygen content in the solution is under 0.05 ppm, indicating a partial oxygen pressure of under 0.6 mm Hg. Finally, the concentration of the hemoglobin moiety

TABLE I
CHARACTERISTICS OF STABILIZED HEMOGLOBIN
SOLUTION

Parameter	Value
Hb concentration	6.3%
Methemoglobin	3.5%
P_{50} (37°)	24.2 mm Hg
Colloidal osmotic pressure	38.0 mm Hg
Viscosity (37°)	2.7 cP
Pyrogenicity	Pass test
Sterility	Pass test

and the pH of the solution are adjusted to 6% and 7.40, respectively. The solution is filtered through 0.22-μm membrane filter for sterilization.

Pyridoxalated PEG–Hb has a P_{50} for oxygen of 21.3 mm Hg in 0.1 M phosphate buffer, pH 7.40, at 37°. It has a plasma half-life in the rat of 12.8 hr.

Stabilization of PEG–Hb

The ester bond between PEG and succinic acid in PEG–Hb is labile to hydrolysis. One approach to increase the stability of the bond between PEG and Hb is to remove the labile ester linkage between the polyethylene moiety and the terminal carboxyls by oxidizing both terminal alcoholic groups of PEG to carboxylic groups through the use of a metal catalyst, to yield α-carboxymethyl-ω-carboxymethoxylpolyoxyethylene, which is activated and coupled to pyridoxalated Hb as in the case of PEG.[15] The resultant conjugate is designated "stabilized hemoglobin." The characteristics of this stabilized hemoglobin solution are given in Table I. Maltose is an effective protectant for the preparation against oxidation to the methemoglobin form during lyophilization and storage, and the solution has been extensively investigated as a blood substitute.[28–32]

[28] M. Hori, in "International Symposium on Stabilized Hemoglobin" (M. Hori and Y. Nose, eds.), p. 1. ISAO Press, Cleveland, OH, 1987.
[29] M. Matsushita, Y. Iwashita, K. Iwasaki, H. Ohki, M. Nasu, T. Horiuchi, J. F. Chen, J. Goldcamp, S. Murabayashi, H. Harasaki, P. S. Malchesky, and Y. Nose, in "International Symposium on Stabilized Hemoglobin" (M. Hori and Y. Nose, eds.), p. 17. ISAO Press, Cleveland, OH, 1987.
[30] K. Nishi, K. Yano, T. Yamakawa, and T. Ohta, in "International Symposium on Stabilized Hemoglobin" (M. Hori and Y. Nose, eds.), p. 37. ISAO Press, Cleveland, OH, 1987.
[31] K. Ito, M. Kobayashi, H. Ikeda, and S. Sekiguchi, in "International Symposium on Stabilized Hemoglobin" (M. Hori and Y. Nose, eds.), p. 55. ISAO Press, Cleveland, OH, 1987.
[32] Y. Sohara, in "International Symposium on Stabilized Hemoglobin" (M. Hori and Y. Nose, eds.), p. 69. ISAO Press, Cleveland, OH, 1987.

Monomethoxypolyoxyethylene–Hemoglobin

PEG has two hydroxyl groups at the two termini. When these are derivatized into functional groups capable of reacting with Hb, the presence of two reactive groups on the same polymer makes possible cross-linking reactions. Such cross-linking is abolished by blocking one of the two termini, as in the case of monomethoxypolyoxyethylene (MPOE).

In one of the methods described for conjugation to Hb,[16] 80 g (4 mmol) of MW 5000 MPOE from Aldrich (Milwaukee, WI) is dissolved in tetrahydrofuran (300 ml) and treated with naphthalene sodium under nitrogen at room temperature for 3 hr. Then $BrCH_2COOC_2H_5$ (1.4 ml; 12 mmol) is added dropwise with stirring. After 4 hr of reaction the ethyl ester obtained is precipitated with ether, dried, dissolved in water, and saponified with 0.1 N NaOH at 55° for 24 hr to yield MPOE-carboxylic acid ($MPOE-O-CH_2COOH$). The solution is then acidified with 1 N HCl down to pH 2.5, and the polymer is taken up with chloroform. After several washings with water, the organic layer is dried over $MgSO_4$ and treated with charcoal. The MPOE-carboxylic acid is precipitated with dry ether, filtered, and dried under vacuum. This run of operations is repeated until the potentiometric titration gives a constant value for the quantity of fixed COOH.

The MPOE-carboxylic acid (5 g; 1 mmol) is dissolved in dry ethyl acetate (60 ml) and activated by N-hydroxysuccinimide (0.15 g; 1.25 mmol) and dicyclohexylcarbodiimide (0.26 g; 1.25 mmol) at 30° for 15 hr. Dicyclohexylurea is removed by filtration and the polymer precipitated with dry ether is taken up with chloroform and crystallized from this solution by dropwise addition of ether at 0°. This procedure is repeated several times until the spectrophotometric analysis of succinimidyl groups[33] gives a constant value.

Coupling to hemoglobin is performed at 5° by diluting 1.5 ml of a 10% Hb solution (2.3 μmol) with 2 ml 0.1 M phosphate buffer, pH 5.8, and 300 mg of MPOE-carboxylic succinimidyl ester (0.95 mol activated ester per mole polymer) is added under stirring. The reaction mixture is stirred at 6° for 2 hr and analyzed by gel permeation chromatography on Ultrogel AcA 34 (linear fractionation range MW 20,000–350,000; exclusion limit 750,000) in 0.05 M phosphate buffer (pH 7.2) at 6°. Pyridoxalated deoxyHb also may be employed in place of Hb, in which case the coupling is carried out in a nitrogen atmosphere. The absence of oxygen is to be checked in all the solutions used.

The P_{50} for oxygen of MPOE–Hb prepared from Hb is 6.2 mm Hg at 25°, and that prepared from pyridoxalated deoxyHb is 9.8 mm Hg, which

[33] T. Miron and M. Wilchek, Anal. Biochem. 126, 433 (1982).

is close to the value of 10 mm Hg for free pyridoxalated Hb under the same conditions.[16]

Polyvinylpyrrolidone–Hemoglobin

Polyvinylpyrrolidone (PVP) is another synthetic polymer that has been conjugated to Hb. The synthesis of PVP–Hb proceeds through the preparation of carboxyl-activated PVP.[12]

Synthesis of Activated PVP

To bring about partial hydrolysis, 50 g of PVP (MW 25,000–35,000) is dissolved in 1 liter of 0.25 N NaOH and heated at 140° for 42 hr under nitrogen in an autoclave. It is then adjusted to pH 5 with concentrated HCl and ultrafiltered through an Amicon UM10 membrane to remove salts. Water is removed through azeotropic distillation with benzene, and the extent of hydrolysis is determined by titration of the secondary amino groups. To blockade these amino groups, 50 g of the partially hydrolyzed PVP is dissolved in 300 ml of dichloromethane/dimethylformamide (1 : 1) and mixed with 0.5 M of acetic acid anhydride. It is left at room temperature for 1 hr and refluxed for 4 hr. Evacuated to about 100 ml, the solution is added dropwise into ethyl ether under strong stirring. The acetylated PVP precipitate is filtered, washed with ether, and dried to constant weight under vacuum over phosphorus pentoxide.

To activate its carboxyl groups, 50 g of acetylated PVP dissolved in 500 ml of dichloromethane/dimethylformamide (1 : 1) is mixed at 0° with 15.47 g of N-hydroxysuccinimide followed with a solution of 27.75 g dicyclohexylcarbodiimide in 50 ml of dichloromethane. The solution is stirred at 0° for 14 hr before centrifugation to remove the dicyclohexylurea. The supernatant solution (about 300 ml) is added dropwise into 5 liters of cold ether under strong stirring. The white precipitate is filtered, washed repeatedly with ether, and dried in the cold over phosphorus pentoxide.

Binding of Hemoglobin to Activated PVP. Hemoglobin (27 g) is dissolved in 1 liter of 5% sodium carbonate and treated at 4° with 40 g of activated PVP for 24 hr with stirring. The preparation is lyophilized and redissolved in 300 ml of distilled water. After a 20-fold volume diafiltration, it is again lyophilized.

PVP–Hb displays a P_{50} for oxygen binding of about 12 mm Hg at 37°, pH 7.4, compared to about 18 mm Hg for free Hb (Fig. 4).[12] Normthermic perfusion of isolated porcine kidneys demonstrates that PVP–Hb provides an adequate oxygen supply to the organ, is devoid of toxic effects on renal function, and enters into urine severalfold more slowly than unmodified Hb.[12]

Polyanion–Hemoglobin

The oxygen affinity of Hb is lowered by the binding of polyanions to its 2,3-diphosphoglycerate effector site. By linking Hb covalently to a polyanionic carrier, an increase of the effective molecular size of Hb and a decrease of its oxygen affinity may be accomplished simultaneously.[34] Polyanionic carriers that have been synthesized for this purpose include dextran-sulfate (SF-Dx), dextran-phosphate (P-Dx), and dextran-benzene hexacarboxylate (Dx-BHC).

Conjugation to SF-Dx and P-Dx

SF-Dx and P-Dx (MW 40,000) are treated with sodium periodate to generate the dialdehydyl derivatives, which are in turn coupled to the amino groups on Hb and are further stabilized by reduction with sodium borohydride, as described above in the synthesis of Dx–Hb from Dx-dialdehyde.[26]

Conjugation to Dx-BHC

Aminopropyl-Dx is prepared[35] by dissolving 5 g of dextran in 7.5 ml of 25% aqueous $Zn(BF_4)_2$ and 5 ml of water. Epichlorohydrin (25 ml) is added with vigorous stirring; the mixture is allowed to react for 3 hr at 80° and subsequently overnight at room temperature. The polymer is precipitated by pouring the solution dropwise into acetone, filtered, and dried under reduced pressure. The resulting dextran derivative has the structure of $Dx-O-CH_2CH(OH)CH_2Cl$. This product (4.1 g containing ~3% Cl) is purified by repeated dissolution in water and precipitation by acetone and methanol. The chlorine atom is subsequently replaced by an amino group by dissolving the compound in 60 ml of H_2O and 20 ml of 14 M aqueous ammonia. The solution is stirred for 20 hr at room temperature and then poured dropwise into 1 liter of methanol. The resulting precipitate of aminopropyl-Dx (3-amino-2-hydroxypropyl ether of dextran) is filtered, washed with acetone, and dried under reduced pressure. The yield at this stage is about 3.5 g.

Benzene hexacarboxylic acid is coupled to aminopropyl-Dx to form Dx-BHC through the use of 1-(3-dimethylaminopropyl)-3-ethylcarbodiimide hydrochloride (EDCI) as condensing agent. Because benzene hexacarboxylic acid has six carboxylic acid groups, reaction with an amino

[34] D. Zygmunt, M. Leonard, F. Bonneaux, D. Sacco, and E. Dellacherie, *Int. J. Biol. Macromol.* **9**, 343 (1987).
[35] P. Hubert, J. Mester, E. Dellacherie, J. Neel, and E. E. Baulieu, *Proc. Natl. Acad. Sci. U.S.A.* **75**, 3143 (1978).

group on aminopropyl-Dx still leaves it with up to five carboxylic acid groups. One of these may be linked to an amino group on Hb through further use of the water-soluble EDCI as condensing agent.

The SF-Dx–Hb conjugate displays an oxygen P_{50} of 6.5 mm Hg in 0.05 M NaCl at pH 7, 25°. It is right-shifted relative to free Hb, which has under the same conditions a P_{50} of only 4.5 mm Hg. The right-shifting of the Dx-BHC–Hb conjugate is far more pronounced; its P_{50} is 28.5 mm Hg.

[21] Structural Characterization of Modified Hemoglobins

By RICHARD T. JONES

Much can be learned about the structure of modified hemoglobins by analysis of their peptide patterns. Many improvements have been made since Ingram first applied this approach to structural studies of sickle cell hemoglobin.[1] With the recent interest in chemical modifications of hemoglobin and genetically engineered human hemoglobins, rapid and sensitive methods for structural characterization of modified hemoglobins have found increasing application. This chapter presents an approach to the structural characterization of chemically or genetically altered hemoglobins at the level of modifications to the primary structure. Most of the procedures used depend on high-performance liquid chromatography (HPLC).

Scheme for Structural Characterization

Because chemical modifications often lead to heterogeneous mixtures of reaction products, strategies for structural characterization necessitate several steps before peptide and amino acid analyses can be done. Figure 1 presents a scheme for structural characterization of modified hemoglobins, outlining the steps utilized in our laboratory.[2–5] The first step is to assess the degree of heterogeneity of the material to be characterized. This is

[1] V. M. Ingram, *Nature (London)* **178,** 792 (1956).
[2] M. P. Kavanaugh, D. T.-B. Shih, and R. T. Jones, *Acta Haematol.* **78,** 99 (1987).
[3] M. P. Kavanaugh, D. T.-B. Shih, and R. T. Jones, *Biochemistry* **27,** 1804 (1988).
[4] R. Kluger, J. Wodzinska, R. T. Jones, C. Head, T. L. Fujita, and D. T. Shih, *Biochemistry* **31,** 7551 (1992).
[5] R. T. Jones, C. Head, T. L. Fujita, D. T.-B. Shih, J. Wodzinska, and R. Kluger, *Biochemistry* **32,** 215 (1993).

Step	Procedure
1. Assessment of heterogeneity	1. Analytical HPLC (a) Anion exchange (b) Cation exchange
2. Isolation of single Hb component	2. Preparative (a) HPLC (i) Anion exchange (ii) Cation exchange (b) Low pressure (i) DEAE - Sephacel (ii) CM - Sephadex
3. Identification of modified globin chains	3. Analytical C_4 reversed-phase HPLC
4. Isolation of single modified globin chain	4. Preparative C_4 reversed-phase HPLC
5. Estimation of molecular weight of modified globin chain	5. (a) SDS-PAGE (b) Electrospray ionization mass spectrometry
6. Modification of SH groups	6. (a) Oxidation → cysteic acid (b) Aminoethylation (c) Other
7. Formation of peptide fragments	7. Enzyme hydrolysis in urea with trypsin followed by GluC
8. Separation and isolation of peptide fragments	8. Peptide pattern analysis using C_{18} reversed-phase HPLC
9. Determination of amino acid composition of modified peptide(s)	9. Amino acid analysis by reversed-phase HPLC
10. Determination of amino acid sequences	10. Automatic sequence analysis

FIG. 1. Scheme for structural characterization of modified hemoglobin.

done by analytical ion-exchange HPLC. The second step, if possible and practical, should be the isolation of single pure hemoglobin components by either preparative ion-exchange HPLC or low-pressure liquid column chromatography. Because of the presence of two different kinds of globin chain subunits in most hemoglobins, the third step is to assess heterogeneity of the reaction mixture and to identify the modified globin chain(s). This is most simply and rapidly accomplished by analytical reversed-phase HPLC using a C_4 column. This third step is also used to follow the progress of purification during the isolation of single Hb components. The fourth step is the actual isolation of the modified globin chain of interest by preparative reversed-phase HPLC. Preferably this is done using a single Hb component, otherwise it can sometimes be done using the original reaction mixture or a partially purified Hb fraction. Because some kinds of modifications lead to cross-linking of two or more chains, while other modifications may affect only a single chain, estimation of the molecular weight of the modified globin chain may be a necessary fifth step. Sodium dodecyl sulfate polyacrylamide gel electrophoresis (SDS–PAGE) analysis of either isolated globin chain or a purified Hb component is a simple and practical way of estimating the molecular weight of denatured globin chains. It is also possible to obtain much more accurate determinations of molecular mass, including that of the modifying groups, by use of electrospray ionization mass spectrometry.

Modification of sulfhydryl (SH) groups of both chains may be necessary as a sixth step in order to prevent formation of disulfides that can compromise the subsequent enzyme hydrolysis, peptide solubilization, and separation of peptide fragments. This can be achieved either by oxidation of the cysteinyl residues to cysteic acid or by reacting the sulfhydryl groups with one of several different blocking agents. The seventh step is the formation of peptide fragments of the entire modified globin. The insolubility of some of the tryptic peptides of both α and β chains can be overcome by aminoethylation of the SH groups and/or by a second enzyme hydrolysis step using an endoproteinase that cleaves the C-terminal peptide bond of many glutamyl residues. The eighth step is the separation and isolation of peptide fragments. This is done at both the analytical and preparative scales by reversed-phase HPLC with a C_{18} column. Most of the peptides can be detected by their absorption of light in the low ultraviolet light range. Analysis of the peptide patterns for missing or reduced amounts of normal, unmodified peptides as well as the detection of new, modified peptides often leads to the identification of residue(s) that have been altered. The ninth step, if necessary, is the determination of the amino acid composition of modified peptide(s) in order to identify the normal tryptic peptide or peptide fragments they contain. In theory, a tenth step of determining the amino acid sequences may be necessary in order to confirm alterations that result from genetically engineered changes or some types of chemical modifications. In our experience, this step is generally not needed for the characterization of chemically modified Hbs.

Reagents

HPLC-grade acetonitrile and water from Mallinckrodt and trifluoro-acetic acid (TFA) from Pierce Chemical Co. (Rockford, IL) have been found to be satisfactory. Trypsin from Worthington Biochemical Co. (Freehold, NJ) and the *Staphylococcus aureus* V8 endoproteinase GluC from Boehringer Mannheim Biochemical (Indianapolis, IN) have been used for enzyme hydrolysis. Reagents for oxidation and aminoethylation of globin chains are obtained from Aldrich Chemical Co. Inc. (Milwaukee, WI) or Sigma Chemical Co. (St. Louis, MO). Reagents for amino acid analysis are supplied by Pierce Chemical Co. and Aldrich Chemical Co., Inc. DEAE-Sephacel and CM-Sephadex are from Pharmacia Fine Chemicals AB (Uppsala, Sweden). Other reagents for preparation of buffers and developers for chromatography have been of analytical grade or better.

Instrumentation

Any HPLC system would be suitable for the procedures described in this chapter. In our laboratory analytical and preparative separations of hemoglobins are done by ion exchange using a Beckman/Altex model 332 (Fullerton, CA) gradient liquid chromatograph with a model 421 micropro-cessor system controller, two model 110A single-piston reciprocation pumps with a dynamically stirred gradient mixing chamber, and a model 210 syringe-loading sample injection valve. The detector is a Schoeffel Instrument Corp./Kratos (Westwood, NJ) Spectroflow monitor model SF770 with a model GM770 monochromator. This is connected to a Linear (Irvine, CA) potentiometric strip chart recorder. The signal from the detec-tor is also recorded and processed by an IBM System (Danbury, CT) 9000 computer (described below). Effluent is collected with a Gilson FC-80K microfractionator.

Analytical and preparative separations of globin chains are done by reversed-phase HPLC using another Altex model 332 gradient liquid chro-matograph like the one described above, but with a model 420 microproces-sor system controller. The detector for this system is a Hitachi model 100-30 variable-wavelength (visible/UV) unit (Tokyo, Japan) equipped with an 8-μl analytical flow cell connected to a Hewlett Packard model 3392A (Avondale, PA) integrating recorder. The detector signal is also recorded and processed by the IBM System 9000 computer. The effluent is collected with a Gilson FC-80K microfractionator (Middleton, WI) in a chemical hood.

Separation and isolation of peptides are done by reversed-phase HPLC using an IBM Instruments Inc. model LC 9533 ternary liquid chromato-graph attached to a model F9522 fixed-UV module and a model F9523 variable-UV module. The detector signals are recorded by an IBM Instruments Inc. System 9000 computer, which also controls the IBM HPLC system. The computer can record and process four signals simultaneously. The column effluent is collected with a Gilson FC-80K microfractionator in an externally vented box.

Amino acid analyses are done by separation and quantitation of phenylthiocarbamylamino acid derivatives using a second IBM model LC 9533 ternary liquid chromatograph equipped with one fixed-UV detector and one variable-UV detector. The signals are recorded and integrated by the IBM System 9000 computer, which controls the HPLC and an IBM Instruments Inc. model LC 9505 automatic sampler. This system is also used for the automatic analytical separation of globin chains by reversed-phase HPLC.

The hemoglobin present in the effluent from large preparative liquid chromatograms run in the cold room is detected with a Beckman Model 153 analytical UV detector connected to a Kipp and Zonen model BD 40 potentiometric strip chart recorder (Delft, Holland).

Procedures

Analytical and Preparative Separations of Hemoglobin Components by Ion-Exchange Chromatography

Both analytical and preparative separations of Hbs can be achieved by ion-exchange HPLC. Several different procedures have been used for assessment of heterogeneity and for the isolation of single Hb components for further analysis. For analytical purposes the cation-exchange HPLC procedure of Ou et al.,[6] with a 200 × 4.6 mm column of 5-μm microparticulate poly(aspartic acid) silica packing (PolyCAT A; Custom LC, Houston, TX) can provide excellent resolution, as illustrated in Fig. 2. This shows the separation of the hemoglobins resulting from treating deoxyHb with trans-4,4'-diisothiocyanostilbene-2,2'-disulfonate (DIDS). The developers and gradient are given in Table I. Although this procedure is useful for assessing heterogeneity, it has not been used for preparative purposes because of the limited load capacity of this ion exchanger. Another cation system in use for both analytical and preparative separations of hemoglobins includes a SynChropak CM300 column (250 × 4.1 mm for analytical and 250 × 10 mm for preparative separations, from SynChrom, Inc., Linden, IN) and developers containing 30 mM bis-Tris, pH 6.4, and various gradients of sodium acetate starting at 30 mM and ending at 300 mM.[7]

[6] C. Ou, G. J. Buffone, and G. L. Reimer, *J. Chromatogr.* **418,** 266 (1983).
[7] T. H. J. Huisman, *J. Chromatogr.* **418,** 277 (1987).

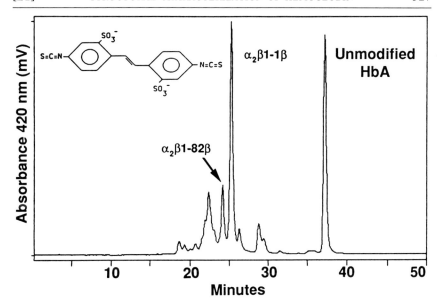

FIG. 2. Analytical cation-exchange HPLC separation of DIDS–deoxyhemoglobin reaction mixture using a poly(aspartic acid)-coated silica column. The gradient and developer conditions are given in Table I.

In our experience, this procedure has been more useful for rechromatography in the isolation of single hemoglobin components than for analytical purposes or as the initial procedure for preparative separation of modified hemoglobins. Separations of hemoglobins by anion-exchange HPLC have been done with a SynChropak AX300 column (250 × 4.1 mm for analytical and 250 × 10 mm for preparative separations) using developers containing 15 mM Tris–acetate at pH 8.0 and various gradients of sodium acetate starting at 10 mM and ending at 150 mM.[7] An example of an analytical separation of Hb modified with bis(3,5-dibromosalicyl) fumarate[8–10] (DBBF) is shown in Fig. 3 using the gradient given in Table I. The hemoglobins are detected in the effluent by monitoring at either 420 or 540 nm. For preparative purposes, 5–10 mg of hemoglobin can be separated on

[8] This sample of $\alpha99$–$99\alpha\beta_2$ cross-linked hemoglobin was supplied to us by Dr. Mario A. Marini of the Letterman Army Institute of Research, Presidio of San Francisco, CA. This was from Batch 11 obtained from Baxter Healthcare Corp. (Round Lake, IL) under contract from the United States Army Medical Research and Development command.

[9] K. D. Vandegriff, F. Medina, M. A. Marini, and R. M. Winslow, *J. Biol. Chem.* **264,** 17824 (1989).

[10] R. Chatterjee, E. V. Welty, R. Y. Walder, S. L. Pruitt, P. H. Rogers, A. Arnone, and J. A. Walder, *J. Biol. Chem.* **261,** 9929 (1986).

TABLE I
GRADIENTS FOR ANALYTICAL ION-EXCHANGE
HPLC PROCEDURES

Time (min)	Pump A (%)	Pump B (%)
Gradient for polyCAT A 200 × 4.6 mm column[a–c]		
0	100	0
25	44	56
35	0	100
50	0	100
52[d]	100	0
67	100	0
Gradient for Analytical AX300 column[e–g]		
0	93.4	6.6
5	90	10
15	60	40
20	34	66
25	0	100
30	0	100
31[h]	93.4	6.6
53	93.4	6.6

[a] Solution for pump A is 40 mM bis-Tris, 4 mM potassium cyanide, in water adjusted to pH 6.5 with hydrochloric acid.

[b] 100% flow rate is 1 ml/min.

[c] Solution for pump B is 40 mM bis-Tris, 4 mM potassium cyanide, and 100 mM sodium chloride, in water adjusted to pH 6.5 with hydrochloric acid.

[d] Program for reequilibration is shown after 50 min.

[e] Solution for pump A is 15 mM Tris in water adjusted to pH 8.0 with acetic acid.

[f] 100% flow rate is 1 ml/min.

[g] Solution for pump B is 15 mM Tris and 150 mM sodium acetate, in water adjusted to pH 8.0 with acetic acid.

[h] Program for equilibration is shown after 30 min.

the preparative columns of SynChropak AX300. Dilute hemoglobin solutions are concentrated by using either a magnetically stirred pressure ultrafiltration chamber with Amicon PM10 Diaflo ultrafilters or with Amicon CF25 25,000 MW Centriflo membrane cones (Amicon Corp., Danvers, MA). The isolation and purification of single hemoglobin components in

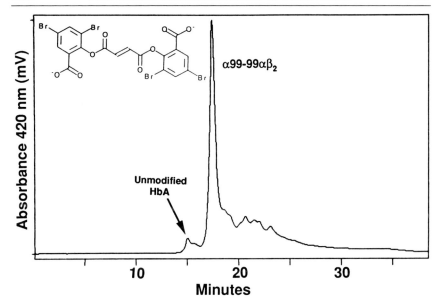

FIG. 3. Analytical anion-exchange HPLC separation of DBBF–hemoglobin sample from Baxter Healthcare Corp. using a SynChropak AX300 column. The gradient and developer conditions are given in Table I.

amounts of more than a few milligrams are best done by standard ion-exchange chromatography using DEAE-Sephacel[11] followed by rechromatography with CM-Sephadex.[12] Purified hemoglobin fractions are generally stored in the CO form on ice to minimize methemoglobin formation and denaturation, particularly if functional or X-ray crystallographic studies are to be made.

Analytical and Preparative Separation of Globin Chains by Reversed-Phase HPLC

Heme and the globin chains can be separated by reversed-phase HPLC using 330-Å pore size C_4 Vydac columns (250 × 4.6 mm for analytical and 250 × 12 mm for preparative separations; The Separations Group, Hesperia, CA) and developers containing 0.1% (v/v) TFA and various gradients of acetonitrile modified after the procedure of Shelton *et al.*[13] Beginning in 1991 the Vydac C_4 columns obtained from The Separations

---

[11] T. H. J. Huisman and A. M. Dozy, *J. Chromatogr.* **19,** 160 (1965).
[12] W. A. Schroeder and T. H. J. Huisman, *Clin. Biochem. Anal.* **9,** 1 (1980).
[13] J. B. Shelton, J. R. Shelton, and W. A. Schroeder, *J. Liq. Chromatogr.* **7.** 1969 (1984).

TABLE II
GRADIENT FOR GLOBIN CHAIN SEPARATION
WITH VYDAC C$_4$ COLUMNS

Time (min)	Pump A^{a,b} (%)	Pump B^c (%)
0	50	50
60	37.5	62.5
80	14	86
82^d	14	86
84	50	50
109	50	50

[a] Solution for pump A is 0.1% TFA, 20% acetonitrile, and 80% water by volumes.
[b] 100% flow rate is 1 ml/min for the 250 × 4.6 mm analytical column and 2 ml/min for the 250 × 12 mm preparative column.
[c] Solution for pump B is 0.1% TFA, 60% acetonitrile, and 40% water by volumes.
[d] Program for reequilibration is shown after 82 min.

Group have been found to have stronger retention properties for globin chains than columns obtained earlier. This is apparently due to changes in the manufacturer's method of preparing the packing material. Elution patterns of globin chains found for columns obtained before 1991 can be achieved with the newer columns by starting at a higher initial concentration of acetonitrile. The gradient conditions now used for analytical separations made with the newer columns are shown in Table II. (The gradient used for older columns is given in the legend of Fig. 5.) Gradients and flow rates for the preparative column are altered according to the increase in cross-sectional area and the elution profile and by trial and error in order to obtain satisfactory resolution of globin components to be isolated. We are currently using a 250 × 10 mm Dynamax-300A 5-μm C$_4$ column with a 50 × 10 mm Dynamax-300A 5-μm C$_4$ guard column (Rainin Instrument Co. Inc., Woburn, MA) for preparative separations. The effluent is monitored at 220 nm and the globin chains are recovered from the column effluent by lyophilization.

Figure 4 shows the globin chains present in the reaction mixture of deoxyHb modified with fumaroyl bis(methyl phosphate).[5,14] The structures of each globin chain identified in this chromatogram were established by subsequent SDS–PAGE and peptide pattern analyses. In our experience,

[14] R. Kluger, A. S. Grant, S. L. Bearne, and M. R. Trachsel, *J. Org. Chem.* **55,** 2864 (1990).

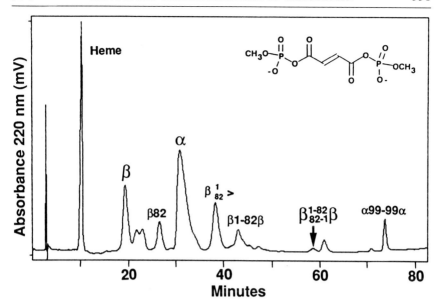

FIG. 4. Analytical separation of globin chains in a reaction mixture of deoxyHb treated with fumaroyl bis(methyl phosphate) by reversed-phase HPLC using a Vydac C_4 column obtained after 1991 and 0.1% TFA in an acetonitrile–water gradient described in Table II.

all of the chemically modified globin chains so far studied elute more slowly from the C_4 column than do their corresponding unmodified chains. Cross-linked chains elute more slowly than un-cross-linked chains. Cross-linked α chains elute the slowest of all modified globins.

Figure 5 is a chromatogram of the chains of the major Hb component isolated from the DBBF–Hb sample shown in Fig. 3 that had been modified by treating deoxyHb with 3,5-dibromosalicyl bisfumarate. This shows the slowly eluting, chemically modified α chain as well as the normal, unmodified β chain and heme.

Estimation of Molecular Weight

The presence of cross-linking of globin chains is determined by polyacrylamide gel electrophoresis in the presence of 0.1% sodium dodecyl sulfate according to the procedure of Laemmli.[15] A 15% polyacrylamide gel that is 2.7% cross-linked is used to distinguish between the 16,000 molecular weight monomers and 32,000 molecular weight cross-linked tetramers. The hemoglobins and globins from HPLC separations are pre-

[15] U. K. Laemmli, *Nature (London)* **227**, 680 (1974).

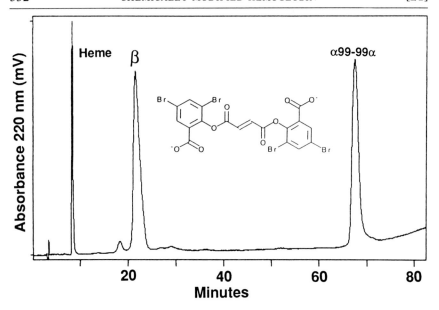

FIG. 5. Analytical separation of globin chains of the major DBBF–Hb component corresponding to the main zone in Fig. 3 by reversed-phase HPLC using a Vydac C_4 column obtained before 1991. Development was with 0.1% TFA in an acetonitrile–water gradient that started at 37.6% acetonitrile and changed to 42.6% by 60 min and then to 52% by 80 min until 82 min, after which it returned to 37.6% by 84 min and reequilibrated until 110 min.

pared by heat denaturation in a buffer containing 65 mM Tris-HCl, pH 6.8, 2% SDS, 10% (v/v) glycerol, and 5% (v/v) 2-mercaptoethanol. Approximately 5–20 μg of protein is applied to the gel and subjected to electrophoresis at 20 mA for about 4–5 hr.

Electrospray ionization mass spectrometry can be used to measure the molecular masses of modified Hbs by a standard method.[16] We have confirmed the exact molecular weights of cross-linked hemoglobins, globin chains, and peptides, including the weights of the modifying groups, by this method. It can also be used to detect the presence of side products with unsuspected structures. We have used this technique to identify a modified Hb in which an ester has been formed between one of the reactive groups of the Hb modifier and a molecule of bis-Tris.[17]

[16] J. B. Fenn, M. Mann, C. K. Meng, S. F. Wong, and C. M. Whitehouse, *Science* **246,** 64 (1989).
[17] R. T. Jones, C. Head, and T. Fujita, unpublished studies (1992).

Chemical Modifications and Enzyme Hydrolysis of Globin Chains

For some structural studies of modified Hbs the cysteinyl residues must be oxidized to cysteic acid with performic acid,[18] aminoethylated according to the procedure of Morrison *et al.*[19] or blocked in some other way. This has been necessary for our characterization of α99–99α cross-linked hemoglobins.[17] However, initial structural studies of a modified globin chain are generally done without modification of the cysteinyl residues. If the first peptide pattern analysis indicates that the globin contains α chains or β chains that are resistant to enzymatic hydrolysis, then oxidation of the cysteinyl residues is done either on another portion of the unhydrolyzed globin chain or on the remaining portion of the enzyme hydrolyzate. The procedure we use is to prepare performic acid by mixing 2.375 ml of 96–100% formic acid with 0.125 ml of 30% hydrogen peroxide and allowing it to stand at room temperature for 2 hr before chilling to −10° in a salted ice bath. Up to 4 mg of globin or lyophilized enzyme hydrolyzate dissolved in 0.2 ml 30% formic acid with one drop of absolute methanol is chilled to −10° and treated with 0.1 ml of the chilled performic acid solution for 3 hr at −5 to −10°. This is diluted with 6 or more ml of ice-cold HPLC-grade or deionized water, then frozen and lyophilized.

Globin chains either with or without their cysteinyl residues modified are hydrolyzed with trypsin (Worthington) carried out at room temperature (25°) in 80 mM ammonium bicarbonate buffer at pH 8.5 for 24 hr with a ratio of trypsin to globin of 1 : 50 by weight. In most of our current studies of chemically modified hemoglobins, the tryptic hydrolysis is heated in boiling water for 2 min to denature the trypsin. This is followed by hydrolysis with endoproteinase GluC from *S. aureus* V8, at room temperature at pH 8.5 for another 24 hr. We have found that more complete hydrolysis of some of the more resistant globin chain preparations can be obtained if the hydrolysis is carried out in the presence of either urea or sodium dodecyl sulfate. Our procedure is to dissolve the globin in 8 *M* urea and allow this to stand at room temperature for 2–4 hr. This is then diluted to 2 *M* urea with 80 mM ammonium bicarbonate buffer at pH 8.5 and hydrolyzed with trypsin (2% by weight) for 24 hr at room temperature. The tryptic hydrolyzate is then heated in boiling water for 2 min, diluted to 1 *M* urea with 80 mM ammonium bicarbonate buffer, and hydrolyzed with endoproteinase GluC (1% by weight) for another 24 hr at room temperature. The hydrolyzate, if not clear, is centrifuged before injection into the HPLC system, which has a guard column before the main C₁₈

[18] N. P. Neuman, this series, Vol. 25, p. 393.
[19] W. T. Morrison, A. D. Pressley, J. G. Adams, III, and W. P. Winter, *Hemoglobin* **5**, 403 (1981).

reversed-phase column. The urea used in this procedure is deionized by filtering through an AG 501-X8 analytical-grade mixed-bed resin (Bio-Rad Laboratories, Richmond, CA) and then through a 13-mm nylon 0.45-μm filter (Micron Separations Inc., Westboro, MA).

Separation of Peptides by Reversed-Phase HPLC

Peptide fragments are separated for both analytical and preparative purposes by HPLC procedures using reversed-phase C_{18} columns (250 × 4.6 mm Vydac; The Separations Group, Hesperia, CA, Cat. # 218TP54.6). Most separations are now made using developers of 0.1% TFA and gradients of acetonitrile starting at 0% and ending at 100%, generated over a period of up to 100 min and pumped at 1 ml/min. This has been modified after the procedure of Shelton et al.[20] As shown in Table III the typical gradient for both α- and β-chain peptides starts at 0% acetonitrile and changes to 12.5% by 10 min, 25% by 60 min, 50% by 100 min, and 100% by 105 min. A second developer system with 10 mM ammonium acetate buffer at pH 6.0 and acetonitrile concentration gradients can be used for the initial separation but is more often used for rechromatography. Its gradient is shown in Table III and was adapted from the procedures of Schroeder et al.[21] and Wilson et al.[22] In both cases, the effluent is monitored at 214 nm, which detects most peptides, and also at either 280 nm to detect tyrosyl- and tryptophanyl-containing peptides or at other wavelengths for other UV-absorbing groups that may be present in the chemical modifiers. The solvents and volatile solutes are removed from the peptides in the effluent by lyophilization or vacuum evaporation.

Figure 6 is a composite of three separate tryptic peptide patterns of the modified α chain (see Fig. 5) from the main Hb component (see Fig. 3) of deoxyHb treated with bis(3,5-dibromosalicyl) fumarate. The sequences of the peptide fragments shown in this and other peptide patterns are listed in Table IV. Figure 6A is the tryptic peptide pattern of performic acid-oxidized, modified globin hydrolyzed with trypsin only and without urea. It reveals all of the normal peptides of unmodified α chains except for the absence of αT-11 and αT-12 and a reduction in the amount of αT-13. In addition, there is a complex of three new peaks that elute near the end of the chromatogram from about 80 to 90 min. Quantitative amino acid analyses of these new peptides indicated that the first peak contains

---

[20] J. B. Shelton, J. R. Shelton, W. A. Schroeder, and D. R. Powars, *Hemoglobin* **9**, 325 (1985).
[21] W. A. Schroeder, J. B. Shelton, J. R. Shelton, and D. R. Powars, *J. Chromatogr.* **174**, 385 (1979).
[22] J. B. Wilson, H. Lam, P. Pravatmuang, and T. H. J. Huisman, *J. Chromatogr.* **179**, 271 (1979).

TABLE III
GRADIENT FOR PEPTIDE SEPARATIONS
WITH 250 × 4.6 mm VYDAC C$_{18}$ COLUMN

Time (min)	Pump A^{a,b} (%)	Pump B^c (%)
TFA/acetonitrile gradient		
0	100	0
10	87.5	12.5
60	75	25
100	50	50
105	0	100
107^d	0	100
110	100	0
130	100	0
Ammonium acetate/acetonitrile gradient^{e-g}		
0	100	0
5	100	0
12	94	6
80	70	30
100	30	70
105^{h,i}	100	0

^{a} Solution for pump A is 0.1% TFA in water by volume.
^{b} 100% flow rate is 1.0 ml/min.
^{c} Solution for pump B is 0.1% TFA in acetonitrile by volume.
^{d} Program for reequilibration is shown after 107 min.
^{e} Solution for pump A is 10 mM ammonium acetate at pH 6.0 (0.77 g sodium acetate per liter of water, adjusted to pH 6.0 with about 4 drops of 5 M acetic acid).
^{f} 100% flow rate is 0.9 ml/min.
^{g} Solution for pump B is 100% acetonitrile.
^{h} The column should be stripped at the end of the day with 0.1% TFA in 100% acetonitrile to remove uneluted core peptides.
^{i} Program for reequilibration is shown after 110 min.

equimolar amounts of αT-11 and αT-12, the second contains 2 mol each of αT-11 and αT-12 plus 1 mol of αT-13, and the third contains equimolar amounts of all three of these tryptic peptides. As shown in Fig. 6B, much greater hydrolysis of the αT-13 tryptic peptide from the modified αT-11,12 is achieved by hydrolyzing with trypsin in the presence of urea. In order

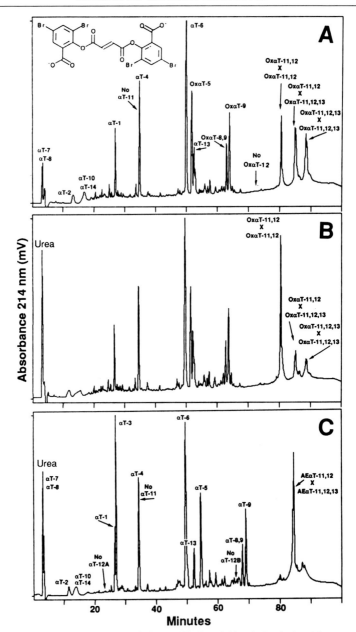

Fig. 6. Tryptic peptide patterns of the modified α-globin chains isolated by preparative chain separation of the major DBBF–Hb (see Figs. 3 and 5). (A) Modified α chains were hydrolyzed with trypsin after performic acid oxidation (Ox). (B) The same as A but the hydrolysis was done in the presence of 2 M urea. (C) Aminoethylated (AE) modified α

to be able to account for all of the normal tryptic peptides of the α chain, an aminoethylated preparation of the major modified globin from the same sample was also hydrolyzed with trypsin in the presence of urea. The peptide pattern is shown in Fig. 6C. The main modified tryptic peptide contains 2 mol of αT-11 and αT-12 to 1 mol of αT-13. This peptide appears to result from the failure of tryptic hydrolysis of both Lys-99(α) residues that connect αT-11 and αT-12 and one of the two Lys-127(α) residues that connect αT-12 and αT-13. The resistance of one Lys-127(α) residue to tryptic hydrolysis of this aminoethylated preparation may be due to the retention of some secondary or tertiary structure in the cross-linked globin that does not permit access of trypsin to an otherwise susceptible bond. The observation of almost complete hydrolysis of this bond in the case of the oxidized globin treated with trypsin in the presence of urea indicates that neither Lys-127(α) bond is blocked by fumarate. On the other hand, the complete resistance of the Lys-99(α) bond to tryptic hydrolysis is consistent with it being blocked by fumarate. These peptide data plus the finding of dimer chains by the SDS–PAGE procedure indicate that this modified Hb is cross-linked between the Lys-99(α) residues of its two α chains. This is consistent with the observation published by Chatterjee et al.[10] that fumarate forms a diamide linkage between the ε-amino side chains of the two α99 lysyl residues.

Figure 7 is a composite of the peptide patterns of the β chains of three different hemoglobins isolated from a reaction mixture obtained by treating deoxyHb with fumaroyl bis(methyl phosphate).[5,14] The hydrolysis was done in the presence of urea, first with trypsin followed by GluC endopro-teinase as described above. Figure 7A shows the peptide pattern of the β chain that elutes at about 38 min in Fig. 4. It lacks any normal βT-1, βT-9, and βT-10a' peptides. (The sequences of the β chain peptide frag-ments are also listed in Table IV.) In addition to the other β-chain tryptic peptides, a modified peptide was found eluting at about 84 min. Its amino acid composition was consistent with equimolar amounts of βT-1, βT-9, and βT-10a' peptides. Because the unhydrolyzed β chain migrates as a globin chain monomer, these data indicate that fumarate forms an internal linkage between the amino group of the Val-1(β) residue and the ε-amino group of the Lys-82(β) residue of the same chain. Both β chains in the modified hemoglobin must be modified but not cross-linked together. Be-

chains hydrolyzed with trypsin in the presence of 2 M urea. Separation was by reversed-phase HPLC using a Vydac C_{18} column and 0.1% TFA in an acetonitrile–water gradient that started at 0% acetonitrile until 5 min and then changed to 13.5% by 20 min, 30% by 60 min, 65% by 90 min, 100% by 95 min, and then returned to 0% after 100 min.

TABLE IV

HUMAN HEMOGLOBIN PEPTIDE FRAGMENTS

GluC peptides, sequences, and chain residue numbers

A. α chain

Tryptic peptide no.	Sequence
1	Val-Leu-Ser-Pro-Ala-Asp-Lys 1 2 3 4 5 6 7
2	Thr-Asn-Val-Lys 8 9 10 11
3	Ala-Ala-Trp-Gly-Lys 12 13 14 15 16
4	<- - - - - - - - - - 4ab - - - - - - - - - - -> <- - - - - - 4a - - - - ->GluC<- - 4b- - - ->GluC<- - - 4c - - -> Val-Gly-Ala-His-Ala-Gly-Glu | Tyr-Gly-Ala-Glu | Ala-Leu-Glu-Arg 17 18 19 20 21 22 23 24 25 26 27 28 29 30 31
5	Met-Phe-Leu-Ser-Phe-Pro-Thr-Thr-Lys 32 33 34 35 36 37 38 39 40
6	GluC Thr-Tyr-Phe-Pro-His-Phe-Asp | Leu-Ser-His-Gly-Ser-Ala-Gln-Val-Lys 41 42 43 44 45 46 47 48 49 50 51 52 53 54 55 56
7	Gly-His-Gly-Lys 57 58 59 60
8	Lys 61
9	Val-Ala-Asp-Ala-Leu-Thr-Asn-Ala-Val-Ala-His-Val-Asp-Asp-Met-Pro-Asn-Ala-Leu-Ser-Ala-Leu-Ser-Asp-Leu-His-Ala-His-Lys 62 63 64 65 66 67 68 69 70 71 72 73 74 75 76 77 78 79 80 81 82 83 84 85 86 87 88 89 90

10 Leu-Arg
    91 92

11 Val-Asp-Pro-Val-Asn-Phe-Lys
    93 94 95 96 97 98 99

12 Leu-Leu-Ser-His-Cys-Leu-Leu-Val-Thr-Leu-Ala-Ala-His-Leu-Pro-Ala-Glu-Phe-Thr-Pro-Ala-Val-His-Ala-Ser-Leu-Asp-Lys
    100 101 102 103 104 105 106 107 108 109 110 111 112 113 114 115 116 117 118 119 120 121 122 123 124 125 126 127

13 Phe-Leu-Ala-Ser-Val-Ser-Thr-Val-Leu-Thr-Ser-Lys
    128 129 130 131 132 133 134 135 136 137 138 139

14 Tyr-Arg
    140 141

B. β chain

1 <- - - - - >1a<- - - - >GluC<-1b->
   Val-His-Leu-Thr-Pro-Glu l Glu-Lys
    1 2 3 4 5 6 7 8

2 Ser-Ala-Val-Thr-Ala-Leu-Trp-Gly-Lys
   9 10 11 12 13 14 15 16 17

3 <- - - - 3a - - ->GluC<- - -3b - ->GluC<- - -3c- - ->
   Val-Asn-Val-Asp-Glu l Val-Gly-Gly-Glu l Ala-Leu-Gly-Arg
    18 19 20 21 22 23 24 25 26 27 28 29 30

4 Leu-Leu-Val-Val-Tyr-Pro-Trp-Thr-Gln-Arg
    31 32 33 34 35 36 37 38 39 40

5 <- -5a- ->GluC<- - - - - - - - - - - - - 5b - - - - - - - ->
   Phe-Phe-Glu l Ser-Phe-Gly-Asp-Leu-Ser-Thr-Pro-Asp-Ala-Val-Met-Gly-Asn-Pro-Lys
    41 42 43 44 45 46 47 48 49 50 51 52 53 54 55 56 57 58 59

6 Val-Lys
   60 61

(continued)

TABLE IV (continued)

GluC peptides, sequences, and chain residue numbers

Tryptic peptide no.	GluC peptides, sequences, and chain residue numbers
7	Ala-His-Gly-Lys 62 63 64 65
8	Lys 66
9	GluC Val-Leu-Gly-Ala-Phe-Ser-Asp-Gly-Leu-Ala-His-Leu-Asp ǀ Asn-Leu-Lys 67 68 69 70 71 72 73 74 75 76 77 78 79 80 81 82
10	<- - - - - -10a' - - - - - - >GluC<- - 10b' - - -> Gly-Thr-Phe-Ala-Thr-Leu-Ser-Glu ǀ Leu-His-Cys-Asp-Lys 83 84 85 86 87 88 89 90 91 92 93 94 95
11	Leu-His-Val-Asp-Pro-Glu-Asn-Phe-Arg 96 97 98 99 100 101 102 103 104
12	Leu-Leu-Gly-Asn-Val-Leu-Val-Cys-Val-Leu-Ala-His-His-Phe-Gly-Lys 105 106 107 108 109 110 111 112 113 114 115 116 117 118 119 120
13	<- - - - - - - - - - - - - - - >"Trypsin" Glu-Phe-Thr-Pro-Pro-Val-Gln-Ala-Ala-Tyr ǀ Gln-Lys 121 122 123 124 125 126 127 128 129 130 131 132
14	Val-Val-Ala-Gly-Val-Ala-Asn-Ala-Leu-Ala-His-Lys 133 134 135 136 137 138 139 140 141 142 143 144
15	Tyr-His 145 146

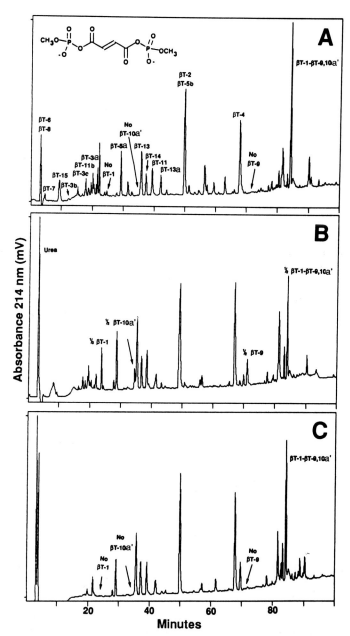

FIG. 7. Tryptic-GluC peptide patterns of modified β-globin chains isolated from three different hemoglobin components purified from a mixture resulting from treating deoxyHb with fumaroyl bis(methyl phosphate). The modified β-globin chains correspond to those in Fig. 4 eluting at (A) 38 min, (B) 42 min, and (C) 59 min. Separation was by reversed-phase HPLC using a Vydac C_{18} column and 0.1% TFA in an acetonitrile–water gradient as described in Table III.

cause its α chains are unmodified, the formula of this modified hemoglobin can be designated as $\alpha_2(\beta \, {}_{82}^{1}\!\!>\!\!F)_2$, where F represents a molecule of fumarate with amide bonds to Val-1(β) and Lys-82(β). Figure 7B is the peptide pattern of the β chain that elutes at about 43 min in Fig. 4. It shows one-half the normal amount of βT-1, βT-9, and βT-10a' and the presence of about half the molar amount of the same modified peptide as found in Fig. 7A. This β chain migrated as a globin chain dimer in SDS–PAGE. Therefore, its parent hemoglobin is concluded to be cross-linked between the Val-1(β) of one β chain and the Lys-82(β) of the other β chain. Its formula is $\alpha_2\beta$1—F—82β. The peptide pattern in Fig. 7C is very similar to that of Fig. 7A. This is from the hydrolyzate of globin that elutes at about 59 min of Fig. 4. It differs from the modified β chain whose peptide pattern is shown in Fig. 7A by migrating as a globin chain dimer rather than as a monomer in SDS–PAGE. We conclude that the hemoglobin from which this came has a double cross-linkage between the Val-1(β) residue of each chain and the Lys-82(β) residue of the opposite chain. As Fig. 7A and C illustrates, the peptide pattern alone may not provide sufficient data to conclude the structural change that is present without additional information about the molecular weight of the modified chain.

Amino Acid Composition and Sequence Analyses

Amino acid composition analysis of peptides may be necessary to confirm the elution position of normal tryptic and GluC fragments as well as to determine the ratios of normal peptides present in the modified peptides. Any method of compositional analysis can be used that is sensitive enough to give accurate quantitative data with the amounts of peptide that can be recovered from the C_{18} reversed-phase column after loading hydrolyzate from about 0.1 to 1 mg of globin chain. The procedure we use is to hydrolyze the peptide in evacuated tubes using 6 M HCl vapor at 110° for 22 hr. In some cases a hydrolysis time of 48 or 72 hr is used if a Val—Val or other resistant bond is known to be present. The amino acids are derivatized with phenyl isothiocyanate and the resultant phenylthiocarbamyl amino acid derivatives are separated by reversed-phase HPLC using an IBM octadecylsilane column (IBM Instruments Inc., Cat. # 8635308).[23] Effluent is monitored at 254 nm and the signal is recorded and integrated with the IBM Systems 9000 computer.

Amino acid sequence analysis is generally not necessary for determining the structural changes of chemically modified hemoglobins at the level of primary structure. A standard automatic method like that of Hunkapiller

[23] R. L. Heinrikson and S. C. Meredith, *Anal. Biochem.* **136**, 65 (1984).

et al.[24] is satisfactory if needed. This procedure cannot be used if the N-terminal amino group is chemically modified, as in the case of the examples illustrated in Fig. 7.

Conclusions

Complete and systematic structural characterizations of chemically and genetically altered hemoglobins can be accomplished by the steps and procedures presented. However, for some purposes, sufficient structural information may be obtained by doing analytical ion-exchange HPLC and reversed-phase HPLC chain separation of the modified hemoglobin mixture, followed by a preparative chain separation and peptide pattern analysis of the altered globin chain of interest. This can be done on as little as 2–3 mg of hemoglobin and within a few days time, provided the HPLC procedures are already in operation. On the other hand, some modified hemoglobins, when present in complicated mixtures, may require much more material, time, and most or all of the steps presented.

Acknowledgments

This work was supported in part by the U.S. Army Medical Research and Development Command under Contract No. DAMD17-89-C-9002 and by Grant HL20142 from the National Institutes of Health. The views and findings contained in this report are those of the author and should not be construed as an official Department of the Army position, policy, or decision unless so designated by other documentation. Charlotte Head and Thomas Fujita have provided valuable technical assistance in the work described. Professor Ronald Kluger supplied the cross-linking reagents and helpful advice through our collaborative studies of chemical modification of human hemoglobins.

[24] M. W. Hunkapiller, R. M. Hewick, W. J. Dreyer, and L. E. Hood, this series, Vol. 91, p. 399.

Section V

Recombinant Hemoglobin

[22] Production of Human Hemoglobin in *Escherichia coli* Using Cleavable Fusion Protein Expression Vector

By Timm-H. Jessen, Noboru H. Komiyama, Jeremy Tame,
Josée Pagnier, Daniel Shih, Ben Luisi, Giulio Fermi,
and Kiyoshi Nagai

Introduction

The first expression of β-globin in *Escherichia coli* was reported in 1980 by Guarante et al.[1] The cDNA encoding rabbit globin was placed under the control of the *lacUV5* promoter and next to the β-galactosidase ribosomal binding sequence to facilitate translation of the β-globin cDNA. The production of β-globin was demonstrated by radiolabeling, the amount of β-globin produced by this method being far too small to do any biochemical experiment. In order to overcome this difficulty Nagai and Thøgersen developed the cleavable fusion protein expression vector pLcIIFXβ-globin.[2] In this vector, human β-globin cDNA was fused to a short coding sequence of the highly expressed λ phage *c*II gene. The hybrid gene was efficiently transcribed from the λ p_L promoter, and the short coding sequence of the *c*II gene ensured efficient translational initiation of the hybrid gene. An oligonucleotide encoding Ile-Glu-Gly-Arg was inserted at the junction between the λ *c*II gene and β-globin cDNA. Blood coagulation factor X_a, a protease in the blood clotting system, recognizes this tetrapeptide sequence in prothrombin and activates it by specifically cleaving the peptide bond immediately to the C-terminal side of the arginine. The β-globin hybrid protein was then cleaved specifically and efficiently by factor X_a at the peptide bond following the tetrapeptide, and thus the authentic β-globin chain was liberated.[2]

Nagai *et al.* folded β-globin synthesized in *E. coli* and reconstituted functional hemoglobin (Hb) from it with heme and native α chain prepared from human blood.[3] They engineered the first manmade mutant hemoglobins by site-directed mutagenesis and reported their oxygen binding properties. Luisi and Nagai crystallized these mutants and solved their crystal structures.[4] These mutants were structurally identical to that of human

[1] L. Guarante, G. Lauer, T. M. Roberts, and M. Ptashne, *Cell (Cambridge, Mass.)* **20,** 543 (1980).
[2] K. Nagai and H. C. Thøgersen, *Nature (London)* **309,** 810 (1984).
[3] K. Nagai, M. F. Perutz, and C. Poyart, *Proc. Natl. Acad. Sci. U.S.A.* **82,** 7252 (1985).
[4] B. F. Luisi and K. Nagai, *Nature (London)* **320,** 555 (1986).

Hb A except in the immediate vicinity of the substituted residues. Subsequently Tame applied the same expression vector to human α-globin and reported the oxygen binding properties and crystal structure of hemoglobins in which the distal residues of the α subunit had been mutated.[5] This made it possible to introduce any mutation at any site in human hemoglobin. The oxygen binding properties and crystal structures of a number of mutants in both α and β subunits have been studied to date.

Our method remained the only method of making human hemoglobin in a microorganism until soluble and functional Hb was produced in *E. coli* by Hoffman *et al.*[6] Unlike our method, which involves the solubilization of inclusion bodies and the purification and refolding of globins, the native functional form of Hb was produced in *E. coli* using a vector coexpressing α- and β-globins. Hb made by this method has, at the N termini, an extra Met residue, which arises from the initiation AUG codon. The oxygen binding of this Hb is only slightly different from that of human hemoglobin A. Subsequently Wagenbach *et al.*[7] and Ogden *et al.*[8] independently produced the soluble and functional form of human hemoglobin by coexpressing α- and β-globins in the yeast *Saccharomyces cerevisiae*. The N termini of both subunits were processed correctly and hence this is the first authentic functional Hb tetramer to be produced in a microorganism.

Despite these developments our method is still valuable because it permits the preparation from *E. coli* of large amounts of Hb that is chemically identical to human Hb. The functional parameters and crystal structure of hemoglobin obtained by our method can be directly compared with those of native and naturally occurring mutant Hbs. We can routinely grow crystals of mutant proteins produced by this method. In this article we describe our method in detail, including some improvements.

Expression Vectors pT7cIIFXα-globin and pT7cIIFXβ-globin

In the original expression system the fusion protein gene cIIFXβ-globin was under the control of the λ p_L promoter, and the *E. coli* strain QY13, a defective λ lysogenic strain with the heat-sensitive cI857 repressor, was used as a host strain.[3] Protein synthesis was induced by heating the culture to 42°. The heat-inducible expression system is not very practical for large-scale work and we therefore recloned the cIIFXβ-globin and

[5] J. R. H. Tame, Ph.D. Thesis, University of Cambridge (1989).
[6] S. J. Hoffman, D. L. Looker, J. M. Roehrich, P. E. Cozart, S. L. Durfee, J. L. Tedesco, and G. L. Stetler, *Proc. Natl. Acad. Sci. U.S.A.* **87**, 8521 (1990).
[7] M. Wagenbach, K. O'Rourke, L. Vitez, A. Wieczorek, S. Hoffman, S. Durfee, J. Tedesco, and G. Stetler, *Bio/Technology* **9**, 57 (1991).
[8] J. E. Ogden, R. Harris, and M. T. Wilson, this volume [24].

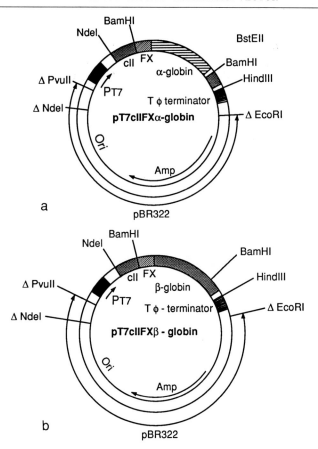

FIG. 1. Physical maps of expression vectors pT7cIIFXα-globin (a) and pT7cIIFXβ-globin (b). The hybrid genes cIIFXα-globin and cIIFXβ-globin are under the control of the T7 bacteriophage RNA polymerase ϕ10 promoter and ribosomal binding sequence. The globin genes are fused to a small segment of λ phase cII gene encoding the N-terminal 31 amino acids. At the junction, oligonucleotides encoding an Ile-Glu-Gly-Arg tetrapeptide sequence were inserted. This is the specific recognition sequence for blood coagulation factor X_a. Reprinted with permission from *Nature* (Ref. 15) Copyright (1987) Macmillan Magazines Limited.

cIIFXα-globin genes into the T7 RNA polymerase expression vectors pRK172[9] and pET3a.[10] Figure 1 shows the physical maps of pT7cIIFXα-globin and pT7cIIFXβ-globin. The cIIFXβ-globin gene was cut out from M13 mp18 cIIFXβ-globin (Fig. 2b) with *Nde*I and *Hind*III and recloned into the pRK172 vector. The *Bam*HI fragment containing the FXα-globin

[9] M. Mcleod, M. Stein, and D. Beach, *EMBO J.* **6,** 729 (1987).
[10] F. W. Studier, A. H. Rosenberg, J. J. Dunn, J. W. Dubendorff, this series, Vol. 185, p. 60.

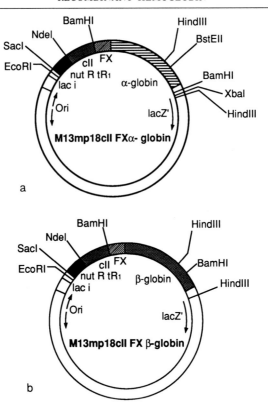

FIG. 2. Physical map of the M13 *mp*18 phage clones containing the hybrid genes *c*IIFXα-globin (a) and *c*IIFXβ-globin (b). Site-directed mutagenesis can be carried out using these vectors, and the hybrid gene can be subcloned into the T7 vector to yield *p*T7*c*IIFXα-globin and *p*T7*c*IIFXβ-globin.

was cut out from M13 *mp*18 *c*IIFXα-globin (Fig. 2a) and cloned into *Bam*HI-cleaved *p*T7*c*IIFXβ-globin. This cloning step should be carried out using a host *E. coli* strain not carrying the T7 RNA polymerase gene, otherwise there would be a selection against recombinant clones expressing foreign protein. The recombinant globin clones are then transformed into the *E. coli* strain BL21(DE3) containing the T7 RNA polymerase gene under the control of *lacUV*5 promoter.[10] Synthesis of T7 RNA polymerase is normally induced by isopropyl β-D-galactoside (IPTG), the T7 RNA polymerase in turn transcribing the *c*IIFX-globin genes. In the case of *p*T7*c*IIFXα-globin and *p*T7*c*IIFXβ-globin, the genes are induced toward the end of the log phase even without addition of IPTG, and large amounts of these fusion proteins were produced (Fig. 3).

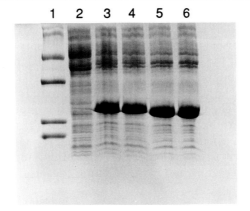

FIG. 3. Expression of hybrid protein cIIFXα-globin and cIIFXβ-globin in BL21(DE3) cells harboring pT7cIIFXα-globin or pT7cIIFXβ-globin. Molecular weight markers (from bottom to top), lane 1: 12,300 (cytochrome *c*), 17,200 (myoglobin), 30,000 (carbonate dehydratase), and 42,600 (ovalbumin); lane 2, BL21(DE3); lanes 3 and 4, BL21(DE3) harboring pT7cIIFXβ-globin; lanes 5 and 6, pT7cIIFXα-globin. Reprinted with permission from *Nature* (Ref. 15) Copyright (1987) Macmillan Magazines Limited.

A single colony of BL21(DE3) harboring pT7cIIFXα-globin or pT7cIIFXβ-globin is picked from a fresh plate and inoculated into 200 ml of 2 × TY medium with 100 μg/ml ampicillin. After culturing for 6 hr at 37°, a 10-ml aliquot is added to 1 liter of 2 × TY medium with 100 μg/ml ampicillin in 2-liter flasks. The culture is harvested after 24 hr and 100–120 g of cells are normally obtained from 10 liters of culture.

Preparation of Inclusion Bodies and Purification of Fusion Proteins

Materials

Lysis buffer: 50 mM Tris-HCl, pH 8.0, 25% sucrose (w/v), 1 mM EDTA
Detergent buffer: 0.2 M NaCl, 1% deoxycholic acid (w/v), 1% Nonidet P-40 (v/v), 20 mM Tris-HCl, pH 7.5, 2 mM EDTA
Deoxyribonuclease I (Sigma, St. Louis, MO; D5025), 10 mg/ml in 0.2 M NaCl

Methods

In this procedure,[3,11] 100 g of harvested cells is suspended in 80 ml of lysis buffer and lysed by addition of 200 mg of lysozyme dissolved in 20

[11] K. Nagai and H. C. Thøgersen, this series, Vol. 153, p. 461.

ml of lysis buffer. After 1 hr incubation at 4°, stock solutions of $MgCl_2$, $MnCl_2$, and DNase I are added to final concentrations of 10 mM, 1 mM, and 10 μg/ml, respectively, and the mixture is stirred gently at room temperature until the viscosity decreases. Then 200 ml of the detergent buffer is added and the suspension centrifuged at 2500 g for 15 min at 4°. The brownish turbid supernatant is sucked off carefully without disturbing the white pellet. The pellet is then carefully resuspended in 1 mM EDTA and 0.5% (v/v) Triton X-100 in order to solubilize the membrane components. The suspension is centrifuged at 2500 g for 15 min at 4° and the milky supernatant is discarded. This procedure is repeated three times until a clear supernatant and tight pellet are obtained.

Purification of Fusion Proteins

Inclusion bodies containing the β-globin fusion protein can be dissolved directly in 100 ml of 8 M urea, 25 mM Tris–acetate buffer, pH 5.0 (dilute 100 mM stock solution), 1 mM EDTA, and 1 mM DTT (dithiothreitol). Insoluble material is removed by centrifugation, and a slightly greenish clear supernatant is obtained. Inclusion bodies containing the α-globin fusion protein are not soluble in the urea buffer and first are dissolved in 100 ml of 6 M guanidinium hydrochloride, 50 mM Tris-HCl, pH 8.0, 1 mM EDTA, and 1 mM DTT. After removing insoluble material by centrifugation, the supernatant is dialyzed against three changes of 5 liters of distilled water, and then against 2 liters of urea buffer. Precipitates formed during dialysis against water should be completely dissolved in the urea buffer.

The urea buffer containing either α- or β-fusion protein is centrifuged at 10,000 g for 15 min at room temperature, and the supernatant is applied to a 4 × 10 cm column of CM-Sepharose CL-6B (Pharmacia, Piscataway, NJ) that has been equilibrated with 25 mM Tris–acetate, pH 5.0, without urea. The column is washed with the urea buffer until the optical density of the effluent is lower than 0.2, and a linear gradient formed with 500 ml of the urea buffer and 500 ml of 350 mM NaCl in the urea buffer is applied. The major peak containing the fusion protein is eluted at about 200 mM NaCl (Fig. 4). The peak fractions are applied to a 18% (w/v) SDS–polyacrylamide gel, and fractions containing large amounts of pure fusion protein are pooled and concentrated down to 10 ml using a 350-ml Amicon apparatus with a Diaflo PM10 membrane (Amicon, Danvers, MA). The sample is then applied to a 5 × 50 cm Sephacryl S-200 column equilibrated with 5 M guanidinium hydrochloride, 50 mM Tris-HCl, pH 8.0, 1 mM EDTA, and 1 mM DTT at a flow rate of 100 ml/hr. This step removes some nonprotein impurities that inhibit refolding (Fig. 5). The pooled fractions containing the pure fusion protein are concentrated to approxi-

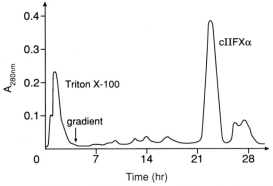

FIG. 4. Elution pattern of *c*IIFXα-globin hybrid protein from the CM-Sepharose column.

mately 2 mg/ml, based on ε_{mM} (280 nm) = 14.5, by ultrafiltration using a PM10 membrane.

Cleavage of Fusion Proteins by Coagulation Factor Xa

The globin fusion proteins tend to precipitate at room temperature, so the following procedures should be carried out on ice. Factor X_a is a stable enzyme but is inactivated by DTT, especially in the presence of even trace amounts of urea or guanidinium hydrochloride. The preparation and activation of factor X_a has been described in detail by Nagai and Thøgersen.[11] Factor X_a is now commercially available, and the enzyme from Boehringer and Stego has worked well (J. Pagnier, unpublished result). Ca^{2+} is essential for the activation of factor X by Russell's viper venom but not for the enzymatic activity toward the globin fusion proteins.

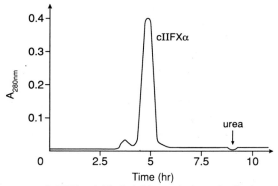

FIG. 5. Elution pattern of *c*IIFXα-globin hybrid protein from the Sephacryl S-200 column.

The pooled fractions from the Sephacryl S-200 column are diluted to 1 mg/ml and first dialyzed against two changes of cold distilled water and then against 8 M urea, 50 mM Tris-HCl, pH 8.0, and 1 mM EDTA. After the globin is completely dissolved in the urea buffer the solution is dialyzed against two changes of cold cleavage buffer (0.5 M urea, 50 mM Tris-HCl, pH 8.0, 1 mM EDTA). The solution should not be stirred or shaken vigorously as this can induce extensive precipitation of the protein. Activated factor X_a is then added to the globin solution at the enzyme–substrate molar ratio of 1 : 200 and 1 : 100 for cIIFXα-globin and cIIFXβ-globin fusion proteins, respectively. The time course of digestion is monitored by analyzing small aliquots on 18% SDS–polyacrylamide gels. Additional enzyme may be added if the cleavage takes place too slowly. A pure preparation of factor X_a should not cleave globins internally, and therefore digestion can be left to proceed longer. When the digestion is at least 80% complete (6 to 12 hr), the solution is then dialyzed against 8 M urea, 50 mM Tris-HCl, pH 8.0, 1 mM DTT, and 1 mM EDTA and concentrated to 5 mg/ml by ultrafiltration using a Diaflo PM10 membrane (Amicon).

Preparation of α and β Chains from Human Blood

The method originally developed by Kilmartin and Rossi-Bernardi[12,13] is used with slight modifications. Red blood cells are washed with cold 0.9% (w/v) sodium chloride saturated with carbon monoxide and recollected by centrifuging at 2500 g for 10 min at 4°. This is repeated three times until the supernatant is no longer turbid. An equal volume of cold distilled water is added to lyse the cells, which are then incubated on ice for 1 hr. Solid NaCl is added to a final concentration of 3% (w/v). The lysed cells are then spun at 10,000 g for 1 hr at 4°. The supernatant containing Hb is carefully poured off, leaving a very soft, jellylike pellet. The Hb solution is respun to remove as much of the cell membrane as possible. The resulting solution of about 6% Hb is bubbled with CO again to protect the protein from oxidation.

DE-52 (1 kg, Whatman, Clifton, NJ) is suspended in water and fines are removed as described by the manufacturer. The suspended resin is adjusted to pH 8.0 with orthophosphoric acid and sodium hydroxide before pouring it into a 8 × 28 cm column. The column is then equilibrated with 10 mM Tris–phosphate, pH 8.0; this is done by washing the column with the buffer until the pH and conductivity of the effluent are the same as those of the affluent. A 4 × 15 cm column of CM-23 (Whatman) is prepared

[12] J. V. Kilmartin, this series, Vol. 76, p. 167.
[13] J. V. Kilmartin and L. Rossi-Bernardi, *Biochem. J.* **124**, 31 (1971).

in the same way and equilibrated with 10 mM sodium phosphate buffer, pH 6.7.

Sodium chloride (2.4 g) is added to 400 ml of 6% carbonmonoxyHb while stirring gently. Then p-hydroxymercuribenzoate (p-HMB; 2 g; Sigma) is dissolved in 100 ml of 2 M glycine–KOH buffer, pH 8.0. KH$_2$PO$_4$ (11 g) is added gradually and the pH is adjusted to 5.8 with 4 M orthophosphoric acid. The solution is then bubbled with CO and left on ice overnight. The p-HMB–Hb solution is spun at 10,000 g for 10 min to remove any precipitate that forms and is passed down a 15 × 60 cm Sephadex G-25 (fine) column equilibrated with 10 mM Tris–phosphate, pH 8.0. The Hb is then applied to the DE-52 column; the β chain remains at the top of the column but the α chain will move down the column. Once all the sample has been loaded onto the column the buffer at the top is bubbled with CO, and the α chain is eluted with the same buffer. After the α chain has been collected, the DE-52 column is washed with 10 liters of CO-saturated 5 mM Tris–phosphate, pH 8.5, containing 50 ml of 2-mercaptoethanol. Meanwhile, 3.5 ml of 2-mercaptoethanol are added to the eluted α chain, which is then bubbled with CO and left on ice for 30 min. The pH of the α-chain solution is lowered to 6.6 with 4 M orthophosphoric acid, and the solution is applied to the CM-23 column. The column is washed with 2 liters of 10 mM sodium phosphate, pH 6.6, containing 6.25 ml of 2-mercaptoethanol, then with 2 liters of the same buffer without 2-mercaptoethanol. The α chain is extremely unstable in 2-mercaptoethanol at low pH so all the buffers must be kept cold and saturated with CO. The α chain is then eluted from the column with 30 mM Tris-HCl, pH 8.5. Once the DE-52 column has been washed with the 2-mercaptoethanol buffer it is washed thoroughly with the same buffer without mercaptoethanol before eluting the β chain with 60 mM potassium phosphate, pH 7.4. The α and β chains are concentrated in Amicon ultrafiltration cells and stored under CO in a conical flask secured with a rubber stopper at 4° for up to 1 year. They can be stored indefinitely in liquid nitrogen.

Reconstitution of Mutant β-Globin into Hemoglobin Tetramers

The recombinant β-globin solution in 8 M urea, 50 mM Tris-HCl, pH 8.0, 1 mM DTT, and 1 mM EDTA is diluted to 5 mg/ml (based on ε_{mM} = 14.5 at 280 nm). A 1.2 M excess of CO-saturated α-globin is diluted into 16 volumes (to the volume of the β-chain solution) of ice-cold 10 mM Tris-HCl, pH 8.0, and 1 mM EDTA, bubbled with CO, and the mutant β-globin is added slowly using a peristaltic pump with gentle stirring. Once all the β-globin has been added, a 1.2 M excess of hemin dicyanide solution is added in the same way. The hemin dicyanide solution is prepared by dissolving 25 mg of bovine hemin (Sigma) in 2.5 ml of 0.1 M KOH, adding 20 ml of water and then 2.5 ml of 1 M KCN. The protein solution is allowed to stand for 15 min and then dialyzed against a large volume of cold 0.1 M potassium phosphate buffer, pH 7.4, 1 mM EDTA, and 50 mM KCN at 4°. After 3 hr the buffer is changed and the Hb is dialyzed overnight. Precipitate formed during dialysis is removed by centrifugation.

Reconstitution of Mutant α Chain into Hemoglobin Tetramer

In this method[5] the cleaved α-globin fusion protein is dialyzed against 8 M urea, 50 mM Tris-HCl, pH 8.0, 1 mM EDTA, and 1 mM DTT and diluted to 5 mg/ml. The globin solution is slowly diluted with gentle stirring into 20 volumes of ice-cold 20 mM potassium phosphate, pH 5.7, 1 mM EDTA, and 1 mM DTT using a peristaltic pump. First, 25 mg of hemin is dissolved in 2.5 ml of 0.1 M KOH and diluted with 20 ml of water and 2.5 ml of 1 M KCN. The hemin solution is then diluted further into 245 ml of water, and finally 5 ml of 1 M potassium phosphate, pH 6.7, is added. A 1.4 M excess hemin solution is added to the diluted α-globin solution, and then 1.2 M excess β chain purified from human blood is added to the refolded α-globin solution. After being bubbled with CO the solution is dialyzed against two changes of 0.1 M potassium phosphate buffer, pH 7.4, 1 mM EDTA, and 0.1 mM KCN.

Reduction and Final Purification of Reconstituted Hbs

The reconstituted Hb[3,5] is concentrated to a final volume of 15–40 ml in a 2.5-liter Amicon concentration cell and transferred to a gas-tight bottle with a rubber septum. The solution is degassed by alternately flushing pure nitrogen and evacuating the gas with a vacuum pump. The protein solution is then saturated with CO; 174 mg of sodium dithionite is placed in a gas-tight tube with a rubber septum, and the air inside is replaced with nitrogen by evacuation and flushing with nitrogen. Deoxygenated water (10 ml) is injected into this test tube to make a 0.1 M solution of dithionite, and a 5 M excess of dithionite over heme is injected into the Hb solution. The color of the protein lightens rapidly as the cyanomet heme of the β subunit is reduced to the ferrous carbonmonoxy form. After 15 min on ice to ensure the reduction has gone to completion, the protein is gel filtered on a 4 × 50 cm Sephadex G-25 (fine) column equilibrated with 10 mM sodium phosphate, pH 6.0. The eluted protein is then applied to a 4 × 10 cm

CM-52 (Whatman) column equilibrated with 10 mM sodium phosphate, pH 6.0, and washed with the same buffer. The protein is eluted with a linear gradient made with 500 ml of 10 mM sodium phosphate, pH 6.0, and 500 ml of 70 mM sodium phosphate, pH 6.9. The reconstituted Hb is eluted first and then any excess α chain is eluted. The fractions containing Hb are pooled and then dialyzed against 10 mM Tris-HCl, pH 8.0. The reconstituted Hb is dialyzed against 10 mM Tris-HCl buffer, pH 8.0, and applied to a 2 × 10 cm DEAE-Sepharose column equilibrated with the same buffer. The reconstituted Hb is eluted with a linear gradient formed with 250 ml of 10 mM Tris-HCl, pH 8.0, and 250 ml of 60 mM NaCl in the same buffer. The main fractions are collected and concentrated by ultrafiltration using a PM10 membrane. Carbon monoxide is removed by the method of Kilmartin and Rossi-Bernardi.[13] The Hb solution of less than 5% concentration is rotated in a small pear-shaped flask with a rotary evaporator, the flask being immersed in ice–water. The sample is illuminated with a lamp under the continuous stream of water-saturated oxygen.

Crystallizations

Using the method of Perutz for crystallization of native deoxy hemoglobin,[14] we have been able to crystallize mutant proteins having one or multiple substitutions in either the α or β subunits.[4,15–18] The hemoglobin solution is first spun down in an Eppendorf centrifuge to remove any particulate matter that may affect crystal growth. The precipitating agent is phosphate-buffered ammonium sulfate, pH 6.5, and ferrous citrate is included as the reducing agent. The salt range used is slightly expanded compared to that of the original recipe, from first surveying crystallization conditions. Also, the crystallization volume is scaled to one-tenth that originally recommended in order to conserve material, and small vials sealed with a rubber stopper are used. Crystallizations are set up in a nitrogen glove box, and the vials are stored in glass jars under nitrogen at room temperature. Crystals grow to optimal size (0.5 × 0.1 × 0.1 mm, typically) in the course of 1–2 weeks.

Crystals are mounted in capillaries under a nitrogen atmosphere for

[14] M. F. Perutz, *J. Cryst. Growth* **2**, 54 (1968).
[15] K. Nagai, B. Luisi, D. Shih, G. Miyazaki, K. Imai, C. Poyart, A. De Young, L. Kwiatkowski, R. W. Noble, S.-H. Lin, and N. T. Yu, *Nature (London)* **329**, 858 (1987).
[16] B. F. Luisi, K. Nagai, and M. F. Perutz, *Acta Haematol.* **78**, 85 (1987).
[17] J. Tame, D. T.-B. Shih, J. Pagnier, G. Fermi, and K. Nagai, *J. Mol. Biol.* **218**, 761 (1991).
[18] T.-H. Jessen, R. E. Weber, G. Fermi, J. Tame, and G. Braunitzer, *Proc. Natl. Acad. Sci. U.S.A.* **88**, 6519 (1991).

data collection. The crystals of mutant Hbs are usually closely isomorphous with native crystals, and belong to the space group $P2_1$ with typical cell dimensions $a = 63$, $b = 83.5$, $c = 53.8$ Å, and $\beta = 99.3°$. One tetramer occupies the asymmetric unit, providing two independent images of the site of mutation. We have also tried growing crystals of mutants in the R state in the presence of CO using the procedure of Perutz.[14] These crystals are more difficult to grow, and we recommend using an expanded range of salt concentration.

Characterization of Reconstituted Hb

Determination of Oxygen Equilibrium Curves

Oxygen equilibrium curves of Hb samples are measured using the automatic method of Imai.[19] The oxygen equilibrium curve of Hb A that we first produced in *E. coli* had a slightly lower Hill coefficient than did human Hb A.[3] Later we introduced a further purification step on a DEAE-Sepharose column, and then the oxygen equilibrium curves of wild-type Hb A generated in *E. coli* were in good agreement with those of human Hb A.[15] The oxygen equilibrium curves are very sensitive to the quality of Hb, and this is the most rigorous test of the authenticity of Hb A produced in *E. coli*. Crystal structures of mutants produced in *E. coli* and refolded *in vitro* are identical to that of human Hb A except in the immediate vicinity of the mutations.[4,14-18] Extensive kinetic ligand binding studies by Olson and colleagues[20-22] have shown that Hb A produced in *E. coli* is identical to human Hb A. Resonance Raman and nuclear magnetic resonance spectra of Hb A produced in *E. coli* have also been reported.[15,23-25] Various aspects of Hb functions have been tested by site-directed mutagenesis.

[19] K. Imai, this series, Vol. 76, p. 438.
[20] J. S. Olson, A. J. Mathews, R. J. Rohlfs, B. A. Springer, K. D. Egeberg, S. G. Sliger, J. Tame, J.-P. Renaud, and K. Nagai, *Nature (London)* **336,** 265 (1988).
[21] A. J. Mathews, R. J. Rohlfs, J. S. Olson, J. Tame, J.-P. Renaud, and K. Nagai, *J. Biol. Chem.* **264,** 16573 (1989).
[22] A. J. Mathews, J. S. Olson, J. Tame, J.-P. Renaud, and K. Nagai, *J. Biol. Chem.* **266,** 21631 (1991).
[23] S.-H. Lin, N. T. Yu, J. Tame, D. Shih, J.-P. Renaud, J. Pagnier, and K. Nagai, *Biochemistry* **29,** 5562 (1990).
[24] K. Ishimori, I. Morishima, K. Imai, K. Fushitani, G. Miyazaki, D. Shih, J. Tame, J. Pagnier, and K. Nagai, *J. Biol. Chem.* **264,** 14624 (1989).
[25] K. Ishimori, K. Imai, G. Miyazaki, T. Kitagawa, Y. Wada, H. Morimoto, and I. Morishima, *Biochemistry* **31,** 3256 (1992).

Role of Distal Residues of Hb

The ability to mutate any residue in the hemoglobin molecule has allowed us to probe the functional roles of the distal histidine and valine that lie in van der Waals contact with bound ligands. From the crystallographic structures of T- and R-state hemoglobin it had been proposed[26] that valine-E11β lowers ligand affinity in the T state but not in the R state. In the 2.1-Å structure of oxyhemoglobin the positions of the distal histidines were seen to be different in the α and β subunits. In the α subunits the imidazole group lies in a suitable position to form a hydrogen bond with the ligand, but no such bond seems to form in the β subunits. The interactions between the distal residues and bound ligands proposed on the basis of the crystallographic structures could be tested only by obtaining suitable mutant hemoglobins in which the distal residues were replaced by other amino acids. Valine-E11 of both subunits was mutated to alanine, leucine, and isoleucine. In the α subunits these mutations had no effect on the oxygen equilibrium curve (Fig. 6a), indicating that the residue plays little steric role in determining oxygen affinity.[17] The equivalent mutation in the β subunit gave very different results. The Val-E11β → Ile mutant (Fig. 6b) shows considerably lower oxygen affinity compared to Hb A at all oxygen pressures, whereas the Val-E11β → Ala mutant shows increased oxygen affinity in the T state. Interestingly, Hb Val-E11β → Leu shows an oxygen equilibrium curve identical to that of Hb A. The crystal structures of the Ile and Leu mutant Hbs showed that this arises because the Leu-E11 can move away from the binding site, while the more sterically restricted Ile side chain cannot (Fig. 7a).

In order to examine the role of the distal histidine, this residue was mutated to glycine. Both the His-E7α → Gly and His-E7β → Gly mutants give large cavities in the oxygen binding pocket (Fig. 7b) and were too unstable for oxygen equilibrium curves to be measured accurately because the heme rapidly oxidized to the ferric state. By measuring the association and dissociation rates of oxygen in stopped-flow experiments, however, we have been able to show that the oxygen affinity of the α chain is lowered significantly by this mutation whereas that of the β chain is unaffected.[20-22] These results agree very well with the functional interpretation of the crystal structures. The functional studies of engineered hemoglobin mutants have highlighted the differences between the α and β subunits particularly in the role of hydrogen bonding to stabilize bound ligand. It is interesting that highly conserved residues such as His-E7 and Val-E11 can have such different properties in two closely related proteins.

[26] M. F. Perutz and G. Fermi, "Hemoglobin and Myoglobin, Atlas of Protein Sequence and Structure," Vol. 2. Oxford Univ. Press, New York, 1981.

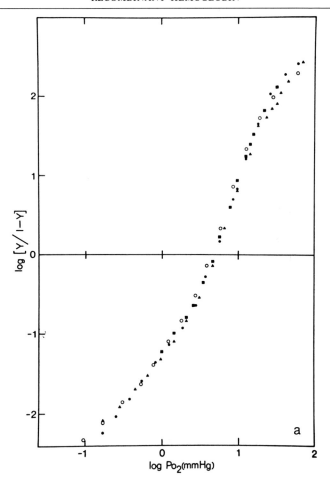

FIG. 6. (a) Oxygen equilibrium curves of the Val-E11α mutants: ■, Val-E11α (Hb A); ●, Ala-E11α; ▲, Ile-E11α; ○, Leu-E11α. (b) Oxygen equilibrium curves of the Val-E11β mutants: ●, Ala-E11β; ▲, Ile-E11β; ○, Leu-E11β; △, Gln-E7β; ×, Hb A control. Reproduced from Nagai *et al.*[15]

Studies on Molecular Evolution of Hemoglobin

All the vertebrate Hbs are tetrameric, consisting of two α and two β subunits, and show cooperative oxygen binding (heme–heme interaction). The amino acid sequences of Hb from over 100 vertebrate species have been determined. Even the most distantly related Hbs show over 40% sequence identity[27]; the crystal structure of antarctic fish Hb confirms that

[27] T. Kleinschmidt and J. G. Sgouros, *Biol. Chem. Hoppe-Seyler* **368,** 579 (1987).

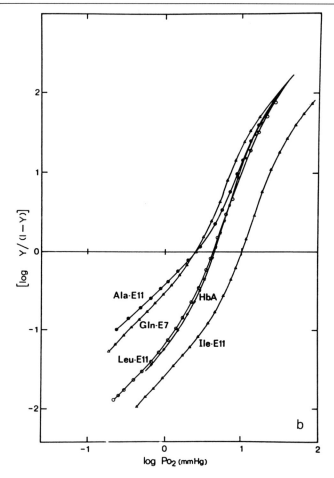

FIG. 6. (*continued*)

the architecture of the Hb molecule is also conserved during vertebrate evolution.[28] In contrast, the heterotropic allosteric effect has diverged, and the oxygen affinity of Hb is modulated by different metabolites in different species. Certain fish Hbs show an enhanced alkaline Bohr effect, referred to as the Root effect; crocodilian Hbs respond to bicarbonate ions instead of DPG or ATP.[29] The molecular mechanisms of these effects have been proposed by examining the effect of amino acid substitutions,

[28] L. Camardella, C. Caruso, R. D'Avino, G. di Prisco, B. Rutigliano, M. Tamburrini, G. Fermi, and M. F. Perutz, *J. Mol. Biol.* **224**, 449 (1992).
[29] M. F. Perutz, *Mol. Biol. Evol.* **1**(1), 1 (1983).

a

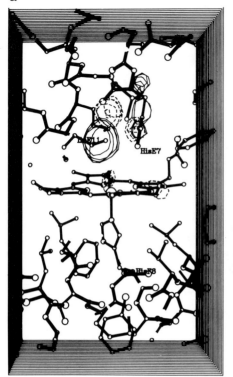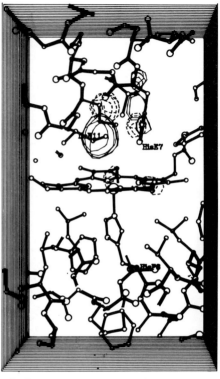

Scale = 12.1318 mm/A   Sections −11 to 12
Fod(E11ILE)−Fon, +/−.09 eA**(−3)

Scale = 12.1318 mm/A   Sections −11 to 12
Fod(E11ILE)−Fon, +/−.09 eA**(−3)

FIG. 7. (a) A difference Fourier map of the Ile-E11β mutant at 2.45 Å resolution. Solid contours, $+0.09$ eÅ^{-3}; broken contours, -0.09 eÅ^{-3}. (b) The $(2F_{obs} - F_{cal})$ map of the Gly-E7β mutant at 2.8 Å. The electron density map contoured at 0.3 Å^{-3}.[15] Reproduced from Nagai et al.[15]

using the atomic model of human Hb.[28] The protein engineering of Hb allows us to introduce multiple mutations into human Hb and to examine their effects experimentally.[5,16]

Molecular Mechanism of Sickle Cell Anemia

Sickle cell anemia is caused by a single amino acid substitution in the β subunit of hemoglobin. This mutation creates a hydrophobic patch on the surface of the molecule, which at low oxygen tension causes the hemoglobin molecules to polymerize into long rigid fibers by hydrophobic interaction. The bundle of hemoglobin fibers within red blood cells distorts the cells into the characteristic sickle form. The roles of the $\beta6$ and $\beta23$

b

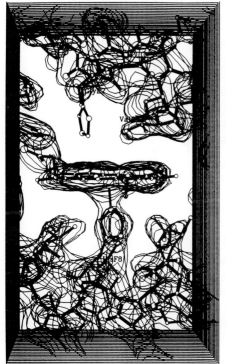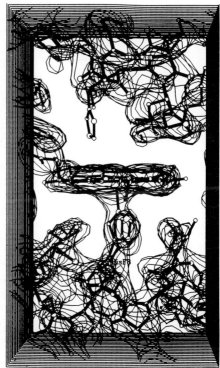

Scale = 12.1318 mm/A   Sections −11 to 12   Scale =12.1318 mm/A   Sections −11 to 12
     2Fod(E7Gly)-Fon, 0.3 eA**(−3)   ·· 2Fod(E7Gly)-Fon, 0.3 eA**(−3)

FIG. 7. (*continued*)

residues in polymerization of Hb molecules have been investigated by protein engineering.[30]

Conclusion

We have shown that engineering of Hb is a powerful method of investigating various functional and structural aspects of Hb. The vectors described in this article are available from the authors.

[30] M. T. Bihoureau, V. Baudin, M. Marden, N. Lacaze, B. Bohn, J. Kister, O. Schaad, A. Dumoulin, S. J. Edelstein, C. Poyart, and J. Pagnier, *Protein Sci.* **1**, 145 (1992).

Acknowledgments

We thank Dr. Max Perutz for suggesting this project and for his continuous encouragement and support. We also thank Colin Young for growing cells and Kiyohiro Imai, John Kilmartin, Peter Esnouf, Steve Anderson, Keith McKenney, Christian Thøgersen, Charlie Craik, and Claude Poyart for their help.

[23] Expression of Recombinant Human Hemoglobin in *Escherichia coli*

By Douglas Looker, Antony J. Mathews, Justin O. Neway, and Gary L. Stetler

Introduction

By utilizing a high-level expression system for production of heterologous proteins in *Escherichia coli,* in combination with controlled fermentation in defined media and efficient, rapid purification, we have developed a system for production and purification of large quantities of recombinant human hemoglobin (rHb) in *E. coli.* Expression is regulated by the strong, inducible *tac* promoter, which is a hybrid promoter composed of the *trp* promoter "−35" and the *lacUV5* promoter "−10" and operator sequences.[1] By constructing the α-globin and β-globin genes using synthetic oligonucleotides we have optimized codon usage for *E. coli*[2] and incorporated a number of convenient restriction enzyme sites for easy manipulation of coding sequences. The genes are transcribed as a polycistronic message. Translational coupler sequences[3] have been included upstream of both the α-globin and β-globin genes to enhance translation initiation. Controlled fermentation in defined medium reduces or eliminates the production of growth inhibitory metabolites often produced in complex media and assures a constant supply of essential nutrients, allowing growth of *E. coli* to high cell densities while maintaining high levels of heterologous protein expression.[4] Crude cell lysate is treated with polyethyleneimine to precipitate nucleic acids and clarified by centrifugation. Recombinant hemoglobin is then extracted and purified in three ion-exchange chromatography steps.

[1] J. Brosius and A. Holy, *Proc. Natl. Acad. Sci. U.S.A.* **81,** 6929 (1984).
[2] M. Gouy and C. Gautier, *Nucleic Acids Res.* **10,** 7055 (1982).
[3] B. E. Schoner, H. M. Hsiung, R. M. Belagaje, N. G. Mayne, and R. G. Schoner, *Proc. Natl. Acad. Sci. U.S.A.* **81,** 5403 (1984).
[4] H. L. MacDonald and J. O. Neway, *Appl. Environ. Microbiol.* **56,** 640 (1990).

Design of Expression System

The expression plasmid pKK223-3 was obtained from Pharmacia LKB (Piscataway, NJ). This plasmid contains the *tac* promoter (Ptac). Expression from this plasmid is inducible by lactose and analogs such as isopropyl β-D-thiogalactoside (IPTG), melibiose, and cellobiose. Immediately downstream of Ptac is a multiple cloning site used for inserting the synthetic α-globin and β-globin genes followed by 5S rRNA coding sequences and the transcriptional terminators T1 and T2.

Each globin gene, in combination with its translational coupler (see below), is constructed from synthetic oligonucleotides. The majority of the codons selected for the synthetic genes are those optimal for *E. coli* expression,[2] except where it is necessary to create convenient restriction endonuclease sites. Both genes were cloned originally into M13 bacteriophage vectors for sequencing. The synthetic α-globin gene was then ligated into pKK223-3, using the *Eco*RI and *Pst*I restriction sites present in the multiple cloning site. The β-globin gene was inserted downstream of the α-globin gene, using the *Pst*I and *Hin*dIII sites present in the multiple cloning site. The expression plasmid constructed in this manner contains a single copy of the α-globin gene, followed by a single copy of the β-globin gene (Fig. 1a).

High-level expression of recombinant proteins requires that both transcriptional and translational rates be maximized. Therefore, in combination with the strong promoter, Ptac, we incorporated translational coupler sequences[3] 5' of the α- and β-globin coding sequences (Fig. 1b).

A more detailed description of the design and construction of the rHb expression plasmid, pDLIII-13e, has been published.[5] The plasmid is available from the authors on written request.

The *E. coli* strain JM109[6] *recA1 endA1 gyrA96 thi hsdR17 supE44 relA1* Δ(*lac-proAB*) {*F' traD36 proAB lacI^qZDM15*} was chosen as the expression host.

Fermentation Protocol (2 liter)

Stock cultures of JM109 cells containing plasmid pDLIII-13e are grown to OD_{600} = 1–2 in LB or 2× YT medium containing 100 μg/ml ampicillin. The culture is aliquoted, dimethyl sulfoxide is added to a final concentration of 7% (v/v), and the aliquots are stored at $-80°C$. To start inocula for 2-liter fermentations, a scraping is taken from the top of a frozen

[5] S. J. Hoffman, D. L. Looker, J. M. Roehrich, P. E. Cozart, S. L. Durfee, J. L. Tedesco, and G. L. Stetler, *Proc. Natl. Acad. Sci. U.S.A.* **87**, 8521 (1990).
[6] C. Yanisch-Perron, J. Vieira, and J. Messing, *Gene* **33**, 103 (1985).

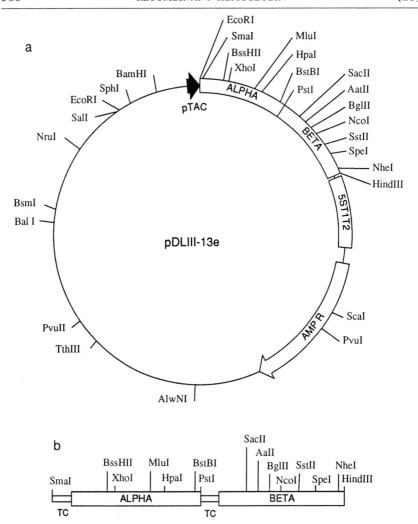

FIG. 1. (a) Map of recombinant human hemoglobin expression plasmid, pDLIII-13e. Unique restriction endonuclease sites are shown. pTAC, *tac* promoter; ALPHA, α-globin gene; BETA, β-globin gene; 5S, 5S ribosomal RNA gene; T1T2, transcriptional terminators; AMP R, β-lactamase gene. (b) Detailed map of α-globin and β-globin coding sequences from pDLIII-13e, showing positions of translational couplers (TC) and unique restriction endonuclease sites.

TABLE I
COMPOSITION OF DM-1 AND DM-4 MEDIA

Medium/component	Amount (per liter)	Procedure
DM-1 inoculum medium		
KH_2PO_4	4.1 g	Autoclave then add the following sterile
K_2HPO_4	7.0 g	components:
$(NH_4)_2SO_4$	2.0 g	20 ml of 10% yeast extract
Trisodium citrate dihydrate	1.0 g	Ampicillin to a final concentration of
Trace metals (see below)	3.0 ml	100 mg/ml
$MgSO_4 \cdot 7H_2O$	0.154 g	Thiamin to a final concentration of
Proline	0.230 g	30 mg/ml
		Glucose to a final concentration of 1%
DM-4 fermentation medium		
KH_2PO_4	2.0 g	Autoclave then add the following sterile
K_2HPO_4	3.6 g	components:
$(NH_4)_2SO_4$	2.0 g	Thiamin to a final concentration of
Trisodium citrate dihydrate	1.5 g	50 mg/ml
Trace metals (see below)	3.0 ml	Ampicillin to a final concentration of
$MgSO_4 \cdot 7H_2O$	0.154 g	100 mg/ml
Proline	0.230 g	Glucose to a final concentration of 1%
Trace metal solution		
$FeCl_3 \cdot H_2O$	27.0 g	
$ZnCl_2$	1.3 g	
$CoCl_2 \cdot 6H_2O$	2.0 g	
$Na_2MoO_4 \cdot 2H_2O$	2.0 g	
$CaCl_2 \cdot 2H_2O$	1.0 g	
$CuSO_4 \cdot 5H_2O$	1.27 g	
H_3BO_3	0.5 g	
HCl	100.0 ml	

aliquot and inoculated into 200 ml of DM-1 medium (see Table I). The culture is incubated with shaking at 37° for 10–12 hr or until the OD_{600} = 1.0–1.5, and is then used to inoculate the 2-liter fermentor. A 2-liter Bioflo III fermentor (New Brunswick, Edison, NJ), containing 1800 ml of DM-4 medium (see Table I) with 1% (w/v) glucose and 100 μg/ml ampicillin is equilibrated to 30° and a dissolved oxygen reading of 100% air saturation at 400 rpm and is then inoculated. The culture is maintained at pH 6.8 with a base feed of 15% (v/v) NH_4OH. A dissolved oxygen minimum of 20% (air saturation) is maintained by increasing agitation as needed. Glucose concentration is monitored using Chemstrip uG urine test strips (Boehringer Mannheim, Indianapolis, IN). When the glucose concentration decreases to less than 0.25%, 40% (w/v) glucose is added to a final concentration of 1% (w/v). At OD_{600} = 20, rHb production is induced by

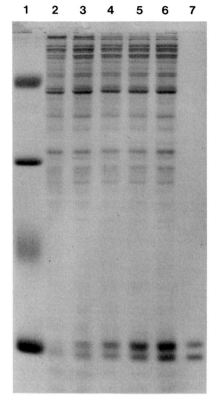

Fig. 2. SDS–PAGE analysis of intracellular accumulation of α-globin and β-globin proteins during an 8-hr induction. Lane 1, molecular weight markers; lane 2, preinduction; lanes 3–6, 2, 4, 6, and 8 hr postinduction; lane 7, human hemoglobin A_0 standard.

addition of IPTG to a final concentration of 30 μM; 2 ml of 25% (v/v) polypropylene glycol (PPG) is added as an antifoaming agent, followed by 10 ml of freshly prepared hemin solution (10 mg/ml in 15% NH_4OH). These quantities of hemin and PPG also are added at 3 and 6 hr postinduction. Induction is continued for 8 hr. At 2-hr intervals during the induction, 1-ml aliquots are removed from the fermentor, microcentrifuged for 2 min, and the cell pellets are stored at $-80°$ for later quantitation of rHb in cell lysates. Cells are harvested by centrifugation at 5000 g for 20 min at $4°$ and stored at $-80°$. Figure 2 shows an SDS–PAGE analysis of the intracellular accumulation of α-globin and β-globin proteins during an 8-hr induction. Each lane was loaded with 50 μg of protein as calculated from the cell density. Accumulation of α-globin and β-globin increased through-

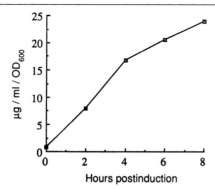

FIG. 3. Recombinant hemoglobin production by *E. coli* strain JM109 during an 8-hr induction of a 2-liter fermentation. Results are expressed as μg rHB/ml/OD_{600}.

out 8 hr of induction (see also Fig. 3). Densitometric analyses indicate that α- and β-globin accumulate to 5–10% of the total cell protein 8 hr postinduction.

Quantitation of rHb in Cell Lysate

Soluble rHb in crude or clarified cell lysate is readily quantitated using standard difference spectroscopy techniques, in either the Soret or visible region of the spectrum. Reduced hemoglobin is determined from ($HbCO-HbO_2$) difference spectra, whereas methemoglobin (metHb) may be determined from (cyanometHb–metHb) difference spectra at pH 8.0. Standard curves and difference extinction coefficients are determined easily using purified hemoglobin standards. The total soluble hemoglobin concentration is then determined from the sum of the reduced and oxidized hemoglobin values.

Purification of Recombinant Hemoglobin from Bacterial Cell Lysate

In general, the purification was carried out at low temperatures and near pH 8 in order to minimize autoxidation.[7] The example given is for the purification of the low-oxygen-affinity mutant rHb Yoshizuka ($\beta108$ Asn → Asp).[8]

Preparation of Cell Lysate

Partially thawed cells (to aid cell breakage, cells should be frozen before use) are resuspended by low-speed blending in 3 ml/g cells of 40

[7] A. Riggs, this series, Vol. 76, p. 5.
[8] T. Imamura, S. Fujita, Y. Ohta, M. Hanada, and T. Yanase, *J. Clin. Invest.* **48,** 2341 (1969).

mM Tris base/1 mM benzamidine hydrochloride (Sigma Chem. Co., St. Louis, MO). The mixture is maintained at 4–10° throughout. Lysozyme (10 mg/ml stock solution in 40 mM Tris-Cl buffer, pH 8.0, prepared within 1 hr of use and stored on ice) is added to give 1 mg lysozyme/g cells. Typically the 2-liter fermentor yields approximately 150–200 g of cells. The mixture is stirred gently to mix (to avoid denaturation of the enzyme) and incubated at 4–10° for 15 min. The mixture becomes extremely viscous during this time. $MgCl_2$, $MnCl_2$, and deoxyribonuclease I (DNase I, Sigma Chem. Co., prepared as 10 mg/ml stock solution) are added to final concentrations of 10 mM, 1 mM, and 30 μg/ml, respectively. These components are mixed gently and incubated at 4–10° with occasional stirring for 30 min. During this time the lysate viscosity decreases dramatically. The lysate is blended briefly and filtered through cheesecloth prior to passage through a Microfluidizer cell disruption device (Microfluidics Corporation, Newton, MA) in accordance with the manufacturer's instructions. Alternatively, the lysate may be blended for several minutes at high speed to achieve thorough cell disruption. The following procedures are performed at 4° or on ice.

Lysate Clarification

The lysate is stirred (magnetic stirrer) and adjusted to pH 8 with 1 M Tris base. Polyethyleneimine [PEI; 10% (v/v) aqueous stock, adjusted to pH 8 with 6 M HCl] is added with stirring to a final concentration of 0.5% (v/v) and incubated at 4–10° for 10 min to precipitate nucleic acids.[9] The PEI-treated crude lysate is centrifuged at 23,000 g for 30 min at 4°. The supernatant is recovered and adjusted to pH 7.4 with 0.5 M HCl or 1 M Tris base as required. The conductivity is reduced until equal to that of the Q1 column equilibration buffer. This is achieved by diafiltration with oxygen-saturated, cold deionized water in a Minitan tangential flow ultrafiltration unit (Millipore, Bedford, MA), using eight stacked membranes (nominal molecular weight cutoff, 10,000). Some turbidity may develop during this process, but it does not interfere with the chromatography and may be removed by further centrifugation if desired. The extract is concentrated using the Minitan apparatus until the volume is equal to 2–3 column bed volumes.

Chromatographic Purification Steps

Three chromatographic steps are employed. The first column captures a large amount of the unwanted bacterial proteins while the hemoglobin

---

[9] R. K. Scopes, "Protein Purification: Principles and Practice," 2nd ed. Springer-Verlag, New York, 1987.

passes through. Although this column can be run at higher pH to capture the rHb, nucleic acids and most bacterial proteins are considerably more acidic than rHb and tend to displace it off the resin. The second and third columns further remove bacterial proteins and resolve oxidized and reduced hemoglobin species. All chromatography was carried out at 4°.

Column Q1

Fast Flow Q-Sepharose (Pharmacia LKB), 5-cm diameter × 14 cm (250-ml bed volume), is equilibrated with 20 mM Tris-HCl/0.1 mM triethyl-enetetraamine hydrochloride (TETA; Aldrich Chem. Co., Milwaukee, WI), pH 7.4, at 4°. The chelator (TETA) is included to scavenge trace copper.[10] The column is loaded at 20 ml/min and the hemoglobin-containing eluate is collected. After loading, the column is washed with 100 ml of equilibration buffer and the eluate is collected and combined with the earlier portion. After adjusting the eluate to pH 8.3 with 1 M Tris base, the conductivity is lowered (Minitan diafiltration as above) until equal to that of the Q2 column equilibration buffer at 4°.

Column Q2

Fast Flow Q-Sepharose anion-exchange resin (Pharmacia LKB), 5-cm diameter × 14 cm (250-ml bed volume), is equilibrated with 20 mM Tris-Cl/1 mM TETA, pH 8.3, at 4°. The column is loaded at a flow rate of 20 ml/min and then washed with 1 bed volume of equilibration buffer. The hemoglobin is eluted at 12 ml/min with a linear gradient (total volume 1 liter) from 0 to 160 mM NaCl in equilibration buffer. Hemoglobin-containing fractions are pooled on the basis of metHb content, concentrated by ultrafiltration in the Minitan apparatus, and adjusted to pH 6.8 with 0.5 M monobasic sodium phosphate. The hemoglobin is then diluted with cold deionized water to yield a conductivity equal to that of the S1 column equilibration buffer. If the total volume exceeds 2 S1 column bed volumes it should be concentrated further before loading.

Column S1

Fast Flow S-Sepharose cation-exchange resin (LKB Pharmacia), 5-cm diameter × 14 cm (250-mL bed volume), is equilibrated with 10 mM sodium phosphate, pH 6.8, at 4°. The column is loaded at 15 ml/min and then washed with 1 bed volume of equilibration buffer. A linear gradient

---

[10] J. M. Rifkind, *Biochemistry* **13**, 2475 (1974).

TABLE II
PURIFICATION AND RECOVERY OF rHb FROM *Escherichia coli* LYSATE

Purification step	Total protein[a] (mg)	Total rHb[b] (mg)	Purification (-fold)	Recovery (%)
Crude lysate[c]	46,483	917	—	100
Clarified lysate[d]	9541	761	4	83
Q1 Flow Through	2228	675	15	74
Q2 peak[e]	720	500	34.5	55
S1 peak	446	459[f]	50	50

[a] Determined by BCA assay (Pierce Chemical Co., Rockford, IL), using hemoglobin as a standard.
[b] Determined by difference spectroscopy.
[c] Measured after lysis step.
[d] Supernatant from centrifugation step.
[e] All peaks were pooled on the basis of metHb content.
[f] Final product contained 1% metHb.

of equilibration buffer versus 20 mM sodium phosphate, pH 8.3, is used to elute the hemoglobin (flow rate 10 ml/min, total volume 800 ml). Fractions again are pooled on the basis of metHb content, and the hemoglobin is concentrated by ultrafiltration to 50 mg/ml and stored in liquid nitrogen. Recovery of hemoglobin and stepwise yields are shown in Table II.

Functional Properties

The functional properties of the purified hemoglobin can be evaluated by measuring the equilibrium oxygen binding properties using a HEMOX analyzer (TCS Medical Products, Southampton, PA) as described by Hoffman *et al.*[5] Figure 4 shows representative equilibrium oxygen binding curves for pure rHb0.0, a recombinant version of adult human hemoglobin bearing $\alpha 1, \beta 1$ Val $\rightarrow$ Met mutations, purified adult human Hb, and purified rHb Yoshizuka. The P_{50} and n_{max} values calculated from these data are shown in Table III. The differences in functional properties are due to the altered N termini and the β-subunit mutation, $\beta 108$ Asn $\rightarrow$ Asp of the recombinant protein. In this case, over the range of partial pressures of oxygen shown in Fig. 4, the mutant hemoglobin does not reach 100% saturation. It does reach saturation at higher oxygen partial pressures.

Conclusions

We have designed an expression system that, in combination with high cell density fermentation and efficient purification methods, is capable of

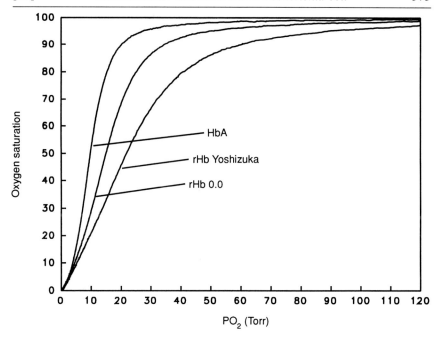

FIG. 4. Representative oxygen affinity curves obtained from the Hemox analyzer expressed as percent oxygen saturation versus oxygen partial pressure in Torr. Conditions: all measurements were carried out at 37° and pH 7.4 in 50 mM HEPES buffer containing 100 mM NaCl.

producing large quantities of soluble rHb. By 8 hr postinduction, rHb accumulates to approximately 25 μg/ml/OD$_{600}$. A final purification recovery of 50% will yield 12–13 mg/liter/OD$_{600}$ (Fig. 3) of purified rHb. Our expression plasmid utilizes synthetic α-globin and β-globin genes with

TABLE III
OXYGEN AFFINITY (P_{50}) AND COOPERATIVITY (n_{max})
OF HEMOGLOBINS[a]

Hemoglobin	P_{50} (Torr)	n_{max}
rHb0.0	14.9	2.6
Adult human HbA	9.7	3.0
rHb Yoshizuka	22.6	2.3
β108 Asn → Asp		

[a] All measurements were carried out at 37° and pH 7.4 in 50 mM HEPES buffer containing 100 mM sodium chloride.

usefully placed restriction enzyme sites. This facilitates cassette insertion of specific mutations into the globin genes and will allow analyses of structure–function relationships in hemoglobin. A comparison of the amount of total globins present in the *E. coli* lysate at harvest (approximately 1–5% of the total cell protein; Fig. 2) with the amount of hemoglobin present in the crude lysate (20–25 μg/ml/OD$_{600}$; Fig. 3) indicates that most of the globin synthesized is assembled into soluble tetrameric hemoglobin.

This system has been used to design and express an engineered form of human hemoglobin that has some of the properties desired in a hemoglobin-based blood substitute.[11]

Acknowledgments

We thank Camille Moore-Einsel and Erin Milne for their technical assistance.

[11] D. L. Looker, D. Abbott-Brown, P. Cozart, S. Durfee, S. Hoffman, A. J. Mathews, J. Miller-Roehrich, S. Shoemaker, S. Trimble, G. Fermi, N. H. Komiyama, K. Nagai, and G. L. Stetler, *Nature (London)* **35**, 258 (1992).

[24] Production of Recombinant Human Hemoglobin A in *Saccharomyces cerevisiae*

By JILL E. OGDEN, ROY HARRIS, and MICHAEL T. WILSON

Introduction

The ability to produce human hemoglobin (Hb) in a genetically engineered host microorganism offers a number of valuable opportunities. First, a recombinant microorganism provides an attractive alternative to outdated stocks of red blood cells as a source of Hb for formulation into a Hb-based red cell substitute. Second, specific mutant Hbs can be synthesized and produced in the host for subsequent detailed structural and functional studies.

Both *Saccharomyces cerevisiae*[1–3] and *Escherichia coli*[4–8] have been

[1] J. E. Ogden, J. R. Woodrow, K. A. Perks, R. Harris, D. Coghlan, and M. T. Wilson, *Biomater., Artif. Cells, Immob. Biotechnol.* **19**, 457 (1991).
[2] M. Wagenbach, K. O'Rourke, L. Vitez, A. Wieczorek, S. Hoffman, S. Durfee, J. Tedesco, and G. Stetler, *Bio/Technology* **9**, 57 (1991).
[3] D. Coghlan, G. Jones, K. A. Denton, M. T. Wilson, B. Chan, R. Harris, J. R. Woodrow, and J. E. Ogden, *Eur. J. Biochem.* **207**, 931 (1992).
[4] K. Nagai and H. C. Thørgersen, *Nature (London)* **309**, 810 (1984).

employed for the production of recombinant human Hb. Nagai and co-workers were the first to describe the synthesis of Hb using *E. coli* as host.[4-6] Their approach required the separate expression of α- and β-globins as insoluble, intracellular fusion proteins, which after solubilization and specific proteolytic cleavage were mixed *in vitro* with exogenous heme to produce fully functional Hb.

A more straightforward method for producing recombinant Hb (rHb) in *S. cerevisiae*[1-3] and *E. coli*[7] has been described that avoids the refolding and reconstitution steps. This entails coexpression of α- and β-globins within the same cell. The α- and β-globin chains fold *in vivo*, complete with endogenous heme, to produce soluble $\alpha_2\beta_2$ tetramers. Recombinant Hb derived from *E. coli* using this method, however, retains the translation-initiating methionine residues at the N termini of the α- and β-globin chains, which affects the functional properties of the molecule.[7] In contrast, recombinant Hb produced in *S. cerevisiae* has correctly processed globin-chain N termini[2,3] and physical and functional characterization demonstrates that the recombinant protein is identical to hemoglobin A (Hb A) purified from erythrocytes.[3]

In this chapter we describe methods for expression of recombinant human Hb A (rHb A) in *S. cerevisiae,* the purification of the protein, and techniques used for subsequent structural and functional characterization.

Coexpression of α- and β-Globins

Coexpression of α- and β-globins to produce rHb A can be achieved either by using a single plasmid carrying both α- and β-globin expression cassettes, or by cotransformation with two plasmids, each carrying α- and β-globin expression cassettes and complementary auxotrophic markers. Here we describe examples of both of these approaches.

Plasmids

Standard recombinant DNA techniques were used for plasmid construction.[9] The α- and β-globin cDNAs were modified for insertion into yeast expression vectors using site-directed *in vitro* mutagenesis (Amer-

[5] K. Nagai, M. F. Perutz, and C. Poyart, *Proc. Natl. Acad. Sci. U.S.A.* **82,** 7252 (1985).
[6] S. J. Hoffman and K. Nagai, U.S. Pat. 5,828,588 (1991).
[7] S. J. Hoffman, D. L. Looker, J. M. Roehrick, P. E. Cozart, S. L. Durfee, J. L. Tedesco, and G. L. Stetler, *Proc. Natl. Acad. Sci. U.S.A.* **87,** 8521 (1990).
[8] D. Looker, A. J. Mathews, J. O. Neway, and G. L. Stetler, this volume [23].
[9] J. Sambrook, E. F. Fritsch, and T. Maniatis, "Molecular Cloning: A Laboratory Manual." Cold Spring Harbor Lab., Cold Spring Harbor, NY, 1989.

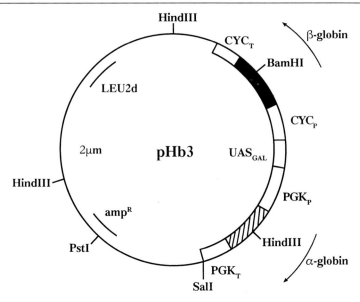

FIG. 1. Structure of plasmid pHb3. The promoter and transcription terminator sequences of *CYC1* (CYC_P, CYC_T) and *PGK* (PGK_P, PGT_T), the UAS_GAL, and the orientation of the α- and β-globin cDNAs are shown. The *E. coli* ampicillin resistance gene, *amp*R, and *S. cerevisiae LEU2d* and 2μm plasmid functions are also indicated.

sham International, UK) and synthetic oligonucleotides (Applied Biosystems, Foster City, CA; 380B DNA synthesizer).[10,11]

Coexpression Plasmid pHb3. The coexpression plasmid, pHb3 (Fig. 1), was constructed by inserting a *Bam*HI α-globin cDNA fragment and a *Bgl*II β-globin cDNA fragment into a pJDB207-based,[12] galactose-inducible bidirectional expression vector pEK30.[13] Expression of α- and β-globin in pHb3 is directed by *PGK*[14] and *CYC1*[15] promoter fragments, respectively, under the control of the galactose-inducible upstream activation sequence of the *GAL1-10* genes (UAS_GAL).[16] This allows regulation of globin expression, so that expression is repressed in the presence of

[10] J. E. Ogden and C. Homer, unpublished results.
[11] B. Chan, unpublished results.
[12] J. D. Beggs, *Alfred Benzon Symp.* **16,** 383 (1981).
[13] E. Kenny and E. Hinchliffe, Eur. Pat. Appl. EP 317,254.
[14] M. J. Dobson, M. F. Tuite, N. Roberts, A. J. Kingsman, and S. M. Kingsman, *Nucleic Acids Res.* **10,** 2625 (1982).
[15] L. Guarente and T. Mason, *Cell (Cambridge, Mass.)* **32,** 1279 (1983).
[16] M. Johnston and R. W. Davis, *Mol. Cell. Biol.* **4,** 1440 (1984).

glucose and induced when the cells are grown on galactose. Control plasmids containing only α-globin or β-globin cDNAs were also constructed. *Plasmids for Cotransformation.* pGLUαPRB and pGTβPRB[3] (Fig. 2) are pJDB207-based plasmids that contain the *LEU2d*[12] and *TRP1*[17] genes, respectively, as yeast-selectable markers. In these plasmids, appropriately modified α- and β-globin cDNAs are inserted downstream of the *PRB1* promoter[18] at a *Hin*dIII site. Similar vectors have been described previously for the expression of other heterologous proteins.[18] Cotransformation of pGLUαPRB and pGTβPRB into leucine- and tryptophan-requiring yeast strains and subsequent growth on selective media allow for maintenance of both plasmids. pGLUαPRB also contains the *URA3* gene[19] as an additional selectable marker. In pGTβPRB the *ADH1* terminator is replaced by the *PGK* terminator.

Media

Yeast minimal medium (MM) is used: 0.67% (w/v) yeast nitrogen base without amino acids (Difco, Detroit, MI), containing 2% (w/v) glucose or 2% (w/v) galactose as carbon source. Growth supplements (final concentration; uracil, 20 μg/ml; adenine 10 μg/ml; tryptophan, 20 μg/ml) are added when necessary.

Yeast Strains and Transformation

We have used DBY745 (α, *ade1, leu2, ura3*) and a protease-deficient strain, DM477 (α, *pra1, prb1, leu2, trp1, ura3*), for the production of recombinant Hb A using pHb3. DM477 is also used for cotransformation with pGLUαPRB and pGTβPRB.

Plasmids are transformed into these strains using the standard spheroplast procedure of Hinnen *et al.*,[20] selecting for complementation of the relevant auxotrophic mutation, *leu2* for pHb3 and *leu2* and *trp1* for pGLUαPRB and pGTβPRB cotransformants.

Small-Scale Analysis of Transformants

To check for expression of α- and β-globins, several transformants are grown on a small scale, and protein extracts are prepared and analyzed

[17] K. Struhl, D. T. Stinchcomb, S. Scherer, and R. W. Davis, *Gene* **8**, 121 (1979).
[18] D. Sleep, G. P. Belfield, D. J. Ballance, J. Steven, S. Jones, L. R. Evans, P. D. Moir, and A. R. Goodey, *Bio/Technology* **9**, 183 (1991).
[19] D. Botstein, S. C. Falco, S. E. Stewart, M. Breenan, S. Scherer, D. T. Stinchcomb, K. Struhl, and R. W. Davis, *Gene* **8**, 17 (1979).
[20] A. Hinnen, J. B. Hicks, and G. R. Fink, *Proc. Natl. Acad. Sci. U.S.A.* **75**, 1929 (1978).

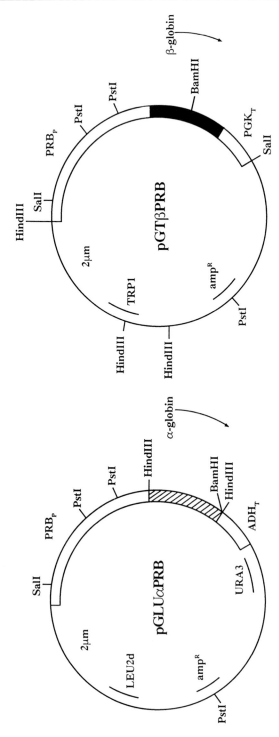

FIG. 2. Structure of plasmids pGLUαPRB and pGTβPRB for cotransformation of yeast. Expression of α- and β-globins is directed by the *PRB1* promoter (PRB$_P$). ADH$_T$ and PGK$_T$ are the transcription terminator sequences of the *ADH1* and *PGK* genes, respectively. The yeast selectable markers are *LEU2d* and *URA3* on pGLUαPRB and *TRP1* on pGTβPRB. Also indicated are the *E. coli* ampicillin resistance gene, *amp*^{R}, and yeast 2μm sequences.

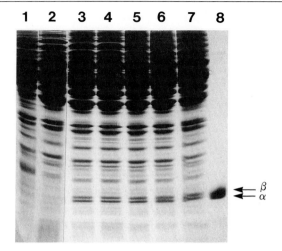

FIG. 3. SDS–PAGE analysis of total soluble proteins from rHb A-producing yeast. Samples (100 μg) of total soluble protein fractions from untransformed DM477 (negative control, lanes 1 and 2) and various rHb A-producing DM477 transformants (lanes 3–7) are resolved on a 15% SDS–PAGE gel with standard Hb A (2 μg) (lane 8) as control. Proteins are visualized by staining with Coomassie blue.

on SDS–PAGE gels. The α- and β-globins are detected either by Coomassie blue staining of the gels (Fig. 3) or by immunoblotting using a rabbit antihuman Hb antibody as probe.

Methods. Transformants are grown overnight at 30° without shaking in 3 ml of MM containing glucose. Appropriate supplements are added throughout. For galactose induction, overnight cultures are diluted 100-fold into 100 ml of MM containing glucose and grown at 30° with shaking until $A_{600\ nm} = 1$ (approximately 1×10^7 cells/ml). The cells are spun down and washed in a small volume of MM containing galactose, resuspended in 100 ml of the same medium, and grown for a further 48 hr. Overnight cultures of DM477 cotransformants are inoculated into 100 ml of MM containing glucose and grown for 72 hr at 30° with shaking.

The cells are harvested in 50-ml aliquots, centrifuged, and washed with 10 ml of 150 mM HEPES, pH 7.0. The cell pellets are finally resuspended in 1.5 ml of the same buffer, and glass beads (0.4 mm diameter, BDH) are added up to the meniscus. The cells are lysed by vortexing for 4×30 sec, with intermittent cooling on ice. A further 0.5 ml of 150 mM HEPES, pH 7.0, is added and the suspension is vortexed for 30 sec. After allowing the glass beads to settle, the cell extract is transferred to a clean Eppendorf tube and centrifuged for 5 min. This separates the soluble protein fractions (supernatant) from the insoluble fractions (pellet). Samples of the soluble protein fractions (50–100 μg total protein) and standard Hb A (200 ng–2 μg, Sigma) in sample buffer [10 mM Tris-HCl, pH 8.0, 2% (w/v) SDS, 5% (w/v) 2-mercaptoethanol, 1 mM EDTA, 0.01% bromphenol blue] are boiled immediately in a water bath and then resolved on 15% reducing SDS–PAGE gels. We use gels of approximately 15 × 18 cm to ensure good separation of the globin chains. Proteins can be detected by staining the gels with Coomassie blue (Fig. 3). Alternatively, proteins can be electroblotted onto Immobilon P membrane (Millipore, Bedford, MA). Nonspecific binding sites on the membrane are blocked by washing in TBS (20 mM Tris-HCl, pH 7.6, 0.15 M NaCl) containing 0.05% (v/v) Nonidet P-40 and 5% (w/v) milk powder for 45 min. The membrane is incubated with rabbit antihuman Hb antibody (1 : 1000 dilution; Sigma) in TBS containing 0.05% Nonidet P-40 and 0.2% milk powder for 1 hr, washed for 3 × 5 min in the same buffer, and then incubated with peroxidase-conjugated goat antirabbit antibody (1 : 1000 dilution; Sigma) for 1 hr. After washing for 3 × 5 min, immunoreactive proteins are visualized using the ECL detection system (Amersham), according to the manufacturer's instructions.

Assays of Hemoglobin in Cell Extracts

There can be considerable variation in levels of rHb A expression between transformants, probably due to differences in plasmid copy number. It is advisable, therefore, to screen several transformants before choosing one for large-scale growth and subsequent purification of rHb A. Analysis of transformants by SDS–PAGE gels gives some indication of differences in expression levels. We also use a spectrophotometric assay in which rHb A in cell extracts is quantitated from the height of the Soret peak in second-derivative spectra, by comparison with standard Hb A of known concentration.[21] It is important to use samples that have been deoxygenated with sodium dithionite to distinguish rHb A in cell extracts (λ_{max} 430 nm) from the indigenous yeast cytochrome c (λ_{max} 415 nm). Because cell extracts can be quite turbid, we find that clarification using polyethyleneimine (PEI) prior to measuring spectra gives better results.

Methods

PEI [20 μl of 5% (v/v) solution in 150 mM HEPES, pH 7.0] is added to 1 ml of cell extract in an Eppendorf tube and the mixture is centrifuged

---

[21] H. van Urk and D. Coghlan, unpublished results.

for 30 sec. The clarified supernatant is removed and 12.5 μl of sodium dithionite [10% (w/v) solution in 150 mM HEPES, pH 7.0] is added. The absorbance spectrum (290–610 nm) of the sample is measured in the second-derivative mode. Dilutions of standard Hb A, treated in the same way, are used to prepare a standard curve. The rHb A content of cell extracts is calculated as a percentage of total soluble protein, determined using Pierce (Rockford, IL) Coomassie protein assay reagent.

Purification of rHb A

We use cation-exchange chromatography for purifying rHb A from yeast cell lysates. To prevent the formation of metHb during purification, all chromatography buffers are purged thoroughly with helium to displace dissolved oxygen. In addition CO is bubbled through the lysate for 3–5 min immediately after cell breakage. The rHb A fractions are maintained in the carbonmonoxy form by repurging with CO at each stage of purification. CarbonmonoxyHb can be converted to the oxy form by flushing the solution on ice with oxygen under bright illumination.

Large-Scale Growth of Yeasts and Preparation of Cell Lysates

Yeast transformants can be grown on a large scale in shake flasks (12 × 2-liter flasks each containing 1 liter of medium). We also grow yeasts by fed batch fermentation using established protocols.[22] For purification of rHb A from DBY745 transformants it is important to incorporate protease inhibitors [2 mM phenylmethylsulfonyl fluoride (PMSF), 2 mM EDTA] in the cell lysis buffer to prevent protease damage of rHb A. This is not necessary when using DM477.

Methods. A suitable transformant is grown overnight in 200 ml of MM containing 2% (w/v) glucose at 30° with shaking. Growth supplements are added throughout as necessary. The overnight culture is diluted 1 : 100 into 12 × 1-liter batches of MM containing 2% (w/v) galactose (for pHb3 transformants) or MM containing 2% (w/v) glucose (for DM477 cotransformants). Cultures are grown at 30° with shaking until in the stationary phase of growth (usually 72 hr).

Yeast cells are harvested by centrifugation at 7000 g for 20 min at 4°. The cells are either used immediately or can be frozen at −70° in thin layers in plastic bags for storage at −20°. The yeast cells are resuspended in 10 mM phosphate buffer, pH 6.5 (containing 2 mM PMSF, 2 mM EDTA if necessary), at a concentration of approximately 60% wet weight/volume

[22] S. H. Collins, *in* "Protein Production by Biotechnology" (T. J. R. Harris, ed.), p. 61. Elsevier, New York, 1990.

(100 g of cells made up to ≈160 ml with buffer). The cells are broken in a Bead Beater (Biospec Products, Bartlesville, OK) using 0.4-mm-diameter glass beads and ice cooling for four 30-sec periods, each separated by a 4-min interval.

The glass beads are removed by vacuum filtration through a glass scinter funnel, and the filtrate is clarified by centrifugation at 30,000 g for 30 min at 4°. The supernatant is removed, taking care not to disturb the surface lipid layer, and is immediately purged with CO for 5 min. The extract is kept on ice throughout. The CO-purged lysate is desalted using a Sephadex G-25 (Pharmacia, Piscataway, NJ) column (5 × 30 cm; ≈600 ml) equilibrated at 4° with 10 mM phosphate buffer, pH 6.5. The column eluate is monitored at $A_{280\ nm}$ and the main protein fraction is collected. The pH of the desalted eluate is adjusted if necessary to pH 6.5 with glacial acetic acid. The conductivity of the eluate should be reduced, if necessary, to less than 2 millisieverts by adding Milli Q water. The eluate is repurged briefly with CO before the next step in purification.

Ion-Exchange Chromatography

Desalted eluates are subjected to three successive ion-exchange chromatography steps[23]: S-Sepharose Fast Flow, CM-Sepharose Fast Flow, and Q Sepharose Fast Flow.

Methods. The desalted extract is applied to a 50-ml (3.5 × 5 cm) S-Sepharose Fast Flow column (Pharmacia) equilibrated at 4° with 10 mM phosphate buffer, pH 6.5, at a flow rate of 2.5 ml/min. The rHb A is eluted using a linear pH gradient to pH 8.0 (10 mM phosphate) over 10 column volumes using a Pharmacia FPLC chromatography system. The buffers are purged with helium before use. The elution of rHb A (red fractions) is monitored at $A_{280\ nm}$ and $A_{405\ nm}$. The pooled fractions containing rHb A are purged with CO and the pH is reduced to 6.5 by the addition of glacial acetic acid.

The S eluate (pH 6.5) is loaded onto a 10-ml (1.2 × 9 cm) CM-Sepharose Fast Flow column (Pharmacia) at 4° equilibrated in 10 mM phosphate, pH 6.5, at a flow rate of 1 ml/min. The same elution procedure (linear pH gradient up to pH 8.0) is used as described here. The rHb A elutes as a single major $A_{405\ nm}$ peak, and fractions collected are pooled.

A 10-ml Q-Sepharose Fast Flow (FF) column (Pharmacia) (1.5 × 6 cm) is equilibrated in 20 mM Tris-HCl, pH 8.5, at 4° and a flow rate of 1 ml/min. The pH of the CM eluate is adjusted to 8.5 by the addition of

---

[23] D. Coghlan, unpublished results.

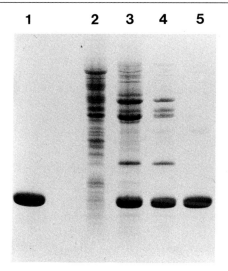

FIG. 4. SDS–PAGE analysis of rHb A purification fractions. Samples are resolved on a 20% homogeneous Phast gel (Pharmacia) and stained with Coomassie blue. All lanes, except lane 2, are loaded with a total of 1 μg Hb. Lane 1, standard Hb A (Sigma); lane 2, yeast cell lysate (2 μg total protein); lane 3, S-Sepharose eluate; lane 4, CM-Sepharose eluate; and lane 5, Q-Sepharose eluate. On these gels α- and β-globins comigrate.

2 M Tris base and is repurged with CO. This is then loaded onto the Q-Sepharose FF column at 1 ml/min. The flow rate is reduced to 0.5 ml/min or less, and the rHb A is eluted stepwise using either 20 mM bis-Tris buffer, pH 6.6, or 20 mM HEPES, pH 6.6. This step not only gives an improvement in purity but also allows buffer exchange and is an important concentration step.

Purity Assessment

The purity of rHb A is monitored during the purification either by spectrophotometry or by using reducing SDS–PAGE. A rapid quantitative estimation of purity can be obtained from UV/Vis scans (250–650 nm) of rHb A fractions and standard Hb A in their carbonmonoxy forms, using the following comparative absorbance ratios: (A_{419}/A_{276})rHb A/(A_{419}/A_{276})Hb A. For SDS–PAGE, fractions containing rHb A at a concentration of approximately 1 mg/ml in sample buffer (see above) are resolved either on 20% homogeneous Phast gels (Pharmacia) (Fig. 4) or on conventional SDS–PAGE gels.

Techniques for Physical Characterization of rHb A

Methods

HPLC Separation of Globin Chains. A 100-μl sample of purified rHb A (50–500 μg) is mixed with 100 μl of 8 M urea and chromatographed on an HPLC Vydac C_4 reverse-phase (RP) column.[24] The column is equilibrated at 1 ml/min in 70% solvent A : 30% solvent B [solvent A is 0.1% trifluoroacetic acid (TFA) in water; solvent B is 90% acetonitrile/0.09% TFA]. Solvent B is increased to 45% over 5 min, followed by a 30-min linear gradient to 55% B. Separation of heme, β-globin, and α-globin is monitored at $A_{214 \text{ nm}}$ and $A_{280 \text{ nm}}$.

Peptide Mapping. Samples (200–500 μg rHb A or isolated α- or β-globin chains) in 500 μl of buffer are precipitated in an Eppendorf tube by adding 1 ml of acetone. After vortexing, the tubes are incubated at $-20°$ for 1 hr and then centrifuged for 10 min. The supernatant is discarded; the pellets are dried briefly by rotary evaporation and are then resuspended in 100 μl of 6 M guanidine hydrochloride in 0.5 M Tris-HCl, pH 8.0, 2 mM EDTA. The samples are reduced [10 μl of 100 mM dithiothreitol (DTT) in water, 37° for 1–2 hr], carboxyamidomethylated [20 μl of 100 mM iodoacetamide (Sigma) in 0.1 M Tris-HCl, pH 8.0, 37° for 2–3 hr in the dark], and then diluted 1 : 3 by adding 400 μl of water. Samples are digested with 3× 5-μl aliquots of trypsin (Sigma, 1 mg/ml in 1 mM HCl, stored at 4°) over 48 hr at 37° with shaking. The digests are separated on a Pharmacia Pep-S RP-column. The column is equilibrated in 95% solvent A : 5% solvent B (solvent A is 0.1% TFA in water; solvent B is 70% acetonitrile/ 0.085% TFA) at 0.5 ml/min. The peptides are eluted using a linear gradient up to 60% B over 50 min with detection at $A_{214 \text{ nm}}$.

Amino Acid and N-Terminal Analysis. The α- and β-globin chains of rHb A and Hb A are separated by RP-HPLC as described above. Aliquots containing 20 μg of each chain are dried separately and resuspended in 20 μl of 0.1% TFA. Each chain (15 μg) is subjected to 10 cycles of automated Edman degradation using an Applied Biosystems 477A protein sequencer and to automated acid hydrolysis and amino acid composition analysis using an Applied Biosystems 420 A amino acid analyzer.

Mass Spectrometric Analysis. Purified rHb A (1 mg) is dialyzed against Milli Q water and analyzed by Electrospray mass spectrometry (ESMS) using a VG Biotech BIO-Q mass spectrometer (M-Scan Ltd., Ascot, UK).

---

[24] S. Rahbar and Y. Asmerom, *Hemoglobin* **13**, 475 (1989).

Functional Studies

The simplest and most direct methods to confirm that purified yeast-derived rHb A is fully functional rely on optical absorption spectroscopy. We have used such an approach to determine the oxygen equilibrium curves and CO combination kinetics for the recombinant protein. These methods have been described extensively elsewhere.[25,26] Only minimal details of the most straightforward techniques will be described here.

Prior to functional studies, rHb A purified in the carbonmonoxy form is converted to the oxy form by exposure on ice to a stream of humidified pure oxygen under illumination (Osram Dulux EL.20W low-heat bulb) with periodic agitation for approximately 1 hr. Control experiments are carried out with standard Hb A purified from erythrocytes using established procedures.[27]

Oxygen Equilibrium Curves

Oxygen equilibrium curves may be obtained conveniently without the requirement for specialized apparatus either by (1) serial addition of air to a modified optical cuvette containing a deoxyHb solution under vacuum or by (2) controlled enzymatic deoxygenation of a solution of oxyHb in an optical cuvette equipped with an oxygen electrode.

Titration Method. The titration method has been described adequately elsewhere.[26] This method is simple to use providing that care is taken to avoid excessive bubble formation and hence denaturation during the evacuation procedures needed to generate deoxyHb, and that full equilibration is achieved after each addition of air. Equilibration is ensured by transfer of the sample to the tonometer bulb (Fig. 5) by tilting the apparatus, which is then agitated in a water bath for 5 min. An example of an oxygen equilibrium curve obtained for rHb A using this method is given in Fig. 6.

Enzymatic Method. Enzymatic depletion of oxygen has the advantage that Hb is not subjected to lengthy evacuation procedures and also that the cuvette remains at all times within the spectrophotometer housing, thus eliminating repositioning errors. The disadvantages are that the substrates and/or products of the enzymes used to consume oxygen may degrade Hb. We have found that a suitable method uses cytochrome *c* oxidase

[25] E. Antonini and M. Brunori, *in* "Hemoglobin and Myoglobin in Their Reactions with Ligands, Frontiers in Biology" (A. Neuberger and E. L. Tatum, eds.). North-Holland Publ., Amsterdam, 1971.

[26] B. Giardina and G. Amiconi, this series, Vol. 76, p. 417.

[27] A. Riggs, this series, Vol. 76, p. 5.

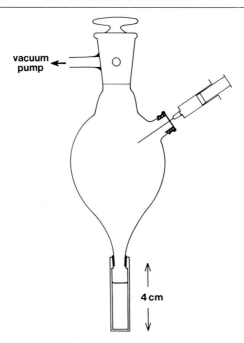

vacuum
pump

4 cm

FIG. 5. Diagram of tonometer bulb used to determine oxygen equilibrium curves. The total combined volume (approximately 150 ml) of the tonometer and cuvette must be determined by weighing the system empty and when completely filled with a liquid of known density. A sample of hemoglobin is deoxygenated by evacuation, and the deoxy spectrum is taken. The volume of the sample following deoxygenation is similarly determined by weight. Additions of 0.5–2 ml of air (21% oxygen) at known atmospheric pressure and humidity are made via the vaccine cap as shown. Tonometers may be modified for use with small volumes.[25]

and ascorbate/cytochrome c. Alternatively, ascorbate oxidase/ascorbate may be used. These systems reduce oxygen to water as opposed to hydrogen peroxide. It is important to remove the oxygen in about 20 min to minimize the risk of ascorbate damage to the heme group.[28]

Methods. A simple system comprises a normal 1-cm pathlength optical cuvette, with ground aperture, into which an oxygen electrode (Yellow Springs Instrument Company, Yellow Springs, OH) can fit tightly. A small gap should remain at one point through which the reactants can be injected using a microsyringe. It is essential that the electrode protrudes into the cuvette and is close (1 cm) to a magnetic follower at the bottom of the cuvette, because the response of the electrode is critically dependent on efficient stirring of the solution close to the membrane. A suitable system

[28] J. C. Docherty and S. B. Brown, *Biochem. J.* **207,** 583 (1982).

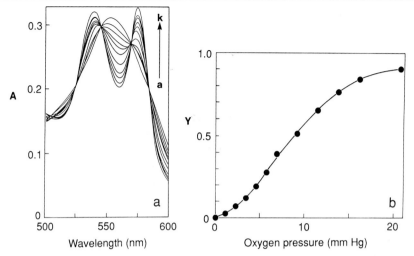

FIG. 6. Determination of oxygen equilibrium curves. (a) Spectral transition from deoxy- to oxyrHb A (25 μM heme in 30 mM HEPES, pH 7.0, 0.2 M NaCl, 25°) ($a \rightarrow k$) on titration with air. The final spectrum was obtained under pure oxygen. The spectra are characteristic of authentic human Hb A. (b) The fractional saturation (Y) determined from $\Delta A_{576-560\ nm}$ is plotted against oxygen pressure (mm Hg). The equilibrium curve is characterized by a P_{50} (the partial pressure at which Hb is 50% saturated) of 9.0 mm Hg, $n_{max} = 2.9$.

comprises Hb at a concentration of 25–100 μM heme, 10 mM sodium ascorbate, 50 μM tetramethylphenylenediamine (TMPD; electron media- tor, Sigma), 0.25 μM cytochrome c (Type VI, Sigma), and 0.5 μM (total heme) bovine cytochrome c oxidase (prepared according to Yonetani[29]) in 0.1 M sodium phosphate buffer, pH 6.8, containing 0.05% (v/v) Tween 80, a nonionic detergent to solubilize the enzyme. Alternatively, 2 units ascorbate oxidase (Sigma) can be substituted for cytochrome c oxidase, and the TMPD, cytochrome c, and Tween 80 can be omitted. In both systems the enzyme should be sufficiently active to deoxygenate the solu- tion within about 20 min.

Oxygen equilibrium curves are obtained by plotting the absorbance at a single wavelength (e.g., 576 nm) against the calibrated output from the oxygen electrode. Alternatively, spectra may be recorded throughout oxygen depletion as in Fig. 6. These must, however, be taken at rapid scan rates to ensure minimal distortion due to oxygen depletion during the course of each scan.[30]

[29] T. Yonetani, *J. Biol. Chem.* **236,** 1680 (1961).
[30] K. D. Vandegriff and R. I. Shrager, this series, Vol. 232.

Carbon Monoxide Binding Kinetics

The autocatalytic time course for CO binding to Hb is a convenient probe for the T to R transition and hence functional integrity. The kinetics of binding may be followed at low protein concentration because deoxyHb tetramers are stable, and dissociation into dimers does not occur in the micromolar concentration range.[31]

Methods. OxyHb (5 to 10 ml of approximately 5 μM heme in 0.1 M phosphate buffer, pH 7.0) solution is partially deoxygenated by three cycles of evacuation and flushing with nitrogen and then drawn into a syringe. Full deoxygenation is achieved by addition of a few grains of solid sodium dithionite or by addition of a 200 mM solution of sodium dithionite made up in anaerobic buffer to a final concentration of 2 mM. This solution is mixed in a standard stopped-flow apparatus (Applied Photophysics, Leatherhead, UK) with solutions of CO in buffer containing 2 mM sodium dithionite. Solutions of CO, at required concentrations (e.g., 25–200 μM), are prepared by diluting a saturated solution of CO in water (equilibration of water at 20° with CO at atmospheric pressure (101 kPa) gives a 1.03 mM solution of CO).

Rapid-mixing devices designed to be used with spectrophotometers (e.g., HiTech, Salisbury, UK) are also suitable for use at low CO concentrations. In this case it is often necessary, because of the slow response of the A/D converter, to use an analog signal from the spectrophotometer both for display on an oscilloscope and for conversion to digital form for analysis. Absorbance changes at 436 nm are recorded as a function of time in the 1-msec to 1-sec time range.

At pH 7.0 in 0.1 M phosphate buffer at 20°, the time course for CO combination to deoxyHb is autocatalytic, reflecting the T to R transition on CO binding. A series of time courses over a range of CO concentrations allow the second-order rate constant (l') to be determined by standard procedures.[25] The rate constant under these conditions increases from approximately $1 \times 10^5\ M^{-1}\ sec^{-1}$ to $2 \times 10^5\ M^{-1}\ sec^{-1}$ throughout the course of binding (see Table I). The time course for CO combination to rHb A should be free of any fast initial component ($l' > 10^6\ M^{-1}\ sec^{-1}$), indicating the presence of free or noncooperative subunits.

Results and Comments

Using the purification methods described, we routinely obtain rHb A of >95% purity, with an overall recovery of 50% even when starting with yeast transformants producing rHb A at as little as 0.5% of the total

[31] B. W. Turner, D. W. Pettigrew, and G. Ackers, this series, Vol. 76, p. 596.

TABLE I
PHYSICAL AND FUNCTIONAL CHARACTERISTICS OF YEAST-DERIVED rHb A AND
STANDARD Hb A FROM ERYTHROCYTES

Parameter	Characteristics
Spectroscopic analysis	Normalized spectra of the oxy, deoxy, and carbonmonoxy derivatives of rHb A and Hb A are superimposable between 350 and 650 nm
RP-HPLC analysis	Stoichiometry of α-globin : β-globin : heme ratio is identical to that of Hb A
Peptide mapping	Tryptic maps of rHb A and recombinant α- and β-globin chains are identical to those of Hb A
N-terminal analysis	Recombinant α- and β-globin chains have correctly processed N termini
Amino acid analysis	Amino acid composition of rHb A is identical to standard Hb A
Electrospray mass spectrometry	Mass: α-globin, 15,126.0 Da; β-globin, 15,866.2 Da. These masses are within 1 Da of those of standard α- and β-globins
Oxygen binding properties	P_{50}: rHb A, 9.0 mm Hg; Hb A, 8.9 mm Hg (25 μM heme, 30 mM HEPES, pH 7.0, 0.2 M NaCl, 25°) n_{max}: rHb A, 2.9; Hb A, 3.0 Bohr effect [Δ log P_{50}/Δ pH (pH 7.0 − 7.6)]: rHb A, −0.37; Hb A, −0.34
CO binding properties	l': rHb A, 2.19 × 0.05 ± 10^5 M^{-1} sec^{-1}; Hb A, 2.02 × 0.03 ± 10^5 M^{-1} sec^{-1} (2 μM heme, 0.1 M sodium phosphate, pH 7.0, 20°)

soluble cell protein. Representative results for the physical and functional characterization of rHb A employing the techniques outlined are given in Table I. Physical analysis demonstrates that rHb A is identical to Hb A purified from erythrocytes. The spectroscopic and chromatographic techniques also confirm that the yeast-derived protein contains a full complement of endogenously produced heme groups (i.e., four heme groups per tetramer). In addition, equilibrium and kinetic measurements of the oxygen and CO binding properties of rHb A demonstrate that the functional characteristics of the recombinant protein are indistinguishable from Hb A.

The ability to synthesize authentic human Hb A in a simple eukaryote, such as *S. cerevisiae,* provides an ideal system for the expression of engineered mutant Hbs for structural and functional studies. The coexpression systems and purification protocol described here can be easily adapted for the production of mutant Hbs for further analysis. Moreover, genetically engineered yeast producing recombinant human Hb A or mutant Hbs offers the opportunity for provision of Hb for development into Hb-based oxygen carriers.

Acknowledgments

We would like to thank many of our colleagues for helpful discussion during the preparation of this manuscript, and especially Bernard Chan, Sarah Gilbert, Katharine Denton, Gareth Jones, David Coghlan, and Henk van Urk for allowing us to refer to their unpublished data. We also thank Kelly Morris for preparing the manuscript.

[25] Purification and Characterization of Recombinant Human Sickle Hemoglobin Expressed in Yeast

By Jose Javier Martin de Llano, Olaf Schneewind, Gary L. Stetler, and James M. Manning

In the genetic disease sickle cell anemia the abnormal hemoglobin S (Hb S) that is expressed has the hydrophobic side chain of Val-6(β) in place of the hydrophilic side chain of Glu-6(β) in hemoglobin A. This substitution on the exterior of the protein causes the aggregation of deoxygenated hemoglobin S tetramers, leading to the distortion of erythrocytes in the venous circulation.[1-8] Although the initial aggregation contact is between Val-6(β) and the Phe-85(β)/Leu-88(β) region of an adjacent Hb S tetramer, there are many subsequent points of contact that strengthen the overall stability of the polymer at low oxygen tensions.[3-5] Although the identity of some of the contact sites in the aggregate is known,[3-5,9] there is a lack of information about other contract sites as well as their relative contribution to the overall strength of the aggregate. The number and nature of such susceptible sites have been limited mainly to those hydrophilic side chains with enhanced reactivity because of their location in the protein.[8,10]

The availability of recombinant DNA technology now permits studies at any site on the hemoglobin S tetramer, heretofore not possible to modify by other methods. For hemoglobin S, the ability to alter hydrophobic sites

---

[1] L. Pauling, H. Itano, S. J. Singer, and I. C. Wells, *Science* **110**, 543 (1949).
[2] V. M. Ingram, *Nature (London)* **178**, 792 (1956).
[3] B. C. Wishner, K. B. Ward, E. E. Lattman, and W. E. Love, *J. Mol. Biol.* **98**, 179 (1975).
[4] S. J. Edelstein and R. H. Crepeau, *J. Mol. Biol.* **134**, 851 (1979).
[5] T. E. Wellems and R. Josephs, *J. Mol. Biol.* **135**, 651 (1979).
[6] J. B. Herrick, *Arch. Intern. Med.* **6**, 517 (1910).
[7] A. Cerami and J. M. Manning, *Proc. Natl. Acad. Sci. U.S.A.* **68**, 1180 (1971).
[8] J. Dean and A. N. Schechter, *N. Engl. J. Med.* **299**, 752, 804, 863 (1978).
[9] R. M. Bookchin, R. L. Nagel, and H. M. Ranney, *J. Biol. Chem.* **242**, 248 (1967).
[10] J. M. Manning, *Adv. Enzymol. Mol. Biol.* **64**, 55 (1991).

can provide relevant information because the initial (and some of the subsequent) stages of aggregation of Hb S involve hydrophobic interactions. Here we describe the expression of recombinant human sickle hemoglobin in a yeast system[11] in which it is possible to obtain hemoglobin tetramers that are assembled *in vivo* using endogenous yeast heme. Another advantage of the yeast system is that the N-terminal residues of proteins are processed so that the hemoglobin expressed has the correct N-terminal residues. Hence, the yeast expression system circumvents the difficulties with fusion proteins encountered in other systems. A more complete description of the advantages of the yeast expression system can be found in Martin de Llano *et al.*[11] and Wagenbach *et al.*[12] Recombinant sickle Hb can be expressed and readily purified by the procedures described in this chapter.

Reagent and Enzymes

The restriction endonucleases, T4 DNA ligase, and alkaline phosphatase were from Boehringer Mannheim (Indianapolis, IN) and T4 polynucleotide kinase was from New England Biolabs (Beverly, MA). The DNA sequencing kit and T7 DNA polymerase (Sequenase version 2.0) were from United States Biochemical (Cleveland, OH). The oligonucleotides were synthesized by Operon Technologies (Alameda, CA). To create the sickle mutation in the β-globin gene, the oligonucleotide 5-CAGACTTCTCCACAGGAGTCAG was used; the underlined base was used for the mutation. All the other reagents were of the highest purity commercially available.

Escherichia coli Strains

Escherichia coli XL1-Blue was used in most of the experiments. The *E. coli* strain BW313 was used in the site-directed mutagenesis step, as described by Kunkel.[13] The *E. coli* cells were grown at 37° in Luria–Bertani (LB) medium (10 g tryptone, 5 g yeast extract, and 10 g NaCl per liter); antibiotics were added when required.

Saccharomyces cerevisiae Strain and Culture Conditions

The strain used was GSY112 *cir⁰ Matα pep*4::*HIS*3 *prb*1-Δ1.6R *his*3-Δ200 *ura*3-52 *leu*2::*hisG can*1. The yeast was grown in rich YPD medium

---

[11] J. M. Martin de Llano, O. Schneewind, G. Stetler, and J. M. Manning, *Proc. Natl. Acad. Sci. U.S.A.* **90**, 918 (1993).

[12] M. Wagenbach, K. O'Rourke, L. Vitez, A. Wieczarek, S. Hoffman, S. Durfee, J. Tedesco, and G. Stetler, *Bio/Technology* **9**, 57 (1991).

[13] T. A. Kunkel, *Proc. Natl. Acad. Sci. U.S.A.* **82**, 488 (1985).

(10 g yeast extract, 20 g peptone, and 20 g dextrose per liter). To select those cells containing the pGS389 plasmid (see below), agar plates of complete minimal medium either without uracil or without uracil and L-leucine were used. To maintain the presence of this plasmid in the yeast, the incubations were in complete minimal liquid media without uracil or L-leucine at 30°. Liquid cultures were shaken at 300 rpm either in 2.8-liter Fernbach flasks or in 2-liter Erlenmeyer flasks (1.2 liters of culture/flask).

Plasmids

pGS389 (15.5 kbp), which is an *E. coli*/yeast shuttle plasmid that contains the human α- and β-globin cDNAs, was used.[12] The expression cassette containing both globin genes and their promoters can be excised as a *Not*I fragment. pGS189 (5.3 kbp) is a derivative of pSK(+) plasmid (Stratagene, La Jolla, CA) and contains the α/β-globin expression cassette named above. This plasmid contains an ampicillin-selectable marker and can replicate only in *E. coli*. The digestion of the plasmid with *Sph*I yields a DNA fragment of 1.2 kbp containing the β-globin cDNA.

Site-Directed Mutagenesis

The plasmid pGS189 was digested with *Sph*I into two DNA fragments of 1.2 and 4.1 kb. The 1.2-kb fragment containing the β-globin cDNA was inserted into the RF form of M13mp18 previously digested with the same restriction enzyme so that the phage contains the sense DNA strand of the β-globin gene.

The *E. coli* strain BW313 was transfected with that recombinant phage and the oligonucleotide described above was used to create the mutation β6 Glu $\rightarrow$ Val following the procedure described by Kunkel.[13] The presence of this mutation was confirmed by sequencing the single-stranded DNA of the mutant phage. The RF form of the mutant phage was digested with *Sph*I and the fragment containing the mutated β-globin gene was recombined with the large *Sph*I fragment of pGS189, creating pGS189ˢ. The correct orientation of the recombinant fragments was confirmed by DNA sequencing. The plasmid pGS189ˢ was treated with *Not*I to release the cassette with the sickle globin gene, which was inserted subsequently into pGS389 digested with the same restriction enzyme. The correct orientation of the DNA fragments was again confirmed by DNA sequencing and the recombinant plasmid was transformed into the yeast GSY112 *cir*⁰ strain using the lithium acetate method.[14] Transformed cells were selected

[14] H. Ito, Y. Fukuda, K. Murata, and A. Kimura, *J. Bacteriol.* **153**, 163 (1983).

on complete minimal media without uracil. To increase the copy number of plasmids per cell the selected clones were restreaked up to four times on agar plates of complete minimal medium lacking uracil and L-leucine.[15]

Growth of Yeast and Induction of Hb S

The procedure to culture the yeast strain GSY112 [pGS389^s] and to induce the synthesis of recombinant hemoglobin S (rHb S) is based on that previously described for rHb A.[12] Typically, an inoculum of yeast was grown in 200 ml of complete minimal medium without uracil or L-leucine; after 24 hr, another 800 ml of the same medium was added to the inoculum and the incubation continued for another 20 hr. This culture was diluted with 9 liters of YP medium and 2% (v/v) ethanol as the carbon source was added. After each increase of 3.0 in A_{600}, an additional 1% (v/v) ethanol was added. When the A_{600} reached 10, hemoglobion expression was induced by the addition of galactose to a final concentration of 2% (20% galactose stock, sterile filtered). After 24 hr, the culture was bubbled with CO and harvested by centrifugation at 7000 g for 10 min at 4°. The cells were washed with the extraction buffer (20 mM sodium phosphate, 20 mM NaCl, 10 mM benzamidine, 0.5 mM EDTA, 0.5 mM EGTA, 0.1% Triton X-100, pH 8.2, bubbled with CO) and collected under the same conditions. If not used immediately, the cell pellets were stored frozen at −80°.

Growth of Yeast Harboring Plasmids for Hb A and Hb S

In order to determine whether sickle Hb was deleterious to yeast, the growth rates and yields of normal and sickle Hb were compared. In general, yeast synthesized about equivalent amounts of Hb A and Hb S at about the same rate (1–2 mg of purified Hb per liter of liquid culture, depending on the amount of cell breakage). Hence, the presence of sickle hemoglobin was not harmful to the yeast, i.e., in the amounts produced, and there was no adverse effect on the metabolism or the morphology of the yeast.

Isolation of Recombinant Hb S

Yeast cells (70 g, wet weight) were resuspended in 100 ml of the extraction buffer at 4°, bubbled with CO, and 200 g of acid-washed glass

---

[15] K. A. Armstrong, T. Som, F. C. Volkert, A. Rose, and J. R. F. Broach, in "Yeast Genetic Engineering" (P. J. Barr, A. J. Brake, and P. Valenzuda, eds.), p. 165. Butterworth, Stoneham, MA, 1989.

beads [425–600 μm, Sigma (St. Louis, MO) or 450–520 μm, Thomas Scientific, Swedesboro, NJ] were added in the 200-ml stainless-steel mixing chamber of the Omni Mixer homogenizer either with a titanium or stainless steel blade (Omni International, Waterbury, CT). A Bead Beater homogenizer (Biospec Products, Burtlesville, OK) can also be used. The vessel was placed in ice–water and the homogenizer was operated at 5000 rpm four times for 5 min each with 3-min intervals; the temperature of the homogenate should not rise above 10°. The suspension, which had a gray hue, was decanted. The beads were washed with 25 ml of the extraction buffer and both suspensions were mixed and centrifuged at 4°, 15,000 g for 20 min. The reddish and cloudy supernatant (~100 ml, pH 6.3–6.5) was bubbled with CO.

Chromatography of rHb S on Carboxymethyl Cellulose

The supernatant was concentrated to about 20 ml with both an ultrafiltration chamber (Amicon, Beverly, MA) and Centriprep tubes (Amicon, Beverly, MA) (membrane cutoff in both cases was 10,000). The sample was dialyzed overnight against three changes of 1 liter each of 10 mM potassium phosphate, 0.5 mM EDTA, 0.5 mM EGTA, pH 5.85, bubbled with CO; the pH of this buffer must not be lower than 5.85 in order to avoid oxidation and precipitation of the rHb S during the dialysis step. After centrifugation at 15,000 g for 30 min at 4°, the red supernatant (~20 ml) was applied to a carboxymethyl cellulose (CM-52) column (0.9 × 20 cm), previously equilibrated in the buffer used in the dialysis step; a linear gradient of 150 ml each of this buffer and 15 mM potassium phosphate, 0.5 mM EDTA, 0.5 mM EGTA, pH 8.0 (both CO bubbled) was applied. The absorbance of the eluent was monitored at 280 and at 540 nm (Fig. 1). The 280-nm wavelength detects protein, nucleic acid, and other ultraviolet-absorbing material. The 540-nm wavelength detects only the presence of heme-containing proteins. Most of the material in the extract did not adhere to the resin and eluted close to the void volume of the column. After application of the gradient, several more nonheme proteins preceded a peak of hemoglobin, which eluted late in the gradient. Those fractions corresponding to peaks containing protein absorbing at 540 nm were bubbled with CO and kept at −20° or −80° for further analysis. The elution position of this heme protein corresponds to that of natural oxyhemoglobin S purified from human sickle erythrocytes.

Alternatively, a batch procedure for adsorption of the recombinant Hb S on CM-52 was sometimes used. The supernatant solutions (~300 ml), obtained from three batches of 70 g each of yeast cells broken by the procedure described above, were mixed and dialyzed overnight against

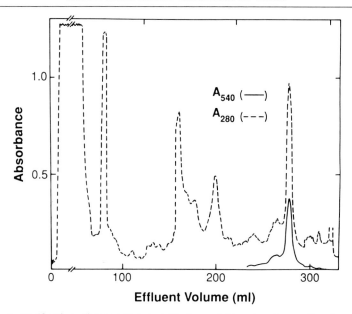

FIG. 1. Purification of recombinant sickle hemoglobin. A column of carboxymethyl cellulose (CM-52) was used as described in the text. The amount of protein in the yeast extract applied was about 850 mg and the yield of purified Hb was 2.7 mg.

three changes of 12 liters each of 10 mM potassium phosphate, 0.5 mM EDTA, 0.5 mM EGTA, pH 5.85, that had been bubbled with CO. After centrifugation (15,000 g for 30 min at 4°), the supernatant containing rHb S was mixed with 1.6 ml of CM-52 resin equilibrated in the same buffer and stirred at low speed for 1 hr at 4° in a vessel. In order to recover the resin, it was passed dropwise through a Millipore (Bedford, MA) fritted glass filter; the wet resin (bright red) was washed with 5–10 ml of buffer and removed from the filter. Another 0.6 ml of the resin was added to the filtrate, and the mix was filtered again as described. The spectra of the original supernatant and the filtrates at ~530–580 nm indicated that more than 80% of the material absorbing at this wavelength range is retained by the resin. After addition of these combined resins (2.2 ml) to the top of the column bed (0.9 × 20 cm) described above, the chromatography on CM-52 was developed as described above.

Purification by HPLC

The rHb S obtained from the CM-52 column is at least 85–90% pure and is useful for most purposes. If necessary to remove the minor contaminants further purification was sometimes employed on a SynChropak CM-300

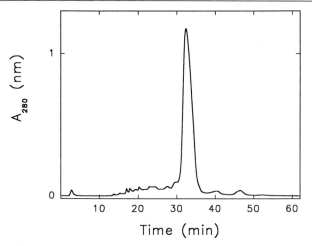

FIG. 2. Purification of rHb S by HPLC. A sample of rHb S purified by CM-52 (4 mg) was chromatographed on a SynChropak CM-300 column as described in the text. The same elution pattern is obtained when monitoring at 540 or 410 nm.

column (250 × 4.6 mm) (SynChrom Inc., Lafayette, IN) on a Shimadzu (Columbia, MD) LC-6A HPLC system at room temperature; a guard column (50 × 4.6 mm) was inserted between the injector and the column. The buffers used in the separation were similar to those previously described[16,17]: 30 mM bis-Tris, 1 mM EDTA, pH 6.4 (buffer A), and 30 mM bis-Tris, 150 mM sodium acetate, 1 mM EDTA, pH 6.4 (buffer B). Both buffers were filtered through a 0.45-μm filter and degassed. The protein sample was dialyzed against 30 mM bis-Tris, 30 mM sodium acetate, 1 mM EDTA, pH 6.4 (bubbled with CO), and concentrated if necessary; 200–400 μl were injected onto the column. The gradient used was from 20 to 70% buffer B over 10 min, 70 to 100% buffer B over 50 min, and then 15 min 100% buffer B, at a flow of 1 ml/min; the elution of the proteins was monitored at 280 nm or at 410 or 540 nm. The fractions comprising the major peak (Fig. 2) were pooled, bubbled with CO, and kept at −80°. With this system it is possible to remove minor components from the rHb S sample obtained after chromatography on CM-52. The elution pattern at 540 or 410 nm is the same as that at 280 nm shown in Fig. 2, indicating that these minor components could be heme proteins. The peaks eluting after the rHb S are probably oxidized forms of this hemoglobin, as indicated by their spectra (data not shown).

[16] T. H. J. Huisman, J. Chromatogr. **418**, 277 (1987).
[17] J. B. Wilson, in "HPLC of Biological Macromolecules: Methods and Applications" (K. M. Gooding and F. E. Regnier, eds.), p. 457. Dekker, New York, 1990.

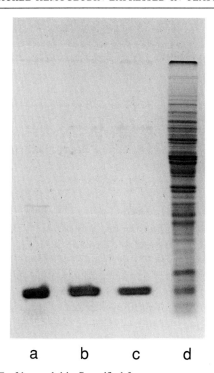

FIG. 3. SDS–PAGE of hemoglobin S purified from yeast extract. Purified, natural Hbs A and S were applied to lanes a and b, respectively. The recombinant Hb S, purified as described in Fig. 1, is shown in lane c. A sample of the yeast extract was applied to lane d.

Molecular Weight of Recombinant Sickle Hemoglobin Subunits

The recombinant sickle hemoglobin isolated on CM-52 (Fig. 1) behaved predominantly as a single band on electrophoresis in SDS–PAGE (Fig. 3, lane c). Natural hemoglobins A and S were included for comparison (Fig. 3, lanes a and b, respectively). The degree of purification by chromatography on CM-52 (Fig. 1) can be appreciated by a comparison of lanes c and d; the latter sample indicates the large number of proteins in the yeast cell extract before fractionation on the CM-52 column. The molecular weight of the recombinant Hb S subunits was in the correct range, approximately 16,000, by comparison with standards.

A more precise determination of the molecular weight of the rHb S was obtained by mass spectrometry. In this analysis the hemoglobin subunits are separated so that it is possible to determine their individual molecular weights. Masses of $15,126.6 \pm 1$ and $15,839.5 \pm 1$ were found for the α and β chains, respectively; these values agree well with the calculated molecular weights of 15,126 and 15,837 for the natural α and β^s chains, respectively.

Spectral Properties

The absorption maxima and extinction coefficients of the recombinant Hb S in the UV–visible range are nearly the same as the corresponding values for natural Hb S. The ratios of these absorbances, which are shown in Table I, indicate that these proteins are indistinguishable with respect to these spectral properties.

Isoelectric Focusing

The sample of recombinant sickle hemoglobin purified as described in Fig. 1 was analyzed by isoelectric focusing with the pH 6–8 Resolve-Hb kit (Isolab, Akron, OH) on a Pharmacia (Piscataway, NJ) Multiphor II system at 15° (Fig. 4). The results indicate that the recombinant Hb S (Fig. 4, lane c) migrated at exactly the same position as natural Hb S (Fig. 4, lane b). A sample of natural human Hb A is shown in lane a to indicate the high resolving ability of this system as well as the absence of the latter in the recombinant Hb S.

HPLC Separation of Recombinant Globin Chains

Separation of the globin chains of natural and recombinant sickle hemoglobin was achieved by reversed-phase HPLC on a Vydac (Hesperia, CA) C_4 column (250 × 4.6 mm) at a flow of 1 ml/min with a gradient from 37.6 to 40.0% acetonitrile with 0.1% trifluoroacetic acid (TFA) over 6 min and

TABLE I

SPECTRAL PROPERTIES OF NATURAL AND RECOMBINANT SICKLE HEMOGLOBIN[a]

Spectral ratio	Recombinant Hb S	Natural Hb S	Published values[b]
568/555	1.22	1.25	1.26
539/555	1.24	1.25	1.27
539/568	1.02	1.00	1.00
420/539	13.71	13.07	13.37
420/568	13.97	13.07	13.42
280/539	2.69	2.53	—
420/280	5.1	5.16	—

[a] The values are for the carbon monoxide derivative of hemoglobin.
[b] From Antonini and Brunori.[18]

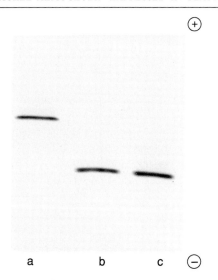

FIG. 4. Isoelectric focusing of recombinant and natural hemoglobins. A pH 6–8 range gel (Isolab, Akron, Ohio) was electrophoresed at 10 W for 45 min. Lane a, natural Hb A; lane b, natural Hb S; lane c, recombinant Hb S.

then to 44.8% in 60 min. In this system there is complete separation of the globin chains from one another (Fig. 5). Moreover, the elution position of the chains of recombinant sickle Hb coincided with those of the natural Hb S chains. These purified recombinant and natural globin chains were used for further characterization by amino acid analysis and N-terminal sequencing as described next.

Amino Acid Analysis

The α and β chains isolated by HPLC on Vydac C_4 as described above were subjected to amino acid analysis after acid hydrolysis (Table II). The results, which were obtained on a Beckman 6300 instrument with a System Gold data handling system, indicate that the amino acid compositions of the recombinant globin chains are in good agreement with analyses of the globin chains from natural Hb S isolated by the same HPLC procedure, as well as the published values.[18]

[18] E. Antonini and M. Brunori, "Hemoglobins and Myoglobins in Their Reaction with Ligands." Elsevier, New York, 1971.

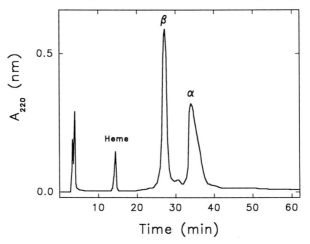

FIG. 5. Separation of recombinant Hb S globin chains by reversed-phase HPLC. A sample of rHb S (150 μg) was chromatographed on a Vydac C_4 column as described in the text. The fractions comprising the α and β chains were pooled and used in further analysis.

TABLE II
AMINO ACID ANALYSIS OF RECOMBINANT α- AND β^s-GLOBINS

Amino acid	α Chain		β^s Chain	
	Theory	Found	Theory	Found
Asp	12	12.5	13	13.5
Thr	9	8.8	7	7.2
Ser	**11**	**10.4**	**5**	**5.5**
Glx	**5**	**5.7**	**10**	**10.6**
Pro	7	6.7	7	6.1
Gly	**7**	**7.2**	**13**	**12.5**
Ala	**21**	**21.1**	**15**	**15.1**
Cys	1	1.0	2	2.1
Val	**13**	**11.8**	**19**	**16.7**
Met	2	1.6	1	1.1
Ile	0	0	0	0
Leu	18	18.8	18	18.4
Tyr	3	2.4	3	3.1
Phe	7	6.7	8	7.9
His	10	10.0	9	9.0
Lys	11	12.1	11	11.9
Arg	3	3.2	3	3.3

[a] The five amino acids shown in boldface type are those for which there are significant differences between the α and β chains.[18]

TABLE III
PARTIAL N-TERMINAL SEQUENCES OF α AND β CHAINS
OF RECOMBINANT Hb S[a]

Chain	Sequence
α	H₂N-Val-Leu-Ser-Pro-Ala-Asp-Lys-Thr--
	(215) (110) (59) (141) (148) (122) (129) (76)
β	H₂N-Val-His-Leu-Thr-Pro-**Val**-Glu-Lys--
	(330) (135) (228) (102) (122) (172) (92) (106)

[a] The sequence was initiated with about 200–300 pmol of each subunit purified by HPLC. The number in parentheses below each amino acid is the picomol amount of PTH-amino acid recovered at each cycle of Edman degradation. The **Val** residue, which is six residues from the N terminus, is the one formed by the mutagenesis.

N-Terminal Sequencing

The α and β chains separated by HPLC on a Vydac C₄ column as described above were subjected to N-terminal analysis by Edman degradation on an Applied Biosystems (Foster City, CA) gas-phase sequencer. The results (Table III) for the first eight amino acid residues of each chain clearly indicate that the sequences, which were obtained in high yield, are correct for the N-terminal segment of each α and β subunit. Furthermore, the presence of a Val residue at the sixth position of the β chain confirms that the mutagenesis was at the correct position.

Functional Studies

After purification of recombinant Hb S the oxygen equilibrium curve can be determined as described below as long as the amount of metHb, which is determined by the presence of a characteristic spectral band at 630 nm, does not exceed 5%. If the procedures described above are performed at 0–4° with all buffers saturated with CO gas, the amount of metHb can be kept to this low, tolerable value. If, however, a more extensive amount of metHb has been formed, it should be removed prior to determination of the oxygen equilibrium curve. The sample is treated with fresh 0.15 M sodium dithionite to remove metHb and layered onto a Sephadex G-25 column (1 × 25 cm) equilibrated in 50 mM bis-Tris–acetate, pH 7.5, containing sodium dithionite; the column was eluted with the same buffer without sodium dithionite. When collecting the hemoglo-

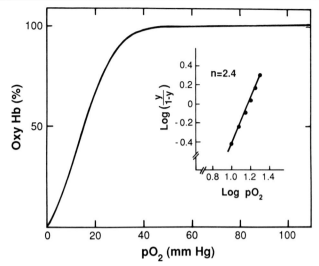

Fig. 6. Oxygen dissociation curve and Hill plot of recombinant sickle hemoglobin. The oxygen equilibrium curve was obtained at 37° on recombinant Hb S (0.5 mM in tetramer) in 50 mM bis-Tris, pH 7.5, and 0.1 M KCl. The inset shows the Hill plot where the log of the fractional oxygen saturation, Y, is plotted against the log of the partial pressure, p, of oxygen.

bin from this column, it is preferable to take a narrow band to ensure the absence of sodium dithionite.

The carbonmonoxy derivative of Hb S was converted to the oxy derivative by two exposures each of 200 W of incandescent light in a round-bottom flask in an atmosphere of pure O_2, as described previously.[19] The Hb sample, in 50 mM bis-Tris–acetate, pH 7.5, was then concentrated to ~0.5 mM (tetramer) using Amicon Centriprep and Centricon (Amicon, Beverly, MA) apparatus just prior to determination of its oxygen equilibrium properties on a modified Aminco Hem-O-Scan (Silver Springs, MD) instrument.

Measurement of the oxygen equilibrium curve for recombinant Hb S at 37° in the presence of 0.1 M KCl indicates that it has a P_{50} of about 15 mm Hg (Fig. 6), close to the value that we find for natural Hb S. These determinations were done below the gelling concentration of Hb S. The cooperativity of recombinant Hb S (Hill coefficient, n = 2.4, Fig. 6, inset) is the same as that for natural Hb S. In order to determine whether the recombinant Hb S was responsive to allosteric regulators, the effect of 2,3-diphosphoglycerate (2,3-DPG) was evaluated. Addition of 2,3-DPG to

[19] J. M. Manning, this series, vol. 76, p. 159.

the recombinant Hb S reduced its oxygen affinity to the same extent as it did for natural Hb S, i.e., with a right shift of 15 mm Hg.

Additional Remarks

Recombinant sickle hemoglobin, with structure and functional properties identical to those of natural sickle hemoglobin, can be readily obtained in reasonable yield with the relatively simple purification procedure described in this chapter. The sickle hemoglobin is expressed in its fully processed state without the need for dealing with fusion proteins. Therefore, it should be feasible to prepare double mutants of hemoglobin, i.e., a sickle hemoglobin combined with a mutation at any other site in the tetramer. Systematic site-directed mutagenesis studies using the yeast expression system described here should provide useful information on the relative contributions of the various contact points in the Hb S aggregate.

Acknowledgments

The authors are grateful to Adelaide Acquaviva and to Laura Manning for their skillful assistance with the typescript. The advice of Dr. Thelma Chen in the initial phases of these studies is gratefully acknowledged. We are indebted to Dr. Klaus Schneider and Dr. Brian Chait for determining the molecular weight of the subunits of the recombinant sickle Hb.

This work was supported in part by NIH Grant HL18819 and by Biomedical Research Support Grant BRSG-507-RR-07065 from the Division of Research Resources, National Institutes of Health, to Rockefeller University.

[26] Preparation of Recombinant Hemoglobin in Transgenic Mice

By Michael P. Reilly, Steven L. McCune, Thomas M. Ryan, Tim M. Townes, Makoto Katsumata, and Toshio Asakura

Introduction: Applications of Transgenic Mice

The production of transgenic animals by the introduction and stable integration of foreign DNA into the germ line has widespread impact in many areas of biological research and has led to significant advances in our understanding of gene regulation, mammalian development, and human diseases. Transgenic animals have also been utilized to produce therapeutically important protein products, such as tissue plasminogen activator, factor IX and α_1-antitrypsin.[1,2] In addition, transgenic technology affords the opportunity to create animal models to study the pathophysiology of genetic diseases under controlled experimental conditions.

There are several important applications of transgenic technology to the study of hemoglobin (Hb), such as the development of blood substitutes, production of abnormal hemoglobins with selected mutations, the investigation of globin gene regulation, and the ability to create animal models to study the consequences of abnormal hemoglobins *in vivo*. Normal human Hb A, produced in a transgenic mouse system, has normal functional properties[3] and represents a means of studying and developing blood substitutes. Similar strategies can be employed to create hemoglobins with specific amino acid substitutions in the α- and/or β-globin chains. The mutated hemoglobins can then be used to study various structural and functional relationships. An advantage of using a transgenic mouse system to produce mutant hemoglobins is that such hemoglobins can be generated in tetrameric forms that need only be purified from endogenous mouse hemoglobins without requiring recombination with heme or reassembly of subunits. The utilization of transgenic techniques has contributed to the investigation of globin gene expression[4] and hemoglobin switching.[5] Transgenic mice may also be utilized to create animal models to study clinically significant human hemoglobinopathies, such as sickle cell disease.[6-9] Such models may be used to investigate the pathophysiology of diseases under controlled experimental conditions, to develop new therapeutic ap-

[1] H. Westphal, *FASEB* **3**, 117 (1989).

[2] A. L. Archibald, M. McClenaghan, V. Hornsey, J. P. Simons, and A. J. Clark, *Proc. Natl. Acad. Sci. U.S.A.* **87**, 5178 (1990).

[3] R. R. Behringer, T. M. Ryan, M. P. Reilly, T. Asakura, R. D. Palmiter, R. L. Brinster, and T. M. Townes, *Science* **245**, 971 (1989).

[4] T. M. Townes and R. R. Behringer, *Trends Genet.* **6**, 219 (1990).

[5] N. Raich, T. Papayannopoulou, G. Stamatoyannopoulos, and T. Enver, *Blood* **79**, 861 (1992).

[6] D. R. Greaves, P. Fraser, M. A. Vidal, M. J. Hedges, D. Ropers, L. Luzzatto, and F. Grosveld, *Nature (London)* **343**, 183 (1990).

[7] T. M. Ryan, T. M. Townes, M. P. Reilly, T. Asakura, R. D. Palmiter, R. L. Brinster, and R. R. Behringer, *Science* **247**, 566 (1990).

[8] E. Rubin, H. E. Witkowska, E. Spangler, P. Curtin, B. Lubin, N. Mohandas, and S. Clift, *J. Clin. Invest.* **87**, 639 (1991).

[9] M. Trudel, N. Saadane, M. C. Garel, J. Bardakdjian-Michau, Y. Blouquit, J. L. Guerquin-Kern, P. Royer-Fessard, D. Vidaud, A. Pachnis, P. H. Romeo, Y. Beuzard, and F. Costantini, *EMBO J.* **10**, 3157 (1991).

proaches, and to test specific agents for treatment of human hemoglobin-opathies.

Outline of Procedure

Transgenic animals may be created by pronuclear microinjection of one-cell embryos, retroviral infection of preimplantation embryos, or selection of embryonic stem cells, after homologous recombination with foreign DNA, and transfer to blastocysts.[10] The common approach to generate transgenic mice, schematically shown in Fig. 1, requires (1) collection of fertilized eggs from hormonally stimulated female mice, (2) microinjection of DNA constructs into the male pronucleus of one-cell embryos, and (3) transfer of two-cell embryos or blastocysts to pseudopregnant females. The resulting progeny can be screened for the presence of the transgene by Southern blotting or polymerase chain reaction (PCR) analysis of tail DNA digests or, as in the case of mice transgenic for hemoglobin, the blood may be screened for the presence of the foreign protein by electrophoresis, high-performance liquid chromatography (HPLC), or monoclonal antibodies.

Figure 2 provides a flow chart and timetable for the production of transgenic mice. Egg donors, fertile studs, and surrogate mothers should be available several days before the beginning of the experiment. Vasectomized males should be prepared at least 1 month prior to the beginning of the experiment to allow for recovery from the surgical procedure and to test for sterility. The first step in the procedure is to induce superovulation in the egg donor. A superovulated female yields 20–70 eggs compared to only 6–10 eggs per naturally ovulating mouse, therefore, reducing the number of animals needed and the time required for collection of enough eggs. Superovulation is induced by injecting each egg donor with pregnant mare's serum (PMS) as a substitute for follicle-stimulating hormone (day -2). Human chorionic gonadotropin (hCG), in place of luteinizing hormone, is administered 40–44 hr later and the superovulated females are placed with fertile stud males (day 0). The following morning (day 1) each female is examined for the presence of a copulation plug. The successfully mated females are sacrificed and fertilized eggs are collected from the ampulla of the oviduct. The male pronucleus of the fertilized egg is injected with the purified DNA preparation and incubated overnight at 37° and 5% CO_2. Pseudopregnant surrogate mothers are prepared by mating with vasectomized stud mice. On day 2, two-cell embryos that survived

[10] R. Jaenisch, *Science* **240**, 1468 (1988).

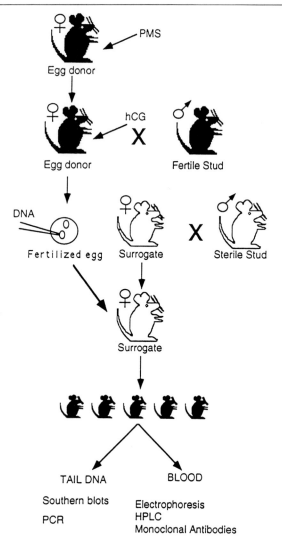

FIG. 1. Schematic diagram of the procedure for generation of transgenic mice by pronuclear microinjection.

the microinjection are transplanted to the oviducts of successfully mated surrogate mothers. Alternatively, blastocysts may be transplanted to the uteri of surrogate mothers on day 5. The estimated date of delivery is 20 days after fertilization of the eggs and also depends on the strain of the egg and sperm donors.

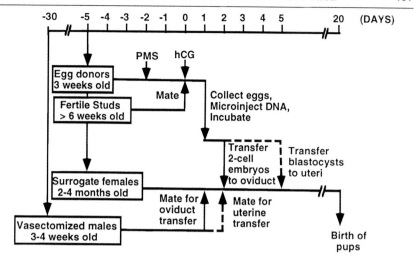

FIG. 2. Timetable for the production of transgenic mice. In this chart the injection of PMS is considered the beginning of the procedure (day 1). Egg donors and surrogate females should be available several days before initiating each microinjection experiment. Fertile and sterile males may be utilized for at least 8 months. Vasectomized males should be prepared at least 1 month before the initial procedure.

Transgenic Expression of Human Hemoglobins

The establishment of transgenic mice that produce human hemoglobins requires the preparation of genes that contain the desired mutation, either by cloning a known mutant or by site-directed mutagenesis. The construct must also contain regulatory sequences that direct high-level, tissue, and temporal-specific expression of the transgene. DNase I-hypersensitive sites (HS)[11,12] that flank the human β-globin locus comprise the β-globin locus control region (LCR). The LCR sequences direct high-level expression of the globin gene in transgenic animals.[13,14] High-level, coordinated, tissue, and temporal-specific expression, of human α- and β-globin genes allow the generation of complete human hemoglobin tetramers in mouse erythrocytes.

[11] D. Tuan, W. Solomon, Q. Li, and I. M. London, Proc. Natl. Acad. Sci. U.S.A. 82, 6384 (1985).

[12] W. C. Forrester, C. Thompson, J. T. Elder, and M. Groudine, Proc. Natl. Acad. Sci. U.S.A. 83, 1359 (1986).

[13] F. Grosveld, G. van Assendelft, D. R. Greaves, and G. Kollias, Cell (Cambridge, Mass.) 51, 75 (1987).

[14] T. M. Ryan, R. R. Behringer, N. C. Martin, T. M. Townes, R. D. Palmiter, and R. L. Brinster, Genes Dev. 3, 314 (1989).

We will describe the preparation of constructs that contain the LCR and human α- or β-globin genes, selection of background strains, production and detection of transgenic animals, and purification of human hemoglobins from transgenic mice. More detailed descriptions of microinjection procedures and microsurgical techniques required for production of transgenic mice have been described elsewhere in this series,[15] and in reviews and manuals.[16-18]

Preparation of DNA Constructs

Preparation of Mutated Hemoglobins

A number of mutations have been introduced into the α- and β-globin chains of human hemoglobin for the study of structure–function relationships in this protein. These mutations have ranged in purpose from altering the oxygen transport characteristics of hemoglobin to changing its tendency to polymerize. A simple and efficient method for site-directed mutagenesis of hemoglobin is described below.

Site-Directed Mutagenesis

Materials

Altered Sites Kit (Promega, Madison, WI)
Ampicillin (Sigma, St. Louis, MO)
Tetracycline (Sigma)
Mutagenesis oligo(s)
Bacterial culture supplies
Agarose gel supplies
Sequenase DNA sequencing kit (US Biochemical, Cleveland, OH)

Procedure. Although many techniques for conducting site-directed mutagenesis have been described, our laboratory has made extensive use of the Altered Sites mutagenesis kit (Promega), which employs a novel technique involving ampicillin selection.[19] This technique permits the successful production of mutants in 10–50% of the bacterial colonies that are generated.

[15] J. W. Gordon and F. H. Ruddle, this series, Vol. 101, p. 411.
[16] R. D. Palmiter and R. L. Brinster, *Annu. Rev. Genet.* **20,** 465 (1986).
[17] B. L. M. Hogan, F. Costantini, and E. Lacy, "Manipulating the Mouse Embryo: A Laboratory Manual." Cold Spring Harbor Press, Cold Spring Harbor, NY, 1986.
[18] M. Monk, ed., "Mammalian Development: A Practical Approach." IRL Press, Washington, DC, 1987.
[19] M. Lewis and D. Thompson, *Nucleic Acids Res.* **18,** 3439 (1990).

The initial step is to clone the globin gene to be mutated into the pAlter plasmid provided in the kit. This plasmid contains a functional tetracycline gene and an ampicillin gene with a four-base deletion. This deletion removes an essential residue and also introduces a frameshift mutation in the rest of the ampicillin gene product, virtually eliminating any chance of a spontaneous reversion to ampicillin resistance. Tetracycline resistance is used to select bacteria containing the pAlter plasmid with the cloned globin gene. Single-stranded DNA is then generated using either the RM408 or M13K07 helper phage. Oligonucleotides encoding the desired mutations are then designed such that they are in the same orientation as a repair oligonucleotide, which introduces the four bases required to restore ampicillin resistance. The oligonucleotides are annealed to the single-stranded DNA, and DNA polymerase and ligase are added to perform synthesis of double-stranded DNA from the single-stranded template. The double-stranded DNA is electroporated or transformed into the BMH-71 strain of *E. coli,* which lacks a mismatch repair system. These bacteria are then grown overnight in ampicillin selection and the DNA from the BMH cells is prepared and electroporated into JM109 cells, which have a functional mismatch repair system. Theoretically, one-half of the BMH-71 cells should be ampicillin resistant, and the other half should be ampicillin sensitive. The JM109 cells should receive DNA only from cells that are able to replicate under ampicillin selection. All of the JM109 cells would be expected to have a plasmid containing not only the functional ampicillin gene, but also any mutations encoded by the other oligonucleotide(s) added to the annealing reaction.

Detailed instructions for site-directed mutagenesis are described in the Altered Sites kit instructions. Antibiotic concentrations were reduced versus those suggested in the kit. Selections were performed using ampicillin at 75 μg/ml and tetracycline at 5 μg/ml.

Preparation of Cosmid Constructs

The globin gene containing the desired mutations is then cloned into a construct containing the β-globin LCR. The construct described below contains all five upstream DNase I-hypersensitive sites of the β-globin LCR. This construct has been shown to drive expression of the human α- and β-globin genes at a level of 100% of mouse β-globin expression on a per gene copy basis.[13,14] The constructs are coinjected into fertilized eggs for the production of transgenic mice that synthesize human hemoglobin with the desired mutations.

A cosmid containing the globin gene and LCR is made via a four-way ligation. After the ligation, the cosmid is packaged using the Packagene

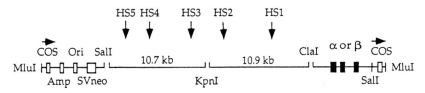

FIG. 3. Preparation of construct containing hypersensitive sites and globin genes. See text for detailed description.

System (Promega), and this packaged DNA is used to infect LE 392 cells. A detailed protocol for the production of LCR/globin gene constructs is presented below.

Cosmid Preparation

  Materials

  Packagene System (Promega)
  Bacterial culture supplies
  SM buffer: 100 mM NaCl, 10 mM MgSO$_4$, 50 mM Tris, 0.01% gelatin
  Procedure. The first step in the production of LCR/globin cosmids is a four-way ligation that unites the globin gene with all five of the upstream DNase I-hypersensitive sites.[14] A 10.7-kb *Sal*I–*Kpn*I fragment containing HS5, HS4, and HS3 is isolated from pTR68, a modified pUC plasmid. A 10.9-kb *Kpn*I–*Cla*I fragment containing HS2 and HS1 is isolated from pTR69, another modified pUC plasmid. These two fragments are ligated together with fragments containing Cos right and left arms and packaged using commercially available λ extracts (Fig. 3).

  The left arm is a 6.6-kb *Mlu*I–*Sal*I fragment from pCV108[20] that contains a Cos site, an ampicillin resistance gene, a *Col*EI origin of replication, and the SVneo gene. The right arm is a 5.1-kb *Cla*I–*Mlu*I fragment from plasmid pTR65 that contains a 4.1-kb *Cla*I–*Sal*I fragment of the β-globin gene and a 1.0-kb *Sal*I–*Mlu*I Cos site framgent from pCV108. The *Cla*I and *Sal*I sites on either side of the β-globin gene replace natural *Hpa*I and *Xba*I sites, respectively. Constructs with an α-globin gene contain a 3.8-kb *Cla*I–*Sal*I fragment from pTR96 instead of the 4.1-kb β-globin fragment described above. The *Cla*I and *Sal*I sites on either side of the α-globin gene replace natural *Xho*II and *Eco*RI sites, respectively.

  1. The four fragments are ligated together in a 2 : 1 : 1 : 2 vector arms-to-inserts ratio and packaged using Promega Packagene extracts. This ligation proceeds overnight at room temperature in a volume of 8 ml and

[20] Y.-F. Lau and Y. W. Kan, *Proc. Natl. Acad. Sci. U.S.A.* **80,** 5225 (1983).

contains approximately 600 ng of DNA. After ligation, Promega Packagene extract is added to the ligation mix. The ligation mix is packaged for 2 hr. Following packaging, 0.5 ml of SM buffer is added to the ligation mixture. Finally, 25 μl of chloroform is added to the ligated DNA and the tube is lightly vortexed.

2. Packaged DNA (100 μl) is used to infect 100 μl of the LE392 cells that are in log phase. This infection is conducted at 37° for 15 min, then LB broth is added and the culture is incubated for 1 hr. This 1-hr culture is then microcentrifuged to pellet the cells. The bacteria are resuspended in 0.1 ml of LB broth and are plated on LB plates with ampicillin at 75 μg/ml. Colonies that grow on these plates are picked and used to inoculate overnight cultures. Plasmid DNA from these cells is miniprepped and digested with *Sal*I, *Kpn*I, and *Cla*I to confirm the integrity of the construct.

DNA, which is cut correctly, is maxiprepped to provide DNA for cloning and microinjection work. For the LCR/globin constructs, the fragments used for microinjection are the 25-kb *Sal–Sal* fragments that contain α- or β-globin genes and all five upstream hypersensitive sites. These fragments can be purified from the 8-kb vector using the techniques described below.

Purification of DNA for Microinjection

DNA for microinjection is routinely purified away from contaminating bacterial plasmid sequences prior to microinjection. Previous work has demonstrated that the expression of mammalian genes is often inhibited by the presence of attached vector DNA.[21] Consequently, DNA for microinjection is digested with a restriction enzyme that cleaves the vector sequences while leaving the desired gene intact. The plasmid sequences can then be removed from the mammalian DNA by the different methods discussed below.

Electrophoretic Purification. DNA for microinjection can be purified from vector sequences by electrophoresis through agarose gels. The plasmid DNA that contains the gene of interest is digested with one or more restriction enzymes. The restriction enzymes used in this digest should be chosen such that the gene of interest is located on a fragment of DNA that differs in size from the fragment containing plasmid vector sequences. For the density gradient techniques discussed below, it is preferable that the injection fragment be larger than the fragment or fragments containing vector sequences.

[21] T. M. Townes, J. B. Lingrel, R. L. Brinster, and R. D. Palmiter, *EMBO J.* **4,** 1715 (1985).

Electrophoretic purification of DNA fragments for microinjection involves the separation of the products of a restriction digest using a gel composed of low-melting-point agarose. The desired DNA fragments are cut out of the gel, and the agarose is melted by warming the sample. The DNA is then purified by extraction with organic solvents. Electrophoresis through low-melting-point agarose gels is the preferred method for purifying relatively small DNA fragments for microinjection, although constructs of over 25 kb can also be purified in this way.

Agarose Gel Electrophoresis

Materials

TE buffer, pH 8.0: 10 mM Tris, 1 mM EDTA
3 M Sodium acetate, pH 5.2
Agarose gel supplies
Low-gelling-point agarose (Seaplaque, FMC Bio Products, Rockland, ME)
Ethidium bromide solution (5 mg/ml)
Microcentrifuge
Water bath
UV light

Procedure

1. A gel of 0.8–1.5% low-melting-point agarose is poured and allowed to solidify at 4° for at least 30 min. The gel should contain ethidium bromide at a final concentration of 0.5 μg/ml. The plasmid containing the gene of interest is digested with appropriate restriction enzymes and an aliquot of the digest is electrophoresed on an analytical gel to confirm that the digest has proceeded to completion. The rest of the sample can then be electrophoresed on the low-melting-point gel, but the current through the gel should be reduced compared with that applied to a standard agarose gel of the same size. For fragments that are large (>10 kb), the current is routinely reduced even further in order to improve band resolution and reduce smearing. The progress of the fragments through the gel should be monitored using a UV light turned to a long-wavelength setting. However, exposure to UV light should be minimized.

2. When individual bands are discernible and well-separated, the band of interest can be cut out of the gel. The gel slice should be placed under a UV light set for long-wavelength emission. Any portion of the gel slice that does not fluoresce due to the presence of DNA should be trimmed away. In general, the lower the volume of the gel slice, the better the

recovery of DNA. The gel slice is then placed in a microcentrifuge tube and weighed. An equal amount of TE buffer is added to the tube, and the sample is placed at 65° for 10 min or until the agarose has melted.

3. The sample is then extracted twice with phenol equilibrated with TE buffer, once with 1 : 1 phenol : chloroform, and once with chloroform. Multiple extractions with *sec*-butanol are performed to concentrate the sample to 400 μl. The *sec*-butanol forms the top phase in these extractions. Once concentrated, the sample is extracted two times with chloroform. One-tenth volume of 3 M sodium acetate is added, mixed, and DNA is precipitated by addition of two volumes of ethanol. The fragment is pelleted, washed with 70% ethanol, and briefly dried.

4. The DNA is then resuspended in 400 μl of TE buffer. Two more phenol extractions are performed, followed by one phenol : chloroform and one chloroform extraction. The sample is again ethanol precipitated, washed twice with 70% ethanol, and then dried. The DNA is then resuspended in a small volume of TE buffer, 10–20 μl. Then 1–2 μl is assayed using a fluorescence spectroscopy technique for determining DNA concentration.[22] Once diluted, the sample is ready for microinjection. Samples for microinjection are routinely electrophoresed on analytical-scale agarose gels to confirm that they are the correct size and that they are free from contaminating vector sequences. Occasionally, a significant amount of the fragment forms an aggregate that does not enter the analytical gel. When 10% or more of the DNA remains in the well, the fragment is not suitable for microinjection and a new preparation of the DNA should be made. Again, exposure of the fragment to UV light during preparation should be minimized. Fragments for microinjection are also digested with several restriction enzymes to assess their integrity and purity. The analytical digests should produce the appropriate fragments and 100% of the sample should be digested. Partial digests indicate that the DNA is not suitable for microinjection.

Density Gradient Purification. Density gradient preparation of DNA for microinjection has several advantages over the low-melting-agarose technique described above. Density gradient ultracentrifugation has been successfully employed for the preparation of DNA ranging in size from a few kilobases up to more than 40 kb. Density gradient ultracentrifugation generally provides a better yield of DNA than do low-melting-agarose methods. The use of gradients also avoids agarose contamination of microinjection fragments, which can lead to clogging of the injection needle. Our laboratory routinely uses both sucrose and potassium acetate gradients in the preparation of microinjection fragments. These methods are adapted

[22] C. Labarca and K. Paigen, *Anal. Biochem.* **102**, 344 (1980).

from the procedures described in Ausubel *et al.*[23] The potassium acetate procedure is less time consuming but sucrose gradient centrifugation provides better resolution of fragments of similar size.

Sucrose Gradient

Materials

Gradient solutions (20 ml each)
   10% sucrose solution: 10% (w/v) sucrose, 1 M NaCl, 20 mM Tris, pH 8.0, 5 mM EDTA
   40% sucrose solution: 40% (w/v) sucrose, 1 M NaCl, 20 mM Tris, pH 8.0, 5 mM EDTA
3 M Sodium acetate solution, pH 5.2
Ethidium bromide solution (5 mg/ml)
TE buffer (12 liters), pH 8.0: 10 mM Tris, 1 mM EDTA
Gradient maker
Magnetic stirrer
Polyallomer ultracentrifuge tubes: 1 × 3 1/2 inch (Beckman)
18-gauge needle and 5- to 10-ml syringe
Dialysis tubing: 3/4 inch (BRL, Gaithersburg, MD)
SW28 Rotor or equivalent (Beckman Instruments, Palo Alto, CA)
UV light
4-liter beaker

Procedure

1. Digest 50 μg of DNA for microinjection with a restriction enzyme that cuts between the mammalian DNA and accompanying plasmid sequences. An aliquot of this reaction mixture is analyzed on an agarose gel to confirm that the digest has proceeded to completion. In practice, samples for microinjection are digested in a reaction volume of 250 μl with a fourfold excess of enzyme for 4–6 hr. The digested DNA is then extracted with a 1 : 1 mixture of phenol : chloroform followed by a chloroform extraction. The DNA is then kept on ice until the gradient has been prepared.

2. Gradients are prepared using a two-chamber gradient maker mounted on a magnetic stirrer. A Teflon-coated stir bar in the gradient maker is responsible for mixing the two sucrose solutions and should be adjusted so that the mixing proceeds rapidly. Gradients can be poured with the

[23] F. M. Ausubel, R. Brent, R. E. Kingston, D. D. Moore, J. G. Seidman, J. A. Smith, and K. Struhl, "Current Protocols in Molecular Biology." Wiley (Interscience), New York, 1989.

denser solution entering the centrifuge tube first and the less dense solution layered on top. Alternatively, the gradient can be made by displacing a less dense solution with a more dense sucrose solution. The latter method permits a more rapid pouring of the gradient and results in an equivalent or perhaps slightly better gradient than the first method. To use this latter approach, a capillary tube is placed so that it touches the bottom of the centrifuge tube and is connected to the gradient maker by Tygon or equivalent tubing, thus permitting the gradient to be loaded from the bottom by displacement of the less dense solution. Sucrose gradients that range from 10 to 40% sucrose permit the separation of a wide size range of DNA fragments. Because sucrose gradients do not reach an equilibrium, the duration of a spin should be based on the size of the fragments to be separated.

To pour gradients for use in an SW28 rotor, 18 ml of each of the sucrose solutions is added to the separate chambers of the gradient maker. Ethidium bromide is added to the sucrose solutions to a final concentration of 0.5 μg/ml. The ethidium permits the visualization of the separated DNA fragments after centrifugation.

Gradients may require from 15 to 30 min each to pour. Once poured, gradients are best stored at 4° until all the desired gradients are ready. The gradient maker should be washed with distilled water after each gradient has been made. Gradients are balanced by addition of the less dense sucrose solution.

3. The DNA restriction digest mixture is warmed to 65° for 5 min prior to loading on the prepared gradients. This loading should be done as gently as possible in order not to disrupt the formed gradient. Samples can be spun in an SW28 or equivalent swinging-bucket rotor. Typically, a spin is conducted at 20° for 16–20 hr with a rotor speed of 25,000 rpm. For example, a 16-hr spin at this speed is sufficient to separate the 25-kb HS 1–5 β-globin LCR from the accompanying 8-kb of vector sequences.

4. After centrifugation, a hand-held UV light, preferably with a long wavelength setting, can be positioned to illuminate the separated fragments in the gradient. A 5- to 10-ml syringe with an 18-gauge needle is used to pierce the side of the centrifuge tube and remove the fragment of interest. Fragments can usually be collected in 3–5 ml of the sucrose gradient solution. These collected samples are then put into 3/4-inch dialysis tubing and dialyzed against 4 liters of TE buffer overnight at 4° with at least three changes of buffer. The volume of sample will increase on dialysis so the tubing should be able to accommodate 2–3 times the volume of the original sample. This dialysis should be done in the dark because the DNA still contains intercalated ethidium bromide.

5. After dialysis, the sample is transferred to a centrifuge tube with a 25- to 50-ml capacity. One-tenth volume of 3 M sodium acetate is added and mixed. The DNA is precipitated overnight at $-20°$ after addition of two volumes of ethanol.

6. Following precipitation, the tubes are centrifuged to pellet the DNA. The pellets are washed with 70% ethanol, air dried, and gently resuspended in 400 μl of TE buffer. The sample is then extracted twice with phenol: chloroform and twice with chloroform. One-tenth volume of 3 M sodium acetate is added and DNA is precipitated with two volumes of ethanol. The sample is washed with 70% ethanol, briefly dried, and resuspended in 25 μl of TE buffer.

7. The integrity and purity of the DNA is assessed as described above for fragments prepared by agarose gel electrophoresis and quantified by the Hoechst dye assay.[22]

Potassium Acetate Gradient

The following protocol is used in the preparation of DNA fragments for microinjection by potassium acetate gradients. Potassium acetate gradients provide a higher yield of purified DNA than do sucrose gradients. If the construct to be microinjected is in short supply, potassium acetate gradients would be the method of choice. Like sucrose gradients, potassium acetate gradients do not reach an equilibrium. Therefore, the duration of the spin must be adjusted according to the sizes of fragments generated by restriction enzyme digestion.

General procedures for pouring gradients and loading samples are described under the procedures section for sucrose gradients. Procedures that specifically apply to potassium acetate gradients are discussed below. As with the sucrose gradients, these instructions are applicable to density gradient centrifugation using an SW28 or equivalent rotor.

Materials

Gradient solutions (20 ml each)
  5% Potassium acetate solution: 5% (w/v) potassium acetate, 10 mM Tris, pH 8.0, 5 mM EDTA
  20% Potassium acetate solution: 20% (w/v) potassium acetate, 10 mM Tris, pH 8.0, 5 mM EDTA
3 M Sodium acetate solution, pH 5.2
Ethidium bromide solution (5 mg/ml)
TE buffer (12 liters), pH 8.0: 10 mM Tris, 1 mM EDTA
Gradient maker
Magnetic stirrer

Polyallomer ultracentrifuge tubes: 1 × 3 1/2 inch (Beckman)
18-gauge needle and 5- to 10-ml syringe
Dialysis tubing: 3/4 inch (BRL)
SW28 rotor or equivalent (Beckman)
UV light
4-liter beaker

Procedure

1. Digest 20–25 μg of the sample for microinjection in a volume of at least 250 μl. A useful estimate is that half of the DNA that would be loaded on a sucrose gradient can be loaded on a potassium acetate gradient. Recovery of DNA from potassium acetate gradients will usually range from 30 to 50%. Extract the digested DNA with 1 : 1 phenol : chloroform followed by a chloroform extraction. Place the sample on ice until gradients are ready for loading.

2. Pour gradients as described in the procedure section for sucrose gradients. The preferred technique is to displace the 5% potassium acetate solution with the 20% solution. Again, ethidium bromide should be added to the potassium acetate solutions to a final concentration of 0.5 μg/ml prior to the pouring of the gradients.

3. The DNA sample is heated at 65° for 5 min prior to loading on the gradients. Standard parameters for centrifugation in an SW28 or equivalent swinging-bucket rotor are 25,000 rpm at 20° for 16–20 hr. With potassium acetate gradients, the DNA of interest is pelleted at the bottom of the centrifuge tube. Centrifuge conditions should be adjusted so that separation is maintained between the pellet and the band composed of vector DNA. Centrifuge times are generally reduced for potassium acetate versus sucrose gradients. For example, using sucrose gradients, the β-globin LCR construct of 25 kb can be separated in 16–18 hr from an 8-kb vector fragment. Using potassium acetate gradients, this same separation is accomplished in 12–14 hr. A 16-hr centrifugation of the same fragments on potassium acetate gradients is less satisfactory because an unacceptable amount of the vector fragment is pelleted with the LCR construct.

4. After centrifugation, a hand-held UV light set to a long wavelength should be used to visualize the DNA in the gradients. A pellet of DNA at the bottom of the tube should fluoresce brightly. A band that represents the vector fragment should also be visible in the gradient. To avoid dislodging the pellet, the gradient solution is not poured away. Rather, a pipette should be used to draw off the liquid above the DNA pellet.

5. Resuspend the pellet in 400 μl of TE buffer and extract twice with isoamyl alcohol to remove the ethidium bromide. Isoamyl alcohol will form the top layer and the bottom phase will contain the DNA. These

extractions are followed by two 1 : 1 phenol : chloroform extractions and two chloroform extractions. One-tenth volume of 3 M sodium acetate is added and mixed and DNA is precipitated with two volumes of ethanol. After centrifugation, the precipitated DNA is washed with 70% ethanol, briefly dried, and resuspended in 25–50 μl of TE buffer. Again, the integrity and purity of the DNA are assessed as described above. After determining the concentration by the Hoechst assay,[22] the fragment is diluted and is ready for microinjection.

The techniques described above can be used to purify DNA fragments that differ greatly in size. Successful microinjections have been accomplished using DNA prepared by each of these techniques.

Establishment of Transgenic Mouse Lines: Selection of Mouse Strains

The selection of appropriate background strains for egg donors and stud males is important for any transgenic experiment. Many experiments require the defined genetic background of an inbred strain in order to interpret the effects of the injected construct. However, the use of F_1 hybrid mating pairs has the advantage of producing fertilized eggs with higher survival rates after microinjection and higher efficiency in retaining injected DNA compared to inbred strains.[24] Other considerations when choosing mouse strains for microinjection experiments are the number of eggs produced after superovulation, the size of the pronuclei, the mating behavior of the stud mice, and the litter-rearing abilities of the surrogate females. Another important consideration, when producing normal or mutant hemoglobins in mice, is the composition of endogenous mouse hemoglobins, as discussed below.

The mouse α-globin complex located on chromosome 11, *Hba* (Fig. 4), consists of three functional genes that code for an embryonic globin (*x*) and adult α-globins (*a1, a2*).[25] At least five different adult α-globin chains have been described in common mouse strains and the *a1* and *a2* alleles may code for identical or for two different α-globin chains. The various α-globin chains differ at one or more positions, resulting in changes in neutral amino acids, and can be distinguished by isoelectric focusing.[26]

Hemoglobin variants of common mouse strains are most often due to β-globin differences and can be electrophoretically distinguished. The mouse β-globin complex located on chromosome 7, *Hbb* (Fig. 4), consists

[24] R. L. Brinster, H. Y. Chen, M. E. Trumbauer, M. K. Yagle, and R. D. Palmiter, *Proc. Natl. Acad. Sci. U.S.A.* **82**, 4438 (1985).

[25] M. F. Lyon, J. E. Barker, and R. A. Popp, *J. Hered.* **79**, 93 (1988).

[26] J. B. Whitney, III, R. R. Cobb, R. A. Popp, and T. W. O'Rourke, *Proc. Natl. Acad. Sci. U.S.A.* **82**, 7646 (1985).

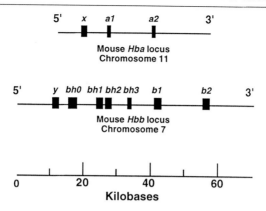

FIG. 4. Organization of mouse globin loci.

of linked genes that code for embryonic β-like globin (*y*), adult β-globins (*b1*, *b2*), and several homologous genes of unknown function (*bh0*, *bh1*, *bh2*, *bh3*).[27] Mouse strains can be distinguished by the type of adult β-globin synthesized. Mice of the single haplotype (*Hbb^s*) produce one β-globin chain, β-single, whereas those of the diffuse haplotype (*Hbb^d*) synthesize two distinct β-globin chains, β-major and β-minor. A deletional mutation that results in the absence of β-major globins and production of only β-minor globins in the homozygous state represents a β-thalassemic model.[28] Mice of a less common haplotype, *Hbb^p*, produce β-major and a variant of β-minor.

For subsequent separation of transgenically produced hemoglobins it is convenient to limit the number of endogenous mouse globins. The use of mice with only β-single globin for transgenic experiments results in the production of endogenous mouse hemoglobin, transgenically produced human hemoglobin, and two mouse–human heterodimer hybrids, on separation by ion-exchange chromatography. The amount of endogenous mouse hemoglobins can be further reduced by mating transgenic animals onto a background that is homozygous for mouse β-thalassemia. This results in a relative increase in human β-globin compared to mouse β-minor, with a subsequent increase in human tetrameric hemoglobin and a decrease in endogenous mouse hemoglobin. The use of animals with β-major and β-minor globins increases the number of possible hybrid

[27] C. J. Jahn, C. A. Hutchinson, III, S. J. Phillips, S. Weaver, N. L. Haigwood, C. F. Voliva, and M. H. Edgell, *Cell* (*Cambridge, Mass.*) **21**, 159 (1980).

[28] L. C. Skow, B. A. Burkhart, F. M. Johnson, R. A. Popp, D. M. Popp, S. Z. Goldberg, W. F. Anderson, L. B. Barnett, and S. E. Lewis, *Cell* (*Cambridge, Mass.*) **34**, 1043 (1983).

TABLE I
MOUSE STRAINS AND β-GLOBIN COMPOSITION

Hb pattern	β-Globins	Inbred strains	F_1 hybrids
Hbb^s	Single	C57BL/6; SEC/1; SJL; SWR; C58	C57BL/6 × SJL
Hbb^d	Major; minor	BALB/c; DBA/2; CBA; 129; AKR	AKR × BALB/c
Hbb^s/Hbb^d	Single; major; minor	—	BALB/c × C57BL/6; BALB/c × SJL; C57BL/6 × CBA

hemoglobins. Table I lists common inbred and F_1 hybrid strains and their β-globin haplotypes.

Egg Donors and Fertile Studs

The selection of F_1 hybrids, such as C57BL/6 × SJL, as egg donors and fertile studs has the advantages of yielding large number of eggs on superovulation and increased survival of injected eggs. In addition, these mice produce only endogenous hemoglobin of the β-single globin type, which is advantageous when producing foreign hemoglobins in mouse red blood cells, as described above. Superovulation of 10 C57BL/6 × SJL females and mating with an equal number of C57BL/6 × SJL males should yield 200–300 injectable fertilized eggs, which is sufficient for one experiment.

Surrogate Mothers and Sterile (Vasectomized) Stud Mice

CD-1 or ICR females, 2–4 months old, are chosen as surrogate mothers because of their good litter-rearing abilities. The estrus cycle of surrogates may be synchronized by hormonal injections as described for superovulation of egg donors. This ensures that a sufficient number of plugged females will be available as recipients of transplanted embryos. Vasectomized stud mice are required to mate with female mice that will serve as embryo recipients and surrogate mothers. SJL or C57BL/6 × SJL males are chosen as sterile stud mice because of their aggressive mating behavior and should be vasectomized at 3–4 weeks of age, before they reach sexual maturity. Vasectomized mice should be tested for sterility, prior to mating with surrogate mothers, by allowing them to plug three females and checking for ensuing pregnancy.

Preparation of Male and Female Mice

All survival surgery should be performed under aseptic conditions with sterilized surgical instruments. Application of a veterinary ophthalmic ointment is necessary to prevent drying of the cornea during surgical procedures.

Materials

Ketamine (100 mg/ml)
Xylazine (20 mg/ml)
1-ml syringe; 27-gauge needle
70% ethanol
Surgical instruments (Roboz Surgical Instrument Co., Washington, DC)
　Microdissecting tweezers, No. 55, No. 5
　Microdissecting forceps, 4 inches, serrated, straight, extra delicate
　Microdissecting forceps, 4 inches, serrated, full curved, extra delicate
　Microdissecting scissors, 4-1/4 inches, straight, blunt
　Hartman mosquito forceps, 3-1/2 inches, curved
　6.0 silk suture

Vasectomized Males

1. Obtain sexually immature male mice, 4 weeks old and weighing approximately 15 g. Prepare a 1 : 1 mixture of ketamine and xylazine and administer 0.02–0.03 ml by intraperitoneal (i.p.) injection using a 1-ml syringe and 27-gauge needle. Final anesthetic concentrations are ketamine (80 mg/kg) and xylazine (16 mg/kg).

2. Sterilize the lower abdomen with 70% ethanol and shave the hair along the lower abdomen with a single-edge razor blade. There is a noticeable color difference under the exposed skin over the bladder area. Lift the skin between the light and dark area with straight forceps, and with the microdissecting scissors make one transverse incision, about 5 mm in length, through the skin at the midline and level with the top of the hind legs. Make a similar incision in the underlying muscle layer, being careful to avoid blood vessels and the bladder.

3. Grasp the fat pad on the right side of the incision with the forceps and gently pull it out of the body cavity; the testes, epididymis, and vas deferens will also come out. The vas deferens is located under the inside aspect of the testes after the tightly coiled, whitish epididymis. The vas deferens is held in a membrane with a prominent blood vessel running alongside. Puncture the membrane with the microdissecting tweezers and separate from the vas deferens. Slip a 5-cm length of 6.0 silk suture under the vas deferens, slide it close to the testis, and tie a double knot. Tie

another suture approximately 5 mm away from the first. Make two cuts between the knots and remove a 5-mm length of the vas deferens. Pull up on the muscle layer with straight forceps and gently push on the fat pad to return the testis to the body cavity. Avoid pushing directly on the testis. Repeat the procedure on the other testis.

4. Place two stitches in the body wall and two more stitches in the skin layer.

5. Test the vasectomized males for sterility by allowing them to plug three females and checking for ensuing pregnancy. A record of plugging performance should be maintained for each vasectomized male. They can be used for 6–8 months.

Egg Donors

In order to obtain a sufficient number of fertilized eggs from a minimum number of animals, superovulation is induced by i.p. injection of 5 IU pregnant mare's serum gonadotropin. An i.p. injection of 10 IU human chorionic gonadotropin is performed 40–44 hr later. Superovulation can yield 20–70 eggs per mouse; however, fertilized egg recovery is notably affected by the mouse strain used, noise, light leakage during the dark cycle, and high animal density in the cage. Sexually immature mice (3–4 weeks old) produce more eggs than older mice on hormonal stimulation.

Materials

Pregnant mare's serum (PMS)
Human chorionic gonadotropin (hCG)
Sterile 0.9% NaCl
1-ml syringe, 27-gauge needle

Induction of Superovulation and Mating

1. Resuspend PMS (50 IU/ml) and hCG (100 IU/ml) in sterile 0.9% NaCl. Aliquots of each suspension may be frozen at $-70°$. Hormones may be thawed and used as needed but should not be refrozen.

2. Administer 0.1 ml (5 IU) of PMS by i.p. injection of each egg donor (3–4 weeks old) with a 1-ml syringe and 27-gauge needle.

3. Administer 0.1 ml (10 IU) of hCG by i.p. injection 40–44 hr later.

4. After the injection of hCG place one or two females with a fertile stud mouse and examine the following morning. Successfully mated females are identified by the presence of a copulation plug in the vaginal opening.

Surrogate Females (Pseudopregnant Females)

Successful mating with a vasectomized male, as evidenced by a copulation plug, ensures that the female's reproductive tract will be receptive to transplanted embryos. Outbred or F_1 hybrid females are the best choice as surrogate mothers and are used at 2–4 months of age to obtain the best mating efficiency and yield of offspring. Females are mated with vasectomized males the night before embryo transfer to the oviducts or two nights before for transfer of blastocysts to the uteri. Because the estrus cycle is 4–5 days and females are receptive to mating only during estrus, a sufficient number of females should be placed with vasectomized males to ensure the required number of recipients for transplanted embryos. Alternatively, the estrus cycles of the surrogates may be synchronized by administration of PMS and hCG as described for the induction of superovulation.

Preparation of Micromanipulation and Microinjection Supplies

Buffer Preparations. Several buffers are required for washing fertilized eggs and maintaining one-celled embryos during the microinjection procedure. Whitten's medium (WM)[29] is used for long-term culture of embryos in a CO_2 incubator. WM containing HEPES buffer (HWM) is used for short-term maintenance of the eggs during manipulation under normal atmospheric conditions. The addition of hyaluronidase to HWM is used to remove cumulus cells from the fertilized eggs.

Materials

Stock solution (g/100 ml): NaCl (5.14), KCl (0.356), KH_2PO_4 (0.162), $Mg_2SO_4 \cdot 7H_2O$ (0.294), dextrose (1.0), calcium lactate pentahydrate (0.527), potassium or sodium penicillin G (0.075), streptomycin sulfate (0.05); sodium lactate, 3.7 ml

Whitten's media (WM): stock solution, 10.0 ml; sodium pyruvate (0.0275 g/ml), 0.10 ml; bovine serum albumin (BSA), 0.40 g; $NaHCO_3$, 0.19 g; cell culture-grade H_2O, 89.9 ml

HEPES–Whitten's media (HWM): stock solution, 10.0 ml; sodium pyruvate (0.0275 g/ml), 0.10 ml; BSA, 0.40 g; HEPES, 0.4766 g; 1% phenol red, 0.05 ml; cell culture-grade H_2O, 89.9 ml

10 mM EDTA–WM: Na_2EDTA, 0.0372 g; WM, 10.0 ml

Hyaluronidase–HWM: hyaluronidase, 500 U; 10 mM EDTA–WM, 0.1 ml; HWM, 10.0 ml

---

[29] W. K. Whitten, *Adv. Biosci.* **6**, 129 (1971).

Procedure. Mix components for stock solution, as indicated above, in culture-grade H_2O. Stock solutions may be prepared in advance and frozen ($-20°$). Working solutions should be made every 2 months or as needed by adding freshly prepared sodium pyruvate, BSA, $NaHCO_3$ for WM; HEPES and phenol red for HWM. Adjust the pH of HWM to 7.4 by adding 7–8 drops of 5 N NaOH to 100 ml of media. Filter sterilize all media preparations through disposable filters (0.2 or 0.45 μm) and store in sterile tissue culture flasks. Sterile technique must be used during the preparation of all media. The use of cell culture-grade water and cell culture-tested reagents is essential. Prior to use add 100 μl of 10 mM EDTA–WM to 10 ml of each buffer solution.

Manipulation Pipettes and Microinjection Needles

Materials

Glass Pasteur pipettes, 9 inch (DNA loading pipette)
Micropipettes, 50 μl (embryo transfer pipettes)
Pipette 1BBL, no filament, 1 mm diameter, 4 inches long (holding pipette and microinjection needles; World Precision Instruments, Inc., Sarasota, FL)
Microbunsen burner
Microforge
Hexamethyldisilazane (HMSD), (Pierce, Rockford, IL)

Procedure

DNA SOLUTION LOADING PIPETTES. DNA solution loading pipettes are needed to fill the microinjection needle. Place the tapered portion of a Pasteur pipette in the pinpoint of the flame of a microbunsen burner. Heat until the glass is soft, quickly remove from the flame, pull the glass approximately 20 cm, and twist to break the glass strand. Set aside the lower part, discard the top. The cut end can be made even by gently squeezing the pipette with forceps, with ground glass in the teeth, and pulling. The pipette should be 6–7 cm long and of <500 μm outer diameter.

Siliconize the pipettes by placing them in a sealed chamber and adding a few drops of HMSD. A simple chamber can be constructed by placing two 15-ml conical culture tubes mouth-to-mouth and sealing with tape. These pipettes may also be used to collect and transfer fertilized eggs during wash and incubation steps. A mouthpiece and aspirator tube commonly supplied with micropipettes is a convenient holding device for the micropipettes.

HOLDING PIPETTES. Holding pipettes are required to hold the fertilized eggs during the microinjection procedure. Place a capillary pipette (1BBL, no filament) in the pinpoint of the flame of a microbunsen burner. Heat until the glass is soft, quickly remove from the flame, pull the glass approximately 10–20 cm, and twist to break the glass strand. The outer diameter of the pulled pipette should be 130–140 μm. The end of the pipette should be even and can be heat polished by bringing the tip close to the glass bead of the microforge. Apply more heat to melt the glass until the inner diameter is 20–25 μm. The pipette is placed over the glass bead of the microforge so that the end extends over the bead; heat is applied until the pipette bends to the desired degree. A bend of approximately 15° is introduced 2–3 mm from the end of the pipette, so that the end can be positioned horizontally in the microinjection chamber.

A pipette filler (#4555, Becton Dickinson, Parsippany, NJ) is attached to the holding pipette by a length of tubing in order to control the suction necessary to hold the eggs. The holding pipettes may be reused if they are cleaned immediately after each use by soaking overnight in 0.2 M HCl. After cleaning the pipettes should be rinsed in deionized H_2O and 95% ethanol, dried, and stored in a tightly sealed 50-ml screw-cap tube.

MICROINJECTION NEEDLES. Clean pipettes with detergent, rinse well with deionized H_2O, rinse with 70% ethanol, then with 95% ethanol, dry, and place in a tightly sealed 50-ml screw-cap tube. Use a computerized pipette puller (Micropipet Puller #P87, Sutter Instr. Co., Novato, CA) to make injection pipettes of suitable shape and pore size. Siliconize the injection pipettes by placing the pipettes on a suitable support, such as clay, in a culture dish with a few drops of HMSD and cover.

EMBRYO AND BLASTOCYST TRANSFER PIPETTES. Transfer pipettes are prepared from 50-μl capillary pipettes by heating and pulling as described above. Embryo transfer pipettes should have an inner diameter of 170–180 μm, whereas the pipettes for blastocyst transfer should be 250–280 μm.

MICROINJECTION CHAMBER. Clean a glass microscope slide with ethanol and air dry. Prepare a microinjection chamber by placing a plastic ring (1 mm high, 20 mm in diameter) on a microscope slide and place a small amount of wax on either side of the ring. The plastic ring can be conveniently obtained by slicing the lip from the snap cap of a disposable culture tube (Fisher #14-956-1J). Secure the ring by heating the slide over a flame; allow the wax to melt and run around the ring. Remove from the flame and allow the wax to cool.

When using the microinjection chamber, add 8 drops of HWM and carefully spread the solution around the ring so that the entire wax seal is covered; add 8 drops of mineral oil on top of the buffer solution to prevent evaporation. The oil will dissolve the seal if the aqueous solution does not completely cover the wax. Prepare a new microinjection chamber if precipitates form in the buffer solution during the course of the microinjection procedure.

Microinjection and Transfer of Embryos

Collection of Fertilized Eggs

1. Successfully mated, superovulated females are sacrificed by cervical dislocation and the oviducts are removed. Sterilize the ventral side of the animal with 70% ethanol. Open the abdominal cavity and avoid contamination with hair.

2. Lift the intestines out of the way and locate each uterine horn. Lift the junction between the uterus and the oviduct with a fine-tip microdissecting forceps and gently pull away from the body cavity. Use another forceps to tear the membrane away from the oviduct and uterus. While maintaining the hold on the junction cut on the uterus side of the junction with a microdissecting scissors.

3. While holding the severed end of the uterus gently pull the oviduct and ovary taut and carefully cut between the oviduct and ovary.

4. Prepare a 100-mm culture dish with several of drops of HWM. Place the dissected tissue in a drop of HWM on a culture dish and wash the tissue in the buffer to remove fat, blood, and hair.

5. Prepare another culture dish with four small drops of hyaluronidase–HWM and one large drop to collect eggs. The eggs are washed in hyaluronidase–HWM to remove the surrounding cumulus cells. Place the washed tissue in the first drop, place the dish under a dissecting scope, hold near the cut end of the uterus with a microdissecting forceps, locate the cluster of eggs, and tear the membrane with forceps near the cluster; fluid pressure inside the bulge should force the eggs out. If a prominent bulge is not observed in the upper end of the oviduct the eggs may be flushed out by inserting a blunt-end 30-gauge needle into the infundibulum and applying gentle pressure from a 5-ml syringe containing HWM.

6. Fill a siliconized transfer pipette with fresh HWM and gently blow the eggs away from the cumulus cells; aspirate the eggs into the pipette, transfer to a fresh drop of HWM, and separate fragmented, shrunken, or unfertilized eggs from fertilized eggs (one or two polar bodies visible). Collect the good fertilized eggs in the large drop of media. The eggs are examined and those that appear healthy and properly fertilized with prominent pronuclei are transferred to fresh WM media.

7. Place WM, equilibrated at 5% CO_2 at 37°, in the inner and outer chambers of an organ culture dish (Falcon #3037). Collect some WM

buffer in the transfer pipette and cover the eggs to dilute the effect of the hyaluronidase–HWM, collect the eggs in the transfer pipette and place them in the outer ring of the organ culture dish, wash the eggs two times by aspirating them and moving them from one area to another in the outer ring, and finally transfer the eggs to the inner chamber. The fertilized eggs are then incubated at 37°, 5% CO_2 for a few hours to allow the male pronuclei to expand.

Microinjection of Embryos

The basic equipment required for the microinjection of fertilized eggs includes an inverted microscope, preferably with Nomarski optics, micromanipulators for the injection needle and the holding pipette, and a microinjector. Specific recommendations of equipment and setups can be found elsewhere.[15-18]

Procedure

1. Prepare a microinjection chamber as previously described. Place 8 drops of HWM in the chamber and cover with mineral oil.
2. Examine fertilized eggs using the dissecting microscope, collect 20–30 eggs with prominent pronuclei in a transfer pipette, and place them in the microinjection chamber.
3. Use a DNA solution loading pipette to load the injection needle.
4. Position the holding pipette next to an egg and apply suction to hold the egg in place. Focus on the larger (male) pronucleus and position the injection needle in the same focal plane.
5. Move the needle to pierce the pronucleus and inject 1–2 pl of DNA solution. The volume of the pronucleus is about 1 pl, so the volume injected can be estimated from the expansion of the pronucleus. The optimal DNA concentration for microinjection is 2–5 ng/μl. Injection of higher concentrations may be detrimental to embryo development. Although there is no relationship between DNA concentration and integrated copy number, at least 100–500 copies of the construct should be injected. For example, there are approximately 500 copies/picoliter of a 5-kb fragment, at a DNA concentration of 2.5 ng/μl.

The use of a sharp needle is essential for smooth microinjection. When the injection needle gets dull or dirty, chromosomes stick to the needle and may be pulled out of the pronucleus when the needle is retracted. The nuclear membrane becomes elastic and difficult to penetrate at advanced developmental states. The injection needle tip is always blunt regardless of how fine it has been pulled. A sharper edge may be obtained by simply hitting the injection needle against the egg-holding pipette. Transfer the injected eggs to the inner well of an organ culture dish containing WM equilibrated at 5% CO_2 and 37°.

Transfer of Embryos to Pseudopregnant Females

Properly developed two-cell embryos can be transferred to the infundibulum of pseudopregnant females the day following microinjection. Alternatively, the embryos can be incubated for 4 days until they reach the blastocyst stage and then transferred to the uteri of pseudopregnant females that have been mated 2 days before. Generally, 50–70% of the embryos develop normally and are delivered.

Procedure

Oviduct transfer

1. Anesthetize the pseudopregnant females by i.p. injection of ketamine and xylazine, as described above. A veterinary ophthalmic ointment should be applied to both eyes as previously described.

2. Place the mouse on the lid of a culture dish and swab the back with 70% ethanol. Shave a small area on the dorsal lateral side of the mouse near the spine at the level of the last rib. Observe the color variation over the liver and the fat pad. Make a small incision through the skin (less than 1 cm) and a similar incision through the muscle layer over the ovary, carefully avoiding blood vessels. Gently pull the fat pad, ovary, oviduct, and uterus outside of the body wall.

3. Place the animal under a stereo dissecting microscope and tear a small hole in the bursa over the infundibulum with microdissecting forceps. Collect the embryos in a transfer pipette, insert them into the infundibulum down to the opening of the ampulla, and gently blow the embryos from the micropipette; 7–10 embryos can be placed in each oviduct. Pull up on the muscle layer with microdissecting tweezers and gently push on the fat pad to return the organs inside the body wall. Place two sutures in the body wall and two in the skin layer. Repeat the process on the other side. The entire procedure should be completed within 15 min. Surrogate females serve as embryo recipients once and are euthanized after the litter is weaned.

Uterine transfer

1. Proceed as indicated above to expose the uterus. Dip a 25-gauge needle in a solution of 5% tryptan blue in 70% ethanol. Pull up on the uterus with the forceps and puncture the uterine wall with the 25-gauge needle. If the needle is placed properly in the lumen of the uterus, it

should move freely in and out, otherwise it is between muscle layers of the uterine wall and must be repositioned.

2. Place the mouse under a dissecting microscope and locate the hole surrounded by the blue dye. Load a transfer pipette with 7–10 blastocysts, insert the pipette into the hole, and expel the embryos into the uterus. Suture the incisions as indicated above.

Detection of Transgenic Mice

Southern Blot analysis

Southern blot analysis of DNA digested with restriction enzymes that cut once within the injected fragment can be used to determine copy number, transgene orientation, and the number of integration sites. Injected DNA fragments are usually integrated in a single chromosomal site in head-to-tail manner. One or more unpredictable size bands, containing either 5' or 3' flanking region, will be detected, depending on the probe used for hybridization and the restriction site in the flanking region. This extra band may provide information for estimation of number of integration sites and can be used as an internal marker for the signal of one copy gene. However, it is necessary to confirm the integration status by using several different restriction enzymes.

Injected DNA is often integrated into two, or possibly more, chromosomal sites and the genotype may segregate in the F_1 generation. Thus, the integration status of injected DNA must be confirmed in animals after the F_1 generation, and an independent transgenic line should be established from each segregated progeny.

Genomic DNA isolated from tail is usually not as clean as that isolated from liver, and certain restriction enzymes, such as *Hind*III and *Xba*I, do not cut well. Protocols for isolation of genomic DNA[17] and Southern blots can be found elsewhere.[30]

Polymerase Chain Reaction Analysis

The progeny can be screened for the presence of the transgene by PCR analysis of tail lysates. Tail tissue (2–3 mm) is removed from each mouse at 1 week of age and is lysed in 200 μl of lysis buffer. Tail lysate can be directly used as a template for the PCR without further purification of genomic DNA. Alternatively, nucleated cells isolated from peripheral blood can be used for the PCR analysis.

---

[30] E. Southern, this series, Vol. 68, p. 152.

Materials

PCR lysis buffer: 50 mM KCl, 10 mM Tris-HCl, 2.5 mM MgCl$_2$
Adjust pH to 8.3, autoclave, and freeze in a small aliquots. Do not add gelatin or BSA because these samples contain a fair amount of protein.
Proteinase K stock: prepare 20 mg/ml of proteinase K in 10 mM Tris-HCl (pH 7.5)
Tail lysis buffer: PCR lysis buffer, 1.000 ml; Tween 20, 0.005 ml; proteinase K, 0.030 ml
K buffer: PCR lysis buffer, 1.000 ml; Tween 20, 0.005 ml; proteinase K, 0.005 ml

Procedure

DIRECT PCR OF TAIL DNA

1. Cut ~5 mm of tail from a 1-week-old mouse and place the tissue in a 1.5-ml microcentrifuge tube. Avoid blood contamination.

2. Add 200 μl PCR tail lysis buffer to the tube and incubate at 55° overnight.

3. Boil the sample for 15 min to inactivate proteinase K.

4. Spin the sample for 5 min and transfer the supernatant to a clean 500-μl tube.

5. In a separate 500-μl tube add (total = 100.0 μl): 39.5 μl of diethyl pyrocarbonate (DEPC)-treated H$_2$O, 10.0 μl of 10× Cetus buffer (Perkin-Elmer Cetus, Norwalk, CT), 3.0 μl of 30 mM MgCl$_2$, 10.0 μl of dNTP (2 mM each), 6.0 μl of 3' primer (0.02 μg/μl), 6.0 μl of 5' primer (0.02 μg/μl), 25.0 μl of lysate, and 0.5 μl of AmpliTaq (5 U/μl) (Perkin-Elmer Cetus).

6. Add 2 drops of mineral oil.

7. Amplify in the PCR machine at 94° for 2 min, 55° for 2 min, and 72° for 2 min for 25–40 cycles.

8. Transfer the lower layer to a clean 500-μl tube or just remove the mineral oil.

9. Run 15 μl of sample (mix with 1.5 μl of dye) in 3% NuSieve (FMC BioProducts), 1% regular agarose (or 4% regular agarose) gel.

BLOOD DNA PCR

1. Mix 50 μl of whole blood (do not use heparinized blood because heparin inhibits PCR) with ~0.5 ml of TE buffer (10 mM Tris–1 mM EDTA, pH 7.5) in a 1.5-ml microcentrifuge tube. Shake or vortex gently so that red blood cells are lysed before clotting. Samples may be kept on ice until the next step (~1 hr).

2. Spin for 10 sec at 15,000 g in a microcentrifuge and then remove the supernatant.

3. Add 0.5 ml of TE and resuspend the white pellet by gentle vortexing.

4. Repeat steps 2 and 3 until the supernatant is clear (2–3 times), then remove the supernatant and resuspend the pellet in 100 μl of K buffer.

5. Incubate for 45 min at 55°, and then boil the sample for 10 min to inactivate proteinase K and to denature DNA for subsequent PCR.

Detection of Expressed Hemoglobin

The expression of human hemoglobins in transgenic mice can be detected relatively simply by assays of hemolysates collected from the pups. Depending on the hemoglobin mutation engineered, the transgenically produced protein can be detected by cellulose acetate electrophoresis, reversed-phase HPLC, or by use of specific monoclonal antibodies.

Electrophoresis

Materials

Cystamine dihydrochloride (1.126 g)
Dithiothreitol (0.10 ml)
10% NH_4OH (5.0 ml)
deionized H_2O (9.0 ml)

Procedure. Cellulose acetate electrophoresis of cystamine-modified hemoglobins has been employed to distinguish differences between mouse hemoglobins of the *Hbb^s* and *Hbb^d* haplotypes.[31] Cystamine is a disulfide reagent that increases the positive charge of proteins that contain cysteine. Because mouse diffuse hemoglobins contain an extra cysteine in each β-chain, not present in the single hemoglobins, they are more positively charged and are readily separated by cellulose acetate electrophoresis.

Prepare a stock solution of cystamine modification solution (CMS) as described above. The stock solution can be stored frozen in 0.5-ml aliquots and diluted 5 times with deionized H_2O prior to use. Collect blood by retroorbital puncture into heparinized 40-μl capillary tubes. Transfer the blood to 1.5-ml microcentrifuge tubes containing 0.5 ml of dilute CMS. Freeze–thaw the samples in dry ice and acetone to ensure complete hemolysis. Hemolysates can then be analyzed by cellulose acetate electrophoresis according to standard techniques. Mice transgenic for human Hb A and Hb S are easily distinguished from mice homozygous for β-single.

Reversed-Phase HPLC

Materials

Vydac C_4 column (0.46 × 25 cm, 300-Å pore size, 5-mm particle size (The Separations Group, Hesperia, CA)

---

[31] J. B. Whitney, III, *Biochem. Genet.* **16,** 667 (1978).

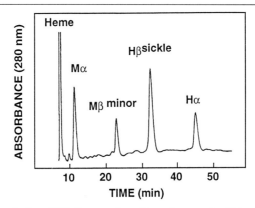

FIG. 5. Reversed-phase HPLC separation of hemolysates of red blood cells from a mouse transgenic for Hb S and homozygous for mouse β-thalassemia. Mouse globins are easily separated from human globins. Mouse β^{single}, β^{major}, and β^{minor} are eluted at similar times but can be resolved by modifications of the gradient.

Eluant A, 20% (v/v) acetonitrile–0.3% TCA
Eluant B, 60% (v/v) acetonitrile–0.3% TCA
Hemolyzing solution: 5 mM potassium phosphate–0.5 mM EDTA, pH 7.5

Procedure. Blood is collected by retroorbital puncture into heparinized capillary tubes. Transfer the blood to 1.5-ml microcentrifuge tubes and wash the red blood cells (RBCs) 3 times by resuspension in saline and centrifugation at 3000 g. Hemolyze the washed RBCs with 3–5 volumes of hemolyzing solution. Centrifuge at 10,000 g to remove membranes. Determine the hemoglobin concentration of the sample and adjust as necessary so that 2.5 μg of sample is injected onto the column. Eluants should be filtered and degassed. A modified gradient was used to separate the globins.[32] Mouse and human globin chains were separated using a linear gradient of 56% B to 62% B over 60 min at a flow rate of 1 ml/min. Detection is at 280 nm. Figure 5 shows an HPLC pattern from a hemolysate of a mouse transgenic for Hb S on a homozygous β-thalassemic background. The mouse α- and β-globins are easily resolved from the human globins. Individual globin chains can be identified by comparison with known samples or collected and subjected to tryptic digestion and amino acid sequencing of peptides.

[32] J. B. Shelton, J. R. Shelton, and W. A. Schroeder, *J. Liq. Chromatogr.* **7,** 1969 (1984).

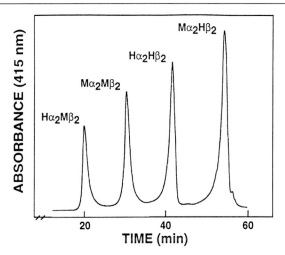

FIG. 6. Anion-exchange HPLC separation of hemolysates of red blood cells from a mouse transgenic for Hb A and homozygous for mouse β-thalassemia. Individual peaks were purified and separated by reversed-phase HPLC to identify the globin composition.

Ion-Exchange HPLC

Materials

SynCropak AX 300 (0.41 × 30 cm; SynChrom, Inc., Lafayette, IN)
Eluant A, 20 mM Tris, 1.5 mM KCN, pH 8.6
Eluant B, 20 mM Tris, 20 mM sodium acetate, 1.5 mM KCN, pH 7.0
Procedure. Sample collection and preparation is as described above. Hemoglobins can be separated by anion-exchange chromatography, as described here, or cation-exchange chromatography.[33] Approximately 50 μg mg of hemoglobin may be loaded onto the column. Mouse, human, and hybrid hemoglobins can be separated by ion-exchange HPLC using a linear gradient of 0–30% B over 60 min. Detection is at 415 nm. Figure 6 shows the HPLC chromatogram of a mouse transgenic for Hb A on a homozygous β-thalassemic background. The major hemoglobin species present include endogenous mouse minor hemoglobin ($M\alpha_2M\beta_2$), human Hb A ($H\alpha_2H\beta_2$), and two mouse human hybrid hemoglobins ($H\alpha_2M\beta_2$, $M\alpha_2H\beta_2$). Each hemoglobin peak can be collected, concentrated, and its globin composition verified by reversed-phase HPLC as described above. Preparative columns may be employed to isolate larger quantities of a

[33] J. B. Wilson, *in* "HPLC of Biological Molecules. Methods and Applications" (R. Goodieng and G. Regnier, eds). p. 457. Dekker, New York, 1990.

particular hemoglobin for further structural and functional characterization.

Monoclonal Antibodies to Specific Human Hemoglobins. A commercially available monoclonal antibody based kit (HemoCard, Isolab, Inc., Akron, OH) is available for the detection of Hb A and Hb S. Briefly, a small amount (2 μl) of whole blood is mixed with a sample conditioner to lyse the cells prior to mixing with reagents which contain monoclonal antibodies to either Hb A or Hb S. The antibodies specifically bind to amino acids at or near position 6 of the β-globin. Antibody–Hb complexes bind to the membrane of the test card, imparting a reddish color indicative of a positive test. This method is convenient for rapid screening of offspring for the presence of expressed human hemoglobin and requires minimal amounts of blood. This method is of course limited to the availability of appropriate antibodies.

Properties of Recombinant Hemoglobins

The structural and functional properties of hemoglobins produced in a transgenic mouse system can be investigated by standard techniques, as described elsewhere in this volume. Mice that produce human hemoglobins may be used to study pathological effects of abnormal hemoglobins under controlled experimental conditions. For example, RBCs from mice transgenic for Hb S sickle under deoxygenated conditions and may be utilized as a readily available blood source to screen possible antisickling agents. Hb S transgenic mice may be used as animal models to study pathophysiological changes *in vivo* under controlled experimental conditions.

Future Considerations

In summary, transgenic techniques provide a powerful approach for further studies of the structural and functional implications of mutated hemoglobins, the pathological consequences of abnormal hemoglobins, and the prospect of producing hemoglobins to be utilized as blood substitutes. Furthermore, investigations utilizing transgenic approaches will allow an increased understanding of expression and regulation of globin genes, which is important for the development of animal models and treatment of hemoglobinopathies. Such investigations of hemoglobin diseases also provide a basis for further refinement of techniques necessary for targeted insertion of genes into human cells, which will increase the prospects of gene therapy as the ultimate cure for certain genetic disorders.

[27] Transgenic Swine as a Recombinant Production System for Human Hemoglobin

By JOHN S. LOGAN and MICHAEL J. MARTIN

Introduction

Human hemoglobin, after appropriate modification, may be potentially useful as a red blood cell substitute for transporting oxygen from the lungs to the tissues. A recombinant source of human hemoglobin separate from the human donor pool has several potential advantages: unlimited supply, controlled and validated source, and use of genetically modified variants. In this chapter, we will describe the steps necessary to produce recombinant human hemoglobin in the red blood cells of pigs and illustrate the separation of human and porcine hemoglobins by ion-exchange chromatography. Although hemoglobin A is the initial molecule, a similar process would be appropriate with hemoglobin variants.

Expression Cassettes

There has been extensive research on the expression and regulation of human globin genes in transgenic mice[1,2] and, although it is not yet known at the same level of detail which of the sequences are necessary for expression in transgenic pigs, several features are likely to be important.

Locus Control Region

The locus control region (LCR) element is necessary for high-level expression of globin genes in transgenic mice.[3] In the human β-globin locus the LCR is contained within a series of DNase I-hypersensitive sites that are located 5′ to the human ε gene. Deletion analysis of this region has revealed important regulatory sites and has defined a minimum sequence thought to be necessary for all functional activity of the LCR. A construct containing this region, the microlocus, is illustrated in

[1] S. H. Orkin, *Cell (Cambridge, Mass.)* **63**, 665 (1990).
[2] T. M. Townes and R. R. Behringer, *Trends Genet.* **6**, 219 (1990).
[3] F. Grosveld, G. Blom van Asendelft, D. R. Greaves, and G. Kollias, *Cell (Cambridge, Mass.)* **51**, 975 (1987).

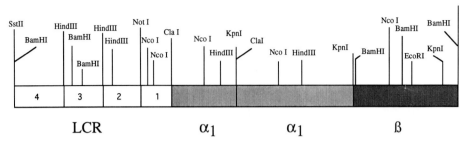

FIG. 1. Diagram of the LCR $\alpha_1\alpha_1\beta_1$ construct. This 16.9-kb DNA vector contains four DNase I-hypersensitive sites cloned from the human β-globin locus control region (LCR), two copies of the human α_1-globin gene, and a single copy of the human adult β-globin gene.[4,5]

Fig. 1.[4,5] This sequence is an essential component of all constructs directed at obtaining expression in cells of the red blood cell lineage. The LCR has several functions that are very interesting and unique. It is thought to confer copy-number dependence and integration-site independence based on transgenic mice experiments.[3] Integration-site independence suggests that the LCR can insulate this region from the surrounding genomic sequences and was established from the observation that all mice containing this sequence expressed the gene. Copy-number dependence implies that as the copy number increases so does the level of expression. To date, in transgenic pigs it is not yet possible to establish whether all of these features are maintained, although all transgenic pigs containing this sequence obtained to date have been shown to express.[6]

Globin Genes

The construct should contain the genes for α- and β-globin and generally genomic sequences are preferred. These genomic sequences contain upstream and downstream regulatory elements and the coding sequences. In addition to having α- and β-globin genes, it is important that they be expressed at approximately equal levels. Unbalanced expression, at least in the case of transgenic mouse experiments, can lead to thalassemia.[4] In initial published experiments in transgenic swine, α-globin was expressed

[4] O. Hanscombe, M. Vidal, J. Kaeda, L. Luzzatto, D. R. Greaves, and F. Grosveld, *Genes Dev.* **3**, 1572 (1989).

[5] D. R. Greaves, P. Fraser, M. A. Vidal, M. J. Hedges, D. Ropers, L. Lazzatto, and F. Grosveld, *Nature (London)* **343**, 183 (1990).

[6] J. S. Logan and M. J. Martin, unpublished observations.

at a higher level than β-globin and it is likely that future constructs should take this into account.[7]

In summary, the preferred construct should contain an LCR and the appropriate α- and β-globin genes. An illustration of such a construct is shown in Fig. 1.

DNA Preparation for Microinjection

In making DNA suitable for microinjection, one should aim for 1 to 5 μg of DNA as a single linear DNA fragment. The DNA fragment should contain no vector sequences, i.e., plasmids or bacteriophages. It is thought that although these sequences have little or no effect on integration frequencies, they may inhibit expression of the transgene. The following steps are suggested.

1. Purify the vector (e.g., plasmid, cosmid, bacteriophage) containing the gene construct by cesium chloride density gradient centrifugation or by ion-exchange column chromatography (see Sambrook et al.,[8] pp. 15, 16).

2. Digest the construct with appropriate restriction endonucleases to remove all or most extraneous (e.g., prokaryotic) sequences. In the construct illustrated in Fig. 1, this would be achieved by digestion with SstII.

3. Isolate the linear DNA fragment by gel electrophoresis and electroelution. As illustrated for the construct in Fig. 1, the two fragments are 16.9 kb, containing the globin sequences, and approximately 3 kb, containing the vector sequences.

4. Following electroelution, purify the linear DNA fragment using ion-exchange column chromatography [e.g., Schleicher and Schuell, (Keene, New Hampshire) Elutip or Qiagen (Chatsworth, CA)] followed by ethanol precipitation. The columns should be prepared and run as described by the manufacturer.

5. The linear DNA fragment should be suspended in Tris–EDTA buffer (10 mM Tris, 0.25 mM EDTA, pH 7.5). The concentration of DNA can be estimated by a number of methods. We generally use agarose gel electrophoresis in the presence of 0.5 μg/ml ethidium bromide and visual comparison of the fluorescent intensity of the band after comparison to known standards.

6. Immediately prior to microinjection the DNA is diluted to 1–2 ng/μl in Tris–EDTA buffer.

---

[7] M. E. Swanson, M. J. Martin, J. K. O'Donnell, K. Hoover, W. Lago, V. Huntress, C. T. Parsons, C. A. Pinkert, S. Pilder, and J. S. Logan, *Bio/Technology* **10**, 556 (1992).

[8] J. Sambrook, E. F. Fritsch, and T. Maniatis, "Molecular Cloning: A Laboratory Manual." Cold Spring Harbor Lab., Cold Spring Harbor, NY, 1989.

Production of Fertilized Eggs for Microinjection

To date, the only proven method for the production of transgenic swine is by nuclear microinjection of fertilized eggs, generally at the one-cell stage. Unfortunately, the efficiency associated with this process is low. Only 0.31 to 1.73% of porcine ova injected with DNA can be expected to develop into integration-positive swine.[9] Because of this inherent inefficiency, it is imperative that performance be maximized throughout all phases of transgenic swine production: ova production (superovulation), breeding management, ova recovery and culture, DNA microinjection, and embryo transfer.

Estrous synchronization can be best achieved by the feeding of an orally active progestin (allyl trenbolone)[10-13] at a dose of 15–18 mg/pig/day to sexually mature gilts for 14 to 21 days. This is followed by superovulation of potential ova donors. Superovulation provides two benefits: it increases the yield of ova and enables one to determine accurately the optimum time at which to recover one- and two-cell ova for subsequent DNA microinjection. Superovulation of sexually mature gilts is achieved through the administration of pregnant mare serum gonadotropin (PMSG: 1250 to 1500 IU) 24 hr after the last feeding of allyl trenbolone, followed 72 to 80 hr later by a single injection of human chorionic gonadotropin (hCG; 500 to 750 IU).[14-16]

The dosage of PMSG that each pig receives depends on the target ovulation rate one chooses and the hormone source. An adequate dose of PMSG and hCG will generate an ovulation response somewhere between 20 and 30 per female. As the ovulatory response to PMSG/hCG approaches or exceeds 40 per female, data indicate that the risk of obtaining immature, degenerate, or otherwise abnormal ova increases as well.[17] Martin et al.[18] have found that the ovulatory response to PMSG in swine varies with the source of the hormone. Hence, it is necessary to character-

[9] V. G. Pursel, D. J. Bolt, K. F. Miller, C. A. Pinkert, and R. E. Hammer, *J. Reprod. Fertil.* **40**, 235 (1990).
[10] D. L. Davis, J. W. Knight, D. B. Killian, and B. N. Day, *J. Anim. Sci.* **49**, 1506 (1979).
[11] D. A. Redmer and B. N. Day, *J. Anim. Sci.* **53**, 1089 (1981).
[12] V. G. Pursel, D. O. Elliot, C. W. Newman, and R. B. Staigmiller, *J. Anim. Sci.* **52**, 130 (1981).
[13] C. J. Ashworth, C. S. Haley, and I. Wilmut, *Theriogenology* **37**, 433 (1992).
[14] R. J. Wall, V. G. Pursel, R. E. Hammer, and R. L. Brinster, *Biol. Reprod.* **32**, 645 (1985).
[15] R. E. Hammer, V. G. Pursel, C. E. Rexroad, Jr., R. J. Wall, D. J. Bolt, K. M. Ebert, R. D. Palmiter, and R. L. Brinster, *Nature (London)* **315**, 680 (1985).
[16] W. F. Pope, M. H. Wilde, and S. Xie, *Biol. Reprod.* **39**, 882 (1989).
[17] W. Holtz and B. Schlieper, *Theriogenology* **35**, 1237 (1991).
[18] M. J. Martin, B. A. Didion, and C. L. Markert, *Theriogenology* **32**, 929 (1989).

ize the ovulatory response to PMSG with each new source. The gilts are bred at 12 and 24 hr after they demonstrate the onset of estrus.

Ova Recovery

Surgical recovery of one- and two-cell ova for microinjection is performed between 60 and 66 hr after hCG administration.[14,15,19] General anesthesia may be induced in swine by administering 0.4 mg acepromazine/kg body weight + 7.5 mg ketamine/kg body weight dissolved in 10 ml 0.9% saline through a peripheral ear vein.

Following anesthetization, the reproductive tract is exposed via a mid-ventral laparotomy. Once the oviduct has been located, a drawn glass cannula (o.d., 5 mm; length, 8 cm) is inserted into the ostium a distance of 3 cm and anchored in place by a single 2-O silk tie through the surrounding mesosalpinx.[20,21] A 20-gauge needle is inserted through the oviduct wall and into the lumen at a point just superior to the uterotubal junction. Dulbecco's phosphate-buffered saline (10 ml) supplemented with 2% bovine serum albumin (BSA) and warmed to 38° is infused through the needle into the oviduct. The oviduct is then stripped from the uterotubal junction toward the ostium.

Medium is collected in 15-ml sterile plastic tubes. Flushings are transferred to sterile 60 × 15-mm Petri dishes and searched at low (50×) power using a stereomicroscope. Ova are washed twice in G medium (Table I)[19] and transferred to microdrops of the same medium that have been overlaid with silicon oil. Ova are stored at 38° under a 90% N_2, 5% O_2, and 5% CO_2 atmosphere.

Ova Culture

Recovery, microinjection, and transfer of ova should take no longer than 4 to 6 hr. Although porcine ova can be cultured much longer than 4 to 6 hr, pregnancy and embryo survival rates will decrease as the period of culture is prolonged.[22–24]

[19] C. A. Pinkert, D. L. Kooyman, A. Baumgartner, and D. H. Keisler, *J. Reprod. Fertil.* **87**, 63 (1989).
[20] C. K. Vincent, O. W. Robison, and L. C. Ulberg, *J. Anim. Sci.* **23**, 1084 (1964).
[21] B. N. Day, *Theriogenology* **11**, 27 (1979).
[22] J. E. James, P. D. Reeser, D. L. Davis, E. C. Straiton, A. C. Talbot, and C. Polge, *Theriogenology* **14**, 463 (1980).
[23] J. E. James, D. M. James, P. A. Martin, D. E. Reed, and D. L. Davis, *J. Vet. Med. Assoc.* **183**, 525 (1983).
[24] B. Blum-Reckow and W. Holtz, *J. Anim. Sci.* **69**, 3335 (1991).

TABLE I
G Medium for Pig DNA Culture

Component	Concentration (g/100 ml)
NaCl	0.7017
KCl	0.036
$CaCl_2 \cdot 2H_2O$	0.025
KH_2PO_4	0.016
$MgSO_4 \cdot 7H_2O$	0.029
$NaHCO_3$	0.211
Dextrose (powder)	0.100
Disodium EDTA	0.0037
BSA	1.50
Phenol red (0.5%)[a]	0.2 ml

[a] From Gibco, #630-5100; (Grand Island, NY).

Ova Preparation

In contrast to murine ova, visualization of nuclear structures in one- and two-cell porcine ova is impossible without the aid of centrifugation.[15] Prior to microinjection, 10 to 15 porcine ova are placed in a 2-ml Eppendorf tube that contains 1 ml of G medium (Table I; with 20.85 mM HEPES substituted for sodium bicarbonate) and centrifuged for 6 min at 15,000 g. Initial centrifugation should reveal pronuclei or nuclei in at least 60% of the ova.[25] A brief period of culture (2 to 4 hr) followed by a second 4-min period of centrifugation can reveal nuclear structures in an additional proportion of the remaining ova.[18]

Microinjection Procedure(s)

A detailed description of the microscope optics and equipment necessary for this process is given in Hogan et al.[26] Preparatory steps include the production of holding and injection pipettes, infusion of the microinjection lines with an appropriate fluid, and attachment of the pipettes to their respective lines. Holding pipettes should be drawn so that the tip o.d. and i.d. are approximately 35 and 10–15 μm, respectively. The injection pipette tip o.d. should not exceed 1–2 μm and is initially closed. The tip can be opened and sharpened by breaking it on the edge of the holding

[25] G. Brem, K. Springham, E. Meier, H. Kraublich, B. Brenig, M. Muller, and E. L. Winnacker, "Transgenic Models in Medicine and Agriculture." 1989. 116:61–72.

[26] B. Hogan, F. Costantini, and E. Lacy, "Manipulating the Mouse Embryo: A Laboratory Manual." Cold Spring Harbor Lab., Cold Spring Harbor, NY, 1986.

pipette or grinding it on a diamond-dust-coated wheel. The microinjection lines are filled with one of the following fluids: silicon oil (100 and 350 viscosity fluids mixed to give a final viscosity of 175) or Fluorinert Electronic Liquid FC77 (3M Company, St. Paul, MN).

One- or two-cell ova are placed in a microdrop of G medium (Table I) containing HEPES buffer (15 ova per drop) on a depression slide. Another 1- to 2-μl drop containing the DNA–buffer solution is placed above or below this drop. Both drops are covered with silicon or paraffin oil.

The injection pipette is loaded by drawing up medium first, followed by DNA. When the DNA solution has stopped flowing into the injection pipette, the syringe plunger is turned down until DNA begins to flow slowly out of the tip. The injection pipette is then transferred to the microdrop containing ova and the magnification increased to 250×. Once an ovum with visible pronuclei or nuclei has been located and immobilized, the injection pipette is inserted through the zona pellucida and into the pronucleus of a one-cell ovum or nucleus of a two-cell ovum. On penetration of the membrane the pronucleus or nucleus should immediately begin to expand. At the instant expansion ceases the injection pipette is withdrawn and the ovum is released. It is estimated that an average of 1 to 2 pl of DNA solution is injected per nuclear structure.[26]

Ova Transfer

Ova injected at the one- or two-cell stage are transferred to the oviducts of sexually mature recipients.[27–29] Recipients are anesthetized and the reproductive tract is exposed as previously described. Ova and 1–2 ml of medium are aspirated into the tubing obtained from a 21-gauge × 3/4 inch infusion set. The tube is passed through the ostium of the oviduct until it reaches the isthmus. Ova are expelled as the tube is slowly withdrawn.

The pregnancy rate among recipients that receive either nonmicroinjected or microinjected ova appears to be directly proportional to the number of ova transferred.[25,30]

An indication of the overall efficiency of the process is given in Table II using a construct capable of producing human hemoglobin.

[27] P. J. Dziuk, C. Polge, and L. E. A. Rowson, *J. Anim. Sci.* **23,** 37 (1964).
[28] C. E. Pope and B. N. Day, *J. Anim. Sci.* **44,** 1036 (1977).
[29] D. H. Segal and R. D. Baker, *J. Anim. Sci.* **37,** 762 (1973).
[30] C. E. Pope, R. K. Christenson, V. A. Zimmerman-Pope, and B. N. Day, *J. Anim. Sci.* **35,** 805 (1972).

TABLE II
Transgenic Pig Production: Construct #263

Trial performance		Efficiency	
Parameter	Number	Parameter	Number[a]
Total donors	129	Total ova collected	1477
Donors used	63	Total fertilized	1326
Recipients used	21	Total injected	710
Farrowings (births)	12	Injected ova transferred	710
Pigs born	69	Control ova transferred	22
Transgenic pigs	5	Recipients used	21
Expression	5	Pigs born (male, female)	33, 36
		Transgenics (male, female)	4, 1 (0.70%)[b]
		Expression	5

[a] Data are from five trials.

[b] Overall efficiency (percentage of transgenic offspring that developed from transferred microinjected ova).

Identification of Transgenics

A biopsy of the tail of the piglet is taken 1 day after birth and a small amount of blood is collected. The tail is processed to make DNA, and integration can be determined by a number of standard molecular biology techniques, including Southern blot analysis at high stringency [final wash conditions, $0.1 \times$ SSC ($1 \times$ SSC is 0.15 M NaCl, 0.015 M sodium citrate, pH 7.0) 0.1% SDS (sodium dodecyl sulfate)] at 65° or polymerase chain reaction using appropriate primers.

The expression of the construct can be determined by isoelectric focusing (IEF) analysis of the blood. About 0.4 μg of total hemoglobin from a hemolysate of human and transgenic pig blood is subjected to isoelectric focusing on a 1% agarose gel containing a gradient from pH 6–8. Electrophoresis is performed exactly as described by the manufacturer (IsoLabs, Inc., Akron, Ohio). Gels are then immediately fixed in 10% trichloroacetic acid without staining and are photographed directly. Quantitation can be achieved by laser densitometry. A representative IEF analysis is shown in Fig. 2. Under these conditions, hemoglobins run as an $\alpha\beta$ dimer. There are four potential dimers, but only three are formed *in vivo*.[7] The three dimers are human α, human β; pig α, pig β; and the hybrid human α, pig β.

Purification of Human Hemoglobin

The first step is the collection of the blood from the pig. This can be accomplished by a number of techniques, the most favored of which is

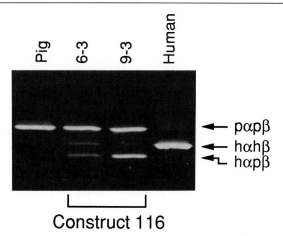

Construct 116

FIG. 2. Analysis of hemoglobin expression. Isoelectric focusing in agarose gels is employed for the detection of human hemoglobin in transgenic swine. A gradient of pH 6–8 is used to separate the isoforms of hemoglobin. Two transgenic samples are shown. In both animals, human and pig–human hybrid hemoglobins as well as the endogenous pig hemoglobin can be detected.

venipuncture of the jugular vein. The blood is washed by standard procedures to obtain packed red blood cells, and stroma-free hemoglobin is prepared by one of a number of techniques listed in this volume of *Methods in Enzymology*. At this stage the primary chromatography step involves the separation of porcine and human hemoglobin. This can be achieved by the use of a number of different ion-exchange resins. A brief description of the conditions for these resins is given.

pH

Because of the closeness of the p*I* values of the human, pig, and hybrid proteins, the choice of pH for each chromatographic resin is critical. Therefore, for each resin a series of pH scouting runs is performed and the optimal pH is chosen. Routinely the pH for the scouting runs range from pH 7.5 up to pH 8.5.

Mono P

Hemolysate (2.6 mg) from a transgenic pig is loaded onto a Mono P 5/20 column (Pharmacia, Piscataway, NJ) equilibrated with buffer A (10 mM Tris, pH 8.10). Hemoglobins are eluted with a 0–75 mM NaCl gradient in buffer A at a flow rate of 1.0 ml/min. With a 4-ml column,

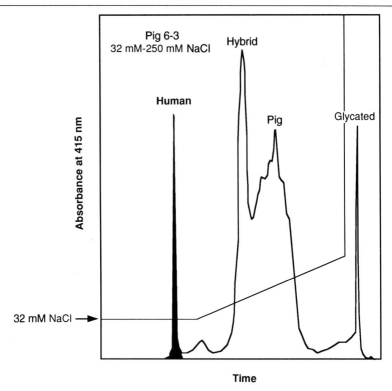

FIG. 3. Purification of human hemoglobin. Ion-exchange chromatography on a Mono P resin was employed for the separation of human hemoglobin in the hemolysate of a transgenic pig. Under the isocratic elution conditions shown, human hemoglobin flows through at 32 mM NaCl. Samples for each peak were analyzed by isoelectric focusing.

a 60-ml gradient is utilized. Detection of the proteins is followed by recording the absorbance at 405 nm. Peak fractions are collected, concentrated (Centricon-10 concentrators, Amicon, Inc., Beverly, MA), and identified by IEF gel analysis (as described previously). Recombinant human Hb elutes first followed by the hybrid and pig hemoglobins. The surprising observation that the human hemoglobin elutes prior to the other hemoglobins led to the possibility that the human hemoglobin could be purified with a batch procedure. To demonstrate the feasibility of this, 2.6 mg of hemolysate was loaded onto the same Mono P column and the human hemoglobin was eluted with buffer A containing 32 mM NaCl at 1.0 ml/min to afford essentially pure recombinant human hemoglobin (Fig. 3).

Toyopearl DEAE

Alternatives to the Mono P resin have been tested. One of these resins is the Toyopearl DEAE resin from Tosohaus, Inc. (Montgomeryville, PA). Total Hb (3.2 mg) is loaded onto an HR 5/20 column (4 ml) packed with Toyopearl DEAE 650M resin that has been equilibrated with buffer B (10 mM Tris, pH 7.9, and 20 mM glycine). The hemoglobins are eluted from the column with a 0–50 mM NaCl gradient in buffer B at a flow rate of 1 ml/min. The total gradient volume is 70 ml.

Q-Sepharose Fast Flow

Hemolysate (100 mg) is loaded onto a Pharmacia HR 16/50 column (100-ml bed volume) packed with Q-Sepharose Fast Flow resin that has been equilibrated with buffer C (10 mM Tris, pH 7.7, and 20 mM glycine). Hemoglobins are eluted with a 2-liter gradient from 0 to 50 mM NaCl in buffer C at a flow rate of 5 ml/min. Using these conditions, the recombinant human hemoglobin again elutes first and can be separated from both the hybrid and pig hemoglobins present in the hemolysate.

Clearly, the amount of hemoglobin loaded, the size of the column, and batch versus gradient elution can be modified as appropriate for the specific aim of the separation.

Section VI

Hemoglobin Stability and Degradation

[28] Detection, Formation, and Relevance of Hemichromes and Hemochromes

By Joseph M. Rifkind, Omoefe Abugo, Abraham Levy, and Jane Heim

Native hemoglobins contain one histidine coordinated to the iron on the proximal side of the heme and another histidine in the ligand pocket on the distal side of the heme. Oxygen located in this pocket in oxyhemoglobin can be replaced by other neutral ligands, such as carbon monoxide, nitric oxide, and alkylisocyanide in the reduced Fe(II) hemoglobins. The iron in deoxyhemoglobin is in a high-spin state, whereas all of the Fe(II) complexes are in a low-spin state. Oxidized Fe(III) hemoglobin ligands bound in this pocket include water, hydroxide, fluoride, azide, thiocyanate, imidazole, and cyanide. The spin state for these complexes depends on the strength of the axial ligand and can be high spin (e.g., fluoride), low spin (e.g., cyanide), or a mixture (e.g., hydroxide).

Low-spin hemoglobin complexes also exist when the exogenous ligand is replaced by an endogenous amino acid side chain.[1] These low-spin complexes define the Fe(II) hemochromes[2] and Fe(III) hemichromes.[3] For many other heme proteins, low-spin complexes involving axial coordination of two amino acid side chains are functional. This is the case for cytochrome *c*, wherein the heme is bound to a histidine imidazole and a sulfur atom of methionine.[4] For cytochrome *P*-450,[5,6] in which oxygen is a substrate for hydroxylation reactions, some of the functional states contain two amino acids bound to the iron, and in other states one of these axial ligands is displaced by an exogenous ligand, i.e., oxygen.

For hemoglobin, the only amino acid side chain in the ligand pocket is the distal histidine. The N^ε-nitrogen of this histidine is located more than 4 Å from the iron in hemoglobin and is therefore not expected to coordinate in native hemoglobin. However, under certain conditions it does bind to the iron, forming low-spin hemi(hemo)chromes. It has been

[1] E. A. Rachmilewitz, J. Peisach, and W. E. Blumberg, *J. Biol. Chem.* **246**, 3356 (1971).
[2] K. H. Mayo, D. Kucheida, F. Parak, and J. C. W. Chien, *Proc. Natl. Acad. Sci. U.S.A.* **80**, 5294 (1983).
[3] J. Peisach, W. E. Blumberg, and A. Adler, *Ann. N.Y. Acad. Sci.* **206**, 310 (1973).
[4] D. L. Brautigan, B. A. Feinberg, B. M. Hoffman, E. Margoliash, J. Peisach, and W. E. Blumberg, *J. Biol. Chem.* **252**, 574 (1977).
[5] M. Chevion, J. Peisach, and W. E. Blumberg, *J. Biol. Chem.* **252**, 3637 (1977).
[6] J. H. Dawson, L. A. Andersson, and M. Sono, *J. Biol. Chem.* **257**, 3606 (1982).

argued that the formation of any hemi(hemo)chrome defines a denatured state of hemoglobin.[7] However, recent evidence indicates that some of these hemi(hemo)chrome states are actually part of the distribution of substates populated, although to a lesser extent, in native hemoglobin.[8,9] This is particularly true for the complexes formed from the more stable Fe(II) hemoglobin.[9]

Because of the connotation of a denatured state when using the term hemi(hemo)chrome, we will also use the term bishistidine when referring to the low-spin complexes formed with the distal histidine, which may be associated with native hemoglobin. These states can be distinguished from those formed under denaturing conditions by reversibility and the absence of subsequent reactions such as precipitation[10] and dissociation of the heme.[7,11,12] Under certain conditions, it has even been suggested that some of the hemichromes involve coordination with the β93 cysteine[13] as well as other more remote amino acids.[3] Such complexes clearly require major rearrangements of the globin conformation and must be considered a form of denatured hemoglobin.

In this review we will first discuss the methods for determining the presence of these low-spin complexes. The characterization of different types of hemichromes, primarily on the basis of electron paramagnetic resonance (EPR), will be reviewed. We will then discuss the different ways of forming hemi(hemo)chromes. This review will range from processes clearly dissociated with denaturation to those clearly indicative of native substates. Finally, the significance of hemi(hemo)chrome formation will be discussed. The information regarding hemoglobin that can be obtained from studies of hemi(hemo)chrome formation will be discussed as well as the possible functional role of bishistidine complex formation.

Detection of Hemi(hemo)chromes

Visible spectra for these low-spin complexes are distinct from the usual high-spin and low-spin complexes. The visible spectra of the Fe(II) hemochromes[1,14] have bands at 529 and 558 nm, with the extinction coefficient at the peak of the shorter wavelength band about one-half of that

[7] E. A. Rachmilewitz, *Semin. Hematol.* **11**, 441 (1974).

[8] A. Levy, J. C. Walker, and J. M. Rifkind, *J. Appl. Phys.* **53**, 2066 (1982).

[9] A. Levy and J. M. Rifkind, *Biochemistry* **24**, 6050 (1985).

[10] V. W. Macdonald and S. Charache, *Biochim. Biophys. Acta* **701**, 39 (1982).

[11] C. C. Winterbourn, *Semin. Hematol.* **27**, 41 (1990).

[12] C. C. Winterbourn and R. W. Carrell, *J. Clin. Invest.* **54**, 678 (1974).

[13] L. M. Neto, M. Tabak, and O. R. Nascimento, *J. Inorg. Biochem.* **40**, 309 (1990).

[14] D. L. Drabkin, *J. Biol. Chem.* **146**, 605 (1942).

of the longer wavelength band. These spectra are clearly different[15] from the low-spin complexes with exogenous ligands and for deoxyhemoglobin without any axial ligand in the sixth position. However, all of the different hemoglobin complexes absorb in the same spectral region and it is therefore difficult to quantitate and even detect low levels of hemochrome by visible spectroscopy. This difficulty is further complicated if all hemochromes do not have identical absorption spectra, in which case even multicomponent fitting will not resolve the problem.

The ideal method to identify, quantitate, and characterize hemochromes in the presence of other components is Mössbauer spectroscopy.[2,9,15a-18] This method, primarily used on iron systems, measures the recoilless absorption. For studies on iron, the ^{57}Fe isotope is used to absorb the 14.4-keV γ-ray emitted from the $I = 3/2$ excited state formed when ^{57}Co decays to ^{57}Fe. In the Mössbauer spectrometer the ^{57}Co source is mounted on a drive unit that moves the source through a range of velocities. The Doppler shift produces small, well-defined changes in the energy of the emitted γ-ray. The ability to detect small changes in energy provides a very sensitive probe of the electronic state of the heme iron. The Mössbauer spectrum is influenced by the oxidation state, the spin state, and the ligands bound to the heme. The parameters that define the spectrum are the isomer shift (the chemical shift) and the quadrupole splitting. The isomer shift, i.e., the shift in the absorption, is determined by the electron density at the nucleus. The quadrupole splitting, which splits the absorption, is determined by the asymmetric electronic charge distribution around the nucleus. For paramagnetic complexes, a contribution from the magnetic field of the nucleus is also obtained. Changes in the spectrum and its intensity with temperature also provide valuable dynamic information in the region of the iron.

The low-spin hemochrome has an isomer shift of 0.45 mm/sec and a quadrupole splitting of 1.08 mm/sec (Fig. 1). Both of these parameters, which are very similar to those of the low-spin complex made by reducing the pyridine complex formed by adding basic pyridine to heme proteins, are clearly distinct from other hemoglobin complexes[9] of both Fe(II) and Fe(III). The ability to detect even low levels of hemochrome in the pres-

[15] E. Antonini and M. Brunori, "Hemoglobin and Myoglobin in Their Reactions with Ligands," p. 19. North-Holland Publ., Amsterdam and London, 1971.

[15a] B. Balko, this series, Vol. 76, p. 329.

[16] W. S. Caughey, W. Y. Fujimoto, A. J. Bearden, and T. H. Moss, *Biochemistry* **5**, 1255 (1966).

[17] R. W. Grant and L. E. Topol, *Biophys. J.* **9**, 1446 (1969).

[18] G. C. Papaefthymiou, B. H. Huynh, C. S. Yen, J. L. Groves, and C. S. Wu, *J. Chem. Phys.* **62**, 2995 (1975).

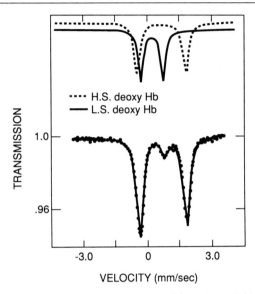

FIG. 1. Mössbauer absorption spectrum at 87 K of deoxyhemoglobin. High-spin (H.S.) deoxyHb is the normal deoxyhemoglobin; low-spin (L.S.) deoxyHb is a hemochrome. Most of this hemochrome was produced during a 14-hr incubation at 210 K. The ability to resolve the low-spin complex (hemochrome) is indicated by the weak absorption at ~1 mm/sec. Adapted with permission from Levy and Rifkind.[9] Copyright 1985 American Chemical Society.

ence of deoxyhemoglobin is further facilitated by the fact that the high-velocity line for the hemochrome is clearly separated from the Mössbauer lines of other Fe(II) complexes.

Nevertheless, the use of Mössbauer spectroscopy to study hemochromes has been very limited. This phenomenon can be attributed to several factors: (1) Mössbauer spectroscopy requires specialized equipment, which is not generally available. (2) The method is very insensitive. Millimolar concentrations (high for most other spectroscopies) are the lower limits for Mössbauer studies. Signal averaging for at least hours, but often days, is required for good signal/noise. (3) The Mössbauer method only detects the ^{57}Fe isotope. Unless substitution of the dominant ^{56}Fe is performed, it is almost impossible to obtain good spectra.

For FE(III) complexes, a visible spectrum with a peak at 535 nm and a shoulder at 565 nm is considered indicative of a hemichrome[1] and can be distinguished from the other high-spin or low-spin complexes of methemoglobin. However, the identification and quantification by visible spectroscopy is even more uncertain for hemichromes than for hemochromes. This conclusion is based on the poorly defined methemoglobin spectrum with mixtures of high-spin and low-spin complexes[19] superimposed on differences for different exogenous ligands. The spectra also depend on pH, temperature, and ionic strength. Furthermore, a number of different classes of hemichromes have been identified by electron paramagnetic resonance (EPR) and it is nevertheless assumed that the visible spectra for all these hemichromes are the same. In organic solvents with simple heme complexes it was in fact shown that different classes of hemichromes can have different visible spectra.[20] The analysis of hemichromes by visible spectroscopy is further complicated by the turbidity that frequently follows hemichrome formation.[10,21,22]

Mössbauer spectroscopy also can detect hemichromes.[8,9] However, the resolution is not nearly as good as for Fe(II) hemoglobin. The isomer shift of 0.77 mm/sec and the quadrupole splitting of 2.12 mm/sec are only 0.04 mm/sec lower and 0.47 mm/sec higher, respectively, than for the low-spin hydroxide.

EPR[22a] is the ideal method to identify, quantitate, and characterize hemichromes.[3,7,8,13,23–29] This method, like Mössbauer spectroscopy, probes the metal center of the hemoglobin and is sensitive to the distribution of electrons around the iron. However, EPR requires an unpaired electron and is therefore unable to detect the diamagnetic hemochromes.

The low-spin hemichromes are clearly distinct from the high-spin Fe(III) complexes (Fig. 2). However, the strength of EPR for studies on hemichromes is associated with the rhombic distortion of the electron distribution of low-spin hemoglobin complexes. This phenomenon is attributed to the fact that the axial ligands, e.g., the proximal and distal

[19] T. Iizuka and M. Kotani, *Biochim. Biophys. Acta* **194**, 351 (1969).

[20] J. Peisach and W. B. Mims, *Biochemistry* **16**, 2795 (1977).

[21] M. Brunori, G. Falcioni, E. Fioretti, B. Giardina, and G. Rotilio, *Eur. J. Biochem.* **53**, 99 (1975).

[22] V. W. MacDonald and S. Charache, *J. Lab. Clin. Med.* **102**, 762 (1983).

[22a] W. E. Blumberg, this series, Vol. 76, p. 312.

[23] A. Levy, P. Kuppusamy, and J. M. Rifkind, *Biochemistry* **29**, 9311 (1990).

[24] H. Rein and O. Ristau, *Biochim. Biophys. Acta* **94**, 516 (1965).

[25] J. Peisach, W. E. Blumberg, B. A. Wittenberg, J. B. Wittenberg, and L. Kampa, *Proc. Natl. Acad. Sci. U.S.A.* **63**, 934 (1969).

[26] K. Gersonde and A. Wollmer, *Eur. J. Biochem.* **15**, 226 (1970).

[27] H. Rein, O. Ristau, G. R. Jänig, and F. Jung, *FEBS Lett.* **15**, 21 (1971).

[28] H. Rein, O. Ristau, and K. Ruckpaul, *Biochim. Biophys. Acta* **393**, 373 (1975).

[29] E. A. Rachmilewitz, J. Peisach, T. B. Bradley, Jr., and W. E. Blumberg, *Nature (London)* **222**, 248 (1969).

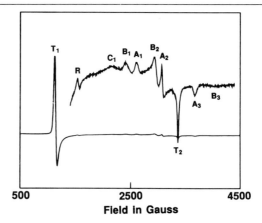

FIG. 2. EPR spectrum of methemoglobin at pH 7.6. Despite the fact that the large majority of the hemoglobin is high spin (T_1, T_2), by a 20-fold expansion of the y axis it is possible to discern three low-spin complexes (A, B, C). Complex A corresponds to the low-spin hydroxide and complexes B and C are two different hemichromes. R is associated with a rhombic high-spin complex. Reprinted with permission from Levy et al.[23] Copyright 1990 American Chemical Society.

histidines, also influence the distribution of electrons in the heme plane. In the derivative EPR spectrum this rhombic distortion is indicated by three different g values.

The only other low-spin heme complex in addition to the hemichromes in the absence of other exogenous ligands is the low-spin hydroxide, which has a well-defined EPR spectrum, with $g_1 = 2.59$, $g_2 = 2.17$, and $g_3 = 1.83$. The wide range of possible g values from above 3 to below 1 makes it possible to characterize and quantitate the hemichrome complexes within one hemoglobin sample even in the pH range where appreciable hydroxide complex is present or in a large excess of EPR-silent Fe(II) hemoglobin.

Characterization of Different Classes of Hemichromes by EPR

The three g values for low-spin ($S = \frac{1}{2}$) heme complexes originate from the distortion of the octahedral symmetry around the Fe(III). The tetragonal (V) and rhombic (Δ) splitting of the ligand field can be calculated from the three g values using the single-hole Griffith model.[30–32] From

[30] J. S. Griffith, Proc. R. Soc. London, Ser. A 235, 23 (1956).
[31] T. L. Bohan, J. Magn. Reson. 26, 109 (1977).
[32] C. P. S. Taylor, Biochim. Biophys. Acta 491, 137 (1977).

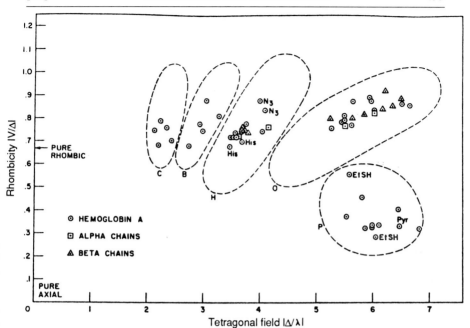

FIG. 3. Crystal field parameters for ferric low-spin compounds of hemoglobin and its isolated subunits. The five different types of hemichromes are indicated by the regions C, B, H, O, and P. Unless indicated otherwise, all points are for compounds formed from ligands endogenous to the hemoglobin molecule. Low-spin compounds can also be formed using the following reagents: EtSH, mercaptoethanol; Pyr, pyridine; His, histidine; N_3, azide. Reprinted with permission from Blumberg and Peisach.[33] Copyright 1971 American Chemical Society.

these parameters[3,5,33] and the spin-orbit coupling parameter (λ), V/Δ and Δ/λ have been defined as a measure of rhombicity and tetragonality, respectively. The rhombicity was considered a geometric term representative of the asymmetric distribution of the electron density, whereas the tetragonality was considered representative of the electron density at the iron. In a plot (Fig. 3) of V/Δ versus Δ/λ, each heme complex was represented by a point, and these points fell into clusters defining five types of low-spin heme complexes (C, B, H, O, and P).[3,33] Many sample complexes as well as heme proteins have EPR spectra that fall into one of these classes.

For heme proteins, all of which contain the same Fe-porphyrin, differences in the ligand field parameters are determined by the two axial ligands.

[33] W. E. Blumberg and J. Peisach, *Adv. Chem. Ser.* **100**, 271–291 (1971).

It was therefore suggested that the different types of complexes identify the nature of the axial ligand and not the particular protein. Thus complexes with a hydroxyl group trans to a histidine fall in the O region with very similar parameters whether from myoglobin, hemoglobin, α chains, or β chains.[3] It was even proposed that in hemoglobin the hemichromes with different EPR parameters correspond to coordination with different amino acid side chains. Attempts were made to identify the ligands responsible for these different hemichromes by using simple heme complexes[20] with known axial ligands and heme proteins,[3] where the axial ligands had been determined by other methods.

Hemichromes B and H were both associated with a bisimidazole heme complex[20] expected for coordination of the distal histidine located in the heme pocket. Similar complexes are obtained with bisimidazole ferric heme and with cytochrome b_5, known to have two histidines coordinated to the iron.[34] Hemichrome C has the same EPR parameters as cytochrome c, where trans to the proximal histidine is a methionine sulfur.[4] Hemichrome P has a spectrum produced by the addition of a mercaptide to Fe(III) hemoglobin, which is similar to cytochrome P-450, where one of the axial ligands is a cysteine residue.[5,6]

It is surprising that all four types of hemichromes are formed with hemoglobin. The closest methionine to the heme is $\alpha32$ (B13), with the other two methionines, $\alpha76$ and $\beta55$, in the EF corner and in the $\alpha_1\beta_2$ interface, far removed from the heme. The enhanced formation of hemichrome C by the unstable hemoglobin Riverdale-Bronx[35] involves the substitution of $\beta24$ (B6) glycine by arginine. This substitution is nowhere in the region of a methionine residue, although gross conformational changes are expected by placing a charged amino acid at this internal site of the β chain. It is even more difficult to explain, in terms of methionine coordination, a type C hemichrome formed by incubating horse hemoglobin in the 200–250 K region.[8] Formation of this hemichrome, together with a B-type hemichrome formed under the same conditions, is completely reversible,[8] with the hemoglobin returning to the normal methemoglobin when the temperature is raised to 295 K.

Hemichrome P, consistent with coordination of a cysteine, may involve a displacement of the proximal histidine by the $\beta93$ cysteine located one residue away in the F helix. However, even this displacement requires a major rearrangement of the F helix.[13] Hemichrome P, originally formed

[34] F. S. Mathews, P. Argos, and M. Levine, *Cold Spring Harbor Symp. Quant. Biol.* **36,** 387 (1972).
[35] H. M. Ranney, A. S. Jacobs, L. Udem, and R. Zalusky, *Biochem. Biophys. Res. Commun.* **33,** 1004 (1968).

during irreversible denaturation, was consistent with such a major rearrangement. Interestingly, it has been shown that hemichrome P can also be formed reversibly by dehydration of Fe(III) hemoglobin,[13] with the usual high-spin forms of hemoglobin returned when the water content is increased above 0.4 (g H_2O/g Hb). However, even in this case involvement of the $\beta93$ cysteine is supported by the inability to form this hemichrome when the sulfhydryl is blocked. It should, however, be noted that a complex with EPR parameters similar to that of hemichrome P is also observed with bleomycin,[36,37] which does not contain a heme and for which sulfur is not considered to be one of the ligands.

The correspondence between EPR of heme proteins and the nature of the axial ligand has been very helpful in classifying a large body of EPR data. The EPR spectra, even for heme complexes, are, however, not only sensitive to the nature of the axial ligands, but also to the location and orientation of the two axial ligands relative to the heme and relative to each other.[38–42]

This conclusion is necessary to explain studies on cytochrome c_3 from *Desulfovibrio vulgaris*,[43,44] which has four heme centers with two axial histidines. Three distinct low-spin EPR components are detected with the highest g values, $g_z = 3.16$, 2.95, and 2.71. The complex with $g_z = 3.16$ is thought to have a 64° dihedral angle between the planes of the two imidazoles. In the other two complexes, the imidazole rings are nearly coplanar, with differences in the Fe–imidazole bond length.

In model heme complexes the EPRs for substituted imidazoles are found with the highest g value ranging from 3.5 to 2.8[39,40] From these studies it has been confirmed that the Fe–N (imidazole) bond length and the relative orientation of the two imidazole rings can have rather dramatic effects on the EPR. The effect of sterically hindered bases is seen by comparing[39] the nonhindered bis-4-methylimidazole complex $g_1 = 2.87$, $g_2 = 2.27$, $g_3 = 1.59$ with the hindered bis-2-methylimidazole with $g_1 = 3.51$, $g_2 = 1.7$, $g_3 = 0.55$. This could partially reflect a greater Fe–N (imidazole) bond distance, i.e., a weaker ligand for the sterically hindered

[36] Y. Sugiura, *J. Am. Chem. Soc.* **102**, 5208 (1980).
[37] A. Levy, P. T. Manoharan, J. M. Rifkind, J. C. Walker, F. C. Haberle, N. G. Kumar, J. D. Glickson, and G. A. Elgavish, *Biochim. Biophys. Acta* **991**, 97 (1989).
[38] G. Palmer, *Biochem. Soc. Trans.* **13**, 548 (1985).
[39] J. C. Salerno and J. S. Leigh, *J. Am. Chem. Soc.* **106**, 2156 (1984).
[40] C. T. Migita and M. Iwaizumi, *J. Am. Chem. Soc.* **103**, 4378 (1981).
[41] F. A. Walker, D. Reis, and V. L. Balke, *J. Am. Chem. Soc.* **106**, 6888 (1984).
[42] D. K. Geiger, Y. J. Lee, and W. R. Scheidt, *J. Am. Chem. Soc.* **106**, 6339 (1984).
[43] D. V. DerVartanian and J. LeGall, *Biochim. Biophys. Acta* **502**, 458 (1978).
[44] Y. Higuichi, M. Kusunoki, Y. Matsura, N. Yasuoka, and M. Kakudo, *J. Mol. Biol.* **172**, 109 (1984).

2-methylimidazole. However, X-ray data also indicate that the sterically hindered hemes prefer a perpendicular orientation[45,46] and a puckered porphyrin. In a protein, in addition to the bond length and the relative orientation of the two imidazole rings, the orientation of the imidazoles with respect to the hemes should also influence the EPR spectrum.

As a result of the multiple factors that can alter the ligand field parameters, it is not generally possible to determine uniquely the nature of the ligands and/or the configuration of the axial ligands from EPR spectra. The different classes of hemichromes, where both axial ligands are part of the globin chain, do, however, represent altered globin conformations in the region of the heme.

The low-temperature EPR studies of methemoglobin provide an example of two classes of hemichromes associated with the coordination of the distal histidine.[23,47] In one class the water is thought to be retained in the ligand pocket, thereby influencing the approach of the distal histidine to the iron. In a second class the water is ejected from the ligand pocket. These two classes of bishistidine complexes have EPR parameters of the earlier described type H and B, respectively.

It has been further shown that the lineshapes of low-spin EPR complexes can reflect a distribution of g values caused by conformational flexibility between closely related states.[48,49] At low temperature in the range of liquid helium, it is sometimes possible by EPR to resolve partially some of these differences in heme configuration. This is the case for the bishistidine complex formed at low temperature[23] for which the lineshape of the highest g value and how it changes with pH have been used to suggest multiple states (Fig. 4). In fact, for horse hemoglobin[8,23] two well-resolved components with all three g values different have been resolved at 11 K.

Formation of Hemichromes

Hemi(hemo)chrome Formation Associated with Disruption of Native Globin Structure

Hemi(hemo)chrome formation requires the coordination of endogenous amino acid side chains not normally bound to the iron. Therefore,

[45] R. G. Little, K. R. Dymock, and J. A. Ibers, *J. Am. Chem. Soc.* **97**, 4530 (1975).
[46] J. R. Kirner, J. L. Hoard, and C. A. Reed, *Abstr. Pap., 175th Natl. Meet. Am. Chem. Soc., Abstr. Inor.* 14 (1978).
[47] A. Levy, V. S. Sharma, L. Zhang, and J. M. Rifkind, *Biophys. J.* **61**, 750 (1992).
[48] J. C. Salerno, *J. Biol. Chem.* **259**, 2331 (1984).
[49] J. C. Salerno, *Biochem. Soc. Trans.* **13**, 611 (1985).

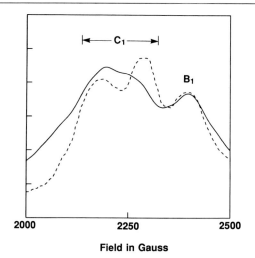

FIG. 4. The multiple class C bishistidine complexes formed during incubation of human (—) and horse (----) methemoglobin at pH 6.4. Reprinted with permission from Levy *et al.*[23] Copyright 1990 American Chemical Society.

it is to be expected that structural changes that disrupt the globin structure[3,7,11] can facilitate hemi(hemo)chrome formation. Consistent with this expectation is the observation that hemichromes are formed from the Fe(III) state much more readily than from the more stable Fe(II) state.[1,10,11,22] Furthermore, normal hemoglobin A is oxidized very slowly and the oxidized Fe(III) hemoglobin is converted to hemichromes very slowly.

Subunit Dissociation. An appreciable enhancement of hemichrome formation takes place when tetrameric hemoglobin is separated into α and β chains,[1,21,29,33,50,51] with hemichrome formation faster for α chains than β chains.[1,33] The relationship between dissociation and hemichromes is the result of conformational changes in the heme pocket that facilitate hemichrome formation when constraints associated with subunit interactions are relieved.

Disruption of just the $\alpha_1\beta_2$ interfaces when tetrameric hemoglobin dissociates into $\alpha\beta$ dimers at low concentrations produces a dramatic increase in the rates of autoxidation[52] without any noticeable increase

[50] R. Banerjee, Y. Alpert, F. Leterrier, and R. J. P. Williams, *Biochemistry* **8,** 2862 (1969).
[51] T. Yonetani, T. Iizuka, and M. R. Waterman, *J. Biol. Chem.* **246,** 7683 (1971).
[52] L. Zhang, A. Levy, and J. M. Rifkind, *J. Biol. Chem.* **266,** 24698 (1991).

in hemichrome formation. However, even in this case the enhanced autoxidation (see below) has been attributed to transient hemichrome formation.[53]

A coupling between subunit dissociation and hemichrome formation is most pronounced in the dimeric and tetrameric hemoglobin from the mollusc *Scapharca inaequivalvis*.[54,55] The dimeric hemoglobin after oxidation dissociates very rapidly in the time range of milliseconds into monomeric hemichromes. For the tetrameric hemoglobin, the oxidized form is first converted to a tetrameric hemichrome, which is subsequently dissociated into dimers and monomers.

Unlike hemoglobin chains, most myoglobins, even though they are monomeric, do not readily form hemichromes.[13,56-58] This has been attributed to a rather rigid distal pocket, at least in sperm whale myoglobin. Protozoan myoglobin,[59] with a number of amino acid deletions relative to sperm whale myoglobin, does, however, readily form hemichromes.

Denaturation. Hemoglobin A, which normally forms hemichromes very slowly, produces hemichromes more rapidly under denaturing conditions.[1,3,7,10,22,26,60-63] Thus hemichromes are produced in the course of denaturation by heat or by the addition of a number of reagents, including H_2O_2, nitrite, menadione, acetylphenylhydrazine, salicylate, benzoate, urea, formamide, phenol, and unbuffered pyridine. This phenomenon is understood in terms of the altered structure that facilitates the new coordination required for hemichrome formation. However, at least in certain cases, it is believed that the hemichrome formed acts as a precursor to the subsequent denaturation,[7,10,12,22] including precipitation and/or heme dissociation.

[53] A. Levy, O. Abugo, M. Franks, and J. M. Rifkind, *J. Inorg. Biochem.* **43**, 327 (1991).
[54] C. Spagnuolo, F. Ascoli, E. Chiancone, P. Vecchini, and E. Antonini, *J. Mol. Biol.* **164**, 627 (1983).
[55] C. Spagnuolo, R. D'Alessandro, and E. Chiancone, *Biochim. Biophys. Acta* **1074**, 270 (1991).
[56] A. Levy, K. Alston, and J. M. Rifkind, *J. Biomol. Struct. Dyn.* **1**, 1299 (1984).
[57] G. B. Ogunmola, A. Zipp, F. Chen, and W. Kauzmann, *Proc. Natl. Acad. Sci. U.S.A.* **74**, 1 (1977).
[58] R. G. Alden, J. D. Satterlee, J. Mintorovitch, I. Constantini, M. R. Ondrias, and B. I. Swanson, *J. Biol. Chem.* **264**, 1933 (1989).
[59] Y. Tsubamoto, A. Matsuoka, K. Yusa, and K. Shikama, *Eur. J. Biochem.* **193**, 55 (1990).
[60] T. C. Hollocher, *J. Biol. Chem.* **241**, 1958 (1966).
[61] A. Tomoda, K. Sugimoto, M. Suhara, M. Takeshita, and Y. Yoneyama, *Biochem. J.* **171**, 329 (1978).
[62] D. L. Drabkin and J. M. Austin, *J. Biol. Chem.* **112**, 89 (1935–1936).
[63] K. Tsushima, *J. Biochem. (Tokyo)* **41**, 215 (1954).

This relationship can be understood in terms of the hemichrome stabilizing a subconformation that is more susceptible to denaturation. The resultant hemichrome conformation may expose previously unexposed hydrophobic groups or weaken heme–globin contacts, facilitating further denaturation.

The level of denaturants necessary to form hemichromes and subsequent precipitation is a very sensitive measure of hemoglobin stability. Thus abnormal, unstable hemoglobins are detected *in vitro* by precipitation of the hemoglobins at 50° for 1 hr and/or the addition of 17% 2-propanol.[64] It was further shown that even the relatively stable abnormal hemoglobin Hb E β26 (B8) Glu → Lys, where the amino acid substitution is on the surface of the molecule, is more susceptible to menadione-induced oxidation and hemichrome formation than is Hb A.[22]

Abnormal Unstable Hemoglobins. Abnormal unstable hemoglobins[7,11,12,64] generally arise from mutations that result in an amino acid substitution that perturbs the structure. In addition, unstable hemoglobins can involve deletions arising from unequal crossovers between genes, and equal crossovers between genes, which can produce multiple amino acid substitutions.

Many of these unstable hemoglobins involve amino acid substitutions in the vicinity of the heme. These include several substitutions of phenylalanine CD1, which is in van der Waals contact and almost parallel to the heme. Hb Louisville (Bucuresti)[65] has the β-chain phenylalanine (CD1) (β42) replaced by leucine and Hb Torino[66] has the α-chain phenylalanine CD1 (α43) replaced by valine. The smaller size of the substitutions increases flexibility in the heme pocket, thereby facilitating hemichrome formation. Hb Hammersmith,[67] in which the hydrophobic phenylalanine at CD1 of the β chain is replaced by the polar serine, produces an even more drastic perturbation, with rapid oxidation and hemichrome formation and a tendency for the heme to dissociate from the globin. Other hemoglobins with altered heme globin contacts that enhance formation of hemichromes[12] are Hb Seattle[68] β70 (E14) Ala → Asp, Hb Sydney[69] β67 (E11)

[64] H. Lehmann and R. G. Huntsman, "Man's Haemoglobins." North-Holland Publ., Amsterdam and Oxford, 1974.
[65] V. Bratu, P. A. Lorkin, H. Lehmann, and C. Predescu, *Biochim. Biophys. Acta* **251**, 1 (1971).
[66] A. Beretta, V. Prato, E. Gallo, and H. Lehmann, *Nature (London)* **217**, 1016 (1968).
[67] J. V. Dacie, N. K. Shinton, P. J. Gaffney, Jr., R. W. Carrell, and H. Lehmann, *Nature (London)* **216**, 663 (1967).
[68] S. Kurachi, M. Hermodson, S. Hornung, and G. Stamatoyannopoulos, *Nature (London), New Biol.* **243**, 275 (1973).
[69] R. W. Carrell, H. Lehmann, P. A. Lorkin, E. Raik, and E. Hunter, *Nature (London)* **215**, 626 (1967).

Val → Ala, Hb Christchurch[70] β71 (E15) Phe → Ser, and Hb Sabine[71] β91 (F7) Leu → Pro. In Hb Köln[72] β98 (FG5) Val → Met the mutated residue is both in contact with the heme and also involved in the $\alpha_1\beta_2$ subunit interface.

Instability and hemichrome formation also have been found for modifications that are not in direct contact with the heme but are expected to alter the heme pocket configuration. Hb Philly[73] β35 (C1) Tyr → Phe is lacking a hydrogen bond with aspartic acid α126 (H9) in the $\alpha_1\beta_1$ interface, resulting in dissociation into monomers with the resultant oxidation and hemichrome formation found for isolated chains.

A number of unstable hemoglobins have also been found for mutations involving the B helix close to where it makes contact with the E helix. These changes are expected to alter the configuration of amino acids in the distal pocket stabilizing hemichrome formation. Replacement of glycine at β24 (B6) is sensitive to even subtle changes. Thus Hb Savannah,[74] with the glycine replaced by valine, which is also hydrophobic but slightly larger, is unstable. Hb Riverdale-Bronx,[35] with glycine replaced by the larger hydrophilic arginine, readily forms hemichromes. Other substitutions in this region include Hb Castilla[75] β32 (B14) Leu → Arg, Hb Freiburg[76] with β23 (B5) Val deleted, and Hb Genova[77] β28 (B10) Leu → Pro. Hb Burke[78] β107 (G9) Gly → Arg involves a surface amino acid and rarely produces hemolytic problems. However, it is unstable to heat and forms hemichromes more readily than does Hb A.

For most of these unstable hemoglobins there is both an enhanced rate of autoxidation and hemichrome formation. There are, however, unstable hemoglobins that are oxidized much more readily than Hb A but do not form hemichromes. This is the case for the hemoglobin M types, for which the alteration involves either the replacement of the distal or proximal histidine by tyrosine,[64,79] or for Hb M Milwaukee,

[70] R. W. Carrell and M. C. Owen, *Biochim. Biophys. Acta* **236,** 507 (1971).
[71] R. G. Schneider, U. Satoshi, J. B. Alperin, B. Brimhall, and R. T. Jones, *N. Eng. J. Med.* **280,** 739 (1969).
[72] R. W. Carrell, H. Lehmann, and H. E. Hutchison, *Nature (London)* **210,** 915 (1966).
[73] R. F. Rieder, F. A. Oski, and J. B. Clegg, *J. Clin. Invest.* **48,** 1627 (1969).
[74] T. H. J. Huisman, A. K. Brown, G. D. Efremov, J. B. Wilson, C. A. Reynolds, R. Uy, and L. L. Smith, *J. Clin. Invest.* **50,** 650 (1971).
[75] J. Thillet, M. C. Garel, R. Bierme, and J. Rosa, *Biochim. Biophys. Acta* **624,** 293 (1980).
[76] R. T. Jones, B. Brimhall, T. H. J. Huisman, E. Kleihauer, and K. Betke, *Science* **154,** 1024 (1966).
[77] G. Sansone, R. W. Carrell, and H. Lehmann, *Nature (London)* **214,** 877 (1967).
[78] S. Kobayashi, T. Nara, Y. Nakano, H. Fukazawa, G. Kawamura, H. Kitajima, M. Sugita, K. Joh, Y. Ohba, Y. Hattori, and T. Miyaji, *Hemoglobin* **10,** 661 (1986).
[79] F. Haurowitz, *J. Biol. Chem.* **193,** 443 (1951).

involving the replacement of valine $\beta67$ (E11) by glutamic acid. In these cases the substituted amino acids are thought to coordinate with the iron, stabilizing the metHb form. By the definition of hemichromes involving coordination with an endogenous amino acid, these complexes may be considered hemichromes. However, they form high-spin complexes and do not satisfy the usual spectroscopic criteria for hemichromes.

An interesting example in which an amino acid substitution also enhances oxidation but not hemichrome formation is Hb Toulouse[75] $\beta66$ (E10) Lys → Glu. This amino acid points toward the surface and no significant conformational change is expected to enhance hemichrome formation. However, oxidation is enhanced, presumably by changing the acidic nature of the heme pocket.

Dehydration. Dehydration is another way of disrupting the native globin structure. Interactions of hydrophilic amino acids with water and the orientation of hydrophobic amino acids away from the aqueous surface of the molecule play a major role in determining the protein conformation. It is therefore to be expected that as one dehydrates a protein, lowering the activity of water, alterations in conformation do take place that can facilitate hemi(hemo)chrome formation.

For deoxyhemoglobin it has been shown by visible[79,80] and Mössbauer[17,18] spectroscopy that a low-spin complex is formed during dehydration. Interestingly, this complex is not formed with deoxymyoglobin,[17] which has a more rigid ligand pocket. The formation of the bishistidine complex does not seem to be simply due to the removal of water from the ligand. Thus, although only one of the hemoglobin chains has a water in the deoxyhemoglobin pocket, both types of chains form the low-spin hemochrome.[18]

For methemoglobin, the formation of hemichromes has also been found during dehydration. This has been detected not only by visible spectroscopy[81] and Mössbauer spectroscopy[16] but also by photoacoustic spectroscopy,[82] which is less sensitive to scattering artifacts than normal visible spectroscopy, and by electron spin resonance.[13] By photoacoustic spectroscopy, the spectral species formed with lyophilized methemoglobin were analyzed. The low-spin complex formed from the aquo methemoglobin was shown to resemble the imidazole complex of methemoglobin. For bovine methemoglobin at pH 7.0, 82% of the hemoglobin contained the low-spin bishistidine complex.

[80] R. Zeynek, *Nov. Lek. Poznan* **38**, 406 (1926).
[81] D. Keilin and E. F. Hartree, *Nature (London)* **170**, 161 (1952).
[82] G. M. Alter, *J. Biol. Chem.* **258**, 14960 (1983).

The formation of this low-spin complex was studied in the presence of different methemoglobin ligands.[82] No imidazole complex was formed when the strong ligand cyanide was bound to the heme. However, binding of weaker ligands such as fluoride and azide only partially inhibited the formation of the bishistidine complex, with the relative amount of this complex following the order $H_2O > F > N_3$. It was found that ammonium sulfate precipitation also produces some bishistidine complex. This process, which does not dehydrate the protein to the same extent as lyophilization, however, produced appreciably less hemichrome.[82]

The effect of dehydration on the EPR spectrum[13] was studied at various water contents by incubating dried samples with salt solutions for 4 days and weighing the equilibrated hemoglobin solution. It was shown that the hemichromes are formed from hemoglobin below a level of 0.4 g H_2O/g protein, whereas for myoglobin the water content had to be decreased to 0.2 g H_2O/g protein. It was also found that for fully dehydrated samples only 5% of the hemoglobin remained as high-spin methemoglobin, whereas 35% of the myoglobin remained as high-spin metmyoglobin.

By EPR it was also possible to identify, in addition to the bishistidine hemichrome, a P-type hemichrome thought to involve coordination with the $\beta93$ cysteine residue. It was suggested that the dehydration causes a rotation of the F helix, breaking the Fe–proximal histidine bond and enabling the cysteine to bind in its place. This complex was not detected with myoglobin lacking this cysteine or when the sulfhydryl group was blocked. Interestingly, these changes were found to be reversible even though formation of P-type hemichromes by other methods are generally not reversible. A related method for forming hemichromes is the use of solvent mixtures,[83,84] which are thought to dehydrate the protein when the mole fraction of the nonaqueous solvent becomes high, decreasing the activity of the water.

Hemi(hemo)chrome Formation under Conditions in which Native Globin Structure Is Not Disrupted

The methods for hemi(hemo)chrome formation involving denaturants, dehydration, and amino acid substitution alter the globin structure, weakening the interactions stabilizing the native conformation, and thereby facilitating hemi(hemo)chrome formation. Increasing the pressure and lowering the temperature, two other methods for producing hemi(hemo)-chromes, should stabilize the native structure. The formation of hemi-

[83] A. C. I. Anusiem, J. G. Beetlestone, J. B. Kushimo, and A. A. Oshodi, *Arch. Biochem. Biophys.* **175**, 138 (1976).
[84] A. C. I. Anusiem and A. A. Oshodi, *Arch. Biochem. Biophys.* **189**, 392 (1978).

chromes under these conditions suggests that the hemi(hemo)chromes can form without disrupting the native conformation. This contention will be supported further by evidence for transient hemi(hemo)chrome formation.

Pressure. The effect of pressure on hemoproteins has been studied by visible spectroscopy,[57,58] magnetic susceptibility,[85] Raman spectroscopy,[58] and nuclear magnetic resonance.[86] These studies demonstrate a stabilization of the low-spin state, which has been interpreted as formation of hemi(hemo)chromes. In a study of metmyoglobin[87] it was shown that these pressure changes are similar to those that occur during temperature, pH, and urea denaturation. In a study of the effect of pressure on different myoglobin and hemoglobin complexes,[57] it was shown that the hemichrome with a spectrum similar to the imidazole complex was obtained for the H_2O, fluoride, and azide complexes of both oxidized myoglobin and hemoglobin. As found in dehydration studies, no hemichrome was formed with cyanide bound to the heme.

In the Fe(II) state, it was shown that at elevated pressures oxyhemoglobin and oxymyoglobin[57] undergo oxidation and formation of hemichrome. No bishistidine complex was found with carbon monoxide bound to the heme. For the deoxygenated proteins, reversible changes in the spectrum were found; however, these looked different from those formed with the oxidized proteins or oxyhemoglobin, and could be attributed to that of an Fe(II) hemochrome.

A more detailed study of the Fe(II) proteins,[58] extending over a wider range of pressure, by both visible spectroscopy and Raman spectroscopy, has demonstrated formation of the low-spin hemochrome complex for the deoxygenated protein. For oxyhemoglobin it was found that the oxygen first came off producing a deoxylike spectrum and subsequently forming a low-spin complex. The initial deoxylike spectrum at only modest pressures was attributed to the O_2 coming off but still retained in the pocket. This may be analogous to the partial displacement of oxygen by the distal histidine observed at low temperature (see below). At still higher pressures a low-spin complex is produced that is clearly distinct from that found with the deoxygenated proteins and may involve oxidation and formation of the hemichrome.

As found for hemichrome formation by other methods, the more rigid ligand pocket in myoglobin than in hemoglobin makes it more difficult to

[85] C. Messana, M. Cerdonio, P. Shenkin, R. W. Noble, G. Fermi, R. N. Perutz, and M. F. Perutz, *Biochemistry* **17**, 3652 (1978).
[86] I. Morishima and M. Hara, *J. Biol. Chem.* **258**, 14428 (1983).
[87] A. Zipp and W. Kauzmann, *Biochemistry* **12**, 4217 (1973).

form the hemichromes and hemochromes at elevated pressure.[57,58] There is also a greater tendency for the myoglobin changes to be reversible.

The role of the distal histidine has been investigated by comparing *Glycera dibranchiata* hemoglobin,[58] a monomeric protein that contains no histidine residue on the distal side of the heme. This protein does, however, possess greater flexibility in the distal pocket than does sperm whale myoglobin. Interestingly, low-spin states are also found with this protein, although greater pressures were required. It is suggested that in the *Glycera* hemoglobin the pressure produces a volume compression that results in proximal forces driving the iron into the heme pocket.[58] The resultant low-spin deoxy state is still coordinated with only five ligands. The different nature of the changes in *Glycera* hemoglobin explains the complete reversibility of all changes even at very high hydrostatic pressures.

Temperature. The temperature-dependent stabilization of low-spin states of aqueous methemoglobin was originally observed in the frozen state.[19] The low-spin component observed by magnetic susceptibility and visible spectroscopy was found to increase in intensity for slowly cooled samples. The visible spectrum of the low-spin component detected in the frozen state, with peaks at 527 and 560 nm, was clearly different from that of the low-spin hydroxide, with bands at 540 and 572 nm.

Electron spin resonance, which is particularly powerful for characterizing low-spin Fe(III) heme complexes, did not identify these low-spin complexes at liquid nitrogen temperature.[51] It was, however, observed that the intensity of the high-spin complex was decreased, suggesting an unresolved low-spin complex. These low-spin complexes were identified as hemichromes with a sixth nitrogen, presumably arising from the distal histidine, by using a high-modulation amplitude or lower temperatures.[24,28,51] Consistent with the magnetic susceptibility studies, slow cooling at a rate of $0.3°/min$ produced an appreciable increase in the total intensity of these low-spin hemichromes.[8]

An understanding of the dependence of hemichrome formation on rate of cooling was provided by studies in which rapidly frozen samples were incubated at various elevated temperatures and then cooled back down to the lower temperatures used for EPR or Mössbauer measurements.[9] It was found that below 210 K there was no additional low-spin bishistidine complex formed, even for long periods of incubation. The rate of formation increased dramatically above this temperature. From these data it was possible to calculate a lower limit for the activation energy associated with formation of bishistidine complexes of 60 kJ mol^{-1}. This large activation energy results in a critical temperaure below which the interconversion to low spin essentially does not take place. Thus, as the temperature is lowered, the equilibrium is shifted toward the bishistidine complexes, while the rate for the spin transition decreases. The anomalous effects associated with freezing are therefore not freezing artifacts as originally suggested, but a temperature-dependent freezing in of the spin equilibrium.

Mössbauer spectroscopy was used to study the temperature-dependent formation of the bishistidine complex from Fe(II) deoxyhemoglobin. In this case again an activation energy ≥ 60 kJ mole^{-1} was detected, resulting in a critical temperature below which the spin equilibrium was frozen.

Different groups[8,23–28,50,51] have reported different g values for the EPR spectra of the low-temperature-stabilized complexes of methemoglobin in the absence of added denaturants. They do, however, tend to fall into three classes, which we have designated as A, B, and C[23] (Fig. 5). The A complexes have the highest g value in the region of 2.6, a middle g value around 2.15, and the lowest g value around 1.85. The intensities of these complexes increase at higher pH and correspond to the hydroxide complexes or other complexes in which the distal histidine is not directly coordinated to the iron. B complexes, with the highest g value in the range of 2.83–2.75 and the lowest g value in the range 1.69–1.63, have ligand field parameters of the reversible hemichrome H. C complexes have the highest g value between 3.08 and 2.89 and the lowest g value ≤ 1.51. Ligand field parameters for these complexes fall within the range of the irreversible B-type hemichrome or the C-type hemichrome.[8]

Unlike the hemichromes formed by the addition of denaturants, all of the different bishistidine complexes formed by lowering the temperature in both the Fe(II) and Fe(III) states are reversible.[8,9,23] Thus the original distribution of species present in a rapidly frozen sample is also found when the sample is thawed and refrozen. The temperature dependencies for the formation of the two classes of bishistidine complexes (B and C), are, however, clearly different. Complex B is the low-spin bishistidine complex present at room temperature. This was demonstrated[23] by comparing the EPR spectra of methemoglobin frozen by submerging the sam-

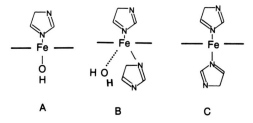

Fig. 5. Proposed structure of the three classes of low-spin complexes detected in the low-temperature EPR of nondenatured hemoglobin.

ple in liquid nitrogen (for which the freezing time is several seconds) with those of methemoglobin prepared by freeze quenching in less than 10 msec. Because the bishistidine complexes are formed very slowly below a critical temperature in the range of 210 K, the freeze quenching should retain the room temperature distribution of species. Interestingly, a negligible amount of complex C is present in the freeze-quenched sample, and complex B in the freeze-quenched sample is always greater than that found in the liquid nitrogen frozen sample.

The level of complex B is increased by lowering the solution temperature of the sample prior to freeze quenching. However, at subzero temperatures complex C becomes the dominant low-spin bishistidine complex, and generally complex B and the other methemoglobin complexes are transformed into complex C. The equilibrium constant for the formation of complex C increases as the temperature is lowered, although the rate constant for the transformation decreases.

In order to explain these results, it has been suggested[23] that the bishistidine complex B may still have water retained in the ligand pocket (Fig. 5) and therefore in solution it is in rapid equilibrium with high-spin methemoglobin. The water in the pocket would also determine the orientation of the histidine as it approaches the heme and explain the distinct spectral properties for this complex. The formation of the bishistidine complex C would therefore not only require the reorientation of the distal histidine but also the egress of the water from the heme pocket (Fig. 5). Interestingly, the critical temperature range below which bishistidine complexes are no longer formed corresponds to the low temperature at which the egress of ligand from the heme pocket becomes extremely low.

It was further shown that the C class of bishistidine complexes contains a number of different components,[8,23] depending on hemoglobin species, pH, and time (Fig. 4). The most dramatic difference is with horse hemoglobin, wherein two distinct components are clearly resolved with $g_1 = 2.9$ and 3.1, respectively. For human hemoglobin, although there are indications of overlapping signals, the bands are broader and show up as a single broad asymmetric band. The differences in these complexes can be understood in terms of subtle differences in the relative orientation of the two histidines coordinated to the iron.[38,41–47] This situation is particularly true when both axial ligands are part of the globin chain and any subtle differences in conformation translate into changes in the relative orientation of the two histidines, which will alter the EPR parameters.

Several authors have suggested the formation of low-spin complexes for which the distal histidine is not directly coordinated with the heme iron, but instead is linked to the iron via a hydrogen-bonded water mole-

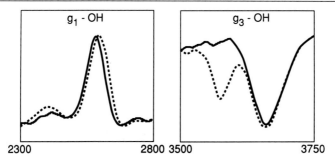

FIG. 6. The two hydroxide-like complexes detected in hemoglobin samples. In the region from 2300 to 2800 gauss the lowest field bands (g_1) are shown. In the region from 3500 to 3750 gauss the highest field bands (g_3) of the same complexes are shown. Spectra are shown for the small percentage of methemoglobin in a sample of deoxyhemoglobin (——) and in (----) oxyhemoglobin.

cule.[28,84,88] It was suggested that a shift of the histidine toward the iron could strengthen the interaction, causing the proton of the hydrogen bond to be shifted closer to the imidazole, resulting in a low-spin complex.

Such a complex may not be distinguished readily from the bishistidine complexes by visible spectroscopy. However, the EPR spectra should clearly indicate a class A hydroxy-type complex. This interaction may explain the origin of two class A complexes with slightly different g values detected at 11 K (Fig. 6). We in fact have shown[89] changes in these two OH-like components, depending on pH, conformation, and incubation. These results are consistent with a low-spin water complex stabilized by the approach of the distal histidine.

Pulse Radiolysis. A particularly interesting method of producing hemochromes is the reduction of methemoglobin by solvated electrons generated by pulse radiolysis.[88,90–93] These Fe(II) bishistidine complexes are intermediate species in the reduction of hemoglobin that rapidly decay to high-spin deoxyhemoglobin in the absence of oxygen, and to oxyhemoglobin in the presence of oxygen. They are formed from the water, hydroxide, azide, and fluoride complexes of hemoglobin. They, however, are not

[88] J. W. van Leeuwen, J. Butler, and A. J. Swallow, *Biochim. Biophys. Acta* **667,** 185 (1981).

[89] J. M. Rifkind, O. Abugo, A. Levy, and M. Franks, unpublished results (1992).

[90] L. A. Blumenfeld, R. M. Davydov, S. N. Magonov, and R. O. Vilu, *FEBS Lett.* **49,** 246 (1974).

[91] A. Raap, J. W. van Leeuwen, H. S. Rollema, and S. H. de Bruin, *FEBS Lett.* **81,** 111 (1977).

[92] A. Raap, J. W. van Leeuwen, H. S. Rollema, and S. H. de Bruin, *Eur. J. Biochem.* **88,** 555 (1978).

[93] Z. Gasyna, *Biochim. Biophys. Acta* **577,** 207 (1979).

detected when the strong ligand cyanide is bound to hemoglobin. The rate of decay of the hemochrome spectrum depends on pH, phosphate concentration, and temperature.[88,92]

At low temperature (below 140 K) in a water–ethylene glycol solvent mixture, the hemochrome spectrum is stable for hours.[90,93] Myoglobin, which has been found to form bishistidine complexes much less readily, does not form this intermediate at room temperature. At 77 K, low-spin intermediates were also produced with myoglobin.[93] However, these myoglobin intermediates begin to decay back to methemoglobin above 100 K and are thought to involve the ligand of oxidized myoglobin retained coordinated to the deoxymyoglobin and not a complex with the distal histidine. The possibility that the hemoglobin intermediate complex, at least at high pH, also retains the hydroxide from methemoglobin bonded to the distal histidine has also been suggested.[93] However, the similarity between the spectra of these transient species and those of the distal histidine hemochrome, as well as the similarity of the spectra produced with different metHb ligands, support a complex with the distal histidine.

Bishistidine Complex as Substate of Native Hemoglobin

Evidence for Such a Substate

Direct evidence for a bishistidine low-spin complex present in room temperature Fe(III) methemoglobin and Fe(II) deoxyhemoglobin comes from rapidly freezing hemoglobin.[8,9,23] In these samples, reflecting the room temperature distribution of states, bishistidine complexes were detected by EPR and Mössbauer spectroscopy. For methemoglobin this contention was further supported by freeze quenching, which freezes the sample within 10 msec.

The possibility that these low-spin complexes originate from a small fraction of denatured hemoglobin is ruled out by the thermal equilibrium involving this complex. Thus, freeze-quenched samples starting from 278 K show an increase in the low-spin bishistidine complex relative to those quenched from an initial temperature of 310 K.[23]

The shift from low-spin to high-spin methemoglobin was also shown by temperature-jump relaxation measurements.[94–96] On the basis of the spectral changes associated with this relaxation process and the large enthalpy and entropy associated with this process, these changes also

[94] U. Dreyer and G. Ilgenfritz, *Biochem. Biophys. Res. Commun.* **87,** 1011 (1979).
[95] A. Bracht, B. R. Eufinger, H. J. Neumann, G. Niephaus, A. Redhardt, and J. Schlitter, *FEBS Lett.* **114,** 157 (1980).
[96] H.-J. Steinhoff, K. Lieutenant, and A. Redhardt, *Biochim. Biophys. Acta* **996,** 49 (1989).

were attributed to enhanced coordination with the distal histidine[95] as the temperature was lowered.

The correspondence between the room temperature transition in solution and the low-temperature processes in the solid state is best shown by a study where EPR and temperature-jump experiments were compared[96] and the equilibrium constant was fit by a single regression line over the temperature range 236–291 K, combining data from both methods. The relaxation times were, however, slower in the frozen state than in solution. This phenomenon is consistent with the lower mobility, analogous to a viscosity effect[97] in the frozen solution.

Further evidence that these complexes cannot be considered denatured forms of hemoglobin comes from the ability to react with exogenous ligands. The Fe(II) bishistidine complex thus disappears when the sample is oxygenated.[9]

For Fe(III) it was shown that even a weak ligand such as fluoride[98] reacts with the bishistidine complex B, present at room temperature, as readily as it does with the other aqueous complexes. This indicates an equilibrium between complex B, high-spin metHb, and the hydroxide complexes.

The bishistidine complex C, formed only at low temperatures, was also shown to react with fluoride. A sample of fluoride rapidly mixed with hemoglobin immediately quenched to low temperature does not react with fluoride until the temperature is raised to 255 K. By incubation at 233 K the bishistidine complex C was formed in this sample. Complex C, normally stable until the temperature is raised above 273 K, however, rapidly reacted with fluoride at 255 K (Fig. 7). Thus both classes of bishistidine complexes readily react with fluoride.

To explain the formation of the transient Fe(II) distal histidine bond during pulse radiolysis it has been suggested[92] that the histidine is the primary acceptor of the solvated electron. The negative imidazole ring could then displace the negative metHb ligand and attack the iron, causing the heme reduction. This hypothesis is, however, not consistent with other evidence that the heme can be reduced directly by solvated electrons via the porphyrin periphery.[99] Furthermore, this mechanism should show a relatively slow reduction of the heme, limited by the approach of the imidazolide anion to the iron, and not the almost instantaneous reduction

[97] A. Ansari, J. Berendzen, D. Braunstein, B. R. Cowen, H. Frauenfelder, M. K. Hong, I. E. T. Iben, J. B. Johnson, P. Ormos, T. B. Sauke, R. Scholl, A. Schulte, P. J. Steinbach, J. Vittitow, and R. D. Young, *Biophys. Chem.* **26,** 337 (1987).

[98] A. Levy, P. Kuppusamy, and J. M. Rifkind, *Int. Biophys. Congr. 9th,* Jerusalem, Israel, 123 (1987).

[99] E. O. Olivas, D. J. A. de Waal, and R. G. Wilkens, *J. Biol. Chem.* **252,** 4038 (1977).

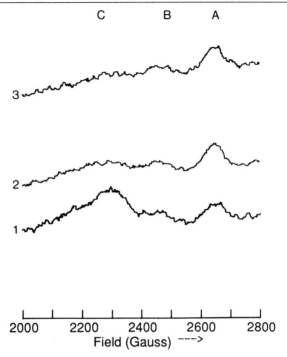

FIG. 7. The binding of fluoride to the bishistidine complex C. Hemoglobin was rapidly mixed with fluoride and quenched within 10 msec by freeze quenching. The sample was then incubated at 233 K for 32 min in order to produce complex C (spectrum 1). This sample was then incubated at 255 K for 8 min (spectrum 2) and for 24 min (spectrum 3). At this temperature the fluoride reacts with complex C.

observed. The transient bishistidine Fe(II) complex during pulse radiolysis is, instead, supportive of a substrate in which the distal histidine can approach the iron even in methemoglobin coordinated with other ligands. This hemichrome state would stabilize the reduced heme formed by direct interaction of the solvated electron with the heme, thereby producing the hemochrome spectrum.

Population of Substate. The X-ray structures of both methemoglobin and deoxyhemoglobin indicate a distance between the iron and the N^ε of the distal histidine of greater than 4 Å. Coordination requires a decrease of this distance by nearly 2 Å. Such a decrease cannot be accomplished by rotation of the histidine side chain and would seem to preclude a distal histidine bond without disruption of the tertiary structure of hemoglobin. The results cited above indicate that the necessary rearrangements for formation of a bond between the iron and the distal histidine do, nevertheless, take place.

The simple potential energy diagram depicted in Fig. 8 explains the formation and relative population of these bishistidine complexes. The minimum of the broad, shallow well, r_0, represents the distance between the iron and the histidine obtained from the energy-minimized X-ray structure. The thermal fluctuations accessible in the heme pocket can result in large variations in this distance. These conformational fluctuations must require not only rotation of bonds but also movement of the E helix relative to the heme.[95] The narrow, deep well, r_1, represents the bishistidine complex, and the barrier height (E_b) represents the conformational rearrangement necessary to bring the histidine close enough to form a bond. Although formation of this complex is energetically favored, the large multiplicity of nonbonded states results in very low levels of bishistidine complex at room temperature. Unlike the temperature changes that redistribute the hemoglobin among the accessible states, the addition of denaturants or amino acid substitutions can be understood in terms of a change in the shape of the potential energy diagram, so that the barrier height E_b is lower and perhaps r_0 decreases. Bishistidine complexes that may be very similar spectroscopically are then formed more readily and perhaps irreversibly.

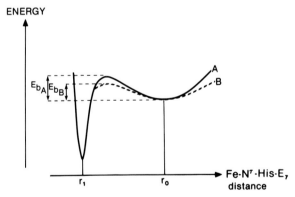

FIG. 8. Potential energy diagram for the formation of the distal histidine complexes. Here r_0 is the distance between the iron and the histidine obtained from the energy-minimized X-ray structure; r_1 is the iron–histidine distance, with the distal histidine coordinated to the iron. E_b represents the barrier height for the formation of the distal histidine complex. The comparision of A and B indicates how changes in pocket dynamics influence the formation of the distal histidine complex. In B, greater flexibility lowers the barrier height, stabilizing the distal histidine complex. Reprinted with permission from Levy and Rifkind. Copyright 1985 American Chemical Society.

Functional Relevance of Hemi(hemo)chrome Formation

The functional relevance of these reactions can be separated into (1) insight regarding function that can be obtained from studies of hemi(hemo)-chrome formation and (2) the possible functional relevance of the hemi (hemo)chrome complexes.

Hemichrome Formation and Distal Pocket

NMR and Raman spectroscopy provide insights into the globin config-uration and dynamics on the proximal side of the heme. However, on the distal side of the heme, where the globin is not directly coordinated to the heme, there are no simple spectroscopic probes of globin dynamics. However, the relevance of globin dynamics in controlling ligand binding as well as heme oxidation is well recognized.

Because of the large barrier associated with bringing the distal histidine close enough to the iron to form a bond (Fig. 8), hemichrome formation, which can readily be detected at low temperature, provides an extremely sensitive probe of distal pocket flexibility.[9]

The sensitivity of the method is indicated by the comparison between myoglobin and hemoglobin. The structures of both proteins as well as the X-ray distances between the iron and the N^ε of the distal histidines are very similar. Nevertheless, a number of studies indicate that hemichrome and hemochrome formation occur much less readily for myoglobin.[13,56-58] These results indicate that there is less distal pocket flexibility in myoglo-bin than in hemoglobin. This contention is supported by X-ray[100] and Raman[101] results.

Electron paramagnetic resonance, which is sensitive to the relative orientation of the bound distal histidine, provides another dimension to probe the distal configuration of the heme. An example of this sensitivity is provided by the different distribution of class C complexes formed at low temperature with horse and human hemoglobins (Fig. 4).

The sensitivity of bishistidine complex formation to hemoglobin con-formation is indicated by a decrease in the low-temperature-dependent hemichrome[27,28] and by the pulse radiolysis transient hemochrome[91,102] when effectors are added to stabilize the T state of hemoglobin. Conforma-tional changes have also been proposed to explain the decreased hemi-chrome when isolated chains are compared with tetrameric hemoglo-

[100] H. Frauenfelder, G. A. Petsko, and D. Tsernoglou, *Nature (London)* **280**, 558 (1979).
[101] E. W. Findsen, T. W. Scott, M. R. Chance, J. M. Friedman, and M. R. Ondrias, *J. Am. Chem. Soc.* **107**, 3355 (1985).
[102] J. Wilting, A. Raap, R. Braams, S. H. de Bruin, H. S. Rollema, and L. H. M. Janssen, *J. Biol. Chem.* **249**, 6325 (1974).

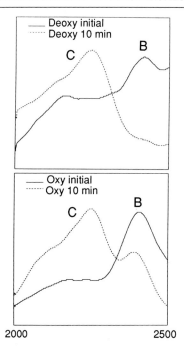

FIG. 9. Comparison of the relative concentrations of bishistidine complexes B and C for the small percentage of methemoglobin present in deoxyhemoglobin and oxyhemoglobin. —, Initially frozen sample; ----, after incubation at 238 K.

bin.[25,50,51] A more direct and detailed comparison of the effect of quaternary structure on the distal pocket dynamics was obtained by comparing the EPR spectra of the small percentage of oxidized chains present in a sample of hemoglobin fully oxygenated and fully deoxygenated[103]. From the EPR spectra and change induced during the incubation at 238 K, it was found that the R state has an increased level of bishistidine complex (B) in the immediately frozen sample. However, during the low-temperature incubation, the transformation from complex B with the water in the pocket to complex C without water is faster and more complete for deoxyhemoglobin than oxyhemoglobin (Fig. 9). In addition, the OH-type complex with water H-bonded to the distal histidine is stabilized for oxyhemoglobin (Fig. 6).

These results suggest that although interactions with the distal histidine are more pronounced in the R state, the actual displacement of the ligand

[103] O. Abugo, A. Levy, and J. M. Rifkind, *36th Annu. Biophys. Soc. Meet.* Abstr. 310 Houston, Texas (1992).

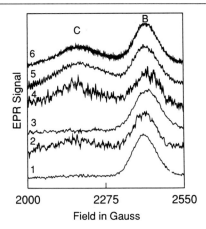

FIG. 10. Bishistidine complexes B and C after incubation at 233 K for different valency hybrids. 1, $(\alpha^+\beta^{CO})(\alpha^+\beta^{CO})$; 2, $(\alpha^+\beta^{O_2})(\alpha^+\beta^{O_2})$; 3, $(\alpha^{CO}\beta^+)(\alpha^{CO}\beta^+)$; 4, $(\alpha^{O_2}\beta^+)(\alpha^{O_2}\beta^+)$; 5, $(\alpha^+\beta^+)(\alpha^{CO}\beta^{CO})$; 6, $(\alpha^+\beta^+)(\alpha^{O_2}\beta^{O_2})$. Reproduced from Levy et al.[47] by copyright permission of the Biophysical Society.

from the heme pocket by the distal histidine occurs more readily in the T state. This must reflect a difference in the approach of the distal histidine to the heme. Additional studies will be required to understand how these differences may relate to the ligand reactions.

These types of interactions have been used further to indicate a new mode for subunit interactions in hemoglobin,[47] transmitted via the distal pocket and across the $\alpha_1\beta_1$ interface. This mode is distinct from the T $\leftrightarrow$ R equilibrium associated with the tilt of the proximal histidine and transmitted through the FG corner across the $\alpha_1\beta_2$ interface. For this purpose the influence of binding ligands to the Fe(II) chain of valency hybrids was studied.

With oxygen (O_2) bound to the reduced chains, the EPR spectra of the Fe(III) chains at low temperature and the changes during low-temperature incubation were similar to those of fully oxidized hemoglobin, with the amount of complex B with H_2O still in the pocket decreasing during low-temperature incubation and complex C, with the H_2O displaced, becoming the dominant bishistidine complex. When O_2 was replaced by CO on the hybrids, with either both α chains or both β chains oxidized, dramatic changes were found (Fig. 10). A high level of complex B was found which actually increased during low temperature incubation. No complex C was formed, suggesting that the distal histidine could no longer force the water out of the ligand pocket. Perturbations were also found for the high-spin aqueous complex.[47]

Both O_2 and CO hybrids are in an R quaternary conformation, with the major structural differences produced by the orientation of CO in the distal pocket.[104] The EPR results indicate that these perturbations are transmitted to the oxidized chains. By comparing different valency hybrids (Fig. 10) and finding that the hybrid with one $\alpha\beta$ dimer oxidized ($\alpha^+\beta^+$) ($\alpha^L\beta^L$) did not show the same perturbation, it was concluded that these perturbations are transmitted across the $\alpha_1\beta_1$ interface. These studies of the distribution and temperature-dependent changes of the bishistidine complexes have thus been used to define a method for transmitting subunit interaction across the $\alpha_1\beta_1$ interface via distal perturbation. It is thought that these interactions may modulate the T $\leftrightarrow$ R transformation and thus help regulate oxygenation.

Possible Functional Role of Bishistidine Complexes

Heinz Body Formation. The first area in which hemichrome formation plays a role, although perhaps a negative role, is in the formation of Heinz bodies[7,11,12,29,64] and the subsequent destruction of red cells.

It has been shown that hemichromes bind to the cytoplasmic portion of band 3 of the erythrocyte membrane.[105] The binding of hemoglobin to band 3 of the erythrocyte membrane involves insertion of the N-terminal peptide of band 3 into the central cavity of hemoglobin between the β chains. This binding is enhanced for deoxyhemoglobin,[106] in which the distance between the β chains is greater, and is blocked by cross-linking the β chains. The binding of hemichromes to band 3 also involves the same N-terminal cytoplasmic peptide.[107-109] However, the binding is much stronger than for hemoglobin and it does not involve insertion into the central cavity. Instead it involves the interaction of two hemoglobin molecules with each band 3 chain.

Hemichromes rapidly copolymerize[108] with the soluble cytoplasmic domain of the erythrocyte membrane band 3 and form an insoluble copolymer. It has been suggested that in the intact cell the same binding of

[104] A. Levy, *Biochemistry* **28**, 7144 (1989).
[105] J. A. Walder, R. Chatterjee, T. L. Steck, P. S. Low, G. F. Musso, E. T. Kaiser, P. H. Rogers, and A. Arnone, *J. Biol. Chem.* **259**, 10238 (1984).
[106] A. Tsuneshige, K. Imai, and I. Tyuma, *J. Biochem. (Tokyo)* **101**, 695 (1987).
[107] S. M. Waugh and P. S. Low, *Biochemistry* **24**, 34 (1985).
[108] S. M. Waugh, J. A. Walder, and P. S. Low, *Biochemistry* **26**, 1777 (1987).
[109] R. Kannan, R. Labotka, and P. S. Low, *J. Biol. Chem.* **263**, 13766 (1988).

hemichromes to band 3 causes a clustering of band 3 and results in the formation of insoluble Heinz bodies.[110-112] This hypothesis is supported by several studies of abnormal hemoglobins that show that the formation of Heinz bodies correlates with the presence of hemichromes bound to the membrane.[7,12,29,71-74]

The cells with Heinz bodies are rapidly taken out of circulation. Thus in splenectomized patients increased levels of Heinz bodies are detected.[111] It is clear, at least with certain abnormal hemoglobins, that the hemichrome formation and subsequent Heinz body formation are part of the etiology of hemolytic anemia.

A possible role for hemichromes in the elimination of normal cells has also been postulated. In normal individuals the most dense, oldest erythrocytes are found to contain Heinz bodies in splenectomized patients.[110,111] It has been reported that erythrocytes with Heinz bodies contain more surface-bound immunoglobin (IgG) molecules, which seem to be clustered around the Heinz body. It has, in fact, been suggested that the clustering of band 3 is responsible for the enhanced binding of IgG molecules. It is thought that these bound antibodies can promote elimination of the damaged or old erythrocytes from circulation. In this way hemichrome formation may actually be associated with triggering the elimination of the older, less functional erythrocytes.

Oxidation. As discussed above, the initial reduced complex resulting from pulse radiolysis is a hemochrome. Results suggest that these hemochromes can be oxidized much faster than normal hemoglobin.[113] In discussing these results it must be recognized that the protein structure in these rapidly reduced species formed by pulse radiolysis has not yet relaxed to a stable reduced form.[113] Therefore, this enhanced oxidation may not be associated with the presence of a hemochrome.

More recent results,[53] however, suggest that a transient hemochrome may actually be involved in autoxidation of hemoglobin. This is based on low-temperature studies of hemoglobin in the solid state that demonstrate the formation of a free radical signal as an intermediate in the oxidation of oxyhemoglobin. This intermediate requires O_2 bound to the heme, and the level of this intermediate formed during low temperature incubation correlates with the rate of oxidation. It was suggested that this signal corresponds to superoxide displaced from oxyhemoglobin by the nucleophilic attack of the distal histidine, but still retained in the ligand pocket

[110] P. S. Low, S. M. Waugh, K. Zinke, and D. Drenckhahn, *Science* **227,** 531 (1985).
[111] K. Schlüter and D. Drenckhahn, *Proc. Natl. Acad. Sci. U.S.A.* **83,** 6137 (1986).
[112] S. M. Waugh, B. M. Willardson, R. Kannan, R. J. Labotka, and P. S. Low, *J. Clin. Invest.* **78,** 1155 (1986).
[113] L. A. Blumenfeld and R. M. Davidov, *Biochim. Biophys. Acta* **549,** 255 (1979).

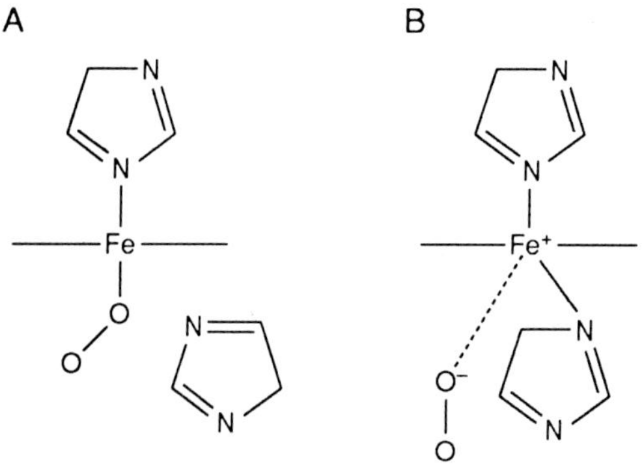

A   B

FIG. 11. The nucleophilic displacement of oxygen as a superoxide ion by the distal histidine. (A) The heme pocket with oxygen bound and the distal histine removed from the heme iron. (B) The oxygen displaced from the iron as a superoxide ion.

(Fig. 11). This transient hemichrome species will decay into oxidized hemoglobin with superoxide released from the heme pocket at elevated temperatures.

A relationship between oxidation and the distal histidine interactions is supported by comparing the transition from class B bishistidine complex to the class C bishistidine complex for different samples. These results suggest a correlation between rates of oxidation and the tendency for the B complex with water still in the pocket to undergo a transition into a class C complex with water displaced from the pocket. Thus oxidation correlates with the ability of the distal histidine to displace the exogenous ligand i.e., the superoxide, from the heme pocket.

Ligand Binding. The question we would like to discuss is what relevance the minor substrate with the distal histidine coordinated to the iron may have on the ligand binding properties of hemoglobin in both the Fe(II) and Fe(III) states.

Most studies that have addressed a role of the distal histidine in ligand biding have emphasized a possible role of blocking or competing with exogenous ligands and thereby lowering the affinity. However, there are several lines of evidence that actually suggest an enhanced affinity associated with hemi(hemo)chrome formation. This was indicated by studying the binding of oxygen to hemoglobin subsequent to pulse radiolysis.[88] At relatively high pH values (pH 8–9) the decay of the hemochrome spectra is actually slower than the binding of oxygen. These results indicate that oxygen is bound to the transient hemochrome structure faster than to the relaxed high-spin deoxyhemoglobin. It has in fact been suggested that an intermediate hemochrome complex may also be related to the highly reactive intermediates produced by flash photolysis.[92] Increased ligand affinity associated with hemochrome formation is also supported by the increased oxygen affinity in the range of 8–12 M ethylene glycol.[114] By EPR studies we have also found that the class C bishistidine complex actually binds fluoride more readily than the low-spin hydroxide (Fig. 7).

These few results are clearly only suggestive. However, they raise a fascinating possibility that the distal histidine may actually facilitate ligand binding. How can this be understood? The formation of class B bishistidine complex with both the distal histidine and the exogenous ligand associated with the iron demonstrates that coordination of the distal histidine need not block access of the heme pocket to the exogenous ligand. It is even possible that rearrangement resulting from distal histidine coordination may facilitate certain pathways for ligand access to the heme pocket.

Furthermore, it has been shown that after the ligand enters the heme pocket there remains an appreciable barrier to the actual coordination of the ligand.[9,97,115] This barrier is attributed to the required energy to produce the rearrangements on the proximal side of the heme, permitting the heme to assume a low-spin state with the iron in the heme plane. The distal histidine already in the ligand pocket may therefore be particularly suited to lower this barrier for coordination of the exogenous ligand.

[114] R. N. Haire and B. E. Hedlund, *Biochemistry* **22**, 327 (1983).
[115] R. H. Austin, K. W. Beeson, L. Eisenstein, H. Frauenfelder, and I. C. Gunsalus, *Biochemistry* **14**, 5355 (1975).

[29] Measuring Relative Rates of Hemoglobin Oxidation and Denaturation*

By Victor W. Macdonald

In aqueous media, without the active methemoglobin reductase system present in intact red blood cells, the ferrous heme groups of hemoglobin can autoxidize to form continually increasing amounts of high-spin ferric heme. Under conditions that promote tetramer dissociation, subsequent tertiary structural distortions in the globin monomers occur; these distortions are associated with iron spin-state transitions and possible geometric distortions of the heme moiety within individual heme pockets. Binding of low-spin ferric heme at the sixth coordinate position of the heme iron

* The opinions or assertions contained herein are the private views of the author and are not to be construed as official or as reflecting the views of the Department of the Army or the Department of Defense (AR 360-5).

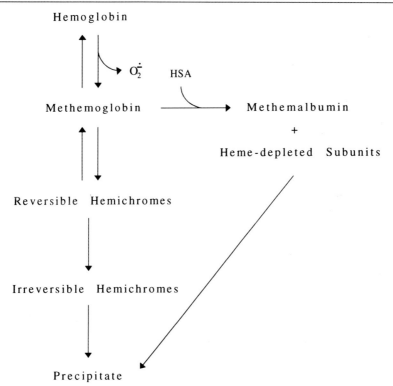

FIG. 1. General scheme of hemoglobin oxidation and subsequent degradation. Aquomethemoglobin is the first reaction product in aqueous media at neutral pH and in the absence of a higher affinity ligand that can bind to the sixth coordinate ferric iron. The heme iron can then undergo a high to low spin-state transition with formation of reversible hemichromes, which if not reduced back to the ferrous form will convert to an irreversible (nonreducible) form with subsequent precipitation. Transfer of ferric heme to human serum albumin (HSA) and the resultant heme depletion bypasses this pathway.

to normally distant amino acid side chains within the heme pocket can then result in the formation and accumulation of soluble and insoluble hemichromes.[1,2] A generalized outline of this pathway is provided in Fig. 1. An alternate degradative path is also shown, wherein the high-spin ferric heme of hemoglobin is transferred to human serum albumin to form methemalbumin.[3] The apoprotein thus formed is insoluble.

[1] E. A. Rachmilewitz, J. Peisach, and W. E. Blumberg, *J. Biol. Chem.* **246**, 3356 (1971).
[2] C. Winterbourn and R. W. Carrell, *J. Clin. Invest.* **54**, 678 (1974).
[3] H. F. Bunn and J. H. Jandl, *J. Biol. Chem.* **243**, 465 (1968).

There is experimental evidence that supports a model of anion-induced autoxidation in which nucleophilic anion displacement of molecular oxygen results in an intermediate ferrous heme/anion complex that acts as an electron donor to displaced oxygen (Scheme I):

$$Hb(Fe^{2+})_n(O_2)_n \rightleftharpoons Hb(Fe^{2+})_n(O_2)_{n-m} + (O_2)_m$$
$$Hb(Fe^{2+})_n(O_2)_{n-m} + (L^-)_m \rightleftharpoons Hb(Fe^{2+})_n(O_2)_{n-m}(L^-)_m$$
$$Hb(Fe^{2+})_n(O_2)_{n-m}(L^-)_m + (O_2)_m \rightarrow$$
$$Hb(Fe^{2+})_{n-m}(O_2)_{n-m}Hb(Fe^{3+})_m(L^-)_m + (O_2^-)_m$$

SCHEME I

where n is number of heme moieties per molecule ($= 4$ for tetrameric Hb), m is the number of released or displaced oxygen molecules ($= 4$ in fully oxidized or deoxyHb), and L^- is a nucleophilic anion. The reaction is accelerated by strong nucleophilic anions and by dissociation of molecular oxygen. At high partial pressure of oxygen (P_{O_2}) and saturating amounts of anion, oxidation is pseudo-first-order with respect to hemoglobin, and anion-methemoglobin is the final ferric heme product.[4] The process in Scheme I predicts the inverse relationship observed between hemoglobin and myoglobin autoxidation rates and dissolved oxygen concentration and predicts the decrease in these rates observed as hemoglobin desaturation progresses and oxygen disappears from solution.[5,6]

Efforts to synthesize chemically modified cell-free hemoglobin preparations with practical value as *in vivo* oxygen-carrying blood substitutes have focused on a variety of chemical agents that serve as intramolecular cross-linkers. The primary goals of these modifications have been to shift hemoglobin oxygen affinity into a physiologically acceptable range and to stabilize hemoglobin tetramers in a manner that will lengthen intravascular retention times. Based on Scheme I, allosteric modifications that shift the equilibrium distribution in favor of deoxyHb have the potential to accelerate anion-induced autoxidation. Because dissociation of the hemoglobin tetramer increases the rate of hemichrome formation,[1] intramolecular cross-linking agents that stabilize the quaternary structure of hemoglobin and impair dissociation into dimers should, at the same time, inhibit formation of hemichrome precipitates. The methods described below are used to assess the susceptibility of structurally modified hemoglobins

[4] W. J. Wallace, R. A. Houtchens, J. C. Maxwell, and W. S. Caughey, *J. Biol. Chem.* **257**, 4966 (1982).
[5] W. J. Wallace and W. S. Caughey, in "Biochemical and Clinical Aspects of Oxygen" (W. S. Caughey, ed.), p. 69. Academic Press, New York, 1979.
[6] A. Levy, L. Zhang, and J. M. Rifkind, in "Oxy-Radicals in Molecular Biology and Pathology" (P. A. Cerutti, E. Fridovich, and J. McCord, eds.), p. 11. Alan R. Liss, New York, 1988.

(as well as naturally occurring hemoglobin variants) to anion-induced autoxidation and to gauge the stability of hemoglobin oxidation products in a hydrophobic milieu designed to induce dimerization.

Hemoglobin Oxidation

Rapid Spectral Scanning

Multiwavelength rapid-scanning spectrophotometry allows reaction sequences involving interconversion of spectrally distinct hemoglobin derivatives to be monitored by recording complete spectra of ongoing reactions. In the examples presented here, spectral scans (500–700 nm in 2-nm steps, 1-sec duration) are collected at fixed intervals over specified time periods in a rapid-scanning diode array spectrophotometer (Hewlett-Packard 9450A, Pleasanton, CA). Reaction rates are slow enough that little or no distortion of individual spectra occurs during the 1-sec scan period. Reaction spectra consist of constantly changing combinations (presumably linear) of spectrally distinct hemoglobin derivatives, individual examples of which are shown in Fig. 2.

Oxygen Binding Measurements

The P_{50} (P_{O_2} at half-saturation) of each hemoglobin is measured separately in the autoxidation reaction buffer. Oxygen dissociation curves are fit to a six-parameter Adair equation from which fractional hemoglobin saturation with oxygen versus P_{O_2} is determined.[7,8]

Reaction Mixtures

Anion-induced autoxidations are carried out with 0.1 g/dl hemoglobin (62 μM heme) in 0.1 M potassium phosphate buffer containing 100 mM azide, pH 7.2, at 27° and ambient P_{O_2} in capped quartz cuvettes. Azide is a strongly nucleophilic anion and in the presence of oxygen serves as an effective promoter of autoxidation. The number of major reaction components is minimized, and azomethemoglobin is the final product,[4] as predicted by the last step in Scheme I. Reaction rates are fast enough for significant autoxidation to occur over a reasonable data collection time period, yet are slow enough that individual spectra accumulated over a 1-sec time period are not distorted by absorbance changes occurring during

---

[7] K. Imai, this series, Vol. 76, p. 438.
[8] K. D. Vandegriff, F. Medina, M. A. Marini, and R. M. Winslow, *J. Biol. Chem.* **264,** 17824 (1989).

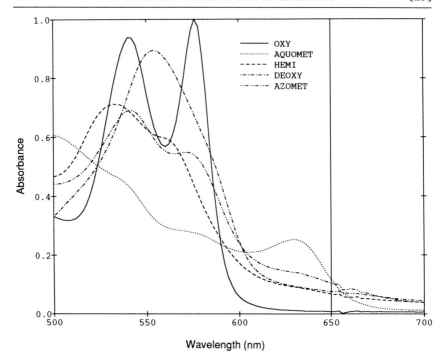

FIG. 2. Absorption spectra of human oxyhemoglobin (OXY), deoxyhemoglobin (DEOXY), aquomethemoglobin (AQUOMET), azomethemoglobin (AZOMET), and hemichromes (HEMI) measured on the Hewlett-Packard diode array spectrophotometer. The buffer is 0.1 M potassium phosphate, pH 7.20, plus excess sodium azide (100 mM) to form azomethemoglobin or 0.5 M salicylate with ferric hemoglobin to form hemichromes. Each scan spans a 200-nm range in 2-nm steps and is the time-averaged signal taken over 1 sec on the diode array spectrophotometer. The spectrum between 500 and 650 nm is used for multicomponent analysis, and the absorbance at 700 nm is used as an index of turbidity. The signal noise above 650 nm is due to the overlapping deuterium line at 656 nm.

the scanning period. A typical series of spectra for Hb A_0 accumulated at 10-min intervals over 4 hr is shown in Fig. 3.

Reactions are initiated by addition of hemoglobin to the reaction mixture, followed by gentle inversion of the cuvette to achieve a uniform mixture. The cuvette is placed immediately in a temperature-controlled cuvette holder in the spectrophotometer, and reactions are run without mechanical stirring. This minimizes optical noise due to artifacts of mechanical mixing, such as erratic spinning of stirring bars or changes in stirring rates. Initial methemoglobin concentrations should be less than 5%, but in practical application with chemically modified hemoglobins, they may be between 5 and 10%.

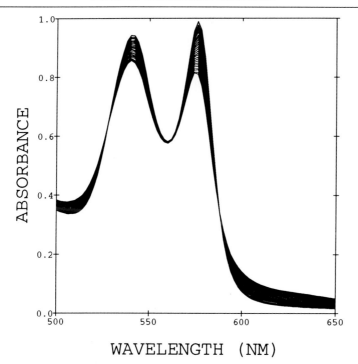

WAVELENGTH (NM)

FIG. 3. Absorption spectra of Hb A_0 during autoxidation in the presence of 100 mM azide. Scans are at 10-min intervals over a period of 4 hr, at which time the reaction mixture contains 36% azomethemoglobin.

Analysis

Spectra are analyzed by least-squares multicomponent analysis,[9-12] solving the matrix equation $A \times X = B$, where A contains absorbances of baseline spectra of pure hemoglobin derivatives, X is a matrix containing the fraction of each component in A at each time point, and B is the matrix of absorbance spectra recorded over time.

Spectral calibration sets are composed of individual hemoglobin derivatives that can be made by standard techniques.[1,13,14] Measurements should be made in the same buffer system in which reactions will be run. The

[9] R. I. Shrager and R. W. Hendler, *Anal. Chem.* **54,** 1147 (1982).
[10] H. Martens and T. Naes, *Trends Anal. Chem.* **3,** 204 (1984).
[11] T. Naes and H. Martens, *Trends Anal. Chem.* **3,** 266 (1984).
[12] K. D. Vandegriff and R. I. Shrager, this series, Vol. 232 [22].
[13] E. E. Di Iorio, this series, Vol. 76, p. 57.
[14] V. W. Macdonald and S. Charache, *Biochim. Biophys. Acta* **701,** 39 (1982).

visible spectrum of aquomethemoglobin is particularly sensitive to changes in pH.[15] Hardware differences among different spectrophotometers require that baseline spectra be generated on the same spectrophotometer in which the experimental spectra will be measured. The absorbance spectra in Fig. 2 were taken on an HP8450A diode array spectrophotometer and are scaled to identical heme concentrations by matching appropriate isosbestic points. Inclusion of all five derivatives in matrix A will yield the fraction of each component in the columns of X, calculated at each time point that spectra were recorded during the reaction.

This type of analysis requires assumptions about the number and identity of distinct spectral components present in the reaction mixture during the oxidation sequence. Reaction intermediates not present as pure component spectra in the calibration set (i.e., the columns of matrix A) will not be detected. Direct multivariate calibration in which the number of wavelengths utilized exceeds the number of absorbing constituents in the reaction mixture is preferred over methods that match the number of wavelengths to the number of absorbing components.[14,16] Increasing the number of instrument variables (e.g., wavelengths and absorbances) and analyzing the data by a least-squares fit to the calibration data set decrease the effect of random noise and increase the precision of the concentration measurements.[10] This analysis is performed best in conjunction with more extensive data reduction and error detection techniques, such as principal component analysis or singular value decomposition, which can be used to determine the total number of spectral transitions, including minor components, present in the original data set. Such methods serve to check assumptions about the reaction sequence and provide information that can be utilized to build and validate reaction models.[12,17,18]

In the presence of 100 mM azide, oxyhemoglobin disappears from the signal set as azomethemoglobin is formed. This reaction can be plotted as a percent decrease in the major component, oxyhemoglobin (Fig. 4). Pseudo-first-order rate constants (k'_1) are determined from single exponential fits[19] to the decrease in oxyhemoglobin concentration with time and are compared to those of Hb A_0 run on the same day. HPLC-purified Hb A_0[20] serves as the control against which different chemically modified hemoglobin derivatives are compared.

[15] S. M. Snell and M. A. Marini, *J. Biochem. Biophys. Methods* **17**, 25 (1988).
[16] R. E. Benesch, R. Benesch, and S. Young, *Anal. Biochem.* **55**, 245 (1973).
[17] R. I. Shrager, *Chemomet. Intel. Lab. Syst.* **1**, 59 (1986).
[18] M. Maeder and A. D. Zuberbuhler, *Anal. Chem.* **62**, 2220 (1990).
[19] W. J. Wallace, J. C. Maxwell, and W. S. Caughey, *Biochem. Biophys. Res. Commun.* **57**, 1104 (1974).
[20] S. M. Christensen, F. Medina, R. M. Winslow, S. M. Snell, A. Zegna, and M. A. Marini, *J. Biochem. Biophys. Methods* **17**, 143 (1988).

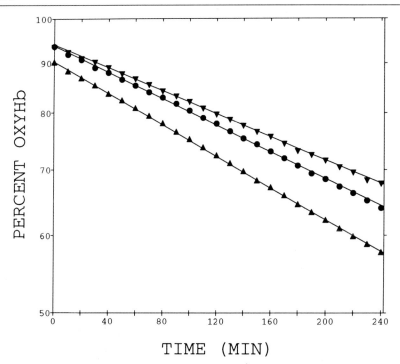

FIG. 4. Semilogarithmic plots of the decrease in percent oxyhemoglobin versus time during autoxidation of Hb A_0 (●), $\alpha\alpha$Hb (▲), and $\beta\beta$Hb (▼) with 100 mM azide (see text for definitions of the chemically modified derivatives). Lines are single exponential fits to the data.

Plots of rate constants (relative to that for Hb A_0) versus P_{50} for five different chemically modified human hemoglobin derivatives are shown in Fig. 5.[21] $\alpha\alpha$Hb is a human hemoglobin derivative covalently cross-linked between Lys-99(α) residues with bis(3,5-dibromosalicyl) fumarate under deoxy conditions.[22] $\beta\beta$Hb is covalently cross-linked primarily between Lys-82(β) residues with the same compound added under aerobic conditions.[23,24] Hb[T] and Hb[R] are the 64-kDa fractions isolated from a heterogeneous polymeric mixture of human hemoglobin reacted with

[21] V. W. Macdonald, K. D. Vandegriff, R. M. Winslow, D. Currell, C. Fronticelli, J. C. Hsia, and J. C. Bakker, *Biomater., Artif. Cells, Immobilization Biotechnol.* **19,** 425 (1991).

[22] S. R. Snyder, E. V. Welty, R. Y. Walder, L. A. Williams, and J. A. Walder, *Proc. Natl. Acad. Sci. U.S.A.* **84,** 7280 (1987).

[23] R. Chatterjee, R. Y. Walder, A. Arnone, and J. A. Walder, *Biochemistry* **21,** 5901 (1982).

[24] N. Shibayama, K. Imai, H. Hirata, H. Hiraiwa, H. Morimoto, and S. Saigo, *Biochemistry* **30,** 8158 (1991).

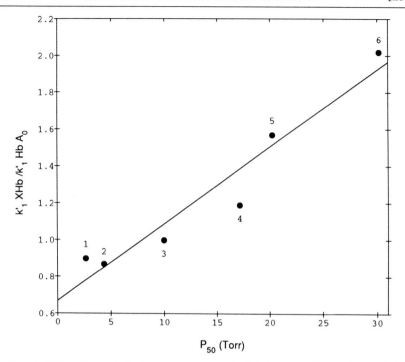

FIG. 5. Ratios of pseudo-first-order rate constants (k_i') for autoxidation of cross-linked human hemoglobins (XHb) to that for Hb A_0 as a function of P_{50}. 1, Hb[R]; 2, $\beta\beta$Hb; 3, HbA$_0$; 4, $\alpha\alpha$Hb; 5, Hb[T]; 6, HbNFPLP (see text for definitions of the chemically modified derivatives). Data from Macdonald et al.[21]

o-raffinose[25] in the presence and absence, respectively, of oxygen. Reacting 2-nor-2-formylpyridoxal 5'-phosphate (NFPLP) with human hemoglobin under deoxygenated conditions[26] produces Hb–NFPLP, which is covalently cross-linked between Lys-82(β) and Val-1(β) residues across the β cleft. The results of this analysis are consistent with the concept that oxygen affinity and rates of anion-induced autoxidation may be inversely proportional in chemically cross-linked native hemoglobin.

Stability of Oxidized Hemoglobin

Drug-induced oxidations are carried out at 27° with 0.1 g/dl hemoglobin (62 μM heme) with 40 mM menadione in 0.1 M potassium phosphate buffer

[25] J. C. Hsia, U.S. Pat. 4,857,636 (1989).
[26] R. E. Benesch and R. Benesch, this series, Vol. 76, p. 147.

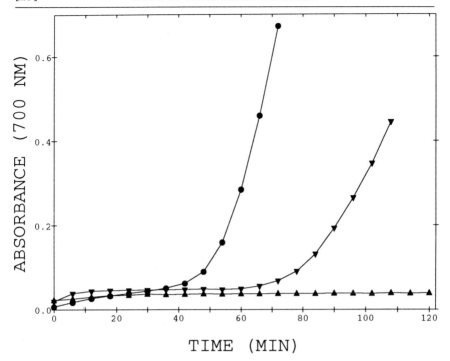

FIG. 6. Absorbance at 700 nm as an index of turbidity to monitor the onset of precipitation during menadione-induced oxidation of Hb A_0 (●), $\alpha\alpha$Hb (▲), and Hb–4-ICBS (▼) (see text for definitions of the chemically modified derivatives). Data from Macdonald et al.[21]

containing 1.1 M 2-propanol, pH 7.2. Superoxide dismutase is present at an appropriate concentration to allow maximal oxidation rates (usually 10 μg/ml).[14] This concentration of 2-propanol is sufficient to destabilize the native conformation of hemoglobin without denaturing the protein. This three-carbon monohydric alcohol is presumed to induce dissociation of the oxidized hemoglobin tetramer, leading to tertiary structural deformations, and is required for significant hemichrome formation[27] and precipitation (see Fig. 1). The reaction is initiated by addition of menadione (in 2-propanol) to the reaction mixture containing all other components, including hemoglobin. Spectra are recorded at 2- or 6-min intervals as described above, and onset of precipitation is indicated by a change in turbidity, monitored as an increase in the apparent absorbance at 700 nm.

This system is ideal for determining the relative stabilities of ferric hemoglobin products because it promotes hemichrome formation and in-

[27] V. W. Macdonald and S. Charache, *Biochim. Biophys. Acta* **705,** 48 (1982).

terconversion to forms in which protein unfolding is so pronounced that the ferric heme can no longer be reduced. Differences in the time of onset of precipitation for three hemoglobin derivatives are shown in Fig. 6. Unmodified Hb A_0 serves as a standard control. The reaction between human oxyhemoglobin and 4-isothiocyanatobenzenesulfonic acid[28] (4-ICBS) produces Hb–ICBS. Two to four amino-terminal residues are modified by this compound in a manner that may allow one or more sulfonic acid groups to interact with more positively charged groups across the β cleft to form an electrostatic bond. This modified hemoglobin tetramer is stabilized but is not covalently cross-linked. It appears that covalent cross-linking of the human hemoglobin tetramer confers the greatest degree of stability in solution for oxidized hemoglobin.

Care must be taken not to overinterpret the rate of apparent increase in absorbance as precipitates form. By running these reactions in the absence of stirring, precipitated hemoglobin will settle through the light path to the bottom of the cuvette. The time of onset of precipitation can be determined, and initial rates of precipitation can be compared, but the signal gradually will cease to reflect the total amount of light-scattering material formed over time. Interpretations must also be tempered by the realization that light scattering will be affected by differences in particle size and shape among different chemically modified hemoglobin preparations. This is most likely to be a factor when comparing the stability of hemoglobin tetramers with that of polymers.

[28] D. L. Currell, B. Law, M. Stevens, P. Murata, C. Ioppolo, and F. Martini, *Biochem. Biophys. Res. Commun.* **102**, 348 (1981).

[30] Hydrogen Peroxide-Mediated Ferrylhemoglobin Generation *in Vitro* and in Red Blood Cells

By CECILIA GIULIVI and KELVIN J. A. DAVIES

The oxidation of methemoglobin by hydrogen peroxide is explained by a heterolytic cleavage of the O—O bond of the coordinated peroxide, yielding a two-electron oxidation product of the hemoprotein, known as ferrylhemoglobin ($\cdot X—Fe^{IV}=O$, where $\cdot X$ denotes an amino acid radical in the apoprotein and Fe is the heme iron). The first oxidation equivalent provided by this reaction is in the form of the oxo–ferryl complex,

$Fe^{IV}=O$, similar to compound II reported for peroxidases,[1] and the second oxidation equivalent is detected as a transient apoprotein radical as reported for myoglobin.[2] Therefore, the formation of the radical form of the ferryl species bears the important implication that it bridges the discrepancy between the retention of one oxidizing equivalent in the ferrylheme complex and the observed catalysis of two-electron oxidations. The overall two-electron oxidation of methemoglobin by hydrogen peroxide could be written as in reaction (1):

$$HX—Fe^{III} + H_2O_2 \longrightarrow \cdot X—Fe^{IV}=O + H_2O \qquad (1)$$

The available experimental data suggest that this protein radical is located on an aromatic residue.[2,3] Current interest in the mechanism by which reactive oxygen species can cause cellular injury in several tissues has focused on the potential cytotoxicity of hemoprotein/hydrogen peroxide interactions,[3] and it is clear that ferrylhemoglobin, like ferrylmyoglobin, is a strong oxidant that can promote oxidation,[4] peroxidation,[4] and epoxidation[5] of various biomolecules *in vitro*.

Because ferrylhemoglobin is regarded as an oxidizing agent capable of promoting oxidation of molecules of biological relevance, it is of some value (1) to obtain ferrylhemoglobin as an oxidizing experimental tool in various systems and (2) to characterize ferrylhemoglobin in systems in which it may be produced, in order to distinguish it from other species with similar spectral profiles.

In this chapter we describe procedures for the preparation and spectral characterization of ferrylhemoglobin.

Preparation of Ferrylhemoglobin

In our studies bovine and human hemoglobin are purchased from Sigma Chemical Co. (St. Louis, MO) and usually a solution containing 1 mM hemoprotein in Krebs–Ringer buffer is prepared. The hemoproteins are dialyzed thoroughly before use (with dialysis membranes of 6000–8000

[1] P. George and D. H. Irvine, *Biochem. J.* **52**, 511 (1952).
[2] K. N. King and M. E. Winfield, *J. Biol. Chem.* **238**, 1520 (1963); T. Yonetani and H. Schleyer, *ibid.* **242**, 1974 (1967); J. F. Gibson, D. J. E. Ingram, and P. Nichols, *Nature (London)* **181**, 1398 (1958); K. Harada, I. Yamazaki, M. Tamura, and H. Watanabe, *Arch. Biochem. Biophys.* **275**, 354 (1989).
[3] D. Galaris, E. Cadenas, and P. Hochstein, *Arch. Biochem. Biophys.* **273**, 497 (1989); J. Kanner and S. Harel, *ibid.* **237**, 314 (1985); M. B. Grisham, *J. Free Radical Biol. Med.* **1**, 227 (1985); J. Kanner and S. Harel, *Lipids* **20**, 625 (1985); H. R. Rice, M. Y. Lee, and W. D. Brown, *Arch. Biochem. Biophys.* **221**, 417 (1983).
[4] E. Cadenas, *Annu. Rev. Biochem.* **58**, 79 (1989).
[5] P. R. Ortiz de Montellano and C. E. Catalano, *J. Biol. Chem.* **260**, 9265 (1985).

molecular weight cutoff, to remove fragments of low molecular weight) against a solution containing 0.1 M EDTA and 0.1 M potassium phosphate buffer (pH 7.4).

The starting material for preparing ferrylhemoglobin can be either methemoglobin or oxyhemoglobin. In the former case, the methemoglobin obtained after dialysis is ready for use. Oxyhemoglobin should be prepared by reducing methemoglobin with dithionite and then oxygenating the deoxyhemoglobin so produced on a Sephadex column. Methemoglobin can be reduced with sodium dithionite (molar ratio, 1 : 1.2), in deoxygenated 0.01 M potassium phosphate buffer (pH 7.8). Reduced hemoglobin (deoxyhemoglobin) should then be chromatographically purified on a Sephadex G-25 column (30 × 1.5 cm) using 0.01 M potassium phosphate buffer (pH 7.8) as the eluant[6] (this step also oxygenates the hemoglobin). A small percentage of methemoglobin (2–3%) always remains in the solutions of oxyhemoglobin so obtained, due to an unavoidable autoxidation. The chromatographed oxyhemoglobin can be concentrated in Centricon microconcentrator tubes (30-kDa cutoff; Amicon Division, W.R. Grace & Co.).

The oxidation of oxyhemoglobin or methemoglobin to ferrylhemoglobin can be achieved either using a flux of hydrogen peroxide or by adding a bolus of peroxide directly to the protein. The system we selected for producing a flux of hydrogen peroxide is glucose plus glucose oxidase (from *Aspergillus niger*, A grade, Calbiochem, La Jolla, CA), which catalyzes the reaction glucose + oxygen → gluconic acid + hydrogen peroxide. The rate of hydrogen peroxide generation can be measured in several ways, including (1) spectrophotometrically by following the formation of the compound horseradish peroxidase–hydrogen peroxide (compound II), as described by Boveris *et al.*,[7] in a dual-wavelength, double-beam spectrophotometer; (2) spectrofluorometrically by the formation of a fluorescent dimer of *p*-hydroxyphenylacetic acid as described by Danner *et al.*[8] and Hinsberg *et al.*[9]; or (3) by monitoring the rate of oxygen consumption in an oxygraph fitted with a Clark-type oxygen electrode.[7]

When using a bolus addition of hydrogen peroxide to oxidize oxyhemoglobin or methemoglobin to ferrylhemoglobin, the concentration of hydrogen peroxide should be checked in the spectrophotometer by measuring the absorbance at 240 nm ($\varepsilon = 40 \ M^{-1} \ cm^{-1}$). Figure 1 shows the percentage of ferrylhemoglobin formed at different ratios of methemoglobin to hydrogen peroxide. Usually a hydrogen peroxide : protein molar ratio of

[6] E. E. Di Iorio, this series, Vol. 76, p. 57.

[7] A. Boveris, E. Martino, and A. O. M. Stoppani, *Anal. Biochem.* **80,** 145 (1977).

[8] D. J. Danner, P. J. Brignac, D. Arceneaux, and V. Patel, *Arch. Biochem. Biophys.* **156,** 754 (1973).

[9] D. W. Hinsberg, III, K. H. Milby, and R. N. Zare, *Anal. Chem.* **53,** 1509 (1981).

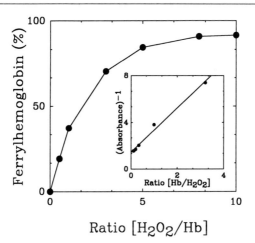

FIG. 1. Ferrylhemoglobin formation at different ratios of hydrogen peroxide to hemoglobin. Bovine methemoglobin (50 μM) in Krebs–Ringer phosphate buffer (pH 7.4) was treated with various concentrations of H_2O_2 to obtain ferrylhemoglobin. Ferrylhemoglobin production was followed spectrophotometrically at 556 nm. The inset shows a double-reciprocal plot of optical absorbance versus the ratio of H_2O_2/Hb, for which the linear correlation coefficient (r) was 0.98. Reprinted, with permission, from Ref. 11.

between 5:1 and 10:1 should be used in order to obtain a 90–100% yield of ferrylhemoglobin. Any excess hydrogen peroxide may be easily removed by adding catalase (500 units/ml) to the reaction mixture.

Spectrophotometric Analysis of Ferrylhemoglobin

Addition of hydrogen peroxide to oxyhemoglobin in a molar ratio of 10:1 causes a decrease in the absorbance at 541.2 and 576.4 nm accompanied by the appearance of two new peaks at 544.6 and 591.6 nm (Fig. 2A). The increase in the absorption at these two wavelengths might indicate a composite of $HX—Fe^{IV}=O$ and $\cdot X—Fe^{IV}=O$, which are spectrophotometrically indistinguishable.

The oxidation of methemoglobin by hydrogen peroxide (molar ratio, 2:1) is shown in Fig. 2B. The absorption spectrum of methemoglobin in the visible range shows a decrease in the absorption maximums at 503.2, 540.4, 578.4, and 634.8 nm along with an increase in the absorption at 544.6 and 591.6 nm, with isosbestic points at 466.2, 516.4, and 609.2 nm. These changes in the absorption spectrum of hemoglobin can be understood as a redox transition from methemoglobin to ferrylhemoglobin.

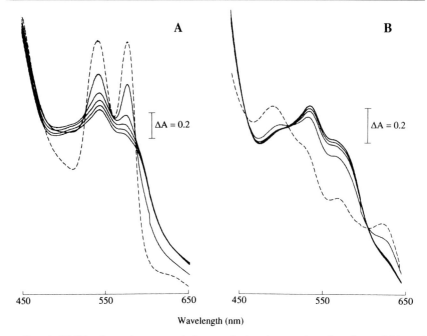

FIG. 2. Visible absorption spectral changes during the reaction of oxyhemoglobin and methemoglobin with hydrogen peroxide. Assay conditions: (A) 50 μM bovine oxyhemoglobin (dashed line) in 0.1 M potassium phosphate buffer (pH 7.4) was treated with 0.5 mM hydrogen peroxide to form ferrylhemoglobin (solid line). (B) 50 μM bovine methemoglobin (dashed line) in 0.1 M potassium phosphate buffer (pH 7.4) was supplemented with 0.1 mM hydrogen peroxide (solid line). Repetitive scans were taken every minute.

It is of interest to determine whether ferrylhemoglobin is formed in a biological system such as intact red blood cells, which contain 5 mM of oxyhemoglobin. Intact red blood cells (from both human and bovine sources) exposed to a continuous flux rate of hydrogen peroxide produced by the glucose–glucose oxidase system produced a spectral profile that can be understood as a transition from oxyhemoglobin to methemoglobin (Fig. 3A). To determine if this oxidation occurred through the intermediate state of ferrylhemoglobin, we performed further spectrophotometric studies in the presence of sodium sulfide (Na$_2$S). This compound adds to a β–β double bond of a pyrrole, forming a chlorin-type structure.[10] Addition of 2 mM Na$_2$S to a mixture of intact red blood cells and glucose–glucose oxidase produced an absorption maximum at 620 nm (Fig. 3B). This spectrum was identical to that obtained on addition of Na$_2$S to ferrylhemoglobin

[10] J. A. Berzofsky, J. Peisach, and W. E. Blumberg, *J. Biol. Chem.* **246**, 3367 (1971).

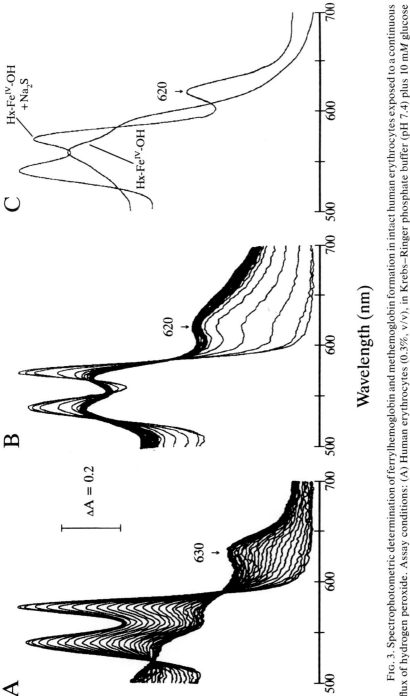

FIG. 3. Spectrophotometric determination of ferrylhemoglobin and methemoglobin formation in intact human erythrocytes exposed to a continuous flux of hydrogen peroxide. Assay conditions: (A) Human erythrocytes (0.3%, v/v), in Krebs–Ringer phosphate buffer (pH 7.4) plus 10 mM glucose and 7.5 μg/ml glucose oxidase, were incubated at 25°. (B) Experimental conditions as in A, except for the addition of 2 mM Na$_2$S. In both A and B, repetitive scans were recorded each minute over a 30-min incubation. (C) Ferrylhemoglobin was obtained by adding H$_2$O$_2$ to methemoglobin in a molar ratio of 5 : 1. Catalase (2 μM) was added to decompose excess H$_2$O$_2$ and Na$_2$S (2 mM) was added in order to obtain the sulfo derivative of ferrylhemoglobin. Reprinted, with permission, from Ref. 11.

(Fig. 3C) and is therefore indicative of the presence of ferrylhemoglobin in red blood cells treated with a continuous flux of hydrogen peroxide. The lack of an observable ferrylhemoglobin spectrum without the addition of Na_2S as an intermediate in the oxidation of oxyhemoglobin by hydrogen peroxide can be explained in terms of a comproportionation reaction between oxyhemoglobin (whose intracellular concentration is 5 mM) and the ferryl species formed.[11]

Acknowledgment

The work reported in this paper was supported by NIH/NIEHS Grant No. ES 03598 to KJAD.

[11] C. Giulivi and K. J. A. Davies, *J. Biol. Chem.* **265**, 19453 (1990).

[31] The Stability of the Heme–Globin Linkage: Measurement of Heme Exchange

By RUTH E. BENESCH

The bonds between heme and globin in normal hemoglobin (Hb) are too strong for direct measurement of the association constant. Therefore this important parameter, as well as its variation as a result of mutations or chemical modification, has been largely ignored. The few published values were all obtained with ferrihemoglobin (Hb$^+$), in which the heme–globin linkage is weaker than in the ferrous compounds, and they all utilized heme transfer to another acceptor. Banerjee[1] first measured heme transfer from ferrimyoglobin to organic bases such as histidylhistidine and pilocarpine and then used apomyoglobin as the acceptor for the hemes of Hb$^+$.[2] The values for the association constant thus obtained ranged from 10^{12} to 10^{16}. Bunn and Jandl[3] measured heme exchange between different ferrihemoglobins and heme transfer to serum albumin after separating the products by ion-exchange chromatography. Finally, Hrkal *et al.*[4] followed heme transfer from ferrihemoglobin to the plasma heme-binding protein, hemopexin.

[1] R. Banerjee, *Biochim. Biophys. Acta* **64**, 368 (1962).
[2] R. Banerjee, *Biochim. Biophys. Acta* **64**, 385 (1962).
[3] H. F. Bunn and J. H. Jandl, *J. Biol. Chem.* **243**, 465 (1967).
[4] Z. Hrkal, Z. Vodrazka, and I. Kalousek, *Eur. J. Biochem.* **43**, 73 (1974).

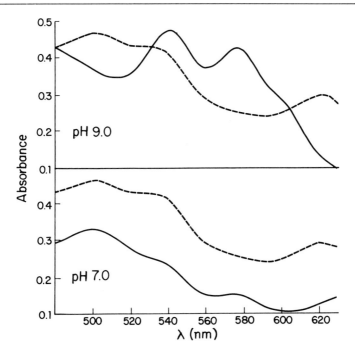

FIG. 1. Spectra of Hb^+ and MHA. Solid lines, Hb^+ A; dashed lines, MHA. Protein concentration: 0.05 mN in 0.25 M Tris buffer. The spectrum of MHA is identical at the two pH values.

The method described here[5] uses heme transfer from Hb^+ to serum albumin under conditions wherein this transfer can be followed spectrophotometrically. It is exceedingly simple and readily yields information on both the equilibrium and the kinetic stability of the heme–globin linkage.

Principle

The measurements are based on the fact[5] that the well-known change in the spectrum of ferrihemoglobin, which accompanies the loss of a proton from the Fe^{3+}-linked water molecule, with a pK of 8.2,[6] is absent in methemalbumin (MHA). Therefore, although the spectra of these two heme proteins are quite similar in neutral and acid solution (Fig. 1, lower curves), they show a large difference with increasing pH (Fig. 1, upper curves). Transfer of heme from globin to albumin can thus be measured

[5] R. E. Benesch and S. Kwong, *J. Biol. Chem.* **265**, 14881 (1990).
[6] D. L. Drabkin, *IUB Symp. Ser.* **19**, 142 (1961).

spectrophotometrically at pH 9, using standard methods for the analysis of two colored components in a mixture.[7,8]

Reagents

Hemolysates are prepared by standard methods.[9] The major component of normal human blood, Hb A_0, is isolated by chromatography on DEAE-Sepharose Fast Flow (Pharmacia-LKB Biotechnology Inc., Piscataway, NJ) at 4°, using a linear gradient of 0.05 M Tris-HCl buffer, pH 8.3–7.3. Hb Rothchild ($\beta^{37\,Trp\rightarrow Arg}$) was a gift from Dr. Helen Ranney of the University of California, San Diego, Hemoglobin cross-linked between the β chains with bis(pyridoxal) tetraphosphate is prepared as described.[10,11]

For oxidation to the ferric form, 1.2 equivalents of potassium ferricyanide are added with stirring to a solution of oxyhemoglobin at pH 7.0 and room temperature (to avoid oxidation of the β93 SH groups), followed by passage through Sephadex G-25 (fine) in 0.05 M bis-Tris buffer pH 7.5, 0.1 M Cl$^-$ to remove ferro- and ferricyanide. The concentration of Hb$^+$ is determined spectrophotometrically after conversion to ferrihemoglobin cyanide ($\varepsilon_{540} = 11.0 \times 10^3\ M^{-1}\ cm^{-1}$, heme basis).

The albumins (Sigma, St. Louis, MO) are dissolved in the bis-Tris buffer to give a 1 mM stock solution. No change has been found in the heme exchange behavior of these solutions on storage at 4° for 5 days.

Hemin chloride (Sigma) is dissolved in a small volume of 0.1 N NaOH and diluted with the pH 7.5 bis-Tris buffer to give a 1 mM stock solution. This is used immediately to prepare the 0.1 mM methemalbumin standard by mixing it with an equal volume of 1 mM albumin and fivefold dilution of the mixture with the same buffer.

Because these reactions involve transfer of heme from a tetramer (Hb$^+$) to a monomer (methemalbumin), all concentrations are expressed *on a heme basis* and are listed as millinormal (mN) or micronormal (μN) to indicate this fact.

[7] A. Zwart, A. Buursma, E. J. van Kampen, B. Oeseburg, P. H. W. van der Ploeg, and W. G. Zijlstra, *J. Clin. Chem. Clin. Biochem.* **19**, 457 (1981).

[8] A. Zwart, A. Buursma, E. J. van Kampen, and W. G. Zijlstra, *Clin. Chem. (Winston-Salem, N.C.)* **30**, 373 (1984).

[9] R. E. Benesch, R. Benesch, R. D. Renthal, and N. Maeda, *Biochemistry* **11**, 3576 (1972).

[10] R. E. Benesch and S. Kwong, *Biochem. Biophys. Res. Commun.* **156**, 9 (1988).

[11] R. E. Benesch and S. Kwong *J. Protein Chem.* **10**, 503 (1991).

Spectrophotometry

We have used a Hewlett-Packard diode array spectrophotometer (Model 8451A) to follow the progress of the heme exchange. This instrument uses reversed optics geometry, i.e., the light is dispersed by the monochromator after, rather than before, it passes through the sample. It is then received by a series of diode detectors spaced at 2-nm intervals. This spectrophotometer is especially convenient for the present purpose, because it has a built-in program for multicomponent analysis in an "over-determined" system,[8] i.e., when the number of data points exceeds the number of components. The wavelength range 470–630 nm, with 0.05 mN Hb$^+$ and 0.05 mN MHA as standards, is suitable for concentrations above about 0.01 mN. For lower concentrations the Soret region (380–420 nm) should be used with appropriately dilute standards. Because the rate of heme exchange is sensitive to temperature, it is obviously important to have an adequately thermostatted cell compartment.

For other spectrophotometers, measurements at two wavelengths and solution of the simultaneous equations, relating the absorbance to the concentrations and extinction coefficients of the two components, can be used. Detailed directions and a simple program for a desktop calculator are given in Zwart et al.[7]

Measurement of the Rate of Heme Transfer

The reactions are started by adding the albumin (in a small volume) to a temperature-equilibrated cuvette containing 1.0 ml of 0.5 M Tris buffer, pH 9.05, and Hb$^+$ in the 0.05 M bisTris buffer, pH 7.5, to give a final volume of 2.0 ml. The final pH is 9.00.

Some typical results with equinormal concentrations of Hb$^+$ and human serum albumin (HSA) are shown in Fig. 2. It is clear that the reaction is strongly biphasic, with about half the hemes exchanging much more rapidly than the rest, confirming previous observations[3,4,12,13] that heme is bound more tightly to the α than to the β chains. The product of the first phase is therefore largely globin with heme bound to the α chain only, and this is soluble under the conditions of the experiments. During this phase the sum of [Hb$^+$] + [MHA] should remain constant and equal to the initial Hb$^+$ concentration. The second phase results in the formation of heme-free globin, which is unstable and precipitates as a fine suspension. We have therefore not studied this part of the reaction.

[12] K. H. Winterhalter, C. Ioppolo, and E. Antonini, *Biochemistry* **10**, 3790 (1971).
[13] R. Cassoly, this series, Vol. 76, p. 121.

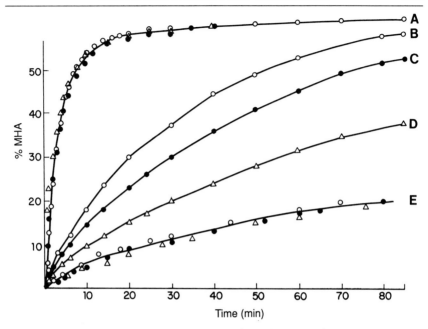

FIG. 2. Time course of heme transfer from Hb$^+$ to HSA. A, Hb$^+$ Rothchild; B, C, and D, Hb$^+$ A; E, Hb$^+$ XL. Initial concentrations: Open circles, [Hb$^+$] = [HSA] = 0.05 mN (O), 0.20 mN (●), and 0.50 mN (△). All were in 0.25 M Tris buffer (pH 9.05), 0.025 M bisTris buffer (pH 7.5), 0.05 M NaCl (final pH 9.00), temperature 20°; % MHA = [MHA]/ ([Hb$^+$] + [MHA]) × 100.

The rate is first-order with respect to [Hb$^+$] up to about 25% MHA and is independent of the albumin concentration over a wide range (data not shown). It can be seen from Fig. 2 that the rate for Hb$^+$ A varies inversely with the hemoglobin concentration, suggesting that dissociation of heme from $\alpha\beta$ dimers is faster than from intact tetramers. This conclusion was borne out by several other observations: (1) The rate was much faster and independent of the Hb$^+$ concentration (Fig. 2, curve A) in the case of Hb Rothchild ($\beta^{37\,Trp\rightarrow Arg}$), a mutant that is completely dissociated into $\alpha\beta$ dimers at the concentrations used for these measurements.[14] (2) Several hemoglobins with a covalent cross-link between the β chains, which therefore cannot dissociate into $\alpha\beta$ dimers, transferred their hemes much more slowly than Hb A and, again, the rate was independent of the Hb$^+$ concentration (Fig. 2, curve E). From these results it could be concluded that heme transfer from $\alpha\beta$ dimers is 50–100 times faster than

[14] V. S. Sharma, G. L. Newton, H. M. Ranney, F. Ahmed, J. W. Harris, and E. H. Danish, *J. Mol. Biol.* **144,** 267 (1980).

TABLE I
EQUILIBRIUM HEME DISTRIBUTION RATIOS[a]

Hemoglobin	Initial			
	$[Hb^+]$ (μN)	$\dfrac{[HSA]}{[Hb^+]}$	$\dfrac{[BSA]}{[Hb^+]}$	R
Hb^+ A				
Effect of total concentration	25	0.4	—	3.8
	50	0.4	—	3.8
	250	0.4	—	3.8
Effect of [Alb]/[Hb$^+$] ratio	50	0.4	—	3.8
	50	0.5	—	3.8
	50	0.7	—	3.4
Transfer to BSA	50	—	2.0	0.0056
	50	—	4.0	0.0046
	50	—	10	0.0045
Hb XL[b]	50	0.8	—	0.29
	50	1.2	—	0.34
	50	1.6	—	0.32
Hb Rothchild	50		1.0	0.044
	50		2.0	0.036
	50		6.0	0.032

[a] R is the equilibrium heme distribution ratio as defined in the text. Constant values were reached in 2 hr or less for all of these experiments except for Hb XL, which required 20 hr.

[b] Hemoglobin cross-linked covalently between the β chains with bis(pyridoxal) tetraphosphate.

from tetramers. Measurement of the rate of heme exchange as a function of the hemoglobin concentration therefore provides a simple and rapid method for detecting changes in the tetramer–dimer dissociation constant of mutant or chemically modified hemoglobins. (3) Finally, the rate of heme transfer from Hb$^+$ A to bovine serum albumin (BSA) is the same as to the human protein, despite the fact that the affinity of BSA for heme is three orders of magnitude lower than that of HSA (see below).

Equilibrium of Heme Transfer Reaction

In order to attain an equilibrium distribution of heme between globin and albumin (Alb), lower ratios of albumin to Hb$^+$ are used. This simultaneously eliminates the contribution of the α-chain hemes with their larger affinity for globin. We have used the equilibrium heme distribution ratio

$$R = \frac{[MHA][apoHb]}{[Hb^+][Alb]}$$

to compare the stability of the heme–β-globin linkage in different hemoglobins. [MHA] and [Hb$^+$] are measured spectrophotometrically as described above and [apoHb] and [Alb] are [Hb$^+$]$_{initial}$ − [MHA]$_{equilib}$ and [Alb]$_{initial}$ − [MHA]$_{equilib}$, respectively.

Some examples are given in Table I. It can be seen that the value of R is independent of the initial ratio, as well as of the total protein concentration. Attainment of equilibrium is, of course, faster at lower protein concentration and very much slower for the cross-linked hemoglobins, which require overnight incubation to reach heme exchange equilibrium.

It should be noted that the spectrum of bovine MHA is somewhat different from that of human MHA,[5] so that bovine MHA must be used as the standard for measurements with the bovine protein. Furthermore, the affinity of BSA for heme is three orders of magnitude lower than that of HSA (Table I), making BSA very suitable as an acceptor for hemoglobin mutants with a defective heme–globin linkage, such as Hb Rothchild.

[32] Release of Iron from Hemoglobin

By S. Scott Panter

In normal physiological states, iron is found almost exclusively bound to protein molecules in specific subcellular compartments, and the concentration of free or labile iron is very tightly controlled.[1] When decompartmentalized, however, iron can contribute to the development of pathophysiological states through various mechanisms.[1,2]

One pathophysiological state that can result from increased concentrations of free iron is bacterial infection. Iron is an element that is essential for bacterial growth, and the normal homeostatic mechanism of combating bacterial infection is to create a state of hypoferremia.[3] In the presence of increased concentrations of free iron, the probability of rapid bacterial growth greatly increases, possibly resulting in sepsis.[3] A second state of pathophysiology that can arise from high concentrations of free iron is related to the ability of iron to catalyze free radical reactions that can

[1] B. Halliwell and J. M. C. Gutteridge, this series, Vol. 186, p. 1.
[2] B. Halliwell and J. M. C. Gutteridge, "Free Radicals in Biology and Medicine." Oxford Univ. Press, Oxford and New York, 1989.
[3] M. J. Kluger and J. J. Bullen, in "Iron and Infection" (J. J. Bullen and E. Griffiths, eds.), p. 243. Wiley, New York, 1987.

cause considerable damage to cellular constituents.[2,4] Iron is a transition element that normally participates in oxidative and reductive reactions, but its reactivity is directed and controlled by protein molecules.[5] The redox activity of iron, however, can continue in solution in a nonspecific or site-specific but uncontrolled fashion, potentially damaging proteins, nucleic acids, and lipids.[6,7] Such cellular damage is mediated by highly reactive oxygen-free radicals, possibly the hydroxyl radical, generated through the iron-catalyzed Haber–Weiss reaction, which is represented by reactions (1)–(3).[8]

$$O_2^- + Fe(III) \leftrightharpoons O_2 + Fe(II) \tag{1}$$
$$2O_2^- + 2H^+ \rightarrow O_2 + H_2O_2 \tag{2}$$
$$Fe(II) + H_2O_2 \rightarrow Fe(III) + OH^- + \cdot OH \tag{3}$$

Hemoglobin comprises the largest reservoir of iron in the human body (70–75%).[9] Ferrous iron, which has six coordination positions, is bound in the heme pocket of hemoglobin by the four pyrrole nitrogen atoms of the protoporphyrin moiety, forming a tetradentate chelate with iron.[10] The imidazole residue of the proximal histidine of the heme pocket occupies the fifth coordination position. The sixth coordination position of iron is open in nonhemichromic hemoglobin and is free to bind ligands distal to the heme moiety.[10]

To function as a redox couple, iron must have at least one free coordination position,[11] which is the case for hemoglobin-bound iron. Consequently, it was hypothesized that hemoglobin-bound iron can catalyze the Haber–Weiss reaction while still bound in the heme pocket.[12,13] However, other studies have suggested that iron cannot act as a biological Fenton reagent in the heme pocket and that the redox-reactive species is either free iron or heme outside the heme pocket.[14–16] Despite the fact that

[4] H. B. Dunford, *Free Radical Biol. Med.* **3**, 405 (1987).
[5] A. Bezkorovainy, in "Iron and Infection" (J. J. Bullen and E. Griffiths, eds.), p. 27. Wiley, New York, 1987.
[6] J. M. Braughler, L. A. Duncan, and R. L. Chase, *J. Biol. Chem.* **261**, 10282 (1986).
[7] G. Minotti and S. D. Aust, *J. Biol. Chem.* **262**, 1098 (1987).
[8] T. P. Ryan and S. D. Aust, *CRC Crit. Rev. Toxicol.* **22**, 119 (1992).
[9] C. J. Gubler, *Science* **123**, 87 (1956).
[10] R. E. Dickerson and I. Geis, "Hemoglobin: Structure, Function, Evolution, and Pathology." Benjamin/Cummings, Menlo Park, CA, 1983.
[11] E. Graf, J. R. Mahoney, R. G. Bryant, and J. W. Eaton, *J. Biol. Chem.* **259**, 3620 (1984).
[12] S. M. H. Sadrzadeh, E. Graf, S. S. Panter, P. E. Hallaway, and J. W. Eaton, *J. Biol. Chem.* **259**, 14354 (1984).
[13] A. I. Alayash, J. C. Fratantoni, C. Bonaventura, J. Bonaventura, and E. Bucci, *Arch. Biochem. Biophys.* **298**, 114 (1992).
[14] J. M. C. Gutteridge, *FEBS Lett.* **201**, 291 (1986).

iron is bound tightly to hemoglobin, it can be liberated under specific circumstances, yielding a source of "free, reactive iron."[2] The term "free, reactive iron" is meant to imply only that iron has been released from heme and ligated to another moiety, perhaps the distal histidine in the heme pocket. Once outside the heme pocket, iron can be translocated to other iron-binding sites, i.e., inorganic chelators, lipids, other proteins, or nucleic acids, where it can catalyze the Haber–Weiss reaction [see Eqs. (1)–(3)].[1,2] Situations in which the heme moiety is released from hemoglobin, instead of iron, are under study by other investigators.[17]

Thus, the measurement of free iron in a variety of solutions will permit prediction and management of potential biological hazards from iron-catalyzed free radical reactions. In addition, the ability to measure iron released by hemoglobin in different experimental paradigms facilitates studies of the effects of structural modifications of the hemoglobin molecule on the stability of iron in heme. And finally, the measurement of free iron concentrations in solutions of hemoglobin is an important part of quality control/quality assurance of large-scale production of hemoglobin-based oxygen carriers and is included as one of the physicochemical variables in the points to consider for hemoglobin-based oxygen carriers.[18]

Preparation of Hemoglobin Solutions

Two different preparations of hemoglobin are used to characterize the assays described in this chapter. The first, designated "stroma-free hemoglobin" (SFHb), is derived from outdated human erythrocytes that are subjected to gentle, hypoosmotic lysis, followed by ultrafiltration, which removes all residual membranes ("stroma"). This preparation contains hemoglobin as well as other components of the red blood cell. The second type of hemoglobin, designated adult human hemoglobin (HbA_0), is produced by subjecting stroma-free hemoglobin to high-performance liquid chromatography, yielding a chromatographically pure hemoglobin solution. Both types of hemoglobin are prepared according to techniques previously described.[19–21] Cyanomethemoglobin is prepared according to

---

[15] A. Puppo and B. Halliwell, *Biochem. J.* **249**, 185 (1988).

[16] J. M. C. Gutteridge and A. Smith, *Biochem. J.* **256**, 861 (1988).

[17] R. E. Benesch, this volume, [31].

[18] J. C. Fratantoni, *Transfusion (Philadelphia)* **31**, 369 (1991).

[19] S. M. Christensen, F. Medina, R. M. Winslow, S. M. Snell, A. Zegna, and M. A. Marini, *J. Biochem. Biophys. Methods* **17**, 143 (1988).

[20] K. W. Chapman, S. M. Snell, R. G. Jesse, J. K. Morano, J. Everse, and R. M. Winslow, *Biomater., Artif. Cells, Immobilization Biotechnol.* **20**, 415 (1992).

[21] R. M. Winslow and K. W. Chapman, this volume [1].

the method of Di Iorio.[22] All hemoglobin samples were supplied by the Letterman Army Institute of Research hemoglobin production facility.

Detection of Free Iron in Solutions of Purified Hemoglobin

Outside the red cell membrane, when hemoglobin is processed or purified, small amounts of free iron can be detected as Fe(II) using the following method. All hemoglobin samples are assayed in duplicate; 250 μl of each hemoglobin solution is added to a 12 × 75-mm disposable borosilicate glass centrifuge tube, to which 250 μl of 20% trichloroacetic acid (TCA) (w/v) is added. The tubes are then centrifuged in a clinical centrifuge (Sero-fuge II, Becton Dickinson, Rutherford, NJ) to pellet the precipitated protein. From the supernatant, 250 μl is transferred to a new 12 × 75-mm test tube containing 250 μl of distilled water. Subsequently, 2.5 ml of Sigma iron buffer reagent [Sigma Chemical Co., St. Louis, MO, Cat. # 565-1; hydroxylamine hydrochloride, 1.5% (w/v) in acetate buffer, pH 4.5, with added surfactant][23] is added to each tube, which is then mixed thoroughly. Finally, 50 μl of Sigma iron color reagent [Cat. # 565-3; ferrozine, 0.85% (w/v) in hydroxylamine hydrochloride][23] is added to each tube, and the color is developed for 30 min at 37° or 60 min at room temperature. The content of each tube is transferred to a disposable 4.0-ml cuvette and read at 560 nm. After correcting for dilution, data are expressed as micrograms of iron per gram of hemoglobin.

A standard curve is constructed using the Sigma iron standard (Cat. # 565-5; 500 μg/dl in hydroxylamine hydrochloride),[23] which is a 500 μg/dl solution of iron in hydroxylamine hydrochloride solution. Six standards, ranging from 0.25 to 2.5 μg, are used routinely and have been found to be reproducible over time (see Fig. 1). The molar extinction coefficient calculated from this standard curve agrees with previously published estimates (approximately 28,000 liter mol^{-1} cm^{-1}).[24-26] A number of different hemoglobin preparations have been tested, including stroma-free red cell hemolysate, purified hemoglobin A_0, and hemoglobin covalently cross-linked between the α subunits.[27] Typical results are shown in Table I. The values for iron content have been corrected for dilution.

[22] E. E. Di Iorio, this series, Vol. 76, p. 57.
[23] Sigma Diagnostics Procedure #565 (1990).
[24] L. L. Stookey, *Anal. Chem.* **42,** 779 (1970).
[25] P. Carter, *Anal. Biochem.* **40,** 450 (1971).
[26] J. D. Artiss, S. Vinogradov, and B. Zak, *Clin. Biochem.* **14,** 311 (1981).
[27] R. Y. Walder, M. E. Andracki, and J. A. Walder, this series [17].

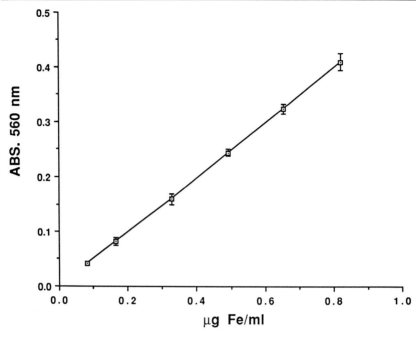

FIG. 1. Standard curve of iron detected spectrophotometrically as a complex with the chromophore ferrozine. The data represent mean values ± standard deviation for eight independent determinations.

TABLE I
ASSAY OF FREE OR LABILE IRON IN DIFFERENT
HEMOGLOBIN SOLUTIONS

Hemoglobin preparation[a]	Concentration (g/dl)	Iron content (μg Fe/g Hb)
SFHb (91098)	8.38	115.6
HbA_0 (91040)	9.90	133.4
HbA_0 (91098)	6.83	136.8
$\alpha\alpha$XLHb (91119)	4.99	164.7

[a] Primary chromatographic component of adult human hemoglobin (HbA_0), stroma-free hemoglobin (SFHb), and hemoglobin cross-linked between the α subunits with diaspirin ($\alpha\alpha$XLHb). The numbers in parentheses represent the Julian date when that specific preparation was started at the Letterman Army Institute of Research production facility.

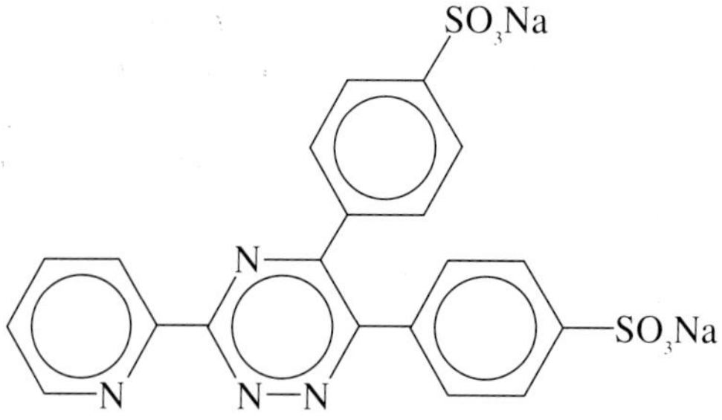

SO₃Na

SCHEME I. Ferrozine.

Comments

Since its initial characterization,[24] ferrozine (Scheme I) has been used extensively as a chromophore to measure iron in a variety of assays.[25,26,28,29] Because ferrozine binds only Fe(II), reducing reagents are required and several different compounds have been utilized, including ascorbic acid, thioglycolic acid, and hydroxylamine. Copper can interfere with ferrozine-dependent iron measurements, and copper-complexing agents have been included routinely in a number of iron assays.[25,26,28,29] Copper has not been found to be a significant contaminant of hemoglobin solutions, and copper-complexing agents have not been included in the present studies. Ferrozine also forms a colored complex with cobalt, which also is not a significant contaminant in our hemoglobin solutions; colored complexes are not formed between ferrozine and calcium or magnesium.[24] The original ferrozine assay for serum iron[25] has proved to be reliable but perhaps less sensitive than the bleomycin assay for free iron; however, numerous substances interfere with the latter assay, thus reducing its utility.[14] There is another assay for free iron based on the iron-binding chromophore Ferene S,[26] but because chromophore solutions have to be made fresh daily, the assay is cumbersome, which contributes to variability in the determinations.

The measurement of "free" or adventitious iron in hemoglobin solutions has been a controversial subject, with some investigators suggesting that nonheme iron might be produced as an artifact of the assay. Using the techniques described in this chapter, between 3.0 and 5.0% of the total iron is detected as nonheme. Significantly lower concentrations of nonheme iron were reported in another study of cross-linked hemoglobin

[28] D. P. Derman, A. Green, T. H. Bothwell, B. Graham, L. McNamara, A. P. MacPhail, and R. D. Baynes, *Ann. Clin. Biochem.* **26,** 144 (1989).
[29] W. W. Fish, this series, Vol. 158, p. 357.

[0.5–1.0 parts per million (ppm) in 10 g% hemoglobin solutions].[30] A major difference between the two assay systems is that the assay described here utilizes an acid precipitation step, but the latter study separates hemoglobin from "free" iron by ultrafiltration in the presence of diethylenetriaminepentaacetic acid. An acid-labile pool of iron associated with hemoglobin ("human blood") has been reported, and despite the fact that this earlier study used considerably more harsh conditions of acid treatment (16 hr at 37° in 0.8% HCl), the amount of nonheme iron recovered (3.4–7.0% of total iron) was in the same range as those reported in this chapter.[31] One additional study using an acid precipitation step reported 3.4% nonheme iron in fresh bovine blood.[32] Consequently, the acid precipitation step used in the assays described in this chapter may facilitate the detection of nonheme iron present in hemoglobin solutions.

To determine whether the iron assay influences the recovery of nonheme iron, the amount of free iron in a hemoglobin solution was determined and defined as the background level. Subsequently, micromolar amounts of reduced and oxidized iron were added to the hemoglobin solution and total recovery was assessed. Fe(II) (as $FeSO_4$) was added to hemoglobin, which was then subjected to the acid denaturation and the iron assay procedures described in the preceding pages of this chapter; all added iron was recovered and detected above the background level. However, if iron was added as Fe(III) (as $FeCl_3$), only the background level of iron was detected; none of the added Fe(III) was recovered. These results suggest that if "nonheme" iron in hemoglobin solutions is present in the oxidized form, it will not be detected by the assay system used for the current experiments. In addition, these data indicate that the labile iron pool in hemoglobin solutions may be present primarily as the reduced species of iron.

The results of the previous series of experiments led us to try to preform the ferrozine–iron complex by adding ferrozine to the hemoglobin solution prior to acid precipitation, predicting that the ferrozine–Fe(II) complex would survive precipitation and could be quantified in the protein-free supernatant. However, no complex was recovered. Hemoglobin A_0 also was added to a preformed complex of ferrozine and Fe(II), after which the hemoglobin was precipitated by TCA or removed by ultrafiltration. None of the ferrozine—Fe(II) complex could be detected in solution after the hemoglobin was removed. These results suggest that either the

[30] T. Marshall, J. Weltzer, T. T. Hai, T. Estep, and M. Farmer, *Biomater., Artif. Cells, Immobilization Biotechnol.* **20,** 453 (1992).
[31] J. W. Legge and R. Lemberg, *Biochem. J.* **35,** 253 (1941).
[32] J. King, S. De Pablo, and F. Montes De Oca, *J. Food Sci.* **55,** 593 (1990).

ferrozine–Fe(II) complex binds to hemoglobin, or it is physically enmeshed in the denatured protein, which carries it into the pellet on centrifugation.

Other experiments were conducted to determine whether iron complexing agents might influence the recovery or detection of nonheme iron in hemoglobin solutions. Ethylenediaminetetraacetic acid (EDTA), histidine, and citrate were added at two different concentrations, 0.1 and 1.0 mM, to the hemoglobin solutions prior to precipitation. The iron recovered when the assay was conducted in the presence of citrate and histidine increased above control levels at both concentrations. EDTA had no effect at the lower concentration, but iron recovery as detected by ferrozine was completely inhibited at the higher concentration. These results indicate that agents that interact with or bind iron may significantly influence the recovery of nonheme iron from hemoglobin solutions.

In summary, hemoglobin solutions seem to contain a pool of labile iron that can be quantified by using ferrozine. Apparently, the exact size of this pool can be influenced by reagents that interact directly with hemoglobin or iron. Consequently, constituents present in the final formulation of a hemoglobin solution may significantly influence the use of this assay system to measure nonheme iron.

Spectrophotometric Measurement of Total Hemoglobin-Bound Iron

In an attempt to release and measure all iron bound by hemoglobin, we modified the technique of Fish,[29] entailing an acid-permanganate-mediated digestion of the protein, which releases all protein-bound iron and converts it to Fe(III). The chromophore is then added in the presence of high concentrations of reducing agents that convert Fe(III) back to Fe(II), which is necessary for the formation of an iron–chromophore complex. A 500-μl aliquot of a solution of 0.6 N HCl in 2.25% (w/v, 0.142 M) KMnO$_4$ is added to a 50-μl sample of a 1.6 mM hemoglobin solution. After digestion for 2 hr at 60°, the sample is subjected to centrifugation (Micro-Fuge II, Becton Dickinson); the supernatant, which tests negative for iron, is removed and the pellet is dried thoroughly. A 2.5-ml aliquot of Sigma iron buffer reagent [Cat. # 565-1; hydroxylamine hydrochloride, 1.5% (w/v) in acetate buffer, pH 4.5, with added surfactant][23] is added to each tube, which is then subjected to thorough mixing. This reagent accomplishes a complete solubilization of the pellet. Finally, 50 μl of Sigma iron color reagent [Cat. # 565-3; ferrozine, 0.85% (w/v) in hydroxylamine hydrochloride][23] is added to each tube, and the color is developed for 30 min at 37° or 60 min at room temperature. The results of such a determina-

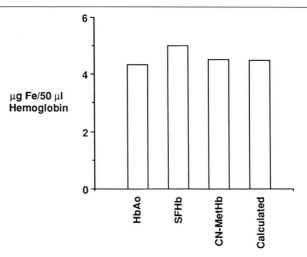

FIG. 2. Measured iron concentration in different hemoglobin solutions compared with the theoretical iron content calculated from the concentration of hemoglobin (1.6 mM).

tion are shown in Fig. 2. The calculated iron content, based on the dilution of all hemoglobin solutions to a starting concentration of 1.6 mM (monomer), was 4.48 μg/50 μl, and the concentrations detected for Hb A_0, SFHb, and cyanomethemoglobin were 4.35, 5.01, and 4.51 μg/50 μl, respectively.

Comments

The total iron measured in hemoglobin solutions with this method corresponds well with theoretical calculations based on the actual concentration of hemoglobin. The spectrophotometric measurement of SFHb iron was higher than estimated by calculation; however, SFHb contains significant amounts of other iron-containing proteins, e.g., catalase and peroxidases, which may have inflated the final measurement. This technique has also been used to quantify the total iron content of different organs. In summary, a rapid and accurate determination of the total iron content of a hemoglobin solution can be made using the chromophore ferrozine.

Effect of Oxidants and Reductants on Release of
  Hemoglobin-Bound Iron

Significant quantities of iron, liberated from hemoglobin by both oxidative and reductive events, can be measured spectrophotometrically. The study of this phenomenon may provide important information about the reactivity of the iron or heme moieties of different hemoglobin molecules under conditions that may be present in disease states or trauma. These experiments may provide insight into the possible toxicity of cell-free hemoglobin solutions.

The loss of iron from oxyhemoglobin through oxidation reactions has been demonstrated previously[14,30,33,34] and can be measured easily on a routine basis. Incubations are conducted in 12 × 75-mm disposable borosilicate glass test tubes in a 1.0-ml final volume with the following constituents at their final concentration: phosphate buffer, pH 7.4, 10.0 mM; hemoglobin (monomer), 160.0 μM; and organic or inorganic peroxides or reducing agents or other effectors at varying concentrations. The reaction is initiated by the addition of the effector reagents, and incubations are conducted for 30–60 min at 37°. At the end of the incubation, protein is precipitated by the addition of 250 μl of TCA (w/v). After centrifugation to pellet the precipitated protein, 500 μl of the supernatant is removed and transferred to another 12 × 75-mm test tube that contains 2.5 ml of Sigma iron buffer reagent (Cat. # 565-1). After a thorough mixing, 50 μl of Sigma iron color reagent (Cat. # 565-3) is added, followed by a second thorough mixing, and the color is allowed to develop for 30 min at 37° or 60 min at room temperature. A standard curve ranging from 0.25 to 2.5 μg iron is included in each assay, and samples are analyzed spectrophotometrically at 560 nm.

The assay procedure can be modified to utilize microtiter plates and a microtiter plate reader, which allows spectrophotometric measurements in 96 wells simultaneously. Hemoglobin incubations are conducted in quadruplicate in a 200-μl final volume in U-bottom 96-well plates (Falcon 3910, Becton Dickinson, Oxnard, CA). When possible, reagents are added using repeat pipettors, which greatly facilitates the preparation of each incubation mixture. In this modified assay, reactions are initiated by the addition of 20 μl of 1.6 mM hemoglobin (per monomer), and after the microtiter plate is covered with a sheet of Parafilm, incubations are conducted at 37° in a tissue culture incubator for 1 hr. At the end of the incubation, 25 μl of 20% TCA is added to each well to precipitate the protein, which is then pelleted by centrifugation at 2500 rpm in a microplate carrier for a tabletop centrifuge (Sorvall RT-6000B, Newton, CT). Following centrifugation, 50 μl of the supernatant from each well is transferred to a second, flat-bottomed microtiter plate (Falcon 3040), in which each

[33] J. M. C. Gutteridge, *Biochim. Biophys. Acta* **834,** 144 (1985).
[34] S. S. Panter, L. J. England, D. M. Hellard, and R. M. Winslow, *Blood* **76,** Suppl 1, 72a (1990).

well contains 200 μl of Sigma iron buffer reagent (Cat. # 565-1). Finally, 10 μl of a 50% solution of Sigma iron color reagent (Cat. # 565-3) is added to each well, and the iron–chromophore complex is allowed to reach equilibrium during a 1-hr incubation at room temperature.

Standard curves for the microtiter plate iron assay are constructed by adding to a separate series of wells 50 μl of Sigma iron standards containing 50.0, 200.0, 500.0, and 1000.0 μg/dl.

The results from a series of experiments are shown in Fig. 3. Using both purified human HbA$_0$ and SFHb, from 3% to over 50% of the total iron, based on heme concentration, was released during a 1-hr incubation at 37°. With the exception of hydrogen peroxide incubated with SFHb, all peroxides, at concentrations 20-fold higher than that of hemoglobin-bound iron, released significant quantities of iron. The release of iron from hemoglobin by ascorbic acid is concentration dependent.

Comments

The limited ability of hydrogen peroxide to release iron from SFHb may be due to the presence of residual red cell catalase in this preparation. The organic peroxides are evidently poor substrates for catalase and are capable of releasing virtually identical amounts of iron from both HbA$_0$ and SFHb. Ascorbic acid liberated iron from both HbA$_0$ and SFHb in a dose-dependent fashion, but there was less total iron released from SFHb than from HbA$_0$. A number of other effectors, particularly reductants, were tested for their ability to release hemoglobin-bound iron, and cysteine, reduced glutathione, and superoxide were particularly effective.[34] Finally, the presence or absence of buffer as well as the type of buffer in incubations of hemoglobin solutions with various effectors can have a pronounced influence on iron release. In the absence of buffering capacity or in the presence of physiological saline, iron release is attenuated, but iron release is increased in the presence of Tris or phosphate buffer. A possible explanation for these results may be that Tris and phosphate are capable of interacting directly with iron, removing it from a hemoglobin-binding site and increasing its solubility.

Previous studies have demonstrated that hemoglobin can serve as a prooxidant, and the results of the iron assays discussed here raise the possibility that the prooxidant effect of hemoglobin may be mediated, in some circumstances, by nonheme iron released from hemoglobin. Although nonheme iron may mediate the prooxidant effects of hemoglobin, the sequence of biochemical reactions resulting in the liberation of hemoglobin-bound iron is not clear. It is not known whether heme dissociation from globin precedes the release of iron or whether iron can be liberated

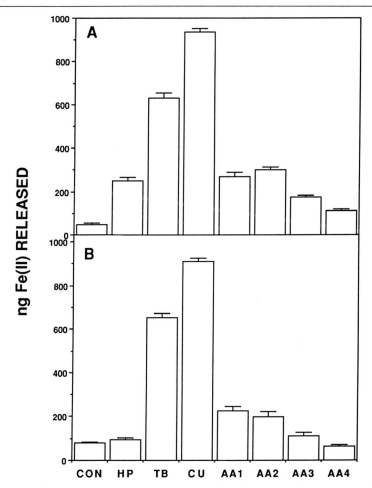

FIG. 3. Iron released by incubation (60 min, 37°) of 0.16 mM Hb A_0 (A) or 0.16 mM SFHb (B) with 10.0 mM hydrogen peroxide (HP), 5.0 mM *tert*-butyl hydroperoxide (TB), 5.0 mM cumene hydroperoxide (CU), or ascorbic acid at concentrations of 15.0 (AA1), 10.0 (AA2), 5.0 (AA3), or 2.0 mM (AA4). The condition "CON" represents the amount of iron released by an incubation of hemoglobin and buffer without additional reagents. The data represent mean values ± standard deviation for eight independent determinations.

directly from globin-bound heme. In fact, the sequence of biochemical reactions may actually differ depending on the reaction conditions. For example, the inclusion of lipids in the incubation might provide a hydrophobic binding site for heme and support heme release, whereas the aqueous conditions of the incubations in the assay described here may increase the liberation of free iron.

In conclusion, as the use of hemoglobin-based oxygen carriers is pondered, more information on the stability, reactivity, and destiny of hemoglobin-bound iron will be required. The iron measurement assays described in this chapter provide a simple, yet reliable, means of accomplishing this goal.

Acknowledgment

The opinions and assertions contained herein are the private views of the author and are not to be construed as official, nor do they reflect the views of the Department of the Army or the Department of Defense (AR 360-5). The author is grateful to J. R. Hess, M. A. Marini, K. D. Vandegriff, and C. Wheeler for helpful discussions, to Mrs. S. Siefert for editorial assistance, and to J. J. Knudsen and D. Profitt for technical assistance.

[33] Thermal Denaturation Procedures for Hemoglobin

By Kenneth W. Olsen

Stability is an important functional property of any protein. In the case of hemoglobin (Hb), qualitative methods, such as the 2-propanol[1] or the heat precipitation[2] tests, have long been used to detect unstable variants. These and other techniques for detecting unstable hemoglobins have been reviewed by Ohba,[3] Carrell,[4] and Huisman and Jonxis.[5] Methods for detecting unstable hemoglobin variants that do not involve thermal denaturation are listed in Table I.[6-9] Quantitative measurements of hemoglobin stability have also been developed. Numerous abnormal hemoglobins are clinically stable, but their stabilities differ from that

[1] A. R. Carrell and R. Kay, *Br. J. Haematol.* **23,** 615 (1972).
[2] J. V. Dacie, A. J. Grimes, A. Meisler, L. Steingold, E. H. Hemsted, G. H. Beaven, and J. C. White, *Br. J. Haematol.* **10,** 388 (1964).
[3] Y. Ohba, *Hemoglobin* **14,** 353 (1990).
[4] R. W. Carrell, *Methods Hematol.* **15,** 109 (1986).
[5] T. H. J. Huisman and J. H. P. Jonxis, "The Hemoglobinopathies: Techniques of Identification." Dekker, New York, 1977.
[6] D. J. Weatherall and J. B. Clegg, "The Thalassaemia Syndromes." Blackwell, Oxford, 1981.
[7] T. Asakura, M. E. Segal, S. Friedman, and E. Schwartz, *JAMA, J. Am. Med. Assoc.* **233,** 156 (1975).
[8] Y. Ohba, Y. Hattori, H. Yoshinaka, M. Matsuoka, T. Miyaji, T. Nakatsuji, and M. Hirano, *Clin. Chim. Acta* **119,** 179 (1981).
[9] R. W. Carrell and H. Lehmann, *J. Clin. Pathol.* **34,** 796 (1981).

TABLE I
NONTHERMAL METHODS OF DETECTING UNSTABLE HEMOGLOBINS

Method	Use	Ref.
Brilliant cresyl blue dye	Test for Hb H inclusions	6
2-Propanol test	Widely used for unstable Hbs but can give false positive if Hb F is elevated	1
Mechanical shaking	Standardized test requires special equipment	7
PCMB precipitation[a]	Usually precipitates unstable subunits, which can be used for structural studies; analysis done with electrophoresis	5, 8
Zinc acetate precipitation	No false positives due to Hb F; Hb A can precipitate if Zn concentration is too high	9

[a] PCMB, p-Chloromercuribenzoate.

of Hb A.[10-13] In addition, the stability of cross-linked hemoglobins is an important aspect in evaluating them as potential blood substitutes.[11,14] Both of these situations require a quantitative measurement of hemoglobin stability that can be used to compare different species. The use of thermal denaturation curves to provide a denaturation temperature (T_m) solves this problem.

Thermal denaturation provides a direct measure of the stability of the hemoglobin. The experiment can be done by two different methods: (1) precipitation as a function of time at a particular temperature[2] or (2) the unfolding of the protein determined from absorbance changes as a function of increasing temperature.[10,11,14]

Rate of Precipitation Procedure

The precipitation test[2] was originally done at 50°, but higher temperatures are needed to detect some abnormal hemoglobins. For example, Hb Hofu [β126 (H4) Val → Glu][15] was identical to adult hemoglobin (Hb A) at 50° but showed an increased denaturation rate at 60°. Although this

[10] P. A. Ockelford, A. Y. Liang, R. M. Wells, M. Vissers, S. O. Brennan, D. Williamson, and R. W. Carrell, Hemoglobin 4, 295 (1980).
[11] T. Yang and K. W. Olsen, Arch. Biochem. Biophys. 261, 283 (1988).
[12] T. Yang and K. W. Olsen, Hemoglobin 13, 147 (1989).
[13] T. Yang and K. W. Olsen, Hemoglobin 14, 641 (1990).
[14] F. L. White and K. W. Olsen, Arch. Biochem. Biophys. 258, 51 (1987).
[15] Y. Ohba, M. Matsuoka, K. Fuyuno, K. Yamamoto, S. Nishijima, and T. Miyaji, Hemoglobin 5, 89 (1981).

experiment has the potential to yield a quantitative rate, most workers using this test have simply reported whether the rate was increased compared to that of a Hb A control.

The precipitation test was originally proposed by Dacie et al.[2] but is normally done by the modified procedure of Huisman and Jonxis,[5] which is described here. The sample (0.2 ml), containing approximately 5 g Hb dl^{-1}, is mixed with 10 ml of 0.1 M sodium phosphate buffer, pH 7.4. The same is done with a control sample of Hb A. The diluted samples are divided into 10 aliquots of 1 ml each in tightly capped microcentrifuge tubes and heated between 60° and 65° in a constant-temperature water bath. The tubes are removed at 0, 2, 4, 6, 8, 10, 15, 20, 25, and 30 min, cooled in an ice bath for 5 min, and centrifuged to remove the precipitate. The absorbance of the supernatant is measured at 523 nm. The percent denatured is calculated by

$$\% \text{ Denatured} = \frac{A(0) - A(t)}{A(0)} \times 100$$

where $A(0)$ is the absorbance of the unheated (i.e., zero time) aliquot and $A(t)$ is the absorbance of the heated sample at time t. The percent denaturation is plotted against time (Fig. 1) and compared with the normal Hb A control. The difference in percent denaturation between the control Hb A and the sample hemoglobin at 10 min is a semiquantitative measure of their relative stabilities.[5] One problem with this method is that it gives an intermediate stability for a sample that is a mixture of two different hemoglobins,[16] as shown in Fig. 1. This is not true for the thermal denaturation procedure given below.

There are several variations on this procedure. The dilution buffer is sometimes 0.2 M sodium phosphate, pH 6.5, but this change does not generally affect the results.[5] Tris–NaCl buffer can also be used, although hemoglobins tend to precipitate more easily in it than in the phosphate buffers.[5] The stabilizing effects of the phosphate may be due to its interactions with hemoglobin. Instead of rate curves, the amount of precipitation after 1 hr at 50° has been used as a quantitative measure of stability for variant hemoglobins. Because this procedure gives less reliable results, it is not recommended.[5]

Because oxyhemoglobin is used in measuring the rate of precipitation, autoxidation is also occurring. The rate of autoxidation can be measured

[16] C. L. Lutcher, J. B. Wilson, M. E. Gravely, P. D. Stevens, C. J. Chen, J. G. Lindeman, S. C. Wong, A. Miller, M. Gottleib, and T. H. J. Huisman, *Blood* **47**, 99 (1976).

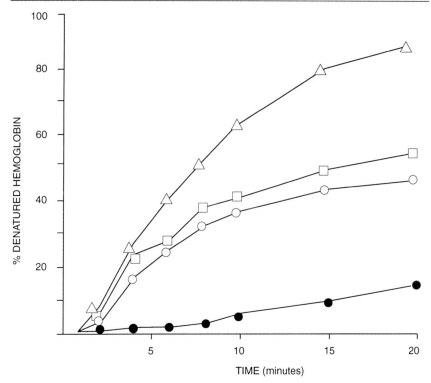

FIG. 1. Thermal precipitation curves for hemoglobins at 62°. These samples were from fresh hemolysates from human subjects. The samples were as follows: △, primarily Hb Leslie [deletion of Glu at $\beta131$ (H9)]; □, a mixture of Hb C [$\beta6$ (A3) Glu replaced by Lys] and Hb Leslie; ○, a mixture of Hb A and Hb Leslie; and ●, Hb A from a normal individual. Small amounts of HbA_2 and Hb F were present in all of these samples. (From Lutcher *et al.*[16] with permission.)

during the precipitation experiment[15] from the absorbance of the supernatant at 577 nm using

$$\% \text{ Autoxidation} = \% \text{ denaturation} + \frac{A_{577}(0) - A_{577}(X)}{A_{577}(HbO_2) - A_{577}(\text{met})}$$

where $A_{577}(0)$ and $A_{577}(X)$ are the absorbencies at 577 nm at zero time and after X min, and $A_{577}(HbO_2)$ and $A_{577}(\text{met})$ are the absorbencies of pure oxy- and methemoglobins, respectively. The $A_{577}(HbO_2)$ can be calculated[15] from the absorbance of cyanomethemoglobin at 540 nm because $A_{577}(HbO_2) = A_{540}(\text{CNmet}) \times 1.398$.

Thermal Denaturation Procedure

Thermal denaturation experiments in which the sample is slowly heated have been more useful in providing quantitative comparisons. Ockelford et al.[10] incubated methemoglobin for 10 min at various temperatures from 37° to 60°. The optical density at 750 nm was used as a measure of the precipitation. The temperature for which A_{750} equaled 0.2 was used to compare different variant hemoglobins. This value depends on the hemoglobin concentration. It is cumbersome to use the large number of temperatures needed to define the point where A_{750} is 0.2. The continuous heating procedure[11,14] described below eliminates these problems.

Denaturation Conditions

The thermal denaturation experiments[11,14] are done in 0.01 M 4-morpholinopropanesulfonic acid (MOPS), pH 7.0, containing 0.9 M guanidine, which prevents hemoglobin precipitation. Normal and cross-linked hemoglobins are not denatured by this concentration of guanidine. The spectrum of Hb A is unperturbed after 24 hr of incubation in this buffer at room temperature. This does not mean, however, that the guanidine does not have an effect. The denaturation temperatures measured in the presence of guanidine are lower than the precipitation temperature measured in its absence. For this reason, it is essential to run a control using normal Hb A with every set of experiments. The most consistent value characterizing the denaturation is the ΔT_m between the experimental sample and the control. It is possible, however, that some unstable variants would be denatured by the guanidine. In this case, it would be omitted, but the T_m would not be directly comparable to ones measured in the presence of guanidine. Due to protein precipitation, the end point of the denaturation is not always obvious when the guanidine is omitted. The MOPS buffer has been used because it has a relatively small change in pH with increasing temperature.[17]

The final hemoglobin concentration in the cuvette should be 7 μM monomer to give an absorbance of approximately 1 for the Soret band. The hemoglobin is oxidized by making the solution 10^{-5} M in $K_3Fe(CN)_6$. The oxidation proceeds for 10–15 min prior to the denaturation and can be monitored by observing the changes in the Soret and other heme absorbance peaks. A diode array spectrophotometer is convenient for following both the oxidation and the denaturation, but a conventional instrument can be used. The advantage of the diode array is the ability to collect data at several wavelengths simultaneously so that the tempera-

---

[17] N. E. Good and S. Izawa, this series, Vol. 24, p. 53.

ture is constant during the measurement. This allows the easy comparison of results at different wavelengths that may shed some light on the unfolding process. In general, the Soret band gives a slightly (2°) lower transition than the protein absorbance at 280 nm,[11,14] indicating that the heme pocket opens up before the core of protein unfolds.

For the denaturation, the sample must be in a sealed cuvette (Teflon stopper) to prevent evaporation. A programmable water circulator (Neslab) is used to increase the temperature from 25° to 70° at 0.3°/min. Data are usually collected at several wavelengths (280, 360, 406, 410, 418, 542, and 576 nm), but one wavelength (406 or 418 nm) is sufficient. Data at higher wavelengths (i.e., 660, 690, 720, 750, and 780 nm) can also be saved if correction for precipitation is needed. The temperature for each spectrum is measured with a thermocouple (Omega). The data are collected every 90 sec. The absorbance at each wavelength and the temperature are saved on a floppy disk. These data are recalled later for analysis.

Due to the sensitivity to experimental conditions and the length of the experiment (3.5 hr), it is best to use a multicell transport so that several samples can be denatured at the same time. If a single-beam instrument, such as a Hewlett-Packard diode array spectrophotometer, is used, this also allows rereferencing of the instrument before each set of readings. A Hb A control should be included with each run. Any slight variation in heating rate, solvent composition, pH, or spectrophotometer performance can be eliminated by reporting the difference in denaturation temperatures between the sample and Hb A (ΔT_m).

The method can be adapted to measure the stability of cyanomet- and carbonmonoxyhemoglobins.[11] The procedure for cyanomethemoglobins is the same as that for methemoglobins except that 2 mM NaCN is added to the denaturation buffer. For carbonmonoxyhemoglobins, wet carbon monoxide gas is bubbled through the oxyhemoglobin solution for 20 min and the Teflon-stoppered cuvette is sealed under CO gas. The visible spectra will demonstrate that the samples are cyanomet- or carbon-monoxyhemoglobins.

Data Analysis

The data analysis can be done using either a two-state model or the first-derivative model. In the two-state model,[18] the protein is considered to be either in the native or the denatured state. The fraction denatured,

---

[18] C. Tanford, *Adv. Protein Chem.* **23**, 121 (1968).

F_D, is given by

$$F_D = \frac{A - A_N}{A_D - A_N}$$

where A, A_N, and A_D are the absorbencies of the experimental mixture, native protein, and denatured protein, respectively. A_N is determined by fitting a linear line to the portion of an absorbance versus temperature plot that corresponds to effect of temperature on the native spectrum (Fig. 2). Similarly, A_D is determined from a linear extrapolation of the temperature effect on the denatured spectrum. The F_D is then the ratio of the appropriate vertical line segments (A/B in Fig. 2). The denaturation temperature (T_m) is defined as the temperature at which F_D is 0.5.

The first-derivative method plots the slope of the absorbance versus temperature curve as a function of temperature. A very simple algorithm calculates the slope (dA/dT) between temperature points $I - 1$ and $I + 1$ and assigns the value to point I. If the data are noisy, which can happen if wavelengths with small absorbencies are being used, the curve can be smoothed using the following algorithm[19]:

$$
\begin{aligned}
D(I) = (1/3059)\{&329D(I) + 324[D(I + 1) + D(I - 1)] + 309[D(I + 2)\\
&+ D(I - 2)] + 284[D(I + 3) + D(I - 3)] + 249[D(I + 4)\\
&+ D(I - 4)] + 204[D(I + 5) + D(I - 5)] + 149[D(I + 6)\\
&+ D(I - 6)] + 84[D(I + 7) + D(I - 7)] + 9[D(I + 8)\\
&+ D(I - 8)] - 76[D(I + 9) + D(I - 9)] + 171[D(I + 10)\\
&+ D(I - 10)]\}
\end{aligned}
$$

where $D(I)$ is the value of the first derivative at temperature point I. This graduation or smoothing formula was obtained[19] by fitting a polynomial of degree 2 or 3 to the data by the method of least squares.

The maxima and minima of the dA/dT versus temperature plot give the temperatures at which the absorbance is changing most rapidly. For an ideal two-state denaturation these maxima and minima would be identical to the T_m values as measured above. For the cross-linked and abnormal hemoglobins that have been studied,[11–14,20,21] when data from the same wavelength are used, this has been the case within the experimental error of measuring the T_m, which is approximately 1°.

Although the two-state model is not absolutely valid for hemoglobin because different values for the T_m have been observed at different wavelengths,[14] the two methods give nearly identical values for the T_m. The major operational advantages of the first-derivative method are that it is

[19] E. T. Whittaker and G. Robinson, "The Calculus of Observations." Blackie, London, 1958.
[20] T. Yang and K. W. Olsen, Biochem. Biophys. Res. Commun. 174, 518 (1991).
[21] R. E. Benesch and S. Kwong, J. Protein Chem. 10, 503 (1991).

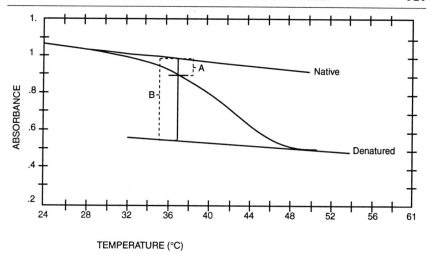

FIG. 2. Plot of A_{418} versus temperature for the denaturation of hemoglobin cross-linked between the $\beta82$ lysines with fumarate. The extrapolated lines show the effects of temperature on the native and denatured spectra. The ratio of the difference between the native line and the experimental curve (A) to the difference between the native and denatured lines (B) is the fraction denatured.

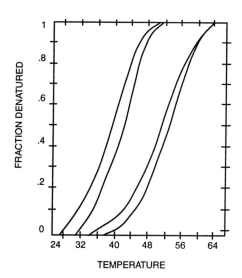

FIG. 3. Comparison of fraction denatured versus temperature for carbonmonoxy- and methemoglobins based on the 418 nm absorbance. From left to right the samples are metHb S, metHb A, carbonmonoxyHb S, and carbonmonoxyHb A. (From Yang and Olsen[11] with permission.)

TABLE II
DENATURATION TEMPERATURES OF VARIANT HEMOGLOBINS

Hemoglobin	Mutation	T_m (°C)[a]	Ref.
Hb A (met)	None	42.2 (0.8)	11, 14
Hb A (CO)	None	53.2 (0.7)	11
Hb A (CN-met)	None	53.6 (0.1)	11
Hb S (met)	$\beta 6$ (A6) Glu → Val	39.2 (0.9)	11
Hb S (CO)	$\beta 6$ (A6) Glu → Val	51.2 (1.2)	11
Hb New York (met)	$\beta 113$ (Gl5) Val → Glu	40.2 (1.2)	12
Hb O-Indonesia (met)	$\beta 116$ (GH4) Glu → Lys	40.5 (1.1)	13
Hb Andrew-Minneapolis (met)	$\beta 144$(HC1) Lys → Asn	42.4 (0.4)	22
Hb A_2 (met)	$\alpha_2 \delta_2$	43.8 (0.5)	22

[a] Values calculated from the absorbance changes at 418 nm by plotting the fraction denatured versus temperature. These values are the average of at least three determinations. The values in parentheses are the standard deviations.

less subjective and that it automatically detects minor transitions due to impurities in the sample. The subjectivity of the two-state model is due to the choice of what portions of the curve represent the effect of temperature on the native and the denatured spectra. In a true two-state situation, this is not generally a problem. However, if the system is multistate or if

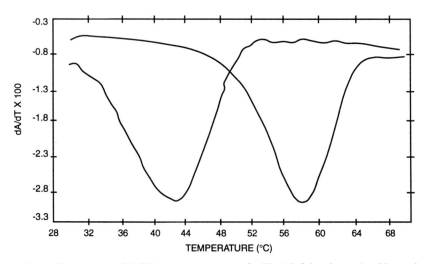

FIG. 4. Comparison of dA/dT versus temperature for Hb A (left-hand curve) and hemoglobin cross-linked between the $\beta 82$ lysines with fumarate (right-hand curve).

TABLE III
DENATURATION TEMPERATURES OF CROSS-LINKED METHEMOGLOBINS

Hemoglobin	Cross-link	T_m (°C)^a	Ref.
Human Hb A	Fumarate between β82 lysines	57.1 (0.7)	14
Human Hb A	Fumarate between α99 lysines	56.5 (1.2)	20
Human Hb A	Fumarate between β chains^b	57.3 (0.3)	20
Hb S	Fumarate between β82 lysines	54.8 (0.8)	11
Hb New York	Fumarate between β82 lysines	53.7 (2.1)	12
Human Hb A	None	44 (NR)	21
Human Hb A	Bis(pyridoxal)tetraphosphate between Val-1(β_1) and Lys-82(β_2)	51 (NR)	21
Bovine Hb A	None	44 (NR)	21
Bovine Hb A	Bis(pyridoxal)tetraphosphate between Val-1(β_1) and Lys-82(β_2)	52 (NR)	21

^{a} Values calculated from the absorbance changes at 418 nm by plotting the fraction denatured versus temperature. These values are the average of at least three determinations. The values in parentheses are the standard deviations. NR, Not reported.
^{b} Minor product isolated from the cross-linking reaction with deoxyhemoglobin.[20]

the sample is impure, the choice can be difficult. Both situations can be true for particular variant or modified hemoglobins.

With some of the modified samples, precipitation can occur during the experiment, apparently due to the cross-link decreasing the solubility of the denatured protein. This effect has been observed with general cross-linkers, such as dimethyl suberimidate (T. D. Corso and K. W. Olsen, unpublished observations, 1988), that cause major changes in the charge and external hydrophobicity of the protein. It has not been observed with singly cross-linked hemoglobins[11,14,20,21] or with abnormal hemoglobins.[11-14] This precipitation causes an increase in the apparent absorbance due to light scattering and masks the unfolding transition. To circumvent this problem, a linear extrapolation of the absorbance from 650 to 800 nm, where there are no absorbance maxima, can be used to correct each data point.

Applications

The thermal denaturation method given above has been applied to a variety of variant and cross-linked hemoglobin studies. The method can be used to detect and quantitate changes in stability for variants that are not clinically unstable. Results for Hb S are shown in Fig. 3. Table II[22] compares the results for several abnormal hemoglobins.

[22] T. Yang, M. S. Thesis, p. 81. Loyola University, Chicago (1986).

This technique has also been used to evaluate the thermal stabilities of cross-linked hemoglobins.[11,12,14,20,21] The interest in cross-linked hemoglobins is due to their potential as blood substitutes.[23] The increased stability that results from these covalent modifications allows them to be pasteurized to kill infectious viruses,[24] such as those for autoimmune deficiency syndrome (AIDS) and hepatitis. Stability also may be an important factor in the long-term storage of the blood substitute. Thermal denaturation easily quantitates the change in stability, as shown in Fig. 4 for oxyhemoglobin cross-linked with bis(3,5-dibromosalicyl) fumarate. Table III compares the results found for several cross-linked hemoglobins.

Acknowledgments

The author thanks Mr. Frank White, Dr. Thao Yang, Dr. Thomas Corso, and Ms. Qun-Ying Zhang for their help in developing the thermal denaturation procedure. This work was supported in part by grants from the American Heart Association of Metropolitan Chicago and from the Research Corporation.

[23] K. D. Vandegriff and R. M. Winslow, *Chem. Ind. (London)*, p. 497 (1991).
[24] T. N. Estep, M. K. Bechtel, T. J. Miller, and A. Bagdasarian, *in* "Blood Substitutes" (T. M. S. Chang and R. P. Geyer, eds.), p. 129. Dekker, New York, 1989.

[34] Photochemical Reduction of Methemoglobin and Methemoglobin Derivatives

By JOHANNES EVERSE

Studies with hemoglobin and myoglobin are often difficult to perform using the pure ferrous proteins, because these proteins oxidize spontaneously and relatively rapidly to the corresponding ferric forms. This spontaneous oxidation makes it difficult to obtain protein preparations that are 100% in the ferrous form. For the same reason, hemoglobin solutions that must be stored for long periods of time must be kept either at very low temperatures or in the complete absence of oxygen, or both, conditions that are not easy to achieve. The spontaneous oxidation to methemoglobin can be especially bothersome when solutions of precious hemoglobin variants or hemoglobin derivatives need to be stored for extended periods of time. One is thus often faced with the problem of having to reduce any methemoglobin that has formed in hemoglobin preparations before valid studies with the hemoglobin can be initiated.

Various procedures for the reduction of methemoglobin present in hemoglobin preparations have been reported in the past, using a variety of reducing agents. Agents that were used for this purpose include dithionite,[1] ascorbate,[2,3] 5-hydroxyanthranilic acid,[4] iron salts and EDTA–metal complexes,[5] ferredoxin,[6] and NADH in the presence of associated enzymes.[7] However, the direct reduction of methemoglobin with a reducing agent sometimes presents difficulties. For example, the reduction with dithionite and EDTA results in the formation of SO_3^{2-} and formaldehyde, respectively. These oxidation products can then react with other parts of the protein, leading to chemical modifications of the protein and a subsequent deterioration of the pigment.[8,9] The reduction of methemoglobin with reducing agents such as ascorbate and cysteine reaches an equilibrium when there are still considerable amounts of methemoglobin present in the solution.[10]

Further research into this problem led to the use of catalysts to mediate the transfer of electrons between an electron donor and methemoglobin. With this approach electron donors can be used that otherwise would react only very slowly or not at all with methemoglobin. Thus, in a preliminary report, Kajita et al.[11] published a method for the regeneration of deoxyhemoglobin from methemoglobin, using phenazine methosulfate as the catalyst and NADH as the electron donor. The reduction was reported to occur even in the presence of oxygen, but the rate was quite slow. The authors did not report whether a complete reduction of the methemoglobin could be achieved with this technique.

In another approach to the problem, several investigations have been reported in which small amounts of methemoglobin were reduced electrochemically, using electrodes coated with methylene blue.[12,13] A similar approach was taken by Durliat and Comtat,[14] who described the electrochemical reduction of methemoglobin at a Pt electrode, using flavin mono-

[1] C. Bauer and B. Pacyna, Anal. Chem. 65, 445 (1975).
[2] A. Tomoda, Y. Yoneyama, and M. Takeshita, Experientia 32, 932 (1976).
[3] A. Tomoda, A. Tsuji, M. Matsukawa, and Y. Yoneyama, J. Biol. Chem. 253, 7420 (1978).
[4] K. Goda, T. Ueda, and Y. Kotake, Biochem. Biophys. Res. Commun. 78, 1198 (1977).
[5] A. G. Mauk and H. B. Gray, Biochem. Biophys. Res. Commun. 86, 206 (1979).
[6] M. Nagai and Y. Yoneyama, J. Biol. Chem. 258, 14379 (1983).
[7] M. Hayashi, T. Suzuki, and M. Shin, Biochim. Biophys. Acta 310, 309 (1973).
[8] K. Dalziel and J. R. P. O'Brien, Biochem. J. 67, 119 (1957).
[9] W. R. Frisell, W. C. Choong, and C. G. MacKenzie, J. Biol. Chem. 234, 1297 (1959).
[10] A. Tomoda, M. Takeshita, and Y. Yoneyama, J. Biol. Chem. 253, 7415 (1978).
[11] A. Kajita, K. Noguchi, and R. Shukuya, Biochem. Biophys. Res. Commun. 39, 1199 (1970).
[12] S. Song and S. Dong, Bioelectrochem. Bioenerg. 19, 337 (1988).
[13] J. Ye and R. P. Baldwin, Anal. Chem. 60, 2263 (1988).
[14] H. Durliat and M. Comtat, J. Biol. Chem. 262, 11497 (1987).

nucleotide (FMN) as the catalyst. Because oxygen is reduced before met-hemoglobin under those conditions, the reduction was performed under anaerobic conditions. The authors reported that 25 ml of a 1 mM hemoglobin solution containing 55% methemoglobin could be completely reduced to deoxyhemoglobin in 25 min, using 0.2 mM FMN.

The reduction of methemoglobin with photochemically reduced FMN as a catalyst was first reported by Yubisui et al.[15] These authors showed that the reduction of methemoglobin by reduced FMN is extremely fast compared to the rates obtained with other agents, and is also quite specific. The photoactivated FMN was reduced using EDTA as the electron donor. McCormick et al.[16] published a list of electron donors that readily react with photoactivated FMN. These include methionine, ascorbate, nicotine, glycine, sarcosine, dimethylglycine, and others, besides EDTA. Most of these agents reduce photoactivated FMN at a rate similar to that of EDTA.

In all studies cited above (with the exception of Durliat and Comtat[14]) the experiments on the reduction of methemoglobin were carried out more or less on an analytical scale, i.e., small reaction vessels were used and only small quantities of methemoglobin were reduced, because the aims were to obtain chemical and/or kinetic information about the reaction. In this chapter we describe a method for the reduction of methemoglobin in hemoglobin preparations that can be used in the laboratory on a preparative scale. The method employs photoactivated FMN to catalyze the reduction of methemoglobin by methionine.

The reduction of methemoglobin by reduced FMN proceeds with a rate constant of $3.3 \times 10^8 \ M^{-1} \ sec^{-1}$, which is very fast compared with the rate with other reducing agents.[15] Moreover, the redox potential ($E^{0\prime}$) of FMN/FMNH$_2$ is -0.219 V, whereas that of Fe^{3+}/Fe^{2+} in hemoglobin is 0.144 V.[17-19] From these values one can calculate that the equilibrium constant for the reduction of methemoglobin by FMNH$_2$ at pH 7.0 is about 10^{12}. The reduction of methemoglobin is thus not only rapid, but also goes to completion even in the presence of catalytic amounts of FMNH$_2$, as long as the FMNH$_2$ is regenerated. The photoactivation of FMN by visible light is also an extremely fast process. Therefore, the rate-limiting step in the overall reaction scheme is most likely the reduction of the photoactivated FMN by the electron donor. The use of FMN as the catalyst thus provides a convenient way to reduce small amounts of methemoglobin that may be present in hemoglobin solutions.

[15] T. Yubisui, S. Matsukawa, and Y. Yoneyama, J. Biol. Chem. 255, 11694 (1980).
[16] D. B. McCormick, J. F. Koster, and C. Veeger, Eur. J. Biochem. 2, 387 (1967).
[17] J. F. Taylor and A. B. Hastings, J. Biol. Chem. 131, 649 (1939).
[18] J. F. Taylor and V. E. Morgan, J. Biol. Chem. 144, 15 (1942).
[19] H. J. Lowe and W. M. Clark, J. Biol. Chem. 221, 983 (1956).

The reduction of photoactivated FMN can be accomplished with a variety of electron donors and we have tested several of these. Of the donors tested, methionine appears to be a superior reductant, because its oxidation products do not react with the protein, and they can be removed easily from the hemoglobin solution, either by column chromatography or by dialysis. In order to do such comparative studies as well as to obtain meaningful kinetic data about the reduction of methemoglobin, we did most of the experiments reported in this chapter starting with 100% methemoglobin preparations.

Materials

Highly purified human hemoglobin A_0 (Hb A_0) is prepared as described elsewhere.[20,21] Riboflavin mononucleotide, DL-methionine, and other chemicals are purchased from Sigma Chemical Co. (St. Louis, MO). FMN immobilized on Sepharose is also obtained from Sigma; this particular lot contained 0.68 μM FMN per milliliter of packed resin.

Oxygen concentrations are monitored with an Instech Laboratories dual oxygen electrode amplifier, connected to a chart recorder. Spectrophotometric measurements are taken with a Hewlett-Packard 8451A UV/ VIS spectrophotometer, equipped with an HP 9121D Micro Disk Drive. As a light source we use a Kodak slide projector, containing a 300-W light source. The optical path of the projector is used to focus as much of the light as possible onto the reaction vessel. The distance between the lens of the projector and the reaction vessel is about 5 cm, and a fan is placed at a right angle from the projector to protect the reaction vessel from any heating effects. Under these conditions the temperature of the hemoglobin solution does not change more than 2° during a 30-min illumination period.

Preparation of Methemoglobins

Pure methemoglobin is prepared from hemoglobin A_0 with the following procedure. A small molar excess (10–20%) of potassium ferricyanide is weighed out and dissolved in a minimal amount of water. The hemoglobin solution is cooled in an ice bath, and the ferricyanide solution is then added, slowly, dropwise, under stirring and in the dark. Stirring is continued in the cold and in the dark for about 30 min. The oxidized hemoglobin solution is then allowed to warm to room temperature and is passed

[20] This volume [1].
[21] S. M. Christensen, F. Medina, R. M. Winslow, S. M. Snell, A. Zegna, and M. A. Marini, *J. Biochem. Biophys. Methods* **17,** 143 (1988).

through a Dowex mixed-bed resin column to remove the reaction products. If, after passage through the column, the conductivity of the solution is more than 10 mhos, the solution again is passed through a fresh mixed-bed resin column. The same procedure is used for the oxidation of the hemoglobin derivatives. Alternatively, the reaction products may be removed by extensive dialysis against cold, distilled water.

Reduction of Methemoglobins

A methemoglobin solution [20 ml of a 0.1% solution (1 mg/ml)] in 10 mM Tris buffer, pH 7.2, is placed in a 20-ml Coulter counting vial. Reductant is added to the solution in solid form to yield a final concentration of 20 mM. The vial is then closed with a rubber stopper, equipped with an oxygen electrode and with two 18-gauge $\times$ $1\frac{1}{2}$ and one 20-gauge $\times$ $3\frac{1}{2}$ hypodermic needles. The long needle is used to add reagents to the reaction mixture and to withdraw samples for analysis. The solution is stirred continuously with a small stirring bar. The solution is deoxygenated by continuously flushing pure nitrogen gas over the solution through one of the 18-gauge needles, with the other needle serving as a vent. Then 200 μl of a 0.01% solution of FMN is added to the hemoglobin solution in the dark (final concentration, 2 μM), and nitrogen flushing is continued until the oxygen pressure is less than 2 Torr. Absolute darkness is not necessary during this period; shielding the reaction vessel from direct exposure to the overhead lights by means of a piece of cardboard proves sufficient. The reduction is started by turning on the light source. At specified time intervals 200-μl samples are removed, diluted to 1 ml with oxygenated buffer, and the absorption spectra of the samples are measured from 450 to 650 nm. The percent MetHb and oxyHb present in the sample are then calculated by computer analysis from the absorbances at 562, 576, and 606 nm. The reaction rates presented in the tables are the rates obtained during the first 2–3 minutes of the reaction. Figure 1 shows a diagram of the reaction vessel.

Effect of Oxygen on Reduction of Methemoglobin

The oxidation–reduction potential for the reduction of molecular oxygen to hydrogen peroxide is 0.295 V at pH 7.0,[22] whereas the redox potential for the reduction of the ferric to the ferrous form of hemoglobin is 0.16 V under the same conditions.[12,17] Acting on a methemoglobin solution at neutral pH, reduced FMN will therefore first reduce any oxygen

---

[22] D. T. Sawyer, *Basic Life Sci.* **49**, 11 (1988).

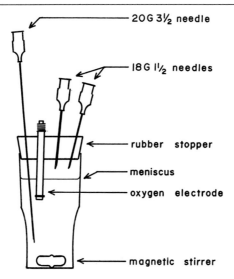

FIG. 1. A schematic diagram of the reaction vessel used in the experiments.

to hydrogen peroxide before it will reduce the methemoglobin. It is well known that hydrogen peroxide in turn reacts with ferrous hemoglobin to form methemoglobin again. Therefore, to achieve a complete reduction of the methemoglobin, it is imperative that the reduction with reduced riboflavin be done in the complete absence of oxygen.

In the experimental procedure described above it will be noticed that on starting the reaction by turning on the light source there will first be a rapid depletion of any remaining oxygen from the reaction mixture. Only after all oxygen is gone will the reduction of methemoglobin become noticeable.

Rate Profile of Methemoglobin Reduction

Figure 2 shows a typical set of data obtained when 0.1% methemoglobin is reduced using 20 mM DL-methionine as the electron donor. The reduction is rapid and virtually linear, but slows down when the reaction is near completion. The results of two independent experiments are shown, illustrating the slight variation in rate that may be observed between individual experiments. Figure 2 also illustrates the increase in reduced hemoglobin present in the reaction mixture, which closely parallels the reduction of the methemoglobin.

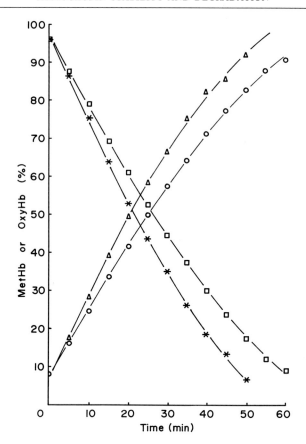

FIG. 2. Time course of the reduction of methemoglobin with photoactivated FMN and of the formation of reduced hemoglobin. Samples were taken at the indicated times and diluted in oxygenated buffer. The results of two independent experiments are shown: decrease in metHb (*) with concomitant increase in reduced Hb ($\triangle$), and decrease in metHb ($\square$) with concomitant increase in reduced Hb ($\bigcirc$). Methemoglobin, 0.1 g/100 ml; FMN, 2 μM; DL-methionine, 20 mM; buffer, 10 mM Tris-HCl, pH 7.2. Total volume, 20 ml.

Effects of Various Catalysts

A comparison of the effectiveness of FMN with that of methylene blue and phenazine methosulfate as a catalyst in the reduction of methemoglobin revealed that the latter two are quite inferior to FMN. Figure 3 shows the results of an experiment in which phenazine methosulfate was used as the catalyst for the reduction of methemoglobin. In the absence of light the reduction is very slow, just a few percent per hour, as was also reported by Kajita et al.[11] Illumination of the reaction mixture results in a large

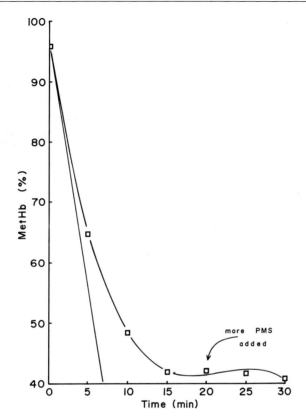

FIG. 3. Time course of the reduction of methemoglobin using photoactivated phenazine methosulfate(PMS). Samples were taken at the indicated time points (□), whereas the straight line indicates the initial rate. Conditions as in Fig. 2, except that FMN was replaced with 33 μM phenazine methosulfate.

increase in the rate, but the rate of reduction was nevertheless much slower than with an equivalent amount of FMN. (Note that 33 μM phenazine methosulfate was used in the experiment illustrated in Fig. 3, whereas it took only 2 μM FMN to obtain the rate shown in Fig. 2.) Moreover, the reaction does not go beyond the reduction of about 50% of the methemoglobin, and adding more phenazine methosulfate does not result in any further reduction. Phenazine methosulfate can therefore not be used as a catalyst when a mixture of hemoglobin and methemoglobin is to be reduced to 100% deoxyhemoglobin. In our experience results obtained with methylene blue as a catalyst were similarly inferior to those obtained with FMN.

TABLE I

Reduction of Methemoglobin with Various Electron Donors[a]

Electron donor	Rate of reduction (% reduced/min)		
(20 mM)	Experiment 1	Experiment 2	Average
Ascorbate	6.71	6.16	6.435
EDTA	5.88	6.05	5.965
DL-Methionine	4.01	4.20	4.105
Nicotine	4.75	3.13	3.940
Dithioerythritol	2.77	2.98	2.875
L-Cysteine	1.49	1.65	1.570

[a] Methemoglobin: 0.075 g/100 ml.

Comparison of Various Electron Donors

Table I tabulates the initial rates obtained for the reduction of methemoglobin with various electron donors. The reduction is fastest in the presence of ascorbate, followed by EDTA and methionine. Unfortunately, oxidation of EDTA by photoactivated FMN leads to the formation of formaldehyde,[9] which is detrimental to proteins. Several other electron donors also form formaldehyde or acetaldehyde as a product, including glycine, dimethylglycine, and sarcosine.[9] Experiments in our laboratory using ascorbate as the electron donor for the reduction of methemoglobin did result in the formation of turbidity toward the end of the reaction. This suggests that undesirable side reactions may take place during the reduction, making ascorbate also a less desirable electron donor for methemoglobin reduction.

Effect of Electron Donor Concentration

The dependence of the reaction rate on the electron donor concentration is illustrated in Table II, where methionine was used as the reductant. The reduction rate increases rapidly with increasing methionine concentration up to 20 mM, where it begins to level off. This could indicate that the oxidation of methionine by photoactivated FMN may be the rate-limiting step in the series of reactions leading to the reduction of methemoglobin.

pH Dependence of the Rate of Reduction

As shown in Table III, the rate of methemoglobin reduction with methionine as the electron donor is independent of pH between pH 6 and

TABLE II
DEPENDENCE OF PHOTOCHEMICAL REDUCTION
OF METHEMOGLOBIN ON METHIONINE
CONCENTRATION[a]

DL-Methionine (mM)	Rate of reduction (% reduced/min)
0	0.43
5	4.65
10	8.16
20	11.92

[a] Methemoglobin: 0.025 g/100 ml.

8. Comparable results are obtained when EDTA is used as the electron donor (data not shown).

Effect of the Optical Density of the Solution

The reduction of methemoglobin proceeds much faster when more dilute solutions of methemoglobin are used, as shown in Table IV. This at first may suggest that the rate of reduction is also dependent on the methemoglobin concentration. In fact, however, it represents an optical problem. The photoactivation of FMN occurs maximally at 450 nm,[23] but at this wavelength the extinction coefficient of metHb is about 37,550 M^{-1} cm^{-1} for the tetramer. Table IV lists the optical densities of reaction

[23] H. Beinert, in "The Enzymes" (P. D. Boyer, H. A. Lardy, and M. Myrbäck, eds.), 2nd ed., p. 339. Academic Press, New York, 1960.

TABLE III
RATE OF REDUCTION OF METHEMOGLOBIN AT
VARIOUS pH VALUES[a]

pH	Rate of reduction (% reduced/min)
5.96	2.482
6.04	2.107
6.95	2.469
7.06	2.013
7.96	2.335
8.01	2.338

[a] Methemoglobin, 0.10 g/100 ml; DL-methionine, 20 mM.

TABLE IV
RATE OF REDUCTION AS FUNCTION OF
METHEMOGLOBIN CONCENTRATION[a]

Methemoglobin (g/100 ml)	Reduction rate (% reduced/min)	OD_{450} in 20-mm light path
0.050	9.690	0.592
0.075	3.846	0.885
0.100	2.235	1.172
0.125	0.999	1.468
0.150	0.740	1.770
0.200	0.281	2.303

[a] DL-Methionine: 20 mM.

mixtures containing various methemoglobin concentrations at 450 nm. In the experimental setup described above, because the total light path of the reaction vessel is 20 mm, at a concentration of 0.1% methemoglobin less than 10% of the light will completely penetrate the reaction mixture, whereas at 0.2% methemoglobin this will be less than 1%. Hence, at the higher methemoglobin concentrations the high absorbancies at 450 nm will prevent the light from completely penetrating the reaction mixture, and consequently the amount of FMN that is photoactivated becomes quite limited. Therefore, the decrease in rate with increasing methemoglobin concentration may reflect primarily a decrease in the steady-state concentration of photoactivated FMN, due to absorbance of the activating light by the other reagents. If this interpretation is correct, then using a reaction vessel with a shorter light path would lead to an increase in the rate under otherwise identical conditions. We found that this was indeed the case (data not shown).

Catalysis by Immobilized FMN

One potentially difficult aspect of the present procedure for the reduction of methemoglobin is the fact that at the completion of the reduction all FMN needs to be removed from the reaction mixture before the solution can be exposed to air (oxygen). This is because FMN in the presence of light will rapidly reduce oxygen to hydrogen peroxide, which in turn will reoxidize the hemoglobin. It is thus essential that the catalyst be removed before the anaerobic conditions are terminated, unless the mixture is kept in complete darkness. Removal of the catalyst from the reaction mixture can be done easily and efficiently if the FMN is immobilized onto some insoluble carrier. A preparation in which FMN is immobilized onto Sepharose beads via the ribose moiety is commercially available, and we have tested this preparation for its potential usefulness in the reduction of methemoglobin. In our experiments 1 ml of packed FMN-containing resin was suspended into the reaction mixture, which also contained 0.1% methemoglobin and 20 mM methionine. The observed rate of methemoglobin reduction was 1.56%/min, which is almost 70% of the rate that we obtained in a parallel experiment with an equivalent amount of soluble FMN. The decrease in rate is probably due to the fact that the presence of the Sepharose may limit the efficiency of the photoactivation process as well as cause some light scattering. Nevertheless, the advantage of easy removal of the catalyst (which can be used again and again) may well outweigh the disadvantage of the rate decrease.

General Comments

For a method for the reduction of methemoglobin in hemoglobin preparations to be useful, several criteria need to be met. First, the reduction of methemoglobin should go to completion under the reaction conditions used. Second, the reagents as well as the reaction products should be inert to the hemoglobin. Third, the reaction products as well as any excess reagents should be removed easily from the hemoglobin solution to prevent contamination of the preparations. These criteria are met with the method described in this section.

Frisell et al.[9] showed, for example, that the oxidation of EDTA (and several other potential electron donors) by photoactivated FMN results in the formation of an almost equimolar amount of formaldehyde as one of the oxidation products. Formaldehyde, however, reacts rapidly with amino groups in proteins, even at low concentrations. In selecting an appropriate electron donor it is therefore important to ascertain the nature of the oxidation product(s) that are formed during the reaction and further to ascertain that these products are inert toward hemoglobin. The above-mentioned characteristic makes EDTA considerably less attractive than methionine as an electron donor to reduce compounds such as methemoglobin, even though EDTA is a considerably more effective reductant.

In this context the interested reader may also consult the work by McGown et al. (this volume [35]), which describes the use of hydrogen gas as the reductant for methemoglobin. Because hydrogen gas is both one of the most inert reductants that can be used for this purpose and is easily removed from the reaction mixture, and its oxidation product (water) is totally harmless, this method is an attractive alternative to the method described herein.

Unfortunately, the redox potentials of most of the potential electron donors and their oxidation products are unknown, including that of methionine. Therefore, the theoretical overall reaction equilibrium of the FMN-catalyzed reduction of methemoglobin cannot be calculated for a number of the potential reductants. Our experimental results nevertheless show that a virtually complete reduction of methemoglobin can be obtained with each of the electron donors mentioned in this chapter, even though the rates may vary considerably. It was nevertheless somewhat surprising to us to find that sulfhydryl-containing compounds such as cysteine and dithioerythritol, which are well-known reducing agents, were less effective than the thioether methionine in reducing photoactivated FMN.

The data presented in Table IV indicate another potential problem that may be encountered when large amounts of methemoglobin are to be reduced with this method. The wavelength at which FMN is optimally activated is about 450 nm. Hemoglobin, however, has a relatively high absorbance at this wavelength. Therefore, FMN present in a solution containing a high concentration of hemoglobin, such as 0.2% and higher, will only be activated at the surface, but little light will penetrate the solution deeply enough to activate the remaining FMN. If highly concentrated hemoglobin solutions are to be used, this difficulty may be overcome by leading the concentrated hemoglobin solutions through a small-diameter glass column that is surrounded by light sources. The ideal diameter of such a column can be calculated readily from the absorbance at 450 nm of the concentrated hemoglobin solution.

Finally, it should be realized that most light sources create a considerable amount of heat. If these light sources are placed close to the reaction vessel, a considerable increase in the temperature of the protein solution may occur during the time of the reaction. Some action needs to be taken in order to circumvent this problem. Ideally, a jacketed reaction vessel should be used, because in this manner the temperature of the reaction can be controlled carefully at all times. Alternatively, a heat shield can be placed between the light source and the reaction vessel, or a fan can be used to provide a cooling effect. Without these precautions, however, some heating of the protein solution will most likely occur during the reaction period, which may result in some denaturation of the hemoglobin.

Acknowledgments

This work was performed at and supported by the Division of Blood Research, Letterman Army Institute of Research, Presidio of San Francisco, CA. The author thanks Col. Robert M. Winslow, Col. John R. Hess, and the staff of the Division for their interest and support of this work.

[35] Regeneration of Functional Hemoglobin from Partially Oxidized Hemoglobin in the Presence of Molecular Hydrogen and a Multicomponent Redox Catalyst*

By Evelyn L. McGown, Mazhar Khan, and Kilian Dill

Methemoglobin can be reduced by a variety of enzymatic systems[1,2] and nonenzymatic photochemical[3] and chemical reducing agents such as dithionite, metabisulfite, cysteine, glutathione, and ascorbic acid.[4–6] All of these systems produce oxidized by-products that may be undesirable and must be removed. Dithionite is the reagent most commonly used, despite side reactions that damage hemoglobin.[7,8] To minimize the undesirable by-products (e.g., H_2O_2), the reduction must be done under strictly anaerobic conditions and the reduced hemoglobin must be separated quickly from the reaction mixture. Separation may be accomplished by gel filtration[9,10] or mixed-bed ion-exchange chromatography.[11] Even then, the dithionite-regenerated hemoglobin may not be functionally identical to the native molecule.[10]

Methemoglobin can be reduced by electrochemical methods.[12,13] This reduction, however, is exceedingly sluggish with conventional solid elec-

---

* This material has been reviewed by Letterman Army Institute of Research and there is no objection to its presentation and/or publication. The opinion or assertions contained herein are the private views of the authors and are not to be construed as official or as reflecting the views of the Department of the Army or the Department of Defense (AR 360-5).

[1] A. Hayashi, T. Suzuki, and M. Shin, *Biochim. Biophys. Acta* **310**, 309 (1973).
[2] T. Suzuki, R. E. Benesch, S. Yung, and R. Benesch, *Anal. Biochem.* **55**, 249 (1973).
[3] J. Everse, this volume [34].
[4] H. F. Bunn and B. G. Forget, "Hemoglobin: Molecular, Genetic and Clinical Aspects," p. 644. Saunders, Philadelphia, 1986.
[5] E. L. McGown, M. F. Lyons, M. A. Marini, and A. Zegna, *Biochim. Biophys. Acta* **1036**, 202 (1990).
[6] C. Bauer and B. Pacyna, *Anal. Biochem.* **65**, 445 (1975).
[7] K. Dalziel and J. R. P. O'Brien, *Biochem. J.* **67**, 119 (1957).
[8] E. E. Di Iorio, this series, Vol. 76, p. 57.
[9] H. B. F. Dixon and R. McIntosh, *Nature (London)* **213**, 99 (1967).
[10] D. Zygmunt, P. Labrude, C. Vigneron, and D. Larcher, *Int. J. Biol. Macromol.* **9**, 197 (1987).
[11] C. Bauer and B. Pacyna, *Anal. Biochem.* **65**, 445 (1975).
[12] S. R. Betso and R. E. Cover, *J. Chem. Soc., Chem. Commun.*, p. 621 (1972).
[13] G. Dryhurst, K. M. Kadish, F. Scheller, and R. Renneberg, "Biological Electrochemistry," pp. 474–477. Academic Press, New York, 1982.

trodes because steric factors prevent close contact between the heme and the electrode. Small molecular weight mediators such as methylene blue[14] or flavin mononucleotide[15] greatly facilitate the reduction, but their use again introduces the problem of removing undesirable by-products.[16]

Molecular hydrogen offers a significant advantage over the reducing agents cited above because it is mild and its oxidation product, H^+, can be buffered easily. The use of hydrogen as a reducing agent has not been practical for proteins to date, because nonspecific protein adsorption to the metal catalyst results in coating of the metal surface and blocking of the catalytic activity. A possible strategy to circumvent the protein adsorption problem would be to protect the metal catalyst with a substance that does not bind protein, but that still affords catalytic activity. Limited success was reported with the use of colloidal platinum to catalyze the reduction of some metalloproteins, but the results with methemoglobin were unsatisfactory.[17]

Polymeric redox catalysts offer a promising approach to the use of molecular hydrogen for methemoglobin reduction. Multicomponent redox catalysts have been synthesized in which atoms of metallic platinum are imbedded in an electroactive polymer synthesized from a monomeric bipyridyl derivative.[18] For convenience, the polymeric catalyst is coated onto an inert material with high surface area (granules of powdered silicon dioxide). With such a catalyst, molecular hydrogen will regenerate functional hemoglobin from partially oxidized hemoglobin.[19] Reversible O_2 binding is restored for regenerated ferrous hemoglobin, with only slight reductions in cooperativity and hemoglobin stability. The advantages of the reduction system include the following considerations: (1) the heterogeneous catalyst avoids the problem of protein adsorption onto bare platinum, (2) catalyst and reducing agent are easily removed from the protein by simple filtration, and (3) the by-product H^+ is buffered easily.

Preparation of Redox Catalyst

This procedure for preparing redox catalyst was adapted from the scheme for Type II catalyst by Chao et al.[18] Potassium tetrachloroplati-

[14] J. Ye and R. P. Baldwin, Anal. Chem. 60, 2263 (1988).
[15] H. Durliat and M. Comtat, J. Biol. Chem. 262, 11497 (1987).
[16] P. Labrune, A. Bergel, and M. Comtat, Biotechnol. Bioeng. 36, 323 (1990).
[17] I. Tabushi and T. Nishiya, Tetrahedron Lett. 24, 5005 (1983).
[18] S. Chao, R. A. Simon, T. E. Mallouk, and M. S. Wrighton, J. Am. Chem. Soc. 110, 2270 (1988).
[19] E. L. McGown, K. Dill, R. J. O'Connor, M. Khan, Y. C. LeTellier, and K. D. Vandegriff, Anal. Biochem. 206, 85 (1992).

nate(II) (99.99%), SiO_2 (325 mesh; 99.99%), trimethyl orthoformate (99%), N,N,N',N'-tetramethyl-1,6-hexanediamine, and 4,4'-bipyridine may be obtained from Aldrich Chemical Co. (Milwaukee, WI). The 3-bromopropyltrichlorosilane is available from Huls Petrarch Systems (Bristol, PA) and trimethoxymethane is available from Fluka Chemika-BioChemika (Ronkonkoma, NY). High-purity gases are available from Air Products (Allentown, PA).

Synthesis of 1-Bromo-3-trimethoxysilylpropane

Trimethoxymethane (40 ml) is put in a 100-ml round-bottom flask, and 17 ml of 3-bromopropyltrichlorosilane is added gradually over a period of 10 min. The mixture is shaken continuously and protected from atmospheric moisture. After allowing the reaction mixture to stand at room temperature overnight, it is distilled under reduced pressure (20 mm Hg). A low-boiling fraction (50–60°) is discarded and the clear liquid (~20.5 g) is collected (b.p. 110°) and stored. *Note:* If the product is stored in a glass-stoppered flask, care must be taken to clean the silicon compound from both the stopper and the inner neck of the flask to prevent fusing, because the silicon compound reacts readily with moisture to form a glassy material.

Synthesis of Compound I

To synthesize N,N'-bis[3-(trimethoxysilyl)propyl]-4,4'-bipyridinium dibromide (**I**, Fig. 1), 4,4'-bipyridine (3.0 g) and 1-bromo-3-trimethoxy-

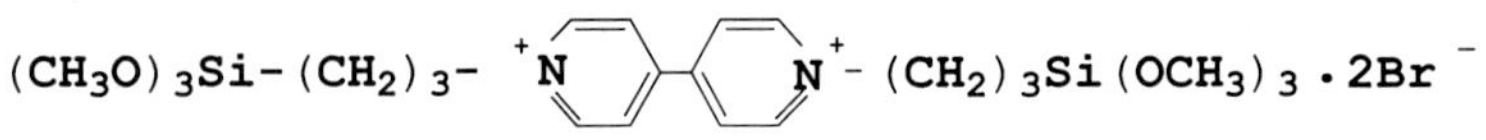

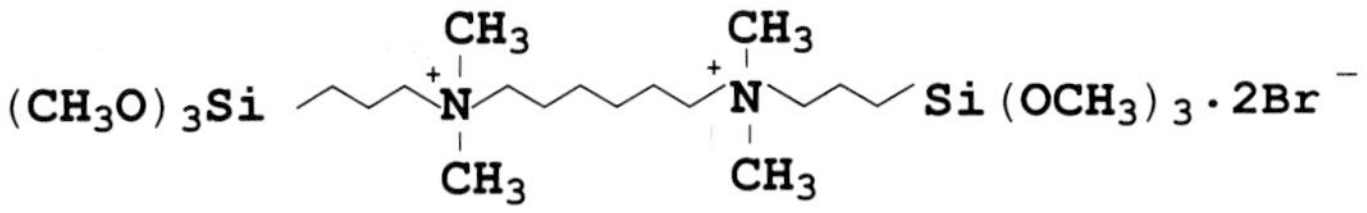

FIG. 1. Structures of intermediates in the synthesis of the redox polymer. (**I**) N,N'-bis[3-(trimethoxysilyl)propyl]-4,4'-bipyridinium dibromide; (**II**) $N,N,N'N'$-tetramethyl-N,N'-bis[3-(trimethoxysilyl)propyl]-1,6-hexanediammonium dibromide. In both structures, the bromide is bound ionically.

silylpropane (18 g) are refluxed in 50 ml of dry acetonitrile for 24 hr. On cooling, the pale yellow crystalline product separates and is filtered. Addition of dry diethyl ether to the mother liquor produces additional solid (combined weight, ~5 g). *Note:* It is important that Compound **I** be prepared immediately before use. When exposed to air, it quickly decomposes and polymerizes.

*Synthesis of Compound **II***

To synthesize N,N,N',N'-tetramethyl-N,N'-bis[3-(trimethoxysilyl) propyl]-1,6-hexanediammonium dibromide (**II**, Fig. 1), 1-bromo-3-methoxysilylpropane (20 g) and N,N,N',N'-tetramethyl-1,6-hexanedia-mine (9 g) are refluxed overnight in 200 ml of dry acetonitrile. The solvent is removed by distillation and the residue is treated with dry ether to allow the oil to separate. The solvent is decanted, another portion of dry ether is added to the oily residue, and the mixture is refluxed for 4 hr, during which time a solid (~23 g) forms (**II**).

*Derivatization of Silicon Dioxide with Compound **II**.* Silicon dioxide (10 g, activated under vacuum at 300° for 48 hr) is suspended in 100 ml of dry acetonitrile with 23 g of **II** and the mixture is refluxed for 48 hr. The solid is filtered, washed with dry acetonitrile, and air dried for 10 min.

Exchange of Charge-Compensating Bromide Ions with $PtCl_4^{2-}$. The derivatized silicon dioxide is suspended in 10 ml of water and mixed with K_2PtCl_4 (3.5 g in 10 ml of water) at room temperature. The mixture is shaken for 15 min and filtered. The slightly red solid is washed with water (3 × 20 ml) to remove excess chloroplatinate.

*Reduction of Surface-Bound $PtCl_4^{2-}$ to Pt Metal and Derivatization with **I**.* The derivatized silicon dioxide with ionically bound $PtCl_4^{2-}$ is resuspended in 0.1 M phosphate buffer (15 ml, pH 7) and hydrogen gas is bubbled gently through the suspension for 15 min. (The solid turns gray within the first few minutes.) The gray solid is filtered and washed with acetonitrile. The washed solid is resuspended in 50 ml of acetonitrile with 3.0 g freshly prepared **I**. A drop of water is added, the mixture is refluxed gently for 48 hr, and the product is filtered and dried. The powder is placed into a glass vial and stored in a desiccator. *Note:* It is important that the storage container be glass because some plastics interact with the catalyst.

Preparation of Partially Oxidized Hemoglobin

We chose to use partially autooxidized hemoglobin because it is a practical problem for the application of hemoglobin as a blood substitute.

A mixture of hemoglobin and methemoglobin can be prepared by autooxidation of sterile human hemoglobin in saline in a sealed flask. The sample is vented through a hypodermic needle attached to a 0.22-μm filter. Each day the sample is flushed with sterile nitrogen (to maintain low oxygen tension and to accelerate autooxidation). After 5 days at 37° the sample contains approximately 50% methemoglobin, as measured by the cyanide addition method.[20] When the desired level of oxidation is reached, the sample is frozen or stored at 4° so that aliquots of hemoglobin can be removed under sterile conditions.

Methemoglobin also can be prepared by oxidation with ferricyanide[8]; however, the latter method has disadvantages: (1) difficulty of removing hexacyanoferrate completely from the reaction mixture and (2) the possibility of increased protein degradation products (e.g., hemichromes) in fully oxidized preparations.

Redox Catalyst and Reduction Process

General

A schematic of the redox catalyst is presented in Fig. 2. The reduction process presumably proceeds according to the following reactions.

Reduction of conducting polymer: $Pt(0) + 2BP^{2+} \rightarrow Pt(II) + 2BP^{+}$
Reduction of one heme site: $BP^{+} + Hb(III) \rightarrow BP^{2+} + Hb(II)$
Regeneration of catalyst: $Pt(II) + H_2 \rightarrow Pt(0) + 2H^{+}$

Because hydrogen ion is a product of the reduction, pH gradients can develop near the surface of the catalyst. This drop in pH must be minimized to avoid damaging the hemoglobin and to minimize reoxidation. It is critical that the solutions be buffered; unbuffered preparations result in significant drops in pH. A 0.1 M sodium phosphate buffer (pH 8) provides a more stable, mild environment, but even under these conditions, a drop of 0.2–0.3 pH units may occur during the reduction process.

It is critical before reducing methemoglobin with H_2 gas to remove all O_2 so that H_2O_2 formation is prevented. Hydrogen peroxide not only will reoxidize hemoglobin, but will damage the catalyst as well. The reduction process is carried out in a sealed, nitrogen-filled glove bag (or box). Oxygen must be excluded until the catalyst and hemoglobin sample have been separated physically.

[20] E. J. van Kampen and W. G. Zijlstra, *Adv. Clin. Chem.* **23,** 199 (1983).

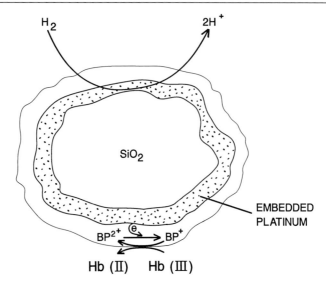

FIG. 2. Stylized schematic of a SiO_2 particle coated with catalyst. The black spots represent elemental platinum embedded in the layer of electroinactive polymer derived from **II**. The surface layer is the electroactive polymer derived from **I**. BP^+/BP^{2+} symbolizes the bipyridyl moiety of the latter.

Other recommended precautions include (1) maintenance of samples at ~4° throughout the reduction process and during subsequent reoxygenation and storage to minimize reoxidation, and (2) use of a fairly high protein concentration (2–3 mM heme) to minimize dissociation of the hemoglobin tetramer into dimers.

Typical Reduction Procedure

Three 25-ml Erlenmeyer flasks, each sealed with a rubber septum, are placed in the glove bag. One contains a prediluted control ferrous hemoglobin solution (3 ml, 9.0 g/dl plus 3 ml of 0.2 M sodium phosphate, pH 8.0), the second contains the autooxidized hemoglobin solution (3 ml, 9.0 g/dl, ~50% methemoglobin), and the third contains the catalyst (about 40 mg) in buffer (3 ml of 0.2 M sodium phosphate, pH 8.0). All flasks contain microstir bars. The glove bag is sealed and purged for 15 min with water-saturated nitrogen. As the bag becomes distended, a tiny opening is made and the air is forced out such that the bag is collapsed and purged of internal gases. It is immediately resealed and reinflated; this process is done three times over the next few hours to ensure complete removal of oxygen. After the last reinflation, the flow of nitrogen is reduced while maintaining a positive pressure of nitrogen to compensate for any small leaks in the bag. While the glove bag is being purged and filled with nitrogen, the three sealed flasks are placed on ice. The flasks are flushed with nitrogen (above the liquid surfaces) for a period of 1.5 hr (with mild stirring) to deoxygenate them.

Next, the catalyst is activated by passing humidified hydrogen through the flask for 0.5 hr (with stirring). At this time, the rapid change from gray to purple confirms the reduction of the bipyridyl moiety of the catalyst. The deoxygenated catalyst and oxidized hemoglobin are combined by removing the rubber septum from each flask and pouring the hemoglobin solution into the catalyst flask. The septum on the latter flask is replaced and the hydrogen atmosphere is maintained for an additional 0.5 hr with stirring. The control also is flushed with hydrogen while being stirred. The reduction is visually apparent within minutes as the solution changes from the brown color of methemoglobin to the violet color of deoxyhemoglobin.

The reaction is stopped after 0.5 hr by filtering the sample through a 0.22-μm filter (Millex-GS; Millipore Products Division; Bedford, MA) into an empty, chilled 25-ml Erlenmeyer flask (containing a stir bar), and the flask is sealed with a septum. The control and the reduced samples are flushed with humidified nitrogen for an additional hour to ensure removal of residual hydrogen. The samples are then removed from the glove bag and placed on ice. At this time, the control and the reduced hemoglobin samples are reoxygenated under a stream of 100% oxygen with mild stirring for 0.5 hr. Aliquots of the samples are then removed for methemoglobin analysis and studies of oxygen equilibrium binding parameters. All samples are kept at 4° during further storage.

Using the procedure detailed above, highly functional hemoglobin can be regenerated from partially oxidized hemoglobin. The reduction reaction is apparent within minutes of mixing methemoglobin with catalyst under a hydrogen atmosphere. The color of the solution changes from brown to purple (deoxyhemoglobin). After reoxygenation, the visible spectrum of the catalytically reduced protein is indistinguishable from that of "native" oxyhemoglobin. Spectroscopic evidence, oxygen equilibrium binding curves, and stability data have been published elsewhere.[19] Catalytically reduced hemoglobin typically shows an apparent 1–4% increase in methemoglobin content during the first few hours. The reason for the small initial reoxidation is unknown. Thereafter, a slower rate of autoxidation is similar to that of the starting material. The oxygen equilibrium binding curves of regenerated hemoglobin are similar to those of the starting material. P_{50} values of 4.2 Torr for regenerated hemoglobin and 3.5 Torr for the controls were measured at pH 7.4 and at 25° in 50 mM bis-Tris with 0.1 M Cl$^-$.[19] The Hill coefficients for regenerated hemoglobin samples are slightly lower than those for the controls (2.7 versus 3.0). The small change in oxygen binding cooperativity may be due to a small percentage of hemes that are altered irreversibly during the autoxidation.

Section VII

Enzymatic Reactions Catalyzed by Hemoglobin

[36] Peroxidative Activities of Hemoglobin and Hemoglobin Derivatives

By JOHANNES EVERSE, MARIA C. JOHNSON, and MARIO A. MARINI

The fact that hemoglobin possesses peroxidative properties was first noticed more than 30 years ago when Tappel showed with *in vitro* studies that methemoglobin and metmyoglobin can react with preexisting lipid hydroperoxides to yield lipid-associated free radicals, which resulted in the propagation of free radical-mediated reactions that produced extensive lipid peroxidation.[1,2] Since then, numerous other investigators have demonstrated that methemoglobin and metmyoglobin can initiate the peroxidation of polyunsaturated fatty acids.[3-7] Szebeni *et al.*[8] have demonstrated substantial membrane lipid peroxidation for model cells composed of hemoglobin entrapped in chemically defined liposomes. Human erythrocytes have also been shown to be susceptible to hemoglobin-catalyzed, H_2O_2-dependent lipid peroxidation.[9-14]

In addition to lipids, proteins and carbohydrates are also susceptible to oxidative degradation by hemoglobin. Thus, the addition of H_2O_2 to erythrocytes or erythrocyte membranes produces high molecular weight membrane protein aggregates,[12,13] whereas deoxyribose is readily degraded by H_2O_2 in the presence of hemoglobin.[15] A review concerning

[1] A. L. Tappel, *Arch. Biochem. Biophys.* **44,** 387 (1953).
[2] A. L. Tappel, *J. Biol. Chem.* **217,** 721 (1955).
[3] J. Kanner and S. Harel, *Arch. Biochem. Biophys.* **237,** 314 (1985).
[4] S. M. H. Sadrzadeh, E. Graf, S. S. Panter, P. E. Halloway, and J. W. Eaton, *J. Biol. Chem.* **259,** 14354 (1984).
[5] J. Kanner and S. Harel, *Lipids* **20,** 625 (1985).
[6] M. B. Grisham, *Free Radical Biol. Med.* **1,** 227 (1985).
[7] S. Harel and J. Kanner, *Free Radical Res. Commun.* **5,** 21 (1988).
[8] J. Szebeni, C. C. Winterbourn, and R. W. Carrell, *Biochem. J.* **220,** 685 (1984).
[9] J. Stocks and T. L. Dormandy, *Br. J. Haematol.* **20,** 95 (1971).
[10] D. Chiu, B. Lubin, and S. Shohet, *in* "Free Radicals in Biology" (W. Pryor, ed.), Vol. 5, p. 115. Academic Press, New York, 1982.
[11] U. Benatti, A. Morelli, G. Damiani, and A. De Flora, *Biochem. Biophys. Res. Commun.* **106,** 1183 (1982).
[12] N. Sauberman, N. L. Fortier, W. Joshi, J. Piotrowski, and L. M. Snyder, *Br. J. Haematol.* **54,** 15 (1983).
[13] J. F. Koster and R. G. Slee, *Biochim. Biophys. Acta* **752,** 233 (1983).
[14] M. R. Clemens, H. Einsele, H. Remmer, and H. D. Waller, *Biochem. Pharmacol.* **34,** 1339 (1985).
[15] A. Puppo and B. Halliwell, *Biochem. J.* **249,** 185 (1988).

the prooxidant activity of hemoglobin and myoglobin has appeared recently.[16]

The mechanism by which hemoglobin exerts its peroxidative activity was long believed to be quite similar to that of the peroxidases, in which hydrogen peroxide first reacts with the native enzyme and withdraws two electrons, generating an intermediary species commonly referred to as Compound I. Compound I then reacts with a suitable substrate and withdraws two electrons from the substrate, which it can do one at a time. The intermediary form of the enzyme that has taken up one electron from a substrate molecule is called Compound II. In both Compound I and Compound II the heme iron is present in the ferryl form, i.e., it carries four positive charges. For a detailed description of the mechanism of various peroxidases, the interested reader is referred to recent review articles on the subject.[17-20]

A detailed study of the interaction of hydrogen peroxide with oxyhemoglobin and methemoglobin has recently been done in Davies' laboratory,[21] and the studies indeed indicate significant similarities between the mechanism of the peroxidases and that of hemoglobin. The results indicate that the reactive intermediates are the ferrylhemoglobins [Hb^+–Fe^{4+} or Hb–Fe^{4+}; see reactions (1)–(6), below], which form on the interaction of hydrogen peroxide with methemoglobin or oxyhemoglobin, respectively, and which, respectively, resemble Compound I and Compound II of the peroxidases. These strong oxidizing agents are the compounds that are capable of withdrawing an electron from a suitable substrate (such as a lipid containing unsaturated fatty acids), resulting in the formation of methemoglobin and a substrate radical as illustrated in reactions (1)–(6).

$$Hb\text{–}Fe^{2+} \cdot O_2 + H_2O_2 + 2H^+ \rightarrow Hb\text{–}Fe^{4+} + O_2 + 2H_2O \qquad (1)$$

$$Hb\text{–}Fe^{3+} + H_2O_2 + 2H^+ \rightarrow Hb^+\text{–}Fe^{4+} + 2H_2O \qquad (2)$$

$$Hb^+\text{–}Fe^{4+} + AH + OH^- \rightarrow Hb\text{–}Fe^{4+} + A\cdot + H_2O \qquad (3)$$

$$Hb\text{–}Fe^{4+} + AH + OH^- \rightarrow Hb\text{–}Fe^{3+} + A\cdot + H_2O \qquad (4)$$

[16] M. B. Grisham and J. Everse, in "Peroxidases in Chemistry and Biology" (J. Everse, K. E. Everse, and M. B. Grisham, eds.), Vol. 1, p. 335. CRC Press, Boca Raton, FL, 1991.

[17] J. K. Hurst, in "Peroxidases in Chemistry and Biology" (J. Everse, K. E. Everse, and M. B. Grisham, eds.), Vol. 1, p. 37. CRC Press, Boca Raton, FL, 1991.

[18] E. L. Thomas, P. M. Bozeman, and D. B. Learn, in "Peroxidases in Chemistry and Biology" (J. Everse, K. E. Everse, and M. B. Grisham, eds.), Vol. 1, p. 123. CRC Press, Boca Raton, FL, 1991.

[19] H. B. Dunford, in "Peroxidases in Chemistry and Biology" (J. Everse, K. E. Everse, and M. B. Grisham, eds.), Vol. 2, p. 1. CRC Press, Boca Raton, FL, 1991.

[20] B. W. Griffin, in "Peroxidases in Chemistry and Biology" (J. Everse, K. E. Everse, M. B. Grisham, eds.), Vol. 2, p. 85. CRC Press, Boca Raton, FL, 1991.

[21] C. Giulivi and K. J. A. Davies, J. Biol. Chem. 265, 19453 (1990); see also this volume [30].

$$A\cdot + A\cdot \rightarrow A_2 \tag{5}$$
$$Hb\text{--}Fe^{4+} + Hb\text{--}Fe^{2+} \cdot O_2 \rightarrow 2Hb\text{--}Fe^{3+} + O_2 \tag{6}$$

where $Hb\text{--}Fe^{2+} \cdot O_2$ and $Hb\text{--}Fe^{3+}$ represent oxyhemoglobin and methemoglobin, respectively, $Hb^{+}\text{--}Fe^{4+}$ and $Hb\text{--}Fe^{4+}$ represent the two forms of ferrylhemoglobin, and AH represents different substrates that can undergo one-electron oxidations to form the radical $A\cdot$.

As already indicated, the peroxidative activity of hemoglobin, like that of the native peroxidases, is very nonspecific toward the substrate that is being oxidized. Indeed, a large variety of compounds can serve as substrates, suggesting that a specific, three-dimensional binding site for the substrate does not exist on the hemoglobin molecule, but that the interaction likely takes place at an "outer edge" of the molecule. For details concerning this type of interaction, the reader is referred to discussions on the mechanism of peroxidases.[17–20]

Preparation of Reagents

Highly purified human and bovine hemoglobin as well as the human hemoglobin derivatives were prepared at the Division of Blood Research, Letterman Army Institute of Research (Presidio of San Francisco, CA). The DBBF-, DBBS-, and DSO-hemoglobins represent modified human hemoglobins in which two lysines of the α chains (Lys-99) were cross-linked using bis(dibromosalicyl) fumarate (DBBF), bis(dibromosalicyl) succinate (DBBS), and di(succinimidyl) oxalate (DSO), respectively, as a cross-linking agent. The presence of this cross-link prevents the dissociation of the tetrameric hemoglobin into $\alpha\beta$ dimers.[22] PLP-hemoglobin is a hemoglobin in which pyridoxal phosphate is covalently bound at the 2,3-diphosphoglycerate-binding site. This also inhibits dissociation of the hemoglobin.[23]

The concentrations of hydrogen peroxide solutions are determined by measuring the absorbance at 230 nm, using a molar extinction coefficient of 81 M^{-1} cm^{-1}.[24] Stock solutions of 176 mM are prepared daily in deionized water and kept in ice. (A 30% hydrogen peroxide solution is 8.8 M; a 1:50 dilution of this yields a 176 mM solution. Using 0.1 ml of the diluted peroxide solution in a 1-ml assay gives a final concentration of 17.6 mM hydrogen peroxide.)

[22] R. Chatterjee, E. V. Welty, R. Y. Walder, S. L. Pruitt, P. H. Rogers, A. Arnone, and J. A. Walder, *J. Biol. Chem.* **261**, 9929 (1986); see also this volume [17].

[23] R. E. Benesch and S. Kwong, *Biochem. Biophys. Res. Commun.* **156**, 9 (1988); see also this volume [16].

[24] J. W. T. Homan-Müller, R. S. Weening, and D. Roos, *J. Lab. Clin. Med.* **85**, 198 (1975).

Most of the peroxidase substrates that can be utilized in the assay are commercially available. These include 2,2′-azinobis(3-ethylbenzothiazolinesulfonic acid) (ABTS), dopamine, guaiacol, *o*-dianisidine, pyrogallol, benzidine and its derivatives, and many others.

The optimal pH for the peroxidase assay of hemoglobin is pH 5.0 to 5.5 and is similar to that of the peroxidases. Because strong oxidizing agents are formed as intermediates of the reaction, the chosen buffer should not be subject to oxidation. Therefore, buffers such as Tris, HEPES, and histidine, should not be used in this assay. An inorganic buffer, such as phosphate or carbonate, would be the buffer of choice, but these may not be at their optimal buffering capacity if assays are to be done at or close to the optimal pH. Organic compounds that do not oxidize readily, such as citrate, can also be used.

Assay Procedures

Assays for peroxidase activity are most conveniently performed in plastic 1-ml cuvettes. We routinely use 0.9 ml of a sodium phosphate–citrate buffer (50 mM each) at pH 5.40, containing the hemoglobin and the substrate. The reaction is started by the addition of 0.1 ml of a 176 mM hydrogen peroxide solution. Adding an equivalent amount of hemoglobin to the blank cuvette allows one to offset the absorbance of the hemoglobin, if any, at the wavelength used.

Assays with the various substrates are done in our laboratory as follows. The monitored changes in absorbance that occur during the assays are the result of the hemoglobin-catalyzed oxidation of the substrate by H_2O_2.

With ABTS

ABTS (100 mg) is dissolved in 20 ml of buffer. This makes a 9.1 mM solution of ABTS. Mix into the assay cuvette 0.85–0.88 ml of the ABTS solution, 0.1 ml of a 176 mM H_2O_2 stock solution, and 20–50 μl of a 0.4 mg/ml hemoglobin solution. Follow the increase in optical density at 410 nm or, preferably, at 740 nm.

With Guaiacol

Add 100 μl of guaiacol to 9.2 ml of buffer and shake well until the guaiacol is dissolved completely. This makes a 0.1 M solution of guaiacol. Mix into the assay cuvette 0.75–0.78 ml of buffer, 100 μl of the 0.1 M guaiacol stock solution, 100 μl of a 176 mM H_2O_2 stock solution, and

20–50 μl of a 1 mg/ml hemoglobin solution. Follow the increase in optical density at 470 nm.

With Pyrogallol

Prepare a 5% stock solution of pyrogallol in buffer. Mix into the assay cuvette 0.75–0.78 ml of buffer, 100 μl of the pyrogallol stock solution, 100 μl of a 176 mM H_2O_2 solution, and 20–50 μl of a 2 mg/ml hemoglobin solution. Follow the increase in optical density at 420 nm.

With Dopamine

Make a stock solution of 38 mg of dopamine in 5 ml of buffer. This constitutes a 40 mM dopamine solution. Mix into the assay cuvette 0.75–0.78 ml of buffer, 100 μl of the dopamine stock solution, 100 μl of a 176 mM H_2O_2 stock solution, and 20–50 μl of a 8 mg/ml hemoglobin solution. Follow the increase in optical density at 475 nm.

With o-Dianisidine

Prepare a 0.2% o-dianisidine solution in methanol. Then make a 10-fold dilution of this solution in buffer. This is the o-dianisidine stock solution, which is 0.82 mM. Note that o-dianisidine is practically insoluble in water, but the HCl salt dissolves readily in water. Mix into the assay cuvette 0.75–0.78 ml of buffer, 100 μl of the o-dianisidine stock solution, 100 μl of a 176 mM H_2O_2 stock solution, and 20–50 μl of a 2 mg/ml hemoglobin solution. Follow the increase in optical density at 450 nm.

In our experiments all reactions are done in triplicate. Rates are monitored for 5 min using a Hewlett-Packard Model 8451A diode array spectrophotometer equipped with a HP9121D microdisk drive. All assay results are stored on a diskette, and the rates obtained in the three assays are averaged subsequently.

Peroxidase Activity of Hemoglobin

We evaluated ABTS, dopamine, o-dianisidine, and guaiacol for their capacity to serve as substrates for the peroxidase activity of the hemoglobins. These four substrates were chosen because of their widely different chemical structures. ABTS is a relatively large molecule and is an azo dye in which two aromatic nitrogens are oxidized. Dopamine is a hydroxyquinone that is oxidized to the corresponding quinone, which eventually forms melanin. Guaiacol is a methoxyphenol that oxidizes to a radical followed by dimerization, whereas o-dianisidine is a methoxybenzidine

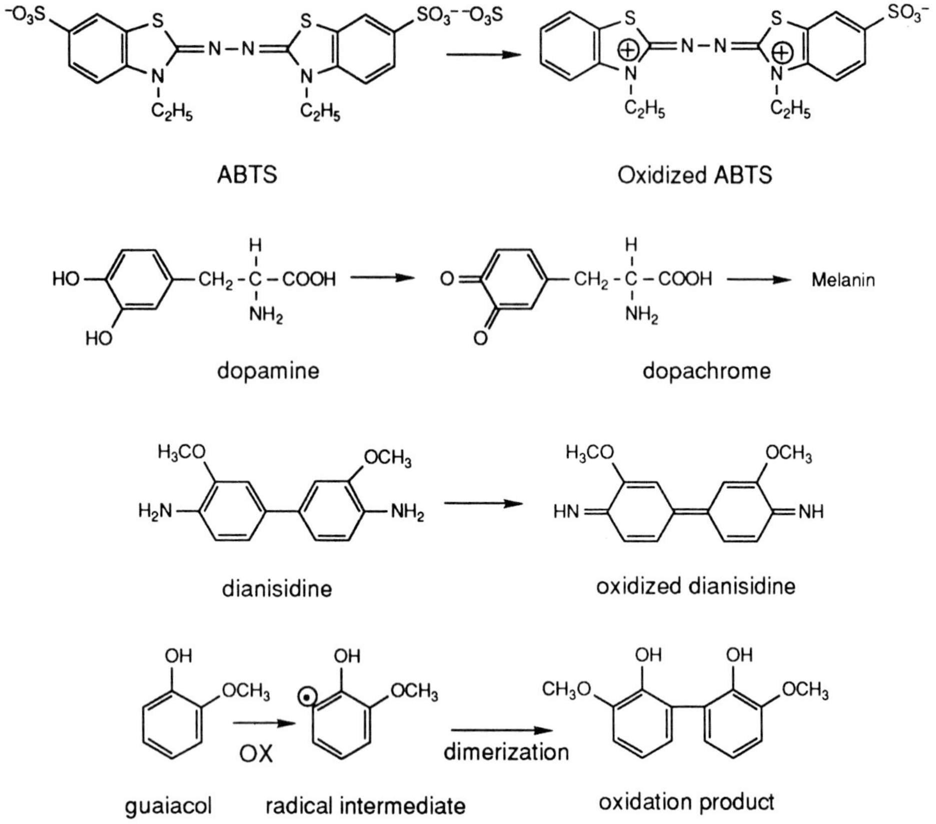

FIG. 1. Chemical structures of the reduced and oxidized products used in this study.

derivative in which the two amino groups are oxidized to imino groups. The structure of these substrates and their oxidation products are presented in Fig. 1. Thus, if any specific type of oxidation were to be expressed preferentially by one or more of the modified hemoglobins, we hoped to detect this by using these chemically unrelated substrates.

Peroxidase assays using ABTS as the substrate are usually monitored at 405–410 nm. Hemoglobin, however, has a rather low peroxidase activity, and therefore the amount of hemoglobin that is required for an assay is 10 to 100 times more than that of ordinary peroxidases. The hemoglobin present in the assay mixture can therefore produce a rather high background absorbance around 410 nm. This may make it necessary to monitor the ABTS reaction at 740 nm, where the absorption of hemoglobin is

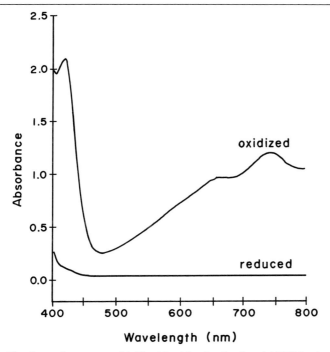

FIG. 2. The absorption spectra of 0.05 mM oxidized and reduced ABTS in phosphate–citrate buffer at pH 5.4. Reduced ABTS is virtually colorless, whereas oxidized ABTS is dark green.

negligible. Figure 2 shows the absorption spectra of the reduced and oxidized forms of ABTS and illustrates the rationale for selecting this wavelength.

The dependence of the rate of oxidation was found to be linear with the hemoglobin concentration over a reasonable range, as illustrated in Fig. 3 for ABTS. Comparable results were obtained with each of the other substrates.

Extinction Coefficients of Substrates

As indicated above, the oxidation of many substrates by heme proteins leads to the formation of free radicals. These in turn can, and often do, give rise to chain reactions leading to the formation of high molecular weight polymers of undefined chemical structures. A typical example is dopamine, which is oxidized to dopachrome. Dopachrome can then react with other dopamine and dopachrome molecules, and eventually form

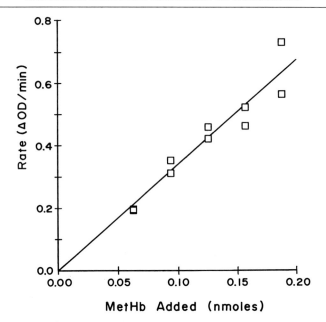

FIG. 3. The rate of ABTS oxidation as a function of methemoglobin concentration. The reaction conditions are described in the text. The concentration of H_2O_2 was 8 mM. The reaction was followed at 740 nm.

the insoluble melanin, the structure of which varies with the reaction conditions. For this reason it is impossible to determine an extinction coefficient for the oxidation product of many of the peroxidase substrates.

Exceptions to this are o-dianisidine and ABTS, which form well-defined oxidation products. The extinction coefficients for oxidized ABTS at various wavelengths have been determined[25] and are listed in Table I. The extinction coefficient of oxidized o-dianisidine at 450 nm is also known $(11,300 \ M^{-1} \ cm^{-1})$.[26]

Determination of K_m and V_{max} Values for H_2O_2

The Michaelis–Menten constants for H_2O_2 were determined with highly purified human and bovine hemoglobins, as well as with several chemically modified hemoglobins. Seven different concentrations of H_2O_2 were used, varying from 0.56 to 36 mM. Rates were determined for both

[25] W. Werner, H. G. Rey, and H. Wielinger, Z. Anal. Chem. **252**, 224 (1970).
[26] L. A. Decker, ed., "Worthington Enzyme Manual," p. 66. Worthington Biochemical Corporation, Freehold, NJ, 1977.

TABLE I

MOLAR EXTINCTION COEFFICIENTS OF OXIDIZED
ABTS AT VARIOUS WAVELENGTHS[a]

Wavelength (nm)	Extinction coefficient (M^{-1} cm^{-1})
405	36,800
420	43,200
436	29,300
578	10,900
623	18,200
660	18,200

[a] Taken from Werner et al.[25]

the initial and the steady-state velocity by linear least squares, using the procedures in the computer program RS1 (Bolt, Beranek, and Newman, Cambridge, MA). (The rate between 0 and 20 sec after the onset of the reaction was used as the initial rate, and the rate between 40 and 60 sec was used as the steady-state rate). The three rates obtained with each peroxide concentration (either initial or steady state) were then averaged and fitted by computer to the Michaelis–Menten equation:

$$v = V_{max}/(1 + K_m/S_0)$$

Typical examples of the computer-fitted curves are shown in Fig. 4. Similar curves are obtained when other substrates are used. Note that there is no evidence for any allosteric effect with regard to the binding of H_2O_2 to the hemoglobin. The values obtained from a set of these curves are tabulated in Table II.

As was illustrated previously in reactions (1)–(6), H_2O_2 binds to and reacts with the hemoglobin, generating the highly reactive ferrylhemoglobin intermediate, which in turn oxidizes the substrate. One might therefore expect that the K_m value for H_2O_2 obtained with the various hemoglobins would be independent of the substrate used. The results in Table II show that such is not the case; variations of more than an order of magnitude are found. It is of interest, though, that the best substrate (ABTS) also yields the highest K_m values.

Bovine hemoglobin differs from human hemoglobin in that it lacks the amino-terminal valine residues of the β chains, as well as a replacement of the $\beta2$ histidine by methionine.[27] Both of these amino acid residues in

[27] W. A. Schroeder, J. R. Shelton, J. B. Shelton, B. Robberson, and D. R. Babin, *Arch. Biochem. Biophys.* **120**, 124 (1967).

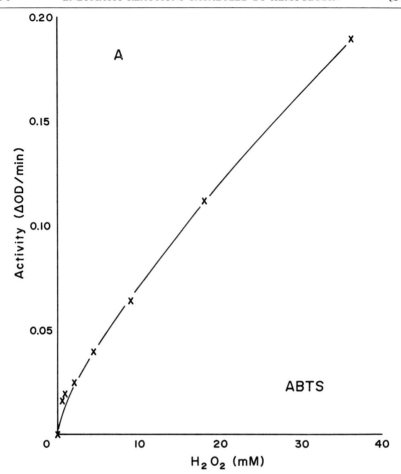

FIG. 4. Peroxidative activity of human hemoglobin A_0 as a function of hydrogen peroxide concentration. (A) With ABTS as the substrate; (B) with dopamine as the substrate.

human hemoglobin are involved in the interaction with 2,3-diphosphoglyc-erate.[28] The P_{50} of bovine hemoglobin is 28 Torr as compared with 13 Torr for the stripped (i.e., DPG-deprived) human hemoglobin.[29] The difference in affinity of the heme iron of these two hemoglobins for oxygen is thus about twofold; it therefore seems reasonable to expect that a significant difference in affinity may also exist in the binding of hydrogen peroxide to the heme irons. However, the only significant differences in K_m

[28] A. Arnone, *Nature (London)* **237,** 146 (1972).
[29] P. M. Breepoel, F. Kreuzer, and M. Hazevoet, *Pfluegers Arch.* **389,** 219 (1981).

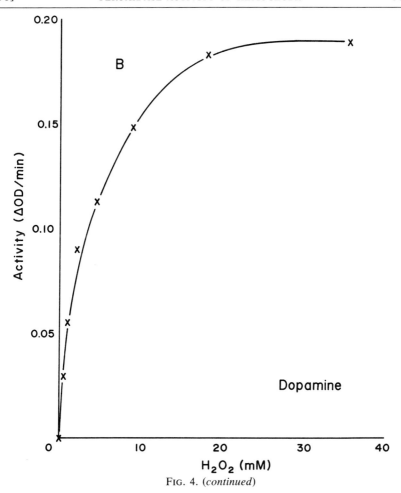

FIG. 4. (*continued*)

and V_{max} values between the human and bovine hemoglobins were found with *o*-dianisidine as the substrate. With all other substrates the values obtained with the two hemoglobins were virtually identical. These results demonstrate that small differences in affinity for H_2O_2 can indeed be demonstrated, but only with selected substrates.

The values obtained with the modified human hemoglobins also are very similar to those of the unmodified human hemoglobin. The largest differences were found in the K_m values using ABTS as the substrate. The K_m value for hemoglobin, 47 mM, is larger than the values for the derivatives, which vary from 11 to 28 mM H_2O_2. Smaller differences are

TABLE II

MICHAELIS CONSTANTS FOR H_2O_2 OF HEMOGLOBIN DERIVATIVES USING
VARIOUS SUBSTRATES[a]

Hemoglobin and parameter	ABTS	o-Dianisidine	Dopamine	Guaiacol
Hb A_0				
V_{max}	2.58	0.23	0.13	0.42
K_m (mM)	47.0	3.51	3.29	8.95
Bovine Hb				
V_{max}	3.00	1.32	0.22	1.08
K_m (mM)	41.9	13.70	2.70	8.99
DBBF-Hb				
V_{max}	1.44	0.18	0.06	0.22
K_m (mM)	11.6	1.62	5.02	3.21
DBBS-Hb				
V_{max}	1.74	0.33	0.17	0.27
K_m (mM)	18.4	4.11	4.55	3.60
DSO-Hb				
V_{max}	1.32	0.28	0.12	0.41
K_m (mM)	13.9	4.25	2.86	2.77
PLP-Hb				
V_{max}	1.92	0.30	0.18	0.40
K_m (mM)	28.6	1.52	4.03	5.68

[a] The values for V_{max} are expressed as the change in optical density per minute.

seen with guaiacol as the substrate, where the K_m values vary from 2.7 to 8.9 mM among the different hemoglobins. Nevertheless, it can be concluded from these data that the heme environments in the various hemoglobin derivatives are not identical to that in the native hemoglobin.

It is noteworthy that the obtained K_m values for H_2O_2 depend strongly on the substrate used. Although the highest values were consistently obtained with ABTS as the substrate, there was no consistent pattern among the different substrates. For example, the K_m value for Hb A_0 obtained with o-dianisidine as the substrate is virtually identical to that obtained with dopamine as the substrate (3.51 versus 3.29), whereas for bovine Hb these values are 13.7 and 2.7, respectively. This clearly shows a difference between the two proteins in the selective oxidation of the various substrates. A comparison of the other values in Table II shows that a similar selectivity for the substrates is observed with the various hemoglobin derivatives.

The V_{max} values in Table II are expressed as changes in optical density per minute because the extinction coefficients for some of the substrates are unknown, as indicated above. Nevertheless, the same relative comparisons can be made, because the extinction coefficients are constants and do not affect a comparison of the results obtained with the different hemoglobins. Thus, the data indicate that at saturating concentrations of H_2O_2, o-dianisidine is oxidized almost six times faster with the bovine hemoglobin than with the human hemoglobin. With the modified human hemoglobins, however, the obtained rates are quite similar to that of the native human hemoglobin.

Use of Substrate Ratios

Small differences at or near the active sites of enzymes can be detected by the determination of substrate and/or coenzyme ratios. This concept was introduced by Kaplan and colleagues in the late 1950s and was used extensively to distinguish enzymes from different sources.[30] Briefly, the concept is that the ratio of the specific activities obtained with two different substrates must always be the same for identical enzymes. However, when a different ratio is obtained using enzymes from different sources, this is virtually conclusive proof that there are differences in the active sites of the two enzymes, either structural or conformational. In addition to electrophoresis, the application of this concept was mainly responsible for the discovery of isoenzymes during the early 1960s.[31,32]

The peroxidative activity of hemoglobin could be used in this manner to test for any structural changes that may have occurred in the hemoglobin molecule as a result of chemical modifications, which in turn may cause an altered oxidative activity of the modified hemoglobin. Although H_2O_2 binds directly to the heme iron, the substrate may donate its electrons at the heme edge or in some cases even interact with the protein portion of the enzyme (in analogy with the peroxidases), because some substrates are too voluminous to penetrate into the heme pocket. Accordingly, any changes in the chemical or structural environment of the heme pocket may be reflected in a change in the rate of oxidation of a given substrate. Four different substrates of unrelated chemical structure were used in this study to maximize the possibility of detecting any changes in oxidative potential. Purified native bovine hemoglobin was used as a positive control, because it is known to have a significantly different affinity for oxygen, and therefore perhaps also for hydrogen peroxide, as compared with human hemoglobin.[29]

[30] N. O. Kaplan, M. M. Ciotti, M. Hamolski, and R. E. Bieber, *Science* **131**, 392 (1960).
[31] N. O. Kaplan and M. M. Ciotti, *Biochim. Biophys. Acta* **49**, 425 (1961).
[32] N. O. Kaplan and M. M. Ciotti, *Ann. N.Y. Acad. Sci.* **94**, 701 (1961).

TABLE III
RATIOS OF RATES WITH ABTS AND o-DIANISIDINE AT VARIOUS
HYDROGEN PEROXIDE CONCENTRATIONS

H_2O_2 concentration (mM)	Human Hb	Bovine Hb	DBBF-Hb	DBBS-Hb	DSO-Hb	PLP-Hb
36.0	5.03	1.41	5.48	3.76	3.38	3.85
18.0	3.97	1.32	4.40	3.16	3.10	2.68
9.0	2.37	1.02	4.01	2.55	2.45	1.75
4.5	1.88	0.77	2.80	1.97	1.91	1.08
2.25	1.70	0.61	2.47	1.62	1.62	0.87
1.12	1.84	0.49	2.10	1.59	1.54	0.84
0.56	2.59	0.91	2.04	1.94	1.52	0.84

At saturating concentrations of the substrates, the increase in optical density obtained with Hb A_0 is more than 10 times faster with ABTS than that obtained with o-dianisidine, whereas this value is much closer to unity for the bovine hemoglobin. This indicates that with respect to ABTS the human hemoglobin expresses a much lower preference for o-dianisidine than does the bovine hemoglobin. Such ratios may vary, however, under nonsaturating conditions. Table III illustrates how the ABTS/o-dianisidine ratio changes for the different hemoglobins when different concentrations of H_2O_2 are used. Nevertheless, at all H_2O_2 concentrations, the difference between the human and the bovine hemoglobin is clearly observable, whereas the values obtained with the modified hemoglobins are much closer to those of the human Hb than to those of the bovine Hb. It is also of interest to note that the ratios obtained with the PLP-Hb show a considerably stronger dependence on the peroxide concentration than, for example, those of the DBBF-Hb.

TABLE IV
RATIOS OF RATES WITH ABTS AND DOPAMINE AT VARIOUS
HYDROGEN PEROXIDE CONCENTRATIONS

H_2O_2 concentration (mM)	Human Hb	Bovine Hb	DBBF-Hb	DBBS-Hb	DSO-Hb	PLP-Hb
36.0	10.00	6.56	15.96	7.45	8.05	6.29
18.0	6.50	5.16	14.45	6.33	6.94	4.70
9.0	4.29	2.90	13.55	5.29	5.29	4.04
4.5	3.56	1.76	9.72	4.48	3.76	2.92
2.25	2.77	1.32	6.34	3.30	2.90	1.99
1.12	3.53	1.28	5.93	2.89	2.83	2.24
0.56	5.42	1.73	8.23	3.40	6.97	2.52

Table IV shows the ratios obtained with ABTS and dopamine. Here the differences between human and bovine hemoglobin are smaller than those shown in Table III, but again the data generally indicate a closer resemblance between the native and modified human hemoglobins than between the human and bovine hemoglobins. It may be noted that the values for the fumaryl α–α cross-linked hemoglobin (DBBF-Hb) in Tables III and IV are higher than those for the other hemoglobin derivatives. This illustrates our observation that, in comparison with the other hemoglobins, the DBBF cross-linked hemoglobin appears to prefer ABTS as a substrate over the other substrates. Although this preference may be interpreted as an indication of the presence of a more extensively altered heme environment in DBBF cross-linked hemoglobin as compared with the other modified hemoglobins, the differences are too small to justify any definite conclusions.

Activity of Liganded Hemoglobins

Oxyhemoglobin is rapidly oxidized to methemoglobin in the presence of low concentrations of hydrogen peroxide.[17] Carbonylhemoglobin is considered more stable to oxidation than oxyhemoglobin, and therefore might alter the results by its presence. We found, however, that hydrogen peroxide appears to react with carbonylhemoglobin as rapidly as it does with oxyhemoglobin, because the oxidative activity of hemoglobin obtained in the presence of CO was essentially the same as in its absence. On the other hand, no oxidation of any of the substrates was observed when cyanomethemoglobin was used, suggesting that hydrogen peroxide may be unable to displace the cyano group in cyanohemoglobin.

Acknowledgments

This work was performed at and supported by the Division of Blood Research, Letterman Army Institute of Research, Presidio of San Francisco, CA. J.E. thanks Col. Robert M. Winslow, Col. John R. Hess, and the staff of the Division for their interest and support of this work.

[37] Cyclooxygenase Activity of Hemoglobin

By Lucilla Zilletti, Mario Ciuffi, Sergio Franchi-Micheli,
Fabio Fusi, Grazia Gentilini, Gloriano Moneti,
Massimo Valoti, and Gian Pietro Sgaragli

Introduction

When hemoglobin (Hb) diffuses into the extracellular space of living tissues, which may occur during hemolysis or when erythrocytes escape from vessels, its catalytic potential toward several endogenous or exogenous compounds can be activated. There has been much attention focused on the capacity of Hb to metabolize polyunsaturated fatty acids (PUFAs), giving rise to the formation of products that may be important in the pathogenesis of systemic (intravascular hemolysis) or local (micro- or macrohemorrhages, inflammation, hemorrhagic infarcts) conditions in which erythrocyte lysis occurs.

At low concentrations Hb catalyzes a quasi-lipoxygenase reaction on linoleic acid with remarkably high substrate specificity.[1] The fatty acid cyclooxygenases [prostaglandin-hydroperoxide (-endoperoxide) synthase; PHS, EC 1.14.99.1], which may be regarded as a special type of lipoxygenase that initiates the biosynthesis of prostaglandins and thromboxanes, require heme for their activity.[2,3] Hb is a good candidate for exerting cyclooxygenase activity owing to its peroxidative properties already described in this volume.[4] A number of reports indicate that PHS shares with peroxidases (PODs, EC 1.11.1.7) the ability to promote peroxidation of xenobiotics and that POD exhibits an incompletely explored oxygenase activity that might have some similarities to the cyclooxygenase activity of PHS.[5]

In a previous paper[5] we have shown that aerobic incubation of 5,8,11,14-eicosatetraenoic acid (arachidonic acid; AA) with POD gives rise to the formation of a number of compounds with prostaglandin-like biological activities. Among these compounds prostaglandins E_2 and $F_{2\alpha}$ (PGE_2 and $PGF_{2\alpha}$) are identified by thin-layer chromatography (TLC) and

[1] H. Kuhn, R. Gotze, T. Schewe, and S. M. Rapoport, Eur. J. Biochem. 120, 161 (1981).
[2] N. Ogino, S. Ohki, S. Yamamoto, and O. Hayaishi, J. Biol. Chem. 253, 5061 (1978).
[3] F. J. Van der Ouderaa, M. Buytenhek, D. H. Nutgeren, and D. A. van Dorp, Biochim. Biophys. Acta 487, 315 (1977).
[4] J. Everse, M. C. Johnson, and M. A. Marini, this volume [36].
[5] L. Zilletti, M. Ciuffi, G. Moneti, S. Franchi-Micheli, M. Valoti, and G. Sgaragli, Biochem. Pharmacol. 38, 2429 (1989).

gas chromatography–mass spectrometry (GC-MS) analysis. Because this catalytic activity is shared by different hemoproteins (cytochrome c, catalase, and Hb) and is inhibited by phenylhydrazine, which is known to inactivate POD irreversibly by complexing heme in the presence of oxygen at doses as low as 1 μM,[6] it is inferred that the heme group and its integrity are of primary importance for this catalytic property of POD.

Here we report an in-depth study of the cyclooxygenase activity of Hb, which has been made possible by the use of either physicochemical or biological analytical methods.

Reagents

Arachidonic acid (AA), human Hb A_0, bovine Hb, bovine N,N-dimethylated Hb, bovine hemin, prostaglandins E_2 and $F_{2\alpha}$, indomethacin, and mepyramine maleate are purchased from Sigma Chemical Co. (St Louis, MO). Methysergide dimaleate is from Sandoz (Basel, Switzerland). Hyoscine sulfate and phenylhydrazine are from BDH (Milan, Italy), D-penicillamine hydrochloride is from Eli-Lilly & Co. (Indianapolis, IN), desferrioxamine methanesulfonate is from Ciba-Geigy SA (Basel, Switzerland), and trimethylsilylimidazole (TMSI) is from Supelco Inc. (Bellefonte, PA). Methemoglobin (metHb) is prepared from human Hb by addition of a slight excess of potassium ferricyanide, which is subsequently removed by passage through a column of Sephadex G-25. Carboxyhemoglobin (HbCO) is obtained by bubbling 20 ml of pure carbon monoxide (CO) into the assay mixture containing human Hb. CO is prepared by the reaction of concentrated sulfuric acid on formic acid. All other reagents are of analytical grade.

Apparatus

A Shimadzu UV 160 spectrophotometer is used for spectrophotometric determinations. Bioassays are performed using a MARB isometric transducer, and contractile responses are recorded on a Linseis Poligraph. Continuous flow–fast atom bombardment–mass spectrometry (CF-FAB-MS) analyses are performed on a VG 7070EQ (VG Analytical, Manchester, England) interfaced to a Digital PDP-8/A Data System and by using xenon atoms at 7 keV kinetic energy. Mass spectra in the negative-ion mode at $M/\Delta M$ 1500 resolution (10% valley definition) are recorded with a run speed of 10 sec/decade. Methanol : water : glycerol (75 : 22 : 3, v : v : v) are used as matrix, at a flow rate of 3 μl/min; the matrix is pumped by a μLC-

[6] O. V. Lebedeva, N. N. Ugarova, and I. V. Berezin, *Biokhimiya* (*Moscow*) **44**, 1766 (1980).

500 Micro Flow Pump (Kontron Instruments, Milan, Italy). The tip probe temperature is maintained at 35° during the analysis and 0.5 μl of the sample is injected by a Rheodyne 7520 valve (Rheodyne, Cotati, CA).

Gas chromatography–mass spectrometry analyses are performed on an HP 5971A mass spectrometry detector coupled to an HP 5890 series II gas chromatograph, equipped with an HP1 capillary column (25 m × 0.2 mm; 0.33-μm film thickness) (Hewlett-Packard Italiana, Cernusco sul Naviglio, Milan, Italy). The carrier gas is helium at a head pressure of 16 psi. The GC oven temperature program is 0.5 min at 90°, then ramped to 250° at a rate of 30°/min; the temperature is then increased at a rate of 6°/min to 310°, where it is held for 5 min. Injector and transfer line temperatures are 285° and 300°, respectively. The injections are made in the splitless mode (0.5-min split valve on) and the injection volume is 2 μl.

Procedure

To assess the cyclooxygenase activity of Hb, the formation of PG-like compounds is detected by means of a classical bioassay method, and identification of true PGs is performed by mass spectrometry analysis.

Incubations: Standard Conditions

Incubations are carried out in unstoppered 25-ml glass tubes under air as gas phase and at 25° in a shaking water bath (60 orbital oscillations/min, 3-cm stroke radius). The composition of the incubation mixture employed routinely is as follows: 0.33 mM AA (to start the reaction), 0.1 μM Hb (Hb concentration is given per tetramer except for spectrophotometric analysis in which is given per heme), 50 mM sodium carbonate buffer, pH 9.5, in a total volume of 4 ml. Incubation time ranges from 2.5 to 40 min (20 min under standard conditions), and the reaction is stopped by acidification to pH 3 with 1 M HCl. To assess the pH dependence of the reaction, 50 mM Tris-HCl buffer and 50 mM sodium carbonate buffer are used. AA is always used soon after opening the vials to avoid the addition of AA oxidation products to the reaction mixture.[5]

Different concentrations (10–300 μM) of AA are incubated with a fixed concentration of Hb in order to define the dependence of the reaction on substrate concentration. The K_m value is determined from double-reciprocal plots as described by Dixon and Webb.[7]

[7] M. Dixon and E. C. Webb, "Enzymes," 3rd ed., p. 550. Longman Group Ltd., London, 1979.

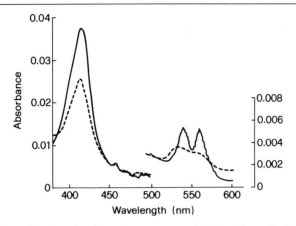

FIG. 1. Effect of AA on the absorption spectrum of HbO_2. Hb nominal concentration, 3×10^{-7} M; AA, 0.15 mM; at 25°. Absorption spectrum of HbO_2 before (continuous line) and 1 min after (dashed line) addition of AA.

Inactivation of Enzyme Activity

In some experiments inactive Hb preparations are used. In order to destroy the enzyme activity, Hb solutions are boiled for 30 min. This treatment abolishes cyclooxygenase activity of Hb, as well as of metHb.

Integrity of Hb during Incubation

The occurrence of degradation or autoxidation of Hb during the incubation with AA is checked by spectrophotometric UV–visible analysis. HbO_2 and metHb spectra are recorded in the region of 350–700 nm. HbO_2 and metHb concentrations are calculated using the extinction coefficients of $\varepsilon = 131$ cm^{-1} mM^{-1} at 425 nm and $\varepsilon = 162$ cm^{-1} M^{-1} at 406 nm,[8,9] respectively. No spectral changes were recorded during a 20-min period in which Hb is maintained under control conditions in the absence of AA. The addition of AA to a solution of HbO_2 causes a rapid decrease in the intensity of the spectrum accompanied by a less appreciable shift of the α and β peaks toward shorter wavelengths (Fig. 1). Similar changes are observed when metHb or HbCO is incubated with AA.

These findings could indicate the formation of a low-spin oxidized form of Hb, namely hemichrome. Other authors have suggested that this could

[8] V. G. Zijlstra, A. Buurmsma, and A. Zwart, *J. Appl. Physiol.* **54,** 1287 (1983).
[9] E. Antonini and M. Brunori, *Front. Biol.* **21,** 16 (1971).

occur when Hb[10] or myoglobin[11] is incubated with myristic or linoleic acid, respectively.

Sample Processing

Extraction of the reaction products is performed with ethyl acetate, according to Piper and Vane,[12] and the aqueous phase is discarded after centrifugation at 1000 g for 10 min. After evaporation of the organic phase to dryness, in a gentle stream of N_2, the residue is suspended directly in Krebs–Henseleit solution for immediate bioassay. Alternatively, the residue is solubilized in ethanol for TLC purification and separation of the reaction products. These are subsequently analyzed by MS.

TLC Purification

The residue obtained after evaporation of the ethyl acetate extract under a stream of N_2 is suspended in 0.2 ml of ethanol and submitted to TLC purification (DC Alufolien Kieselgel 60 F254, Merck), using an eluting system of n-hexane : diethyl ether : acetic acid (15 : 85 : 0.1, v : v : v).[13] PGs do not migrate, as demonstrated by the $R_f = 0$ of standard PGs cochromatographed on the same plate and developed with anisaldehyde. The gel areas of the undeveloped lanes where the residues are applied, and corresponding to $R_f = 0$, are scraped off, suspended in ethyl alcohol, centrifuged, and analyzed by mass spectrometry.

Assessment of PG-like Activity by Bioassay

The products of the reaction are bioassayed for their PGE_2-like activity on strips of rat stomach fundus, according to Vane,[14] and for their $PGF_{2\alpha}$-like activity on rat colon, according to Regoli and Vane.[15] Bioassays are performed in the presence of 0.56 mM methysergide, 0.33 mM hyoscine, 0.35 mM mepyramine, and 2.8 mM indomethacin. The assays are done at least in duplicate, and the biological content is determined at three dose levels. Authentic PGE_2 (1–2 ng) causes a good contractile response of rat stomach and is not effective in rat colon. On the contrary, $PGF_{2\alpha}$ (20–40

[10] A. A. Akhrem, G. M. Andreyuk, M. A. Kisel, and P. A. Kiselev, *Biochim. Biophys. Acta* **992**, 191 (1989).
[11] D. Galaris, A. Sevanian, E. Cadenas, and P. Hochstein, *Arch. Biochem. Biophys.* **281**, 163 (1990).
[12] P. J. Piper and J. R. Vane, *Nature (London)* **223**, 29 (1969).
[13] K. C. Srivastava and K. K. Awasthi, *J. Chromatogr.* **275**, 61 (1983).
[14] J. R. Vane, *Br. J. Pharmacol.* **12**, 344 (1957).
[15] D. Regoli and J. R. Vane, *Br. J. Pharmacol.* **23**, 351 (1964).

ng) elicits a good contraction of rat colon and a weak contraction of rat stomach fundus.

Characterization of Hb-Catalyzed Conversion of AA to PGE₂-like Substances. Hb aerobically incubated with AA gives rise to the formation of substances identified by bioassay as PGE_2-like compounds. The formation rate of these substances (Fig. 2) is well sustained and linear in the first 10 min and then decreases to 0 after 20 min. There is a pH dependence of the reaction with an apparent optimum between 9 and 10, although formation of PGE_2-like products is also sustained at pH values of 7 and 8 (Table I). The initial rate of the reaction is proportional to the concentration of added Hb up to 1 μM, as shown in Fig. 3. A double-reciprocal plot of the oxidation of AA is shown in Fig. 4. The kinetic parameters determined from nonlinear regression analysis of the initial rate data are apparent $K_m = 46$ μM and $V_{max} = 26$ ng of PGE_2-like substances min^{-1} nmol Hb^{-1}. The amount of PGE_2-like substances formed is found in the range of 180–304 ng/sample ($n = 5$) under standard conditions with the maximal concentration of hemoglobin (0.1 μM).

Conversion of AA into PGF$_{2\alpha}$-like Activity Catalyzed by Hb. PGF$_{2\alpha}$-like activity is also present in single samples in amounts close to the sensitivity threshold of rat colon. Pooling dried extracts makes the products of reaction detectable. In this way average amounts of 90–120 ng/sample are found. Alternatively, enrichment of single samples with a

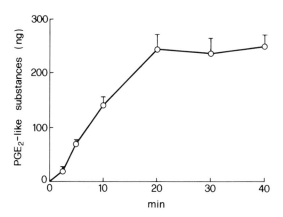

FIG. 2. Time course of the Hb-catalyzed formation of PGE_2-like substances from AA. Formation of PGE_2-like substances is performed at 25° in air in 4 ml of a reaction mixture containing (final concentrations) 50 mM sodium carbonate buffer, pH 9.5, 0.33 mM AA (starter of the reaction), and 0.1 μM Hb. The reaction is blocked with addition of 1 M HCl. PGE_2-like substances are determined by bioassay in the ethyl acetate extracts of the reaction mixture. Points represent the mean ± SEM of three experiments.

TABLE I
HEMOGLOBIN-CATALYZED CONVERSION OF AA
TO PGE$_2$-LIKE COMPOUNDS AT DIFFERENT pH
VALUES[a]

pH	ng/sample
6.0	176 ± 8.8
7.0	240 ± 7
8.0	243 ± 20
9.0	265 ± 17
10.0	284 ± 34
11.0	181 ± 12

[a] Experimental conditions are those reported in
Fig. 2. Different pH values are obtained by
using Tris-HCl buffer and 50 mM sodium carbonate buffer. Figures are means ± SEM of
three experiments.

small amount (10 ng) of authentic PGF$_{2\alpha}$ makes the bioassay possible. By
following this procedure, amounts in the range of 80–115 ng/sample of
PGF$_{2\alpha}$-like substances are found.

Inhibition of Hb-Catalyzed Conversion of AA to PGE$_2$-like Substances

To study the effects of possible inhibitors of the reaction, inhibitors are
preincubated with Hb for 5 min before the addition of AA. Hemoglobin-

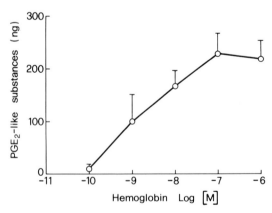

FIG. 3. Dependence of formation rate of PGE$_2$-like substances from AA on Hb concentration in otherwise standard conditions. Each point represents the mean ± SEM of three experiments.

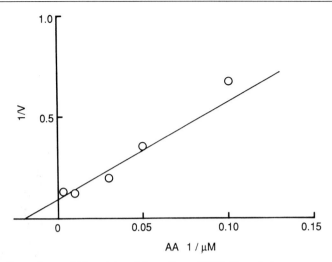

Fig. 4. Dependence of Hb-catalyzed formation of PGE$_2$-like substances on arachidonate concentration. Experimental conditions are those reported in Fig. 2. Arachidonate concentrations ranged between 10 and 300 μM. Each point represents the mean of five experiments.

catalyzed conversion to PGE$_2$-like substances is inhibited by phenylhydrazine in a concentration-related manner with a calculated IC$_{50}$ = 0.17 μM (nonlinear regression analysis) (Table II). Interestingly, this value is similar to that observed when phenylydrazine was used to inhibit cyclooxygenase activity of POD.[5] D-Penicillamine and desferrioxamine inhibit this catalytic activity of Hb to a lesser extent, whereas indomethacin at

TABLE II
EFFECT OF DIFFERENT COMPOUNDS ON HEMOGLOBIN-
CATALYZED FORMATION OF PGE$_2$-LIKE SUBSTANCES FROM AA

Compound	Concentration (μM)	PGE$_2$-like substances[a] (%)
Phenylhydrazine	0.01	92.8
	0.1	78.5
	1	46.7
	10	39.7
Indomethacin	1	100
D-Penicilliamine	1	65
Desferrioxamine	1	61.5

[a] Numbers represent percentage of the amounts obtained without inhibitor, ranging from 190 to 320 ng.

the concentration used lacked any inhibitory effect. This is not surprising because indomethacin does not inhibit the quasi-lipoxygenase activity of Hb.[1]

Conversion of AA to PGE₂-like Substances Catalyzed by Different Forms of Hb

The property of Hb to catalyze aerobically the conversion of AA to PGE_2-like substances is shared by different forms of Hb, as shown in Table III. Although the oxidation state of Fe does not seem to influence the reaction to any extent, the steric hindrance sustained by N-methylation of two pyrroles in the heme moiety does significantly inhibit the reaction. Furthermore, the molecular activity of free hemin is 17% of that exhibited by Hb when the latter is used in comparison, thus suggesting the importance of the protein moiety for this catalytic activity.

Identification of PGs by Mass Spectrometry Analysis

The CF-FAB ionization technique in the negative-ion mode, when performed on standard PGs, gives rise to spectra that are characterized by the almost exclusive presence of the quasi-molecular ions $(M–H)^-$. With this technique, coproduction of fragment ions is negligible or totally absent, thus the method has a high detection sensitivity.

Authentic PGE_2 and $PGF_{2\alpha}$, in fact, give mostly peaks corresponding to $(M–H)^-$ ions at m/z 351 and 353, respectively. Identical spectra are

TABLE III
CONVERSION OF AA TO PGE₂-LIKE SUBSTANCES BY
DIFFERENT FORMS OF Hb[a]

Hemoglobin	PGE₂-like substances[b] (ng)
Hb A_0	275 ± 18.4
Methemoglobin	230 ± 15.6
Bovine Hb	260 ± 14.6
Bovine N,N-dimethylated Hb	192 ± 12.4[c]
Hemin	39 ± 3.7[d]

[a] At 0.1 μM hemoglobin.
[b] Numbers represent mean values ± SEM from four experiments.
[c] Statistically different ($p < 0.05$; Student's t-test) versus bovine Hb.
[d] In the same experiments bovine Hb yields 226 ± 12.5 ng of PGE₂-like substances.

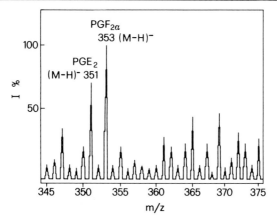

FIG. 5. FAB negative-ion mas spectrum of products formed from arachidonate in the presence of Hb. The ethyl acetate extract of the incubation mixture is submitted to TLC (see Procedures). MS analysis is performed in CF-FAB on the ethanol eluate (0.5 μl injected) of the TLC areas containing PGs.

obtained when products of the reaction, purified on TLC, are submitted to CF-FAB-MS in a glycerol matrix (Fig. 5). The identification of PGs is confirmed by GC-MS analysis with electron impact ionization. For GC-MS, the extracts purified by TLC are methylated with ethereal diazomethane and silylated with TMSI and piperidine (1 : 1, v : v) according to a previously published procedure.[16] The estimation at nanogram levels of PGs is carried out by selecting ion monitoring (SIM); the ions selected for PGE_2 and $PGF_{2\alpha}$ are m/z 321, 349, and 420, and m/z 307, 423, and 494, respectively (Fig. 6).

Comments

The findings reported here demonstrate the conversion of AA into PGE_2 and $PGF_{2\alpha}$ brought about by Hb. In order for Hb to have cyclooxygenase activity, the integrity of heme is essential; furthermore, the protein moiety of Hb plays an important role. Our evidence for the cyclooxygenase activity of Hb permits us to propose a hypothesis for explaining the observation in that, following subarachnoid hemorrhages in humans, the cerebrospinal fluid contains elevated levels of $PGF_{2\alpha}$ and possibly other

[16] S. Nicosia and G. Galli, *Anal. Biochem.* **61**, 336 (1975).

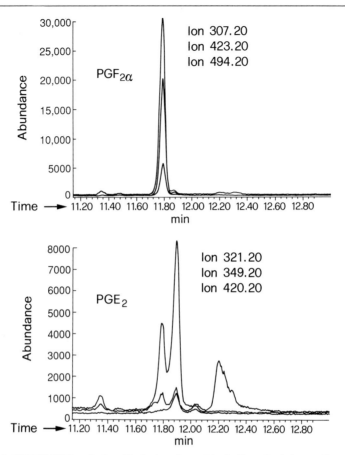

FIG. 6. GC-MS-SIM analysis with electron impact ionization of substances formed from arachidonate incubated with Hb. The compounds present in the ethyl acetate extract of the gel area of the TLC are methylated with diazomethane and silylated with TMSI and piperidine (1 : 1, v : v). Chromatographic peaks were attributed to different PGs by the criterion of the R_f identity.

PGs.[17–20] However, direct experiments are now necessary in order to assess whether this pathway is also operant *in vivo,* when Hb has access to the extracellular space.

It is of note that in previous work[1] it was observed that in *in vitro*

[17] V. Walker and J. D. Pickard, *Adv. Tech. Stand. Neurosurg.* **12,** 3 (1985).

[18] R. Rodriguez y Baena, P. Gaetani, V. Silvani, T. Vigano, M. T. Crivellari, and P. Paoletti, *Acta Neurochir.* **84,** 129 (1987).

[19] V. Seifert, D. Stolke, V. Kaever, and H. Dietz, *Surg. Neurol.* **27,** 243 (1987).

[20] B. B. Cherhazi, S. Giri, and R. M. Joy, *Stroke* **20,** 217 (1989).

experiments "AA is not attacked at all by Hb." This finding might be due to the methods used in that study, which do not allow the detection of the biological activities or the very small amounts of the compounds formed. That study, in fact, was based on the measurement of oxygen consumption during the reaction and of conjugated dienes as oxygenation products.

Finally, our demonstration of the cyclooxygenase activity of Hb suggests the need for caution when interpreting results obtained from experiments in which Hb has been used as a chemical tool without any consideration of the consequences that may arise because of its cyclooxygenase activity. This is, for example, the case when it is used in *in vitro* experiments on the binding and subsequent inactivation of NO that is eventually formed and released into the extracellular spaces in living tissues.[21-24]

Acknowledgments

This research is supported by grants from the Ministry of University and Scientific and Technological Research and from the National Research Council (CNR).
The authors thank Daniela Bindi and Dr. Giuseppe Pieraccini (Centre of Mass Spectrometry, University of Florence) for excellent technical assistance.

[21] W. Martin, G. M. Villani, D. Jothianandan, and R. F. Furchgott, *J. Pharmacol. Exp. Ther.* **232,** 708 (1985).
[22] J. R. Vane, E. E. Anggard, and R. Botting, *N. Engl. J. Med.* **323,** 27 (1990).
[23] R. M. J. Palmer, A. G. Ferrige, and S. Moncada, *Nature (London)* **327,** 524 (1987).
[24] E. M. Schuman and D. V. Madison, *Science* **254,** 1503 (1992).

[38] Hydroxylation and Dealkylation Reactions Catalyzed by Hemoglobin

By J. J. Mieyal and D. W. Starke

Introduction

Hemoglobin (Hb) exhibits monooxygenase-like activity *in vitro* with a variety of substrates in a system containing O_2, NADPH, and *P*-450 reductase.[1-3] The reactions can be catalyzed by Hb with NAD(P)H also

[1] J. J. Mieyal, R. S. Ackerman, J. L. Blumer, and L. S. Freeman, *J. Biol. Chem.* **251,** 3436 (1976).
[2] D. W. Starke, K. S. Blisard, and J. J. Mieyal, *Mol. Pharmacol.* **25,** 467 (1984).
[3] I. Golly and P. Hlavica, *Biochim. Biophys. Acta* **760,** 69 (1983).

in the absence of a reductase enzyme,[1,4–6] but then product formation is coupled less tightly to consumption of NADPH and O_2 and it is inhibited by superoxide dismutase (see below).[1,4] The hemoglobin catalytic system can be reconstituted also with the methemoglobin reductases, albeit not as efficiently as with P-450 reductase.[3] In this regard, Kokubo et al. prepared a flavohemoglobin derivative that displayed aniline hydroxylase activity comparable to that of the Hb plus the P-450 reductase system.[7]

The catalytic activity of Hb, with NADPH and P-450 reductase, includes aromatic p-hydroxylations of aniline and congeners,[1,2] O-demethylations of aromatic ethers,[2] N-demethylations of N-methyl-aniline and benzphetamine,[2] N-hydroxylation of p-chloroaniline,[3] and aliphatic hydroxylation of cyclohexane.[8] The mechanism and the versatility of the monooxygenase activity of Hb are qualitatively similar to those of the liver microsomal cytochromes P-450, including the prototype substrates aniline, p-nitroanisole, benzphetamine, and cyclohexane. Overall the similarities and differences between Hb and P-450 provide an opportunity to examine differential monooxygenase or peroxidase/peroxygenase activities in terms of structure, heme environment, and interaction of the hemoproteins with other components of the oxygenase systems. This may be especially useful in delineating the early events in the reaction scheme, because Hb can be examined in a stable oxyferrous form.[4] Figure 1 shows a general reaction scheme that typifies the reactions catalyzed by Hb and cytochrome P-450.

Within the broader perspective of comparisons of the activities of Hb to those of P-450 and the peroxidases, a number of distinctions have been made.[4,9–12] For example, different patterns of kinetic isotope effects have suggested that the initial electron abstraction in N-dealkylation reactions may occur differently for Hb- and P-450-catalyzed reactions.[9] In a study wherein styrene and H_2O_2 were cosubstrates for Hb,[10] evidence was reported for an unusual cooxidation mechanism that involved an intermediate radical on the Hb protein reacting with O_2, in addition to the familiar ferryl oxygen transfer mechanism associated with P-450 reactions. (Dis-

[4] J. J. Mieyal, Rev. Biochem. Toxicol. 7, 1 (1985).
[5] M. R. Juchau and K. G. Symms, Biochem. Pharmacol. 21, 2053 (1972).
[6] H. G. Jonen, R. Kahl, and G. F. Kahl, Xenobiotica 6, 307 (1976).
[7] T. Kokubo, S. Sassa, and E. T. Kaiser, J. Am. Chem. Soc. 109, 606 (1987).
[8] D. W. Starke and J. J. Mieyal, Biochem. Pharmacol. 38, 201 (1989).
[9] G. T. Miwa, J. S. Walsh, G. L. Kedderis, and P. F. Hollenberg, J. Biol. Chem. 258, 14445 (1983).
[10] P. R. Ortiz de Montellano and C. E. Catalano, J. Biol. Chem. 260, 9265 (1985).
[11] P. R. Kelder, N. J. De Mol, and L. H. M. Janssen, Biochem. Pharmacol. 38, 3593 (1989).
[12] J. J. Mieyal, Bioorg. Chem. 4, 315 (1978).

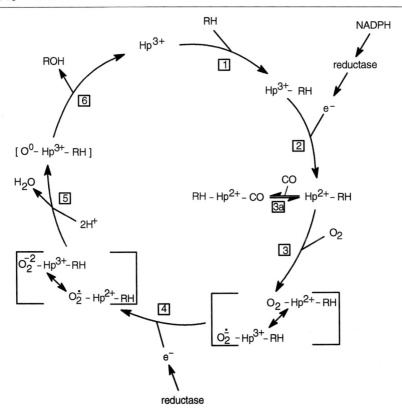

FIG. 1. Schematic representation of the monooxygenase catalytic cycle. RH represents substrate, and Hp represents hemoprotein (hemoglobin or cytochrome P-450). At steps 2 and 4, e^- represents the two electrons from NADPH donated sequentially to the hemoprotein from the NADPH-dependent reductase flavoprotein. The charge indicated on Hp represents the formal charge on the heme iron atom as either Fe^{3+} or Fe^{2+}; the actual valence state of the ferryl iron in the intermediate complex shown after step 5, however, is more accurately represented by Fe^{4+} or Fe^{5+}.

cussions of the mechanistic implications of Hb catalysis versus P-450 and peroxidases have been given in other reviews.[4,12,13]

From a practical point of view, Hb bound to solid supports has been considered for use in industrial catalysis of hydroxylation reactions.[14] In addition, monitoring the monooxygenase activity of hemoglobin is being

[13] L. J. Marnett, P. Weller, and J. R. Battista, in "Cytochrome P-450: Structure, Mechanism, and Biochemistry" (P. R. Ortiz de Montellano, ed.), p. 29. Plenum, New York, 1986.
[14] D. Guillochon, E. Laurent, B. Cambou, and D. Thomas, Ann. N.Y. Acad. Sci. 434, (Enzyme Eng.), 214 (1984).

developed as a potentially sensitive means of assessing subtle changes in the conformational integrity of Hb after molecular modifications by cross-linking agents.[15] The modified hemoglobins could serve as special catalysts or artificial blood substitutes.

In this report, methods are described for measuring hemoglobin-cata-lyzed reactions using the typical reconstituted system containing purified hemoglobin, NADPH, and purified cytochrome P-450 reductase. Hydroxylation of the prototype aromatic substrate aniline and aliphatic substrate cyclohexane, and N- and O-demethylation of the prototype sub-strates N-methylaniline and p-nitroanisole, respectively, are described in detail. Related methods for analogs of these prototype substrates are described briefly and the corresponding original literature reports are cited. In a separate section, methods are described for assessing hemoglobin-dependent hydroxylation and demethylation reactions in intact erythro-cytes.

Materials

Reagent Chemicals

NADPH is from Boehringer-Mannheim Biochemicals (Indianapolis, IN) or ICN Nutritional Biochemicals (Irvine, CA). Aniline, phenol, potas-sium phosphate (mono- and dibasic), trichloroacetic acid, ammonium ace-tate, sodium hydroxide, sodium carbonate, potassium cyanide, potassium ferricyanide, acetonitrile, p-dimethylaminobenzaldehyde, acetic acid, and diethyl ether are purchased from Fisher Scientific (Pittsburgh, PA). The ether is treated with an equal volume of 5% ferrous sulfate solution to deplete peroxides, and then it is washed three times with water. p-Anisidine, o-methoxyphenol, and o-toluidine are obtained from Eastman Organic Chemical Co (Rochester, NY). m-Toluidine and acetylacetone are from Aldrich (Milwaukee, WI). Cytochrome c (from horse heart) and Trizma base are purchased from Sigma (St. Louis, MO). Most substrates are purified by recrystallization or fractional distillation in order to obtain products that display <2° melting or boiling point ranges. Cyclohexane, cyclohexanol, and cycloheptanol are purchased from Aldrich and are used directly; gas chromatographic analysis indicated no detectable impurities.

Preparation of Erythrocytes

Erythrocytes from human blood, obtained from normal volunteers and anticoagulated with heparin, are isolated by low-speed centrifugation in

[15] R. H. Kluger, University of Toronto, Department of Chemistry, personal communica-tion (1992).

a clinical centrifuge. The packed erythrocytes, after removal of the supernatant plasma, are resuspended with an equal volume of isotonic saline and recentrifuged. This washing procedure is repeated four times.

Preparation of Hemoglobin

Usually Hb is purified from washed human erythrocytes by the method of Riggs.[16] Ferric Hb is prepared from oxyHb in 0.1 M potassium phosphate, pH 7, by addition of a 1.2 M excess (relative to heme) of potassium ferricyanide under a nitrogen atmosphere. The mixture is then passed sequentially over Sephadex G-25 and Bio-Rad AG1-X8 (chloride form) (Bio-Rad, Richmond, CA) columns, each equilibrated with 0.1 M potassium phosphate, pH 7, to remove residual ferricyanide and ferrocyanide.

Hemoglobin concentrations are determined by the method of Van Kampen and Zijlstra[17] based on the cyanoferriheme absorbance at 541 nm, $\varepsilon = 11$ mM^{-1} cm^{-1}.

Preparation of P-450 Reductase

Highly purified *P*-450 reductase is prepared from mouse or rat liver microsomes by affinity chromatography according to the method of Shephard *et al.*[18] Reductase preparations usually display specific activities of 29–35 units/mg for reduction of cytochrome *c* at room temperature. This specific activity corresponds to a single band on SDS–polyacrylamide gel electrophoresis. It is confirmed separately that each preparation of *P*-450 reductase also is capable of reducing cytochrome *P*-450 by measuring reductase-dependent benzphetamine demethylation in a complete reconstituted *P*-450 system.[1,19] The solution of purified reductase is purged with N_2, and aliquots are stored frozen at $-70°$. Partially purified *P*-450 reductase can also be used in the reconstituted system with hemoglobin; however, in all cases it is important to verify that the reductase preparations alone have neither contaminating cytochrome *P*-450 (assessed by carbon monoxide difference spectroscopy[20]) nor monooxygenase activity (assessed by testing for aniline hydroxylase activity in the absence of hemoglobin).

[16] A. Riggs, this series, Vol. 76, p. 5.
[17] E. J. Van Kampen and W. G. Zijlstra, *Clin. Chim. Acta* **6**, 538 (1961).
[18] E. A. Shephard, S. F. Pike, B. R. Rabin, and I. R. Phillips, *Anal. Biochem.* **129**, 430 (1983).
[19] M. J. Coon, J. L. Vermillion, K. P. Vatsis, J. S. French, W. L. Dean, and D. A. Haugen, *ACS Symp. Ser.* **44**, 46 (1976).
[20] T. Omura and R. Sato, *J. Biol. Chem.* **239**, 2370 (1964).

Optimal Conditions for Hemoglobin-Catalyzed Reactions with P-450 Reductase

As with any characterization of optimal conditions for an enzymatic system, the process is iterative and requires reconfirmation of the optimum for each parameter after other optima are found. Because reactivity was observed in the absence of the reductase enzyme, most experiments were set up as paired comparisons, i.e., in the presence and absence of reductase. The overall prototype reaction for characterization of the catalytic system was conversion of aniline to p-aminophenol. All other substrates were studied under the conditions established for aniline, except where indicated otherwise (below). (Some comparative studies with myoglobin substituted for Hb also have been done.[1,4]) The complete catalytic system consisted of purified human hemoglobin added to the reaction mixture in the oxidized state (described above), i.e., ferrihemoglobin (Hb^{3+}), NADPH added in excess, purified liver microsomal cytochrome P-450 reductase (described above), substrate, and atmospheric O_2.

Typical Reaction Mixtures

A total volume of 1 ml contains the following components at the final concentrations indicated: 20 mM potassium phosphate buffer, pH 6.8, 1 μM Hb^{3+}, 0.02 units P-450 reductase, the appropriate concentration of substrate (e.g., aniline), and 0.2 mM NADPH. Mixtures (0.9 ml) are prepared without NADPH and equilibrated for 2–3 min at 37° before adding NADPH (0.1 ml of 2 mM) to initiate the reaction. The mixtures are then incubated for an appropriate time (usually 30 min) within the linear portion of the time course of product formation. Then, 0.3 ml of 20% trichloroacetic acid (TCA) is added to terminate each reaction. Zero-time controls are prepared by adding the trichloroacetic acid before the NADPH. The amounts of product formed are assayed as described below for each different substrate.

Time Dependence

The initial rate of hydroxylation of the prototype substrate aniline by this system is linear for at least 30 min at 37°. The reactions of all other substrates reported below display similar time courses in the standard reaction system with Hb^{3+}, except for O-demethylation of p-anisidine, which displays a much shorter linear time course. Consequently, product formation from p-anisidine is assayed after terminating the reactions at 5 min. Certain substrates display a lag phase when HbO_2 is substituted for Hb^{3+} as the initial form of Hb added to the reaction mixture (see Dependence on Redox/Ligand State, below).

Temperature Dependence

The temperature of 37° is chosen to be compatible with the physiological state. The rate of the aniline hydroxylation reaction is directly proportional to temperature, and measurements over the range 30°–40° yield a typical Q_{10} value of 1.9.

Dependence on Hb Concentration and Native Structure

Hb^{3+} concentration was varied over the range 0–2 μM and a linear dependence was observed up to at least 1.2 μM. Hence in all subsequent studies 1.0 μM Hb (4 μM heme) was used. This linear concentration range is similar to that of the cytochromes P-450 in corresponding reconstituted systems. It was shown also that the catalytic activity of Hb is dependent on its native structure, i.e., no catalysis is observed with heat-denatured hemoglobin[1] (see also Dependence on Redox/Ligand State, below).

Dependence on O_2 Concentration and Inhibition by CO

Usually all reactions are carried out with the normal amount of dissolved O_2 associated with atmospheric O_2. With the prototype substrates aniline and cyclohexane,[1,4,8] reaction mixtures were studied that were bubbled with various gas combinations of O_2 plus N_2 or O_2 plus CO. With both substrates, greater product formation is observed (up to 70% increase) as the O_2 content is increased from 20% O_2/80% N_2 to 100% O_2. In contrast, both reactions are inhibited when CO, rather than N_2, is combined with the O_2; e.g., the reaction rates in O_2/CO (20/80) are >60% inhibited (<40% activity) compared to those in O_2/N_2 (20/80).

Dependence on NADPH Concentration

The hydroxylation of aniline by hemoglobin occurs to some extent with NADPH alone in the absence of the reductase. This phenomenon is observed also with the other substrates. When the Hb-catalyzed reactions in the presence of the reductase are studied over a broad range of NADPH concentrations, biphasic kinetic plots (v versus v/NADPH) are observed.[2] The biphasic nature of the curves is interpreted to reflect a high-efficiency (low K_m for NADPH) phase ascribable to reductase-mediated electron transfer and a low-efficiency phase due to direct action of NADPH. Consistent with the K_m of the reductase for NADPH < 10 μM,[21] the reductase-mediated component of the reactions appears to be saturated near 200 μM (0.2 mM) NADPH. The nonreductase contribution at 0.2 mM NADPH

---

[21] J. D. Dignam and H. W. Strobel, *Biochemistry* **16**, 1116 (1977).

is different for different substrates. For example, in the absence of reductase no product is detectable from aniline whereas 0.9 nmol/min/ml of *p*-aminophenol is produced from *p*-anisidine. For this reason product formation in the absence of reductase is subtracted from total product formation for *p*-anisidine at 0.2 mM NADPH and for N-methylaniline and benzphetamine at 1 mM NADPH (see Table II). In all cases the K_m for substrates observed for the reactions in the absence of the reductase is found to be identical to the substrate K_m for the reductase-coupled reaction. This observation indicates that the substrate K_m values reflect the interactions between the substrates and hemoglobin, and the reductase serves to catalyze electron transfer from NADPH to the hemoglobin during the course of the reactions analogous to its role in the *P*-450 system.

Dependence on Reductase Concentration

Two series of typical reaction mixtures are set up, except for the following. For one series the reductase enzyme is omitted; for the other the reductase content is varied over the range 0–0.05 units. The data for the difference in product formation (± reductase) gives a typical linear relationship on reductase content up to a maximum (≥ 0.015 units); more reductase does not increase the rate further. On this basis, 0.02 units of reductase are added to all reaction mixtures. (*Note:* The activity of *P*-450 reductase is sensitive to changes in ionic strength. The units reported in the above description correspond to micromoles of cytochrome *c* reduced per minute by the enzyme assayed in 0.3 M potassium phosphate buffer, pH 7.7[18]).

Dependence on pH

The effects of varying pH on the initial rates of aniline hydroxylation in the presence and absence of the reductase enzyme are somewhat different. Both relationships reflect an increased reactivity with decreased pH in the range pH 5–8. These results suggest that proton transfer may be an important component of the overall reaction scheme. Buffer composition and concentration are not critical. Thus, at the low end of the pH range, sodium acetate buffer is tested as well as potassium phosphate. At the high end, Tris-HCl is tested in addition to potassium phosphate. Changes in either buffer composition or concentration (20–100 mM) cause little, if any, deviation. All subsequent reaction mixtures are buffered at pH 6.8, because this value represents the point at which the effect of the reductase is maximal.[1] The activity at pH 7.4 is about 75% of that at pH 6.8.

Dependence on Redox/Ligand State of Hemoglobin

Hydroxylation and demethylation reactions catalyzed by Hb apparently differ mechanistically according to studies of the prototype substrates aniline and p-nitroanisole.[2,4,22] Two-substrate kinetic patterns were developed from studies of the variation in product formation as a function of aniline and oxygen concentrations (Fig. 2) or of p-nitroanisole and oxygen concentrations (Fig. 3).

The family of lines in Fig. 2A, when aniline concentration is varied at several fixed concentrations of O_2, intersects in the second quadrant between the negative x and the positive y axes. When O_2 is the variable substrate (Fig. 2B), the lines intersect on the y axis. These patterns taken together indicate that the aniline hydroxylation reaction occurs via a rapid equilibrium ordered-bisubstrate mechanism whereby aniline must bind to Hb before O_2.[22,23] The nitroanisole kinetic patterns (Fig. 3) are distinctly different from those for aniline, and they are consistent instead with a noninteracting random addition of O_2 and nitroanisole to Hb.[22,23]

One of the distinct advantages of studying monooxygenase catalysis with Hb is its stability in both the ferric and oxyferrous states, unlike cytochrome *P*-450 for which the unstable oxyferrous form cannot be isolated. Therefore, the interpretation of the two-substrate kinetic studies of the Hb reactions could be tested by examining the relative catalytic efficiencies of Hb^{3+} and HbO_2 in the complete system with NADPH and *P*-450 reductase. With aniline as substrate, there is a distinct lag phase with HbO_2 and the later linear portion of the time course parallels that with Hb^{3+}, which does not display a lag phase.[4] These data confirm that O_2 cannot be bound prior to aniline in the catalytic scheme. In contrast, no difference between Hb^{3+} and HbO_2 is observed when p-nitroanisole is the substrate, consistent with a random order of binding of O_2 and substrate in that case.[4,22]

Subunit Selectivity of Hb Catalysis

To evaluate the individual hydroxylase activities of the subunits within the intact Hb tetramer, a pair of converse valency hybrids [α^{3+}_2 (β^{2+}-CO)$_2$ and β^{3+}_2 (α^{2+}-CO)$_2$] are prepared in which the inhibitor carbon monoxide is present on either the α- or the β subunits, thereby blocking monooxygenase activity of those subunits. The hybrids are prepared by recombination and chromatographic purification of the appropriately ligated α and β subunits, and they are characterized by visible and NMR

---

[22] D. W. Starke, M.S. Thesis, Case Western Reserve University, Cleveland, OH (1984).
[23] I. H. Segal, "Enzyme Kinetics," p. 320. Wiley (Interscience), New York, 1975.

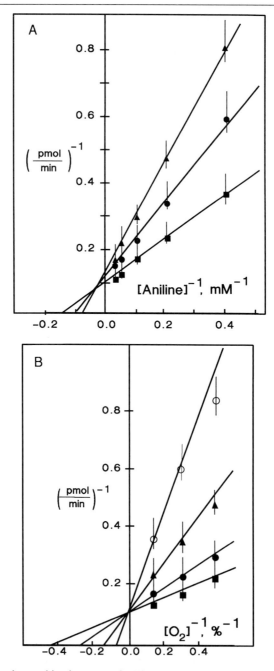

Fig. 2. Two-substrate kinetic patterns for Hb-catalyzed aniline hydroxylation (see legend, p. 583).

spectroscopy and isoelectric focusing. As shown in Table I,[24] the β sub-units in Hb are the much more efficient catalyst of aniline hydroxylation, i.e., V_{max}/K_m is more than 10-fold larger for the β-active hybrid than for the hybrid in which only the α subunit is available for catalysis. The basis for this subunit selectivity appears to be due to better access of the aniline molecule to the vicinity of the heme moiety on the β subunits compared with the α subunits, as shown by ^{1}H NMR T_1 relaxation studies.[25]

Assays of Specific Products

Quantitation of para-Hydroxylation Reaction Products

For aniline and aniline analogs, assays of para-hydroxylation are done according to a modification of the procedure of Brodie and Axelrod[26] in which the p-aminophenol congeners are converted to the corresponding indophenol chromophores for colorimetric assay. This method has been shown to be specific for p-aminophenol, i.e., o- and m-aminophenol do not give rise to the same blue color as does p-aminophenol.[27] We confirmed those observations and also showed that o,p-dihydroxyaniline does not react like p-aminophenol. In addition, p-aminophenol was confirmed as the predominant product in our reaction mixtures by combination gas chromatographic/mass spectroscopic analysis.[1]

Each reaction (constituted as described above) is terminated by adding 0.3 ml of 20% trichloroacetic acid to the 1-ml reaction mixture containing the p-aminophenol. The resultant suspension is centrifuged to remove precipitated protein, and 1 ml of the supernatant is transferred and combined with 0.1 ml of 5% phenol in 2.5 N NaOH (w/v); after mixing, 0.2

[24] B. L. Ferraiolo, G. M. Onady, and J. J. Mieyal, *Biochemistry* **23**, 5528 (1984).
[25] G. M. Onady, B. L. Ferraiolo, and J. J. Mieyal, *Biochemistry* **23**, 5534 (1984).
[26] B. B. Brodie and J. Axelrod, *J. Pharmacol. Exp. Ther.* **94**, 22 (1948).
[27] B. G. Bray and W. V. Thorepe, *Methods Biochem. Anal.* **1**, 27 (1954).

FIG. 2. Two-substrate kinetic patterns for Hb-catalyzed aniline hydroxylation. (A) Dependence of p-aminophenol formation on aniline concentration at various fixed O_2 concentrations. Samples were prepared as described in the text (see Two-Substrate Kinetic Studies) and contained from 2.5 to 40 mM aniline with O_2 at 20% (▲), 33% (●), or 66% (■) relative saturation. Product p-aminophenol was assayed as described in the text. Each symbol represents the mean ± S.E. for at least three determinations. (B) Dependence of p-aminophenol formation on O_2 concentration at various fixed aniline concentrations. Samples contained O_2 at percent saturations from 20 to 66% with aniline concentrations at 5 mM (○), 10 mM (▲), 20 mM (●), and 40 mM (■). Product p-aminophenol was assayed as described in the text. Each symbol represents the mean ± S.E. for at least three determinations.

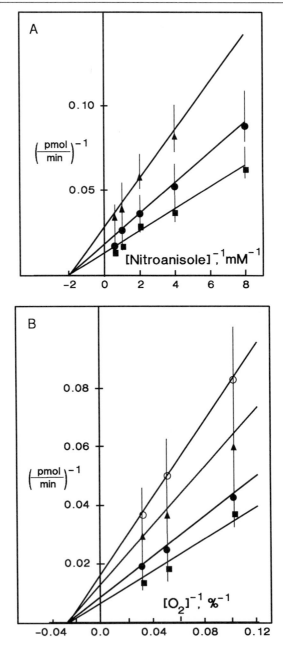

FIG. 3. Two-substrate kinetic patterns for Hb-catalyzed p-nitroanisole O-demethylation (see legend, p. 585).

ml of 2.5 N Na_2CO_3 is added. The resultant mixture is allowed to stand at room temperature for at least 30 min. Extinction coefficients of 14.8 mM^{-1} cm^{-1} at 630 nm for p-aminophenol, 18.0 mM^{-1} cm^{-1} at 630 nm for 4-hydroxy-N-methylaniline, and 8.8 mM^{-1} cm^{-1} at 640 nm for 4-hydroxy-2-methylaniline are obtained from standard curves of the individual products. The indophenol derivative of the product of hydroxylation of m-toluidine absorbs maximally at 620 nm. If an authentic sample of 4-hydroxy-3-methylaniline is not available, the extinction coefficient is estimated as the arithmetic mean of the values for p-aminophenol and 4-hydroxy-2-methylaniline, i.e., 11.8 mM^{-1} cm^{-1}.

Quantitation of Demethylation Reaction Products

Assays of formaldehyde formed via the demethylation of benzphetamine, N-methylaniline, or the various anisole derivatives are performed by a modification of the assay of Nash.[28] In all but the benzphetamine assay the buffer is 20 mM potassium phosphate, pH 6.8. To improve the solubility range for benzphetamine, 0.1 M Tris-HCl, pH 6.8, is used. It is confirmed separately that equivalent results are obtained for demethylation of lower concentrations of benzphetamine in potassium phosphate buffer. After obtaining the 1 ml of supernatant of the terminated reaction mixture, 0.4 ml of a solution containing 0.4% acetylacetone in 4 M ammonium phosphate is added. Samples are mixed and kept at 37° for 45 min (the color develops more rapidly at higher temperatures, but it also decays more rapidly). After the samples are cooled, their absorbances at 412 nm are measured. From standard curves, the extinction coefficient for the formaldehyde–acetylacetone derivative is determined to be 3.8 mM^{-1} cm^{-1}.

[28] T. Nash, *Biochem. J.* **55**, 416 (1953).

FIG. 3. Two-substrate kinetic patterns for Hb-catalyzed p-nitroanisole O-demethylation. (A) Dependence of formaldehyde formation on p-nitroanisole concentration at various fixed O_2 concentrations. Samples were prepared as described in the text (see Two-Substrate Kinetic Studies) and contained from 0.1 to 1.6 mM p-nitroanisole with O_2 at 10% (▲), 20% (●), or 33% (■) relative saturation. Product formaldehyde was assayed as described in the text. Each symbol represents the mean ± S.E. for at least three determinations. (B) Dependence of formaldehyde formation on O_2 concentration at various fixed p-nitroanisole concentrations. Samples contained O_2 at percent saturations from 10 to 33% with p-nitroanisole concentrations at 0.25 mM (○), 0.5 mM (▲), 1.0 mM (●), and 2.0 mM (■). Product formaldehyde was assayed as described in the text. Each symbol represents the mean ± S.E. for at least three determinations.

TABLE I
KINETIC VALUES FOR ANILINE HYDROXYLATION BY
HEMOGLOBIN DERIVATIVES[a]

Hemoglobin derivative[b]	V_{max} (pmol/min/ml)	K_m (mM)	V_{max}/K_m
$\alpha^{3+}_2 \beta^{3+}_2$	77	6	12.8
$(\alpha^{2+}\text{-CO})_2 \beta^{3+}_2$	89	5	17.8
$\alpha^{3+}_2 (\beta^{2+}\text{-CO})_2$	128	102	1.3

[a] The linear portions of double-reciprocal plots ($1/v$ vs. $1/$[aniline]) were extrapolated by linear regression analysis to obtain the V_{max} and K_m values.[24]

[b] All hemoglobin derivatives were purified and characterized as described,[24] and tested at 4 μM concentration with respect to total heme.

Demethylation of N-methylaniline by erythrocytes (see below) is measured by assaying for aniline. The samples and standards (prepared by adding a known amount of authentic aniline to reaction mixtures at zero time) are processed in the same fashion as described for other assays except that the reagent for development of color is a modified Ehrlich's reagent,[29] consisting of 0.5% p-dimethylaminobenzaldehyde in a 5 M sodium formate buffer, pH 3.75. Aniline has a much higher extinction coefficient at 445 nm ($\varepsilon = 10.0$ mM^{-1} cm^{-1}) than N-methylaniline ($\varepsilon = 0.02$ mM^{-1} cm^{-1}) and can easily be detected in the concentration range of interest by subtracting out the absorbance due to the N-methylaniline (i.e., the absorbance of a control sample incubated at 0°). The resulting absorbance difference is then corrected for the average recovery of aniline (85%), which is determined separately. N-Methylaminophenol at a concentration of 20 μM does not give a measurable reaction with this reagent.

Quantitation of Cyclohexane Hydroxylation

Each sample contains 20 mM potassium phosphate buffer, pH 6.8, 1 mM NADPH, 0.2–8 mM cyclohexane, and 0.02 units P-450 reductase in a 1-ml total volume. Mixtures are preincubated for 5 min at 37°, and then the reactions are initiated by addition of Hb^{3+} (final concentration, 1 μM). Controls omitting NADPH are incubated along with the samples. After 25 min of reaction time, a 20-μl aliquot of 250 μM cyclopentanol (internal standard) is added to the solution; 5 min later each reaction is terminated by shaking with 1 ml of ice-cold ether and immersion of the

[29] G. S. Estus and J. J. Mieyal, *Drug Metab. Dispos.* **11**, 471 (1983).

test tube into ice. NADPH is then added to the controls. Each sample is vortexed twice for 30 sec and then centrifuged at top speed in a clinical centrifuge for about 1 min. The ether layer is removed and placed on ice, then evaporated to about 50 μl by a gentle stream of nitrogen. Aliquots (5 μl) of the concentrated ether solutions are then analyzed by gas chromatography [10% Carbowax 20M/Diatoport S column (24 × 0.25 inches) with the injector and detector ports maintained at 250° and column temperature programmed from 40° to 200° at 10° per min; flow rates of He, H_2, and air are 30, 30, and 200 ml/min, respectively]. The extraction efficiencies for cyclohexanol and cycloheptanol are determined to be about 95%. Although controls lacking NADPH are used routinely (described above) and show no cyclohexanol formation, it has been confirmed separately that no cyclohexanol product is detectable in alternative controls in which Hb or cyclohexane is omitted instead.

Two-Substrate Kinetic Studies of Aniline Hydroxylation and P-Nitroanisole O-Demethylation

Reaction mixtures (1 ml) containing 2.5–40 mM aniline or 0.05–2.0 mM p-nitroanisole, 20 mM potassium phosphate (pH 6.8), and 0.02 units of P-450 reductase are prepared in degassed water. The samples are then injected into individual vacutainer evacuated tubes. Samples are flushed for an appropriate time with an appropriate mixture of O_2 and N_2 (controlled by Matheson No. 610 flowmeters) to achieve the desired percent O_2 saturation (analyzed as described below). After O_2 content is adjusted, 50 μl of 20 μM Hb^{3+} (in degassed water) is injected into the samples and they are equilibrated for 5 min at 37°. Then reactions are initiated by injection of 50 μl of 4 mM NADPH (in degassed water). After 30 min reactions are terminated by injection of 0.3 ml of 20% trichloroacetic acid and centrifugation. The tops of the vacutainer tubes are removed, and 1 ml of each sample supernatant is transferred to a clean tube and assayed for p-aminophenol or formaldehyde as described above.

Measurement of O_2 Content of Reaction Mixtures

For each set of samples that are gassed with different O_2/N_2 mixtures, an additional duplicate sample with a midrange substrate concentration is prepared in parallel with the other samples and analyzed for percent O_2 saturation as follows. The oxygen electrode (Yellow Springs Instruments, Yellow Springs, OH) is calibrated by introducing air-saturated 20 mM potassium phosphate buffer into the sample chamber and adjusting the meter to 20% oxygen saturation. The readings of the representative samples are then made relative to this standard. Percent oxygen saturation readings are stable for all of the representative samples, transferred from their vacutainer tubes, except for those that have been flushed with pure O_2. For these the percent O_2 reading decreases slowly, and an extrapolated estimate of % O_2 saturation is necessary.

Relative Efficiency of Hemoglobin-Catalyzed Oxidation of Various Substrates

Hb in the reconstituted system with P-450 reductase can catalyze the full range of monooxygenase-like reactions, including aromatic and aliphatic hydroxylation as well as heteroatom oxidation and N- and O-dealkylation. Complete kinetic analysis has been carried out for various Hb-catalyzed reactions of a number of typical monooxygenase substrates and structural analogs (Table II).[1,3,8] The K_m and V_{max} values vary broadly so that the efficiencies of the reactions (V_{max}/K_m) encompass a range greater than 40,000.

p-Hydroxylation Reactions

With regard to the toluidines, positioning of the methyl group on the aniline ring showed a varied effect on the p-hydroxylation reaction relative to unsubstituted aniline (Table II). A methyl group ortho to the amine increased the K_m more than 10-fold without an effect on V_{max}. A m-methyl group apparently increased K_m further along with a large decrease in maximal activity. With N-methylaniline two reactions occur, p-hydroxylation and N-demethylation, and these display separate K_m and V_{max} values (see below). N-Methyl substitution affected the apparent K_m for p-hydroxylation slightly but decreased V_{max} approximately 2.5-fold relative to aniline. This may be due in part to the competing N-demethylation reaction, which proceeds much more efficiently.

Because phenol resembles aniline in size and electronic properties, it was also tested as a substrate, but complete and reliable kinetic analysis could not be accomplished, because phenol caused denaturative precipitation of hemoglobin. The estimated maximal rate of p-hydroquinone formation (measured by HPLC[22]) appeared to be similar to p-aminophenol formation from aniline, but the K_m for phenol appeared to be >30 mM (data not shown).

N-Demethylation Reactions

Benzphetamine and n-methylaniline have some molecular features in common, but they are quite different in molecular size and the nature of the nitrogen atom involved in the reaction. Benzphetamine is a bulky molecule with two aromatic rings situated at opposite ends of the molecule, whereas N-methylaniline is more compact. Benzphetamine is a tertiary aliphatic amine, whereas N-methylaniline is a secondary aryl amine. In spite of these differences, the K_m and V_{max} values are similar (Table II), and both reactions are highly efficient relative to the other reactions.

Dual Metabolism of N-Methylaniline

The fact that separate K_m values were obtained for the p-hydroxylation and N-demethylation reactions of the single substrate N-methylaniline is consistent with different ternary O_2–Hb–substrate complexes mediating the two different reactions.[2] At least two explanations for this phenomenon are conceivable. (1) There may be a slow conformational equilibrium between forms of the hemoprotein that favor different modes of substrate binding. One mode would favor N-demethylation and the other would favor p-hydroxylation. (2) There may be two separate preexisting noninteracting substrate-binding sites on the hemoprotein molecule.

O-Demethylation Reactions

The O-demethylation of p-anisidine exhibited the highest maximal activity. Its high K_m value, however, limits the reaction efficiency. p-Nitroanisole, a prototypic P-450 substrate, gave the lowest K_m of all the substrates and displayed the greatest efficiency. A Hammet-type free energy analysis of the kinetic data for anisole homologs (p-NH$_2$, p-OH, p-H, p-NO$_2$) showed a linear relationship to the electronic inductive effects of the para substituents.[2] The V_{max} of the O-demethylation reaction is not sensitive to the inductive effect, whereas the K_m appears to be influenced markedly (Table II).

N-Hydroxylation of p-Chloroaniline

Although the study of p-chloroaniline is not strictly comparable to the others because rabbit rather than human Hb was used and at a higher concentration, it did demonstrate that Hb can catalyze the N-hydroxylation reaction when coupled with P-450 reductase or with the methemoglobin reductases.[3] The efficiency of this reaction, however, is quite low.

Substrate-Accelerated Autoxidation of HbO$_2$

Previously we observed that pure aniline accelerated the autoxidation of HbO$_2$ to Hb^{3+} in a concentration-dependent manner that was coincident with the concentration dependence on aniline for p-aminophenol formation in the complete monooxygenase catalytic system, i.e., K_m (autoxida-

TABLE II
KINETICS OF MONOOXYGENASE-LIKE REACTIONS OF HEMOGLOBIN[a]

Substrate	Reaction	V_{max} (min^{-1})	K_m (M)	V_{max}/K_m
Aniline	p-Hydroxylation	0.064	0.0057	11.2
o-Toluidine	p-Hydroxylation	0.068	0.0654	1.0
m-Toluidine	p-Hydroxylation	0.004[b]	0.087[b]	0.05
N-Methylaniline[c]	p-Hydroxylation	0.025	0.0036	6.9
	N-demethylation	0.164	0.00052	315.0
Benzphetamine[c]	N-Demethylation	0.115	0.00042	268.0
p-Anisidine	O-Demethylation	0.658	0.0364	18.1
Anisole	O-Demethylation	0.439	0.00089	493.0

(continued)

TABLE II (continued)

Substrate	Reaction	V_{max} (min^{-1})	K_m (M)	V_{max}/K_m
NO_2—◯—OCH_3 p-Nitroanisole	O-Demethylation	0.211	0.000094	2244.0
CH_3O—◯—OH p-Methoxyphenol	O-Demethylation	0.320	0.0045	71.1
Cl—◯—NH_2 p-Chloroaniline[d]	N-Hydroxylation	0.003	0.0059	0.51
⬡ Cyclohexane[c]	C-Hydroxylation	0.052	0.001	52.0

[a] Reaction conditions for all substrates were the same as those described in the text, except as noted herein.

[b] These values are estimates, because the limited reactivity and solubility of this substrate precluded more accurate determination.

[c] Only 1 mM NADPH was used in the reaction mixtures in these cases. In other cases in which both 0.2 mM and 1.0 mM NADPH were tested, the amount of product formation associated with the reductase-coupled reaction was the same; i.e., only the component of the total rate corresponding to that observed in the absence of reductase was proportionately higher with 1 mM NADPH.

[d] p-Chloroaniline N-hydroxylation was studied by Golly and Hlavica[3] under substantially different conditions from those used by us[1,8] for all of the other substrates. They used rabbit Hb^{3+} at 3.8 μM, 1.2 mM NADP with an NADPH generating system, and 0.15 M potassium phosphate buffer, pH 7.4. In this context, we previously found that the aniline hydroxylase activity of rabbit hemolysates was comparable to that of human hemolysates.[30]

tion) = K_m (metabolism).[31] In separate experiments a number of the other substrates also were found to accelerate HbO_2 autoxidation with concentration dependences parallel to the corresponding K_m values for their metabolic reactions.[2] The phenomenon, however, is not a prerequisite for metabolism by the Hb monooxygenase system, because a number

[30] B. L. Ferraiolo and J. J. Mieyal, Mol. Pharmacol. 21, 1 (1981).
[31] J. J. Mieyal and J. L. Blumer, J. Biol. Chem. 251, 3442 (1976).

of substrates (e.g., benzphetamine, anisole, p-nitroanisole), did not accelerate HbO_2 autoxidation, yet they were metabolized efficiently (Table II). The simplest view of the interaction of Hb with substrates involves the intermediate formation of one or more ternary substrate–Hb–O_2 complexes. Such complexes may then proceed to react in two ways formally; namely, via transfer of electrons from the ferrous iron atom to oxygen (autoxidation) or via the acceptance of an exogenous electron (from NADPH and reductase) in order to complete the monooxygenase process of formation of hydroxylated product (metabolism). The kinetic data presented for the various substrates suggest that distinct and noninterconvertable ternary complexes are formed with different properties for autoxidation and different modes of metabolism.[2] The enhancement of autoxidation of HbO_2 within the catalytic system would be akin to the uncoupling of O_2 and NADPH utilization from substrate oxidation observed for various substrates in the analogous cytochrome P-450 monooxygenase systems.[32]

C-Hydroxylation of Cyclohexane

All of the substrates for the various Hb-catalyzed reactions so far described have in common the presence of unsaturated bonds or heteroatoms that may facilitate the reactions either by serving as oxidation sites or by stabilizing the intermediates that lead to the oxidized product. Cyclohexane is a prototype aliphatic P-450 substrate that has no double bonds or adjacent heteroatoms, and therefore its hydroxylation may be expected to require a more active oxidant. In fact, it was suggested that catalysis of aliphatic hydroxylation might be limited among hemoproteins to cytochrome P-450 due to mechanistic differences in the activation of oxygen.[32] As shown in Table II, however, we found that Hb can catalyze the hydroxylation of cyclohexane in an *in vitro* system analogous to reconstituted cytochrome P-450 systems.[8] In separate experiments, it was ascertained that the cyclohexane hydroxylation reaction was not inhibited either by 100 mM dimethyl sulfoxide, a known hydroxyl radical scavenging agent, or by 1 mM desferrioxamine, a potent chelator of iron.[8] Thus, the possibility of nonenzymatic formation of cyclohexanol in this system by hydroxyl radicals generated in solution was obviated. The K_m and V_{max} values for the cyclohexane reaction fall within the range observed for the other substrates of Hb. Thus, the qualitative similarities between Hb

---

[32] R. E. White and M. J. Coon, *Annu. Rev. Biochem.* **49,** 315 (1980).

TABLE III
PROTOTYPE MONOOXYGENASE REACTIONS OF HUMAN HEMOGLOBINS AND RABBIT LIVER
CYTOCHROMES P-450

	Turnover numbers (min^{-1})				
	Hemoglobin		Cytochrome P-450		Ratio of P-450 to Hb
Reaction	Hb A	Hb F	IIB4 (LM$_2$)	IA2 (LM$_4$)	
p-Hydroxylation of aniline	0.06[a]	1.5[b]	1.0[c]	0.4[c]	0.3–17
N-Demethylation of benzphetamine	0.12[d]		56.8[c]	4.9[c]	41–473
O-Demethylation of p-nitroanisole	0.21[d]		5.7[c]	3.4[c]	16–27
Hydroxylation of cyclohexane	0.05[e]		30.0[f]		600

[a] See Mieyal et al.[1]
[b] See Blisard and Mieyal.[33]
[c] See Coon et al.[19]
[d] See Starke et al.[2]
[e] See Starke and Mieyal.[8]
[f] Reported values for the turnover number for cyclohexanol formation by a number of preparations of the well-characterized rabbit liver P-450 isoenzyme (IIB4, previously known as LM$_2$) range from 9 to 50 min^{-1}, but 30 min^{-1} is considered typical, and a K_m of 0.3 mM is considered typical.[34]

and cytochrome P-450 activities extend to aliphatic hydroxylation, but quantitative differences are evident (Table III).

Comparison of Hemoglobin and P-450 Activities

Table III shows a direct comparison of the turnover numbers for a number of reconstituted Hb and cytochrome P-450 systems using several prototype substrates. The relative activities of Hb A and Hb F are compared with two forms of P-450 from rabbit liver that have been studied extensively.

This comparison shows that the P-450s display the catalytic advantage in most cases, except for aniline hydroxylation by fetal Hb. The most pronounced differences are seen with benzphetamine demethylation and cyclohexane oxidation, where the differences in V_{max} values approach 500-fold or greater. These intrinsic activity differences are likely due to one or both of the following possibilities: (1) The closeness of approach

[33] K. S. Blisard and J. J. Mieyal, Biochem. Biophys. Res. Commun. **96**, 1261 (1980).
[34] G. D. Nordbloom and M. J. Coon, Arch. Biochem. Biophys. **180**, 343 (1977); M. McCarthy and R. E. White, J. Biol. Chem. **258**, 9153 (1983); D. R. Koop, personal communication (1989).

or relative orientation of the bound substrate to the bound active oxygen creates a more reactive situation near the heme site on P-450 compared to Hb, i.e., the "active site" of Hb may not be optimal for monooxygenase function. (2) These two hemoproteins differ in their intrinsic ability to activate oxygen. White and Coon[32] postulated that the thiolate proximal heme ligand of P-450 plays a primary role in the activation of oxygen. Accordingly, the lower activities of Hb would be predicted based on its histidine ligand. Although both histidinyl and thiolate ligands have electron pairs, the sulfur moiety is more polarizable and may be expected to stabilize a higher iron oxidation state more effectively.

Hydroxylation and Demethylation Reactions by Hemoglobin in Intact Erythrocytes

Reaction Mixtures and Product Extraction Procedures

Hydroxylation and demethylation reactions in intact erythrocytes[2] are performed as developed in detail for aniline hydroxylation.[35] Oxyhemoglobin (HbO_2) is shown to be the catalyst by experiments in which the intraerythrocytic HbO_2/Hb^{3+} ratio is varied.[35] Reaction mixtures contain a quantity of erythrocytes corresponding to 1 mM concentration with respect to HbO_2. Each substrate is present at a concentration of 20 mM. This concentration is chosen both because of solubility limitations and to keep the amount of substrate-induced hemolysis at less than 5%. Incubations are carried out for 60 min at 37°, and the reactions are terminated by cooling in ice (trichloroacetic acid precipitation is not used, because the oxidative burst associated with rapid denaturation of the 1 mM HbO_2 can cause artifactual product formation). Products are extracted with purified diethyl ether and back extracted into dilute HCl solutions as described previously.[35] The products of p-hydroxylation of the aniline derivatives are assayed colorimetrically and quantitated relative to standard curves as described above, except that each extinction coefficient is multiplied by an appropriate recovery percentage. Average recovery for each compound is calculated from the ratio of the absorbances of the indophenol derivative of a known amount of the compound that had been carried through the extraction procedure and the same concentration of the compound that had been assayed directly. These values are determined over a wide range of concentrations, and are as follows: p-aminophenol, 50%; 4-hydroxy-2-methylaniline, 60%; recovery of 4-hydroxy-3-methylaniline is taken as the average for p-aminophenol and 4-hydroxy-2-methylaniline,

[35] K. S. Blisard and J. J. Mieyal, *J. Biol. Chem.* **254**, 5104 (1979).

i.e., 55%. Recoveries are found to be reproducible to within 6%. Because N-methylaminophenol is observed to decay relatively rapidly in the presence of erythrocytes, its "recovery" is estimated to be approximately 60% by the following procedure. Sequential additions whereby a known amount of N-methylaminophenol is added to the incubation mixture in five equal aliquots at 0, 15, 30, 45, and 60 min are utilized to simulate product formation from N-methylaniline. After 60 min the reaction mixture is cooled, extracted, and assayed as usual. Decay of N-methylaminophenol is not a significant problem in assays involving the reconstituted system because a shorter time course is monitored.

Verification of Reaction Products

High-performance liquid chromatography is used to confirm that the products of hydroxylation of aniline, o-toluidine, N-methylaniline, and phenol coelute with authentic samples of the expected products.[2,36] A μBondapak C_{18} column (Waters Associates, Inc., Milford, MA) is eluted with 20% acetonitrile in water at a flow rate of 1 ml/min. The samples for analysis are extracted twice with ether and the ether fractions are evaporated to dryness (except for N-methylaniline, for which a second extraction into 1% acetic acid is used for injection). The residues are redissolved in the mobile-phase solution and a 25-μl aliquot is injected. Retention times are as follows: o-toluidine, 8.0 min; 4-hydroxy-2-methylaniline, 5.9 min; N-methylaniline, 30 min; N-methylaminophenol, 6.6 min; phenol, 5.7 min; hydroquinone, 2.8 min. Standard curves are constructed for these assays in the same way as for the colorimetric assays. Aniline, o-toluidine, and N-methylaniline cochromatograph with authentic standards, and quantitative agreement with the colormetric assays is obtained. Gas chromatography is used to confirm that the metabolism of N-methylaniline by erythrocytes gives a product that coelutes with authentic aniline.[2,36] Aniline produced by erythrocyte metabolism of N-methylaniline cochromatographs with authentic aniline, and the rate of metabolism agrees with that measured by the colorimetric assay.

Substrate Specificity of Hemoglobin within Erythrocytes

A number of substrates were tested for metabolism by hemoglobin within intact erythrocytes. The relative rates of metabolism at the same concentration (20 mM) of each of the substrates are shown in Table IV.

---

[36] K. S. Blisard, Ph.D. Thesis, Case Western Reserve University, Cleveland, OH (1980).

TABLE IV
MONOOXYGENASE-LIKE ACTIVITIES
OF HUMAN ERYTHROCYTES[a]

Substrate	Reaction type	Activity (pmol/min/ml)
Aniline	p-Hydroxylation	60.0 ± 3
o-Toluidine	p-Hydroxylation	50.0 ± 3
m-Toluidine	p-Hydroxylation	26.6 ± 3
Phenol	p-Hydroxylation	49.0 ± 4
N-Methylaniline	p-Hydroxylation	16.7 ± 2
	N-Demethylation	640.0 ± 0

[a] Each sample contained 20 mM substrate and erythrocytes in isotonic saline at 1 mM with respect to HbO$_2$ concentration; incubations were for 60 min at 37°; reactions were stopped by cooling on ice and extracting products into ether.[35,36]

For those substrates that were metabolized by the erythrocytes (Table IV) the turnover numbers are obviously much lower than those for the reconstituted system with isolated hemoglobin, because here the Hb concentration was 1 mM rather than 1 μM. Furthermore, unlike the reconstituted system, conversion of HbO$_2$ to Hb^{3+} within the red blood cells inhibited the monooxygenase-like activity.[35] Nevertheless, several similarities in the metabolism of the substrates are evident from Table IV. The relative rates of p-hydroxylation reflect an order comparable to the reconstituted system, i.e., aniline $\sim$ o-toluidine > N-methylaniline. Also, N-methylaniline is dually metabolized and a similar regiospecificity is displayed, N-demethylation > p-hydroxylation. In contrast, two anisole derivatives, p-anisidine (20 mM) and p-nitroanisole (1 mM), give little (if any) of the corresponding demethylated products (data not shown), even when the system was set up for maximal stimulation of monooxygenase activity by the addition of methylene blue and glucose.[35] This result demonstrated another distinct qualitative difference between the reactivity of hemoglobin in its native environment compared with isolated hemoglobin in the reconstituted system with P-450 reductase. The differences in catalytic efficiency of Hb in erythrocytes are likely due to the inefficient NADPH electron transport system in erythrocytes,[37] thereby probably requiring superoxide as a mediator of electrons for the metabolic reactions. Thus the erythrocyte system may behave more like the isolated Hb system when P-450 reductase is omitted (see below).

[37] K. S. Blisard and J. J. Mieyal, Arch. Biochem. Biophys. 210, 762 (1981).

Mechanism of NADPH-Mediated Metabolism by Hb
in Absence of Reductase

The finding that NADPH supports Hb-catalyzed monooxygenase-like reactions at all in the absence of the reductase enzyme indicates that it can directly reduce methemoglobin. It is known that NADPH reduces ferricyanide,[38] and it has been reported that it reduces methemoglobin.[39] We have confirmed the latter observation for our conditions and preparations of Hb, and have shown that reduction of methemoglobin is enhanced by the P-450 reductase enzyme in the quantities used for the reconstituted catalytic system.[1] It is generally believed that direct hydride ($2e^-$) transfer from NADPH is involved in dehydrogenase reactions.[38] It seems likely, however, that reduction of $1e^-$ acceptors such as methemoglobin would occur via $1e^-$ transfer from NADPH, and this necessarily means that unstable free radicals would be formed in these systems, e.g., $NADPH^+$, O_2^-, OH·. It follows that one of these radical species, probably O_2^-, which is the most stable, donates the second electron to the substrate–Hb–O_2 ternary complex to activate it [see step 4 in the monooxygenase cycle (Fig. 1)]. This conclusion is consistent with inhibition data. Thus, the NADPH-mediated reaction is very sensitive to superoxide dismutase, but much less sensitive to scavengers of other radicals.[1] In contrast, the reductase-coupled reactions are not sensitive to superoxide dismutase or to other radical scavengers or iron chelators.[1,8]

Summary

Red blood cells contain many enzymes that are akin to those that catalyze xenobiotic metabolism in liver and other tissues. An obvious exception is the cytochrome P-450 system that is found in virtually all other tissues. *In vitro* studies, however, have shown that hemoglobin can be a broad monooxygenase catalyst, exhibiting the properties of a monooxygenase enzyme. Thus, catalysis by Hb displays typical Michaelis–Menten kinetics, dependence on the native protein, coupling to NADPH-dependent flavoprotein reductases, and inhibition by carbon monoxide. The reconstituted system containing Hb along with P-450 reductase utilizes NADPH and O_2 to catalyze typical monooxygenase reactions, including O- and N-demethylations as well as aromatic and aliphatic hydroxylations, and the catalytic cycle appears to mimic the typical P-450 mechanism. Turnover numbers for aniline hydroxylation are similar for Hb and P-450 reconstituted systems, whereas P-450 systems are more

[38] S. P. Collowick, J. van Eys, and J. H. Park, *Compr. Biochem.* **14**, 1 (1966).
[39] W. D. Brown and H. E. Snyder, *J. Biol. Chem.* **244**, 6702 (1969).

effective for other reactions. Catalysis by Hb seems to be restricted to the β-heme sites of the tetramer, reflecting more facile substrate access. Overall the similarities and differences between Hb and P-450 provide an opportunity to examine the basis for their differential monooxygenase or peroxidase/peroxygenase activities in a comparative manner. Hb may be especially useful in delineating the early events in the respective reaction schemes, because it can be studied in various stable redox/ligand states, including the oxyferrous form.

Similar hemoglobin-catalyzed oxidative biotransformations occur within intact erythrocytes, but apparent turnover numbers are much lower than those with the reconstituted Hb system, suggesting different mechanisms of catalysis. Although Hb-mediated oxidase activity in erythrocytes is low relative to other sites of xenobiotic metabolism, it may contribute to *in situ* activation of xenobiotics leading to oxidative stress, disruption of sulfhydryl homeostasis in the erythrocytes, covalent modification of Hb, and hemolysis.

Acknowledgments

Besides citations of their coauthorship in the reference list, grateful recognition is given to our former co-workers who contributed importantly to the work reviewed in this article, especially Karen S. Blisard, Jeffrey L. Blumer, Bobbe L. Ferraiolo, and Gary M. Onady. The work was supported in part by grants from NIH (R01 GM 20050 and GM 24076), from the American Heart Association (881186 and 91014450), and from the Rockefeller Foundation.

[39] Oxidation of Olefins Catalyzed by Hemoglobin

By Giorgio Belvedere and Michele Samaja

Besides its properties as an O_2 carrier, hemoglobin (Hb) has several enzymatic activities.[1-3] For example, it catalyzes the hydroxylation and the demethylation of various xenobiotics,[4-6] the binding of benzo[a]py-

[1] J. Everse, this volume [34].
[2] J. J. Mieyal and D. W. Starke, this volume [38].
[3] L. Zilletti, M. Cuiffi, S. Franchi-Micheli, F. Fusi, G. Gentilini, G. Maneti, and G. P. Sgaragli, this volume [37].
[4] B. L. Ferraiolo and J. J. Mieyal, *Mol. Pharmacol.* **21,** 1 (1982).
[5] D. W. Starke, K. S. Blisard, and J. J. Mieyal, *Mol. Pharmacol.* **1,** 25 (1984).
[6] O. Augusto, K. L. Kunze, and P. R. Ortiz de Montellano, *J. Biol. Chem.* **257,** 6231 (1982).

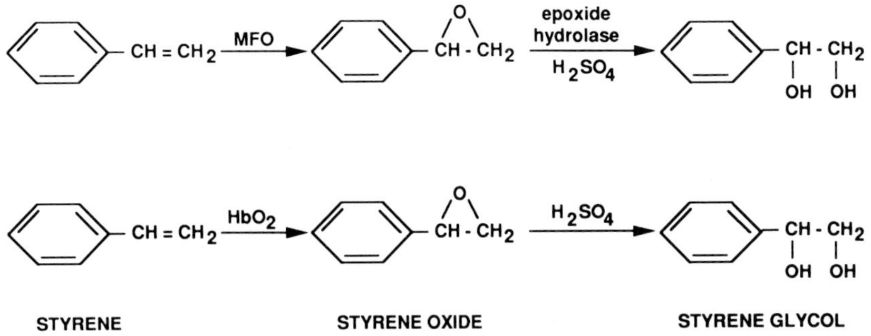

FIG. 1. Reaction of styrene catalyzed by mixed-function oxidases (MFO) and oxygenated hemoglobin (HbO$_2$). The *in vivo* metabolism of styrene oxide includes conversion to styrene glycol catalyzed by epoxide hydrolase. The *in vitro* assay was performed measuring styrene glycol after quantitative hydrolysis with H$_2$SO$_4$.

rene,[7] the oxidation of *trans*-7,8-dihydroxy-7,8-dihydrobenzo[a]pyrene to anti-*trans*-7,8,9,10-tetrahydroxy-7,8,9,10-tetrahydrobenzo[a]pyrene,[8] as well as the oxidation of olefins to olefin oxides.[8-11] Styrene is an olefin monomer used widely in the production of plastics, resins, and synthetic rubber. The first step of styrene metabolism in the organism, the conversion to styrene oxide by mixed-function oxidases (MFOs), occurs mainly in the liver and is followed by further conversion of styrene oxide to styrene glycol, which is catalyzed by epoxide hydrolase.[12] To characterize the properties of the various cytochrome *P*-450-dependent MFOs involved in the oxidation of styrene to styrene oxide, a method was developed to determine styrene oxide produced during incubation of styrene with rat liver microsomes in the form of styrene glycol (Fig. 1).[13,14]

In a study on the oxidation of styrene catalyzed by human lymphocyte MFOs, it was observed that the production of styrene oxide was greatly increased when cells were contaminated with blood.[9] This

[7] R. H. Rice, Y. M. Lee, and W. D. Brown, *Arch. Biochem. Biophys.* **221,** 417 (1983).

[8] C. E. Catalano and P. R. Ortiz de Montellano, *Biochemistry* **26,** 8373 (1987).

[9] G. Belvedere and F. Tursi, *Res. Commun. Chem. Pathol. Pharmacol.* **33,** 273 (1981).

[10] F. Tursi, M. Samaja, M. Salmona, and G. Belvedere, *Experientia* **39,** 593 (1983).

[11] P. R. Ortiz de Montellano and C. E. Catalano, *J. Biol. Chem.* **260,** 9265 (1985).

[12] K. C. Leibman and E. Ortiz, *J. Pharmacol. Exp. Ther.* **173,** 242 (1970).

[13] G. Belvedere, J. Pachecka, L. Cantoni, E. Mussini, and M. Salmona, *J. Chromatogr.* **118,** 387 (1976).

[14] M. Salmona, J. Pachecka, L. Cantoni, G. Belvedere, E. Mussini, and S. Garattini, *Xenobiotica* **6,** 585 (1976).

necessitated a characterization of the enzymatic reaction catalyzed by erythrocytes.

Styrene Oxidation to Styrene Oxide Catalyzed by Oxygenated Erythrocytes

Methods

Preparation of Erythrocytes and Lysates. Human venous blood suitable for transfusion [with 0.68% (w/v) citrate as anticoagulant] is obtained from local blood banks. Blood is centrifuged (100 g) for 10 min at room temperature. The buffy coat is removed and the erythrocyte pellet is resuspended in twice the volume of 154 mM NaCl and then centrifuged at 400 g for 20 min. Erythrocytes are then washed once with phosphate-buffered saline (PBS) containing no Ca^{2+} or Mg^{2+}, centrifuged, and counted with standard methods. The suspension is adjusted to 500×10^6 cells/ml with PBS.

To obtain the lysate, washed erythrocytes are frozen rapidly in acetone/CO_2 and thawed four times at room temperature; the resulting samples are then centrifuged at 10,000 g for 10 min and the supernatant is used for the experiments.

Tonometry. Tonometry is an operation aimed at equilibrating a liquid sample with a gas of known composition to establish a known partial pressure of a gas in a liquid. The gas usually contains O_2 and N_2 in various amounts to obtain the desired partial pressure of O_2 (P_{O_2}) in the liquid. The composition depends on temperature, the kind of preparation (erythrocytes or Hb), and the desired percentage of HbO_2 (S_{O_2}). For the purposes of this work, P_{O_2} values in the range 0–150 mm Hg are achieved.

Obtaining a given gas mixture from cylinders with the composition certified by the manufacturer is accurate but not practical because each desired mixture requires a separate cylinder. We therefore use two gas cylinders, containing either air or N_2. The gas flows are controlled by two high-precision flow meters (Sho-Rate Brooks, SIAD, Milano, Italy) and are then mixed in a chamber containing glass wool. Most of the gas is used for tonometry (see below), and roughly 50 ml/min is directed to a paramagnetic O_2 analyzer (OA 273 Taylor Servomex, Crowborough, Sussex, England; nominal accuracy ±0.01%). We thus obtain virtually any gas repartition in the desired P_{O_2} range using two cylinders. High reproducibility is ensured by calibrating the flow meters in advance.

Tonometry is performed in either an open or a closed system. The first (IL237, Instrumentation Laboratory, Paderno Dugnano, Italy) is particularly suitable when complete deoxygenation of the sample is required.

Gas humidified at 37° (to avoid evaporation of the sample) flows over liquid that is stirred to provide maximal surface area for gas exchange, allowing full equilibration of up to 2-ml samples in 20–25 min.

When intermediate P_{O_2} values are required, the closed system is used. This system is based on anaerobically sealed flasks that are washed with gas of known composition. After this operation, a sample (0.2–0.7 ml) of erythrocytes or Hb is injected into the flask, which is placed into a temperature-controlled rotating device (7 rpm) for 20 min. The flask is designed so that when rotated at this speed the solution or suspension forms a thin layer of liquid, exposing the largest surface area to exchange gas. The advantages of this system are the small sample volume and the low stress to which the sample is exposed, with consequent low risk of Hb denaturation, metHb formation, and depletion of erythrocytic metabolic intermediates such as 2,3-diphosphoglycerate. On the other hand, the closed system cannot completely deoxygenate the sample (see below).

The final P_{O_2} is calculated theoretically from the fraction of O_2 in the gas phase (x_{O_2}), the barometric pressure (P_B), and the water vapor pressure (P_{H_2O}):

$$P_{O_2} = (P_B - P_{H_2O})x_{O_2}$$

This equation does not yield the true P_{O_2} in the closed system. First, the water vapor correction does not apply because the gas used to wash the flask is dry (humidified gas forms water condensation inside the flask). Another adjustment accounts for washing flasks at <37° with consequent increase of total pressure during tonometry:

$$P_{O_2}(t_{ton}) = P_{O_2}(t_w)[(273 + t_{ton})/(273 + t_w)]$$

where t_{ton} and t_w are temperatures (in degrees Celsius) of tonometry and of flask washing, respectively. Total pressure inside the flask is also increased for the addition of incompressible liquid into a finite volume:

$$P_{O_2} = P_{O_2}[V_f/(V_f - V_s)]$$

where V_f and V_s are flask and sample volumes, respectively. Finally, a third correction accounts for the increase of total O_2 in the flask from the O_2 already dissolved in the sample before tonometry. The following formula relates the final P_{O_2} with sample [Hb] (g/liter), its S_{O_2} before tonometry ($S_{O_{2in}}$), final S_{O_2} ($S_{O_{2fin}}$), tonometry temperature (t_{ton}), volume of Hb or erythrocytes (V_s), and barometric pressure (P_B):

$$P_{O_2} = P_{O_2} + [(S_{O_{2in}} - S_{O_{2fin}})/S_{O_{2in}}][Hb] \times V_s P_B(273 + t_{ton})(1.213 \times 10^{-4})$$

Closed tonometry requires care when using samples with a heavy bacterial contamination or with an increased number of leukocytes, because these factors substantially increase the overall O_2 consumption. Under normal conditions, the P_{O_2} decrease for this reason is no more than 0.09 mm Hg over a period of 30 min.[15] Temperature is critical because it influences the water vapor pressure,[16] the Hb O_2 affinity,[17] and the solubility of O_2 in blood or aqueous solutions,[18] and is to be controlled within ±0.1°.

Measurement of S_{O_2}. S_{O_2} is defined as:

$$S_{O_2} = [HbO_2]/([HbO_2] + [Hb])$$

The classical procedures to determine S_{O_2} depend on spectrophotometric methods because the absorption spectra of the various Hb derivatives are well defined,[19] allowing determination of HbO_2 and other Hb derivatives under practically any condition. The drawback of these methods in the absence of sophisticated devices, such as fiber optics equipment, is the need to get the sample from the test tube or the tonometer into the spectrophotometer cuvette without appreciable contamination from atmospheric air.

To avoid these problems, we used a simple and inexpensive yet accurate method to measure S_{O_2} suitable for small volumes of Hb or erythrocytes. This method is based on the fact that Hb becomes oxygenated quickly when diluted and placed in alkaline buffer and requires a 1-ml stoppered aluminum cuvette equipped with a Clark-type O_2 electrode to continuously monitor the P_{O_2} of the buffer and a magnetically operated stirrer (Fig. 2). No temperature control is needed.

Figure 2 illustrates a practical example of this method. Assume we have to analyze a sample of erythrocytes with 50% S_{O_2} and that the cuvette contains 2.5 mM sodium tetraborate and 5 mM Na_2HPO_4 buffer at pH 9.1, previously equilibrated with air (P_{O_2} = 152 mm Hg). When Hb is packed into the erythrocyte, it has a relatively low affinity for O_2, and the P_{O_2} at which S_{O_2} = 50% usually corresponds to roughly 26–28 mm Hg (Fig. 2A). When 10 μl of the erythrocyte suspension is added to 1 ml of buffer, which also contains 1 ml/liter Sterox SE (Baker Chemicals B.V.,

[15] M. Samaja, A. Mosca, M. Luzzana, L. Rossi Bernardi, and R. M. Winslow, *Clin. Chem. (Winston-Salem, N.C.)* **27**, 1856 (1981).

[16] J. R. Hall and R. G. Brouillard, *J. Appl. Physiol.* **58**, 2090 (1985).

[17] M. Samaja, D. Melotti, E. Rovida, and L. Rossi Bernardi, *Clin. Chem. (Winston-Salem, N.C.)* **29**, 110 (1983).

[18] C. Christoforides and J. Hedley-White, *J. Appl. Physiol.* **27**, 592 (1969).

[19] O. W. Van Assendelft, *in* "Spectrophotometry of Haemoglobin Derivatives." Royal Vangorcum Ltd., Assen, The Netherlands, 1970.

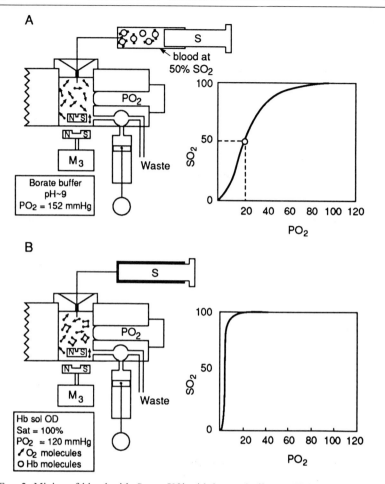

FIG. 2. Mixing of blood with $S_{O_2} = 50\%$ with borate buffer at pH 9.1. Insets in A and B show the O_2 dissociation curve of normal human blood or diluted normal human Hb, respectively.

The Netherlands) as hemolyzing agent, and 0.1 ml/liter silicone antifoam (Wacker Chemie, GMBH, Germany), the erythrocyte membrane is disrupted and Hb becomes diluted rapidly. Under these conditions of pH and protein concentration, the affinity of Hb for O_2 increases greatly, and Hb becomes fully saturated with O_2 at P_{O_2} values as low as 20 mm Hg (Fig. 2B). Because the cuvette in which the reaction occurs is sealed anaerobically, the total O_2 content does not change. Therefore, the oxygenation process of deoxyHb, which usually occurs in less than 20 sec, causes the P_{O_2} of the buffer to decrease in proportion to the amount of deoxyHb present in the original blood sample ($\Delta P_{O_2}'$). The exact amount of deoxyHb in the sample thus can be calculated from this decrease of P_{O_2}, the O_2 solubility,[20] and the volumes of the sample and of the cuvette[21] (see below).

To calculate S_{O_2}, however, the total Hb concentration in the sample used to determine deoxyHb is required. This is done by two independent procedures:

1. By spectrophotometric methods with cuvettes equipped with the O_2 electrode and a fixed-wavelength spectrophotometer (interference filters at 497, 565, and 620 nm) to determine the concentrations of HbO_2, HbCO, and metHb.[22]
2. By the same P_{O_2} electrode, forcing HbO_2 to release O_2 on oxidation by strong oxidant and measuring the increase of P_{O_2} ($\Delta P_{O_2}''$). For this purpose, 5 μl of a solution containing 0.5 M $K_3Fe(CN)_6$ in 0.7 M lactic acid is added because the more acidic pH induces faster (<20 sec) and complete ($>99\%$) oxidation of HbO_2.[21]

With the latter method, the two parameters needed to calculate S_{O_2} are measured simultaneously in the same sample. S_{O_2} is roughly estimated by the following equation:

$$S_{O_2} = [HbO_2]/([deoxyHb] + [HbO_2]) = (\Delta P_{O_2}'' - \Delta P_{O_2}')/\Delta P_{O_2}''$$

A more precise calculation takes into account factors such as the variation of the solubility for O_2, volume displacements, and contributions of dissolved O_2 from addition of the reactants, and these are discussed in Samaja and Rovida.[21] The accuracy of this method is $\pm1\%$. Figure 3 shows how a $\pm10\%$ error in one variable reflects on the final S_{O_2} measurement.

Incubation of Erythrocytes with Styrene. The incubation medium is composed of erythrocytes (500×10^6 cells/ml), 50 μl/ml of 1 M styrene dissolved in acetonitrile, and PBS (pH 7.4) to yield a final Hb concentration of 0.25 mM heme. Blank samples without erythrocytes are also prepared. The incubation is carried out in a Dubnoff incubator for 30 min at 37°. Sealed flasks under N_2 atmosphere are used for deoxygenated samples. The incubation is stopped by adding 0.4 ml of 0.6 N H_2SO_4 per milliliter of incubation medium.

Assay of Styrene Oxide. The samples resulting from the above incubation (1.4 ml) are made alkaline with 0.8 ml of 0.6 N NaOH and then

[20] F. J. W. Roughton and J. W. Severinghaus, *J. Appl. Physiol.* **35**, 861 (1973).
[21] M. Samaja and E. Rovida, *J. Biochem. Biophys. Methods* **7**, 143 (1983).
[22] L. Rossi Bernardi, M. Perrella, M. Luzzana, M. Samaja, and I. Raffaele, *Clin. Chem.* (*Winston-Salem, N.C.*) **23**, 1215 (1977).

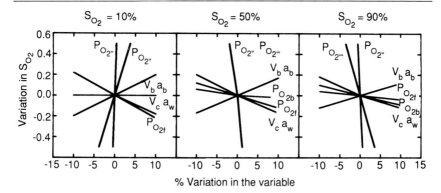

FIG. 3. Effects of the variation of a single variable ($\pm 10\%$ max) on the value of S_{O_2} in three standard cases. Abbreviations: a_b, O_2 solubility coefficient in the erythrocyte sample; a_w, O_2 solubility coefficient in the aqueous buffer; $P_{O_2''}$, P_{O_2} of the buffer after addition of the erythrocyte sample; $P_{O_2'''}$, P_{O_2} of the buffer after addition of the oxidant; $P_{O_{2b}}$, P_{O_2} of the erythrocyte sample before injection into the cuvette; $P_{O_{2f}}$, P_{O_2} of the ferricyanide solution before injection into the cuvette; V_b, volume of the erythrocyte sample; V_c, volume of the cuvette.

extracted twice with 3 ml of ethylacetate on a rotating shaker; 2.5 ml of this solution is transferred to conical test tubes, and the pooled extracts are dried under a gentle stream of N_2 at 37°. The extraction residue is dissolved in 1 ml of toluene, to which 200 μl of 0.05 M trimethylamine in toluene and 100 μl of pure trifluoroacetic anhydride are added. Derivatization is carried out keeping the samples at 60° for 30 min.[23] Samples are cooled at room temperature and then 1 ml of distilled water and 1 ml of 5% ammonia solution (after 1 min) are added. The samples are shaken for 5 min and centrifuged at 3500 g. 1-Bromo-1-phenylethane (50 μl; 28 μg/ml in toluene) is used as a marker and is added to 50 μl of the organic phase. Then 1 μl of this solution is injected into a gas chromatograph (Carlo Erba Strumentazione, Milano, Italy) equipped with a ^{63}Ni electron capture detector and a glass tube column (2 m × 4 mm i.d.) packed with 3% OV-17 on 100-120 mesh Gas-chrom Q (Supelco, Bellefonte, PA). The temperatures of the column, the injector port, and the detector are 140°C, 250°C, and 275°C, respectively. The carrier gas is N_2 at 30 ml/min flow rate.

The calibaration curve was obtained using known amounts of styrene glycol (0.125, 0.25, 0.5, 0.75, 1, 1.5 and 2 μg/ml PBS) processed as described above. The calibration curve is plotted by reporting on the X axis the amount of styrene glycol and on the Y axis the ratio of the area of

[23] G. Gazzotti, E. Garattini, and M. Salmona, *J. Chromatogr.* **188**, 400 (1980).

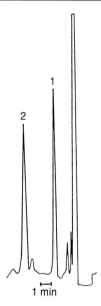

FIG. 4. Gas chromatogram of the styrene glycol trifluoroacetyl derivative. Peak 1 corresponds to the derivatized glycol and peak 2 to 1-bromo-2-phenylethane (marker). The amount of styrene oxide present in blank samples was subtracted from that found in those containing the erythrocytes.

the peak of styrene glycol to that of the marker 1-bromo-2-phenylethane (peaks 1 and 2 in Fig. 4).

Results

Figure 4 is a typical chromatogram of the derivatized styrene glycol extracted from the reaction mixture of erythrocytes and styrene (peak 1) and its marker 1-bromo-2-phenylethane (peak 2). The sensitivity of the method was 100 ng/ml with a linearity range between 125 and 2000 ng/ml.[23]

The formation of styrene oxide at various styrene concentrations in intact erythrocytes and in lysates is shown in Fig. 5. With 50 mM styrene, a concentration that causes complete cell lysis within 5 min, the time courses overlapped (Fig. 5A). At a styrene concentration of 0.8 mM (Fig. 5B), which does not cause lysis, the erythrocytes were more active than the lysates. This was explained on the basis of higher Hb concentration inside the erythrocytes than in the lysate, although the total amount of Hb was the same in both samples.

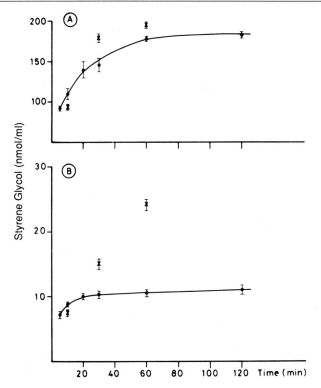

FIG. 5. Time course of styrene oxidation to styrene oxide in human erythrocytes (×) and cell lysate (●). Erythrocytes and lysate preparation were described in text (see Preparation of Erythrocytes and Lysates). Intact cells and lysates were incubated with 50 mM styrene (A) or 0.8 mM styrene (B). Mean ± SEM.

Table I shows some characteristics of the enzymatic oxidation of styrene. The reaction was almost completely inhibited by CO, indicating the involvement of Fe^{II} hemoprotein in the process.[24] In addition, it appears that the O_2-derived free radicals such as $O_2^{\bar{\ }}$, H_2O_2, and OH· do not appear to be involved directly, although they originate from the autoxidation of Hb.[25] In fact, addition of various free radicals scavengers, i.e., superoxide dismutase (which converts $O_2^{\bar{\ }}$ to H_2O_2 and O_2),[25,26] catalase (which re-

[24] I. Golly and P. Hlavica, *Biochim. Biophys. Acta* **760,** 69 (1983).
[25] H. P. Misra and I. Fridovich, *J. Biol. Chem.* **247,** 6960 (1972).
[26] J. M. McCord and I. Fridovich, *J. Biol. Chem.* **244,** 6049 (1969).

TABLE I
STYRENE OXIDATION TO STYRENE OXIDE IN
HUMAN ERYTHROCYTES[a]

System	Styrene glycol[b] (nmol/30 min/ml)
Erythrocytes	130.0 ± 6.0
+ CO[c]	29.2 ± 1.0
+ Superoxide dismutase (50 units/ml)	117.0 ± 1.4
+ Catalase (1750 units/ml)	158.3 ± 16.2
+ Tryptophan (2 mM)	104.0 ± 5.2
+ Mannitol (20 mM)	126.0 ± 6.3
+ Dimethyl sulfoxide (280 mM)	150.0 ± 8.0

[a] Styrene concentration in the incubation mixture was 50 mM. Data are mean ± SEM.
[b] The styrene oxide formed enzymatically was chemically converted to styrene glycol (see text, Incubation with Styrene).
[c] CO was bubbled into the incubation mixture for 1 min before addition of styrene.

moves H_2O_2),[27] and mannitol, dimethyl sulfoxide,[28,29] and tryptophan[28] (which are scavengers of the hydroxyl radical, OH·) did not inhibit styrene oxide formation. Some superoxide dismutase and catalase are contained in the erythrocyte.[25,26] However, it is possible that hemolysis dilutes these enzymes, thus preventing their protective action; therefore, they were added to the medium in saturating amounts.

Based on these experiments, the formation of styrene oxide was attributed to the O_2 bound to Hb. This O_2 may be in a partially reactive form in HbO_2.[30,31] The involvement of HbO_2 in the oxidation of styrene was confirmed by the linear relationship between the fraction of HbO_2 and the amount of styrene oxide produced (Fig. 6).

Oxidation of Olefins to Olefin Oxides Catalyzed by metHb and H_2O_2

Ortiz de Montellano and Catalano have studied extensively the oxidation of styrene and cis- and trans-stilbene catalyzed by bovine metHb and hydrogen peroxide.[8,11]

[27] G. E. Gaetani, S. Galiano, L. Canepa, A. M. Ferraris, and H. N. Kirkman, Blood 73, 334 (1989).
[28] R. C. Armstrong and A. J. Swallow, Radiat. Res. 40, 563 (1969).
[29] J. E. Repine, J. W. Eaton, M. W. Anders, J. R. Hoidal, and R. B. Fox, J. Clin. Invest. 64, 1642 (1979).
[30] J. J. Weiss, Nature (London) 202, 83 (1964).
[31] J. Peisach, W. E. Blumberg, B. A. Wittenberg, and J. B. Wittenberg, J. Biol. Chem. 243, 1871 (1968).

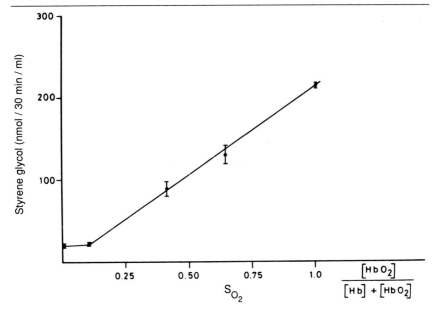

FIG. 6. Relationship between the fraction of HbO_2 and styrene oxidation to styrene oxide in human erythrocytes. Preparation of partially deoxygenated erythrocytes is described in text (see Tonometry).

Methods

Incubations of Hb with Styrene. Duplicate 10-ml incubations, containing bovine Hb (40 μM heme), styrene (10 mM) or *cis*- or *trans*-stilbene (50 μM) in 0.2 M phosphate buffer, pH 7.4, are cooled to 0° for 15 min. An acetonitrile solution of styrene and stilbenes is employed in the preparation of this mixture, but the acetonitrile concentration in the final volume did not exceed 1%. A precooled solution of H_2O_2, to minimize decomposition of the hemoprotein, in 0.2 M phosphate buffer (0.6 mM final concentration) is then added to start the reaction, and the mixtures are incubated at 0° with occasional stirring for 90 min. 2-Undecanone (approximately 5 μg/ml) is added as an internal standard at the end of the incubation to quantitate product formation. For the analysis of stilbene oxides, 1 μg of diphenylmethane is added to the incubation mixture. The samples are then extracted with diethyl ether (2× 5 ml) and the combined extracts are washed with saturated NaCl solution. The organic phase is filtered to remove precipitated protein and then dried over anhydrous K_2CO_3.

The ether solutions are reduced to volumes less than 0.5 ml and the concentrations of styrene metabolites are determined by gas chromatography on a Varian 2100 flame ionization instrument fitted with a 6-ft glass column packed with 3% OV-225 on 100/200 mesh Supelcoport. The injector and detector are held at 250° while the oven temperature is programmed to rise from 80° to 140° at 15°/min. The *cis*- and *trans*-stilbene oxides are analyzed by gas chromatography on a Hewlett-Packard 5890A instrument equipped with a 0.5 mm × 30 m DB-5 capillary column at an isothermal temperature of 180°.

A set of eight incubations to which acetonitrile but not styrene is added are carried through the incubation procedure, and authentic styrene oxide or benzaldehyde (0, 10, 20, 40, 80, and 160 μg) is added before the 2-undecanone standard. For the analysis of stilbene oxides a standard curve is plotted by substituting known amounts (0.1, 0.5, 1, and 5 μg) of the *cis*- and *trans*-stilbene oxides for the stilbenes.

Identification of the Metabolites of Styrene. The ether extracts from 100-ml incubations are freed of styrene by low-pressure liquid chromatography on a Lichroprep Si-60 size B silica gel column (Merck) with 40% acetonitrile/water as the elution solvent. The eluent is monitored at 265 nm with a Hitachi Model 100 variable-wavelength detector. Metabolites are analyzed by high-performance liquid chromatography on an Altech Spherisorb S-5-ODS 5-μm C_{18} reversed-phase column with 40% acetonitrile/water as the elution solvent. The eluent is monitored at 265 nm with the Hitachi Model 100 variable-wavelength detector. High-performance liquid chromatography is employed to avoid the formation of phenylacetaldehyde generated by decomposition of styrene oxide in the gas chromatograph. The amount of phenylacetaldehyde generated this way depends on the history of the chromatographic column. The least decomposition is observed when a gas chromatography column is freshly packed and baked at 200° for 30 min prior to use.

The structures of the metabolites detected in the incubations are confirmed by coinjection and mass spectrometric comparison with authentic samples.

Electron-impact (70-eV) mass spectra are obtained on a Kratos AEI-MS 25 mass spectrometer coupled to a Varian 3700 gas chromatograph fitted with a 30-μm SE 52 column programmed to rise from 50° to 150° at 5°/min.

Incubations with $^{18}O_2$. Styrene (100 μl of a 1 M solution in acetonitrile) is added to a solution of metHb (6.5 mg, 0.1 μmol) in 8.9 ml of 0.2 M phosphate buffer (pH 7.4). The mixture is cooled in an ice bath and is taken six times through a cycle in which the flask is first evacuated to <1 Torr and then brought to atmospheric pressure with nitrogen. In a final cycle 100 ml of $^{18}O_2$ is introduced into the evacuated flask before nitrogen is used to bring the system to slightly above atmospheric pressure (a balloon attached to the flask is used as a pressure indicator and reservoir). A N_2-flushed syringe is then used to transfer 1 ml of a 6 mM H_2O_2 solution that had been similarly cooled and deoxygenated. After a 90-min incubation at 0°, 50 ml of diethyl ether is added by syringe. The mixture is then worked up normally, and the styrene oxide and benzaldehyde obtained are analyzed by gas chromatography–mass spectrometry.

Incubations with $H_2{}^{18}O_2$. Labeled hydrogen peroxide is prepared from $^{18}O_2$ as described by Sawaki and Foote.[32] The hydrogen peroxide concentration is determined by iodometric assay.[33] The ^{18}O content of the peroxide is determined by a new procedure based on alkaline oxidation of menadione to the corresponding epoxide by hydrogen peroxide[34] followed by mass spectrometric analysis. In brief, 35 mg of sodium carbonate is added to 50 μl of the peroxide in 1 ml of double glass-distilled water. A warm solution of menadione (430 μg, 2.5 μmol) in 1 ml of ethanol is then added, and 5 min later the ethanol is removed by rotary evaporation. Ether extraction of the residual mixture after addition of brine, followed by drying over $CaCO_3$, yields a solution that is analyzed by gas chromatography–mass spectrometry. The sample used in the present studies contained 42% labeled O_2, as shown by 42% incorporation of label into menadione epoxide. The incubation of labeled peroxide with styrene and the hemoprotein (40 μM heme concentration) is carried out as already described.

O_2 Evolution. The free O_2 concentration is measured by polarographic methods with the Clark O_2 electrode in a jacketed cell maintained at 0°. The electrode was calibrated against air-saturated water at the same temperature. The appropriate amount of Hb in 0.2 M phosphate buffer (pH 7.4) is mixed with 15-fold excess H_2O_2 (heme basis) into the cell, and O_2 evolution is monitored for 90 min.

Purification and Chemical Epoxidation of Cis-Stilbene. Commercial preparations of *cis*-stilbene are usually contaminated with 2–3% *trans*-stilbene that must be eliminated to avoid interferences. *cis*-Stilbene is purified by low-pressure chromatography on a Lichroprep Si-60 size B silica gel column (Merck). The column is eluted with 10% diethyl ether/ hexane at 6 ml/min. The retention times of *cis*- and *trans*-stilbene are 18.1 and 20.4 min, respectively. The collected 5-ml fractions are examined by isothermal (180°) gas–liquid chromatography (Hewlett-Packard 5890A) equipped with 0.5 × 30 mm DB-5 coated column. The retention times of

[32] Y. Sawaki and C. Foote, *J. Am. Chem. Soc.* **101,** 6292 (1979).

[33] M. Pesez and J. Bartos, *in* "Colorimetric and Fluorimetric Analysis of Organic Compounds and Drugs" (M. K. Schwartz, ed.), p. 329. Dekker, New York, 1974.

[34] L. Fieser and M. Fieser, "Reagents for Organic Synthesis," p. 466. Wiley, New York, 1967.

the cis and trans isomers under these conditions are 9.3 and 15.3 min, respectively. Fractions with low content of the trans isomer are pooled and concentrated on a rotary evaporator. The *cis*-stilbene product after this procedure is contaminated with less than 0.2% of the trans isomer.

Authentic *cis*-stilbene oxide is prepared by vigorously stirring 1 mmol *cis*-stilbene with 2 mmol *m*-chloroperbenzoic acid that has been washed previously with 0.2 M phosphate buffer and dried overnight under vacuum in 10 ml of CH_2Cl_2. The organic layer is washed with 1 N NaOH, water, and saturated NaCl, and is dried over K_2CO_3. The residue gives a peak with a retention time of 3.4 min on analysis by gas–liquid chromatography on a 6-ft column packed with 3% OV 225 on 100/120 mesh Supelcoport that was programmed to rise from 100 to 200° at 12°/min.

Results

Oxidation of Styrene by metHb and H_2O_2. Addition of H_2O_2 to a solution of styrene and Hb at 0° causes an immediate shift of the Soret band peak from 408 to 412 nm, a position characteristic of the ferryl (Fe^{IV}) complex. The shift is followed by a decrease in absorbance at 412 nm (Fig. 7), presumably an index of time-dependent irreversible loss of the heme.[11] The loss is independent of H_2O_2 concentration for H_2O_2/heme ratios ranging from 5 to 40, but when the H_2O_2/heme ratio approaches unity the Soret band shifts back from 412 to 408 nm after 60 min, presumably because of exhaustion of H_2O_2 in the medium and metHb regeneration. The formation of metabolites is not dependent on temperature, although the Soret band is lost more rapidly at 37° than at 0°. Therefore incubations were carried out at 0°.

Two styrene metabolites were detected by gas chromatography after incubation of styrene with metHb and H_2O_2 (Fig. 7, Table II). These metabolites were identified as styrene oxide and benzaldehyde. Benzaldehyde, however, is not a product of secondary oxidation of styrene oxide or phenylethylene glycol because these compounds did not yield benzaldehyde when incubated with Hb and H_2O_2. Unlike styrene oxide, the formation of benzaldehyde is not significant before 30 min of incubation (Fig. 7).

The production of styrene oxide from styrene and metHb was characterized using various scavengers of the reactive O_2 intermediates and anaerobic conditions. The low inhibition by CO suggests that iron does not cycle extensively through the ferrous state because it is not trapped as complex with CO.[24]

Product formation decreased in incubations carried out anaerobically but was never completely suppressed despite extensive efforts to ensure

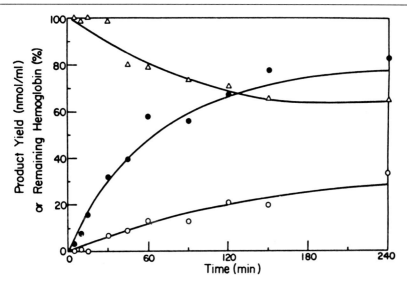

FIG. 7. Loss of heme and formation of styrene oxide and benzaldehyde as a function of time. The absorbance loss at 408 nm in standard incubations of styrene (10 mM) with metHb (10 μM) and H_2O_2 (0.6 mM) at 0°, as a percentage of the absorbance immediately after addition of the peroxide (△), is plotted as a function of time of incubation. The yields of benzaldehyde (○) and styrene oxide (●) in the same incubations are also plotted as a function of time.

complete removal of O_2 before addition of H_2O_2. This is because the reaction of metHb with H_2O_2 results in formation of O_2 (Fig. 8). The amount of O_2 produced per mole of heme in 90 min rises linearly with Hb concentration from 2 mol/mol to 5.5 mol/mol as heme concentration increases from 20 to 160 μM. Approximately 2–3 mol of O_2 are thus produced per mole of heme in standard incubations with 40 μM heme. The O_2 concentration, therefore, can reach values in the range of 100 μM in incubations that are virtually anaerobic.

The H_2O_2-supported, metHb-catalyzed oxidation of styrene under an atmosphere of $^{18}O_2$ resulted in incorporation of the label into 38% of the styrene oxide. The incubation of metHb and $^{18}O_2$ was repeated three times to ensure that the fractional incorporation of label did not result from incomplete removal of unlabeled O_2 from the incubation system. However, the result was approximately the same in all three experiments. It is, therefore, clear that a fraction of styrene oxide is formed in an O_2-dependent process. These results require a parallel oxidative mechanism to incorporate O_2 from a source other than molecular O_2 into styrene oxide. The finding that a fraction of the styrene oxide (70%), roughly complemen-

TABLE II
FACTORS AFFECTING FORMATION OF STYRENE OXIDE
AND BENZALDEHYDE[a]

Incubation	Styrene oxide	Benzaldehyde
Normal incubation[b]	100	100
− H$_2$O$_2$	8	6
− Hb	1	5
+ KCN (50 mM)	15	14
+ Mannitol (50 mM)	104	114
+ Ascorbic acid (50 mM)	3	28
+ POBN (50 mM)	57	101
+ BHT (50 mM)	74	70
+ DABCO (50 mM)	125	60
+ Catalase (60 units/ml)	ND[c]	16
+ Superoxide dismutase (60 units/ml)	107	100
+ CO (1 : 1 with O$_2$)	95	85
+ Argon atmosphere	82	88

[a] Data expressed as percentage of controls. POBN, α-(4-Pyridyl-1-oxide)-N-tert-butyl nitrone; BHT, butylated hydroxytoluene; DABCO, diazabicyclo[2.2.1]octane.

[b] The results are averages of duplicate incubations except for the original controls and the values in the absence of O$_2$, which are averages of six incubations. Styrene oxide and benzaldehyde formations in the control incubations were 54 and 29 nmol/ml, respectively.

[c] ND, Not detectable.

tary to the fraction labeled by molecular O$_2$ (38%), was labeled in incubations with ^{18}O$_2$-labeled H$_2$O$_2$ helps to identify the source of O$_2$ in this second mechanism. The sum of O$_2$- and H$_2$O$_2$-derived product is more than 100%, but this discrepancy is readily accounted for by the fact that O$_2$ is generated from H$_2$O$_2$ by the hemoprotein (Fig. 8). The incorporation of label from molecular O$_2$ into benzaldehyde in the reaction supported by metHb was also examined, but the 17% incorporation observed cannot be taken as a measure of the actual incorporation because the aldehyde O$_2$ exchanges readily with the medium. Mannitol, an OH· scavenger, had no effect on the reaction but butylated hydroxytoluene (BHT) and α-(pyridyl-1-oxide)-N-tert-butyl nitrone (POBN), two radical scavengers of relatively low specificity, partially inhibited metabolite formation (Table II). Ascorbic acid markedly inhibited metabolite formation, but this could reflect the reduction of radical intermediates as the direct reduction of metHb. The superoxide radical is ruled out as the oxidant because superoxide dismutase stimulated rather than inhibited metabolite formation;

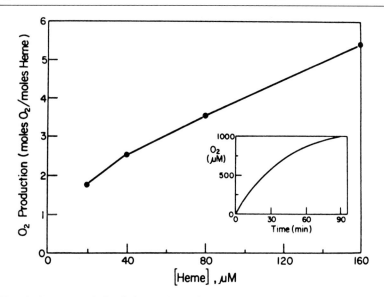

FIG. 8. Oxygen evolution in incubations of metHb with H_2O_2 after 90 min. The O_2 that evolved in standard incubations of metHb with H_2O_2 in a closed cell at 0° is plotted against the hemoprotein concentration. The response of the Clark electrode, expressed as O_2 evolved, is given as a function of time in the inset for the incubation with 160 μM heme (40 μM metHb). No O_2 is evolved in the absence of either component of the reaction mixture.

diazabicyclo[2.2.1]octane (DABCO), a singlet O_2 trap, stimulated styrene oxide formation but somewhat inhibited benzaldehyde formation.

Oxidation of cis- and trans-Stilbene to Stilbene Oxides Catalyzed by metHb and H_2O_2. Incubation of *trans*-stilbene with bovine metHb and H_2O_2, followed by gas–liquid chromatographic analysis of extracts of the incubation mixture, indicates that a single major product is formed with a retention time of 15.3 min (Fig. 9). This product was identified as *trans*-stilbene oxide by cochromatography with authentic material and by the identity of its mass spectrum with a reference sample. A minor product with a retention time of 12.0 min is also detected by gas chromatography, but its identity is unknown (Fig. 9). The formation of both *trans*-stilbene oxide and the unidentified product is strictly dependent on the presence of Hb and H_2O_2. Incubation of *trans*-stilbene with metHb and $H_2^{18}O_2$ (77 atom %) yields *trans*-stilbene oxide in which the O_2 derives quantitatively from the peroxide.

Two major products are found by gas–liquid chromatography in incubations of *cis*-stilbene with Hb and H_2O_2 (Fig. 10, peaks 1 and 3). These were identified by retention time and mass spectrometric comparison with

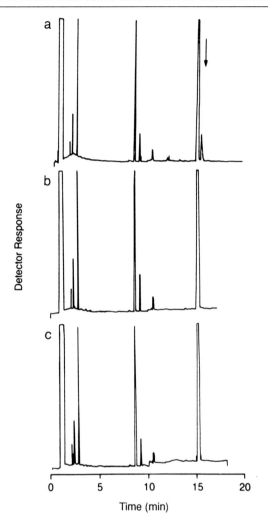

FIG. 9. Gas chromatographic analysis of the products from oxidation of *trans*-stilbene by the Hb/H₂O₂ system: (a) complete incubation system (the arrow marks the position of the major metabolite); (b) incubation with Hb omitted; (c) incubation with H₂O₂ omitted. The incubation and chromatographic conditions are discussed in text (see Incubation of Hb with Styrene). Reprinted with permission from Catalano and Ortiz de Montellano.[8] Copyright 1987 American Chemical Society.

authentic standards as *cis*- and *trans*-stilbene oxide. The *cis*- and *trans*-stilbene oxides are formed in a 3:1 ratio (Table III), but neither product is formed in the absence of Hb or H_2O_2 (Fig. 10). As with *trans*-stilbene, a minor unidentified product (retention time 12.1 min) is also formed.

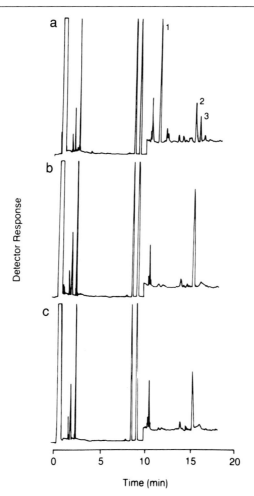

Fig. 10. Gas chromatographic analysis of the products from oxidation of *cis*-stilbene by the Hb/H$_2$O$_2$ system: (a) complete incubation system (peaks 1, 2, and 3, respectively, are *cis*-stilbene oxide, *trans*-stilbene, and *trans*-stilbene oxide); (b) incubation with Hb omitted; (c) incubation with H$_2$O$_2$ omitted. The attenuation of the detector was decreased at approximately 10 min. The incubation and chromatographic conditions are discussed in text (see Incubations of Hb with Styrene). Reprinted with permission from Catalano and Ortiz de Montellano.[8] Copyright 1987 American Chemical Society.

The purified *cis*-stilbene employed in these experiments contains a trace (<0.2%) of *trans*-stilbene, but the proportion of the trans isomer is not measurably increased in the stilbene recovered from incubations with Hb and H$_2$O$_2$.

TABLE III
PRODUCTS AND YIELDS IN H_2O_2-DEPENDENT
REACTIONS CATALYZED BY Hb[a]

Substrate	Product	Yield nmol/ml
Styrene	Styrene oxide	35
trans-Stilbene	*trans*-Stilbene oxide	1.0
cis-Stilbene	*cis*-Stilbene oxide	0.7
	trans-Stilbene oxide	0.2

[a] Adapted with permission from Catalano and Ortiz de Montellano.[8] Copyright 1987 American Chemical Society.

The *cis*- and *trans*-stilbene oxides obtained in incubations of *cis*-stilbene with Hb and $H_2^{18}O_2$ (77 atom %) were analyzed by gas chromatography–mass spectrometry. The $^{18}O_2$ content of the *cis*-stilbene oxide (77%) is essentially identical with that of the H_2O_2 employed in the incubations. The O_2 in the cis epoxide thus derives exclusively from the peroxide. In contrast, no $^{18}O_2$ is detected in the *trans*-stilbene oxide produced from *cis*-stilbene by Hb, so the O_2 in this isomer derives from a source other than the peroxide. When *cis*-stilbene is incubated under $^{18}O_2$, approximately 32% of the *trans*-stilbene oxide product is labeled. It has not been possible to carry out the incubation in $H_2^{18}O$ to see whether O_2 from water is incorporated into the epoxide because the large incubation volume makes the cost of the experiment prohibitive. It is clear, however, that the *trans*-stilbene oxide derived from *cis*-stilbene incorporates O_2 from the atmosphere and probably from water but not from the peroxide.

Discussion

To investigate the oxidation of styrene catalyzed by Hb, we have used washed human erythrocytes and metHb plus H_2O_2. In both systems, the O_2-derived free radicals do not appear involved in styrene oxidation. Indeed, OH· and O_2^- scavengers have no effect. However, the mechanism of styrene oxide formation does not appear to be the same in the two systems because the presence of H_2O_2 is required in the system containing metHb but not in the system with erythrocytes. The strong inhibition by CO in the latter system and the poor effect in the former can be correlated with the occurrence of the Fe^{II} and Fe^{III} states in the two systems, respectively.

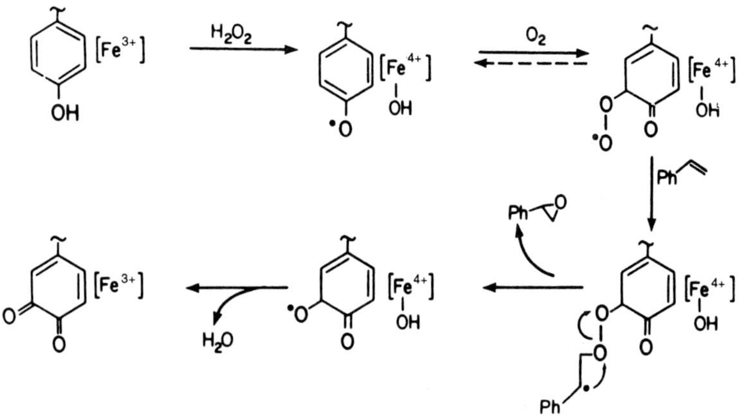

SCHEME I. Proposed mechanism for the incorporation of molecular O_2 into styrene oxide in the H_2O_2-dependent oxidation of styrene by metHb. The prosthetic heme, represented by the brackets around the iron, and the putative action site Tyr are shown. Ph, Phenyl ring in styrene. Alternative fates for the hydroxylated Tyr radical at the end of the sequence can be envisioned.

Strong experimental evidence indicates that in the system with metHb and H_2O_2, the formation of the ferryl complex is intimately connected with epoxidation of styrene. This evidence includes (1) the requirement of H_2O_2, (2) the persistence of metHb in the ferryl state throughout the reaction, (3) the time dependence of the heme loss, and (4) the strong inhibition by cyanide. However this is in contrast with the observation that the ferryl O_2 is found in only 38% of the styrene oxide produced in the presence of $^{18}O_2$. The inability of mannitol and superoxide dismutase to inhibit the reaction (Table II) suggests that neither O_2^- nor OH· acts as oxidant in that reaction, although the modest inhibition by POBN and BHT does not prevent some minor participation of other unidentified radicals. Our data thus point to an oxidative mechanism triggered by a reaction with H_2O_2 but primarily mediated by species other than the ferryl complex or a diffusible O_2 species.

Removal of one electron from Fe^{III} and the protein in the reaction of metHb with H_2O_2 generates the ferryl complex and a protein radical likely centered on a Phe or Tyr residue, as suggested by EPR studies.[35] A computer graphic analysis showed that the Tyr-42 residue is very close to the heme vinyl groups in bovine metHb.[36] This residue is part of the

[35] T. Yonetani and H. Schleyer, *J. Biol. Chem.* **242,** 1979 (1967).
[36] R. E. Dickerson and I. Geis, "Hemoglobin: Structure, Function, Evolution, and Pathology." Benjamin-Cummings, Menlo Park, CA, 1983.

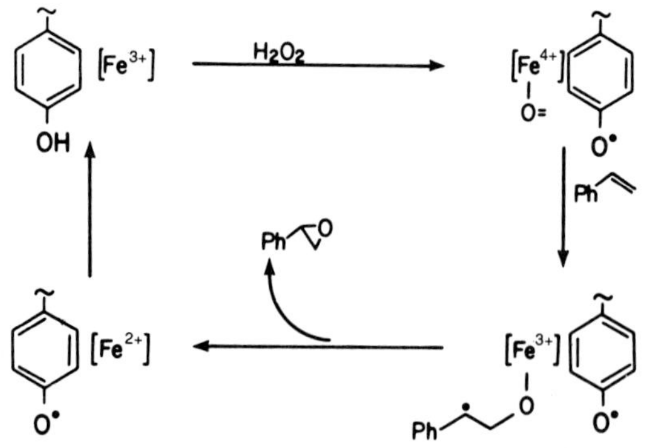

SCHEME II. Mechanism for the incorporation of O_2 from H_2O_2 into styrene oxide in the oxidation of styrene by metHb. The active site of metHb is abbreviated as in Scheme I. Ph, Phenyl ring in styrene. An alternative mechanism employs molecular O_2 generated from H_2O_2 in a sequence similar to that shown in Scheme I.

outer surface of the hemoproteins and thus is in contact with the surrounding medium containing styrene and the heme.

A proposed mechanism (Scheme I) for the activation of molecular O_2 is based on the oxidation of a Tyr residue or another amino acid and is similar to the epoxidation of olefins by lipid peroxy radicals.[37] The peroxy radical generated by the reaction of the protein radical with molecular O_2 would be expected to react with the double bond of styrene to give a transient species with the impaired electron adjacent to the aromatic ring. Rupture of the peroxide bond in a second step would yield styrene oxide and a protein-bound alkoxy radical that could be quenched by electron transfer to or from the heme. This mechanism is consistent with the incorporation of an atom of molecular O_2 into the styrene oxide.

A prerequisite for the above mechanism is that the protein radical involved in styrene oxidation is readily accessible to molecular O_2 and to styrene. Phenylhydrazines, substrates comparable in size to styrene, do bind the heme iron atom and, therefore, can enter the heme pocket.[38,39] However, their motion inside the heme pocket is severely restricted as shown by NMR studies.[40] The discrepancy between the need for an unconstrained reaction environment and the congested nature of the heme binding sites is readily resolved by the proposed mechanism because the

[37] G. A. Reed, E. A. Brooks, and T. E. Eling, J. Biol. Chem. 259, 5591 (1984).
[38] K. L. Kunze and P. R. Ortiz de Montellano, J. Am. Chem. Soc. 105, 1380 (1983).
[39] D. Ringe, G. A. Petsko, D. E. Kerr, and P. R. Ortiz de Montellano, Biochemistry 23, 2 (1984).
[40] P. R. Ortiz de Montellano and D. E. Kerr, Biochemistry 24, 1147 (1985).

reaction with styrene is expected to occur outside the heme pocket. The Tyr radical on the outer protein surface may, therefore, react with O_2 and styrene outside the protein. The Tyr residue can even pick up O_2 inside the heme pocket and deliver it, by a rotation, to styrene outside the protein. This latter mechanism can rationalize preferential incorporation of O_2 from H_2O_2 into styrene oxide.

The incorporation of O_2 from H_2O_2 may stem from a direct reaction of the ferryl O_2 with styrene in the active site (Scheme II), but it is likely that the fraction of O_2 that derives from H_2O_2 and that is incorporated into the epoxide reflects an initial conversion of H_2O_2 to molecular O_2, which is then utilized by the protein radical as in Scheme I to oxidize styrene. The possibility that benzaldehyde arises from reaction of styrene with singlet O_2 generated *in situ* is supported by the partial inhibition caused by DABCO (Table II). The aldehyde could, alternatively, arise from a reaction of the peroxystyrene radical intermediate with a second molecule of O_2.

Styrene is better substrate than *trans*- and *cis*-stilbene in the H_2O_2-dependent oxidation to olefins oxides catalyzed by Hb (Table III). Experiments with $H_2{}^{18}O_2$ and $^{18}O_2$ have shown that O_2 produced from *trans*- and *cis*-stilbene derives exclusively from H_2O_2. On the other side, in styrene oxide and in *trans*-stilbene oxide produced from *cis*-stilbene, O_2 derives partly from $^{18}O_2$.

These experiments indicate that, as far as the oxidation of olefins to olefins oxides is concerned, the behavior of Hb is much like that of an enzyme. This reaction may be relevant *in vivo* in the activation of styrene to its toxic metabolite styrene oxide, as shown by experiments with oxygenated erythrocytes. It was in fact observed that more styrene oxide was formed when perfusing rat livers in the presence of erythrocytes than in a physiological medium.[41] Moreover, sister chromatid exchanges, one of the toxic effects of styrene, were more frequent in cultures of human lymphocytes contaminated with erythrocytes,[42] Although the liver seems able to detoxify to styrene glycol all of the styrene oxide produced by erythrocytes and the liver,[41] the metabolic activation of styrene and possibly other olefins to reactive metabolites by blood may have some relevance in the production of toxic effects.

[41] G. Belvedere, E. Elovaara, and H. Vainio, *Toxicol. Lett.* **23**, 157 (1984).
[42] H. Norppa, H. Vainio, M. Sorsa and G. Belvedere, *Arch. Toxicol., Suppl.* **7**, 286 (1984).

Section VIII

Xenobiotic Adducts of Human Hemoglobin

[40] Quantitative Analysis of Hemoglobin–Xenobiotic Adducts

By Steven R. Tannenbaum and Paul L. Skipper

Hemoglobin (Hb) adducts of chemical carcinogens have found application as biomarkers of exposure and internal dose in occupational settings as well as in cross-sectional epidemiological studies.[1–3] Future application as biomarkers associated with clinical end points is supported by the results of several studies.[4–6] Key to the success of these studies has been the development of precise as well as reasonably accurate methods of quantitative analysis, which are described in several of the following chapters. In this chapter we will discuss some of the more general aspects of the challenge of separating and quantifying hemoglobin adducts. We will also describe how the techniques may be or have been validated. Last, we will try to indicate how the development of a technique for one compound could be used to identify adducts of related compounds.

Properties of Hemoglobin Adducts

In Vivo

Hemoglobin adducts may exhibit the same kinetics of disappearance as does normal hemoglobin, in which case it may be assumed that they are not chemically unstable under physiological conditions. More rapid disappearance than predicted by the normal turnover of hemoglobin has also been observed.[7–9] In these cases the increased rate of adduct disap-

[1] P. L. Skipper and S. R. Tannenbaum, *Carcinogenesis (London)* **11**, 507 (1990).

[2] S. Osterman-Golkar, L. Ehrenberg, D. Segerback, and I. Hallström, *Mutat. Res.* **34**, 1 (1976).

[3] H.-G. Neumann, *Arch. Toxicol.* **56**, 1 (1984).

[4] P. DelSanto, G. Moneti, M. Salvadori, C. Saltutti, A. DelleRose, and P. Dolara, *Cancer Lett.* **60**, 245 (1991).

[5] H. Bartsch, N. Caporaso, M. Coda, F. Kadlubar, C. Malaveille, P. Skipper, G. Talaska, S. R. Tannenbaum, and P. Vineis, *J. Natl. Cancer. Inst.* **82**, 1826 (1990).

[6] G. Talaska, M. Schamer, P. Skipper, S. Tannenbaum, N. Caparoso, L. Unruh, F. F. Kadlubar, H. Bartsch, C. Malaveille, and P. Vineis, *Cancer Epidemiol. Biomarkers Prev.* **1**, 61 (1991).

[7] S. Naylor, L.-S. Gan, B. W. Day, R. Pastorelli, P. L. Skipper, and S. R. Tannenbaum, *Chem. Res. Toxicol.* **3**, 111 (1990).

[8] S. G. Carmella and S. S. Hecht, *Cancer Res.* **47**, 2626 (1987).

pearance may be caused by chemical instability leading to removal of adduct without the loss of hemoglobin, or it may be caused by accelerated protein turnover.[9] Mathematical models relating adduct levels to exposure have been developed to deal with both causes.[10] No evidence exists for organism-mediated repair of adducted hemoglobin.

Chemical Nature

All the known adducts of hemoglobin arise through the formation of a covalent bond between the electrophilic carcinogen or its electrophilic metabolic product and a nucleophilic center of the protein. Adducts formed by the relatively weakly basic amino group of the N-terminal valine residue, cysteine, histidine, and aspartic acid have been characterized. The driving force for the reaction appears to depend in part on the nature of the electrophile. Small molecules, whose number of carbon atoms is less than about 10, react with many centers in proportion to the nucleophilicity of those centers. Of great value in terms of analysis is the fact that the N-terminal valine is generally a reaction center. A second driving force may be the Swain–Scott complementarity of the nucleophilic center with the electrophilic reactant, leading to, for example, the relatively high fraction of carboxylic esters by *N*-nitrosamines, which are metabolized to produce diazohydroxides or diazonium ions.

It is much more difficult to predict where higher molecular weight carcinogens become bound. It may well prove to be the case that the primary driving force is a receptor protein–ligand type of interaction followed by covalent bond formation with the most propitiously situated nucleophilic amino acid. It is well known that hemoglobin binds drugs and other xenobiotics without forming covalent bonds. When hemoglobin is cocrystallized with these compounds and the complexes are examined by X-ray crystallography, it becomes apparent that there are multiple binding sites within the protein.[11] Studies of carcinogen binding to another protein, serum albumin, also support the premise that stereochemical constraints control the reaction to a large degree.[12,13]

[9] M. Maclure, M. S. Bryant, P. L. Skipper, and S. R. Tannenbaum, *Cancer Res.* **50**, 181 (1990).

[10] T. R. Fennel, S. C. J. Sumner, and V. E. Walker, *Cancer Epidemiol. Biomarkers Prev.* **2**, 213 (1992).

[11] M. F. Perutz, G. Fermi, D. J. Abraham, C. Poyart, and E. Bursaux, *J. Am. Chem. Soc.* **108**, 1064 (1986).

[12] P. L. Skipper, M. W. Obiedzinski, S. R. Tannenbaum, D. W. Miller, R. K. Mitchum, and F. F. Kadlubar, *Cancer Res.* **45**, 5122 (1985).

[13] B. W. Day, P. L. Skipper, J. Zaia, and S. R. Tannenbaum, *J. Am. Chem. Soc.* **113**, 8505 (1991).

TABLE I
OBSERVED CONCENTRATIONS OF ADDUCTS IN HUMAN HEMOGLOBIN SPECIMENS

Compound	Determined as	Adduct level[a]		Nature of exposure
		Unexposed[b]	Exposed	
Ethylene oxide[c]	N-Hydroxyethylvaline	10–100[g]	100–500 2400	Cigarette smoking Workplace (1 ppm)
Acrylonitrile[c]	N-Cyanoethylvaline	N.D.	90	Cigarette smoking
4-Amino biphenyl[d]	4-Aminobiphenyl	0.05–0.5	0.25–2.5	Cigarette smoking
3-Amino biphenyl[d]	3-Aminobiphenyl	N.D.–0.01	0.005–0.1	Cigarette smoking
NNN/NNK[e]	4-Hydroxy-1-(3-pyridyl)-1-butanone	N.D.–0.5	N.D.–1.5 N.D.–2.0	Cigarette smoking Snuff dipping
Benzo[a]pyrene[f]	7,8,9,10-Tetra hydroxy-7,8,9,10-tetrahydrobenzo[a] pyrene	0.5–1.5	—	—

[a] Expressed as picomoles per gram of hemoglobin. N.D., Not determined.
[b] Nonsmokers with no known occupational and undetermined environmental exposure.
[c] M. S. Bryant and S. M. Osterman-Golkar, *CIIT Act.* **11**(10), 1 (1991).
[d] Data accumulated from all studies in this laboratory since 1986.
[e] NNN, N'-Nitrosonornicotine; NNK, 4(methylnitrosamino)-1-(3-pyridyl)-1-butanone. S. S. Hecht, S. G. Carmella, and S. E. Murphy, this volume [44].
[f] Unpublished results, this laboratory.
[g] More recent studies indicate that these values may be too high. A range of 8–25 is currently indicated (M. Törnqvist).

Concentration

Ordinary human environmental exposures to the carcinogens that have been studied thus far, or to chemically similar compounds, result in a wide range of hemoglobin adduct levels (Table I). For example, we have found 3-aminobiphenyl adducts arising from exposure to environmental tobacco smoke to be 0.005–0.02 pmol/g Hb,[14] and adducts of acrylonitrile[15] or ethylene oxide[16] produced by active smoking are on the order of 100 pmol/g. The lowest concentrations correspond to hardly more than one part per trillion.

[14] M. Maclure, R. Ben-Abraham Katz, M. S. Bryant, P. L. Skipper, and S. R. Tannenbaum, *Am. J. Public Health* **79**, 1381 (1989).
[15] M. S. Bryant and S. M. Osterman-Golkar, *CIIT Act.* **11**(10), 1 (1991).
[16] E. Bailey, A. G. F. Brooks, C. T. Dollery, P. B. Farmer, B. J. Passingham, M. A. Sleightholm, and D. W. Yates, *Arch. Toxicol.* **62**, 247 (1988).

Methods of Analysis

The method of choice for the analysis of hemoglobin adducts has been gas chromatography combined with mass selective detection (GC-MS). This may not seem to be an obvious choice for something that might be regarded as a substructure of a protein or as a protein with somewhat modified physical properties. Nevertheless, the superior resolving power of GC-MS using capillary columns and the stringent demands of specificity and sensitivity have combined to direct research efforts to find ways to render adducts amenable to gas chromatography. Also, mass selective detection makes it possible to use isotopically labeled analogs ("isotopomers") as internal standards in a procedure known as isotope-dilution mass spectrometry.[17] Four specific examples of this technique, which is generally recognized as one of the most accurate for quantitative analysis, are described in this volume [42–45]. An alternative technique potentially applicable to whole protein, which is still in the developmental stage, is laser-induced fluorescence spectrometry. Its application to the study of adducts of polynuclear aromatic hydrocarbons (PAHs) is also described in this volume [46].

GC-MS can only be applied after adducts have been converted to a low molecular weight form and has been rendered volatile. Two approaches to accomplish the first step have been used successfully. One is hydrolytic cleavage of the carcinogen moiety from the polypeptide chain, which is used for aromatic amine, PAH, and tobacco-specific nitrosamine (TSN) adducts. The aromatic amines form adducts through a sulfinamide bond that is readily hydrolyzed to regenerate the amines. PAH epoxides and the diazonium ions generated by the TSN alkylate carboxylate groups of hemoglobin to form carboxylic esters, which are similarly hydrolyzed under relatively mild conditions. Adducts of the N-terminal valine amino group, on the other hand, are cleaved by a selective Edman procedure that results in an adduct incorporating one amino acid. The net effect, though, is still the same, namely, to form a low molecular weight analyte while leaving the protein intact.

A collateral benefit of the carcinogen residue cleavage approach is that it effects a substantially greater sample cleanup than might otherwise be realized. It is unlikely that an adduct would confer sufficient change in the properties of hemoglobin so that an effective means could be devised to separate adducted from nonadducted protein. In contrast, separation on the basis of molecular weight or on the other physical properties of

---

[17] A. P. De Leenheer, M. F. Lefevere, W. E. Lambert, and E. S. Colinet, *Adv. Clin. Chem.* **24,** 111 (1985).

hemoglobin, which are vastly different from those of many chemical carcinogens, can be very efficient.

Nevertheless, sample purification prior to GC-MS analysis remains a major challenge. One very effective solution to the sample concentration and purification problem is antibody affinity chromatography. Although this technique requires the production of antibodies, which involves a considerable commitment of resources, its power is such that it should be seriously considered where applicable. An example of its use is in the analysis of benzo[a]pyrene (BaP) adducts as described in this volume [45].

Validation of Analytical Methods

Validation of the quantitative analytical techniques may be undertaken on one of several levels. The most rigorous is to demonstrate that analysis of a hemoglobin sample containing a known amount of adduct as determined by independent means yields the correct value over an appropriate range of values. This has only been accomplished in a limited number of cases owing largely to the fact that independent means of determining the adduct content of a hemoglobin sample have been quite difficult to establish. A representative example is that of ethylene oxide–Val-1 adducts. N-Hydroxyethylvaline is stable to the acid hydrolysis conditions used to degrade protein to amino acids. Thus it is possible to alkylate hemoglobin with [14C]ethylene oxide, to hydrolyze to amino acids, and confidently to quantify N-hydroxyethylvaline by radiochromatography.[18]

When adducts are not so stable as to survive protein hydrolysis the difficulty of determining adduct concentration increases considerably. The major obstacle involves knowing what is adduct and what is not. Aromatic amine adducts regenerate aromatic amines if they react with water and thus are indistinguishable from free amines that might arise from other sources. Similarly, diol epoxide and diazonium ion adducts are hydrolyzed to products that are indistinguishable from the respective diol epoxide or diazonium ion hydrolysis products. Thus, it is clear that all nonadducted material that could be determined as adduct must be removed prior to analysis.

Removal of all noncovalently bound material from hemoglobin, though, is not at all simple because the protein is capable of binding a broad range of compounds with great avidity. If this were the only difficulty the problem might be tractable, but not all adducts are stable under the conditions to which they might be exposed in the process of attempted

[18] P. B. Farmer, E. Bailey, S. M. Gorf, M. Tornqvist, S. Osterman-Golkar, A. Kautiainen, and D. P. Lewis-Enright, *Carcinogenesis* (*London*) **7**, 637 (1986).

cleanup. We have found this to be particularly true for the adducts of BaP diol epoxide, which are determined as the tetrahydrotetrols. Although apparently quite stable in tetrameric hemoglobin, the adducts become labile if the protein no more than dissociates into dimers. As a consequence, we have been unable to find conditions under which free tetrahydrotetrols can be separated from hemoglobin without compromising adduct integrity. Similar remarks apply to aromatic amine adducts.

An alternative approach to validation is to determine the dose response for adduct formation in dosed experimental animals using the analytical technique to be validated.[19] If a linear relationship is observed, as it should be at sufficiently low doses, then the assay may be considered to give results that are at least proportional to the true values. The proportionality constant can then be estimated to within a certain range by other means, such as determining the maximum adduct level by measuring total radioactivity in the hemoglobin.

There are two potential drawbacks to this approach. One arises when carcinogen clearance is not complete within a short time frame, such as 1 day (e.g., BaP[20]), which creates an inherent nonlinearity in the dose–response relationship, which may not be distinguishable from nonlinearity of the assay. With clear understanding of the pharmacokinetics of the particular carcinogen, this potential confounder can be finessed.

The second difficulty, though, has no easy resolution. The hemoglobins of different species are strongly conserved, but they are all different, and this can lead to the formation of different adducts by different hemoglobins. The most notable example is that of the rat, which possesses a cysteine-125(β). This highly reactive residue is replaced by proline in human hemoglobin and by alanine in mouse hemoglobin, neither of which is capable of forming adducts. The interspecies difference in hemoglobins can also be more subtle. For example, both mouse and human hemoglobins form ester adducts with BaP diol epoxide, and although the human adducts appear to be stable, the mouse adducts are lost *in vivo* at a relatively rapid rate.[7] Thus, animal models may only be appropriate if the adducts formed are the same as those formed by human hemoglobin and have similar stabilities as well.

The ultimate validation of any method, finally, will be in human population studies. Here one is concerned not so much with determining the true values of adduct levels, but with obtaining reproducible results with good precision. This is not to minimize the worth of accuracy, but for most types of biomonitoring, precision and reproducibility are far more important.

[19] S. E. Murphy, A. Palomino, S. S. Hecht, and D. Hoffman, *Cancer Res.* **50**, 5446 (1990).
[20] M. Uziel and R. Haglund, *Carcinogenesis (London)* **9**, 233 (1988).

Within this context a number of assays may be said to have been validated. *N*-Hydroxyethylvaline, for example, exhibits a reasonable dose–response relationship with cigarette consumption,[16] as do adducts of 4-aminobiphenyl.[21] Adducts of TSN are far higher in snuff dippers and smokers than in nonsmokers.[22] Cyanoethylvaline, formed by reaction of Val-1 with acrylonitrile, is high in smokers and undetectable in nonsmokers.[15] *N*-Hydroxyethylvaline and hemoglobin–aniline adducts are greatly elevated in workers occupationally exposed to ethylene oxide[23] and aniline,[24] respectively.

Conversely, within groups of people with similar exposures to a carcinogen, it has been possible to demonstrate the effect of differences in metabolic phenotype on adduct levels. Specifically, slow acetylators would be expected to exhibit higher adducts of aromatic amines than fast acetylators exposed to the same amount of amine. This is, in fact, what was observed in three groups of subjects stratified by exposure.[25] The difference between the two phenotypes was not large, though, ranging from only 1.3 to 1.6.

Clearly, without good precision such differences would be impossible to detect without the statistical power of very large groups. Thus, it has been a major goal of each of the methods described in the following chapters to achieve high precision and reproducibility.

Qualitative Analysis

The methods described in the following chapters can be used not only for quantitative analysis of the particular adducts for which they were developed, but also for generating evidence for the existence of related adducts. No sample workup procedure will be 100% selective for a chosen compound, so it may be expected that related compounds will also be isolated. Recoveries will of course vary, depending on the degree of relatedness. Nevertheless, it has been fruitful to examine samples in a qualitative manner.

[21] M. S. Bryant, P. Vineis, P. L. Skipper, and S. R. Tannenbaum, *Proc. Natl. Acad. Sci. U.S.A.* **85,** 9788 (1988).
[22] S. G. Carmella, S. S. Kagan, M. Kagan, P. G. Foiles, G. Palladino, A. M. Quart, E. Quart, and S. S. Hecht, *Cancer Res.* **50,** 5438 (1990).
[23] U. Duus, S. Osterman-Golkar, M. Tornqvist, J. Mowrer, S. Holm, and L. Ehrenberg, *in* "Proceedings of the Symposium: Management of Risk from Genotoxic Substances in the Environment" (L. Frey, ed.), p. 141. Swedish National Chemical Inspectorate, Solna, 1989.
[24] J. Lewalter and U. Korallus, *Int. Arch. Occup. Environ. Health* **56,** 179 (1985).
[25] P. Vineis, N. Caporaso, S. R. Tannenbaum, P. L. Skipper, J. Glogowski, H. Bartsch, M. Coda, G. Talaska, and F. Kadlubar, *Cancer Res.* **50,** 3002 (1990).

GC-MS is a particularly powerful technique for qualitative analysis. In GC-MS it is only necessary to monitor a different portion of the chromatogram to detect structural analogs or to change the selected ion in order to detect compounds of different molecular weights. With this approach, we have detected a wide range of aromatic amine adducts in human bloods.[21] We have also detected adducts of chrysene diol epoxide in addition to those formed by benzo[a]pyrene diol epoxide.[26]

It must be emphasized, though, that these types of qualitative analyses should only be used as a means of generating hypotheses about possible exposures. The potential existence of any adduct should be confirmed by additional evidence such as relatedness to known exposure or any of a variety of physical–chemical analyses. Quantitative estimates would require the further development of assays such as those described in this volume.

Acknowledgments

This work was supported by grants from NIEHS (Nos. ES05622, ES02109, and ES04675).

[26] B. W. Day, P. L. Skipper, J. S. Wishnok, J. Coghlin, S. K. Hammond, P. Gann, and S. R. Tannenbaum, *Chem. Res. Toxicol.* **3,** 340 (1990).

[41] Quantitative Mass Spectrometry of Hemoglobin Adducts

By JOHN S. WISHNOK

The focus of this chapter is on quantitative analysis of hemoglobin–carcinogen adducts in epidemiological samples rather than on adducts formed *in vitro* or by dosing experimental animals. In the latter cases there is often a good deal of control over the amounts of material and the identities of the products; in the former there is virtually none except in the selection of the populations to be studied and the amount of blood that can ethically be obtained from the participating individuals. Despite these constraints, adduct quantitation can be done with most high-quality modern gas chromatography–mass spectrometry (GC-MS) equipment capable of splitless or on-column injections and chemical ionization, espeically electron-capture negative-ion chemical ionization (NICI or ECNI). The major differences from more routine GC-MS experiments arise from the fact that adduct analyses typically involve ultratrace quantitation of components present in unavoidably complex mixtures. In addition, especially for epidemiological studies, the absolute amount of material may be limited, thus precluding reanalysis or additional concentration or cleanup.

Electron ionization is the classical method of ionization in which the sample molecules interact with a beam of electrons, generally with an energy of 70 eV. The most common result is loss of an electron from the sample molecules to form a set of singly charged cations. The molecular ions typically contain excess energy and undergo subsequent bond ruptures to generate a characteristic set of fragment ions that constitute the mass spectrum. In many cases, the molecular ions are sufficiently unstable that their relative abundances are low or even absent. Chemical ionization is a less energetic process in which the sample molecules interact in the gas phase with preformed ions or with electrons of energy less than 70 eV. In practice, this is done most commonly by adding a gas, e.g., methane, to the ion source at sufficient pressure that the ionizing electrons interact primarily with the gas rather than with the sample. In the case of methane, the result is a set of cations, i.e., CH_5^+, $C_2H_5^+$, and $C_3H_5^+$, along with thermal electrons. If the sample is relatively basic, it will then be ionized by proton transfer and adduct formation to yield a set of ions with masses of $M + 1$ (from protonation), $M + 29$ (from the $C_2H_5^+$ adduct, and $M + 41$ (from the $C_3H_5^+$ adduct), where M is the molecular weight of the sample. This process is called positive chemical ionization. If the sample molecules have a high electron affinity, they may interact with the thermal electrons to form, typically, singly charged anions; this is most commonly designated negative-ion chemical ionization. Both types of chemical ionization generate less energetic sample ions than is the case with electron ionization, and there is consequently less fragmentation. Electron ionization and chemical ionization are complementary in that chemical ionization often confirms the molecular weight of the sample whereas electron ionization generates the fragmentation patterns that offer clues to the molecular structure.

Electron ionization (EI) or positive chemical ionization (PCI) are often useful in cases in which the adduct is plentiful or additional cleanup steps are incorporated into the workup, but NICI analysis of fluorinated derivatives appears to give the best combination of sensitivity and selectivity for the widest range of compound types. Each new analysis, however, should nonetheless be developed individually by assessing a variety of derivatives in all three ionization modes, both with standards and with actual samples or at least spiked media. Because of possible interference from coeluting components, the best ionization mode for a standard may not be best for the actual analyses. In addition, a given ionization mode may not be best for all members of a series. NICI, for example, is a good ionization mode for most of an extensive list of trimethylsilyl (TMS) derivatives of polycyclic aromatic hydrocarbon (PAH) diols and tetrols; EI, however, is better for derivatized styrene glycol.[1]

These considerations represent a worst-case situation in which the adducts of interest are present at levels near the realistic detection limits of the mass spectrometer. There are cases, e.g., the hydroxyethylvaline adducts (see [43] in this volume), in which the adduct levels are relatively high and the instrumental constraints therefore are not so severe.[2,3]

The data system should have software that can be configured conveniently for manual integrations; the concentrations of analytes are often too low to be dealt with effectively with the automatic integration routines built into some of the commercial GC-MS software.

Despite the high intrinsic sensitivity of GC-MS, adequate operating sensitivity requires attention to detail in every aspect of the methods, including solvent purity, sample preparation, and chromatography. Maximum sensitivity is usually obtained only with a manually tuned, newly cleaned ion source, a new GC column with appropriate coating, and a clean silanized GC injector liner; the sensitivity will then gradually decline as these conditions inevitably deteriorate.

The following sections of this chapter cover most of these considerations in detail.

Mass Spectrometry Tuning

Tuning of the mass spectrometer for highest sensitivity may require optimization of both source temperature and moderator/reagent gas pressure in chemical ionization in addition to optimization of the focusing-electrode voltages. If the highest sensitivity is obtained at lower source temperatures (e.g., 100°–150° versus 200°–250°), this may then require a tradeoff with respect to frequency of source cleaning because of the increased likelihood of material condensing on the cooler source. Negative-ion tuning is relatively straightforward for most modern instruments but somewhat less so for older instruments. If only moderate sensitivity is required, the autotune (automatic tuning) routines, especially those on newer instruments, may be adequate. The autotune algorithms, however, are usually designed for a specific purpose, e.g., to produce a particular

[1] B. W. Day, S. Naylor, L.-S. Gan, Y. Sahali, T. T. Nguyen, P. L. Skipper, J. S. Wishnok, and S. R. Tannenbaum, *J. Chromatogr. Biomed. Appl.* **562,** 563 (1991).

[2] M. Törnqvist, J. Mowrer, S. Jensen, and L. Ehrenberg, *Anal. Biochem.* **154,** 255 (1986).

[3] M. Törnqvist, S. Osterman-Golkar, A. Kautiainen, S. Jensen, P. B. Farmer, and L. Ehrenberg, *Carcinogenesis (London)* **7,** 1519 (1986).

set of relative abundances for the ions in a tuning standard or to allow the instrument to meet specifications on a given test compound, and may therefore not generate the best tune for other analytes.

For maximum sensitivity, especially as the source gets dirty, manual tuning will probably be necessary. This involves adjusting the electron energy and the voltages on the repeller and various focusing lenses (the number and geometry of the lenses vary from manufacturer to manufacturer and sometimes from model to model, so it is difficult to discuss this specifically) to obtain maximum signal and good peak shape from a standard tuning compound, typically perfluorotributylamine (PFTBA). Adduct quantitation is most often done by selectively monitoring a few individual ions, so "correct" relative intensities of the tuning peaks, i.e., given relative abundances for ions in the full mass spectra of the tuning standards, are usually less important than is signal strength in the m/z (mass-to-charge ratio) range expected for the analytes under investigation. For operators inexperienced with manual tuning, it is important to record the values of the autotune parameters before beginning a manual tune; it will then always be possible to set the instrument back to the starting point. The best tuning procedure for a given instrument usually is determined by trial and error. A good starting point is the manual tuning sequence outlined in the operator's manual, but it is often necessary to go beyond the suggested ranges. This may be especially true for source temperature, moderating gas pressure, electron energy, and X-ray voltage, which may not be included in the manual tuning sequences.

Suggested methane pressure for CI experiments, for example, is typically $(1.5-3) \times 10^{-4}$ Torr in operator's manuals. On one of the instruments in our laboratory, however, optimum NICI performance is usually obtained at this pressure, whereas pressures around 5×10^{-5} Torr are better on another. Note also, that the pressure indicated on the gauge or dial will depend on the actual location of the sensor inside the instrument. Optimum pressure for PCI and NICI may not be identical. Figure 1 shows the effect of methane pressure on the relative abundances of ions in the PFTBA NICI spectrum.

Standard electron energies are 70 eV for EI and 200 eV for chemical ionization. Little can usually be gained by changing this parameter in EI, but optimum sensitivity in chemical ionization may occur at values between 100 and 200 eV.

If exceptional sensitivity is required for a single compound, and adequate amounts of a standard are available, it may be possible to introduce the standard into the instrument via the direct insertion probe and then tune the instrument specifically on the ion(s) being used for quantitation.

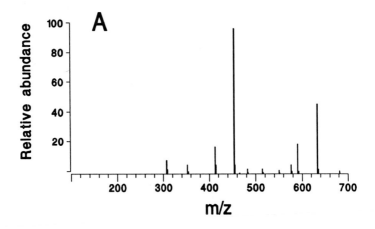

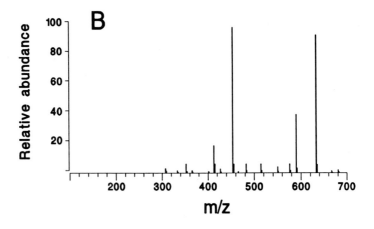

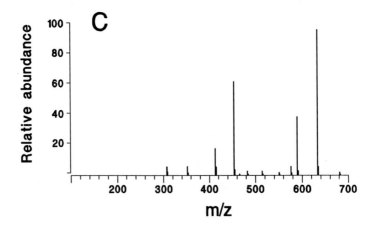

A detailed example of nonconventional NICI tuning is given in Abdel-Baky and Giese.[4]

Performance in NICI is generally more sensitive than EI or PCI to the condition of the ion source. As the source gets dirty, the instrument may perform well in electron ionization (especially at lower masses) and positive chemical ionization and yet be of little use in NICI. It is therefore important to evaluate routinely the instrument with a familiar standard in the appropriate ionization mode.

Qualitative Analysis

There is no generally successful GC-MS technique for identifying completely unknown adducts, although some adducts have been identified in individual cases—usually by a combination of techniques.[5-7] Because of the complexity of the systems, even after extensive cleanup, and the low concentrations of adducts, most carcinogen–hemoglobin adduct research with biological samples has been based on analysis of suspected adducts, i.e., a target compound has been selected and its presence then confirmed by comparison of chromatographic retention times with those of a standard. The chemical ionization mass spectra, NICI spectra in particular, are often simple, thus usually providing only a few useful ions for generation of selected ion chromatograms. As with any target compound analysis, the fewer ions monitored, the less certain is the identity of the compound. Wherever possible, pooled samples should be concentrated and full spectra obtained in order to confirm the identity of the analyte.

[4] S. Abdel-Baky and R. W. Giese, *Anal. Chem.* **63**, 2986 (1991).

[5] P. B. Farmer, *Biomed. Environ. Mass Spectrom.* **17**, 143 (1988).

[6] A. Weston, M. L. Rowe, D. K. Manchester, P. B. Farmer, D. L. Mann, and C. C. Harris, *Carcinogenesis (London)* **10**, 251 (1989).

[7] B. W. Day, P. L. Skipper, J. S. Wishnok, J. Coghlin, S. K. Hammond, P. Gann, and S. R. Tannenbaum, *Chemical Res. Toxicol.* **3**, 340 (1990).

FIG. 1. Effect of methane pressure on the negative-ion chemical ionization mass spectrum of perfluorotributylamine. The methane pressure was varied from about 1×10^{-4} Torr (A) to about 3×10^{-4} Torr (C) as indicated on the manifold pressure gauge (Hewlett-Packard 5989). The instrument was then manually tuned for maximum abundance at $m/z = 633$. The source temperature was 200° for B and C and 204° for A. The spectra were redrawn from manual-tune reports.

Quantitative Analysis

Internal Standards

Successful quantitation by GC-MS depends not only on instrumental sensitivity but also on chromatography, sample workup, and quantitation methods, the most important of which is probably a good internal standard, i.e., a compound present at a known concentration in the solution along with the compound being quantitated, and which can thus be measured in the same run as the analyte. The analyte concentration is then determined as a ratio—by comparing its detector response to that of the internal standard—rather than as an absolute value. At the simplest level this technique will correct for variations in injection volume or in the detector response. Depending on the point in the protocol at which the internal standard is added, and on the degree to which its properties resemble that of the analyte, it may also correct for losses during the workup, e.g., losses due to incomplete recoveries or to decomposition. Despite intensive solvent, sample, and instrument preparations, the samples are complex and the analyte concentrations are often near the limits of detection. In addition, the characteristics of both the gas chromatograph and the mass spectrometer usually change during a series of analyses due to buildup of material on the inlet and column of the mass spectrometer and on the ion source of the mass spectrometer. The amount of material getting into the mass spectrometer and the limit of detection by the mass spectrometer may therefore decrease, sometimes dramatically, over a day or sometimes even a few hours. The effects of these variations can be minimized by using a good internal standard.

The major operational requirements for an internal standard are stability, good chromatographic behavior (i.e., good peak shape and a convenient retention time), and overall properties that will reflect the behavior of the analyte throughout the course of the workup; the best internal standards for GC-MS are therefore stable isotope-labeled analogs ("isotopomers") of the analytes. Except for slight differences in retention time, these compounds will behave virtually identically with the analytes with respect to extraction efficiencies and derivatization yields. The near coelution of the standard with the analyte is also useful as a retention-time marker in instances when retention times are affected by changing column conditions. As noted by Farmer and co-workers,[8] it is important to assure that the label does not exchange during the workup procedure and that

---

[8] P. B. Farmer, E. Bailey, and J. B. Campbell, *IARC Sci. Publ.* **59**, 189 (1984).

it is present in useful fragment ions in the mass spectra. The molecular weight difference between analyte and internal standard should be as large as practical because of possible complications due to natural abundances or to interfering ions in the mass spectra, e.g., those arising from protonation–deprotonation reactions in the ion source. The contribution of natural abundance ^{13}C to the $M + 1$ is 1.1% per carbon and can therefore be substantial for compounds with high carbon numbers. With aminobiphenyl, for example, the internal standard is often fully deuterated[9–11] and consequently has a mass 9 Da higher than that of aminobiphenyl alone. There are no contributions of either compound to the mass spectra of the other. If the internal standard, however, contains, say, a single deuterium or ^{15}N per molecule, then the $M + 1$ ion of the analyte will have the same mass as the molecular ion of the internal standard and corrections may be necessary either by calculation[12–14] or by construction of standard curves. The corrections are additionally complicated by the fact that the observed relative intensities of the natural abundance isotope peaks may vary from instrument to instrument and from day to day on a given instrument. For example, the theoretical relative abundance of $M + 1$ for nitrobenzene is about 6.6%; on one of our instruments this is typical whereas on another the observed value is persistently 6.2%. If stable isotope internal standards are unavailable or impractical, fluorinated or chlorinated analogs often work well, and have been used in a number of cases, e.g., in analyses of 4-aminobiphenyl adducts[15–17] and 4,4'-methylene-bis(2-chloroaniline).[18]

[9] A. Weston, N. E. Caporaso, K. Taghizadeh, R. N. Hoover, S. R. Tannenbaum, P. L. Skipper, J. H. Resau, B. F. Trumpond, and C. C. Harris, *Cancer Res.* **51,** 5219 (1991).

[10] M. S. Bryant, P. Vineis, P. L. Skipper, and S. R. Tannenbaum, *in* "Methods for Detecting DNA-Damaging Agents in Humans" (H. Bartsch, K. Hemminki, and I. K. O'Neill, eds.), p. 133. IARC, Lyon, 1988.

[11] S. R. Tannenbaum, M. S. Bryant, P. L. Skipper, and M. Maclure, *Banbury Rep.* **23,** 63 (1986).

[12] K. Biemann, "Mass Spectrometry: Organic Chemistry Applications." McGraw-Hill, New York, 1962.

[13] L. C. Green, D. A. Wagner, J. Glogowski, P. L. Skipper, J. S. Wishnok, and S. R. Tannenbaum, *Anal. Biochem.* **126,** 131 (1982).

[14] W. Schramm, T. Louton, and W. Schill, *Fresnesius' Z. Anal. Chem.* **294,** 107 (1979).

[15] L. C. Green, P. L. Skipper, R. J. Turesky, M. S. Bryant, and S. R. Tannenbaum, *Cancer Res.* **44,** 4254 (1984).

[16] P. L. Skipper, L. C. Green, M. S. Bryant, S. R. Tannenbaum, and F. F. Kadlubar, *IARC Sci. Publ.* **59,** 143 (1984).

[17] M. S. Bryant, P. L. Skipper, S. R. Tannenbaum, and M. Maclure, *Cancer Res.* **47,** 602 (1987).

[18] G. Sabbioni and H.-G. Neumann, *Arch. Toxicol.* **64,** 451 (1990).

Integration

Most modern mass spectrometer data systems contain good automatic integration routines. These are designed for complex mixtures and generally work well even with partially resolved peaks, drifting baselines, and large dynamic ranges (i.e., mixtures containing components present both at very high and very low concentrations). If the chromatograms are relatively clean in the retention region of interest, if the analyte peaks are symmetrical, if there are no shoulders, and the signal-to-noise characteristics of the analyte signal are good (e.g., signal-to-noise ratio greater than about 2 or 3), then the autointegration routines should probably work well with appropriate parameters. The details of the autointegration setups vary from data system to data system, even those from a given manufacturer, but in general require choosing retention windows, abundance thresholds, and parameters to distinguish peaks from baseline drift. An example of the effects of interaction between chromatography and integration threshold is given in Table I; the precision of integration is notably better when the chromatography is good.

If the criteria of adequate analyte concentration and good chromatography are not met, integrations can be done manually. Manual integration routines also vary from instrument to instrument but usually require the operator to enter the retention times for the integration to start and stop and may also require one or more thresholds. The thresholds can be set

TABLE I
EFFECT OF THRESHOLD SETTING AND PEAK SHAPE ON
AUTOMATIC INTEGRATION[a]

Threshold	Peak area (tailing)	Peak area (no tailing)
8	1170194	1131051
10	1227421	1129873
12	1242520	1126589
14	1077938	1111594
16	905066	1109478
Mean	1124638	1121717
Standard deviation	138689	1032
Range[b]	30	2

[a] The numbers were generated by successive integrations of two chromatograms, one with good peak shape and one with pronounced tailing, using the automatic integration routine with thresholds set as in column one. The compound is nitrobenzene. The integrations were done on a Hewlett-Packard DOS-based data system for a 5971 mass-selective detector.
[b] Range expressed as percentage of the mean.

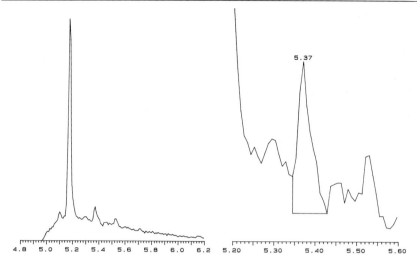

FIG. 2. Graphic manual integration output showing the details of the integrated area for the peak eluting at 5.37 min.

much lower in these cases because only the peaks of interest will be integrated. The most convenient manual integrations allow "zooming in" on the peak(s) of interest, i.e., magnifying that area of the chromatogram, and selecting the start and stop times for integration by visual positioning of a cursor with a mouse or from the keyboard. The routine should then redraw the chromatogram with the integrated area highlighted so the operator can decide whether the integration is appropriate (e.g., Fig. 2). This is especially useful if peaks are tailing or contain shoulders, although, as noted above, it is always easier to quantitate sharp, symmetrical peaks. Although most data systems have the ability to do all of these operations, it is often necessary for the operator to write custom routines linking them smoothly together.

Quality Control

As implied above, if performance appears to have deteriorated, there are many possible causes. If a chromatographic peak becomes asymmetric it may be due to column or injector contamination, to a leak, or to an unexpected component in the sample. If the limit of detection appears lower than expected, it may be due to these chromatographic problems or to mass spectrometer problems such as a dirty ion source, a failing electron multiplier, or changing tune parameters. This may also originate in the sample, most simply by incomplete derivatization or by decomposition of the analyte or the derivative. More subtle problems, e.g., lower apparent analyte levels because of a contaminant coeluting with the internal standard also sometimes occur. It is thus important to develop a system to monitor performance and to help discover and diagnose problems as they occur. The simplest routine procedure is to check the mass spectrometer tuning regularly for total and relative ion abundance. When sufficient material is available, a useful technique is to randomly reanalyze recoded samples.[6] If the analyte—or especially the derivative of choice—is stable, a solution containing both the analyte and the internal standard or the respective derivatives can be made up in relatively large volume and used either routinely or diagnostically to evaluate the overall performance of the GC-MS system.

Pitfalls

The major pitfall is probably the uncertainty that the quantitation is being done on the correct compound. Because of the low concentrations of the analytes and the simplicity of chemical ionization mass spectra, quantitation is often done on single-ion chromatograms. There is rarely enough material for a complete mass spectrum or for structure confirmation by other types of spectroscopy.[7,8] The qualitative information thus typically consists of GC retention times and the fact that the chromatograms are generated by characteristic ions. Although the ions are at least unequivocal (but not sufficient), the retention times can change as the column gets dirty or shorter (as portions are broken off to remove contaminated lengths), so there is always some doubt that the data are meaningful. In a study of aminobiphenyl levels in umbilical cord blood, for example, several samples appeared to have very high levels of 3-aminobiphenyl as indicated by peaks at the appropriate retention time in the selected ion chromatograms at an m/z of 295 (the base peak in the NICI mass spectrum of the pentafluoropropionyl derivative of 3-aminophenyl). In order to confirm this, several samples were pooled and concentrated in order to obtain a full mass spectrum. This spectrum, although containing a prominent ion at an m/z of 295, bore little resemblance to that of the authentic 3-aminobiphenyl derivative.[19]

Thus, although GC-MS is clearly one of the most powerful tools in hemoglobin adduct research, successful use of this technique requires continual intellectual as well as technical rigor.

[19] K. Taghizadeh, unpublished data.

Acknowledgments

This work was supported by DHHS Shared Instrument Grant 1-S10-RR1901 and NIH Grants ES01640, ES02109, ES04675, CA44306, and CA26731. Thanks are due to Paul Skipper for helpful comments on the manuscript.

[42] Aromatic Amine–Hemoglobin Adducts

By Paul L. Skipper and W. G. Stillwell

Human exposure to aromatic amines has multiple documented and potential sources, including tobacco smoke inhalation,[1] workplace contamination,[2] pesticide residues in food,[3] airborne nitroaromatic compounds,[4] and enterobacterial reduction of dyes.[5] Many aromatic amines, perhaps most of the ones with one- and two-ring structures, form hemoglobin adducts (see Scheme I) through a process of hepatic metabolic oxidation to an aryl hydroxylamine followed by heme-mediated oxidation to a nitrosoarene and reaction with a cysteine sulfhydryl group.[6] In human hemoglobin, 4-aminobiphenyl forms a single stable adduct with β93(F9) cysteine, which has been fully characterized as a sulfinic amide.[7] Adduct levels of 4-aminobiphenyl in persons undergoing smoking withdrawal decline nearly as expected on the basis of normal hemoglobin turnover.[8] Other studies with experimental animals, though, indicate that amine adducts or amine-adducted hemoglobin can be removed from circulation more rapidly than normal hemoglobin.[9,10]

[1] M. S. Bryant, P. L. Skipper, S. R. Tannenbaum, and M. Maclure, *Cancer Res.* **47,** 602 (1987).
[2] L. Lewalter and U. Korallus, *Int. Arch. Occup. Environ. Health* **56,** 179 (1985).
[3] G. Sabbioni and H.-G. Neumann, *Carcinogenesis (London)* **11,** 111 (1990).
[4] J. Arey, B. Zielinska, R. Atkinson, and A. M. Winer, *Atmos. Environ.* **21,** 1437 (1987).
[5] B. W. Manning, C. E. Cerniglia, and T. W. Federle, *Appl. Environ. Microbiol.* **50,** 10 (1985).
[6] H.-G. Neumann, *in* "Molecular Dosimetry and Human Cancer" (J. D. Groopman and P. L. Skipper, eds.), p. 363. CRC Press, Boca Raton, FL, 1991.
[7] D. Ringe, R. J. Turesky, P. L. Skipper, and S. R. Tannenbaum, *Chem. Res. Toxicol.* **1,** 22 (1988).
[8] M. Maclure, M. S. Bryant, P. L. Skipper, and S. R. Tannenbaum, *Cancer Res.* **50,** 181 (1990).
[9] H.-G. Neumann, *IARC Sci. Publ.* **59,** 115 (1984).
[10] K. L. Cheever, D. E. Richards, W. W. Weigel, K. B. Begley, D. G. DeBord, T. F. Swearengin, and R. E. Savage, Jr., *Fundam. Appl. Toxicol.* **14,** 273 (1990).

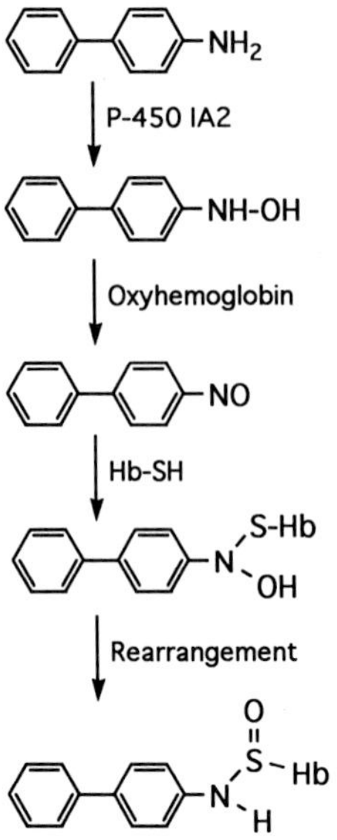

Ph—C6H4—NH2

↓ P-450 1A2

Ph—C6H4—NH-OH

↓ Oxyhemoglobin

Ph—C6H4—NO

↓ Hb-SH

Ph—C6H4—N(S-Hb)(OH)

↓ Rearrangement

Ph—C6H4—N(H)—S(=O)—Hb

SCHEME I

Overall conversion of aromatic amines to adducts can be very high, as much as 10% of intake, making these compounds very amenable to biomonitoring. The adducts are readily hydrolyzed to regenerate free aromatic amines, which can be determined by gas chromatographic analysis.[1,11] The most sensitive and selective detection, which is necessary for studies such as those involving tobacco smoke exposure, is negative-ion chemical ionization mass spectrometry, and this technique is described in detail in [41]. Other exposure scenarios might not demand as sophisticated and expensive equipment, and the interested researcher should consider if alternatives such as electron capture or nitrogen-specific detection would be satisfactory.

Many of the recommendations and procedures that are described below are based on the assumption that the studies will involve extremely low levels of adduct, on the order of 0.01–1 pmol/g hemoglobin. The degree of stringency can undoubtedly be reduced if higher adduct levels are involved. However, it cannot be overemphasized that appropriate controls be employed: the determination of adducts is through analysis of aromatic amines, which are no different from amines that produced the adduct in the first place, and if these are significant environmental contaminants,

[11] L. C. Green, P. L. Skipper, R. J. Turesky, M. S. Bryant, and S. R. Tannenbaum, *Cancer Res.* **44,** 4254 (1984).

then the results can be compromised if the environmental contamination extends to the analytical equipment or laboratory.

Reagents and Solvents

Hexane of a grade suitable for organic trace analysis such as Baker Resi-analyzed (Phillipsburg, NJ) is distilled sequentially through two high efficiency fractionating stills. An effective column for the stills can be made by filling a 50-cm jacketed Vigreux column with small glass helices. Water is distilled from a dilute $KMnO_4/KOH$ solution. Trimethylamine solution in hexane is prepared by adding 1 ml 10 M NaOH to an aqueous solution of 1 g trimethylamine hydrochloride and extracting with 5 ml hexane. It should be stored in a tightly stoppered bottle and not kept more than 1 week. Derivatization grade pentafluoropropionic anhydride (PFPA) should be used.

Equipment

Ideally, all equipment should be dedicated. This applies especially to items such as rotary evaporators and associated glassware, stills, and dialysis equipment. All glassware should be cleaned by soaking in a bath of ethanolic (200 proof) KOH and then rinsed with distilled H_2O and oven dried.

Isolation of Hemoglobin

Red blood cells are separated from whole blood (10 ml) by centrifugation and washed three times with phosphate-buffered normal saline (PBS). They may then be used directly or frozen for storage. Analysis after as long as 2 years in storage at $-20°$ has indicated that there is no loss of adducts over this length of time. Freezing is also convenient in that it facilitates cell lysis, which is accomplished by adding 2–3 volumes of distilled H_2O per volume of packed red blood cells and 2 ml toluene, which has previously been extracted with 1 N HCl. The mixture is shaken vigorously and allowed to stand for 30 min, after which it is centrifuged at 10,000 g for 20 min. The aqueous phase is then dialyzed against distilled water for 3 days, the water being changed each day.

Adduct Hydrolysis

The dialyzed hemoglobin solutions are transferred to 50-ml screw-cap tared centrifuge tubes and the weight of each is determined. The hemoglobin concentration is determined by Drabkin's assay. Appropriate internal standards are added with complete mixing and the solutions are then allowed to stand for $\frac{1}{2}$ to 1 hr. Following this a solution of 10 M NaOH is added, using 0.01 volume per volume of hemoglobin solution. The color of the solution should change from red to brown rather quickly. If the hemoglobin concentrations are particularly high (>50 mg/ml), more NaOH may be needed. After the addition of NaOH, the solutions are held for 2–3 hr at room temperature.

Extraction of the amines is performed twice with a volume of hexane sufficient to nearly fill the centrifuge tube. The tubes are shaken gently for several minutes and then frozen to break the resultant emulsion. Too vigorous extraction results in the formation of unbreakable emulsions. The hexane layer is removed by pipette and passed through a column of drying agents (Na_2SO_4 and $MgSO_4$) into a 50-ml pear-shaped flask. A convenient column can be made from Kimble disposable serological 10-ml glass pipettes as follows. First, a glass bead with diameter slightly greater than the pipette opening is introduced into the pipette. Anhydrous Na_2SO_4 is then added to form a layer 2 cm deep. Finally, anhydrous $MgSO_4$ is added to form an upper layer of about 2 cm. The Na_2SO_4, being granular, serves to facilitate flow around the glass bead while blocking the passage of any of the powdery $MgSO_4$ into the receiving flask.

We have found hexane to be the optimum solvent for amine extractions because it extracts so little else from the hydrolyzates. It is clearly suitable for the relatively nonpolar unsubstituted or alkyl-substituted aromatic monoamines. Other have also used hexane for extraction of chlorinated anilines.[12] One laboratory, though, has found hexane to be ineffective and uses dichloromethane instead.[13]

Derivatization

The dried, combined hexane extracts are treated with 5–10 μl of trimethylamine in hexane and 3–5 μl of PFPA. After 10 min or longer the solvent is removed with a rotary evaporator until the volume has been reduced to 1–2 ml. The solution is transferred to a $\frac{1}{2}$ dram vial and the remainder of the solvent is removed. This can be performed conveniently with a rotary evaporator using a container for the vial made from a standard 24/40 ground glass outer joint. These joints are supplied with an integral length of glass tubing that has a suitable inside diameter to contain the

[12] G. Sabbioni, *Chem.-Biol. Interact.* **81,** 91 (1992).
[13] P. DelSanto, G. Moneti, M. Salvadori, C. Saltutti, A. DelleRose, and P. Dolara, *Cancer Lett.* **60,** 245 (1991).

vials. The tubing is cut to an appropriate length, and the end is flame-sealed. When analyzing particularly volatile amines, such as aniline or alkylanilines, it is important not to leave the vials on the evaporator any longer than necessary to remove the hexane. The samples are reconstituted in a minimal volume, typically 20 μl, of hexane or heptane for injection onto the capillary GC column. A Teflon liner must be used in the caps of these vials because the rubber liners leach contaminants that interfere with the analysis.

Internal Standards

The best internal standard is hemoglobin that has been adducted by the amine of interest labeled with a stable isotope. Adducted hemoglobin can be made by reacting freshly isolated human hemoglobin with the appropriate aromatic hydroxylamine. Details of the synthesis of the hydroxylamines will vary depending on the structure, but in general they can be formed by reduction of the corresponding nitroarenes. After reaction of hemoglobin with the aromatic hydroxylamine, the mixture is dialyzed to remove unbound material. It can then be diluted to a suitable concentration of adduct, divided into aliquots sufficient for a batch of samples to be analyzed, and stored frozen at $-20°$. The concentration of adduct is determined by using the normal isotope form of the amine as an internal standard.

An alternative but less satisfactory internal standard is isotopically labeled amine, which is prepared as a 0.1 N HCl solution. Most aromatic amines are stable in acidic aqueous solution, so these can be stored in the refrigerator for extended periods of time.

Isotopically labeled internal standards are, of course, unsuitable unless mass selective detection is used. If it is desired to use an electron capture or flame ionization detector, then monofluoroaromatic amines or amine adducts make good internal standards. The substitution of fluorine for one hydrogen atom has only a slight effect on the physical properties of the amine, just enough to render it separable by capillary GC from the unsubstituted amine.

Chromatographic Analysis

Gas chromatography is best performed using high-resolution bonded stationary-phase capillary columns. When using stable isotope-labeled internal standards and mass spectrometric detection, nonpolar stationary phases such as DB-1 (J & W Scientific, Folsom, CA) are to be preferred because they are extremely durable and performance changes little with time. Separation of monofluoro internal standards, however, requires a polar stationary phase such as Supelcowax 10 (Supelco, Bellefonte, PA). These phases are far more temperature sensitive and degrade slowly when heated to the temperatures required to elute naphthyl- and biphenyl-amines, so care must be taken in their use.

Simple temperature programs are usually sufficient to achieve good resolution of almost all aromatic amines in one run. The sample can be injected at 60°–100°. Temperature is then ramped linearly at 15°–20°/min to the chosen final temperature. A postrun temperature hold of 2–5 min at a temperature 20° higher than that required to elute the last amine of interest is advisable to avoid carryover from one run to the next.

Splitless injection is essential in the analysis of trace levels of adducts, such as those produced by tobacco smoke exposure, in order to obtain high sample loading of capillary columns. On-column injection is an alternative that achieves even higher sample loading. In our 4-aminobiphenyl adduct studies we routinely inject 20% of the entire sample (4 μl out of 20 μl) in the splitless mode.

Mass Spectrometry

This section will only deal with some specifics that are not addressed in the more general discussion of [41]. It is assumed that the mass spectrometer is equipped for negative-ion chemical ionization.

The PFPA derivatives of aromatic amines produce a very simple negative-ion mass spectrum[14] generally composed of a very weak ion at 1 mass unit less than the molecular weight, the base peak produced by loss of the elements of HF, and several other weak ions, including one at 38 mass units less than the molecular weight, probably resulting from loss of F_2. Each amine is detected by monitoring the ion current at the mass/charge ratio corresponding to the mass of the base peak, which is 20 mass units less than the molecular weight. It should be realized that a monofluoramine will produce significant ion current at 38 mass units below its molecular weight, which is also the same value as the monitored base peak ion produced by the nonfluorinated amine. Thus, if a monofluoramine is used as internal standard, it is necessary that a GC column be used that can separate the two amines.

[14] W. G. Stillwell, M. S. Bryant, and J. S. Wishnok, *Biomed. Environ. Mass Spectrom.* **14**, 221 (1987).

There is no significant mass defect for most of the PFPA derivatives. Thus, the m/z values that are monitored should be adjusted only for the offset observed in the mass axis calibration.

We have not observed any significant difference in mass spectrometric detector response to the pairs of isomers 4-aminobiphenyl and 4-aminobiphenyl-d_9 or aniline and aniline-d_5. The unadjusted peak area ratio for quantitation is therefore applicable, but it is recommended that this be verified for a given instrument or if other isotopes such as ^{13}C or ^{15}N are used.

Comments

It is critically important in these analyses to maintain cleanliness of materials and workplace and to ensure that samples contact only inert materials. This advice is pertinent, of course, to all trace level analysis, but is repeated here because it is easy to overlook if one is used to part per million levels rather than part per trillion, and one part per trillion is in fact possible with the techniques described in this chapter.

Rubber is one of the most significant potential sources of contamination. We can routinely find dibenzylamine in samples, and it is possible that this amine comes from the rubber stoppers used in blood collection tubes.[15] We have also detected other aromatic amines with antioxidant-type structure and suspect that these too arise from contact with rubber at some point. Thus, every effort should be made to use only glass, stainless steel, or Teflon materials.

The precision of this assay can be such that a coefficient of variation (CV) of 5–10% can be obtained. When analyzing for 4-aminobiphenyl adducts with 4'-fluoro-4-aminobiphenyl as standard we obtain a CV near 10%. This is reduced to near 5% when hemoglobin adducted with perdeutero 4-ABP is used. The use of an amine with a different structure can lead to a considerably greater CV.

Acknowledgment

This work was supported by grants from NIEHS (Nos. ES04675 and ES02109).

[15] J. W. Danielson, G. S. Oxborrow, and A. M. Placencia, *J. Parenter. Sci. Technol.* **38,** 90 (1984).

[43] Epoxide Adducts to N-Terminal Valine of Hemoglobin

By MARGARETA TÖRNQVIST

N-Alkyl Edman Method for Hemoglobin Adduct Determination

Measurement of hemoglobin (Hb) adduct levels, for the purpose of identification and dose monitoring of reactive compounds *in vivo*,[1] by analysis of substituted amino acids (histidines, cysteines)[2,3] following total acid hydrolysis of the protein, is insensitive, time-consuming, and susceptible to artifact formation,[4] whereas enzymatic hydrolysis is often incomplete at modified amino acids. In the search for methods permitting the determination of adducts without hydrolysis, the Edman degradation for protein sequencing[5] offered a potential solution because it was found that N-substituted N termini (valines) were cleaved off without acidification following treatment with phenyl isothiocyanates.[6]

The procedure comprises derivatization of globin with pentafluorophenyl isothiocyanate (PFPITC) and analysis by gas chromatography–mass spectrometry (GC/MS) of the pentafluorophenylthiohydantoins (PFPTHs) of *N*-alkylvalines originating from the reaction of N termini with alkylators[7] (Fig. 1). Current studies indicate that the PFPTHs of *N*-alkylvalines are formed without the intermediate formation of thiazolinones, as observed in the corresponding reaction of unsubstituted N termini[5] (R = H in Fig. 1). For understanding the reaction conditions it should be noted that the first step of the reaction, the coupling of the reagent to the amino group, has a higher pH optimum than does the second step, the cleavage of the derivative.

The method has been applied, for example, to the determination of *N*-(2-hydroxyethyl)valine (HOEtVal), *N*-(2-hydroxypropyl)valine, and *N*-hydroxyphenethylvaline, resulting from exposure of humans and ani-

[1] L. Ehrenberg and S. Osterman-Golkar, *Teratog., Carcinog., Mutagen.* **1**, 105 (1980).
[2] C. J. Calleman, L. Ehrenberg, B. Jansson, S. Osterman-Golkar, D. Segerbäck, K. Svensson, and C. A. Wachtmeister, *J. Environ. Pathol. Toxicol.* **2**, 427 (1978).
[3] E. Bailey, T. A. Connors, P. B. Farmer, S. M. Gorf, and J. Rickard, *Cancer Res.* **41**, 2514 (1981).
[4] C. J. Calleman, L. Ehrenberg, S. Osterman-Golkar, and D. Segerbäck, *Acta Chem. Scand., Ser. B* **33**, 488 (1979).
[5] P. Edman and A. Henschen, in "Protein Sequence Determination" (S. B. Needleman, ed.), p. 232. Springer-Verlag, Berlin and New York, 1975.
[6] S. Jensen, M. Törnqvist, and L. Ehrenberg, *Environ. Sci. Res.* **30**, 315 (1984).
[7] M. Törnqvist, J. Mowrer, S. Jensen, and L. Ehrenberg, *Anal. Biochem.* **154**, 255 (1986).

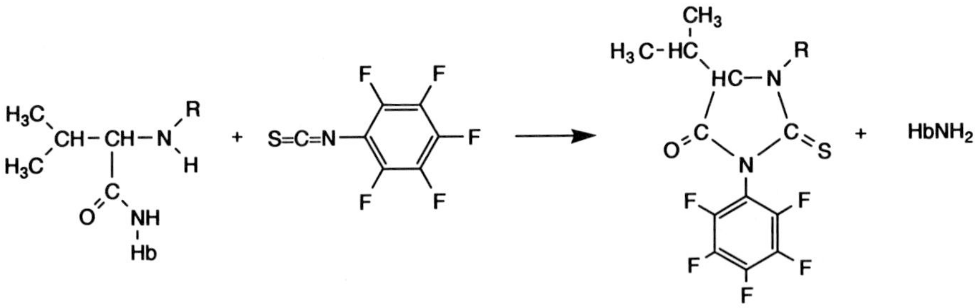

HOEtVal: R = -CH$_2$-CH$_2$-OH

FIG. 1. Edman degradation of N-substituted N-terminal valine.

mals to ethylene oxide, propylene oxide, and styrene oxide, respectively, or to their precursors (ethylene, propylene, and styrene, respectively).[8,9] It is also applicable for the measurements of adducts from other alkylators (e.g., methylvaline) and of Schiff bases from aldehydes, reduced by NaBH$_4$ to alkylvalines.[9,10] Of most low molecular weight adducts studied, background levels have been found, mainly of endogenous origin.[8-11] The procedure is exemplified here by the analysis of HOEtVal–PFPTH [1-(2-hydroxyethyl)-L-5-isopropyl-3-pentafluorophenyl-2-thiohydantoin]. The development of the method has aimed at simplicity, with several refinements and changes being possible.

Chemicals and Equipment

To avoid contamination, the workup of samples for analyses should be done with dedicated glassware in a dedicated laboratory area, separated from the handling of larger amounts of related compounds as used in the synthesis of standards, etc. Before reuse, glassware should be washed with acetone. For analysis of HOEtVal-PFPTH, silanization of glassware is not required.

Analytical-grade solvents are used and, when considered necessary, their purity is checked by electron capture detection–gas chromatography.

[8] M. Törnqvist, in "Human Carcinogen Exposure: Biomonitoring and Risk Assessment" (R. C. Garner, P. B. Farmer, G. T. Steel, and A. S. Wright, eds.), p. 411. Oxford Univ. Press, Oxford and New York, 1991.
[9] M. Törnqvist, in "Use of Biomarkers in Assessing Health and Environmental Impacts of Chemical Pollutants" (C. C. Travis, ed.), p. 17. Plenum, New York, 1993.
[10] A. Kautiainen, M. Törnqvist, K. Svensson, and S. Osterman-Golkar, *Carcinogenesis* (*London*) **10**, 2123 (1989).
[11] M. Törnqvist and A. Kautiainen, *Environ. Health Perspect.* **99**, 39 (1993).

Diethyl ether is tested for peroxides. Formamide is extracted twice with pentane. PFPITC (purum, Fluka, Buchs, Switzerland) is purified on a Sep-Pak Plus silica cartridge (Millipore, Milford, MA), washed with toluene followed by pentane. PFPITC (≤200 μl) is added to the cartridge and eluted with 6 ml pentane, which is then carefully evaporated in a stream of N_2 at room temperature.[12] Vacuum distillation is usually not required. The PFPITC should be handled with care in the hood due to its volatility and supposed toxicity.

Standards

HOEtVal (racemic) is synthesized according to Rydberg *et al.*[13] by a modification of previously described procedures. 2-Bromoisovaleric acid (7.5 g) is added to 2-aminoethanol (20 g) and water (5 ml) and the mixture is refluxed for 15 hr and cooled to room temperature. Following precipitation with acetone (250 ml) and filtration, the product is dissolved in 1 M HCl (50 ml) and applied to a Dowex 50 ion exchanger, which is washed with water (400 ml) and eluted with 2 M aqueous ammonia (300 ml). After evaporation to dryness, the residue is diluted with water (25 ml) and crystallized from acetone–ethanol [2 : 1 (v/v), 70 ml]. White crystals of HOEtVal are obtained in a yield of about 50%. HOEtVal and related compounds have been characterized with respect to melting point, pK_a and NMR.[13]

The HOEtVal–PFPTH is synthesized with more than 90% yield in 1.5 ml 1-propanol : 0.5 M KHCO$_3$ [1 : 2 (v/v)], by reacting HOEtVal (2 mg) with PFPITC (5 μl) at 45° for 2 hr. The product is extracted twice with heptane (2 ml), which is then evaporated under a stream of N_2, and the residue is dissolved in toluene.[7] The derivative could be washed with 0.1 M Na$_2$CO$_3$ as described for globin samples below. Physicochemical properties for a few substituted valines and their PTHs and PFPTHs, after synthesis in a larger scale, have been determined.[13] For HOEtVal–PFPTH in ethanol, λ_{max} − 266 nm and ε = 20.0 × 10^3 M^{-1} cm^{-1}.

Globin alkylated with deuterium substituted ethylene oxide [(^{2}H$_4$)EO] or ethylene oxide (EO) to be used as an internal standard and calibration standard, respectively, can be prepared as follows: To 5 ml red cells suspended in a saline solution with 0.01 M phosphate buffer, pH 7.4, a cold aqueous saline solution is added to a final volume of 25 ml and to a concentration of EO of 40 mM. Because the rate constant for HOEtVal

[12] M. Törnqvist, A. Kautiainen, R. N. Gatz, and L. Ehrenberg, *J. Appl. Toxicol.* **8**(3), 159 (1988).

[13] P. Rydberg, B. Lüning, C. A. Wachtmeister, and M. Törnqvist, *Acta Chem. Scand.* **47**, 813 (1993).

formation[14] at 37° is ~5 × 10^{-5} liter g^{-1} hr^{-1}, this would give ~2 nmol HOEtVal/mg globin in 1 hr at 37°. If the reaction goes to completion (~24 hr), the HOEtVal level will be about four times higher. The reaction is finalized by washing the red cells with cold saline. The exact concentration of EO, which easily evaporates, in the saline solution or in the reaction mixture after spinning down the red cells, can be determined by a modification of a fluorometric method.[15] Globin is precipitated as described below. The determination of the degree of alkylation in the globin is carried out as described earlier[16] or, more simply, in principle, as follows: A mixture of ($^2H_4$)EO-alkylated globin [e.g., 10 mg containing about 2 nmol ($^2H_4$)HOEtVal/mg] together with the amino acid HOEtVal (20 nmol) is hydrolyzed with 6 M HCl (2 ml) at 120° overnight. The HCl is evaporated in a rotoevaporator. The hydrolyzate is dissolved in 6 ml 1-propanol : 0.5 M KHCO$_3$ [1 : 2 (v/v)] and the pH is adjusted to approximately 8.6 with 1 M NaOH. Half of the amount is derivatized with PFPITC (10 μl) as described above. The degree of alkylation is then determined by GC-MS analysis (cf. below). The sensitivity in the analysis of samples prepared in this manner is less than in the analysis of globin samples as described below, but it is sufficient at the high adduct levels in the standard globins. The adduct level in the EO-alkylated standard globin can then be determined through derivatization of a mixture of EO-alkylated and ($^2H_4$)EO-alkylated globin according to derivatization and analysis of globin as described below.

Preparation and Storage of Globin Samples

Blood samples are collected in heparinized vacutainer tubes. If determination of HOEtVal at low levels (e.g., background levels, about 10 pmol/g globin) is intended, blood samples should preferably be drawn using gamma-sterilized equipment, be stored at +4°, and precipitated within 24 hr to avoid artefact formation. Red cells are isolated by centrifugation at approximately 1000 g and washed three times with saline. For isolation of globin,[17] 1 ml hemolysate (red cells diluted with 1.5 volumes of distilled water) is added to 6 ml of concentrated HCl (50 mM) in cold 2-propanol. After centrifugation (approximately 3000 g, 10 min) and removal of cell membranes, the acidified globin is precipitated with cold ethyl acetate (5 ml), added slowly with mixing. The precipitate is washed

[14] D. Segerbäck, Carcinogenesis (London) 11, 307 (1990).
[15] H. J. C. F. Nelis and J. E. Sinsheimer, Anal. Biochem. 115, 151 (1981).
[16] P. B. Farmer, E. Bailey, S. M. Gorf, M. Törnqvist, S. Osterman-Golkar, A. Kautiainen, and D. P. Lewis-Enright, Carcinogenesis (London) 7, 637 (1986).
[17] J. Mowrer, M. Törnqvist, S. Jensen, and L. Ehrenberg, Toxicol. Environ. Chem. 11, 215 (1986).

twice with ethyl acetate, followed by pentane, dried under a stream of N_2, and then dried overnight in a desiccator (~150 mg globin is obtained). Dialysis (e.g., in microcollodion bags, with a vacuum system in a glass holder; from Sartorius, Göttingen) of globin samples and reprecipitation give improved signal to noise ratios in the GC-MS analysis at low adduct levels. Lyophilization of hemoglobin or globin samples should be avoided because of the risk of artifact formation of certain adducts.[18] Normally neither dialysis nor separation of hemoglobin or globin from other proteins has been considered necessary. In special cases enrichment of α- and β-globin chains with alkylated N termini could be used to increase the analytical sensitivity.[19]

Derivatization and Analysis of Globin Samples

Dissolve 50-mg samples of precipitated globin in 1.5 ml formamide in 10-ml Pyrex tubes. Samples are neutralized by addition of 50 μl 1 M NaOH (to a recorded pH of about 6.8). A small amount (50 μl) of the internal standard, $(^2H_4)$EO-alkylated globin, dissolved in formamide (with carrier globin; see below) is added to give a predetermined amount of $(^2H_4)$HOEtVal per gram of globin in the sample, e.g., corresponding to 100 pmol. After dissolution of globin samples through vortexing (about 1 hr), PFPITC (7 μl) is added. In this procedure the yield of HOEtVal–PFPTH is about 70%. Addition of a larger amount of reagent would slightly increase the yield, but may lead to increased formation of by-products that violate the sensitivity of the GC-MS analysis, especially for HOEtVal–PFPTH.

For practical reasons, the derivatization is carried out overnight at room temperature, with vortexing (this is important because the reagent is not immediately dissolved). To reach a better yield the samples are finally warmed at 45° for 1.5 hr.

Small alterations in pH at the beginning of the reaction do not influence the analytical results, and it is normally not necessary to check the pH. During the course of the reaction the pH decreases approximately one unit, which favors the splitting off of the derivative.

The samples are extracted three times with diethyl ether (2 + 2 + 1 ml), the combined ether extracts are evaporated under a stream of N_2, and the residues are redissolved in toluene (1 ml). Phase separation in the ether extraction as well as the extraction of derivatives more hydrophilic than HOEtVal–PFPTH may be facilitated by addition of water to the formamide solution. The toluene extracts are washed twice with water (2 ml), twice with freshly prepared 0.1 M sodium carbonate in water (3 ml)

[18] M. Törnqvist, *Carcinogenesis (London)* **11,** 51 (1990).
[19] E. Bergmark, M. Belew, and S. Osterman-Golkar, *Acta Chem. Scand.* **44,** 630 (1990).

for 10 min or longer (HOEtVal–PFPTH is stable for hours), and again with water. This carbonate wash has to be done with care for other PFPTHs; especially more hydrophilic ones may be less stable. The toluene and volatile by-products, formed by hydrolysis in the treatment with alkaline solution, are carefully evaporated under a stream of N_2 at 60° (more volatile compounds, such as the PFPTH of methylvaline, may be subject to losses). If purification has been obtained, no visible material should remain in the tubes. At this stage further purification procedures could be added, but this should not be required for the analysis of HOEtVal by GC-MS.

In a GC version of the method,[20] for the determination of high adduct levels, a purification on Sep-Pak C_{18} cartridge (Millipore) is added. Such treatment also gives an improved GC-MS analysis at low levels of adducts (about 10 pmol HOEtVal/g globin).

Details in GC-MS analysis depend on the available instrument, etc.[21] With the following instrumentation and experimental parameters high sensitivity and reproducibility have been achieved.[12] The samples are analyzed by GC-MS (Finnigan 4500) using chemical ionization in the negative-ion (NICI) mode with methane as reagent gas and monitoring characteristic ions. The derivatized samples are dissolved in toluene (~50 μl) and ~1 μl is injected via on-column or splitless injector (Program Temperature Vaporizer) on a DB-5 fused silica capillary column (30 m, 0.33 mm i.d., 1 μm phase thickness, J&W Scientific, Inc., San Jose, CA), coupled to an uncoated deactivated precolumn. Typical mass spectrometric settings during analysis are CH_4-moderating bath gas with an ion source pressure of 0.45 Torr (60 Pa), an ion source temperature of 100°, and an ionization energy of 125 eV. The pressure of the He carrier gas is 10 psi (67 kPa) and the GC temperature is programmed 15°/min from 100° to 240°, then 5°/min to 300°. HOEtVal–PFPTH is monitored at m/z 348 (M^- – HF) and m/z 318 (additional loss of CH_2O), and the internal standard ($^2H_4$)HOEtVal–PFPTH is monitored at m/z 352 and 320. Quantification is based on a comparison of peak areas of the major fragments, m/z 348 and 352.

For the MS analysis of PFPTHs of other N-substituted valines by NICI it may be noted that alkylvalines (R = methyl, ethyl in Fig. 1) have major fragments m/z M – 1 (loss of H) and M – 28, and 2-hydroxyalkylvalines lose HF giving the major fragment M – 20. For adducts of oxirane homologs, e.g., propylene oxide, the quantitation is done by summation of the peak areas for the diastereomers of the PFPTH

[20] A. Kautiainen and M. Törnqvist, *Int. Arch. Occup. Environ. Health* **63**, 27 (1991).
[21] M. Törnqvist, A.-L. Magnusson, P. B. Farmer, Y.-S. Tang, A. M. Jeffrey, L. Wazneh, G. D. T. Beulink, H. van der Waal, and N. J. van Sittert, *Anal. Biochem.* **203**, 357 (1992).

of N-(R,S-2-hydroxypropyl)-L-valine. (Current mechanistic studies of the N-alkyl Edman method indicate that racemization of the amino acid occurs.)

An important point is the choice of GC column and, further, for the analysis of hydroxyalkylvaline-PFPTHs, that the use of the column for other purposes should be avoided. Silylation of OH groups of PFPTs has been used.[21,22]

Derivatized samples have been shown to give reproducible results in analysis after years of storage at $-20°$, protected from light, at least with respect to measurement of HOEtVal–PFPTH at levels ≥ 100 pmol/g.

Calibration

It is essential to establish calibration curves from samples with a composition as similar as possible to those of the globin samples to be measured.[12] In the case of HOEtVal–PFPTH calibration, the samples to be derivatized have consisted of 50 mg of low-background globin (from a nonsmoker) with addition of different amounts of the calibration-standard globin and of the internal-standard globin dissolved in formamide. However, when measurement of background levels is intended, other proteins available in pure form may be used together with standard globins.[12]

It has been found that in solutions of too low globin concentrations the alkylated standard globins are subjected to losses, which could show up as nonlinear calibration curves. This may be counteracted by adding carrier globin with low background levels to the standard globin solutions to a concentration of ~ 5 mg globin/ml formamide.[12]

Because calibration curves are linear and samples can be reanalyzed with reproducible results for long periods of time, a check of the reproducibility is obtained by including one calibration sample in every new sample series to be derivatized and analyzed.

If free ($^2H_4$)HOEtVal is used as an internal standard it has to be considered that, because of the higher pK_a, it reacts more slowly than HOEtVal in globin.[21] For the same reason derivatization of free amino acids is done in more alkaline medium (1-propanol : 0.5 M KHCO$_3$, 1 : 2) (pH 8.6). It seems that ($^2H_4$)EO-alkylated globin may be used as an internal standard for other hydroxyalkylvalines and that, e.g., ($^2H_3$)-methylated globin could be used as an internal standard for other alkylvalines.[23] However, for exact quantitation proper calibration is required to allow for differences

[22] E. Bailey, P. B. Farmer, Y.-S. Tang, H. Vangikar, A. Gray, D. Slee, R. M. J. Ings, D. B. Campbell, J. G. McVie, and R. Dubbelman, *Chem. Res. Toxicol.* **4**, 462 (1991).

[23] M. Törnqvist, S. Osterman-Golkar, A. Kautiainen, M. Näslund, C. J. Calleman, and L. Ehrenberg, *Mutat. Res.* **204**, 521 (1988).

in response factors, fragmentation patterns in MS analysis, and possibly in yield.

Sensitivity and Accuracy

The contributions to the uncertainty under the instrumental conditions described have been determined in a few series of background level HOEt-Val measurements.[11] The coefficients of variation (CV) for derivatization, injection, and reading were found to be 4.7, 7.8, and 4.5%, respectively, corresponding to a total CV of 10% in measurements close to the detection level, ~10 pmol/g. In a ring test of HOEtVal-PFPTH determinations different laboratories achieved results that, considering the low level (10–100 pmol/g), were in acceptable agreement (interlaboratory CV, 28%).[21]

By GC-MS-MS analysis of samples prepared as described above the sensitivity was increased by a factor of 10, thus reaching a detection level below 1 pmol/g Hb, which corresponds to an injected amount of about 1 fmol (Törnqvist *et al.*, in preparation). Since in this case the background noise is eliminated a further increase of the sensitivity is possible by using larger samples.

Acknowledgment

The author thanks Anna-Lena Magnusson for excellent technical assistance. The Swedish Environmental Protection Agency has been the main sponsor of the work described.

[44] Tobacco-Specific Nitrosamine–Hemoglobin Adducts

By Stephen S. Hecht, Steven G. Carmella,
and Sharon E. Murphy

Tobacco-specific nitrosamines are a group of carcinogens formed by nitrosation of nicotine and other tobacco alkaloids.[1,2] Two of these compounds, N'-nitrosonornicotine (NNN) and 4-(methylnitrosamino)-1-(3-pyridyl)-1-butanone (NNK), have been implicated in the etiology of cancers of the oral cavity, lung, pancreas, or esophagus caused by use of tobacco products.[3,4] Based on the amounts of NNN and NNK in tobacco

[1] D. Hoffmann and S. S. Hecht, *Cancer Res.* **45**, 935 (1985).
[2] S. S. Hecht and D. Hoffmann, *Carcinogenesis (London)* **9**, 875 (1988).
[3] S. S. Hecht and D. Hoffmann, *Cancer Surv.* **8**, 273 (1989).
[4] D. Hoffmann and S. S. Hecht, *in* "Handbook of Experimental Pharmacology" (C. S. Cooper and P. L. Grover, eds.), p. 63. Springer-Verlag, Heidelberg, 1990.

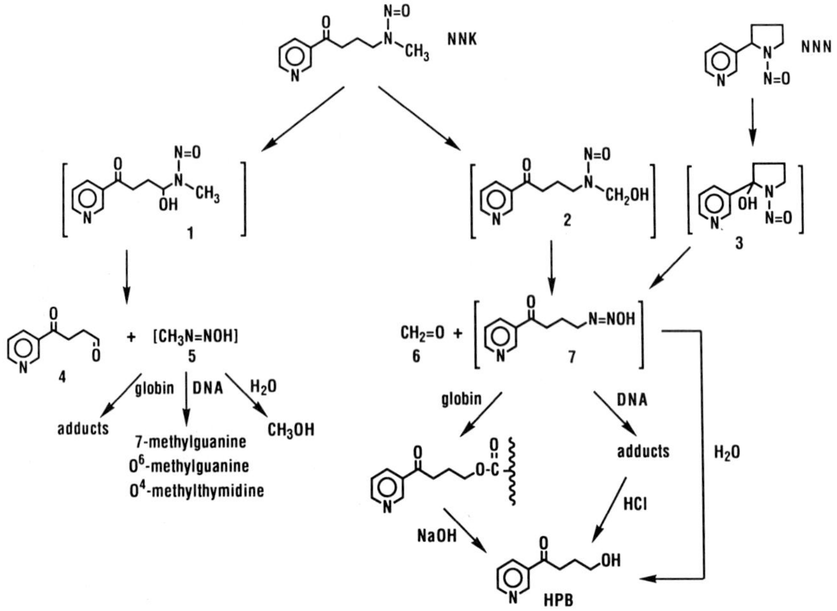

FIG. 1. Metabolic pathways leading to the formation of DNA and globin adducts of NNK and NNN.

products, human exposure to these carcinogens is similar to the doses that cause tumors in laboratory animals. The method described here can potentially provide information not only on an individual's exposure to NNN and NNK, but also on the metabolic activation of these carcinogens to intermediates that bind to cellular macromolecules.

NNN and NNK are converted to highly electrophilic intermediates by enzymatic hydroxylation of the carbons α- to the N-nitroso group. In the case of NNK, α-methylene hydroxylation yields methane diazohydroxide (**5**, Fig. 1), which methylates DNA and hemoglobin. α-Methyl hydroxylation produces 4-(3-pyridyl)-4-oxobutane diazohydroxide (**7**, Fig. 1), which pyridyloxobutylates DNA and hemoglobin. Hydroxylation of NNN at its 2'-position, adjacent to the pyridine ring and N-nitroso groups, yields the same pyridyloxobutylating agent that is produced by methyl hydroxylation of NNK. These reactions are illustrated in Fig. 1 (**1–7**). Extensive studies on the metabolism and DNA binding of NNK have implicated both DNA methylation and pyridyloxobutylation as important factors in carcinogenesis by this nitrosamine.[1,2,5-11] DNA pyridyloxobutylation is also likely to play an important role in NNN carcinogenesis.[1,2,12,13]

Pyridyloxobutylation of hemoglobin was first observed in rats treated with NNK and NNN.[14] Alkylation occurs on the protein, not the protoporphyrin. Treatment of hemoglobin or globin with NaOH at room temperature releases 20–40% of the bound material as 4-hydroxy-1-(3-pyridyl)-1-butanone (HPB).[14-16] The adduct that releases HPB on this mild base hydrolysis is an ester, formed by reaction of either aspartate, glutamate, or the terminal carboxyl group of globin with the diazohydroxide.[17] Other studies have shown that HPB does not react with globin to produce an adduct that releases HPB on mild base hydrolysis.[16] Thus, the release of HPB from globin is a measure of the extent of metabolic activation of NNK or NNN to 4-(3-pyridyl)-4-oxobutane diazohydroxide.

Dose–response studies carried out with NNK in rats have demonstrated that the formation of the ester adduct that releases HPB is linear over a dose range of four orders of magnitude.[15] Comparative studies of globin and DNA pyridyloxobutylation in rats treated with NNK have shown that both parameters increase with dose. However, DNA pyridyloxobutylation in liver and lung does not vary in a linear fashion with dose, and the relationship between DNA and globin pyridyloxobutylation by NNK is thus complex.[15]

The methodology described here was developed to quantify levels of HPB released on mild base hydrolysis of human hemoglobin.[18] It is known

[5] S. S. Hecht, K. G. Jordan, C.-I. Choi, and N. Trushin, *Carcinogenesis (London)* **11**, 1017 (1990).

[6] L. A. Peterson and S. S. Hecht, *Cancer Res.* **51**, 5557 (1991).

[7] S. A. Belinsky, T. R. Devereux, R. R. Maronpot, G. D. Stoner, and M. W. Anderson, *Cancer Res.* **49**, 5305 (1989).

[8] S. A. Belinsky, J. F. Foley, C. M. White, M. W. Anderson, and R. R. Maronpot, *Cancer Res.* **50**, 3772 (1990).

[9] L. A. Peterson, R. Mathew, S. E. Murphy, N. Trushin, and S. S. Hecht, *Carcinogenesis (London)* **12**, 2069 (1991).

[10] S. S. Hecht, D. Lin, and A. Castonguay, *Carcinogenesis (London)* **4**, 305 (1983).

[11] J. Jiao, M. Zielenska, M. W. Anderson, and B. W. Glickman, *Carcinogenesis (London)* **12**, 221 (1991).

[12] S. S. Hecht, B. Reiss, D. Lin, and G. M. Williams, *Carcinogenesis (London)* **3**, 453 (1982).

[13] S. S. Hecht and D. Lin, *Carcinogenesis (London)* **7**, 611 (1986).

[14] S. G. Carmella and S. S. Hecht, *Cancer Res.* **47**, 2626 (1987).

[15] S. E. Murphy, A. Palomino, S. S. Hecht, and D. Hoffmann, *Cancer Res.* **50**, 5446 (1990).

[16] L. A. Peterson, S. G. Carmella, and S. S. Hecht, *Carcinogenesis (London)* **11**, 1329 (1990).

[17] S. G. Carmella, S. S. Kagan, and S. S. Hecht, *Chem. Res. Toxicol.* **5**, 76 (1992).

[18] S. G. Carmella, S. S. Kagan, M. Kagan, P. G. Foiles, G. Palladino, A. M. Quart, E. Quart, and S. S. Hecht, *Cancer Res.* **50**, 5438 (1990).

that individuals vary widely in their abilities to enzymatically activate or detoxify carcinogens such as NNN and NNK.[19] Presumably, those individuals who activate these carcinogens extensively would have a higher risk for cancer development than those who were more proficient in detoxification. Quantitation of hemoglobin adducts of NNN and NNK could thus provide an index of their uptake and metabolic activation.

The method consists of four major steps: (1) isolation of hemoglobin and treatment with base to release HPB; (2) solvent partitions to enrich HPB; (3) derivatization of HPB to its pentafluorobenzoate, which can be detected with great sensitivity and selectivity by capillary column gas chromatography–mass spectrometry with negative-ion chemical ionization detection and selected ion monitoring (GC-MS-NICI-SIM); and (4) quantitation by GC-MS-NICI-SIM.

Materials and Methods

Chemicals

HPB, [4,4-D$_2$]HPB (internal standard), HPB-pentafluorobenzoate, [4,4-D$_2$]HPB-pentafluorobenzoate, and HPB-tetrafluorobenzoate (external standard) are synthesized,[18] or can be obtained from the authors on request. Pentafluorobenzoyl chloride is obtained from Aldrich Chemical Co. (Milwaukee, WI). Pentanophenone and hexanophenone are purchased from Pierce Chemical Co. (Rockford, IL). Trimethylamine (TMA) hydrochloride and dichlorodimethylsilane are obtained from Sigma Chemical Co. (St. Louis, MO).

Instrumentation

Solvent concentration is performed with a Model SVC200H SpeedVac centrifugal concentrator (Savant Instruments, Farmingdale, NY). With this model, heat (43°) may be applied to the sample. All centrifugations are performed using the SpeedVac ($v = 1750$ rpm). HPLC is carried out with a system consisting of two Model 510 pumps, a Model 680 automated gradient controller, a Model 440 fixed-wavelength UV/visible detector, all from Waters Division of Millipore (Milford, MA), and a Model 7125 injector from Rheodyne (Cotati, CA). GC-MS-NICI-SIM is performed with a Hewlett-Packard Model 5988A instrument (Hewlett-Packard Co., Palo Alto, CA).

[19] A. Castonguay, G. D. Stoner, H. A. J. Schut, and S. S. Hecht, *Proc. Natl. Acad. Sci. U.S.A.* **80,** 6694 (1983).

Procedures to Minimize Contamination and Low Recovery

The levels of HPB released from hemoglobin are typically less than one part per billion, and the actual amounts quantified may range from 1 to 100 pg/sample. Therefore, special precautions are necessary to avoid contamination and loss of analyte. Analyses are carried out in a dedicated area or laboratory, with dedicated glassware and HPLC equipment. Organic solvents are HPLC or GC-MS grade and are used without further purification.[20] Sodium hydroxide solution (1 N) is freshly prepared from NaOH pellets and HPLC-grade H_2O. Reagent-grade concentrated HCl is diluted with purified H_2O to 1 N and stored in glass bottles.

Only silanized glassware is used; 24 50-ml disposable borosilicate centrifuge tubes (Kimble, Toledo, OH) are silanized as follows. The tubes are filled with 5% (v/v) dichlorodimethylsilane in CH_2Cl_2 and incubated for 30 min. The silanizing reagent is decanted and the tubes are filled with methanol and allowed to stand for 30 min. The tubes are then rinsed twice with methanol, covered with foil, and allowed to dry overnight.

Silanize 4-ml borosilicate screw cap vials (Baxter, McGaw Park, IL) and 250-μl microvials ("limited-volume inserts," National Scientific Co., Lawrenceville, GA) as follows. The vials are placed in a vacuum desiccator along with a 20-ml scintillation vial containing 2 ml of dichlorodimethylsilane. The desiccator is flushed with N_2 for 3 min. A vacuum is applied for 1 min, and then the desiccator is flushed again with N_2 for 1 min to distribute the silanizing reagent. The vacuum is reapplied and the sealed desiccator is allowed to stand for at least 1 hr. The desiccator is then flushed with N_2. The vials are removed, rinsed twice with methanol, once with CH_2Cl_2, then covered with foil and dried at 180° for 2–3 hr.

All glassware is rinsed with the solvent to be used before each procedure.

Collection of Blood and Isolation of Hemoglobin

Blood is collected in 10-ml heparinized Vacutainers (Baxter, McGaw Park, IL). It is recommended to use at least 5–10 ml human blood for each assay. The blood is centrifuged at 900 g for 10 min at 4°. The supernatant is discarded and the packed red blood cells (RBCs) are washed three times

---

[20] It is critical that solvents be of highest purity. In our experience one batch of HPLC-grade CH_2Cl_2 contained 1% methanol. This resulted in 0% recovery of [4,4-D_2]HPB, because the methanol reacted with all the derivatizing agent.

by centrifugation with 0.9% saline (w/v). The RBC are lysed by mixing with 3 volumes of ice-cold H_2O on a vortex mixer for 5 min. After 15 min at $0°$, KH_2PO_4 buffer, 0.67 M, pH 6.5, is added to make the final buffer concentration 0.2 M. The resulting mixture is centrifuged at 25,000 g for 30 min at $4°$. The supernatant is removed and dialyzed 20 to 24 hr against 10–20 volumes of distilled H_2O, using Spectra/Por 4 cellulose dialysis tubing, with a molecular weight cutoff of 12,000–14,000 (Baxter, McGaw Park, IL). The dialysis H_2O is changed four times.

Treatment of Hemoglobin Solutions with Base

Hemoglobin solution (5–10 ml) is poured into a 50-ml glass centrifuge tube, and 1 N NaOH is added to a final concentration of 0.15 N. The mixture is stirred on a vortex mixer for 5 min and then placed in a sonicator (Branson 1200, Baxter) for 1 hr at room temperature. After sonication, the globin is precipitated by neutralizing the solution with 1 N HCl. Then 50 pg (299 fmol) [4,4-D_2]HPB is added from an aqueous solution (5 pg/μl). This solution is prepared by the addition of 5 μl of a stock solution of [4,4-D_2]HPB (1.2 ng/μl in isopropanol) to 1.2 ml H_2O. The globin and [4,4-D_2]HPB are mixed by inverting the capped tube several times. The globin is pelleted by centrifugation for 10 min. The supernatant, containing HPB, is transferred to a fresh 50-ml glass centrifuge tube.

Solvent Partitioning to Enrich HPB

The supernatant is adjusted to pH 2 within HCl. Under these conditions, HPB is protonated.[21] The supernatant is extracted with an equal volume of CH_2Cl_2 by shaking for 5 min in an automatic shaker. The CH_2Cl_2 layer is separated by centrifugation and discarded. This procedure is repeated. Then, the aqueous layer containing protonated HPB is extracted twice with equal volumes of hexane, using the same procedure. The hexane extracts are discarded. The pH of the aqueous phase is adjusted to 7, which deprotonates the pyridine ring of HPB. It is extracted three times with equal volumes of CH_2Cl_2 as described above. It is important in this step not to carry over any H_2O. The pooled CH_2Cl_2 layers are then

---

[21] We are investigating the use of PrepSep C_{18} extraction columns (Fisher Scientific) in place of partitioning. The supernatant containing HPB is applied to the column, washed with 5 ml H_2O, then HPB eluted with 1 ml methanol. The methanol is added to 9 ml H_2O and HPB is extracted with an equal volume of CH_2Cl_2 once. The CH_2Cl_2 is concentrated to dryness on a SpeedVac.

concentrated to dryness in the SpeedVac concentrator. The concentration step requires approximately 90 min, with heat.

Derivatization of HPB to HPB-Pentafluorobenzoate

A solution of TMA in hexane is prepared as follows. TMA hydrochloride (240 mg) is placed in a 50-ml glass centrifuge tube and to this is added 2.4 ml of 1.0 N NaOH and 17.6 ml of H_2O, followed by 20 ml of hexane. The mixture is shaken for 1 min and allowed to stand for 5 min. The hexane layer is transferred to a 20-ml scintillation vial and dried by addition of 2–3 g of Na_2SO_4. It is shaken and allowed to stand for 5 min, and the hexane solution of TMA is decanted.

For derivatization, reagents are added sequentially with vortexing to the 50-ml centrifuge tube containing the fraction enriched with HPB, as follows: 0.5 ml of CH_2Cl_2, 1 ml of TMA in hexane, and 0.5 ml of 14 mM pentafluorobenzoyl chloride in CH_2Cl_2. The tube is agitated on a mechanical shaker for 30 min at room temperature. The solution is then transferred to a silanized 4-ml vial. The tube is washed with 0.5 ml of CH_2Cl_2 and the washings are added to the solution in the 4-ml vial. The CH_2Cl_2 is removed by evaporation in the SpeedVac, without heat, for 30 min.

The derivatized sample is then purified by HPLC. A 10 cm × 4.6 mm Partisil 5 ODS 3 cartridge column (Whatman Co., Clifton, NJ) is used. The column is washed successively with H_2O, methanol, and 65 : 35 (H_2O : methanol) (v/v) for 1 hr each at 1 ml/min, and then overnight in 65 : 35 (H_2O : methanol) at 0.2 ml/min.

The retention time of HPB-pentafluorobenzoate is marked with pentanophenone and hexanophenone. This marker mixture is prepared by mixing 1 ml each of tetrahydrofuran (THF) and methanol with 5 μl each of pentanophenone (10 mg/ml methanol) and hexanophenone (10 mg/ml methanol). Of this mixture, 70 μl is injected on the column with a solvent program as follows: 65 : 35 (H_2O : methanol), 10 min; then linear to 30 : 70 in 15 min; hold at 30 : 70 for 9 min; then 0 : 100 for 9 min; then return to 65 : 35 in preparation for the next sample. The flow rate is 1 ml/min. Under these conditions, the elution times of the pentanophenone and hexanophenone markers are 29 and 31 min, respectively, and the elution time of HPB-pentafluorobenzoate is 30 min.

The derivatized sample is mixed with 70 μl of the marker mix, vortexed, and allowed to stand for 10 min prior to injection. Standing for 10 min will allow time for methanol to scavenge excess pentafluorobenzoyl chloride. The resulting methyl ester elutes at 25 min. The region eluting between the apices of the pentanophenone and hexanophenone peaks is collected in 4-ml vials. The samples are concentrated to dryness in the SpeedVac, with heat, for 2.5 hr. The dried samples are stored at $-80°$ until analysis by GC-MS-NICI-SIM. Each sample is transferred to a microvial by rinsing the 4-ml vial three times with 6 drops of THF, stirring 2 min each time on a vortex mixer.

Analysis by GC-MS-NICI-SIM

Solutions of HPB-pentafluorobenzoate, [4,4-D_2]HPB-pentafluorobenzoate (internal standard), and HPB-tetrafluorobenzoate (external standard) are prepared in ether, immediately prior to analysis. They are used to generate a standard curve. The concentration of the external standard is 30 fmol/μl. The concentrations of HPB-pentafluorobenzoate and [4,4-D_2]HPB-pentafluorobenzoate range from 5 to 30 fmol/μl. The inclusion of an external standard provides a measure of the sensitivity of detection by GC-MS for a particular sample. We have found that the sensitivity varies between standards and samples, and among samples.

A 0.25-mm i.d. $\times$ 30-m 50% methylphenylsilicone (DB-17) bonded phase column (0.15-μm film thickness), connected to a 0.32-mm i.d. $\times$ 2-m fused silica retention gap is used. The column is connected to the retention gap with a silica press-fit union (Restek Corp., Bellefonte, PA). The seal is reinforced with polyimide resin. Injections are carried out with a septumless on-column injector (J&W Scientific, Folsom, CA). Injections are made with a 10-μl syringe equipped with a 5-inch fused silica needle. The needle is placed into the retention gap such that 2 inches extend into the GC oven. The sample is deposited at an injection speed of 1 μl/sec onto the retention gap with the oven equilibrated at 35°. The injection port temperature is 35° and, after injection, the oven is programmed as follows: 35° for 1.0 min; then 20°/min to 150°; then 4°/min to 215°; then hold for 8 min. The flow rate is 1 ml/min helium measured at 35°.

Ether (10 μl) containing 300 fmol of the external standard, HPB-tetrafluorobenzoate, is added to the conical microvial containing the derivatized sample. Then, the walls of the vial are washed with approximately 3 μl of ether. The volume of the ether solution is then measured by taking it up in a 10-μl syringe. The solution is placed back into the vial, and 50% of it is taken up with a fused silica needle syringe containing enough ether to make the final injection volume 6 μl. It is injected on the column and the temperature program is initiated. The mass spectrometer is operated in the NICI mode with a methane pressure of 0.8 Torr, ionizing energy of 120 eV, and a source temperature of 150°. The molecular ions of HPB-

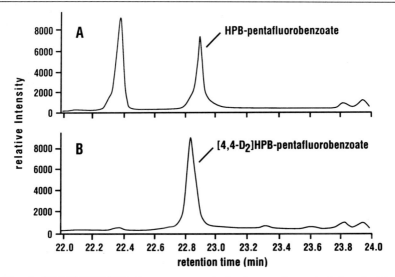

FIG. 2. GC-MS-NICI-SIM chromatograms obtained using the method described here, applied to hemoglobin from a smoker; (A) SIM at m/z 359 (molecular ion of HPB-pentafluorobenzoate) and (B) SIM at m/z 361 (molecular ion of internal standard, [4,4-D$_2$]HPB-pentafluorobenzoate).

pentafluorobenzoate (m/z 359), [4,4-D$_2$]HPB-pentafluorobenzoate (m/z 361), and HPB-tetrafluorobenzoate (m/z 341) are monitored.

Calculations

The amount of HPB in each sample is determined by multiplying the amount of [4,4-D$_2$]HPB added by the ratio of the area of the peaks at 359 and 361 mass units. A correction is made for any difference in GC-MS detector response between HPB and [4,4-D$_2$]HPB. The level of HPB is expressed as femtomoles/gram of Hb. Hemoglobin is analyzed as cyanomethemoglobin (Sigma kit No. 525A). Percent recovery may be calculated using the external standard in conjunction with these data.

Results and Comments

This method for the analysis of HPB is both sensitive and specific. The sensitivity results from the electron capture properties of HPB-pentafluorobenzoate, which make it readily detectable by NICI detection. The detection limit for standard HPB-pentafluorobenzoate, monitored by SIM at its molecular ion of m/z 359, is approximately 100 amol. The

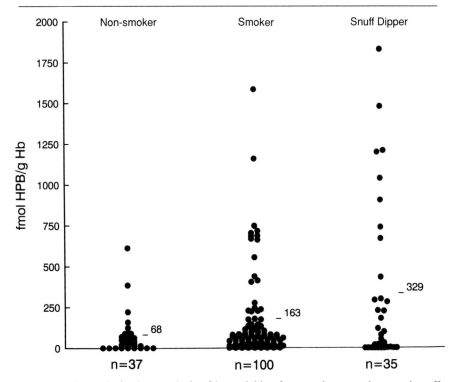

FIG. 3. Data obtained on analysis of hemoglobin of nonsmokers, smokers, and snuff dippers for HPB released on base hydrolysis. Data are from ongoing studies conducted in our laboratory by P. G. Foiles and S. Akerkar, as well as from published studies.[18]

specificity results from the outstanding chromatographic resolution achieved with capillary column GC and the SIM of the particular molecular ion of the analyte. The internal standard, [4,4-D₂]HPB, not only allows quantitation of HPB in the sample, but also provides a reliable GC retention time marker for HPB. We have found that the peak for the molecular ion of [4,4-D₂]HPB, monitored by SIM at m/z 361, generally elutes 0.04 min (typically at 22.85 min) before the peak for the molecular ion of HPB, monitored by SIM at m/z 359 (typically eluting at 22.89 min). This is illustrated in Fig. 2.

The high sensitivity of the method results in detection of some HPB-pentafluorobenzoate in H_2O blanks. The source of this apparent contamination is unknown, but it typically amounts to approximately 60 fmol per 10 ml H_2O (1 ppt). Samples are analyzed in sets of 8–10. Each set includes an H_2O blank as well as a positive control sample, which is a dilution of hemoglobin from a rat treated with [5-³H]NNK. The level of HPB was determined by analysis of [5-^{3}H]HPB to be 500 fmol/g Hb. These negative and positive controls serve as quality control monitors for the analysis. The amount of HPB in the negative control is subtracted from the amount in a given sample. We do not consider a sample to be positive unless the amount of HPB in that sample is at least twice background.

The precision of the method has been described.[18] Replicate values of a given sample run in the same set of samples on the same day have a relative standard deviation of less than 5%. Replicate values of a given sample run in different sets of samples on different days have a relative standard deviation of approximately 25%. The higher relative standard deviation obtained in multiple sets probably results in part from differences in GC-MS responses.

Some representative values for HPB released from human hemoglobin are summarized in Fig. 3. The levels of HPB releasing adducts detected in nonsmokers are generally lower than those of hemoglobin adducts of any other carcinogen examined to date, which is probably a reflection of the tobacco-specific nature of NNK and NNN.[18] Most of the HPB detected in the blood of nonsmokers is likely associated with background from unknown sources, but some may be due to exposure to NNK and NNN in environmental tobacco smoke. In smokers, we have consistently observed elevated levels of HPB-releasing adducts in about 20% of the samples analyzed. The reasons for this intriguing observation are presently under investigation. Preliminary data suggest that levels of HPB-releasing adducts in smokers are not a reflection of the amount of NNK and NNN in a cigarette, but rather may be influenced mainly by metabolic activation and detoxification parameters. Levels of the HPB-releasing adducts in snuff dippers generally appear to be higher than those in smokers. This finding requires further study.

Acknowledgments

This research was supported by Grants No. CA-29580 and CA-32617 from the National Cancer Institute.

[45] Benzo[a]pyrene Diol Epoxide and Related Polynuclear Aromatic Hydrocarbon Adducts of Hemoglobin

By KOLI TAGHIZADEH *and* PAUL L. SKIPPER

Benzo[a]pyrene (BaP) is metabolized by several isozymes of the cytochrome *P*-450 superfamily of enzymes to an array of oxygenated products. Prominent among these in terms of biological activity are two diastereoisomeric (syn or anti) 7,8-dihydroxy-9,10-epoxy-7,8,9,10-tetrahydrobenzopyrenes, the so-called bay region diol epoxides.[1-3] These metabolites alkylate carboxylic acid groups of hemoglobin,[4] and there is evidence that this is the only group alkylated by the anti isomer.[5] Other polynuclear aromatic hydrocarbons are also metabolized to diol epoxides, and it appears that alkylation of hemoglobin carboxyl groups by diol epoxides is a general reaction.[6]

anti - BaPDE   *syn* - BaPDE

The resulting esters are quite labile at neutral pH as a consequence of their benzylic nature, and in the absence of some stabilizing interactions with the protein would not persist *in vivo*. Clearly, though, in some cases there do exist sufficient stabilizing forces to preserve the ester adducts. The lability of the adducts outside the environment of the protein served as the basis for a fluorimetric assay in which globin was enzymatically

[1] A. Dipple, R. C. Moschel, and C. A. H. Bigger, *ACS Monogr.* **182,** 41 (1984).

[2] R. G. Harvey, "Polycyclic Hydrocarbons and Carcinogens," ACS Symp. Ser. 283. American Chemical Society, Washington, DC, 1985.

[3] M. R. Osborne and N. T. Crosby, "Benzopyrenes." Cambridge Univ. Press, Cambridge, UK, 1987.

[4] P. L. Skipper, S. Naylor, L.-S. Gan, B. W. Day, R. Pastorelli, and S. R. Tannenbaum, *Chem. Res. Toxicol.* **2,** 280 (1989).

[5] B. W. Day, P. L. Skipper, R. H. Rich, S. Naylor, and S. R. Tannenbaum, *Chem. Res. Toxicol.* **4,** 359 (1991).

[6] B. W. Day, S. Naylor, L.-S. Gan, Y. Sahali, T. T. Nguyen, P. L. Skipper, J. S. Wishnok, and S. R. Tannenbaum, *Cancer Res.* **50,** 4611 (1990).

METHODS IN ENZYMOLOGY, VOL. 231

digested to destroy its adduct-stabilizing effect and the exposed adducts spontaneously hydrolyzed to tetrahydrotetrols.[6]

This procedure was originally pursued because digestion of the protein not only resulted in hydrolysis of adducts, but also eliminated the protein as a factor in subsequent isolation steps. However, enzymatic digestion required the production of globin from hemoglobin, and when we attempted to introduce GC-MS as the method of quantitation, it became clear that this step was causing unacceptable variance. Thus, we reinvestigated acid- and base-catalyzed hydrolysis of the ester adducts and developed the procedure described below.

Sample Preparation

The procedures described in this section were developed to minimize contact of polyhydroxy-polynuclear aromatic hydrocarbons (PAHs) with glass because these compounds adhere strongly to glass surfaces. It is advised that any glassware that contacts the samples should be silanized before use. Adsorption to glass is most pronounced with aqueous solutions, but also occurs with organic solutions at low temperatures. Glassware is silanized by treatment with a 10% solution of $(CH_3)_2SiCl_2$ in toluene, then rinsed successively with toluene, methanol, and distilled H_2O.

Red blood cells (2 ml) washed three times with 0.9% NaCl are lysed with 2–3 volumes of H_2O that has been redistilled from $KMnO_4$. The lysate is then centrifuged at 14,000 g for 30 min. The supernatant is decanted into tared 50-ml polypropylene centrifuge tubes and the hemoglobin concentration is determined by Drabkin's assay. Internal standard (2 pmol) is added and allowed to equilibrate for 30 min. HCl (1 ml of 6 N) is added to achieve a final concentration of 0.3 N HCl after the sample volume has been made up to 20 ml by further addition of H_2O. The solution is then heated at 80° for 75 min. After cooling, the pH is adjusted to between pH 10 and 11 (by pH paper) with 6 M NaOH and the mixture is extracted three times with ethyl acetate (20 ml).[7] During pH adjustment a precipitate may form if the pH remains near 7 longer than briefly. This precipitate will generally redissolve as the pH is increased further, but it is preferable to pass the critical point quickly.

Extraction is performed by capping the tube and shaking gently. Emulsion often results, but it can be broken by centrifugation if it is not too

[7] This procedure for hydrolysis and extraction is similar to that described in A. Weston, M. L. Rowe, D. K. Manchester, P. B. Farmer, D. L. Mann, and C. C. Harris [*Carcinogenesis (London)* **10,** 251 (1989)]. The critical difference is that by raising pH after hydrolysis, it becomes possible to extract with ethyl acetate, which evaporates more readily and yields a cleaner extract than the more polar isoamyl alcohol.

extensive. The organic phase is removed by aspiration with a plastic pipette. The ethyl acetate extracts are combined in a 50-ml polypropylene beaker and air-dried by placing them in the air flow of a fume hood. The residue is redissolved in 5 ml of phosphate-buffered normal saline (PBS) using a sonicator to ensure complete solution. The solution is then applied to an immunoaffinity column, which is washed with PBS (5 ml) to remove unbound material. Excess PBS is displaced from the column with air before the bound fraction is eluted with a 9:1 (v/v) mixture of CH_3OH and H_2O (10 ml). The eluent is collected in a 20-ml polypropylene vial. The contents of the vial are taken to relative dryness in a SpeedVac concentrator (Savant, Farmingdale, NY). The residue is redissolved in CH_3OH (1 ml) and transferred to a 1.5-ml microcentrifuge tube. The sample is again taken to dryness with a SpeedVac, dissolved in CH_3CN, and redried. The final dissolution and evaporation removes traces of H_2O and is critical for successful subsequent derivatization.

Internal Standard

A suitable standard of $(\pm)$-[7-^{14}C]-r-7,t-8,t-9,c-10-tetrahydroxy-7,8,9,10-tetrahydrobenzo[a]pyrene can be prepared by hydrolysis of $(\pm)$-[7-^{14}C]-r-7,t-8,-dihydroxy-t-9,t-10-epoxy-7,8,9,10-tetrahydrobenzo [a]pyrene, which is commercially available from the NCI chemical carcinogen repository maintained by Chemsyn Science Laboratories (Lenexa, KS). A tetrahydrofuran (THF) solution of the diol epoxide is treated with an equal volume of 1 N HCl for 1 hr at room temperature to effect hydrolysis and the product is isolated by preparative HPLC. Acidic hydrolysis produces predominantly the desired 9,10-trans isomer, but further cleanup is necessary. The following gradient program has been found to be effective with Waters (Bedford, MA) μBondapack C_{18} columns: initial conditions of 85% H_2O (solvent A), 15% CH_3CN (solvent B), changing to 75% A, 25% B at 10 min, then to 72% A, 28% B at 35 min. The 9,10-*trans*-tetrol is the first component to elute at about 25 min, followed by the 9,10-*cis* isomer. Retention times are very sensitive to the condition of the column as well as to how well the column has been equilibrated under the initial elution conditions.

Before use, it is necessary to record the mass spectrum of the [7-^{14}C]tetrol to determine the contribution of all the ^{12}C isomer that may be present. We have used diol epoxide with stated specific activity of 53 mCi/mmol to produce tetrol and have found very little stable isotope contribution to the mass spectrum. This was unexpected because complete substitution of one ^{12}C by ^{14}C should lead to a specific activity of 64 mCi/mmol.

This internal standard has the valuable property that not only is it a heavy atom isotope of the analyte and thus can be used with mass selective detection, but it is also radioactive. The amount used for the assay, 2 pmol, can thus be confirmed by liquid scintillation counting because activity of 2 pmol is about 100 cpm (counts per minute), depending on the exact specific activity. It is important to confirm the concentration of solutions of standard because when stored at $-20°$ a solution of BaP tetrol may decrease in concentration as the result of adsorption to the container.

Immunoaffinity Chromatography Columns

Immunoaffinity chromatography media are prepared by coupling monoclonal antibodies that recognize BaP tetrols to CNBr-activated Sepharose 4B. Effective media can be prepared from relatively impure antibody preparations, such as ammonium sulfate fractions from ascites fluid. Typical protein loading is in the range of 5 mg/ml of swollen gel.

The selection of antibody will determine the range of other PAH tetrols and related compounds that will be detected. Monoclonal antibody 8E11,[8] for example, recognizes at least 10 structures similar to BaP tetrols[6] and has been used to isolate chrysene tetrol from human blood specimens.[9]

GC-MS Analysis

Trimethylsilylation of the polyhydroxy-PAH yields derivatives with excellent properties for both gas chromatography and mass spectrometry.[6] These derivatives, though, are extremely unstable and for trace level analysis must be prepared immediately before injection. Even so, we have found derivatization followed by injection to be unreliable in that a good standard curve cannot be obtained. A successful procedure, instead, is to coinject the silylating reagent, trimethylsilylimidazole in pyridine (Tri-sil "Z"; Pierce, Rockford, IL), and the sample so that derivatization occurs in the injection port or on-column.[10] For this procedure to be completely successful, the sample needs to be "sandwiched" between reagent in the syringe. The technique is to draw up 1 μl of reagent, 2 μl of sample, and finally another 1 μl of reagent. It is not necessary to achieve any mixing within the syringe.

[8] R. M. Santella, C. D. Lin, W. L. Cleveland, and I. B. Weinstein, *Carcinogenesis (London)* **5**, 373 (1984).

[9] B. W. Day, P. L. Skipper, J. S. Wishnok, J. Coghlin, S. K. Hammond, P. Gann, and S. R. Tannenbaum, *Chem. Res. Toxicol.* **3**, 340 (1990).

[10] K. Taghizadeh, B. W. Day, P. L. Skipper, J. S. Wishnok, and S. R. Tannenbaum, *Proc. 39th Am. Soc. Mass Spectrom.*, p. 973 (1991).

Several other precautions are also necessary. Injection port liners must, of course, be silanized. Similarly, it is important to silanize syringes to reduce "ghosting," the appearance of peaks in blank samples. Injection of CH_3CN with reagent between runs likewise reduces carryover. Carryover is a major concern with these PAH alcohols because they have such a high affinity for glass, and it should be assessed regularly.

Gas chromatography conditions are fairly standard. A 15 m × 0.25 mm DB17 stationary-phase (0.25-μm film thickness) fused-silica capillary column (J&W Scientific, Rancho Cordova, CA) is used with helium as carrier gas. Samples are introduced via a splitless injector maintained at 250°. Initial oven temperature is 150°, which is held for 1 min, then ramped to 330° at a rate of 25°/min. A final isothermal phase is maintained for 1 min.

Mass spectrometer operating parameters are also unremarkable. We use a Hewlett-Packard HP5989A MS Engine in the negative-ion chemical ionization mode with CH_4 (99.999%; Matheson, Gloucester, MA) as the moderating bath gas. The source pressure is maintained in the range (1.3–2.0) × 10^{-4} Torr, which is optimal for these compounds, and the source temperature is 200°. The instrument is manually tuned using perfluorotributylamine, which generates ions at 302, 452, and 633. Optimization of the 452 ion is favored.

Tetrahydrotetrols derived from PAH generally produce a single abundant fragment ion under negative-ion chemical ionization corresponding to the loss of a neutral fragment with mass 162 from the molecular ion.[6,11] This neutral fragment is believed to be $(CH_3)_3Si-O-Si(CH_3)_3$. Thus, m/z 446 is monitored to detect tetrols derived from adducts, and m/z 448 is monitored to detect the internal standard.

Quantitation

Two factors render quantitation based on the simple ratio of peak areas incorrect. One is the diastereomer ratios (9,10-cis : 9,10-trans) observed for the [^{14}C]tetrols and the tetrols obtained by adduct hydrolysis, which are different. It is evident that some epimerization of the tetrols occurs under the conditions of hydrolysis, because 5–10% of the added 9,10-*trans*-tetrol standard ends up as the cis isomer. Between 2 and 5% of the stable isotope tetrol is detected in the cis form. The difference is believed to arise because the standard is present as tetrol from the beginning of hydrolysis, whereas the adduct is relatively protected from epimerization

[11] B. W. Day, S. Naylor, L.-S. Gan, Y. Sahali, T. T. Ngyuen, P. L. Skipper, J. S. Wishnok, and S. R. Tannenbaum, *J. Chromatogr. Biomed. Appl.* **562,** 563 (1991).

until it has been hydrolyzed and thereby released from the hydrophobic environment of the protein.

It is necessary, therefore, to determine the abundance of both diastereomers and use the sum to calculate the isotope ratio. This assumes, of course, that no other reactions occur besides epimerization. Recovery of the standard, though, is good, usually 40–50%, so the assumption that tetrols are not degraded during hydrolysis is probably accurate.

The other factor is that the standard tetrol and adduct tetrol differ by only 2 mass units in molecular weight and that the mass spectrum of each contains ions produced by the other. This situation has been analyzed previously, and an equation to calculate the ratio of the components has been derived.[12] In the present case this equation may be written as follows:

$$\text{Adduct} = \frac{A_{446} - (fA_{448})}{A_{448} - (f'A_{446})} \times \text{standard}$$

where A_{446} and A_{448} are the measured peak areas of ions m/z 446 and m/z 448, respectively, f is the 446/448 ratio in the ^{14}C-labeled standard tetrol, and f' is the 448/446 ratio in all the [12C]tetrol.

It is important in these calculations to use the experimentally determined abundance of m/z 448 in the spectrum of all the [12C]tetrol and m/z 446 in the spectrum of the ^{14}C-labeled standard determined with the same instrument that is used for quantitation. The theoretical values based on natural isotope abundances may not be the same and different instruments will not necessarily yield the same results. As indicated earlier, it is also most important to use empirically determined values because the relative amount of all the [12C]tetrol in the ^{14}C-labeled tetrol may not be known with accuracy.

Comments

The present procedure has not yet been thoroughly validated in the way that assays for aromatic amines, alkyl epoxides, and tobacco-specific nitrosamines have been. As discussed in [40], there is no really suitable animal model for human hemoglobin–BaP adduct formation and we have not had access to human blood specimens from individuals with unequivocally elevated exposures to BaP. Cigarette smoke is not a major contributor to human exposure to BaP by most environmental analyses.

Nevertheless, this procedure does give satisfactory reproducibility with the same sample of hemoglobin (coefficient of variation = 0.22) and it gives essentially the same results as the enzymatic digestion procedure.

[12] M. G. Rinsler and F. L. Mitchell, *Z. Klin. Chem. Klin. Biochem.* **12,** 558 (1974).

Adduct levels in hemoglobin from individuals with typical environmental exposure to BaP average around 0.8 pmol/g Hb. The limit of detection is about 0.01 pmol/g Hb, which is an order of magnitude below the lowest observed values.

Acknowledgment

This work was supported by grants from NIEHS (Nos. ES01640 and ES02109).

[46] Analysis of Hemoglobin Adducts by Fluorescence Spectroscopy

By BILLY W. DAY and KULDIP SINGH

Fluorescence spectroscopy has only recently been applied to the study of human hemoglobin adducts. The entirety of the few reports on this method for analysis has consisted of the qualitative and quantitative measurement of polycyclic aromatic hydrocarbon (PAH) metabolite adducts of the protein.

Room Temperature Methods

The technique most used in the quantitation of hemoglobin adducts by room temperature fluorescence has been synchronous scanning fluorescence spectroscopy (SFS). In this method the excitation and emission monochromators are scanned simultaneously at a fixed-wavelength difference, which best corresponds to a Stokes shift of the chromophore of interest. For example, the adducts arising from reaction of the *anti*-diol epoxide of benzo[*a*]pyrene with hemoglobin can be quantified after their hydrolysis, which yields the tetrahydrotetrols of the diol epoxide, by use of a monochromator difference of 34 nm. This corresponds to the energy difference between the S_2 and S_0 electronic states of pyrene, the remaining chromophore in this PAH metabolite.[1-4]

[1] B. W. Day, S. Naylor, L-S. Gan, Y. Sahali, T. T. Nguyen, P. L. Skipper, J. S. Wishnok, and S. R. Tannenbaum, *Cancer Res.* **50**, 4611 (1990).

[2] A. Weston, J. C. Willey, D. K. Manchester, V. L. Wilson, B. R. Brooks, J. S. Choi, M. C. Poirier, G. E. Triviers, M. J. Newman, D. L. Mann, and C. C. Harris, *IARC Sci. Publ.* **89**, 181 (1988).

[3] M. L. Rowe, D. K. Manchester, P. B. Farmer, D. L. Mann, and C. C. Harris, *Carcinogenesis (London)* **10**, 251 (1989).

A plot of fluorescence intensity versus wavelength yields a curve whose maximum is centered on the emission wavelength corresponding to the electronic origin transition. The intensity of this peak is used to quantify the chromophore, as there is a linear correspondence between the hydrolysis products of hemoglobin–benzo[a]pyrene diol epoxide adducts and the intensity of the SFS signal above that of background. These spectra are often cluttered due to interference from other chromophores in the solution, and techniques such as first- and second-derivative SFS can be useful when fluorescing contaminants are present. Because of the low abundance of carcinogen adducts in the presence of unadducted protein, these more precise methods are only reproducibly applicable to large (>20 ml whole blood) samples. Also, unless coupled with other separation and/or spectroscopic/spectrometric techniques, room temperature fluorescence will yield structural information on the adduct only on rare occasions and may in fact be a sum of emissions and quenching effects from closely related chromophores. Another polycyclic aromatic hydrocarbon diol epoxide adduct, that from chrysene, has also been found in fetal hemoglobin, although its identification after hydrolysis to its 1,2,3,4-tetrahydrotetrol was by a combination of UV, room temperature excitation/emission fluorescence, and mass spectral methods.[5] Reiterating what has been put forward in earlier chapters in this section, the exact structure of an adduct must be ascertained before a measurement of its corresponding chromophore in the biological system will carry any meaning as to the potential biochemical outcome of the particular xenobiotic metabolite.

The most sensitive, common, room temperature fluorescence systems consist of two double monochromators, preferably of ≥ 0.22 m pathlength, a photomultiplier tube cooled to $\leq -30°$ by a thermoelectric crystal, a 450-W xenon lamp as the excitation source, and personal computer with software designed for various fluorescence measurements and background subtraction routines. Samples are dissolved in the minimum volume of an appropriate solvent (final volume of 100–500 μl). Phosphate-buffered normal saline has been the most useful solvent for the polar compounds, which are invariably present in biologically derived samples. Less polar solvents can cause precipitation of materials that may scatter light. The samples are placed in a high-quality quartz microcuvette held by an anodized cell holder, both of which are available from several vendors. Dark

---

[4] A. Weston, M. Rowe, M. Poirer, G. Trivers, K. Vahakangas, M. Newman, A. Haugen, D. Manchester, D. Mann, and C. Harris, Int. Arch. Occup. Environ. Health. **60,** 157 (1988).
[5] B. W. Day, P. L. Skipper, J. S. Wishnok, J. Coghlin, S. K. Hammond, P. Gann, and S. R. Tannenbaum, Chem. Res. Toxicol. **3,** 340 (1990).

current and solvent blank background spectra are recorded and subtracted from the sample spectrum. Interference from solvent Raman bands is thus avoided.

Cryogenic Methods

More sensitive measurements, both qualitatively and quantitatively, have been undertaken on hemoglobin–polycyclic aromatic hydrocarbon adducts using laser-induced cryogenic fluorescence spectrometric techniques, one of which is known as fluorescence line-narrowing spectroscopy (FLN).[6] The use of cryogenic temperatures dramatically decreases or abolishes thermal broadening. Coupling cryogenics with the narrow excitation line of a laser to yield site-selective excitation energy, a chromophore imbedded in an inhomogeneous host, such as a protein or glass solvent, undergoes photon absorption and emission transitions corresponding to very narrow ranges of energy. At cryogenic temperatures the chromophores (= impurities = guests) dispersed in the heterogeneous solvents (= glasses = hosts) used for FLN reside in a number of energetically inequivalent microenvironments or sites. Because of the relative orientation of the impurity within the glass, each slightly different microenvironment imparts a specific energy to the chromophore. A change of substituents on the same chromophore imparts additional differences in microenvironment. The result is a collection of identical or very similar guests that undergo the same nominal transition, but only at specific excitation energies corresponding to their various relationships to the host matrix. The spectra of the adducts can be analyzed with respect to chromophore via origin excitation to yield classical zero phonon line/ phonon wing-type spectra. The origin spectra reveal the energies of the S_1–S_0 origin electronic transition, and ground-state vibrational frequencies. These spectra are generally more difficult to obtain due to interference from scattered laser light, and often do not correlate well with the electronic and ground state vibrational energies noted from the corresponding nonpolar parent chromophores in the more easily obtained Shpol'skii spectra. Thus, spectra obtained by the FLN method are typically vibronic in nature. That is, the excitation wavelength is selected to be high enough in energy to be well into the S_1 manifold of the analyte. For the PAH metabolites these energy ranges correspond to wavelengths of ~300 to 450 nm. For the adducts of the *anti*-diol epoxide of benzo[a]- pyrene, these energies are in the 353–368 nm range. Clues as to substitution patterns of the chromophore are obtained via vibronic excitation to reveal

[6] R. Jankowiak and G. J. Small, *Chem. Res. Toxicol.* **4**, 256 (1991).

excited state vibrational frequencies and their corresponding Franck–Condon factors, which provide insight into the structure of the adducts after comparison to synthetic analogs.[7,8]

An example of the utility of vibronic excitation FLN spectroscopy for determination of human hemoglobin adduct structure is shown in Fig. 1. By comparing the FLN spectra of low molecular weight synthetic adducts of the *anti*-diol epoxide of benzo[*a*]pyrene with the spectra obtained from globin from a nonsmoking individual exposed to baseline (Boston area) levels of the parent hydrocarbon, it is apparent that the major, and perhaps only, adduct formed by the *anti*-diol epoxide of benzo[*a*]pyrene with human hemoglobin is a carboxylic ester.[7] Later experiments proved this reaction to take place on the side chain of aspartate-47 on the α chain of human hemoglobin.[9] Through careful design of this experiment, it was determined that the only other possible sites of ester adduct formation could be the C termini of the α and β chains of human hemoglobin.

Quantitation of adducts and their hydrolysis products can often be made more readily by excitation into higher electronic levels of the chromophore. The usefulness of this depends on the populations of the various electronic states. The benzo[*a*]pyrene–7,8,9,10-tetrahydrotetrols, as stated above, have a large cross-section of absorption in their S_2 origin region. Quantitation of these tetrahydrotetrols can be done by examination of their non-line-narrowed, S_2 region excited emission spectra. These results are shown in Fig. 2. We have made an attempt to determine the effect of noncovalent protein binding to the FLN spectra of these tetrols. Tetrols were mixed with an excess of the 8E11 monoclonal antibody, which has high binding affinity for benzo[*a*]pyrene *anti*-diol epoxide adducts and their hydrolysis products. Quantitative analysis of an antibody (8E11)–benzo[*a*]pyrene tetrol complex carried out at femtomole concentration levels is shown in Fig. 3.

Instrumentation and Sample Preparation for Cryogenic Fluorescence Measurements

Instrumental arrangements for laser-induced fluorescence and FLN use as their excitation sources a pumped dye laser with a resolution of <0.1 cm^{-1} and a pulsewidth of 5–10 nsec. Lasers utilized have included

[7] R. Jankowiak, B. W. Day, P.-Q. Lu, M. M. Doxtader, P. L. Skipper, S. R. Tannenbaum, and G. J. Small, *J. Am. Chem. Soc.* **112,** 5866 (1990).

[8] B. W. Day, M. M. Doxtader, R. H. Rich, P. L. Skipper, K. Singh, R. R. Dasari, and S. R. Tannenbaum, *Chem. Res. Toxicol.* **5,** 71 (1992).

[9] B. W. Day, P. L. Skipper, R. H. Rich, S. Naylor, and S. R. Tannenbaum, *Chem. Res. Toxicol.* **4,** 359 (1991).

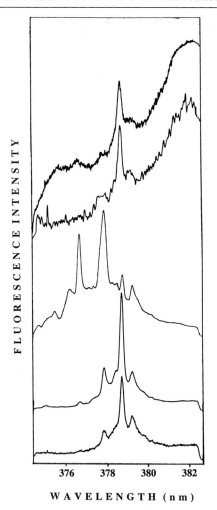

WAVELENGTH (nm)

FIG. 1. Comparison of the vibrationally excited (excitation at 363.4 nm, temperature, 4.2 K) fluorescence line-narrowing spectra of (top to bottom) 7,8,9-trihydroxy-*r*-7,*t*-8,*t*-9,*c*-10-tetrahydrobenzo[*a*]pyren-10-yl *N*-*t*-BOC-L-alaninate ester, a 2-mg sample of intact human globin, *N*-(2-aminoethyl)-*N*-(7,8,9-trihydroxy-*r*-7,*t*-8,*t*-9,*c*-10-tetrahydrobenzo[*a*]pyren-10-yl)amine,   *S*-(7,8,9-trihydroxy-*r*-7,*t*-8,*t*-9,*c*-10-tetrahydrobenzo[*a*]pyren-10-yl)-2-thioethylamine,   and   *S*-(7,8,9-trihydroxy-*r*-7,*t*-8,*t*-9,*c*-10-tetrahydrobenzo[*a*]pyren-10-yl)glutathione thioether.

Nd^{3+}/YAG (yttrium–aluminum–garnet) and excimer (XeCl gas). For example, using the 532-nm laser line from the YAG and pumping various dyes (e.g., rhodamine 640, LDS 698), tunable excitation from 300 to 400 nm is generated by mixing and doubling with KDP crystals. A more

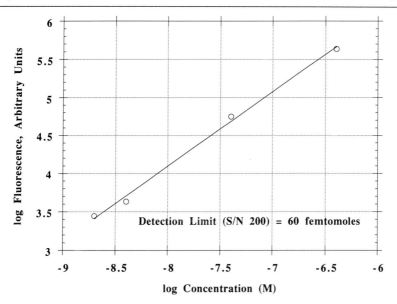

FIG. 2. Quantitation of benzo[a]pyrene-7,8,9,10-tetrahydrotetrols by cryogenic laser-induced fluorescence. Excitation, 343.75 nm; temperature, 4 K. Samples were dissolved in 5:4:1 glycerol–water–ethanol.

expensive but task-simplifying doubling/mixing crystal system is an ultraviolet wavelength extender (UV/WEX), which may be used to obtain laser pulses in the desired range. Useful pulse energies range from 0.2 to 15 mJ/pulse, and occur at a repetition rate of 10–30 Hz, depending on the laser used. Higher pulse energies often lead to problems via photochemical and nonphotochemical hole burning, although the latter has been found to be quite useful when studying systems that emit a broad vibronic envelope at cryogenic temperatures. Pulse energies must be carefully measured in quantitative studies in order to correct for energy fluctuations.

Subsequent to filtering the laser fundamentals, the excitation beam is focused by a cylindrical lens onto the sample compartment of a liquid helium exchange gas cryostat fitted with quartz windows. The temperature of the cryostat is controlled from 4 K to ambient temperatures. Alternatively, a double-nested glass dewar fitted with fused quartz optical windows, which can be filled with liquid helium, may be used. The samples are dissolved in 5:4:1 glycerol–H_2O–ethanol and transferred to 3-mm o.d. × 2-mm i.d. quartz tubes to give a final volume of 30–80 μl. Prior to insertion into the cryostat the samples are thoroughly degassed by several freeze–pump–thaw cycles or by sonication for several minutes at

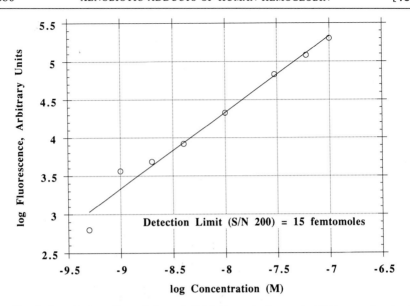

FIG. 3. Quantitation of benzo[a]pyrene-7,8,9,10-tetrahydrotetrols–8E11 monoclonal antibody complex by cryogenic laser-induced fluorescence. Excitation, 343.75 nm; temperature, 4 K. Samples were dissolved in 5 : 4 : 1 glycerol–water–ethanol.

<25°. Problems with photochemical reactions and hole burning are usually reduced if one or both of these degassing/homogenizing procedures are used. The sample tubes are then sealed either by torch under vacuum or with paraffin film. The tubes are enclosed in an aluminum block holder designed to reduce scattered light. For qualitative work the samples may be cooled rapidly to liquid helium temperatures. Our experience in attempts to achieve quantitative analyses has shown that it is best to slowly cool the samples by partial immersion into the headspace of either the cryostat or dewar for about 10 min. This is done in order to form clear glasses, which decreases the specular reflections that an imperfect glass may give rise to. Fluorescence is collected at right angles to the excitation by a 5-cm focal length lens ($F/\# = 1$) and directed to the entrance slit of a monochromator. Installing gratings of 3600 grooves/mm on a 0.32-m pathlength monochromator yields an instrumental resolution of ~12 cm^{-1}. Similarly, a 1-m monochromator with 2400 grooves/mm grating yields spectral resolution of ~3 cm^{-1} in this energy range.

Fluorescence spectra are recorded using a gated intensified photodiode array mounted at the exit plane to give a spectral range of ~9–14 nm, depending on the monochromator/gratings employed. The intensifier can be gated to 100-nsec pulsewidths, and is ordinarily delayed in the range of 20–100 nsec after the laser pulse. Adjustable delay and width of the detector temporal observation window can be controlled via use of a zero-drift controller to trigger a high-voltage pulse generator. This latter system allows gating as low as 5 nsec. Non-line-narrowed spectra can be obtained at 77 K by replacing the photodiode array with a photomultiplier tube, the signal from which is sent through a boxcar averager. A personal computer is used to run the optical multichannel analyzer software, which controls data acquisition. Typical acquisition times are from 0.5 to 5 min. This same computer is used to accumulate signal from the boxcar averager.

Samples can be reproducibly aligned with respect to the monochromator by way of ascertaining the proper reflection angle of an alignment laser, such as a small, inexpensive He/Ne system, which strikes a polished face of the aluminum block holder. Proper imaging and focusing of the sample fluorescence is also facilitated by aiming this alignment laser through the sample. The photodiode array–monochromator system is standardized and calibrated by response in wavelength and intensity to a multiphoton source, such as an elemental hollow cathode lamp.

Acknowledgment

This work was supported by grants from NIEHS (Nos. ES02109 and ES04765).

Addendum

Addendum to Article [11]

By EISHUN TSUCHIDA and TERUYUKI KOMATSU

On the other hand an amphiphilic tetraphenylporphinato-iron derivative having four dialkylphosphocholine groups on one side of the ring plane [new lipid–heme (21)] forms spherical unilamellar vesicles in water.[61] 21 is easily dispersed in deionized water by a sonication method to give a transparent red solution. The homogeneous dispersion did not change for over a year. 21 forms a spherical unilamellar vesicle with a diameter of ~100 nm in dilute aqueous solution (Fig. 7a). This is the first successful example of formation of a porphyrinato-iron vesicle. The thickness of the membrane is ~8 nm. Since the molecular length of 21 is ~3.5 nm, a bilayer structure is conceivable (Fig. 8).

21

The vesicle is obtained in a similar manner in the 3-fold molar excess coexistence of 1-laurylimidazole as an axial base for the O_2 adduct. The visible absorption spectrum of the deoxy complex of the lipid–heme vesicle (λ_{max}: 434, 536, and 558 nm) changed to that of its O_2 adduct on exposure to O_2 (λ_{max}: 432 and 540 nm). The spectrum changed reversibly in response to O_2 pressures. The O_2 adduct changed to the corresponding CO adduct upon bubbling CO gas through the solution (λ_{max}: 431 and 544 nm). The O_2 binding affinity [$P_{1/2}(O_2)$, the O_2 partial pressure at half O_2 binding for the heme] of the lipid–heme vesicle was estimated to be 43 Torr at 37°.

[61] E. Tsuchida, T. Komatsu, K. Arai, and H. Nishide, *J. Chem. Soc., Chem. Commun.* **730** (1993).

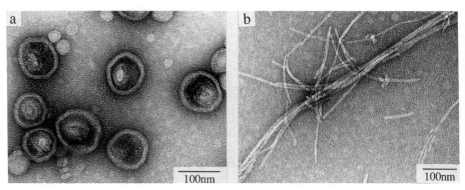

FIG. 7. Transmission electron micrographs of heme assemblies. (a) **21** vesicle; (b) **22** fiber.

Protoporphyrin IX derivative having two alkylphosphocholine groups (**22**) itself produces highly organized rod-like fibers (~7 nm width) in dilute aqueous solution (Fig. 7b).[62] The length of the fibers varies from 0.2 to 1.5 μm. Since the molecular length of **22** is ~4 nm, it may be presumed that the fibers arise from the stacked lipid–porphyrin aggregate. **22** was dispersed with 1-lauryl-2-methylimidazole and DPPC (**22**)/1-lauryl-2-meth-ylimidazole/DPPC [1 : 20 : 100 (molar ratio)] by a sonication method in 30 mM phosphate buffer (pH 7.4) to give a spherical unilamellar vesicle with

[62] E. Tsuchida, T. Komatsu, N. Toyano, S. Kumamoto, and H. Nishide, *J. Chem. Soc., Chem. Commun.*, 1731 (1993).

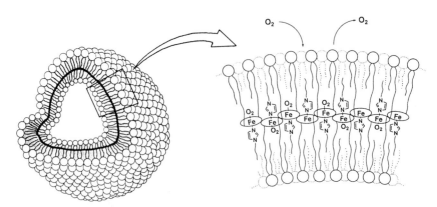

FIG. 8. Lipid–heme vesicle.

22

a diameter of ~50 nm, which is a new hemoglobin model consisting of protoheme derivative. The $P_{1/2}(O_2)$ of the DPPC vesicle embedded **22** was estimated using laser flash photolysis; the $P_{1/2}(O_2)$ was 140 Torr at 25°. These heme assemblies would also be useful as a new molecular architecture in biomimetic reactions.

Author Index

Numbers in parentheses are footnote reference numbers and indicate that an author's work is referred to although the name is not cited in the text.

A

Abbott-Brown, D., 374
Abdel-Baky, S., 637
Abraham, A., 69
Abraham, D. J., 626
Abraham, E. C., 69
Abugo, O., 449, 460, 469, 475, 478(53)
Acharya, A. S., 66, 68, 68(3), 69, 69(3), 194, 197–198, 204, 229, 230(5), 231(5), 232(5), 238, 239(17–19), 240(19), 241(17, 19), 242(19), 245(17–19), 246–248, 251(7–9), 253(10), 261(10), 263(11), 266–267, 290
Achby, M. D., 129
Ackerman, R. S., 573, 574(1), 578(1), 579(1), 583(1), 588(1), 593(1), 597(1)
Ackers, G., 388
Ackers, G. K., 274, 283
Adachi, K., 207
Adams, J. G. III, 333
Addato, E., 44
Adler, A., 449, 453(3), 455(3), 456(3), 459(3), 460(3)
Adler, J., 313
Aebischer, L., 150, 163(2)
Ahmed, F., 500
Aiba, K., 181
Ainscough, R., 114
Air, G. M., 205
Ajisaka, K., 309, 316(11)
Akama, K., 86, 88(27)
Akhrem, A. A., 566
Alard, P., 115
Alayash, A. I., 503
Alden, R. G., 460, 465(58), 474(58)
Ali, M.A.M., 32

Allen, D. W., 67
Alperin, J. B., 462, 478(71)
Alpert, Y., 459, 467(50), 475(50)
Alston, K., 460, 474(56)
Altay, Ç, 22, 27, 28(27), 36(27), 39, 39(27)
Altemüller, A. G., 95, 97(14)
Alter, B. P., 40, 41(57)
Alter, G. M., 463, 464(82)
Amberson, W. R., 82
Ames, B. N., 131, 271
Amiconi, G., 385
Amione, C., 22
Anders, M. W., 608
Anderson, A. J., 288
Anderson, M. W., 659
Anderson, R. R., 95, 98, 99(13), 107(13)
Anderson, W. F., 419
Andersson, L. A., 449, 456(6)
Andracki, M. E., 274, 505
Andreyuk, G. M., 566
Anfinsen, C. B., 196
Angeloni, P., 150, 163(2)
Anggard, E. E., 573
Anjaneyulu, P.S.R., 256
Ansari, A., 471, 480
Antonini, E., 25, 119, 124, 128, 130, 133, 138, 216, 385, 386(25), 388(25), 398(18), 399, 400(18), 451, 460, 499, 565
Anusiem, A.C.I., 464, 469(84)
Arai, K., 168, 171(13), 176(13), 177(13), 189
Arceneaux, D., 492
Archibald, A. L., 404
Arey, J., 643
Argos, P., 456
Ariani, I., 138
Arishima, K., 193
Armarego, W., 114

712 AUTHOR INDEX

Sharma, P. K., 112–113, 120(14), 121–122, 122(67), 123

Sharma, V. S., 458, 468(47), 476(47), 477(47), 500(14)

Shaw, D., 114

Shea, E., 71, 75(20)

Shelton, J. B., 22, 26, 27(25), 29(22), 41(23), 48, 150, 155(5), 156, 157(5), 329, 334, 432, 555

Shelton, J. R., 22, 26, 27(25), 29(22), 41(23), 48, 150, 155(5), 156, 157(5), 329, 334, 432, 555

Shenkin, P., 465

Shephard, E. A., 577

Shibayama, N., 215, 221(2), 274, 280, 288, 487

Shih, D., 207, 212, 347, 357, 358(15)

Shih, D. T., 322

Shih, D. T.-B., 242, 246, 283, 287(8), 322, 330(5), 337(5), 357, 358(17), 359(17)

Shikama, K., 114, 460

Shilton, B., 68

Shimomura, S., 269

Shimuzu, S., 69

Shin, M., 130, 210, 283, 525, 537

Shinton, N. K., 461

Shirai, T., 181, 183(52)

Shishikura, F., 114, 118(26), 119, 121(26), 123(26), 123(77), 124

Shlom, J. M., 116–118, 118(48, 54)

Shoemaker, S., 374

Shohet, S. B., 14, 17, 547

Shrager, R. I., 387, 485–486, 486(12)

Shukuya, R., 525, 530(11)

Sick, H., 95, 97, 97(6, 7, 10, 11), 98, 98(7, 11), 99, 99(7, 11), 100(7), 107(10, 11)

Sigma Diagnostics Procedure #565, 505

Silvani, V., 572

Simmons, J., 121

Simon, M., 87

Simon, R. A., 538

Simoni, J., 79(30), 82, 89

Simoni, S. C., 3

Simons, J. P., 404

Simons, P., 114

Singer, S. J., 390

Singh, K., 674, 677

Sinsheimer, J. E., 653

Skipper, P. L., 625–627, 626(9), 630(7), 631–632, 632(21), 637, 639, 642(7), 643–644, 644(1), 668, 671, 671(6), 672, 672(6), 674–675, 677

Skow, L. C., 419

Slee, D., 656

Slee, R. G., 547

Sleep, D., 377

Sleightholm, M. A., 627, 631(16)

Sliger, S. G., 358, 359(20)

Slightom, J. L., 27

Small, G. J., 676–677

Smith, A., 504

Smith, D. J., 16, 82–83

Smith, F. R., 274

Smith, J. A., 414

Smith, K. M., 95, 97(7), 98, 98(7), 99, 99(7), 100(7)

Smith, L. L., 462, 478(74)

Smith, M., 215

Smith, R. D., 45

Smithies, O., 27, 139

Snell, S. M., 7, 13, 15(12, 27), 288, 289(13), 293(13), 294(13), 295(13), 486, 504, 527

Snow, J. W., 114, 118(26), 121(26), 123(26), 123(77), 124

Snyder, H. E., 597

Snyder, L. M., 547

Snyder, S. R., 10, 276, 487

Sodja, A., 196

Sohara, Y., 318

Sokolovsky, M., 250

Solomon, W., 407

Solway, J., 67

Som, T., 393

Somero, G. N., 126

Song, S., 525, 528(12)

Sono, M., 449, 456(6)

Sorsa, M., 621

Southern, E., 429

Spagnuolo, C., 460

Spangler, E., 404

Sparapany, J. W., 168, 171(11), 178(11), 179

Spartalian, K., 168, 171(10), 177(10), 178(10)

Spector, J. I., 82

Spiteri, D., 46

Springer, B. A., 358, 359(20, 21)

Springham, K., 440

Srivastava, K. C., 566

Staden, R., 114

Staehelin, T., 18

Staigmiller, R. B., 438

Usuki, I., 114
Uy, R., 462, 478(74)
Uziel, M., 629

V

Vaccaro, A. M., 139
Vaccaro-Torracca, A. M., 143
Vahakangas, K., 674
Vainio, H., 621
Vallee, B. L., 250
Valoti, M., 562, 564(5), 569(5)
van Assendelft, G., 407, 409(13)
Van Assendelft, O. W., 141, 602
van Beek, G. U., 285
Van Beeumen, J., 115
Vandegriff, K. D., 12, 327, 387, 483, 485, 486(12), 487, 488(21), 489(21), 524, 538, 543(19), 544(19)
Van der Ouderaa, F. J., 562
van der Ploeg, P.H.W., 498, 499(7)
van der Waal, H., 655, 656(21), 657(21)
van Dorp, D. A., 562
Vane, J. R., 566, 573
van Eys, J., 597
Vangikar, H., 656
Van Hauwaert, M. L., 115
van Kampen, E. J., 498, 499(7, 8), 541, 576
van Leeuwen, J. W., 469, 470(88, 92), 471(92), 474(91), 479(88, 92)
van Noordwijk, J., 81
van Sittert, N. J., 655, 656(21), 657(21)
van Urk, H., 380
Vasudevan, S., 114
Vatsis, K. P., 577, 593(19)
Vecchini, P., 119, 460
Veeger, C., 525
Vermillion, J. L., 577, 593(19)
Verpooten, G., 115
Vestri, R., 43
Vidal, G., 195
Vidal, M., 436
Vidal, M. A., 404, 436
Vidaud, D., 404
Vieira, J., 365
Vigano, T., 572
Vigneron, C., 314, 537
Villani, G. M., 573
Villegas, A., 27, 28(27), 36(27), 39(27)

Vilu, R. O., 469
Vincent, C. K., 439
Vineis, P., 625, 631, 632(21), 639
Vinogradov, S., 114, 505, 507(26)
Vinogradov, S. N., 112–113, 113(7), 114–117, 117(39, 40), 118, 118(26, 48), 119, 120(14, 39, 40, 62), 121, 121(26, 39), 122, 122(40, 67), 123, 123(7, 26, 40), 123(77), 124
Viscio, D. B., 99
Visser, J. P., 195
Vissers, M., 515, 518(10)
Vitale, S., 133, 143
Vitez, L., 374, 375(2), 391, 392(12)
Vithayathil, P. J., 195, 248
Vitrano, E., 161, 162(21)
Vittitow, J., 471, 480(97)
Vlahakes, G. J., 4
Vodrazka, Z., 496, 499(4)
Voliva, C. F., 419
Volkert, F. C., 393

W

Wachtmeister, C. A., 650, 652
Wada, Y., 38, 39(52), 212, 358
Wagenbach, M., 374, 375(2), 391, 392(12)
Wagland, B. M., 114
Wagner, H. N., Jr., 77–78, 81(4)
Waheed, A., 87
Wainwright, N. R., 87
Wajcman, H., 22
Wakabayashi, S., 114
Waks, M., 207
Walder, J., 267
Walder, J. A., 10, 222, 274–276, 276(6), 277(6), 278(5, 6), 279(7), 288, 327, 337(10), 477, 487, 505, 549
Walder, R. Y., 10, 222, 274–276, 276(6), 277(6), 278(5, 6), 279(7), 288, 327, 337(10), 487, 505, 549
Walker, B., 115
Walker, E. L. III, 39
Walker, F. A., 457, 468(41)
Walker, J. C., 450, 453(8), 457, 458(8), 466(8), 467(8), 468(8), 470(8)
Walker, V., 572
Walker, V. E., 626

Subject Index

A

ISBN 0-12-182132-3

90018